Air Pollution Modeling and Its Application XI

NATO • Challenges of Modern Society

A series of volumes comprising multifaceted studies of contemporary problems facing our society, assembled in cooperation with NATO Committee on the Challenges of Modern Society.

Recent volumes in this series:

Volume 5 AIR POLLUTION MODELING AND ITS APPLICATION III
Edited by C. De Wispelaere

Volume 6 REMOTE SENSING FOR THE CONTROL OF MARINE POLLUTION
Edited by Jean-Marie Massin

Volume 7 AIR POLLUTION MODELING AND ITS APPLICATION IV
Edited by C. De Wispelaere

Volume 8 CONTAMINATED LAND: Reclamation and Treatment
Edited by Michael A. Smith

Volume 9 INTERREGIONAL AIR POLLUTION MODELING: The State of the Art
Edited by S. Zwerver and J. van Ham

Volume 10 AIR POLLUTION MODELING AND ITS APPLICATION V
Edited by C. De Wispelaere, Francis A. Schiermeier, and Noor V. Gillani

Volume 11 AIR POLLUTION MODELING AND ITS APPLICATION VI
Edited by Han van Dop

Volume 12 RISK MANAGEMENT OF CHEMICALS IN THE ENVIRONMENT
Edited by Hans M. Seip and Anders B. Heiberg

Volume 13 AIR POLLUTION MODELING AND ITS APPLICATION VII
Edited by Han van Dop

Volume 14 HEALTH AND MEDICAL ASPECTS OF DISASTER PREPAREDNESS
Edited by John C. Duffy

Volume 15 AIR POLLUTION MODELING AND ITS APPLICATION VIII
Edited by Han van Dop and Douw G. Steyn

Volume 16 DIOXIN PERSPECTIVES: A Pilot Study on International Information Exchange on Dioxins and Related Compounds
Edited by Erich W. Bretthauer, Heinrich W. Kraus, and Alessandro di Domenico

Volume 17 AIR POLLUTION MODELING AND ITS APPLICATION IX
Edited by Han van Dop and George Kallos

Volume 18 AIR POLLUTION MODELING AND ITS APPLICATION X
Edited by Sven-Erik Gryning and Millán M. Millán

Volume 19 METHODS OF PESTICIDE EXPOSURE ASSESSMENT
Edited by Patricia B. Curry, Sesh Iyengar, Pamela A. Maloney, and Marco Maroni

Volume 20 PLANNING ESTUARIES
Cees-Jan van Westen and Reinier Jan Scheele

Volume 21 AIR POLLUTION MODELING AND ITS APPLICATION XI
Edited by Sven-Erik Gryning and Francis A. Schiermeier

Air Pollution Modeling and Its Application XI

Edited by

Sven-Erik Gryning

Risø National Laboratory
Roskilde, Denmark

and

Francis A. Schiermeier

U.S. Environmental Protection Agency
Research Triangle Park, North Carolina

Published in cooperation with
NATO Committee on the Challenges of Modern Society

SPRINGER SCIENCE+BUSINESS MEDIA, LLC

Library of Congress Cataloging-in-Publication Data

Air pollution modeling and its application XI / edited by Sven-Erik Gryning and Francis A. Schiermeier.
p. cm. -- (NATO challenges of modern society ; v. 21)
"Published in cooperation with NATO Committee on the Challenges of Modern Society."
"Proceedings of the Twenty-first NATO/CCMS International Technical Meeting on Air Pollution Modeling and Its Application, held November 6-10, 1995, in Baltimore, Maryland"--T.p. verso
Includes bibliographical references and index.
ISBN 978-1-4613-7678-1 ISBN 978-1-4615-5841-5 (eBook)
DOI 10.1007/978-1-4615-5841-5
1. Air--Pollution--Mathematical models--Congresses.
2. Atmospheric diffusion--Mathematical models--Congresses.
I. Gryning, Sven-Erik. II. Schiermeier, Francis A. III. North Atlantic Treaty Organization. Committee on the Challenges of Modern Society. IV. NATO/CCMS International Technical Meeting on Air Pollution Modeling and Its Application (21st : 1995 : Baltimore, Md.) V. Series.
TD881.A475 1996
628.5'3'015118--cc20 96-28437
CIP

Proceedings of the Twenty-first NATO/CCMS International Technical Meeting on Air Pollution Modeling and Its Application, held November 6–10, 1995, in Baltimore, Maryland

ISBN 978-1-4613-7678-1

Originally published by Plenum Press, New York in 1996
Softcover reprint of the hardcover 1st edition 1996

10 9 8 7 6 5 4 3 2 1

PREFACE

In 1969 the North Atlantic Treaty Organization (NATO) established the Committee on Challenges of Modern Society (CCMS). The subject of air pollution was from the start one of the priority problems under study within the framework of various pilot studies undertaken by this committee. The organization of a periodic conference dealing with air pollution modelling and its application has become one of the main activities within the pilot study relating to air pollution. The first five international conferences were organized by the United States as the pilot country; the second five by the Federal Republic of Germany; the third five by Belgium; the next four by The Netherlands; and the two most recent by Denmark.

This volume contains the papers presented at the 21st NATO/CCMS International Technical Meeting (ITM) on Air Pollution Modelling and Its Application, held in Baltimore, Maryland, during November 6-10, 1995. This ITM was jointly organized by the Riso National Laboratory of Denmark (Pilot Country); by the EPA National Exposure Research Laboratory, United States (Host Country); and by the American Meteorological Society, United States (Host Organization).

The ITM was attended by 137 participants representing 29 countries from North and South America, Europe, Asia and Australia. The members of the NATO/CCMS Scientific Committee for the 21st ITM were R.M. van Aalst (The Netherlands), D. Anfossi (Italy), C. Borrego (Portugal), N. Chaumerliac (France), H. van Dop (The Netherlands), S.-E. Gryning, Chairman (Denmark), T. Iversen (Norway), G. Kallos (Greece), W. Klug (Germany), M.M. Millan (Spain), H.R. Olesen (Denmark), G. Schayes (Belgium), F.A. Schiermeier (United States), S. Ulug (Turkey), J.L. Walmsley (Canada), and M.L. Williams (United Kingdom).

The main topics of this ITM were: the Role of Atmospheric Models in Air Pollution Policy and Abatement Strategies; Integrated Regional Modelling; Global and Long-Range Transport; New Developments; Accidental Releases; and Model Assessment and Verification. Invited papers were presented by R. Dennis of the U.S. National Oceanic and Atmospheric Administration (Multi-Pollutant Modeling and Policy: Dressing the Emperor and Avoiding False Paradigms); G. Carmichael of the University of Iowa (Sulfur Deposition in Asia: An Assessment of the Present Situation and the Future Scenarios); A. Flossmann of the Université Blaise Pascal (Mesoscale Modeling of Clouds and Aerosol Particles); and D. Steyn of the University of British Columbia (Air Pollution in Coastal Cities).

As part of the conference, two special sessions were organized on Emerging Air Quality Modeling Technology for High Performance Computing and Communication, and on the NATO/CCMS Pilot Study on Air Pollution Transport and Diffusion over Coastal Urban Areas.

On behalf of the ITM Scientific Committee and as organizers and editors, we should like to record our gratitude to all participants who made the meeting so successful, resulting in publication of this book. Among the participants, we particularly recognize the efforts of the chairpersons and rapporteurs of the various sessions. Thanks are also extended to the American Meteorological Society, the local host organization, who so capably assisted us in organizing this 21st ITM.

Finally, we wish to express our gratitude to the sponsors that made this conference possible. In addition to financial support from NATO/CCMS, the conference received contributions from the United States Environmental Protection Agency, Centro de Estudios Ambientales del Mediterraneo, the European Association for the Science of Air Pollution, the Danish Center for Air Research, and Risø National Laboratory. A special grant was given by NATO/CCMS to facilitate attendance of scientists from Central and Eastern Europe. The next conference in this series will be held in 1997 in Clermont-Ferrand, France.

S.-E. Gryning
F.A. Schiermeier

February 1996

CONTENTS

ROLE OF ATMOSPHERIC MODELS IN AIR POLLUTION POLICY AND ABATEMENT STRATEGIES

Multi-Pollutant Modeling and Policy: Dressing the Emperor and Avoiding False Paradigms 3
Robin L. Dennis

Meteorological Aspects of Ozone Episodes in Southern Ontario - A Mesoscale Model Investigation 13
Rama Seshu Tangirala

Optimization of Ozone Abatement Strategies Using Atmospheric Modeling 17
Marion Wichmann-Fiebig and Thomas Schoenemeyer

Numerical Simulations of the Wind Flow and Ozone Formation during the Heilbronn Ozone Experiment 27
N. Moussiopoulos, P. Sahm, C. Kessler, R. Kunz, C. Schneider, and T. Vögele

Modeling Anthropogenic Sulfur Transport to the Arctic 37
A.P. Dastoor

On the Assessment of Ozone Control Policies for the Northeastern United States 41
S.T. Rao, G. Sistla, W. Hao, K. John, and J. Biswas

INTEGRATED REGIONAL MODELLING

Meteorological Modeling in Sydney Australia: Case Study of a Smog Event 55
Peter J. Hurley, Julie A. Noonan, and Peter C. Manins

Lake Breezes and Their Impact on Ground-Level Ozone Transport over Southwestern Ontario and Southeastern Michigan 63
David M.L. Sills and Peter A. Taylor

Impact of Different Clouds on Tropospheric Chemistry 73
Nadine Chaumerliac, Nicole Audiffren, and Sylvie Cautenet

Transport and Chemistry of Aircraft Emissions in the Vicinity of an Airport 81
Jutta Graf and Hans G. Tremmel

Mass Balances and Interaction of Budget Components of Chemically Reactive Air Pollutants over the Federal State of Baden-Württemberg (Germany) 89
Hans-Jürgen Panitz, Klaus Nester, and Franz Fiedler

Possible Mechanisms for Long Range Transport in the Eastern Mediterranean 99
G. Kallos, V. Kotroni, K. Lagouvardos, M. Varinou, and A. Papadopoulos

Model-Calculations of Transboundary Episodes of High SO_2-Concentrations due to Emissions in the Kola Peninsula 109
Trond Iversen, Anne K. Brenden, Marit H. Jensen, Jan Erik Haugen, and Bjarne Sivertsen

The Impact of the El Niño–Southern Oscillation on the Air Quality over Portugal: The Lisbon Case 119
N. Barros, A. Rocha, and C. Borrego

Air Pollution Assessment Studies for Sweden Based on the Match Model and Air Pollution Measurements 127
Christer Persson, Joakim Langner, and Lennart Robertson

Preliminary Results: Modeling Fine Particulate Mass for the Eastern United States Using the EPA Regional Particulate Model 135
Jason Ching, Francis A. Binkowski, and Jonathan E. Pleim

Uncertainties Associated with Source-Receptor Relationships for Visibility Reducing Species Derived from a Grid Model 145
Prakash Karamchandani and Prasad Pai

Effects of Emission Controls and Stratospheric Ozone Loss on Tropospheric Ozone Production 155
Kevin C. Crist, Gregory R. Carmichael, Brian T. Hotchkiss, and Kuruvilla John

GLOBAL AND LONG-RANGE TRANSPORT

Sulfur Deposition in Asia: An Assessment of the Present Situation and Future Scenarios 167
Gregory R. Carmichael and Richard L. Arndt

The European Long Range Tracer Experiment (ETEX). Preliminary Evaluation of Model Intercomparison Exercise 181
Graeme Archer, Franco Girardi, Giovanni Graziani, Werner Klug, Sonia Mosca, and Katrin Nodop

Long Range Transport of NO_x, SO_x and O_3 over East Asia and the Northern Pacific Ocean Caused by Typhoons 191
Toshihiro Kitada, Seiji Isogawa, and Yutaka Kondo

Deposition of Persistent Organic Pollutants over Europe 203
W.A.J. van Pul, J.A. van Jaarsveld, and C.M.J. Jacobs

Budgets of Ozone on an European and Global Scale 213
Michiel Roemer, Ger Boersen, Peter Builtjes, and Paul Esser

The Effect of Historical Emission Changes and Meteorological Variability on Sulphur Loads in Europe 221
Sophia Mylona

Ozone Modeling over a Large City by Using a Mesoscale Eulerian Meteorological and Transport Model: Madrid Case Study 231
Roberto San José, Luis M. Marcelo, Belen Moreno, and Arturo Ramírez-Montesinos

The Role of Mineral Aerosols in Tropospheric Chemistry 239
Yang Zhang, Li-Ling Chen, Gregory R. Carmichael, and Frank Dentener

The Danish Eulerian Hemispheric Model 249
Jesper Christensen

NEW DEVELOPMENTS

Mesoscale Modeling of Clouds and Aerosol Particles 259
Andrea I. Flossmann

A New MC2/ADOM Modeling System and Its Applications to the Study of the Oxidants in the Canadian Southern Atlantic Region 275
Xiude Lin, Wanmin Gong, and Sylvain Menard

Test of a New Concentration Fluctuation Model for Decision-Makers 285
Per Løfstrøm, Hans Jørgensen, Erik Lyck, and Torben Mikkelsen

Model of Internal Boundary Layer Formation over Inhomogeneous Terrain 295
Ekaterina Batchvarova and Sven-Erik Gryning

High Resolution Dispersion Modeling near Tunnel Exits 303
Roland Ries

Performance of a Flux Conserving and a Semi-Lagrangian Advection Scheme in Simulating a Photochemical Episode 311
Prasad Pai, Prakash Karamchandani, and Akula Venkatram

SO_2 Dry Deposition Parameterization in a Chemistry-General Circulation Model: Model Description and Development 325
Laurens Ganzeveld, Jos Lelieveld, and Geert-Jan Roelofs

The Dynamic Exchange of Pollutants at the Air-Soil Interface and Its Impact on Long Range Transport 333
J.A. van Jaarsveld

Modeling of Cloud Effects on Ozone over Europe 345
Jan Matthijsen, Peter J. H. Builtjes, and Ernst W. Meyer

Turbulent Mixing of Reactive Gases in the Convective Boundary Layer 353
Gé Verver

Comparison between Experimental Data and a Langevin Particle Dispersion Model Including Dry Deposition 363
Edvard Karlsson, Erik Näslund, and Sven-Erik Gryning

PRIME - An Improved Downwash Model 373
Richard H. Osa and Charles Hakkarinen

Improved Algorithms for Estimating the Effects of Pollution Impacts from Area and Open Pit Sources 379
William B. Petersen and Steven G. Perry

New Developments in the CALPUFF Non-Steady-State Modeling System 389
Joseph S. Scire, David G. Strimaitis, and Mark E. Fernau

Local-Similarity Description of Trajectories of Plumes and Jets in Neutrally Stratified Turbulent Shear Flow 399
Eugene Genikhovich

Breakdowns of the Nighttime Planetary Boundary Layer over Complex Terrain 407
Carmen J. Nappo and Richard M. Eckman

A Numerical Simulation of the Eddy Motion and Cloud Distributions over the Santa Barbara Channel 417
Ted Yamada

SCIPUFF - A Generalized Dispersion Model 425
R.I. Sykes, S.F. Parker, D.S. Henn, and R.S. Gabruk

A Modeling Study of Vertical Diffusion of Passive and Reactive Tracers Using Local- and Nonlocal-Closure Boundary Layer Schemes 433
Kiran Alapaty, Rohit Mathur, and Daewon W. Byun

Advanced Mesoscale Dispersion Modeling Using Kinematic Simulation 443
Robert Yamartino, David Strimaitis, and Arno Graff

Dispersion in Idealized Urban Street Canyons 451
Robert N. Meroney, Stillianos Rafailidis, and Michel Pavageau

ACCIDENTAL RELEASES

Validation of a Combination of Two Models for Long-Range Tracer Simulations 461
Jørgen Brandt, Thomas Ellermann, Erik Lyck, Torben Mikkelsen, Søren Thykier-Nielsen, and Zahari Zlatev

Severe Nuclear Accident Program (SNAP) - A Real Time Dispersion Model 471
Jørgen Saltbones, Anstein Foss, and Jerzy Bartnicki

Studies of Dense Gas Dispersion from Short-Duration Transient Releases over Rough Surfaces During Stable Conditions 481
Steven R. Hanna and Kenneth W. Steinberg

Emerging Air Quality Modeling Technologies for High Performance Computing and Communication Environments 491
Daewon W. Byun, Donald Dabdub, Steve Fine, Adel F. Hanna, Rohit Mathur, M. Talat Odman, Armistead Russell, Edward J. Segall, John H. Seinfeld, Peter Steenkiste, and Jeffrey Young

MODEL ASSESSMENT AND VERIFICATION

Air Pollution in Coastal Cities 505
D.G. Steyn

Toward the Establishment of a Common Framework for Model Evaluation 519
H.R. Olesen

Visibility Episodes in Southwestern British Columbia - Evaluation of a Gas-to-Particle Scheme Using the REVEAL Data Set 529
S.C. Pryor, R.J. Barthelmie, and O. Hertel

Validation of a Microscale Pollution Dispersal Model 539
Joachim Eichhorn

Numerical Study of Dispersion over the Greater Athens Area during Two Days of Heavy Pollution Using the Urban Airshed Model 549
Ioannis C. Ziomas, Paraskevi Tzoumaka, Dimitrios Balis, Dimitrios Melas, and Christos S. Zerefos

Intercomparison of 3-D Flow and Particle Models with TRANSALP 1989 Meteorological and Tracer Data 559
E. Ferrero, F. Desiato, G. Brusasca, D. Anfossi, G. Tinarelli, M.G. Morselli, S. Finardi, and D. Sacchetti

Quantitative Evaluation of a Mesoscale Numerical Model Simulation Using Four-Dimensional Data Assimilation of Complex Airflow over the Kanto Region of Japan 569
Itsushi Uno

Wind Flow over Rolling Terrain: A Comparison of Three Recent Models and Full-Scale Observations 583
J.M. Santabárbara, J.F. Hernández, J. Calbó, L. Cremades, T. Mikkelsen, and J.M. Baldasano

A Comparison Study of Air-Quality Model Simulation Results with Aircraft Data 593
Gunilla Svensson and Otto Klemm

A Comparison of CALPUFF Modeling Results with CAPTEX Field Data Results 603
John S. Irwin, Joseph S. Scire, and David G. Strimaitis

Comparison of Measured and Modeled Surface Fluxes of Heat, Moisture, and Chemical Dry Deposition 613
Jonathan E. Pleim, John F. Clarke, Peter L. Finkelstein, Ellen J. Cooter, Thomas G. Ellestad, Aijun Xiu, and Wayne M. Angevine

AERMOD - The Developmental Evaluation 623
R.F. Lee, S.G. Perry, A.J. Cimorelli, R.J. Paine, A. Venkatram, J.C. Weil, and R.B. Wilson

POSTER SESSION

Use of the UB/NMC Model for the Simulation of the Life-Cycle of Chernobyl Radioactive Cloud 635
Milan Dacić and Boško Telenta

Air Pollutant Diffusion-Deposition from a Continuous Point Source 639
Ana G. Ulke and Nicolás A. Mazzeo

IFDM Modeling for Optimal Siting of Air Quality Monitoring Stations around Five Oil Refineries 641
G. Cosemans, G. Dumont, E. Roekens, and J.G. Kretzschmar

An Analytical Formulation for Mixing Length Scales and Eddy Diffusivities in a Convective PBL 643
G.A. Degrazia, A.P. de Oliveira, and O.L.L. Moraes

On a PC-Oriented Eulerian Multi-Level Model for Long-Term Calculations of the Regional Sulphur Deposition 645
Dimiter E. Syrakov

Air Pollution Modeling in the Capital of Prague 647
Jan Pretel, Josef Brechler, Václav Pisa, and Stanislav Böhm

Model Predictions on Urban Air Quality and Their Comparison to Measurements 649
A. Karppinen, J. Kukkonen, E. Valkonen, J. Härkönen, E. Rantakrans, L. Jalkanen, S. Haarala, J. Walden, T. Koskentalo, T. Elolähde, P. Aarnio, and J. Laurikko

Physical Modeling of Car Exhaust Dispersion in Urban Street Canyons - The Effect of Slanted Roofs 653
Stilianos Rafailidis and Michael Schatzmann

An Implicit-Explicit Algorithm for Chemistry-Transport Models 655
Ralf Wolke, Oswald Knoth, and Eberhard Renner

Modeling Studies on the Atmospheric Long-Range Transport of Toxic Metals to the Lake Balaton, Hungary 657
László Bozó, Richard Artz, Joseph Pinto, Joseph Scudlark, and Krisztina Labancz

Long-Term Atmospheric Transport and Deposition of Heavy Metals in Central Europe 659
Jerzy Bartnicki, Krzysztof Olendrzyński, Jozef Pacyna, Stefan Anderberg, and William Stigliani

Complex Mathematical Model Concerning Pollutants Dispersion in the Atmosphere Emitted from Interfering Sources 661
Traian Pop, Livia-Mihaela Pop, Adrian Jelev, and Octavian Olteanu

Numerical Modeling of Three-Dimensional Flow and Pollution Transport
over Complex Terrain During Stable Stratification 665
Alexander Baklanov and Jan Burman

Emergency Warning System in Slovenia for Accidental Radioactive Releases 667
Marija Božnar, Rafael Martinčič, and Primož Mlakar

A Numerical Study about the Transport and Diffusion of Sulfur
Dioxide during the Breeze Cycle in the Island of Mallorca (Spain) 669
Romualdo Romero, Sergio Alonso, and Clemente Ramis

Numerical Simulation of the Mesometeorological Circulations in the Iberian Peninsula .. 671
R. Salvador, E. Mantilla, M.J. Salazar, and M. Millán

The Participation of the Swiss Meteorological Institute in the European
Tracer Experiment (ETEX) 677
D. Schneiter, H. Berger, and S. Galmarini

Analysis of Air Pollutants in Istanbul: A Preliminary Study 679
Mehmet Karaca, Arslan Saral, Mete Tayanç, and Ferruh Ertürk

Validation of ADMS for Dispersion in Convective Conditions 683
Belinda M. Davies

Dense Gas Plume Field Measurements at the Nevada Test Site 687
Gary A. Briggs

PARTICIPANTS 689

AUTHOR INDEX 705

SUBJECT INDEX 707

ROLE OF ATMOSPHERIC MODELS IN AIR POLLUTION POLICY AND ABATEMENT STRATEGIES

chairman: F.A. Schiermeier

rapporteur: D. Melas

MULTI-POLLUTANT MODELING AND POLICY: DRESSING THE EMPEROR AND AVOIDING FALSE PARADIGMS

Robin L. Dennis*

Atmospheric Sciences Modeling Division
Air Resources Laboratory
National Oceanic and Atmospheric Administration
Research Triangle Park, NC 27711
*on assignment to the National Exposure Research Laboratory,
U.S. Environmental Protection Agency

WHERE WE SEEM TO BE HEADED

Multi-pollutant Perspective is Growing in Importance

The problem landscape we know has multi-pollutant interactions. There is a growing recognition of multi-pollutant interactions and a desire to take them into account for assessment and policy purposes.

- SO_2 reductions reduce SO_4^{-2} (improving visibility), but leave more available ammonia to produce NO_3^-, mitigating the visibility improvements from sulfur reductions.
- SO_2 reductions reduce SO_4^{-2}, thereby changing the radiative forcing and the radiative balance with the greenhouse gases, reducing mitigating influences on global warming.
- NO_X reductions help reduce regional ozone and help reduce eutrophication of coastal waters, but may increase urban ozone.
- NO_X reductions decrease the available OH, thus increasing the lifetimes of greenhouse gases, such as CH_4, and exacerbating global warming.
- It appears that whatever we do to mitigate tropospheric problems will/could exacerbate global (stratospheric) problems.

Problems Have a Large Geographic Scale

There is a growing recognition that we have had an artificial separation between global, regional and local issues. We need to include the regional scales in order to deal with local and global problems.

Examples:

Acid rain (now with both sulfur and nitrogen)

Transport distances of importance are the order of 1,000 kilometers. New studies show that geographically-focused reductions still need to include several states and there is no economic advantage to geographically-focused strategies compared to continental ones.

Coastal waters and estuaries (nitrogen)

The airsheds affecting coastal estuaries are large relative to the watershed end points, the order of one million square kilometers. Local emissions contribute only a fraction of the deposition to the watersheds.

Ozone

Ozone airsheds are regional, because O_3 begets O_3 (note establishment of the Ozone Transport Commission); airsheds are yet to be adequately defined.

Toxics (Persistent Organic Pollutants /Heavy Metals)

The airsheds of these pollutants are subcontinental to continental since their lifetimes are longer than that of sulfur.

Greenhouse gases

The effect of NO_X emissions on OH levels (affecting lifetimes of CH_4) is sub-continental to continental in scale.

ADDRESSING MULTI-POLLUTANT AND MULTI-SCALE MEANS WE HAVE TO ADDRESS COMPLEXITY AND UNCERTAINTY

Complexity

Complexity has to be more directly addressed now because the variety of air pollution perturbations, their magnitude, and their geographic scope is so large that we are affecting the fabric of the system, not its margins. To deal effectively with emerging issues, it is becoming necessary to take pollutant interactions into account. Therefore, we have to understand the more complete working of the air quality system with all of its dynamics, feedbacks and nonlinearities.

Complexity is addressed through simplification. But the right level of parsimony is difficult to ascertain and quantify and there is a tension between over-parameterization and over-simpification. It is often said that added complexity makes models harder to use and evaluate, more difficult to understand, yet provides little or not improvement in performance (when compared to historical measurements) or even degrades performance, reinforcing the arguments that complexity is more of a bane than a boon.

Often, because understanding is not the objective of modeling, the realism of the model is not of interest, only calibrated performance. We have tended to do the minimum and increment when there is major policy failure. If realism is not important, then a model can continue to be used in the face of poor performance.

Complexity has to be addressed for today's multipollutant and cross-media problems, because realism is important and the models need to represent the system as well as we can underdtand it. Examples of areas where lack of realism is expected to affect model analyses are:

- Acidic deposition and aerosol loading. For developing regions like the Pacific Rim where SO_2 emissions are anticipated to increase to densities greater than those in the U.S., nonlinearities should have an impact on deposition projections and sulfur lifetimes. Cloud processing, which is only now being considered in regional and global models, is very important to the production and geographic distribution of sulfate aerosols.
- Regional photochemistry. Urban and regional exchange, biogenically driven chemistry, and NO_X plume interactions affect control strategy effectiveness.
- Seasonal photochemical loading. Cloud and precipitation processes, stratospheric ozone influences (O_3 column density and O_3 exchange), and low-concentration photochemistry affect integrated exposures and control strategy effectiveness.
- Hemispheric photochemistry. Feedbacks on OH due to changes in NO_X emissions and to O_3 exchange with the mixed layer and stratosphere affect gas-phase lifetimes of NO_X, SO_2, CH_4 and other greenhouse gases.
- Regional aerosols. Aerosol size distributions are influenced by clouds and there are feedbacks between secondary aerosol loadings and their rate of formation.

We have been moving in the direction of dealing with complexity. We have gone through two phases (generations) of modeling and are now entering a third phase (generation).

First generation
- highly simplified parameterizations for inert pollutants (simple plume dispersion)

Second generation
- single pollutant systems with simplified mechanisms for chemically reacting pollutants and highly simplified meteorology (conceptual extension of boxes or puffs moving and expanding)

Third generation
- multipollutant models with condensed, coupled nonlinear multipollutant descriptions integrated (linked or coupled) with equally complex meteorological process descriptions.

Uncertainty

Uncertainty has to be addressed because we do not know enough and cannot measure enough, ever. Also, our ability to simplify is constrained. Thus, perfect agreement between model predictions and measurements is impossible to achieve. We need to establish the degree of uncertainty to be able to judge model performance. Yet estimates of uncertainty are themselves highly uncertain. On the model development side, there is too much information with too little knowledge about how it all fits together. On the data interpretation side, the signals of system tendency are muddied by stochastic variability and signals that aren't yet resolved that appear to be noise.

Uncertainty is addressed through interrogation between models and the real world. But our current pattern of interrogation is inadequate or incomplete relative to multi-pollutant modeling because it has been schooled by statisticians and regression modelers. Our interrogation has only been concerned with "history matching" of outcome variables.

We have also tended to take a Leggo Block approach to modeling, thinking that once it fits together and gives something that is within a factor of 2 we are done. Any new aspect of the problem, if needed, can be added at will as a new piece. This comes from an inadequate understanding of the importance of model-structure validity to the evaluation of models and from the perspective that often understanding was not a critical objective of model development.

A Key Element of Uncertainty is Ignorance

Ignorance is more critical than is parameter estimate uncertainty -- ignorance of the complete process, ignorance of the complete set of processes operating. Incomplete knowledge has led to choices, scientific judgments, that later proved to be wrong when we gained more knowledge. This is science and it is to be expected at certain stages of development of our understanding.

Examples:

- Isoprene emissions
  - In late 1970's we thought isoprene emissions were unimportant. Ozone response for comparisons of NO_X versus VOC emissions reductions is very different when biogenic emissions are included compared to when they are ignored.
  - Model performance for ozone of some regional models appears to degrade with better science (better isoprene estimates).
- S-wet- to S-dry-deposition ratio
  - Originally we thought wet deposition equaled dry and simpler models were parameterized to mimic this assumption. Later we found this assumption to be wrong through both studies with complex models and measurements.
- S and N lifetimes
  - Newer estimates of sulfur lifetimes are nearly half those of 10 years ago as the influence of clouds on SO_2 conversion is better understood and included in models.

Also, because of the primary concern to "history match," we have tended to compensate for our ignorance by artificially balancing the errors in the modeling system to provide a "reasonable" outcome, without recognizing the part ignorance was playing. This can seriously distort the realism we are trying to portray with the modeling system.

A KEY NEED IS PROJECTION/WHAT-IF ANALYSIS AND EXPLANATORY ANALYSIS AS WELL AS POLICY TOOLS

There are lots of questions in search of answers. Many activities in support of decision making require the use of models and model analysis. These analyses cannot be accomplished only with data.

Dissecting Problems

Who is doing what to whom?
How big are the airsheds? Where are they?
What interactions and feedbacks are there? Are they critical?
Which way NO_X and VOC controls for O_3 mitigation?
Can we allow economic approaches like trading to cut control costs?
Can we spatially/geographically optimize to reduce costs?
What are the relative benefits among controls.
What about chronic versus acute exposures?

Exploring Scientific Issues, Uncertainty and Ignorance

How large are the nonlinearities in the system?
What interactions and feedbacks are there? Are they represented?

How might our ignorance be affecting the answers from the models?
What is the level of uncertainty regarding answers from the models?
What sorts of errors do we get through use of simplified models?
What represents good modeling practice?

Addressing Regulatory Projections

What are the unintended consequences from control of a target pollutant on the other pollutants in the system?
What are the projected episodic/acute exposures to humans and ecosystems ?
What are the projected seasonal/chronic exposures to humans and ecosystems?

These different uses all require an understanding of the system. Thus, the realism or intrinsic truth of the model is of importance to the applications.

WHAT DOES THIS MEAN FOR SCIENCE, MODELING AND POLICY?

Models are critical to the decision making process and are very important to problem solving, yet they are imperfect. The decision process is set up to want or expect perfect answers, and quickly. We need to help modify this process to one that treats models as being associated with scientific understanding (embodying imperfect and incomplete knowledge), not as engineering constructs that are of whole cloth. We have to find a way to be brutally honest, yet constructive and helpful and work with the decision making process. As part of this quest for "honesty with relevancy," we have to find a way to more actively include the scientific community and the scientific peer review process in our more mission-oriented modeling activities and in our interactions with the decision process.

Model development efforts need to be matched with the problem needs. We can only develop adequate realism and attack ignorance and incomplete knowledge through development, testing and field evaluation of the more comprehensive models. We need good models and good data together in concert to do this. Advanced, complex models are the best vehicles for hypothesis testing and testing for our ignorance. They are also the best vehicles for interesting the scientific community and developing a base of credibility for all of the models. The credibility of the simpler models is coming more and more to depend on the science tested in the more complex models. Thus, it becomes increasingly important that the advanced science models be properly evaluated.

a. We need a strong, healthy research program in advanced model development to advance the tools, address uncertainty, and look for ignorance. We should rethink support for development of one-issue and quick-fix models. We need to develop a core capability to handle multipollutant systems and build on that core. We need to continue to attack the basic scientific issues. We need to use the models as investigative tools to help teach us what our collective knowledge is regarding the system. We should also beware of reduced-form models as stand-alones without thorough testing against the full model.

b. We need a more adequate, process-oriented approach to model evaluation. We need to move from outcome testing to process and rate testing, to move from limited episode testing to include the full-range of meteorology. Evaluation needs to be a serious probing of our ignorance. We need to get away from the regression model testing paradigm of "fitting" outcomes. We need to understand the state of our science. We also need a more adequate, continual external scientific review of the models and the science in them. The RADM experience shows how valuable this is, even though it takes time and effort. It is not easy for either the scientist or the policy maker.

c. We need to treat models as scientific tools with strengths and weaknesses, not as yes/no engineering solutions, when using them. We should get away from quick-fix, instant answer attitudes; get away from nonreactive pollutant paradigm; get away from fitting the models (tuning) to outcome levels.

   Current model application guidelines implicitly condone and encourage tuning. Yet, for complex, interacting systems this is a flawed approach to modeling. It totally ignores the fact of compensating errors and ignorance. If we look at common practice, we have obviously failed to move beyond the inert pollutant (first generation) paradigm.

   The tools are only as good as our understanding, our honesty, and our insightfulness (judgment). We need to have the scientific and the decision making communities interact more effectively to change the way models are used in specialized regulatory applications that need to bring our most advanced understanding to bear. We need to work for a more flexible attitude towards the application of models in decision making. We need to be honest about the limitations of our knowledge and our models, yet recognize the power and insight that our collection of knowledge has when embodied in these same models.

d. We need to take advantage of new computer technology for more insightful and easier model interrogation, what-if analyses, and communication of results to the experts and nonexperts. We need to make interrogation easy. We need to make communication more intuitive.

Disclaimer: The information in this document has been funded wholly or in part by the United States Environmental Protection Agency. It has been subjected to Agency review and approved for publication. Mention of trade names or commercial products does not constitute endorsement or recommendation for use.

DISCUSSION

M. WICHMANN-FIEBIG: Concerning uncertainties and ignorance. What is the uncertainty of the emission inventory in the U.S., especially referring to the VOC-split?

R.L. DENNIS: The sulfur dioxide inventory is considered have a reasonable level of uncertainty of less than plus/minus 25 %. The available evidence, including field study comparisons in the middle of the mixed layer, suggests that the uncertainty of nitrogen dioxide emissions is less than plus/minus 50 %. The most recent evidence regarding biogenic isoprene suggests the older estimates (BEIS1) are low by a factor of 3-4 (according to BEIS2). There are still measurement uncertainties to be resolved, but most recent evidence supports the new BEIS2 isoprene emissions estimates for the continental U.S.. There are still concerns about the anthropogenic VOC emissions estimates, with most attention being focused on mobile sources. Evidence of many types, from tunnel studies to driving behavior studies, is still conflicting, leading to a sense that these emissions still have an uncertainty of a factor of two.

R. YAMARTINO: Given that getting the right answer (i.e., ozone predictions) for the wrong reasons has led to past errors in ozone control policies, how do you see your recent RADM findings (i.e., role of hydrogen peroxide, clouds) influencing the current ozone modeling efforts on the OTAG domain?

R.L. DENNIS: I do not expect the recent RADM findings to directly influence the efforts of OTAG (expected to be completed by the end of 1996). The modeling systems and approaches for OTAG have already been defined by the regulatory community. Our goal is for the recent RADM findings and other new findings to influence the modeling and decision making process for the round of modeling expected to follow the OTAG efforts. This next round is termed "mid-course correction modeling," and is expected to occur in the years 1999-2000. The means of influence will be a synthesis of modeling results and implications prepared for the 1998 North American Research Strategy for Tropospheric Ozone (NARSTO) Assessment. Planning for that assessment is under way. Communication links with many principle participants of the OTAG process are being established at this time.

T. IVERSEN: My question is linked to what to me appears as a paradox connected with horizontal and vertical scales of transport. Can you explain why you need a vertical coverage up to the tropopause for the problems you address? And furthermore, if you need such a deep vertical coverage, how can you defend using your quite modest horizontal coverage (compared to hemispheric)?

R.L. DENNIS: The question is a good one. There are two answers depending on whether one takes a modeling-for-applications or a modeling-for-understanding point of view. Because here I am primarily dealing with modeling for understanding I will answer from this perspective. The regional model structure needs to be capable of incorporating the important processes of the system, and in the future to be able to better couple or link to global and hemispheric models. Here we are only talking about the air pollution side of the modeling. With respect to the vertical, clouds are an important exchange mechanisms between the mixed layer and the free troposphere. Deep convection can penetrate well into the free troposphere (especially for the eastern United States), injecting aerosol mass that is radiatively important into this transport regime. We expect a majority of the mass generated in the mixed layer to end up in the free troposphere. Deep convection also means that both the mixed layer and the free troposphere are sources of pollutants (that are then mixed) for precipitation and aqueous processes. Next, the ozone column and clouds are very important relative to actinic flux driving photochemistry. We need to resolve the interactions with clouds throughout the free troposphere. It may be that assessment models can simplify interactions, but the more complete models are needed to test the simplifications and to test the feedbacks between different pollutant problems. With respect to the horizontal, I agree that the horizontal coverage is too modest. We need to be at least at the continental scale. Some important photochemical species have lifetimes at about the continental scale, many important lifetimes are less. Many interactions can be studied at the continental scale, including the influence of boundary conditions that may be driven by hemispheric processes or species involving long, free-troposphere chemical lifetimes to test their importance. Our regional model structure is not yet where we want it to be, but we have to move in increments.

What is the next desirable domain coverage? I believe that continuing with the present vertical extent, but with an explicit coupling with the stratosphere, and expanding our domain to continental scale is where we need to move next with regional modeling. Discussion of errors in the hemispheric models suggest important issues regarding horizontal grid resolution relative to process descriptions and simplification of vertical interactions, especially clouds. These can all be studied with the expansion I suggest and still fit within the current computational constraints. We will then be in a better position, to my mind, to collaborate with hemispheric models and move to the next level of domain scale, as computational constraints recede, to continue to improve our understanding.

METEOROLOGICAL ASPECTS OF OZONE EPISODES IN SOUTHERN ONTARIO: A MESOSCALE MODEL INVESTIGATION

Rama Seshu Tangirala

Air Quality and Meteorology Section
Environmental Monitoring and Reporting Branch
Ontario Ministry of Environment and Energy
125 Resources Road - East Wing
Etobicoke, Ontario, Canada M9P 3V6

INTRODUCTION

Occurrence of high ground level ozone concentrations is of concern in southern Ontario during summer time. Presently, there is a network of about 30 monitoring sites in southern Ontario operated by the Ontario Ministry of Environment and Energy (OMEE) to monitor ground level ozone concentrations for assessing the effects on human health and the damage caused to agriculture, and also to investigate the role of long-range transport on high ozone levels in southern Ontario. Results from this monitoring network have revealed that the Ontario's air quality criterion for ozone (80 ppb 1 hour average) is frequently exceeded during the crop growing season (Yap *et al.*, 1988). Even though there are local sources for ozone precursors in the region it is essential to investigate the relative contributions of long-range transport and local production to the total ground level ozone observed in southern Ontario. This information is necessary for the development and implementation of effective abatement strategies in the region. It is important to note that southern Ontario is bordered by lakes at least on three sides. Interestingly enough, observations generally indicate that the number of days per year on which ground-level ozone over the AAQC (ambient air quality criteria) are more along the lakes in southern Ontario. Hence, any abatement strategy should take into consideration the role of local/mesoscale circulations which would strongly influence the transport and diffusion of contaminants within the boundary layer.

Yap *et al.* (1988), Heidorn and Yap (1986) and several others examined the general meteorological conditions associated with high surface ozone concentrations in southern Ontario. In their analysis with six years of data, Yap *et al.* found an association between primarily three weather regimes and ozone episodes - post high pressure

Air Pollution Modeling and Its Application XI, Edited by
Gryning and Schiermeier, Plenum Press, New York, 1996

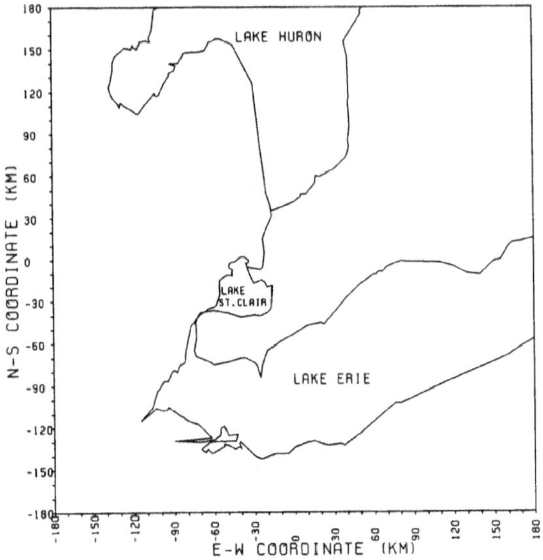

Figure 1: BLFMESO model domain

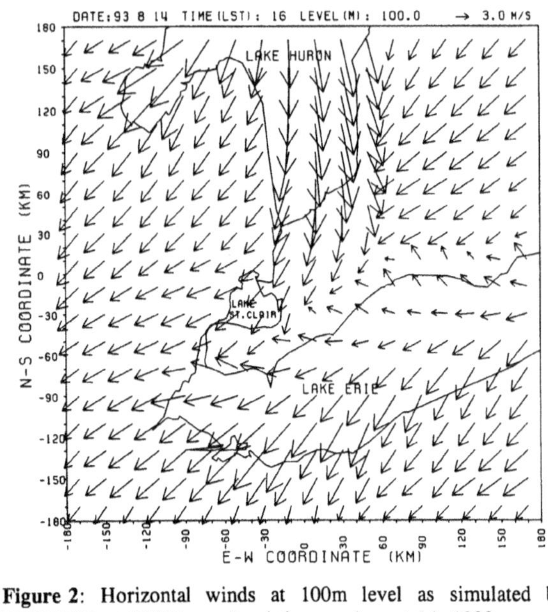

Figure 2: Horizontal winds at 100m level as simulated by BLFMESO at 1600 hours local time on August 14, 1993

ridge/pre-warm front situations, high pressure situations, and weak pressure gradients. Coincidentally, these meteorological regimes can also be very favourable for the development of local lake/land breeze circulations. The OMEE sponsored a mesoscale meteorological observational field study, named the Southern Ontario Oxidant Study - Meteorological Measurements (SOMOS) during the summer of 1993. The SOMOS data, together with the meteorological data from Southeastern Michigan Ozone Study (SEMOS), provide a good data base on lake effects for testing meteorological and oxidant models. Preliminary analysis of the SOMOS dataset indicate the occurrence of local circulations on approximately fifty percent of the days covered by the study period (Sills *et al.*, 1994).

SUMMARY

Daggupaty *et al.* (1994) developed a micro-computer based hydrostatic mesoscale boundary layer meterological model, BLFMESO, for air pollution applications in coastal regions. This hydrostatic mesoscale is capable of simulating the local lake/land circulations reasonably well.

The BLFMESO model is implemented for southern Ontario covering Lake St.Clair, and parts of Lake Erie and Lake Huron in North America (Figure 1) and the model performance is being evaluated against SOMOS dataset (Tangirala and Sills, 1995). The model simulated winds at 100m above ground level at 1600 hours local time are shown in Figure 2. One of the main advantages of the BLFMESO model is that it is operational on a personal computer and takes only about 30 minutes for a 24 hour model simulation. Work is also underway for model intercomparison of BLFMESO with a non-hydrostatic mesoscale model.

In this work the BLFMESO model will be applied to simulate the meteorological aspects of high ground-level ozone concentrations in southern Ontario. This study provides a better understanding of the role of lake/land circulations in ground-level ozone episodes in southern Ontario. Additionally, tentative conclusions can be drawn on the occurrence of relatively higher number of ozone AAQC exceedances along the lakes in southern Ontario.

ACKNOWLEDGEMENTS

Dr. P.K. Misra, Dr. David Yap, and Mr. Duncan Fraser of Environmental Monitoring and Reporting Branch, Ontario Ministry of Environment and Energy, and Prof. Peter Taylor of York University are acknowledged for their useful comments and suggestions during the course of this work.

REFERENCES

Daggupaty S.M., Tangirala R.S., Sahota H., 1994, BLFMESO - A 3-dimensional mesoscale boundary layer meteorological model for microcomputers. *Boundary-Layer Meteorology* **71**:81-107.

Heidorn K. and Yap D., 1986, A synoptic climatology for surface ozone concentrations in southern Ontario, 1976-81. *Atmospheric Environment* **20**:695-703.

Sills D., Taylor P.A., Salmon J., 1994, Southern Ontario Oxidant Study - Meteorological measurements. Final data report. York University and Zephyr North, Ontario, Canada.

Tangirala R.S. and Sills D.M.L., 1995, Application of a mesoscale model to investigate the lake breeze effects on high ozone and oxidant episodes. 11th Symposium on Boundary Layers and Turbulence, March 27-31, 1995, Charlotte, N.C.,U.S.A.

Yap D., Ning T., Dong W., 1988, An assessment of source contributions to the ozone concentrations in southern Ontario 1979-1985. *Atmospheric Environment* **22**: 1161-1168.

OPTIMIZATION OF OZONE ABATEMENT STRATEGIES USING ATMOSPHERIC MODELING

Marion Wichmann-Fiebig[1] and Thomas Schoenemeyer[2]

[1]Northrhine-Westphalia State Environment Agency
Wallneyer Str. 6
D-45133 Essen
[2]Fraunhofer-Institute for Atmospheric Environmental Research
Kreuzeckbahnstr. 19
D-82467 Garmisch-Partenkirchen

INTRODUCTION

According to EU directive 92/72/EEC[1] ozone concentrations of more than 180 µg/m³ should be prevented to protect health and vegetation. However this threshold is frequently exceeded in Germany and other European countries. Peek values in Northrhine-Westphalia in 1994 and 1995 even exceeded 300 µg/m³. As the first abatement strategy for the whole of Germany the German Ozone Act[2] is legally binding since July 26th, 1995. It prescribes measures for the next day if concentration values exceed 240 µg/m³ at three sites at least. Use of vehicles without catalytic converters is prohibited in this case, yet several exceptions for commuters, delivering of important goods etc. are admitted.

Up till now, however, it is not known what will be the effects of this measure and if there are other perhaps more efficient abatement strategies. Especially the effect of speed limits is broadly discussed in Germany. Judging emission reduction effects on ozone concentrations cannot be done in a simple way because of the numerous non-linear chemical reactions involved. Even results from large scale experiments are equivocal because no "zero-case" information is available, i.e. it is not known what concentrations would have been observed if no measures had taken place. Therefore the only way to compare different abatement strategies under identical conditions is modelling.

CONCEPT OF THE NORTHRHINE-WESTPHALIAN OZONE STUDY

To study the effects of short term ozone abatement strategies in general as well as of the German ozone act in special ozone-concentrations were calculated for an episode of four days where high ozone-concentrations were observed. For the simulation of the meteorological parameters the non hydrostatic model METRAS[3] was used. Initialisation took place every 24 hours based on rawinsonde data. Ozone formation due to photochemical reactions was calculated by the CIT-photochemical-model[4] thereby using the gasphase-mechanism of RADM II[5] and an improved algorithm to consider biogenic emissions[6]. The model domain consisted of 112*117 grid cells with a horizontal resolution of 2 km. It included Northrhine-Westphalia and parts of the neighbouring countries. The lowest model layer in CIT had a depth of 70 m. Concentrations were calculated as half an hour mean values. The episode considered lasted from July 24th to 27th 1994. In its beginning and its end concentrations were dominated by advection of ozone and its precursors while in between regional emissions mainly affected the ozone values.

To get information about the transport of ozone and its precursors over the boundaries of the model domain a double-nesting was performed. For the large scale runs which where done using the hydrostatic European Community model EURAD[7] a resolution of 36 km respectively 9 km was chosen. The EMEP emission inventory was used for this large scale runs.

Topography and land use data were aggregated from a detailed inventory[9] with a resolution of 110*150 m^2. Concerning the emission inventory on the 2 km scale we distinguished between emissions from road traffic, from facilities subject to license and other sources like small trade and private consumption. Total VOC and NO_X emissions of these groups are shown in Table 1. Due to the industrial and population structure in Northrhine-Westphalian emissions concentrate in the Rhine-Ruhr area (Figure 1). Because traffic is the main source other relevant emission areas are cities with more than 100,000 inhabitants. With the exception of highways VOC to NO_X-ratios on a mass base are in the order of one to five in urban and above five in rural areas. Highways show ratios below one. Due to these ratios reduction of VOC emissions is expected to be more efficient than reduction of NO_X emissions.

Table 1. The emission inventory of Northrhine-Westphalia for VOC and NO_X for July 1994.

	VOC in kg/month	NO_X in kg/month
Road traffic	1.7 10^7	2.2 10^7
Facilities subject to license	0.3 10^7	2.0 10^7
Small trade/ private consumption	1.2 10^7	0.1 10^7
Sum	3.2 10^7	4.3 10^7

Because no detailed emission inventory was available for the part of the model domain which does not belong to Northrhine-Westphalia EMEP data were also used for those areas. However, comparison with the detailed data base showed the EMEP data to be higher up to a factor of ten. As no other data were available the EMEP inventory had to be used nevertheless knowing some artificial influences on concentrations might appear in our results.

The following scenarios for the state of Northrhine-Westphalia were examined concerning their ozone abatement efficiency:

- *Scenario 1:* Use of cars without catalytic converters is prohibited everywhere in Northrhine-Westphalia.
- *Scenario 2:* Use of cars without catalytic converters is prohibited in the Rhine-Ruhr area and in cities with more than 100,000 inhabitants.
- *Scenario 3:* Speed limit of 90 km/h on highways and 80 km/h outside of cities.

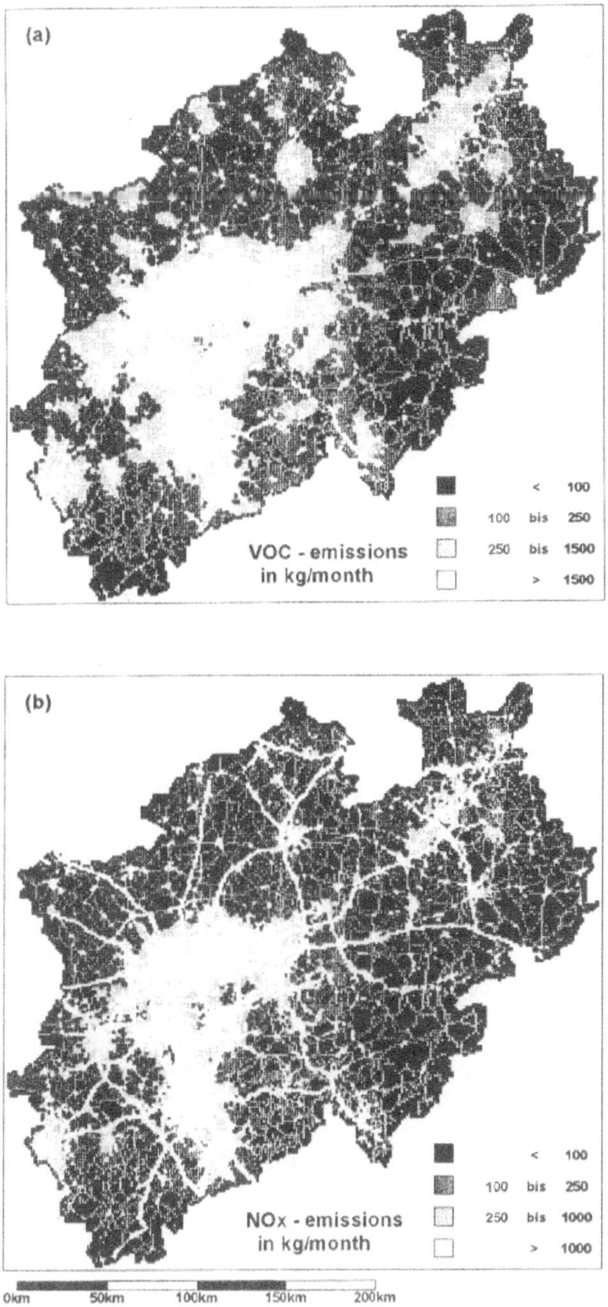

Figure 1. Spatial distribution of (a) VOC and (b) NO_X emissions in Northrhine-Westphalia for July 1994.

The total emission reduction of each scenario can be seen from Figure 2. The German ozone act is supposed to correspond closest to scenario 2 provided that the number of commuters is not too high. In addition to these scenarios the effects of emission reductions concerning industry and small trade are considered which correspond to 12% VOC and 14% NO_x with respective to the total emissions.

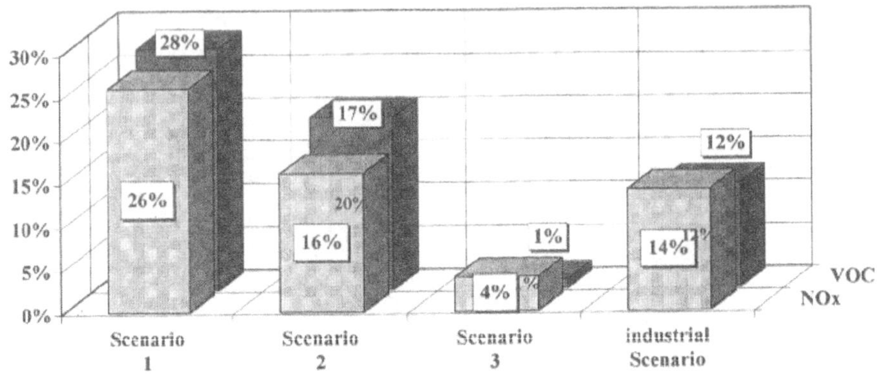

Figure 2. Reduction effects of the scenarios which were examined on the VOC respectively NO_x emissions in per cent of total emissions.

VALIDATION OF MODEL RESULTS

Concerning validation of the model results it has to be distinguished between results of the meteorological and of the photochemical model. As mentioned above the database for the meteorological model was rather poor. Therefore on some days significant differences were found between the observed and the calculated wind field. A mass consistent interpolation of the observed wind data might have resulted in better agreement with the observations. However, usually poor turbulence data result from those wind fields because no equilibrium between turbulence and mean wind is considered. Another shortcoming of the poor database were the insufficient details of the nightly inversion structure in the model results. Agreement with observations might have been improved by fitting the input parameter continuously to the observed structures which however would have taken much computer time and was not the main point of the study.

Observed and calculated immission data were compared at two sites, namely one with urban and another with rural characteristics (Figure 3). Poor agreement on July 25th is caused by severe precipitation at both sites which is not included in the model. On July 26th and 27th dynamics of ozone formation and peak values in general are well reproduced by the model. However, ozone destruction during night-time at the rural site is overestimated by the model. Modelled maximum values on July 26th were found somewhat downwind the urban observation site because of transport effects.

Comparison of spatial distribution in the afternoon of July 26th and 27th reflects the deficiencies of the model approach. Because of the rather high emissions outside of Northrhine-Westphalia ozone is destroyed near this artificial source region. Farther downwind from those regions ozone production is overestimated. Because the wind direction differs in model results and observations the region of maximum concentration does not correspond. However, as peak values and daily dynamics are performed well by the model, results are suitable for a comparison of the scenario results with the base run.

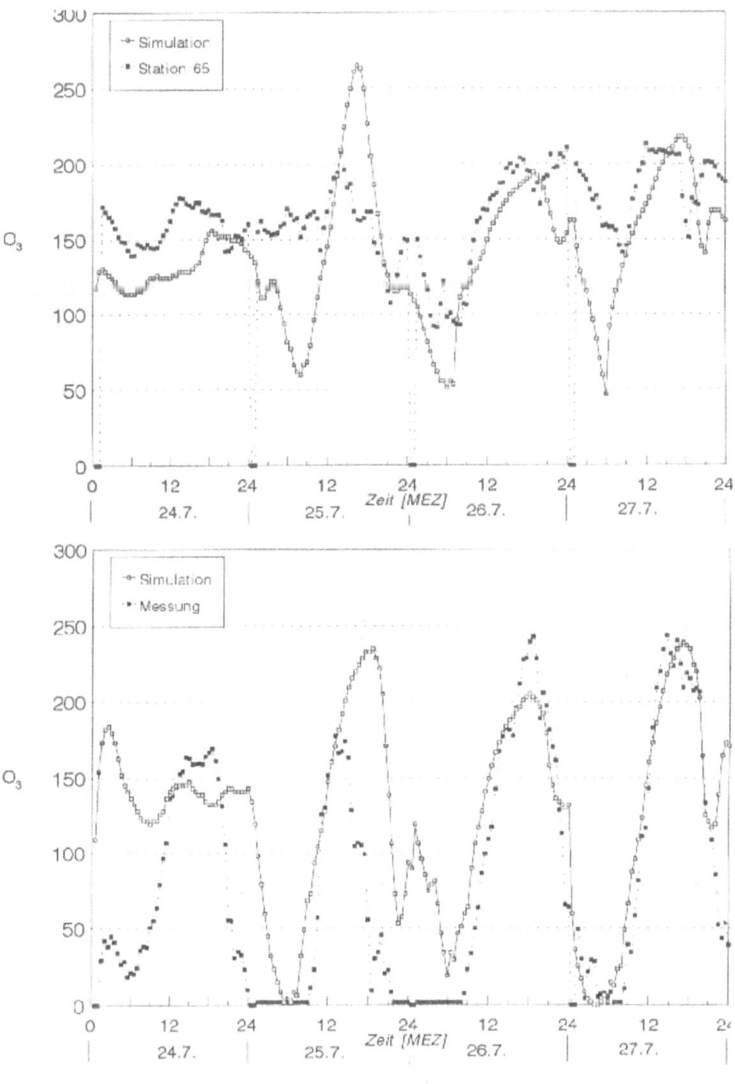

Figure 3. Time series of modelled and observed ozone concentrations in µg/m³ at (a) a rural and (b) an urban site.

COMPARISON OF DIFFERENT OZONE ABATEMENT STRATEGIES

The concept of judging the scenario results is threefold. First, model results have been evaluated with respect to the daily frequency of exceeding a threshold of 180 µg/m³ respectively 240 µg/m³ as half an hour mean values with reference to all model grid cells each of a size of 2x2 km². Second, daily maximum values for different abatement strategies were compared. Third, the spatial ozone distributions were related.

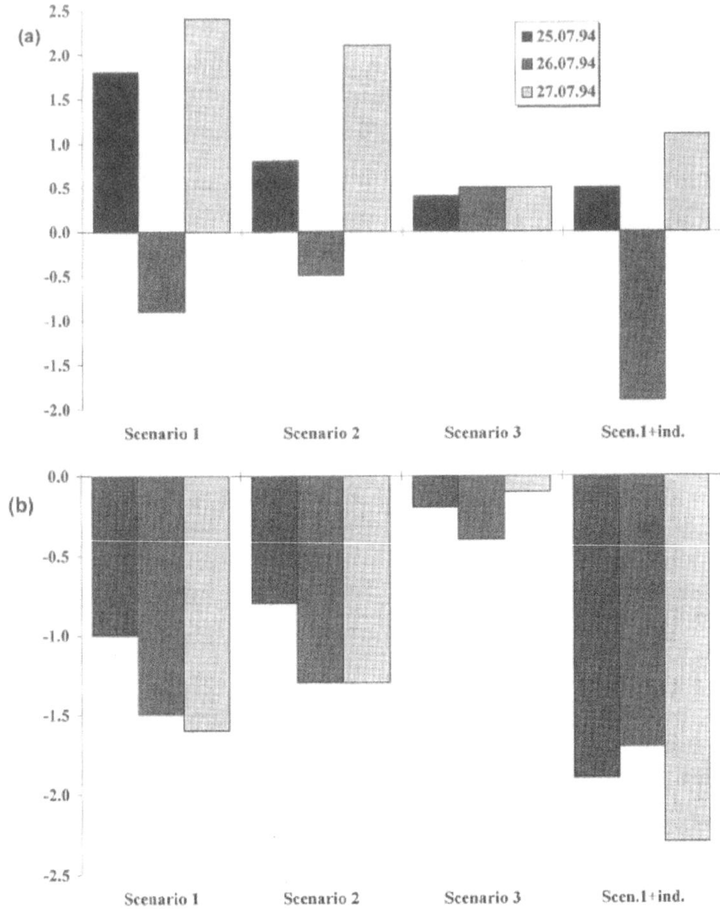

Figure 4. Reduction of hours during which (a) 180 µg/m³ and (b) 240 µg/m³ are exceeded on July 25th, 26th and 27th, respectively, according to the different scenarios considered.

Concerning the exceeding of the 180 µg/m³ threshold only small differences to the base run could be found from the model results (Figure 4a). The reductions even lead to growing frequencies because of partly NO_X determined chemics. Scenario 3 also shows negligible abatement effects with respect to the 240 µg/m³ threshold (Figure 4b). While the area-hours during which 240 µg/m³ ozone are exceeded amount to 5,3% of the total area-hours on July 27th in the base run they are reduced to 3,7% respectively 4,0% by Scenario 1 and 2. Including also industrial emission reductions into scenario 1 nearly doubles the effect. As can be seen from Figure 4 effects differ form day to day. This is caused by the changing meteorological situation. Details still have to be analysed.

Peak values of the base run and the scenarios were compared at three sites. Maximum abatement effects of 17% could again be achieved by scenario 1 in combination with industrial measures. Due to scenario 2 peak values are lowered by approximately 10%. Scenario 3 showed no effect. Looking at the spatial ozone distribution at 5 p.m. on July 26th the base run shows large areas where ozone concentrations exceeded 240 µg/m³ (Figure 5a). At the time considered this threshold was neither exceeded in scenario 1 nor in scenario 2. The combination of scenario 1 with industrial measures (Figure 5b) also lowered the exceeding of 180 µg/m³ significantly. Scenario 3 did not affect the spatial distribution.

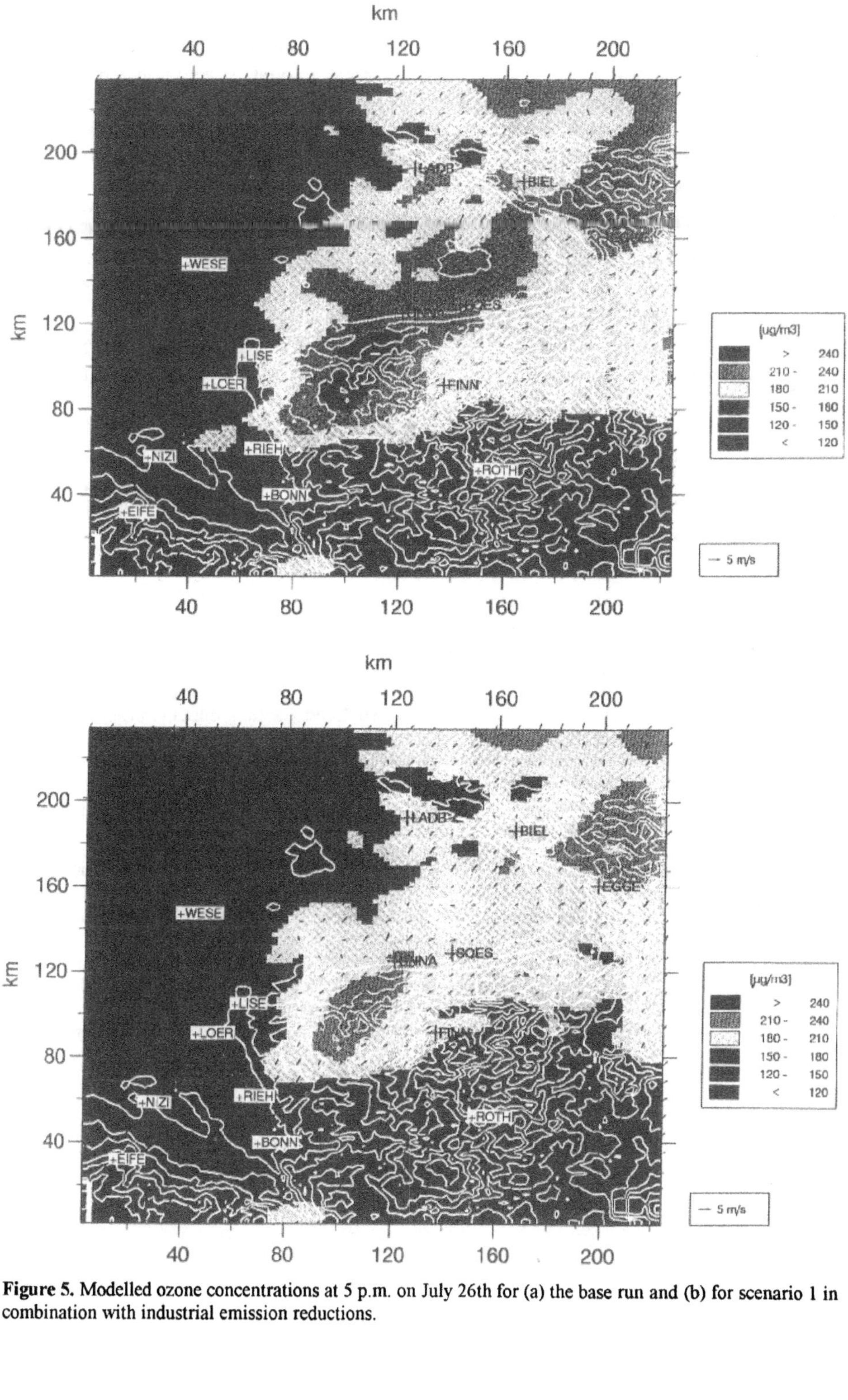

Figure 5. Modelled ozone concentrations at 5 p.m. on July 26th for (a) the base run and (b) for scenario 1 in combination with industrial emission reductions.

RESULTS AND DISCUSSION

The model studies performed showed to be a suitable tool to scrutinise different ozone abatement strategies. As to be expected from other model studies effects are largest for high concentrations and become negligible for concentration values near 180 $\mu g/m^3$. From the different scenarios it was found that not only the total amount of emission reductions is essential for the resulting ozone abatement but that also the spatial distribution of measures is important. This can be seen by comparing scenario 2 to scenario 1. Abatement effects of scenario 2 did not differ significantly from those of scenario 1 though the emission reduction was lowered by 10 % with respect to the total emissions. However, measures of scenario 2 are concentrated in urban areas which showed a high ozone formation potential. As among the scenarios considered scenario 2 corresponds closest to the German Ozone Act it may be concluded that thereby a suitable measure for the abatement of high ozone concentrations is prescribed. Considering industrialised regions like Northrhine-Westphalia lager effects can however be achieved if industrial emission reductions also were performed.

The often discussed speed limit of 90 km/h on highways does not affect ozone concentrations in a notable way. The effect of a more rigorous speed limit have not been studied up till now.

To prevent high ozone concentrations measures should be taken as early as possible. To this purpose a real time prognostic model based on observations and meteorological forecast is needed. Moreover to perform most efficient industrial measures those VOC-components which show the highest ozone formation potential should be identified and purposefully be removed.

REFERENCES

1. EEC, *Information Document concerning Air Pollution by Ozone. Overview of the situation in Europe during summer 1994.* Expert Group on Photochemical Pollution, working document.
2. German Ozone Act, BImSchG in der Fassung vom 14.05.90 (BGBI. I S.880), zuletzt geändert durch ÄndG vom 19.07.1995 (BGBI. I S.930), § 40 a-e.
3. K.H. Schlünzen, *Das mesoskalige Transport- und Strömungsmodell METRAS - Grundlagen, Validierung, Anwendung* - , Thesis, Hamb.Geophys.Einzelschrift. A88, 139pp, 1988.
4. G.J. McRae, J.H. Seinfeld, Development of a second-generation mathematical model for urban pollution, II, model performance evaluation, *Atmos.Environ.*, 17, 501-523, 1984.
5. W.R. Stockwell, P. Middleton, J.S. Chang, The second gereration regional acid deposition model chemical mechanism for regional air quality modeling, *J.Geophys.Res.*, 95, 16343-16367, 1990.
6. R. Steinbrecher, R.Rabong, *Isoprenemission von Fichte: Freilandexperimente im montanen Ökosystem*, Tätigkeitsbericht des Fraunhofer Institutes für Atmosphärische Umweltforschung in Garmisch-Partenkirchen, 1993.
7. H.Hass, H.J. Jakobs, M. Memmensheimer, G. Piekorz, A. Ebel, *Ermittlung des Eintrags von Ozon und Ozonvorläuferstoffen nach NRW.* Bericht im Auftrag des Ministeriums für Umwelt, Raumordnung und Landwirtschaft des Landes Nordrhein-Westfalen, Köln 1994.
8. G. Manier, Topographiekataster für die alten Bundesländer, Institut für Meteorologie der TH Darmstadt, 1994.

DISCUSSION

S.T. RAO: Since the initial conditions affect the model simulations, are the model results for the first two days discarded?

M. WICHMANN-FIEBIG: Model simulations started on July 24th, while results were interpreted for July 25th to 27th.

S.T. RAO: How are the boundary conditions changed when emissions are changed? If the boundary conditions are not varied, the concentrations predicted by the model near the boundaries (also in the interior) are affected. Therefore, the relative efficiency of regional versus local controls cannot be really discerned.

M. WICHMANN-FIEBIG: The aim of the study was to examine effects of emission reductions confined to the State of Northrhine-Westphalia (NRW). As NRW is completely included in the model domain no change of boundary conditions was made. However, the scenarios affected regions of different extension inside of NRW so regional (scale 250 km) versus local (scale 50 km or less) controls could be judged.

B. BORNSTEIN: While model tuning (of formulation parameters) from run to run is discouraged the adjustment of the external (large scale and soil) parameters required for IC (initial conditions) and upper BC (boundary conditions) is necessary as such parameters are frequently "unknown" during the slow speed anticyclonic conditions during which mesoscale forcing are most important.

M. WICHMANN-FIEBIG: In the current study agreement with observed ozone concentrations was significantly improved by using additional meteorological information from the German Weather Service. Instead of only one rawinsonde data with a spatial resolution of about 30 km were used every six hours to update the meteorological parameters. Thereby local transport features gained influence on the ozone distribution. No further tuning - especially none concerning the photochemical modelling - was performed.

D. STEYN: It is disturbing to note that you conclude no effect on episodes having low concentrations ($180 \mu g/m^3$ in your case) since recent epidemiological studies show measurable effects due to ozone at concentrations as low as $10\ \mu g/m^3$.

M. WICHMANN-FIEBIG: Concentrations of the air masses transported into the model domain from outside were already higher than $180\ \mu g/m^3$. As emission control was assumed to be restricted to an area not larger than 250 x 250 km^2 no effects concerning those "background" concentrations could be expected. However, emissions reduction for a larger area, for example for the whole of Germany, should affect lower concentration levels as well as higher ones.

NUMERICAL SIMULATIONS OF THE WIND FLOW AND OZONE FORMATION DURING THE HEILBRONN OZONE EXPERIMENT

N. Moussiopoulos[1], P. Sahm[1],
C. Kessler[2], R. Kunz[2], C. Schneider[2] and T. Vögele[2]

[1] Laboratory of Heat Transfer and Environmental Engineering, Aristotle University Thessaloniki, 54006 Thessaloniki, Greece
[2] Institut für Technische Thermodynamik, Universität Karlsruhe, 76128 Karlsruhe, Germany

ABSTRACT

The "Heilbronn Ozone Experiment" was performed in the period 23-26 June 1994 with the intention to investigate if peak ozone concentrations during summer smog periods may be reduced by the aid of short term local scale interventions. The measures taken were estimated to lead to reductions of road traffic emissions in the study area by up to 40%. With regard to industrial emissions only smaller reductions were achieved.

Extensive meteorological and air quality measurements carried out during the experiment provide valuable information on the situation prevailing. Model simulations with the EUMAC Zooming Model (EZM) were performed to facilitate assessing the effectiveness of the emission reduction interventions.

The evaluation of both measurements and simulation results reveals that the interventions lead to air quality improvements with regard to the primary pollutant concentrations but only to minor changes in peak ozone levels. The main conclusion from the Heilbronn ozone experiment is that concerted large scale interventions to the primary pollutant sources are needed for decisive reductions of peak ozone concentrations in Central Europe.

THE HEILBRONN OZONE EXPERIMENT

Ozone episodes with exceedances of 180 $\mu g/m^3$ frequently occur in the period between March and September in the state of Baden-Württemberg, Germany. Moreover, the background ozone levels at remote areas of this state were found to increase by 5 $\mu g/m^3$ per year between 1984 and 1991 [1]. The "Heilbronn Ozone Experiment" was performed in the period 23-26 June 1994 in the industrialized area of Heilbronn with the intention to investigate, to what extent ozone levels may be reduced by the aid of short term local scale interventions.

With a total population of more than 130,000 inhabitants, Heilbronn and its neighbour Neckarsulm are located in the Neckar valley. Both cities are a densely populated area between Stuttgart, the capital of Baden-Württemberg in the south, and Heidelberg and Mannheim in the northwest where the Neckar joins the river Rhein. The Heilbronn area is traversed by major motorways, both in the north-south and in the west-east directions. The marked smog region Heilbronn/Neckarsulm was considered as suitable for the experiment, as it is an urban agglomeration within a basin where ozone precursors are emitted because of both dense traffic and manifold industrial activities.

Traffic restrictions were applied to passenger cars not classified as "clean" or low-emitting in an area of approx. 45 km^2 comprising the urban agglomerations of Heilbronn and Neckarsulm ("restriction area"). Other emission reduction interventions (speed limits, voluntary actions by the industry) were implemented in an area of approx. 400 km^2 ("speed limit area"). In this area all emissions of traffic, industry and small consumers were inventorized. Extensive meteorological and air quality measurements were performed in a considerably larger area extending over approx. 910 km^2 ("experimental area").

In order to estimate the reduction of the emissions during the experiment, weekday emissions on comparable days without interventions were calculated. Reference weekend emissions were estimated using relevant ratios of weekday/weekend emissions obtained independently of the experiment.

The main findings of the emission calculations may be summarized as follows: Traffic related NO_x and VOC emissions in the restriction area were reduced by up to 40% during the whole period. The NO_x reductions may be attributed to both traffic prohibition within the restriction area and the speed limit of 60 km/h on motorways. They were mainly caused by the reduced exhaust of light and heavy duty trucks. Reduced exhaust of passenger cars was the main reason for the VOC reduction which was achieved to almost 100% by traffic restrictions. The total NO_x and VOC emissions in the speed limit area could be reduced by 15 to 20% during weekdays and up to 25% during the weekend.

Attempts to estimate the air quality changes related to the experiment were done using a regression analysis of measurements from stations within and in the surroundings of the experimental area. By restricting the analysis to comparable periods (i.e. high pressure situations), meteorological influences on the concentrations as well as differences in the pollution load of advected air masses could be minimized. It is evident, however, that this approach is inherently approximative. As the main result of the regression analysis, it appears that the traffic restrictions led to NO_x concentration reductions by 30% on weekdays and by 15% on Sunday.

The experimental evidence reveals that peak ozone concentrations in the restriction area remained unchanged during the experiment. Three-dimensional budget calculations based on the measurements show even a slight increase of the ozone concentrations in the experimental area. However, only a small fraction of the urban plume is within the experimental area and additional measurements at more distant locations would be needed for a complete picture of the ozone formation in this plume.

The meteorological and air quality measurements carried out during the experimental period provide valuable information on the situation prevailing during the experiment. Yet, they do not allow completely assessing the effectiveness of the emissions reduction interventions. This was the reason for performing accompanying model simulations with the EUMAC Zooming Model (EZM).

MODEL SIMULATIONS

The EZM was applied for the period 23-27 June 1994, i.e. the four day period of the "Heilbronn Ozone Experiment" and the following day (measurements are available for this day as well). In the present paper results are presented for the experimental period only. A summery high pressure system in the first days of the considered time period favoured the occurrence of a photosmog episode.

The wind field was computed with the nonhydrostatic mesoscale model MEMO [2,3]. For the computation of the concentration fields the dispersion model MARS [4] was used in conjunction with the chemical reaction mechanism KOREM. The latter represents a modified version of the compact reaction mechanism of Bottenheim and Strausz [5].

In order to take into account all relevant orographic influences on the flow field, a triply nested system (coarse, medium and fine grid simulations) based on the expanded radiation boundary condition was applied. The coarse grid simulations covered an area of 200 × 200 km^2 including most of the Black Forest and the Schwäbische Alb in the south of the domain and the Rhine valley and the Odenwald in the northwest. Initial and lateral boundary values for the coarse grid were derived from upper air soundings at Heilbronn-Wertwiesenpark. The high resolution fine grid covered an area of 20 × 20 km^2 with a horizontal grid spacing of 500 m.

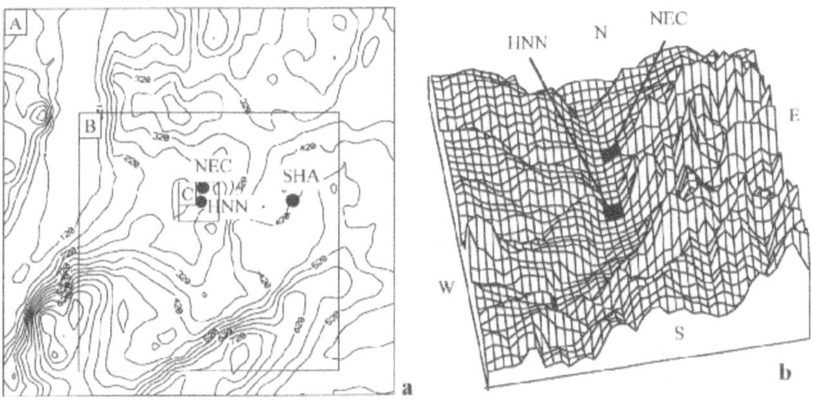

Figure 1: (a) Coarse grid domains for the wind field (A) and dispersion (B) calculations, fine grid domain (C) and locations of Heilbronn (HNN), Neckarsulm (NEC) and Schwäbisch Hall (SHA). (b) Perspective view of fine grid domain and locations of Heilbronn (HNN) and Neckarsulm (NEC).

Lateral boundary concentrations for the dispersion simulations on the fine grid were generated from an additional dispersion simulation on a coarse grid covering an area of 130 × 130 km^2 at a resolution of 5 km. The large scale boundary concentrations for NO_2 and ozone on the coarse grid were derived from results of simulations with the LOTOS model [6]. The necessary emission input data for the coarse and the fine grid dispersion simulations were derived from suitable inventories prepared at the Universities of Stuttgart and Karlsruhe, respectively [7,8]. The coarse grid domains for both wind field and pollutant dispersion calculations, the fine grid domain and the locations of the cities of Heilbronn (HNN), Neckarsulm (NEC) and Schwäbisch Hall (SHA) are shown in Figure 1.

Additional dispersion simulations were performed for the hypothetical case of no emission reductions during the period of the experiment in an attempt to quantify the effectiveness of the emission reduction interventions.

SIMULATION RESULTS

In the following, selected results of the simulations with the EZM are presented and compared with observations. A detailed discussion of the simulation results can be found elsewhere [9]. In view of the limitations of any comparison of this kind, the predictions for both the wind field and the concentrations are in satisfactory agreement with observations. At most stations, deviations between predicted ozone peaks and corresponding observations do not exceed 10 ppb.

The comparison of measured and computed wind speed and direction at Heilbronn (cf. Figure 2) shows a good agreement for all days except for the third one. This deviation is associated with a front which passed through the experimental area during the night between June 24 and 25 without being adequately resolved in the model simulation. The agreement between observations and model results increases from the coarse to the fine grid simulation. This proves that by enhancing the spatial resolution in the model, the representation of physical processes may be significantly refined.

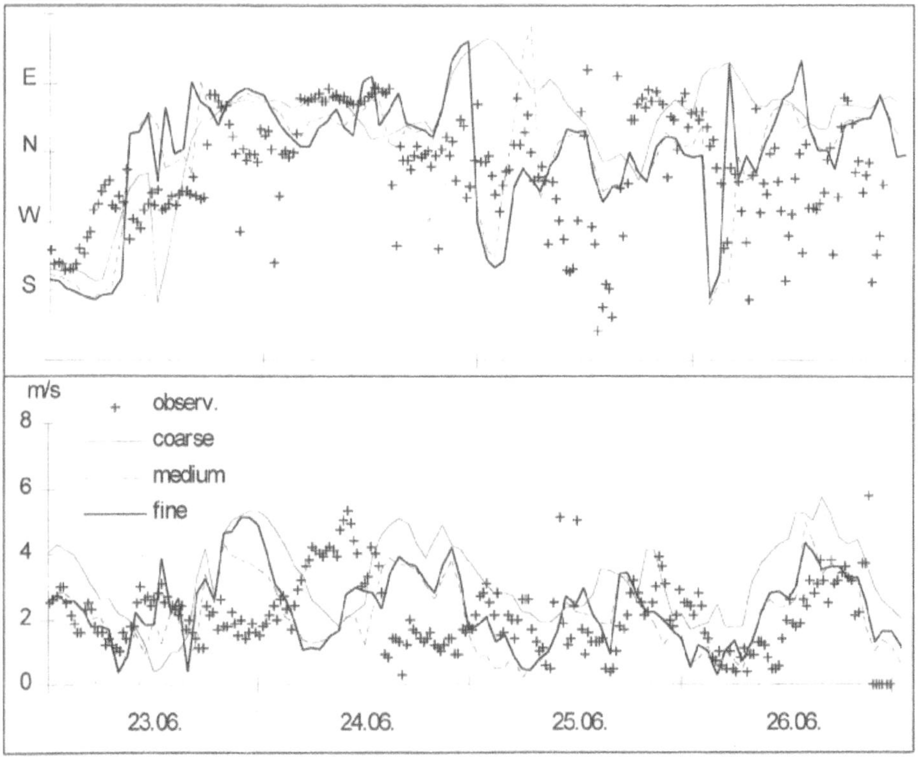

Figure 2: Predicted and observed wind direction and wind speed at Heilbronn.

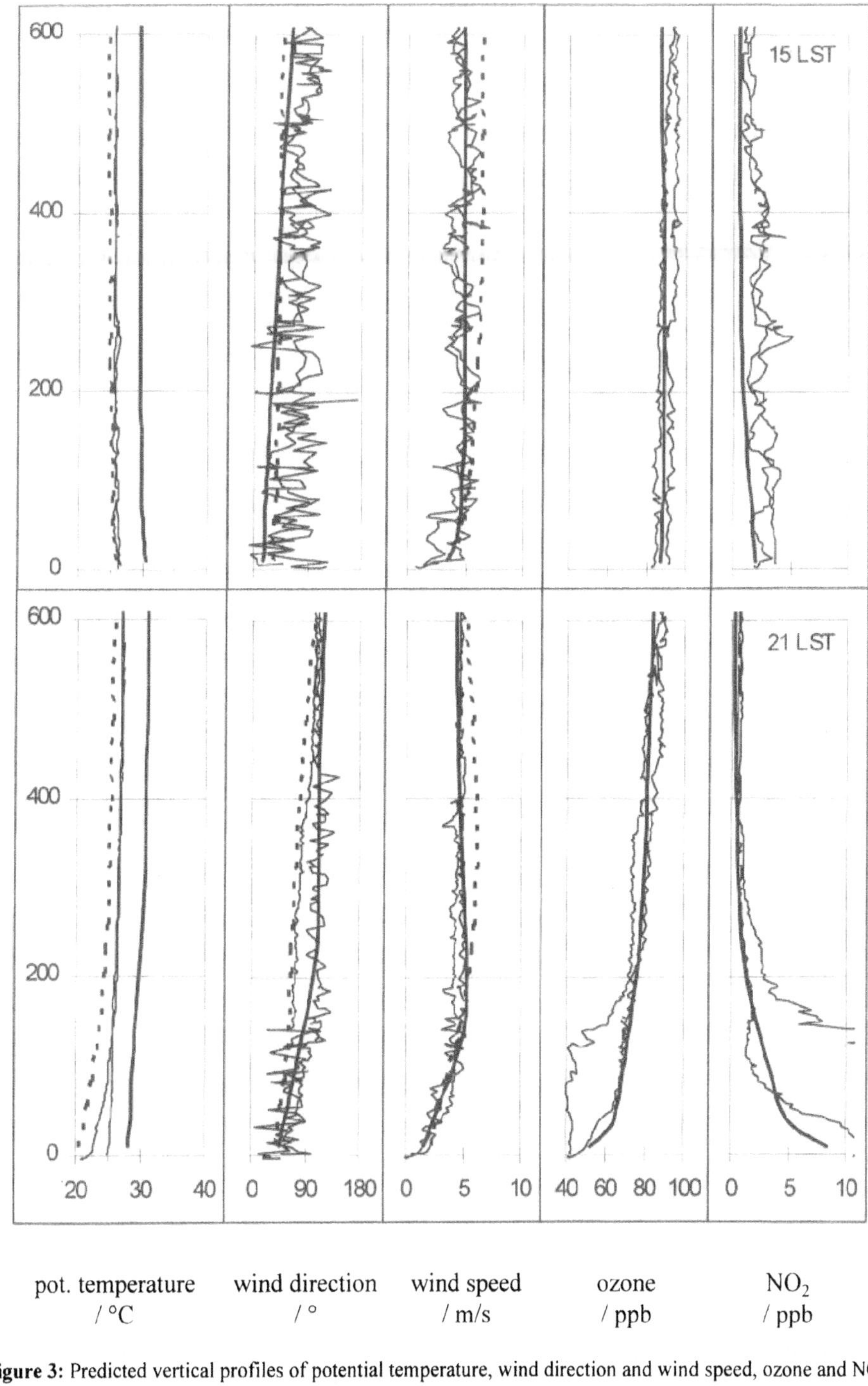

Figure 3: Predicted vertical profiles of potential temperature, wind direction and wind speed, ozone and NO_2 concentrations at Heilbronn Wertwiesenpark at 15 LST (top) and 21 LST (bottom) on July 24, 1994 compared with observations (height in meters). Solid thick lines: model results, dotted thick lines: model results with modified soil parameters, thin lines: observations.

In Figure 3 predicted vertical profiles of potential temperature, wind speed and wind direction as well as the ozone and NO_2 concentrations are compared with corresponding measured values obtained by tethered balloon ascents at Heilbronn Wertwiesenpark at 15 and 21 LST on June 24. Wind as well as concentrations resulting from the fine grid application of the EZM are in very good agreement with the measured profiles. The potential temperature is systematically overpredicted, even at higher elevations (solid line). Sensitivity studies revealed that the air temperature is rather sensitive to changes in the temperature and the moisture of the soil. Using modified values for the latter quantities, the simulation result for the air temperature (dashed line) approaches very closely the observation thus also leading to an improvement of the wind field prediction (no over-prediction of the near ground wind speed). The slight increase of the upper air speed in this simulation is a consequence of minor modifications in the series of upper air soundings used to derive boundary conditions (see above).

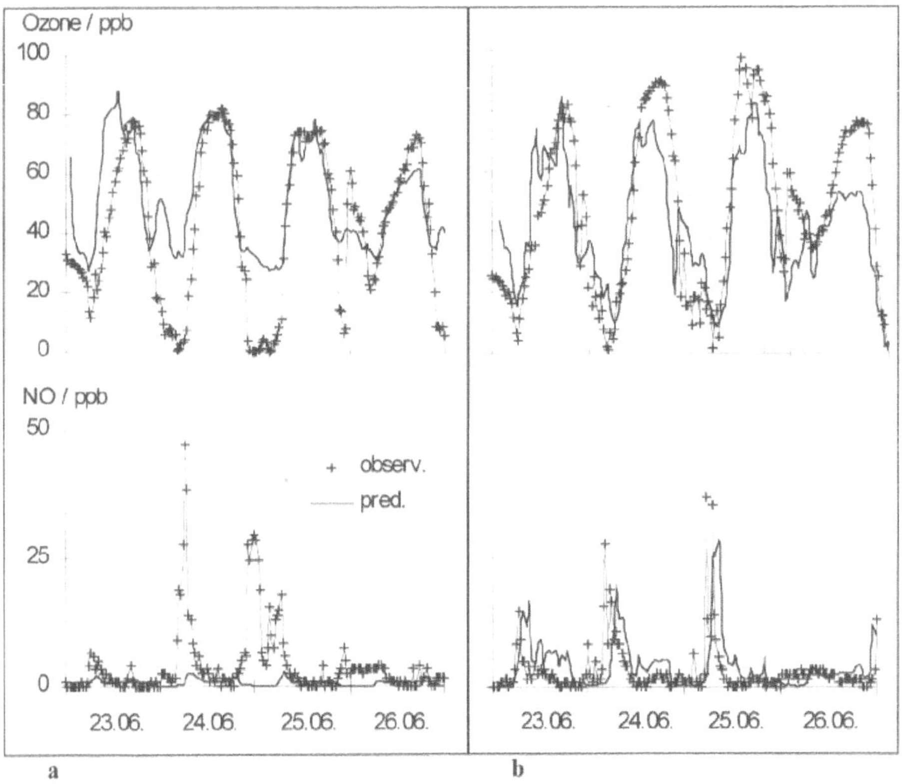

Figure 4: Predicted concentrations of ozone and NO at Schwäbisch Hall (a) and Heilbronn (b) compared with observations.

In Figure 4 the diurnal variations of the ozone and the NO concentrations are given for Schwäbisch Hall (a) and Heilbronn (b). At both locations predictions for the diurnal cycle and peak value of the ozone concentration are in very good agreement with corresponding observations for June 23 to 26. The observed ozone depletion early in the morning is, however, predicted only at Heilbronn. This depletion is caused by NO peak emissions during the morning rush hour. Given the high reactivity and short lifetime of NO, a high spatial resolution is a prerequisite for a reliable prediction of its dispersion. The displayed simulation results for the NO concentration at Heilbronn originate from the fine grid simulation (horizontal grid spacing: 500 m). This explains why the observed NO peaks during the morning hours are satisfactorily reproduced by the model. Only coarse grid simulation results are available for Schwäbisch Hall (spatial resolution of 5 km) and therefore the NO concentration peaks could not be adequately resolved. As a consequence, ozone concentrations at Schwäbisch Hall are overestimated in the morning hours. In the light of the above, it can be stated that the EZM allows satisfactorily predicting the ozone concentrations, provided that an adequate spatial resolution is used.

During the experiment ozone concentrations reached their maximum at about 16 LST of each day. Figure 5a gives an impression of the spatial distribution of the coarse grid results for the ozone concentration at 16 LST on June 24 (the fine grid domain is marked by a square). At that time north-northeasterly winds were prevailing (cf. Fig. 2). Peak ozone concentrations occur about 60 km downwind of the restriction area. In spite of the considerable emission reductions, NO emissions still cause a depletion of ozone near the motorways and within the densely populated restriction area.

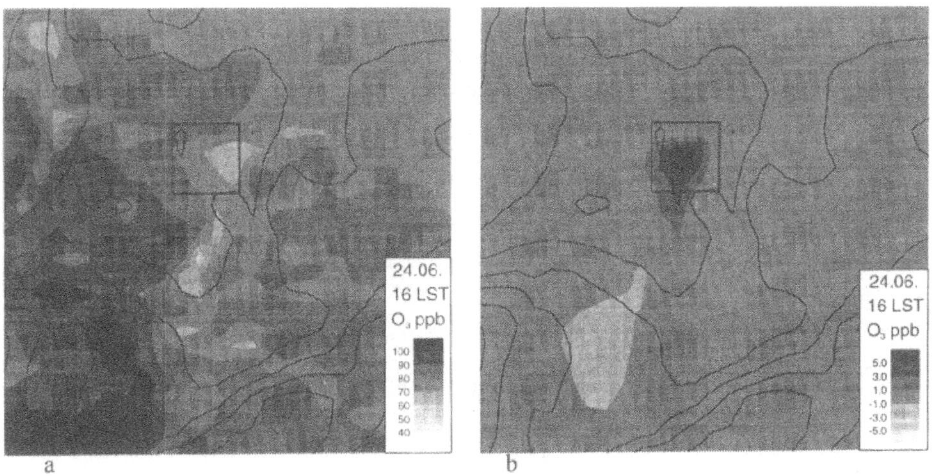

Figure 5: Coarse grid results for the ozone concentration (a) and the difference of ozone concentration with and without interventions (b) at 16 LST on June 24. The frame indicates the location of the fine grid.

As mentioned above, additional dispersion simulations were performed for the hypothetical case of no emission reductions during the experiment. A comparison of the results obtained in the two simulations allows estimating the changes in air pollution levels due to the implementation of the short term local scale interventions. As an example, Figure 5b shows the changes in ozone concentration at 16 LST on June 24 caused by the interventions. In the restriction area and in its vicinity, reduced NO emissions lead to increased ozone concentrations. Downwind of the restriction area, however, the situation is reversed: At this distance, NO has been already converted into NO_2. As a consequence, the emission reductions lead to a decrease of ozone concentrations far away from the source.

CONCLUSIONS

The presented results of the EUMAC Zooming Model for the wind flow and the dispersion and transformation of air pollutants in the Heilbronn area during the period 23-27 June 1994 agree fairly well with measurements performed during the "Heilbronn Ozone Experiment".

Sensitivity studies reveal that the predicted air temperature and consequently the wind speed near the ground are rather sensitive to even minor changes of the temperature and the moisture within the soil.

For a proper prediction of the ozone concentrations it is essential to model the dispersion of precursor substances at an adequate spatial and temporal resolution. Especially NO, which is a highly reactive species with a short lifetime, requires a high spatial resolution.

Additional dispersion calculations performed for the hypothetical case of no emission reductions lead to the conclusion that concerted large scale interventions are necessary to decisively reduce peak ozone concentrations in Central Europe.

ACKNOWLEDGMENT

Major parts of the studies related to the Heilbronn ozone experiment were performed in a collaboration of the Aristotle University Thessaloniki with the Universität Karlsruhe. The authors acknowledge the financial support granted by the Ministry of the Environment of Baden-Württemberg. They also wish to thank Dr. U. Neu, Bern, for his courageous work in editing the Final Report on the scientific evaluation of the experiment.

REFERENCES

1. H. Mayer and J. Schmidt, Trendanalyse von Immissionsdaten in Baden-Württemberg, LFU Karlsruhe, (1992).
2. N. Moussiopoulos, P. Sahm, Th. Flassak and D. Berlowitz, Simulations of the wind field in Athens with the nonhydrostatic mesoscale model MEMO, *Environmental Software 8* 29:42 (1993).
3. R. Kunz and N. Moussiopoulos, Simulation of the wind field in Athens using refined boundary conditions, *Atmos. Env.* in print (1995).
4. N. Moussiopoulos, P. Sahm and Ch. Kessler, Numerical simulation of the photochemical smog formation in Athens, Greece - A case study, *Atmos. Env.* in print (1995).
5. J.W. Bottenheim and O.P. Strausz, Modelling study of a chemically reactive power plant plume, *Atmos. Env.16* 85:97 (1982).
6. P.J.H. Builtjes, The LOTOS-Long Term Ozone Simulation-project, Summary report, TNO-report IMW-R 92/240, Delft, (1992).
7. R. Friedrich, personal communication, (1994).
8. Ch. Schneider, Aufbereitung und Überprüfung der Emissionsdaten der Episode 22.6.-29.6.1994 und Erstellung eines Gesamtemissionskatasters, Abschlußbericht des Teilprojektes im Rahmen des Modellversuchs Regionale Ozonminderungsmaßnahmen, (1995).
9. U. Neu, Ozonversuch Neckarsulm/Heilbronn - Wissenschaftliche Auswertung, Umweltministerium Baden-Württemberg, (1995).

DISCUSSION

G. VERVER: If I am only interested in results on scales of your coarse resolution, would I need calculations on the high resolution and area average the results to the coarse grid?

R. KUNZ: No, in order to simulate processes for a scale corresponding to the coarse grid simulation it is usually sufficient to consider the coarse grid results only. The ozone depletion near pollutant sources during the morning hours, however, might be underpredicted.

S.T. RAO: Are model input variables resolved to the 500 m resolution used in the model?

R. KUNZ: Yes, especially the emissions inventory has a spatial resolution of 500 m.

S.T. RAO: How do you explain better agreement between predicted and observed peak O_3 values in the 5 km resolved domain and poor agreement between measured and predicted peak O_3 values within the 500 m domain?

R. KUNZ: Differences in peak ozone concentrations for Heilbronn are in the range of the observational uncertainties except for the last day when an frontal passage could not be resolved adequately by the model.

R. SAN JOSE: We found in Madrid (Spain) that biogenic emissions are 5 times higher than the anthropogenic emission. Did you include the biogenic emissions and if so in which rate are they when comparing with the anthropogenic emissions.

R. KUNZ: Biogenic emissions were taken into account. Their contribution to the total VOC emissions is about 30 %.

J. WALMSLEY: You hypothesized that increased horizontal resolution improved the results, particularly for nighttime ozone concentration. Did you not verify this hypothesis by moving the high-resolution region to the area of Schwaebisch-Hall?

R. KUNZ: A high-resolution emissions inventory was only available for the region around Heilbronn. Hence, high-resolution simulation was restricted to this area.

S. RAFAILIDIS: Regarding the comparison between numerical predictions and the field data, what was the averaging time for the latter ones? Do we know that these averaging times are sufficient, in view of changing meteorology, possible local effects in the vicinity of the field station,...?

R. KUNZ: We assume that the averaging time of 30 minutes, which was used for the observations, is sufficient to filter out local effects.

P. PAI: Did your field study include measurements of species besides O_3 and NO_x such as VOC and oxidants? If so, did you evaluate your model against these measurements?

R. KUNZ: Measurements of VOC were not available for the model evaluation. The evaluation of O_3 and NO yields a good agreement between measured and calculated oxidant concentrations in the fine grid domain.

MODELING ANTHROPOGENIC SULFUR TRANSPORT TO THE ARCTIC

A. P. Dastoor

Atmospheric Environment Service
2121, Route Trans-canadienne
Dorval, Quebec
CANADA H9P 1J3

A numerical model aimed at simulating Arctic air quality must be able to reproduce the following three main characteristics: the seasonal cycle, the episodic character of the Arctic pollution events and the vertical distribution of the pollutant in polar regions. Chemical transport models used in the past are capable of simulating pollution transport quite well, but they either depend on meteorological preprocessing or climatological data to provide the parameters for the physical processes in the atmosphere. Some models include the effects of clouds as stochastic processes in space-time.

A dynamic global atmospheric sulfur transport model has been constructed to address the issue of long range atmospheric sulfur transport to the Arctic. The global model includes the dynamics of meteorological and tracer fields, thermodynamics, cloud processes, radiation, turbulent boundary layer mixing, multiple 3-dimensional anthropogenic sulfur emission sources, dry and aqueous-phase chemical processes for sulfur, dry deposition and the precipitation scavenging of sulfur. The advantage of this model is that the physical processes in the atmosphere, advection and chemical conversions are calculated within one consistent modeling system. In contrast to an off-line chemical transport model the dynamic global tracer model is capable of studying the interaction between pollutant, radiative and hydrological budgets and atmospheric circulation.

Clouds play an important role in the formation of sulfate from SO_2. Liquid-phase oxidation often makes clouds more efficient in forming sulfate than gas-phase processes involving OH radicals. The incomplete description of clouds and precipitation in most models has represented a major limitation to the modeling of wet chemical processes on a global scale. Most GCMs include simple cloud parameterization schemes. One of the main features of our study is an attempt to a realistic representation of the interaction between clouds and chemical reactions. The model includes a detailed sub-grid scale convective and stratiform condensation scheme which includes cloud liquid water content as a predictive variable. The cloud parameterization predicts the cloud liquid water content, precipitation formation, precipitation flux through each layer, evaporation and the cloud fraction present in a grid cell. These predicted cloud fields are used to compute the in-cloud chemistry, wet scavenging and precipitation evaporative processes for sulfur.

The global sulfur model is built using the Canadian global spectral model. The model is based on the primitive equations and uses the hydrostatic approximation. It has been modified to include cloud water and tracer transport conservation equations. It uses a terrain following coordinate in the vertical. The time integration is performed using a semi-Lagrangian semi-implicit approach. Using this formulation the complete set of model equations are the horizontal momentum equation, thermodynamic equation, continuity equation and moisture, cloud water and tracer conservation equations along with the hydrostatic relation. To avoid excessive damping arising from interpolating rapidly varying fields in the vertical in the vicinity of the tropopause, a 3-dimensional scheme combining a 2-dimensional semi-Lagrangian scheme in the horizontal and a non-interpolating semi-Lagrangian scheme in the vertical is used. The presence of strong gradients and discontinuities in mass fields such as specific humidity, cloud water mixing ratio and tracers make the conventional high order interpolating semi-Lagrangian schemes unacceptable due to their difficulties in preserving the shape of the solutions. We have adapted a three dimensional extension of the quasi-monotone semi-Lagrangain shape preserving scheme for the moisture, cloud water and tracer transport in the model. This scheme is a conventional semi-Lagrangian scheme coupled with a limiting procedure. The scheme is of high order in the domain of sufficiently smooth solution and is able to suppress the wiggles generated by the high-order interpolation in the neighborhood of the strong gradients of the solution.

For the chemical processes the model includes dry phase oxidation of SO_2, in-cloud oxidation of SO_2, uptake of sulfate cloud condensation nuclei (CCN), in-cloud scavenging of SO_2 and SO_4 by precipitation formation, sub-cloud scavenging of SO_2 and SO_4 by precipitation and release of sulfate aerosol by dissipating clouds.

Twelve integrations of one month each were performed for the year 1993. The model was integrated with a resolution of 2.8^o lat./long. in the horizontal and 21 sigma levels in the vertical. Figs. 1a-f show the sulfate distribution in the surface air for three days in winter/spring (January 1, February 1 & March 31) and three days in summer (June 1, July 1 & August 1) as simulated by the model. High levels of pollution are evident in the Arctic in winter months from all sources. The figures illustrate that there is a strong meridional transport in winter, which leads to high concentrations of sulfur in high latitudes, particularly over the Arctic, on the other hand, with the more zonal atmospheric flow in summer the sulfur is mostly confined to mid-latitudes. On February 1 strong transport of sulfur from NE American sources which is related to the strong meridional flow produced by a block situated over Europe. Air pollution in winter is observed mostly over the continents due to the meridional character of the atmospheric flow. Also the maximum level of sulfate concentration found in winter (56 mg/m^3 on March 31) is much higher than in summer (14 mg/m^3 on June 1). On the other hand, in summer, the sulfate concentration over the oceans is quite high, as seen over the Atlantic in July (Fig. 1e). The structure of the distribution is evidence of a dominant zonal flow, which transports the sulfate over the oceans. In fact this result is strong evidence for continental exchange of pollution. It is likely that there is considerable potential for continental pollution exchange in summer.

The model is able to reproduce the most important aspects of the transport of sulfur from source regions to the Arctic. The model is able to simulate anomalous circulations of the atmosphere such as blockings which are important to the meridional sulfur transport. The model is also able to reproduce the difference in concentration levels and distributions of atmospheric sulfur between the warm and the cold seasons. The characteristic annual cycle in sulfur concentrations over the Arctic is well simulated. Realistic parameterization of clouds and wet chemistry provides physically consistent estimates of wet deposition fluxes of sulfur in different seasons.

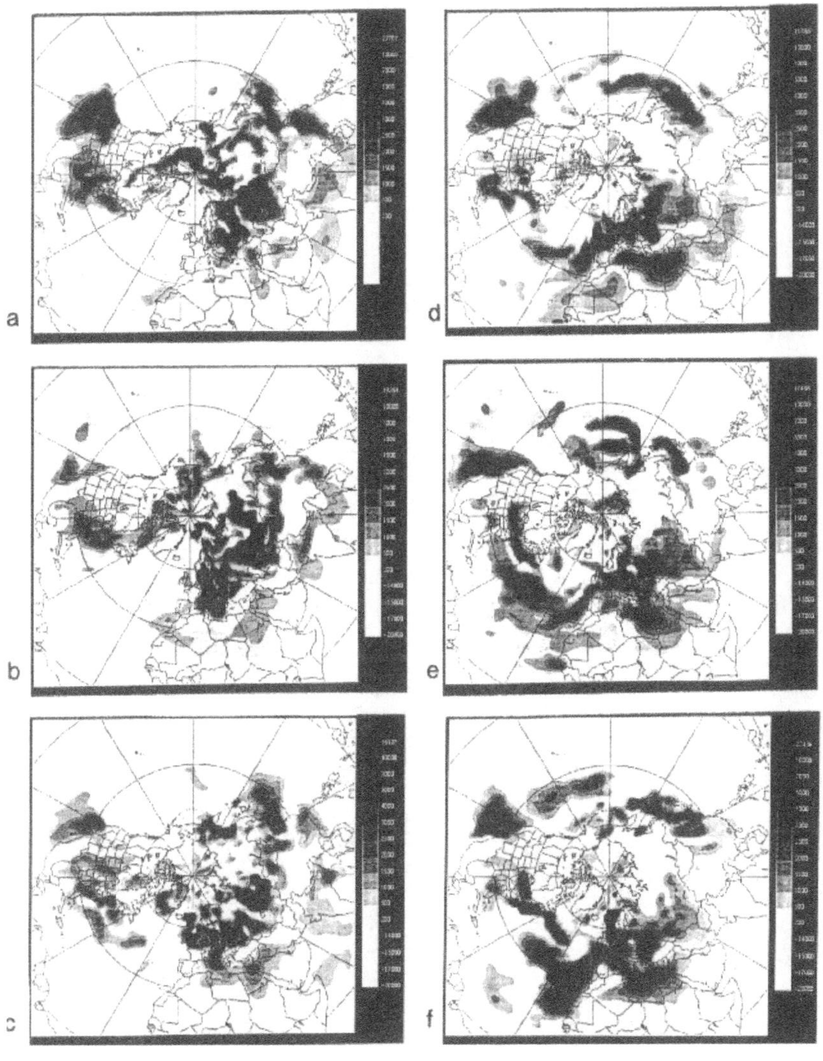

Figure 1. Simulation of sulfate distribution in surface air for three days in winter/spring (*a:* Jan. 1, *b:* Feb. 1, and *c:* Mar. 31) and three days in summer (*d:* June 1, *e:* July 1, and *f:* Aug. 1) in 1993.

DISCUSSION

P. PAI: Since clouds play such an important role in the aqueous phase production of sulfates, how did you evaluate the cloud parameters estimated by your model?

A.P. DASTOOR: Cloud liquid water path, cloud cover, cloud top height, precipitation rates and integrated water vapor simulated by the model are evaluated using satellite visible (GOES-7) and microwave imagery (SSM/I).

E. GENIKHOVICH: It seems to me that your method of the "min-max" correction prevents you from having of point sources of pollutant located inside the grid cell. Is it correct?

A.P. DASTOOR: The limiting procedure is coupled to the semi-Lagrangian time integration scheme to preserve the shape of the solutions. There are no point sources considered within the cells in the model.

ON THE ASSESSMENT OF OZONE CONTROL POLICIES FOR THE NORTHEASTERN UNITED STATES

S. T. Rao[†], G. Sistla[‡], W. Hao[‡], K. John[†] and J. Biswas[†]

†Department of Atmospheric Science
University at Albany
Albany, NY 12222

‡Division of Air Resources
New York State Department of Environmental Conservation
Albany, NY, 12233

INTRODUCTION

Currently, increased attention is being paid to the nation's smog problem in major urban areas because of the lack of a sustained improvement in meeting and maintaining the National Ambient Air Quality Standard (NAAQS) for ozone. The efficacy of various emission control policies at the urban and regional scale can be examined through the application of grid-based photochemical models (NAS, 1991). In this paper, we report on the use of the Regional Oxidant Model (ROM) in the eastern United States for two ozone conducive events in examining the response of ozone concentrations to NOx- and VOC- focussed emission control strategies. As expected, the results indicate that the predicted peak ozone levels decrease as emissions are reduced, but the amount of improvement in ozone concentration levels is strongly dependent upon the prevailing meteorological conditions. ROM simulations reveal that NOx-focussed controls are more effective in reducing the areal extent of the ozone exceedances than VOC-focussed controls. Further, the improvement in the peak ozone levels over urban areas from the base case for the VOC-focussed scenario is found to be comparable to that of the NOx-focussed scenario. However, the peak ozone in rural areas are found to respond more favorably to NOx-focussed controls rather than to VOC-focussed controls. These results demonstrate the need for region-wide NOx and urban-specific VOC control strategies to attain the ozone NAAQS throughout the eastern United States.

Model, Modeling Domain, and Data Bases

Briefly, the Regional Oxidant Model (Lamb 1983) is a meso-scale photochemical grid model with Carbon Bond IV chemical kinetic mechanism (Gery et al.,1989). The modeling domain (Figure 1) commonly called as the SUPROXA domain, extends from east of Mississippi covering eastern United States and portions of Canada with a horizontal grid resolution of 1/6° by 1/4° on a latitude-longitude scale, and has three time-dependent vertical layers. Detailed description of the model design and application can be found in Young, et al. (1989).

Air Pollution Modeling and Its Application XI, Edited by
Gryning and Schiermeier, Plenum Press, New York, 1996

Meteorological Episodes

Two meteorological episodes reflecting persistent high ozone concentrations in the Northeast form the basis for this study. The ozone episodes modeled cover the period July 1 to 9, 1988 and July 13 to 21, 1991. During these episodes, measured ozone levels exceeded 200 ppb on some of the days with a large region of the northeastern U.S. exceeding the ozone NAAQS. Both episodes are associated with such ozone conducive synoptic meteorological features as high surface temperatures ranging from the upper 80's to mid 90's°F, low surface winds, and a high pressure system centered off of the North Carolina coast at the beginning of the episode, and with occasional showers and thunderstorms over some parts of the modeling domain. Meteorological data available from a sparse network of observations were processed using the ROM preprocessor system to generate the relevant input fields for the model for the two episodes.

Emissions

Anthropogenic emissions data for this study were obtained from EPA's interim emissions inventory (EPA, 1993). Interim 1990 emissions inventory for the July '88 episode, and interim 1991 emissions inventory for the July '91 episode were processed using the procedures of Young, et al. (1989). The biogenic emission estimates for these episodes were based on relevant daily meteorological data. Table 1 lists emissions for a typical day over the SUPROXA domain. In general, the two emissions data sets differ by a few percent only, with emissions from the mobile source category contributing about 30% of the total VOCs and NOx. The emissions were processed using the ROM emissions preprocessor system to generate hourly speciated and gridded data. In addition to the base case emissions, a matrix of model runs was developed reflecting several across-the-board reductions in emissions. The subsets consisted of reduction in NOx and VOCs emissions in increments of 25% and are labeled as M2550, M2575, M5025, M7525, where Mxy corresponds to x% reduction in NOx and y% reduction in VOCs, respectively. For example, M2550 is a scenario that includes a 25% reduction in NOx emissions and 50% reduction in VOC emissions from the base case. These emission reductions were applied to the anthropogenic portion of the emissions inventory only.

Simulations

ROM simulations were performed for the two episodes, for the base case and for emission reduction scenarios; the predicted peak ozone concentrations are listed in Table 2. In the case of the July '88 episode, the days July 1 to 4 were considered as 'ramp-up' days, and, therefore, were not considered in the analysis. Similarly, in case of July '91episode, the days July 13 to 15 were not included in the analysis. In addition to the ozone maxima over the SUPROXA domain, the ozone maxima for the urban corridor are listed for areas outlined in Figure 1 are presented in Table 2. Distinct and extended areas of ozone plumes were predicted over the SUPROXA domain for the two episodes. For the July '88 meteorology episode, there were four regions or corridors:

a) industrial/urban regions of Chicago, IL, and Milwaukee, MN, and extending over Lake Michigan
b) industrial/urban areas of Toronto, Canada, Lake Erie, and portions of western New York
c) industrial/urban areas along the eastern seaboard extending from Richmond, VA to portions of southern Maine

Figure 1. SUPROXA Modeling domain. (a) Rural Pennsylvania; (b) rural Georgia; shaded area indicates urban area.

Table 1. Total Emissions (tons/day) over the SUPROXA Domain.

Year	Pollutant	Point	Area	Mobil
1990	NOx	24218.3	11548.2	16124.7
	VOC	9536.7	30174.6	18495.8
	CO	14214.6	41774.8	141842.7
1991	NOx	24113.2	11478.9	16040.9
	VOC	9472.4	29927.8	18814.7
	CO	13989.3	40568.0	134982.0

Table 2. Predicted Maximum Ozone Concentration (ppb) for July 5–8, 1988, and July 16–20, 1991.

Over SUPROXA Domain

DAY	BASE	M2550	M2575	M5025	M7525
July 5	221	171	154	182	160
July 6	218	162	140	178	153
July 7	248	189	147	207	154
July 8	236	184	154	179	130
July 16	167	125	106	151	125
July 17	269	193	122	201	144
July 18	235	173	132	177	130
July 19	202	157	132	154	111
July 20	182	140	120	134	129

Over Northeast Urban Area

DAY	BASE	M2550	M2575	M5025	M7525
July 5	162	127	115	168	131
July 6	218	162	124	178	132
July 7	248	189	147	198	146
July 8	236	184	154	179	127
July 16	154	118	106	151	125
July 17	149	122	110	120	90
July 18	188	151	132	149	110
July 19	202	157	132	154	111
July 20	182	140	120	161	129

Table 3. Number of Grid Cells with Ozone Concentration >124 ppb for July 5–8, 1988, and July 16–20, 1991.

over SUPROXA Domain consisting of 16,128 grids

DAY	BASE	M2550	M2575	M5025	M7525
July 5	379	60	19	65	14
July 6	676	158	57	132	10
July 7	843	212	84	134	20
July 8	839	177	124	79	5
July 16	98	1	0	21	1
July 17	116	11	0	13	4
July 18	218	60	12	51	1
July 19	224	41	5	30	0
July 20	174	25	0	26	2

Over Northeast Urban Area consisting of 396 grids

DAY	BASE	M2550	M2575	M5025	M7525
July 5	74	4	0	22	5
July 6	127	29	0	34	6
July 7	171	88	47	64	14
July 8	189	76	41	47	4
July 16	40	0	0	17	1
July 17	26	0	0	0	0
July 18	69	29	12	20	0
July 19	80	40	5	28	0
July 20	85	24	0	21	2

Table 4. Predicted Peak Ozone Concentration (ppb).

Episode: July 5 to 8, 1988

Area	Base	M2550	M2575	M5025	M7525
Penn. Rural	151	123	114	116	88
Georgia Rural	139	117	112	104	69

Episode: July 16 to 20, 1991

Area	Base	M2550	M2575	M5025	M7525
Penn. Rural	135	114	95	107	82
Georgia Rural	86	69	64	64	48

Table 5. Predicted Episode Average Ozone Concentration (ppb).

Episode: July 5 to 8, 1988

Area	Base	M2550	M2575	M5025	M7525
Urban	83	71	71	63	45
Penn. Rural	79	68	69	58	41
Georgia Rural	49	45	46	40	34

Episode: July 16 to 20, 1991

Area	Base	M2550	M2575	M5025	M7525
Urban	76	65	62	59	45
Penn. Rural	72	63	61	57	44
Georgia Rural	42	37	37	34	28

d) and, a plume extending from Detroit, MI, and portions of West Virginia covering the urban centers of Pittsburgh, PA, and Cleveland, OH.

Although not so pronounced, a similar pattern was seen for the July '91 episode with a patchy ozone plume extending from Detroit, MI to West Virginia.

Results and Discussion

In the case of the July '88 episode, the predicted maxima for the SUPROXA domain for three of the four days are found over the northeastern urban area while in the case of July '91episode only two of the five days fell into that category indicating the influence of day-to-day variations in meteorology on the predicted ozone concentrations. Again similar differences are evident in the predicted maxima between the SUPROXA and the urban corridor area for the control scenarios. In general, the improvement in the predicted peak ozone level from the base case for the matrix of emission control scenarios ranged from 5-10 % depending upon the episode day, with both emissions reduction options showing similar results. Even in the case of urban areas in the Northeast neither emission reduction option seems to be favored overwhelmingly in reducing the peak ozone unlike that found in from the urban airshed model (UAM) applications to the New York airshed (Sistla, et al., 1995). It is also interesting to note that the improvement in the predicted peak ozone is of the order of 10% for both episodes for a reduction of 25% in VOC (for M2550 and M2575 scenarios) or a 25% reduction in NOx emissions (for M5025 and M7525 scenarios). However, when the number of grid cells exceeding the ozone NAAQS (see Table 3) are considered, the ROM results indicate that the M7525 scenario is more favorable than the other scenarios in improving ozone in both the SUPROXA and the northeastern urban region. This is evident from Figure 2 in which a box-plot of distribution of ozone concentrations is shown for one of the episode days for the base and control cases for the SUPROXA and the northeast urban corridor area. The M7525 scenario yields a higher level of overall improvement than the other scenarios both for the SUPROXA and the urban corridor area.

To examine if the effects of emission reductions in rural areas are similar to those found for urban areas, we considered two extended rural areas, as identified in Figure 1, for rural Georgia and rural Pennsylvania, for analysis. The predicted daily maximum for these two areas are listed in Table 4. In case of the rural area in Pennsylvania the predicted maxima are above the ozone NAAQS for both episodes whereas for the area in Georgia exceedances are found only for the July '88 episode. These high ozone levels are indicative of the transported component, since there are no local anthropogenic emissions in these areas except for biogenic emissions. The peak ozone concentrations show an improvement ranging from about 15%- 50% depending upon the emission control scenario, with significant improvement resulting under the M7525 scenario. It is interesting to note that for a change of 25% in VOC emissions (M2550 and M2575 scenarios), the change in peak ozone levels is less than 10%, while for a change of 25% in NOx emissions (M5025 and M7525 scenarios) the change is greater than 25%. This demonstrates the need for the reduction in the overall NOx emissions budget for improving ozone levels over rural areas. Figure 4 displays the box-plot representation of ozone for these two areas for one of the episode days. An overall improvement in the ozone levels from the base case for these emissions reduction simulations is evident in Figure 4.

Another way of assessing the efficacy of the emission reduction scenarios is to examine the average ozone concentrations over urban and rural areas. Table 5 lists the episode-based average concentration for the base and emission reduction scenarios. The largest improvement in the average ozone levels is associated with the M7525 scenario for both urban and rural areas. In particular, the average ozone concentrations for the rural area in Pennsylvania are comparable

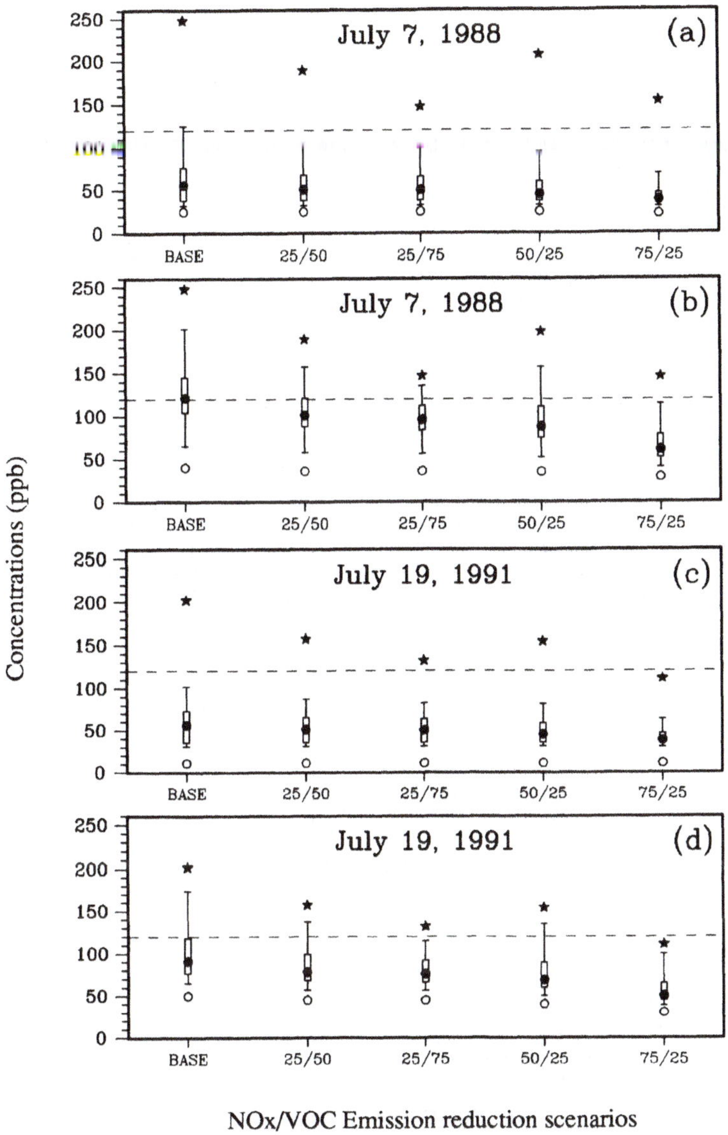

Figure 2. Box-plot distribution of maximum ozone for July 7, 1988 and July 19, 1991 for the SUPROXA (a and c), and the urban area (b and d) for the base and control scenarios.

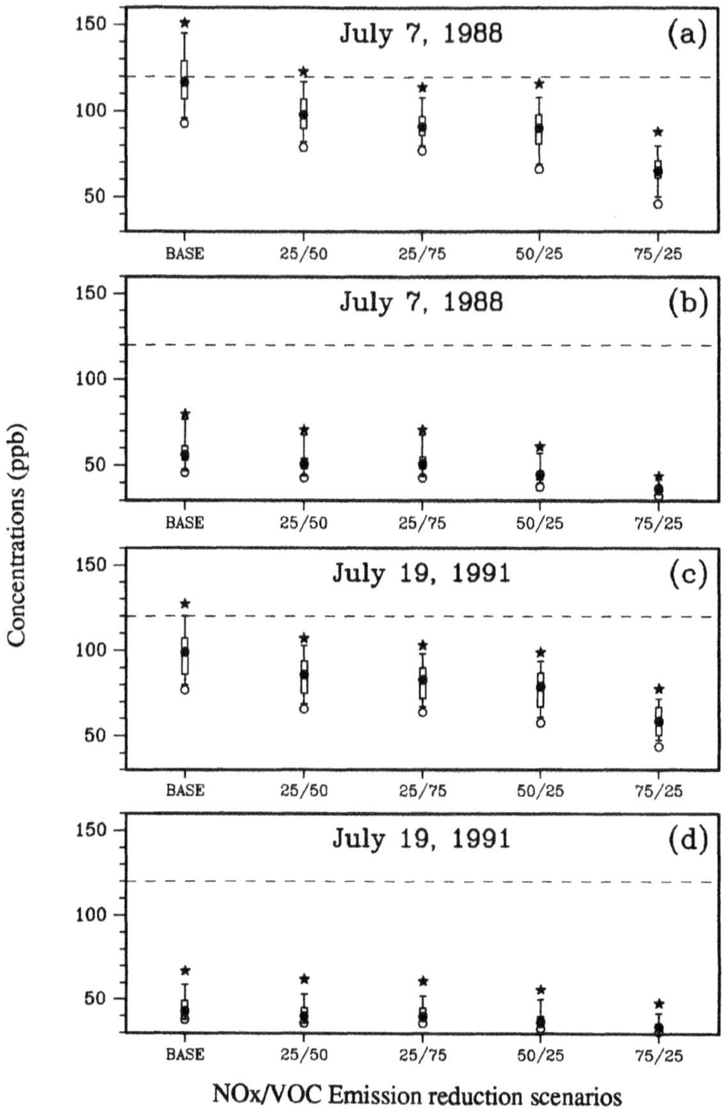

Figure 3. Box-plot distribution of maximum ozone for July 7, 1988 and July 19, 1991 for the rural Pennsylvania (a, c), and rural Georgia (b, d) for the base and control scenarios.

to those in the urban area for both the base case and emission reduction scenario, with VOC-focussed reductions yielding about 15% improvement as compared to 25% or more for NOx-focussed reductions. However, in the case of the rural area in Georgia, VOC-focussed scenarios yield an improvement of about 10% compared to the 20 to 30% improvement from NOx-focussed scenarios. Thus, these simulations suggest the need for designing emission reduction plans to reduce ozone levels both in the rural and in the urban areas. Whereas either VOC- or NOx-focussed controls provide relatively similar level of reduction in the peak ozone concentrations, NOx-focussed controls are more effective in reducing the areal extent of the ozone non-attainment problem than that provided by VOC-focussed controls.

Summary

In this study, a series of simulations was performed with the ROM for two ozone-conducive meteorological episodes over the eastern United States. Examination of the relative efficacy of the NOx- and VOC-focussed emissions control scenarios reveals that NOx-focussed controls are far more effective than VOC-focussed controls in reducing ozone levels both in urban and rural regions. The modeling simulations suggest that it is important to consider emission control policies that would not only reduce the peak ozone concentrations but also the areal extent of the ozone non-attainment problem. Thus, region-wide NOx controls coupled with urban-specific VOC controls would be needed for ozone attainment in the northeastern United States.

References

Gery, M. W.,G. Z. Whitten, J. P. Killus, and M. C. Dodge, (1989), " A photochemical kinetics mechanism for urban and regional scale computer modeling", J. Geophys. Res., 94, 12925-12956.

Lamb, R. G., (1983),"A regional scale (1000 km) model of photochemical air pollution, Part I: Theoretical formulation", EPA-600/3-83-035

NAS, (1991), "Rethinking the ozone problem in urban and regional air pollution", National Academy Press, Washington, DC, 489 pp.

Sistla, G., N. Zhou, J. Y. Ku, S. T. Rao, R. Bornstein, F. Freedman, and P. Thunis, (1995), "Effect of uncertainties in meteorological inputs on the urban airshed model predictions and ozone control strategies", Accepted for publication in Atmos. Env.

Young, J., M.Aissa, T. Boehm, (1989),"Development of the regional oxidant model version 2.1", EPA-600/3-89-044

DISCUSSION

M. PIRINGER: The influence contours of the urban ozone concentrations seemed to be more circular than one would expect from the knowledge of the windroses. Did you take the local meteorology into account in your calculations, especially the local windfield? Is the density of the data sufficient?

S.T. RAO: The isolines of correlation coefficient are drawn using the information among the short-term component of the ozone time series at various monitoring locations. A correlation matrix was computed using ten years worth of data at each site, and then the footprint for the ozone cloud at a selected site is developed. The shape of the ozone cloud is more elliptical than circular as you note; the major axis of the ozone cloud is oriented along the southwest to northeast direction, reflecting the prevailing meteorology. The spatial coverage of ozone monitoring sites is not as dense as we would like. There are very few monitors at rural locations in the eastern United States.

J. WALMSLEY: You mentioned that control strategies must focus on reducing the baseline (long-term) signal. Does this mean that episodes should be ignored and that abatement strategies should not be implemented to reduce episodic ozone levels? If so, is it politically feasible to ignore episodes?

S.T. RAO: Currently, photochemical models are being used to simulate episodic ozone events to determine the maximum ozone concentrations during a given episode. Assuming that everything else in the model remains the same, the model is then applied to determine the reduction in the emissions needed to attain the ozone standard. We have shown that seasonal and long-term variations (the baseline) are the only deterministic components in ozone data, and that ambient ozone concentrations can be effectively modeled as the sum of baseline and white noise. Hence, attainment strategies must be aimed at reducing the baseline. To this end, models need to be applied to simulate the whole ozone season. From this modeled concentration data, the baseline can then be extracted. The amount of reduction in the emissions required to bring the baseline down to the level at which the ozone standard can be met can then be readily determined. As you now, confidence intervals for extreme values are very wide. Thus, this approach reduces the uncertainties associated with the predicted peak ozone levels under extreme events and the control requirements based on extreme values using the modeling information from a few ozone episodes.

INTEGRATED REGIONAL MODELLING

chairmen:	S.E. Gryning C. Nappo
rapporteurs:	E. Batchvarova A.G. Straume

METEOROLOGICAL MODELING IN SYDNEY AUSTRALIA: CASE STUDY OF A SMOG EVENT

Peter J. Hurley, Julie A. Noonan, and Peter C. Manins

CSIRO Division of Atmospheric Research
PMB 1, Mordialloc, Victoria, 3195
Australia

INTRODUCTION

Australia's large coastal cities experience the most intractable form of air pollution — photochemical smog characterised by the presence in the air of excessive concentrations of ozone gas, a respiratory irritant (Manins *et al.*, 1994) — commonly exacerbated at coastal sites by sea breeze recirculation of pollutants.

Sydney is the capital city of New South Wales (NSW), and is Australia's largest city with a population of 3.6 million people. The city and suburbs are contained in a large basin bounded on the east by the sea, and to the north, west and south by hills and mountains. Following concern about photochemical smog levels in Sydney in the 1970's, the State Pollution Control Commission (SPCC) (now the Environment Protection Authority of NSW (EPAN)) established the Sydney Oxidant Study. As part of this study, Hyde *et al.* (1983) investigated the meteorology of high ozone days and found that ozone formed predominantly within complex mesoscale flows and recirculation patterns. This study led to tighter controls on emissions which then seemingly produced a reduction in smog levels. However, Hyde and Johnson (1990) found that the photochemical smog problem had merely shifted to the west beyond the limits of the SPCC monitoring network.

In 1993, the EPAN initiated the Metropolitan Air Quality Study (MAQS), based on the above predictions that urban and industrial expansion in the western part of Sydney could lead to a deterioration of air quality. The study has involved the enhancement of existing air quality measurement sites, and included intensive upper-air field campaigns. One component of the study was contracted out to a team managed by Coffey Partners International, who were to develop and validate an emissions inventory and a set of modelling tools for the region which would ultimately be passed on to the EPAN to assist in future assessment of urban and industrial planning issues. The MAQS consultancy consisted of four major tasks:

Air Pollution Modeling and Its Application XI, Edited by
Gryning and Schiermeier, Plenum Press, New York, 1996

1. Air Emissions Inventory (EPA of Victoria);
2. Meteorology - Air Movements (Maquarie University and CSIRO Division of Atmospheric Research);
3. Air Chemistry (CSIRO Division of Coal and Energy Technology);
4. Development of Airshed Models (EPA of Victoria).

In this paper we describe the meteorological modelling component of the Meteorology - Air Movements Task. The Environmental Consulting and Research Unit (ECRU) of the CSIRO Division of Atmospheric Research conducted numerical meteorological simulations for the MAQS regions, based on high pollution days categorised by Macquarie University using historical data. The meteorological predictions were used as input to MAQS tasks 3 and 4. In Section 2 we will briefly describe the modelling system used for the study, and the configuration of the model for MAQS. In Section 3 we present modelling results and model validation for a detailed case study day in Sydney.

MAQS METEOROLOGICAL MODELLING

Meteorological modelling for MAQS was performed using the Lagrangian Atmospheric Dispersion Model (LADM), described by Physick *et al.* (1994) and Physick (1993). LADM is a two-part model:

- the first part is a mesoscale meteorological model which solves the primitive equations for fluid flow, with parameterised radiation and surface characteristics, and forecasts the local winds, turbulence, temperature and mixing heights;
- the second part is a Lagrangian particle dispersion model, which uses mean flow and turbulence predictions from the meteorological model to disperse pollutant emissions from point, line, or area sources, and predict pollutant concentrations.

For the MAQS region, LADM was run using a series of nested grids with horizontal dimensions of 40x40 points and a vertical dimension of 20 points. Figure 1 shows the

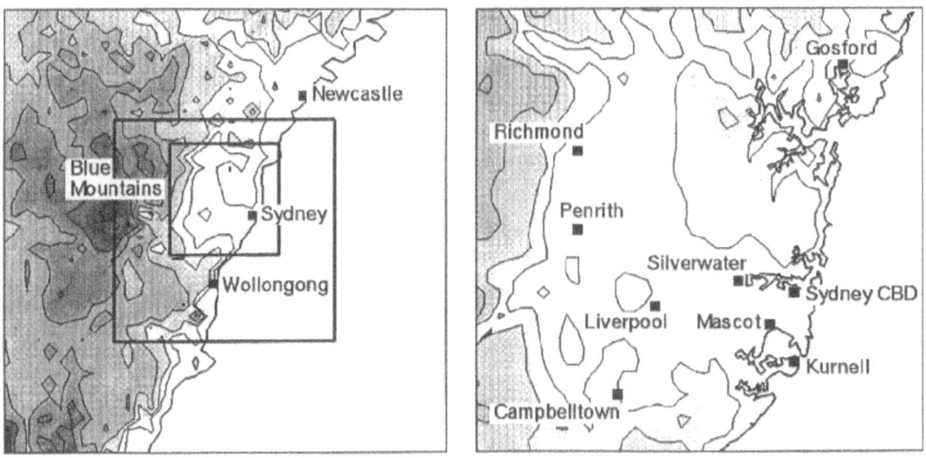

Figure 1. The Sydney 10 km spaced grid domain (left) showing the region covered by the 5 km and 2.5 km spaced domains as dark square outlines, and the Sydney 2.5 km spaced grid domain (right) which covers an area of 100x100 km^2. The regions show terrain contours and shading, with darker shading indicating higher terrain, and contour levels: 0, 100, 250, 500, 750, 1000, 1250, and 1500 m.

10 km, 5 km and 2.5 km spaced grid domains for the Sydney sub-region. The terrain data of Hutchinson and Dowling (1991) were used on these grids.

Two different approaches to meteorological modelling have been used for MAQS:

- Generic modelling: performed to help identify major mechanisms characterising the type of days being modelled and to enhance the understanding of the data;
- Case study modelling: performed in order to describe as much detail as possible about a particular day, and involves first using the generic modelling approach, and then refining the model input (usually to a much greater degree than the accuracy available in the geostrophic wind measurements), until good agreement is obtained between predictions and observations. The assimilation of wind observations (data nudging) into predictions is then performed using the approach of Stauffer and Seaman (1994).

MAQS CASE STUDY SIMULATION (9-10 FEBRUARY 1994)

Surface (10 m) measurements were available on 9-10 February 1994, providing wind speed and direction at seventeen sites, and temperature data at eight sites. Wind direction at the coast remained north-northeasterly throughout the day, while light southerly flow was apparent in the western half of the basin overnight, with northeasterly to easterly winds in the afternoon sea breeze. Flow at the coast was stronger than at inland sites, particularly overnight. An analysis of the synoptic charts showed a high pressure system off the NSW coast directing northerly gradient winds over the Sydney region. Upper air data were available from Mascot Airport, including six-hourly profiles of winds, data from the 0500 hr and 1400 hr radiosonde ascents, and some temperature and wind data from commercial aircraft ascents. Profiles at this location showed that north-northeasterly flow occurred in the poorly resolved lowest 1000 m, with winds backing with height to be westerly by 2000-3000 m. However, the aircraft ascent data showed that the winds between the surface and 500 m had significant north-northwesterly to northerly components.

The model was run for three days with results from simulation days 2 and 3 corresponding to 9 and 10 February 1994 respectively. Initial profiles of synoptic wind speed, wind direction, potential temperature and mixing ratio are given in Table 1. Sea surface temperature was 295 K, and the mean sea level pressure was 1015 hPa. Geostrophic winds remained constant in time. A surface roughness length of 1 m, and a soil moisture of 0.20 $m^3 m^{-3}$ was used, based on an analysis of the previous 12 months rainfall data. The nudged simulation required the input of the hourly averaged measured wind data. Associated variables specified were a 10 km radius of influence and a nudging coefficient of 0.002 s^{-1}. Only surface winds were nudged, due to the lack of detailed wind profiles.

Table 1. Profiles of initial and geostrophic wind speed and direction, and initial potential temperature and mixing ratio for 9-10 February 1994.

height (m)	wind speed ($m s^{-1}$)	wind direction (°)	potential temperature (K)	mixing ratio ($g kg^{-1}$)
0.	13	15	298	15.0
500.	13	20	303	13.0
1000.	13	20	306	5.5
1500.	13	330	309	5.5
3000.	13	240	313	4.0
5000.	13	210	325	0.1

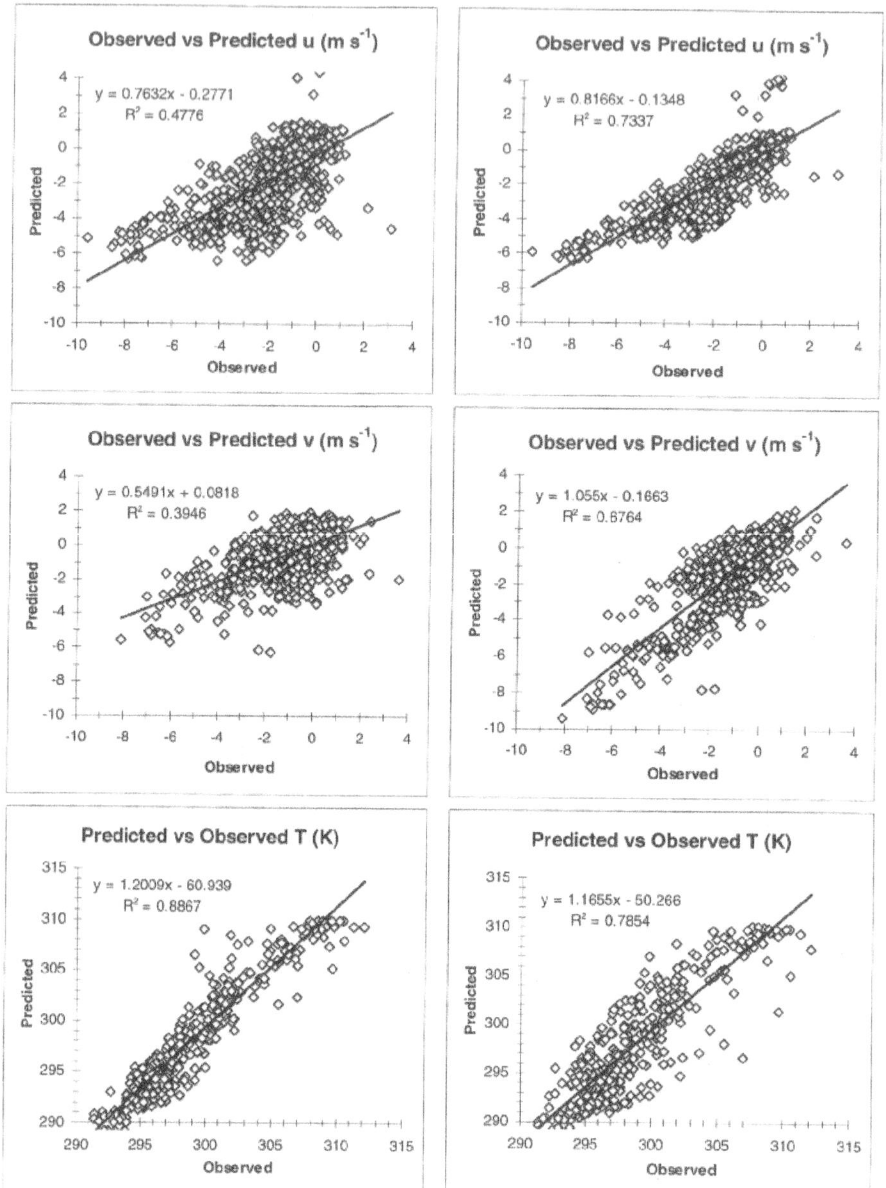

Figure 2. Predicted vs observed wind components (u and v) and temperature (T) for the case study simulation without nudging (left) and with nudging (right). Regression lines and correlation measures are also shown.

Prediction of 10 m agl wind and temperature at observation sites

Figure 2 shows scatter plots of predicted versus observed u- and v-components of the wind and temperature (T), without and with data nudging. The plots also show linear regression lines of best fit and quote the equation of the line and the R^2 correlation measure.

The results without nudging show that R^2 values of 0.48, 0.39, and 0.89 were obtained for u, v, and T respectively, with an over-prediction of low wind speeds, and an under-prediction of high wind speeds and minimum temperatures. As expected, the changes in the solution due to nudging were most pronounced at the observation sites, as indicated by the improvement in regression fits. The results show that R^2 values of 0.73, 0.68, and 0.79 were obtained for u, v, and T respectively, showing an improvement in the winds, but a slight deterioration in the temperatures. When the constraint of exact time matching was relaxed (allowing predictions to be shifted by one hour from the measurements), the degree of scatter reduced and resulted in R^2 values of 0.65, 0.55, and 0.93 without nudging, and 0.84, 0.75, and 0.91 with nudging, for u, v, and T respectively.

The strength of the wind flows was predicted well at almost all sites except for an overnight over-prediction on the western slopes of the basin, and a daytime over-prediction at some central basin sites. The wind direction was generally predicted well when the wind speed was significant. Temperature results show that the model has done well, except for a slight under-prediction of the overnight temperatures at the inland sites.

Wind flow patterns

Figure 3 shows 10 m winds at selected hours for the simulation with data nudging. Before sunrise, winds at the coast were moderate northeasterly, while with increasing distance inland the winds were much weaker and turned from northeasterly to be southerly at the foot of the Blue Mountains. Drainage flows occurred into the Sydney basin from surrounding slopes. In the morning after sunrise, surface heating reversed overnight drainage flows, and the sea breeze began to form at the coast. These effects produced light winds in the basin area and easterly flow on the surrounding slopes and at the coast. By late-afternoon the sea breeze had travelled across the grid producing northeasterlies at the coast and easterlies inland. After sunset, the easterly winds in the western half of the basin collapsed, and light southerly flow again formed. Overnight along the south Sydney coast, the effect of the data nudging can be seen, with winds being a little stronger and having more northerly component compared to the winds on the edge of the grid over the sea.

At upper levels during the morning, a distinctive southerly flow (~1-2 m s^{-1}) occurred up to about 600 m in the western part of the basin, while an easterly flow was found in the central basin and northeasterly flow at the coast. Above about 1000 m, the flow was more northerly turning to northwesterly by 2000 m. The deep southerly flow is thought to be a blocking of the easterly component synoptic wind by the mountains to the west and southwest of the Sydney basin. A strong vertical potential temperature gradient existed in the basin due mainly to the advection of marine air by the previous afternoon's sea breeze. Simple Froude number (the ratio of kinetic and potential energy) calculations suggested that the incoming air had an overnight Froude number less than 1, which is the condition needed for flow blocking by terrain (see Manins and Sawford, 1982). Morning convective mixing gradually removed the strong stability in the lowest layers of the atmosphere, with mixing heights ahead of the sea breeze reaching 600 m by 1200 hr and 2000 m by mid-afternoon.

Trajectories and Particle Positions

Lagrangian particle dispersion calculations were performed for releases from three notional sources (see Fig. 1 for locations) to assist the interpretation of wind flow patterns:

- Sydney central business district (CBD) surface area source;
- Silverwater 200 m high point source;
- Kurnell 200 m high point source;

Particles were continuously released from 1800 hr on 8 February 1994.

Figure 4 shows particle positions on 9 February 1994. Overnight surface emissions from the CBD were caught in light northeasterly flow near the coast and travelled inland in easterly flow and then in light southerly flow in the western part of the basin. By sunrise, CBD pollutants were spread over most of the basin floor covering the south-north extent of the grid domain and the west-east extent in southern Sydney, forming a broad L-shape pattern, although the majority of the particles were in the southern half of the basin. This pattern persisted until late morning when the light easterlies ahead of the sea breeze began to clear away the CBD pollutants in the north. By mid-afternoon, pollutants covered the southern half of the grid domain. Pollutants from the notional Kurnell source travelled down the coast and off the edge of the grid domain throughout the simulation. Pollutants from the notional Silverwater source followed a similar path to the surface CBD emissions.

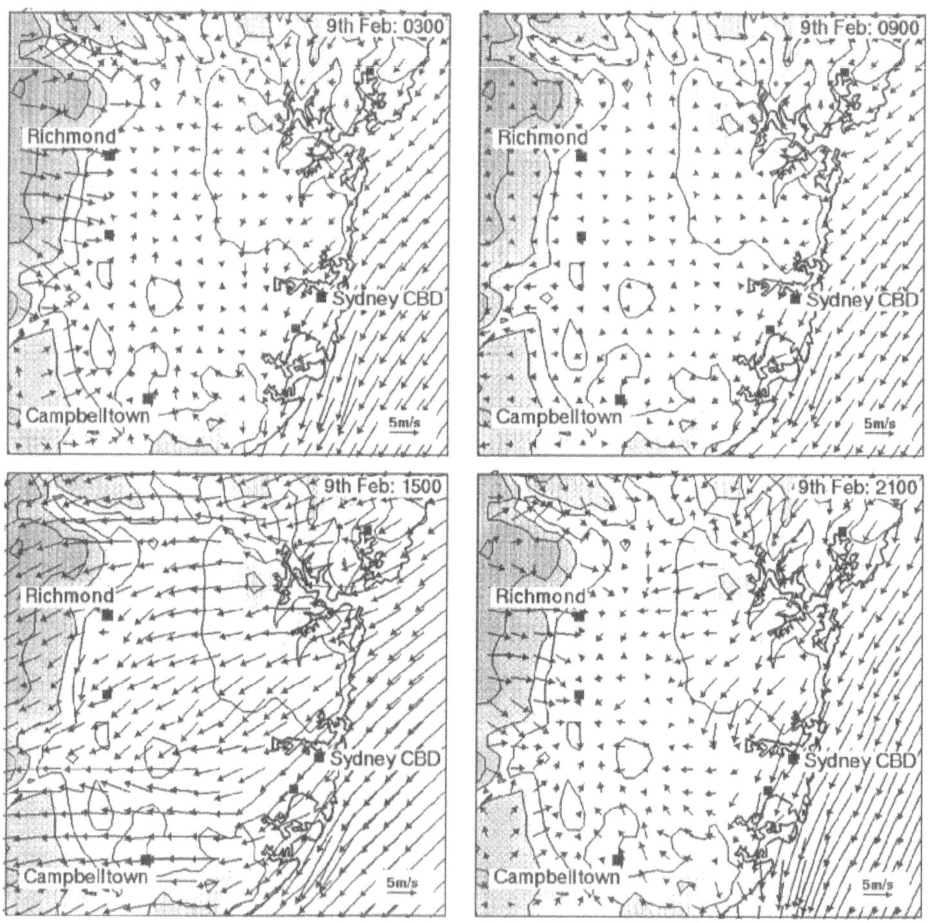

Figure 3. Predicted wind flow patterns at 10 m above ground for the case study simulation with data nudging.

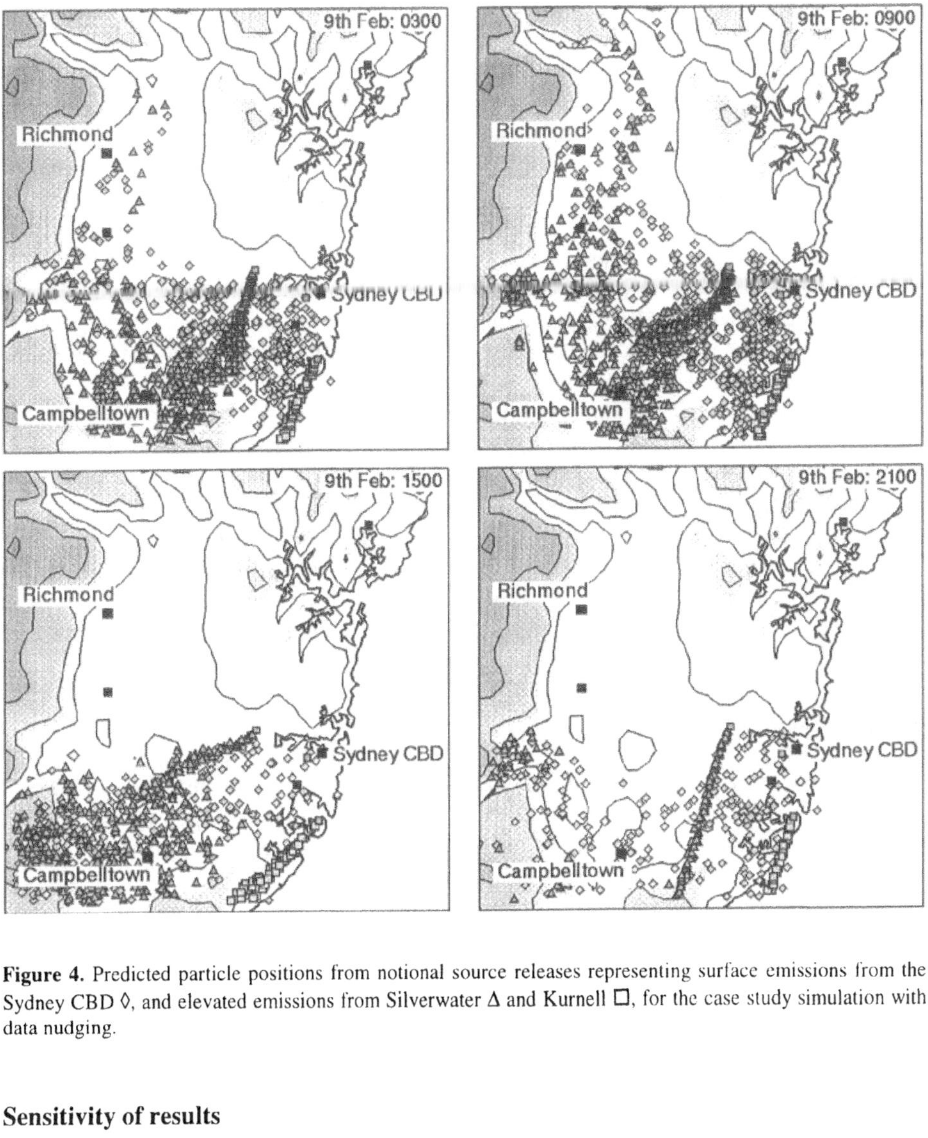

Figure 4. Predicted particle positions from notional source releases representing surface emissions from the Sydney CBD ◊, and elevated emissions from Silverwater Δ and Kurnell □, for the case study simulation with data nudging.

Sensitivity of results

The specification of initial model profiles of synoptic winds is not very accurate due to the difficulty in interpreting synoptic pressure charts, and the need to eliminate mesoscale effects from the observations. The case study simulation was the end result of a multitude of runs, with geostrophic wind directions being varied between 10 and 30° and wind speeds between 7 and 15 m s^{-1}. The final values tended to give better agreement with near coastal sites. The use of initial temperature profiles taken from Mascot observations resulted in an underprediction of daytime temperatures, and so the profile below 3000 m was increased by 3 K. The need for this increase may be due to Mascot being a coastal site.

A major reason for the selection of the reference run rather than one of the sensitivity runs, was that the slightly more northerly wind direction in the lowest 500 m produced more northerly drift in the eastern part of the basin at around sunrise (consistent with aircraft ascent data at Mascot). This was a critical feature for the positioning of the urban plume to the south of the CBD, and enabled the sea breeze to transport the plume to the observed southern region by mid-afternoon.

For the nudging simulation, variations of the radius of influence between 5-15 km, and the nudging coefficient between 0.0003-0.003 s^{-1} were made. The final values of radius 10 km and coefficient 0.002 s^{-1} produced good correlations, and influenced the solution at several surrounding sites without detrimentally affecting the overall result.

CONCLUSIONS

Meteorological predictions for the MAQS case study day 9-10 February 1994 generally reproduced the observed wind and temperatures patterns near the surface, while the use of data nudging further improved the predictions near the observation sites. Upper level wind predictions reproduced the observed profiles at the coast, but no validation data were available at inland sites.

Results showed that morning emissions from the Sydney CBD and surrounding areas travel to the south-west in the afternoon sea breeze flow, and then overnight in the southerly drainage/blocked flow. These characteristics, along with urban emissions and high temperatures, explain the conditions leading to the build-up of smog levels during the morning, and the inland peak values in the afternoon sea breeze.

While good comparisons of winds were not too difficult to obtain using LADM, the critical finding, emphasised in the phase of the study predicting photochemical smog, was that wind trajectories must be accurately obtained so that air masses are correctly positioned throughout the day. This is a far from trivial extra step.

ACKNOWLEDGMENTS

This work was done as part of the MAQS consultancy led by Coffey Partners International, and was funded by the EPAN. LADM modelling was performed in close association with, and depended on, data analysis and characterisation of event days by Dr. Robert Hyde and Ms. Margaret Young of Macquarie University.

REFERENCES

Hyde , R., Malfroy, H.R., Heggie, A.C., and Hawke, G.S., 1983, Nocturnal wind flow across the Sydney Basin, *The Urban Atmosphere - Sydney, a case study*, ed. J.N. Carras and G.M. Johnson, CSIRO, Melbourne, pp.39-60.

Hutchinson, M.F., and Dowling, T.I., 1991, A continental hydrological assessment of a new grid-based digital elevation model of Australia, *Hydrological Processes*, **5**, 45-58.

Hyde, R., and Johnson, G.M., 1990, Pilot Study: Evaluation of air quality issues for the development of Macarthur South and South Creek Valley regions of Sydney, *CSIRO MRL Investigation Report RIR 1885R*, 182 pp.

Manins, P.C., and Sawford, B.L., 1982, Mesoscale observations of upstream blocking, *Quart. J. R. Met. Soc.*, **108**, pp.427-434.

Manins, P.C., Physick, W.L., Hurley, P.J., and Noonan, J.A., 1994, The role of coastal terrain in the dispersion of pollutants from Australia's major cities, In *Clean Air '94: proceedings of the* Clean *Air Society of Australia and New Zealand 12th International Conference & 6th Regional Conference of the IUAPPA*, **2**, Perth. 179-188. Canning Bridge, W. A.: Promaco Conventions for the Society.

Physick, W.L., Noonan, J.A., McGregor, J.L., Hurley, P.J., Abbs, D.J., and Manins, P.C., 1994, LADM: A Lagrangian Atmospheric Dispersion Model, *CSIRO Division of Atmospheric Research Technical Report No. 24.*

Physick, W.L., 1993, LADM (Lagrangian Atmospheric Dispersion Model), *Clean Air Australia.* **27**, 172-175.

Stauffer, D.R., and Seaman, N.L., 1994, Multiscale four dimensional data assimilation, *Journal of Applied Meteorology*, **33**, pp.416-434.

LAKE BREEZES AND THEIR IMPACT ON GROUND-LEVEL OZONE TRANSPORT OVER SOUTHWESTERN ONTARIO AND SOUTHEASTERN MICHIGAN

David M.L. Sills and Peter A. Taylor

Department of Earth and Atmospheric Science
York University
North York, Ontario
CANADA
M3J 1P3

INTRODUCTION

Frequent high concentrations of ground-level ozone over much of eastern North America during the summer of 1988 led to intensified efforts to identify the sources of the precursors of this pollutant and address issues of emissions abatement. The Canadian Council of Ministers of the Environment named the Windsor-Quebec City corridor as the area in Canada where higher than acceptable concentrations of ground-level ozone occur most often and for the longest periods of time (CCME, 1990). In response, a field study including intensive atmospheric chemistry and meteorological measurements along the south-western half of the Windsor-Quebec City Corridor was carried out during the summers of 1992 and 1993 under the auspices of the Canadian Institute of Research in Atmospheric Chemistry (CIRAC) (see Melo, 1993). The observation program was named the Southern Ontario Oxidant Study (SONTOS). In 1993, SONTOS was conducted simultaneously with a similar program, called the Southeastern Michigan Ozone Study (SEMOS), in adjacent southeastern Michigan (see Bowne, 1994). These field studies were further augmented by a surface mesoscale meteorological observation program in southwestern Ontario (SOMOS, Southern Ontario Oxidant Study - Meteorological Measurements) carried out by York University and Zephyr North (see Sills et al., 1994).

The focus of our research at York University has been on the analysis of the consolidated field data and mesoscale meteorological modelling in the SOMOS/SEMOS study area to elucidate the roles of lake and land breezes in the transport of ground-level ozone. Data from the month of August 1993 provided excellent examples of interacting lake effects and enhanced ground-level ozone production.

Air Pollution Modeling and Its Application XI, Edited by
Gryning and Schiermeier, Plenum Press, New York, 1996

FIELD MEASUREMENTS AND ANALYSIS

Mesoscale surface meteorological observations were made in southwestern Ontario (SOMOS) and southeastern Michigan (SEMOS) to supplement operational hourly meteorological observations from Environment Canada's Atmospheric Environment Service (AES), Ontario Ministry of Environment and Energy (OMEE) and the U.S. National

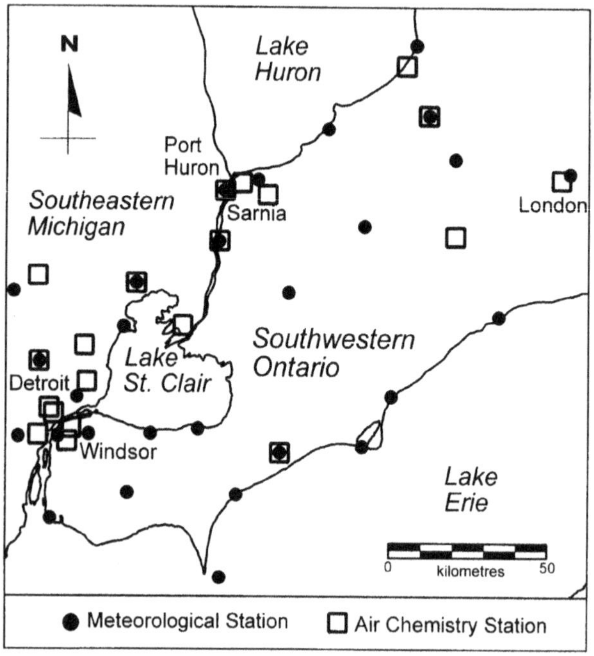

Figure 1. Map of study region showing the locations of 28 surface meteorological stations and 20 surface air chemistry stations.

Weather Service (NWS). Hourly air chemistry measurements of O_3, NO_X, SO_2, etc. from monitoring stations in southwestern Ontario were obtained from OMEE. Similar measurements made in southeastern Michigan were obtained from SEMOS. In total, 28 surface meteorological stations and 20 air chemistry stations provided data in the study region, shown in Figure 1. NWS upper air meteorological data from Flint, Michigan at 12 hour intervals were also obtained for selected days in August 1993.

Visible satellite images were obtained from AES for most of the days in the study period. These high resolution images helped to reveal the positions of lake breeze fronts and other important mesoscale features. Composite images of maximum reflectivity radar for the study region were also archived.

Lake and land breezes were frequently observed in southwestern Ontario during the summer 1993 field campaign. Occurrences of lake and land breezes were initially identified by visual inspection of time series and surface station plots of all station data. In most cases, significant on/offshore components of wind speed distinct from the gradient flow designated lake/land breezes. Lake breezes were then verified using the high-resolution visible satellite imagery. Infrared satellite images were not sufficiently detailed to verify nocturnal land breezes. Table 1 summarizes lake and land breeze occurrences in the study region for the month of August in 1993. Note that lake breezes developed on more than half of the days in the month while land breezes occurred only slightly less frequently. For comparison, other studies have found that Great Lakes lake breezes occur on 25-45% of summer days (Comer and McKendry, 1993). On 32% of August days, well-organized lake breezes from Lake Huron, Lake Erie and Lake St. Clair interacted to produce elongated areas of convergence between the Lakes. In many cases, visible satellite imagery shows enhanced convection, and in some cases thunderstorm development, in these 'convergence zones'. Conversely, well-organized nocturnal land breezes blowing offshore along each Lake occasionally led to the development of weak 'divergence zones' between the Lakes, implying some subsidence over land. This occurred on only 13% of August days owing to the weaker nature of these circulations.

CASE STUDY - AUGUST 14, 1993

An excellent example of lake and land breeze development and interaction from SOMOS was observed on August 14, 1993. Weak synoptic-scale pressure gradients, warm

Table 1. Statistical table for lake breezes in August 1993. A lake or land breeze 'day' refers to a day on which a lake or land breeze was initiated. CON / DIV events are explained in the text.

August 1993 Statistics	Lake breeze		Land breeze	
	Days	%	Days	%
Any of the three shores	19	61	15	48
Lake Erie shore	19	61	14	45
Lake Huron shore	18	58	12	39
Lake St. Clair shore	16	52	12	39
All three shores	16	52	11	35
CON / DIV events	10	32	4	13

temperatures and clear to partly clear skies allowed well-developed lake breezes during the day and land breezes early in the morning and late at night. Hourly ozone maxima were greater than 50ppb at all monitoring stations in the study region, with stations reporting one hour average values up to 147ppb. Afternoon thunderstorms, including one severe storm, developed along lake breeze convergence zones.

Lyons and Olsson (1973), among others, have shown that high levels of pollutants can be introduced behind lake breeze fronts. Figure 2 shows time series for surface ozone and wind direction at Windsor. Ozone reached 89ppb by 13:00 EST after the nocturnal inversion had broken up. After the passage of the Lake St. Clair front at 16:00 EST, ozone climbed to a peak of 105ppb at 17:00 EST. A small 15ppb increase in ozone was also observed at 23:00 EST as the remnants of the Lake Huron front passed the station. Several stations in the Windsor-Detroit area experienced this 'double shot' of ozone. Figure 3a shows the observed wind field and hourly ozone values in the study region at 13:00 EST. The positions of the lake breeze fronts were estimated using surface meteorological data and satellite imagery. Ground-level ozone concentrations increased rapidly behind the Lake Huron lake breeze front as it moved inland at the Lake's extreme southern shore. Port Huron recorded a hourly average of 147ppb while Sarnia observed an hourly average of 130ppb. Ozone also quickly increased as the Lake St. Clair lake breeze moved onshore over the Lake's western coast. The East 7 Mile station in Detroit reported an hourly average of 124pbb at 13:00 EST. Later at 15:00 EST, the lake breezes had converged triggering thunderstorms in the areas of increased uplift (Figure 3b). The thunderstorm at the point of collision of all three lake breeze fronts produced funnel clouds and damaging winds. Note also that ground-level ozone concentrations near Sarnia had decreased significantly well behind the Lake Huron and Lake St. Clair lake breeze fronts.

Lake breeze fronts did not deliver high ground-level ozone to all locations in the study region. Ozone concentrations *decreased* as the Lake Huron lake breeze front moved inland over its southeastern shore toward London. The Lake Erie lake breeze front seemed to have little or no effect on ozone concentrations as it penetrated inland.

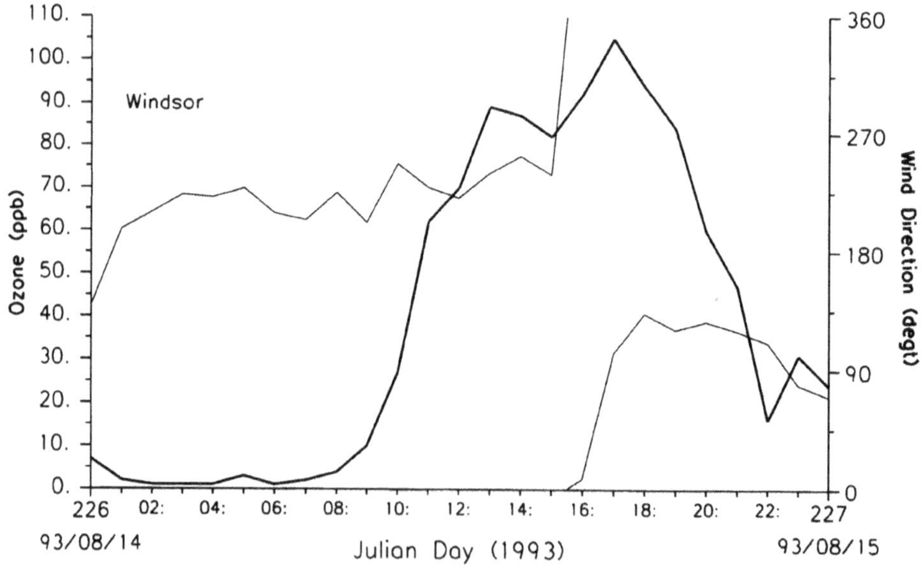

Figure 2. Time series plot showing surface ozone (thick line) and wind direction (thin line) at Windsor. Note the peaks in ozone in the wake of the Lake St. Clair front at 17:00 EST and the Lake Huron front at 23:00 EST.

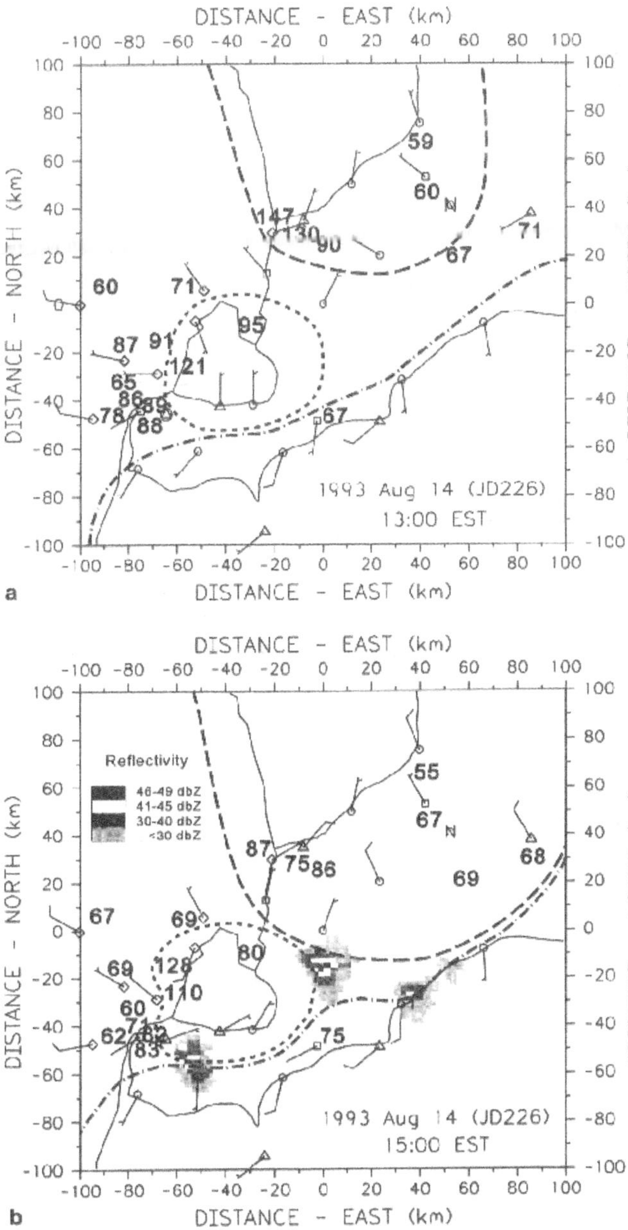

Figures 3a and 3b. Plots of hourly averaged ozone concentrations and the observed wind field at 13:00 EST and 15:00 EST. The broken lines mark the estimated positions of lake breeze fronts; dashed for Lake Huron, dotted for Lake St. Clair, and dashed-dotted for Lake Erie. Maximum reflectivity radar is also included in Figure 3b valid at 15:15 EST.

MESOSCALE METEOROLOGICAL MODELLING

The Mesoscale Compressible Community (MC2) Model, developed by Environment Canada, has been used to simulate the conditions on August 14, 1993. The MC2 model is based on a fully-elastic, non-hydrostatic model developed by Robert, Tanguay and Laprise (1990). The model solves a full set of Euler equations in a three-dimensional terrain-following coordinate system with time-dependent nesting of the lateral boundary conditions supplied by a hemispheric or global model . The model's efficiency is increased through the use of a semi-Lagrangian, semi-implicit numerical scheme that allows a relatively large time step. No chemistry is included in the model at this time but the addition of passive chemical tracers is in progress at York.

The model was initialized using zero hour analyses from the Regional Finite Element (RFE) hemispheric model used operationally by Environment Canada as well as high-resolution topography and climate files. An initial model run at 100km grid spacing and an intermediate run at 25km grid spacing were required to provide initial and boundary conditions for a 100x100 node grid with 5km grid spacing. Twenty-eight variably spaced vertical levels were used for all runs with a model lid near 24km and greatest resolution near the surface.

Output on the 5km grid is shown in Figure 4. By 15:00 EST, lake breezes produced by the model were well-developed with convergence zones located along the north shore of Lake Erie and over Lake St. Clair. We used the observed surface meteorological data to assess the model's accuracy. The model faithfully simulated many observed parameters on this day. However, the model failed to allow the Lake Erie and Lake St. Clair breeze fronts to push northward as seen in the observations. Several attempts were made to correct this situation by improving the quality of the geophysical data with little success. We think that the coarse resolution (100km) of the initialization data from the RFE model near the Great Lakes tends

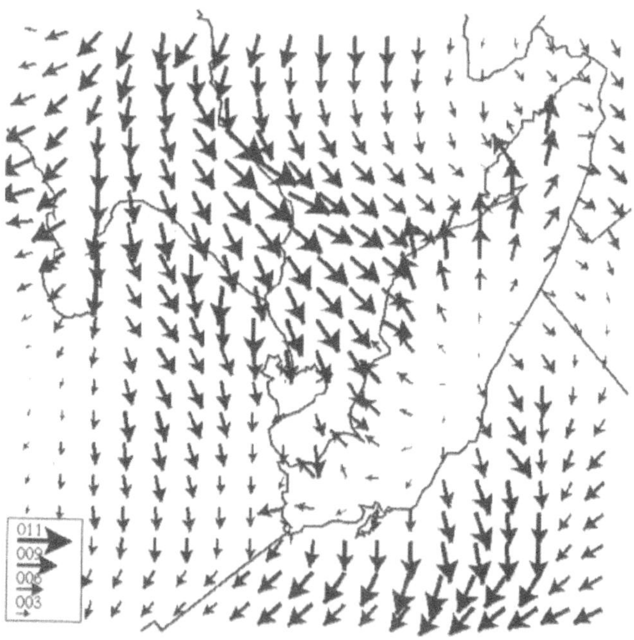

Figure 4. Output from the MC2 model on a grid with 5km spacing valid at 15:00 EST. A linear scale is used for the arrows with the largest arrow representing 11knots.

to cause an over-estimation of the strength of the northwesterly gradient wind in this case. This results in the lake breeze fronts being located more southerly than observed. Work is now underway to improve the accuracy of the initialization data.

CONCLUSIONS

Analysis of meteorological and chemical field data from August 14, 1993 suggests that lake breezes were responsible for rapid increases in ground-level ozone as they reached localities in southwestern Ontario and southeastern Michigan. A light northwesterly gradient wind allowed the Lake Huron lake breeze front to penetrate over 100km to the south, causing several stations near Lake St. Clair to experience two lake breeze related ozone increases through the day. The accuracy of our mesoscale modelling must be improved to more closely match observations so that detailed ozone transport analyses can be conducted. Thus, initialization data more accurately representing the Great Lakes region may be needed. The planned use of passive chemical tracers in the model will likely contribute greatly to the transport analyses.

ACKNOWLEDGEMENTS

The Ontario Ministry of Environment and Energy (OMEE) has provided funding for this continuing project through the Environmental Research Programme. Patrick King at the Atmospheric Environment Service (AES) provided visible satellite images and assisted in their interpretation. Thanks also to Donald Hastie at York for valuable input.

REFERENCES

Bowne, N.E., 1994, *Southeast Michigan Ozone Study Data Information*, ENSR Consulting and Engineering, Glastonbury, CT, 58pp.

Canadian Council of Ministers of the Environment (CCME), *Management Plan for Nitrogen Oxides (NO_x) and Volatile Organic Compounds (VOCs). Phase I*, Nov. 1990.

Melo, O.T., Ed., 1993, *The Southern Ontario Oxidant Study: The Right Atmosphere,* CIRAC, 9.

Tanguay, M., Robert, A., et Laprise, R., 1990, A semi-implicit semi-lagrangian fully compressible regional forecast model, *Mon. Wea. Rev.*. 118:1970.

Lyons, W.A. and Olsson, L.E., 1973, Detailed mesometeorological studies of air pollution dispersion in the chicago lake breeze, *Mon. Wea. Rev.*. 101:387.

Sills, D.M.L., Taylor, P.A., and Salmon, J.R., 1994, *Southern Ontario Oxidant Study - Meteorological Measurements. Final Data Report*, York University and Zephyr North, 87 pp.

DISCUSSION

D. STEYN: Is it possible that the increase in ozone after the passage of the lake breeze front is not due to vertical advection (as you hypothesize), but rather due to acceleration of photochemistry by oxidants isolated aloft and brought to the surface by entrainment in a growing boundary layer?

D. SILLS: In my presentation, I showed that if a sheet of passive tracer at 500 m is present before lake breezes develop in the MC2 simulation, the downward vertical motion behind the 'head' of the lake breeze front and the general subsidence within the lake breeze circulation can act to bring significant amounts of the tracer to the surface. This mechanism could explain observed increases in ozone with the passage of lake breeze fronts in certain regions of the study area. Your suggestion that the surface photochemistry could be accelerated by 'seeding' from oxidants of the previous day is interesting and will be explored.

R. KUNZ: Were all simulations performed with a model in non-hydrostatic mode?

D. SILLS: Yes.

R. KUNZ: Was the nesting one-way or two-way?

D. SILLS: Nesting was one-way. The MC2 model uses a 'sponge zone' at the lateral boundaries in which variables from the integration of the regional model are blended with those of the driving model using an attenuation function. Thus, the values along the outside edge of the sponge zone are those of the driving model while the values along the inside edge of the sponge zone are those of the regional model.

L. GANZEVELD: How is the dry deposition of ozone incorporated? I realize you want to illustrate the mechanism but I thought that if you are using ozone surface layer concentration to illustrate mechanisms you might also see a signal of differences in dry deposition over land/sea and with time.

D. SILLS: The use of the passive tracer was kept extremely simple. It was used only to see if the downward motion behind the lake breeze front would be strong enough to advect a significant amount of tracer to the surface. Thus, the addition of dry deposition was unnecessary for this experiment, However, it might be an interesting feature to include in future simulations.

IMPACT OF DIFFERENT CLOUDS ON TROPOSPHERIC CHEMISTRY

Nadine Chaumerliac, Nicole Audiffren, and Sylvie Cautenet

Laboratoire de Météorologie Physique / OPGC
Université Blaise Pascal / CNRS
24, Avenue des Landais
63177 AUBIERE Cedex FRANCE

INTRODUCTION

There are some experimental evidences that clouds can play an important role on the chemical composition of the troposphere through the modification of photolysis rates, the vertical transport of species and their removal by wet deposition and finally by the aqueous phase chemical reactions within cloud water or rainwater. Chemical analyses in several rainwater samples have demonstrated that it is necessary to know the type of cloud which led to the precipitations (Lefeivre, 1993). This is due to the fact that chemistry in presence of cloud depends on the microphysical features of the cloud which depend the rate of transfer toward aqueous phase. Here, we seek to appreciate the impact of a maritime cloud on tropospheric chemistry in a case of a remote atmosphere. We will make some comparisons with a case of a non-precipitating cloud appearing in the same conditions of temperature and humidity.

MODEL DESCRIPTION

A chemical module (Gregoire et al, 1994) describing the tropospheric photochemistry of ozone precursors in both gaseous and aqueous phases has been included in a mesoscale model with detailed microphysics. A bidimensional mountain wave scenario allows the formation of the orographic cloud. The half-width of the mountain (Chaumerliac et al, 1990) is of 25 km and its height is of is 1km. The horizontal uniform cloud is of 20m/s, and the relative humidity is of 80% until the altitude of 3km. Vertical coordinates of the model follow the terrain shape. The domain of simulation is of 400km large and sixteen levels are used. The timestep for dynamics is of 10s.

The microphysic module comes from Berry-Reinhardt (1974) scheme. This scheme includes the process of self-collection, at the difference of the Kessler's scheme. It assumes a log-normal law for the distribution of both raindrops and cloud droplets. For rainwater, there are two prognostic variables : the number of raindrops and the mixing ratio.

By the way of the autoconversion rate of cloud droplets to raindrops it is possible to choose the type of the cloud which is generated. Here, a maritime cloud (figure 1) is generated in a first run (Audiffren et al, 1995). A second run corresponds to the case of a nonprecipitating cloud : continental cloud (figure 7). Hence, the maritime cloud is composed of small drops of 20microns of diameter and raindrops of increasing size from 100 microns on the upwind side to 240 microns on the downwind side.

Air Pollution Modeling and Its Application XI, Edited by
Gryning and Schiermeier, Plenum Press, New York, 1996

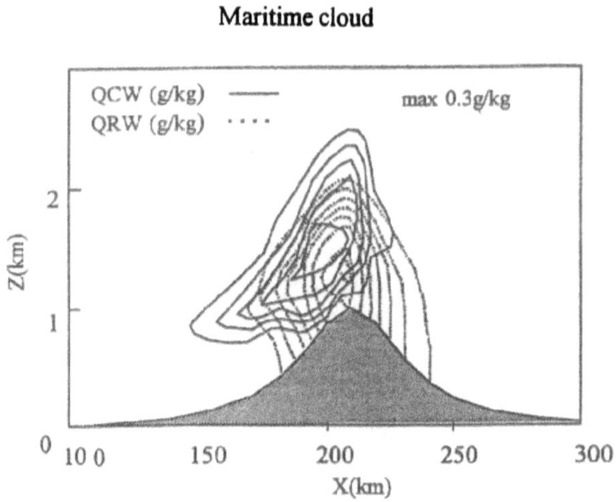

Figure 1: Maritime cloud : cloud water (solid lines) rainwater (dashed lines). Maxima of each water content are of 0.3g/kg.

Chemical reactions of twelve species, in both phases are treated at each timestep into the model. In the presence of cloud, the timestep is reduced to 0.5s. Initial concentrations are taken vertically constant (mol/cm^3) and are shown on table I. The photolysis rates are held constant during the whole simulation in order to focus exclusively on the sensitivity to the granulometry features of the clouds. The complete chemical scheme takes the following form:

$$\frac{d\,C_g}{dt} = P_g - D_g C_g - xwc\; k_{t_l} (C_g - \frac{C_{cl}}{xwc\; H_{eff} RT}) - xwr\; k_{t_r} (C_g - \frac{C_r}{xwr\; LH_{eff} RT})$$

$$\frac{d\,C_{cl}}{dt} = P_{cl} - D_{cl} C_{cl} + xwc\; k_{t_l} (C_g - \frac{C_{cl}}{xwc\; H_{eff} RT})$$

$$\frac{d\,C_r}{dt} = P_r - D_r C_r + xwr\; k_{t_r} (C_g - \frac{C_r}{xwr H_{eff} RT})$$

where C_g and C_{cl}, Cr are respectively the gaseous, cloud water and rainwater concentrations, expressed in molecules per cm^3 of air. P_g, P_{cl} Pr and D_g and D_{cl} D_r are the product and destruction rates respectively for gaseous and aqueous phases. xwc and xwr are the cloud water content and rainwater content (v/v) ; k_{t_l} and kt_r are the mass transfer coefficient (Schwartz, 1986) to cloud water and rainwater, respectively.

Table I Initial concentrations of the chemical species

O_3	30 ppb
H_2O_2	2 ppb
CH_3O_2H	0 ppb
CH_2O	0 ppb
CO	100 ppb
HNO_3	100 ppt
HO_2	0 ppt
OH	0 ppt
NO_2	5 ppt
NO	5 ppt
CH_3O_2	0 ppt
HCOOH	0 ppt

RESULTS

The results are presented at the end of the simulation which is of three hours. The evolution of OH is shown on the following figures. OH is submited to a depletion up to 50% and the radical CH_3O_2 (not shown) up to 80% into the heart of the cloud.

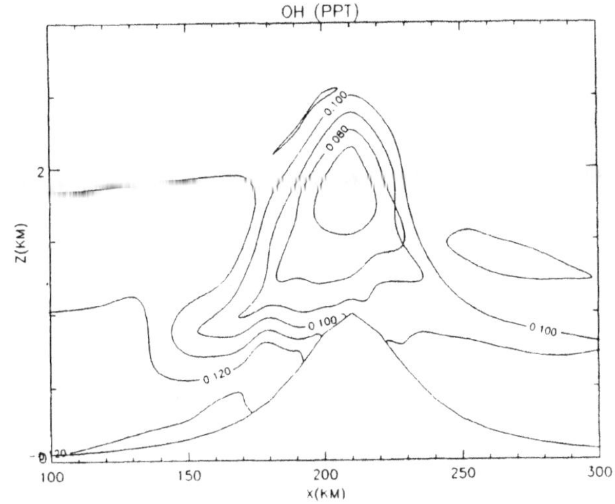

Figure 2: Concentrations of OH in the case of maritime cloud after three hours of simulation

On figure 3 are reported the gaseous production rate P_g/C_g for OH and its gaseous chemical destruction rate Dg following a streamline passing through the heart of the cloud (X=210km) where there are maxima of cloud water and rainwater. One can see that they are very close each other. Hence, the depletion of OH is almost exclusively due to the scavenging of OH as we can see on figure 4 by cloud droplets (term xwc kt_l) and by raindrops (term xwr kt_r). On the contrary, the radical CH_3O_2 shows a strong difference between P_g/Cg and D_g. Moreover, the rate of transfer to cloud water is the main sink for this species until the upper part of the downslope (X=230km), since it is greater than D_g (figure 6) of a factor 100. The other radical HO_2 (not shown) follows the same behaviour as CH_3O_2.

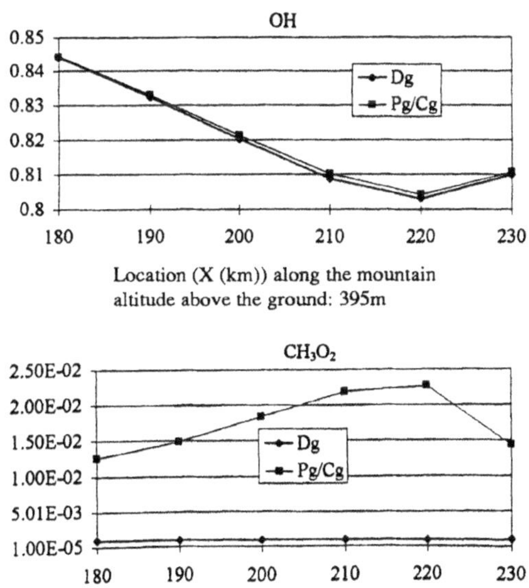

Figure 3: Gaseous production rate P_g/C_g for OH and its gaseous chemical destruction rate Dg

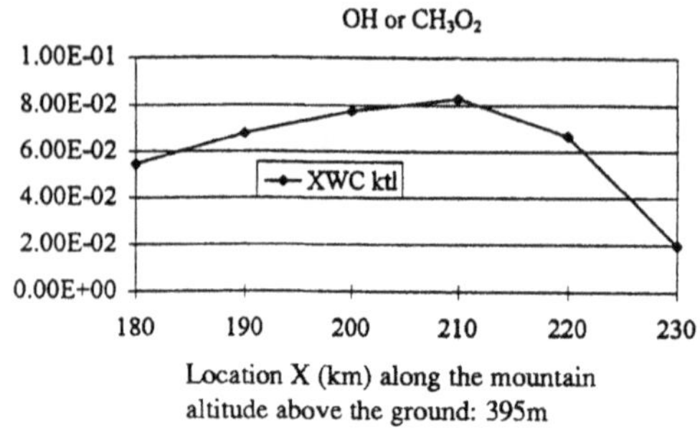

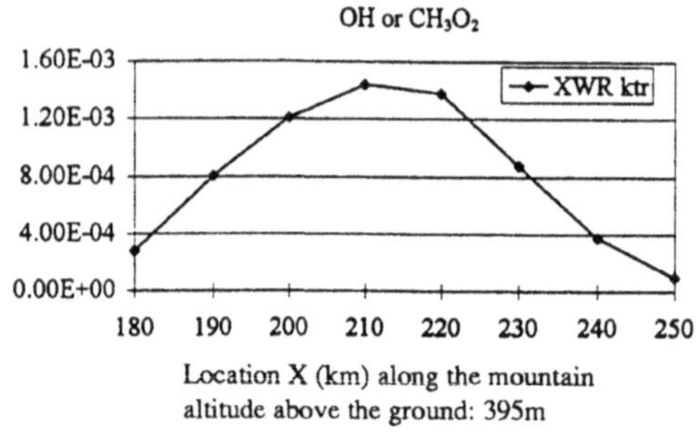

Figure 4: Rate of transfer toward cloud water and rainwater of OH and CH_3O_2 (same sticking coefficient)

The scavenging of the radicals has strong consequences on the ratio of the concentration of the nonsoluble species NO and NO_2. The following figure shows the modification of this ratio in the case of the maritime cloud. The ratio which is of 0.5 far upwind increases to the value 0.65 in the middle of the cloud and decreases to find again its initial value. That means a slowing down of the reactions of conversion of NO into NO_2 :

$$NO + HO_2 \text{------}> NO_2 + OH$$

$$NO + CH_3O_2 \text{---}> NO_2 + CH_2O + HO_2$$

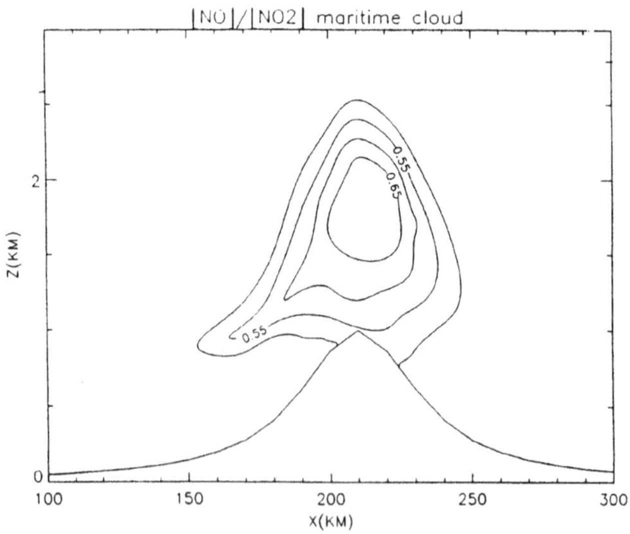

Figure 5: Ratio of the concentrations of NO and NO_2 in the case of maritime cloud

This slowing down contributes to the decrease of the chemical gaseous destruction rates D_g of these two radicals as can be seen on the following figure. These rates decrease from upstream (X=180km) to the middle of the cloud (X=210km) of 14% for HO_2 and 10% for CH_3O_2.

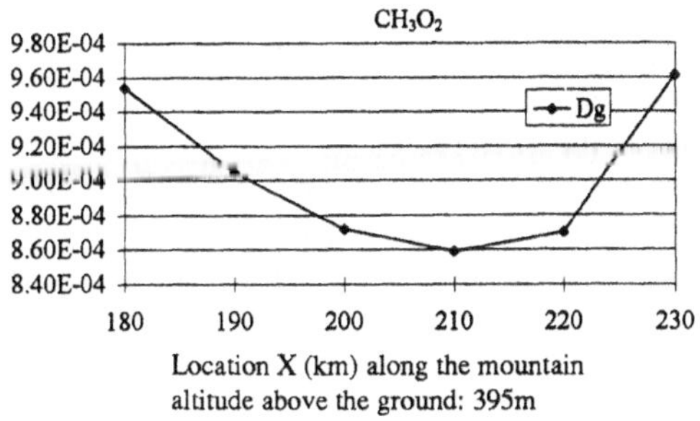

Figure 6: rate of destruction of CH_3O_2 by chemical reactions in gas phase

The differences with the continental case which is composed of droplets are located on the downwind side of the mountain. In this region, the ratio of the concentrations of NO and NO_2 are higher in the continental case with a difference up to 20% with the maritime case. This is due to the different gramulometry of the clouds. The maritime cloud is only formed of big raindrops, the population of cloud water drops is not present. The transfer rates depend on the size of the drop and are more rapid (factor of 100) toward droplets of the continental cloud. Hence, the continental cloud still scavenges radicals more efficiently. On the upwind side of the mountain, differences on this ratio are slight, because the maritime cloud has still in this region a population of droplets of the same size as the continental cloud.

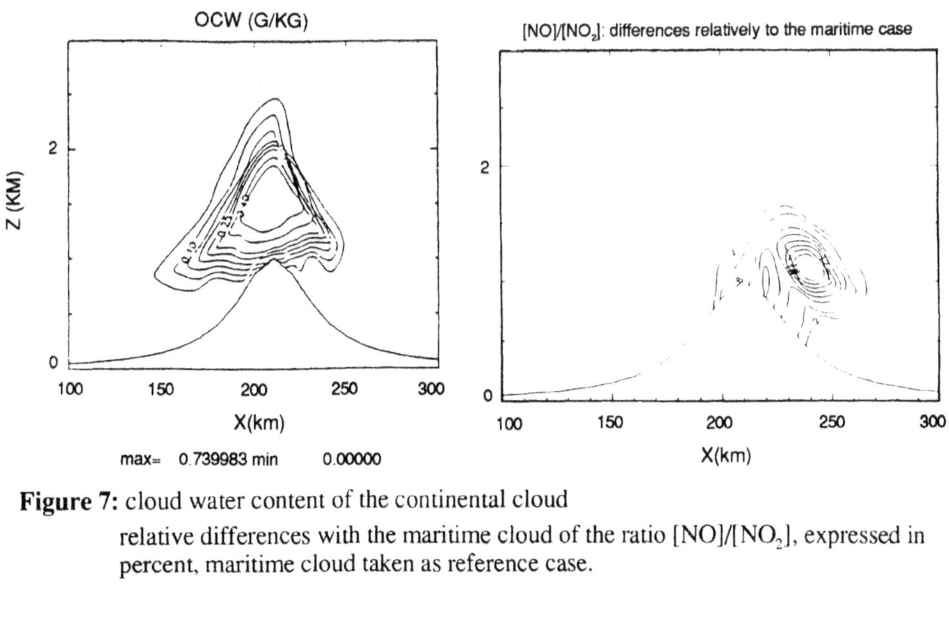

Figure 7: cloud water content of the continental cloud
relative differences with the maritime cloud of the ratio [NO]/[NO_2], expressed in percent, maritime cloud taken as reference case.

CONCLUSION

The scavenging of the radicals leads to a significant modification of the ratio [NO]/[NO_2] in the presence of cloud. However, in the case of a remote atmosphere the level of ozone is just slightly affected (1%).

The impact of the type of ganulometry of the cloud is pointed out on the rate of transfer toward aqueous phases. The rate of transfer has to be compared with the reactions rates in gas phase. Even, the destruction of OH by chemical reactions is of the same degee of the rate of transfer to rainwater and actually beneath the one toward cloud water, the whole contribution (production and destruction) by chemical reactions is quite zero. So for OH, the scavenging is almost the main reason for the depletion into cloud. On the contrary, the scavenging of the radicals is higher than their gas-phase chemical destruction in the case of cloud water and tends to be lower in case of rainwater.

Hence, for a given species the competition between chemical reactions and scavenging from one cloud to another one depends strongly on its microphysical features.

REFERENCES

Audiffren N., Chaumerliac N., and Renard M., 1995, Effects of a polydisperse cloud on tropospheric chemistry, submitted to *J. Geophys. Res.*

Berry E.X. and Reinhardt R.L., 1974, An analysis of cloud drop growth by collection. Part II. Single intitial distributions, *J. Atmos. Sci.*, 31: 1825-1831.

Chaumerliac N., Richard E., Rosset R., and Nickerson E.C., 1990, Impact of two microphysical schemes upon gas scavenging and deposition in a mesoscale meteorological model. *J.Applied Meteorol.*, 30: 88-97.

Gregoire P.J., Cheumerliac N., and Nickerson E.C., Impact of cloud dynamics on tropospheric chemistry: advances in modelling the interactions between microphysical and chemical processes, *J. Atmos. Chem.*, 18, 247-266

Lefeivre B., 1993, Influence of the rain water content on the tropical precipitation chemistry, *EGS*.

Schwartz S., 1986, Mass-transport considerations pertinent to aqueous phase reactions of gases in liquid water clouds, *Chemistry of multiphase systems*, W. Jaechke, Ed., Springer, pp. 415-471.

ACKNOWLEDGMENTS

EDF (Electricité de FRANCE) and IDRIS (Institut du développement et des ressources en informatique scientique) are greatly acknowledged for their financial support and computer time on CRAY C90.

DISCUSSION

C. NAPPO: Were the effects of the release of the latent heat included in the dynamics? Also, was the latent heat included in the cloud chemistry?

A. FLOSSMANN: The effects of the release of latent heat are included in the equation of temperature in the dynamic model. Consequently, the additional buoyancy resulting from condensation is taken into account for the development of the cloud. As the reaction rate and Henry's law coefficients are dependent on temperature, changes of temperature due to release of latent heat also enter into the cloud chemistry module.

TRANSPORT AND CHEMISTRY OF AIRCRAFT EMISSIONS IN THE VICINITY OF AN AIRPORT

Jutta Graf and Hans G. Tremmel

Institut für Physik der Atmosphäre
Deutsche Forschungsanstalt für Luft- und Raumfahrt (DLR)
Oberpfaffenhofen, Germany

INTRODUCTION

During the last few years the investigation of the environmental effects resulting from aircraft exhaust emissions have been intensified due to a growing awareness of climatic changes of the atmosphere. Emissions from aircraft engines include carbonmonoxide (CO), nitrogin oxides (NO_x), organic compounds (HCs), water vapor and various other gases and particles (Schumann, 1994). Under favourable meteorological conditions, the production of ozone (O_3) is mainly controlled by the concentration level of NO_x and HC. Aircraft emissions may have an effect on these ozone precursors and thus on the formation of ozone.

This study estimates the immission concentrations of various gases resulting from aircraft- and motor vehicle emissions in the vicinity of Munich Airport. A hydrostatic mesoscale model in combination with a chemical model is used to calculate the temporal and spatial distribution of the primary emitted and secondary produced pollutants. The emission inventory is derived from air traffic statistics and measured road traffic densities. Based on a climatological statistic of the synoptic forcing and temperature profiles, many simulations are realized, with the goal to achive seasonally and yearly averaged values. In addition, specific episodes are simulated and the results compared with data collected in the vicinty of the airport since 1991. This paper focusses on the preparation of the emission data and the episodic simulation.

MODEL DESCRIPTION

The three-dimensional, hydrostatic model REWIH3D (Heimann, 1990) is used to simulate the meteorological fields and the dispersion and chemical transformation of the emitted pollutants. The model is forced by an outer scale (synoptic) flow, and calculates the mesoscale modifications of the prescribed fields due to the influence of topography and thermal effects. A chemistry submodel, using the RADM-mechanism (Stockwell, 1986) is coupled with REWIH3D. The RADM-mechanism considers 40 pollutant species; 15 are assumed to be in quasi steady state, and 25 are integrated in time. The chemistry equations are solved with the Quasi Steady State Analysis (QSSA) (Hesstvedt et al., 1978).

Air Pollution Modeling and Its Application XI, Edited by
Gryning and Schiermeier, Plenum Press, New York, 1996

The required photolysis frequencies are obtained from a vertical one-dimensional model (Ruggaber, 1994), and calculated every hour assuming horizontal homogenity.

For the present application the model uses an orthogonal horizontal grid with a grid-spacing of 1 km. The atmosphere is devided vertically into 20 non-equidistant layers. The resolution is higher in the mixing layer, and decreases with height. The model domain covers an area of 30 x 30 km^2, the Munich airport is located in the middle of the domain.

EMISSION SOURCES

In addition to air traffic, highways and Munich Airport destination traffic are considerd as emission sources.

Aircraft emissions

For every box of the model domain, the average aircraft emissions E of the primarily emitted species NO_x, CO and (the sum of) hydrocarbons (HC) is calculated according to the following equation:

$$E = (N_{TO} \cdot t_{TO} \cdot e_{TO} + N_A \cdot t_A \cdot e_A + N_I \cdot t_I \cdot e_I) \cdot n \quad (1)$$

where the indices TO, A, and I mean take-off, approach (landing), and idle (taxi). N is the number of aircraft movements (of the respective aircraft engine mode) crossing one grid cell; t is the duration of the aircraft movement in this box; e is the average emission factor (mass per time) per aircraft engine in the respective engine mode, and n is the average number of engines per aircraft.

The total number of aircraft movements N amount to 524 per average day in 1993 and is derived from data of the Munich Airport GmbH. From these data we can calculate an emission inventory for an average day in 1993 as well as inventories for every single day in 1993. The number of aircraft movements are divided equally into approaches and departures, evenly distributed between northern and southern runways. For the air traffic, 36% is to or from the east, 64 % is to or from the west. The preference for the westerly direction is the result of the prevailing westerly wind direction in this region. The distribution of the number of aircraft movements in the taxi-in and taxi-out mode and in the departure and approach mode into the different grid cells could also be derived from these data and also from the flight route model of the BFS (Bundesanstalt für Flugsicherung) and the map of the Munich Airport. The diurnal variation of N (derived from flight schedules), shown in Fig. 1, enables us to calculate hourly emissions.

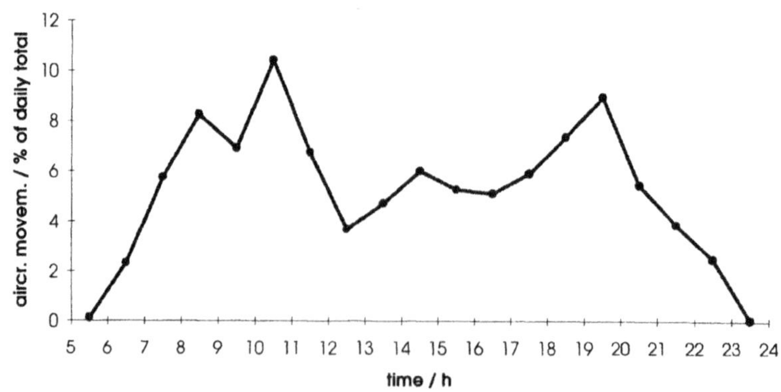

Fig. 1: Diurnal variation of the number of aicraft movements at Munich Airport given in % of the daily total.

The duration, t, of an aircraft movement across one grid cell was calculated as the ratio of path length and aircraft speed, or was inferred from the average duration of taxiing-in (5 min.) and taxiing-out (11 min.) for Munich Airport, respectively.

The average emission factors, e, per aircraft engine were derived from the frequency distribution of aircraft types for the Munich Airport 1993 (the 12 most frequent types are plotted in Fig. 2). Considering the frequency distribution of airlines at Munich Airport, where Lufthansa contributes with 40 % of all aircraft movements, we derived a frequency distribution of aircraft engine types (e.g. B737 at Munich Airport subdivides to: 60% CFM56-3-B1 engine, 26% JT8D-15, 9% CFM56-3C-1, and 5% JT8D-15A). The average emission factors, e, per aircraft engine at Munich Aiport 1993 were then calculated by averaging the specific emission factors of the different engine types (ICAO, 1993) according to their frequency of occurence. The resulting emission factors are listed in Table 1. The average number of engines per aircraft, n, for Munich Airport 1993 amounts to 2.065.

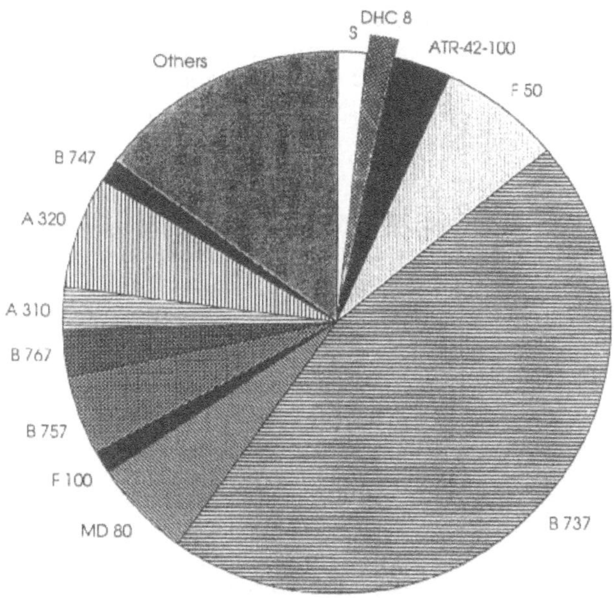

aircraft	%
B 737	45,73
F 50	7,33
A 320	6,59
MD 80	5,82
B 757	4,74
ATR-42-100	3,25
B 767	2,85
A 310	2,42
Saab 340	1,85
DHC 8	1,59
B 747	1,58
F 100	1,51

Fig. 2: Frequency distribution of aircraft types for Munich Airport 1993 (portions in % of all aircraft movements in 1993).

Table 1: Average emission factors, e, in kg/h per aircraft engine for Munich Aiport 1993.

	Take-off	Approach	Idle
e (NO_x)	102.05	9.92	1.70
e (CO)	3.49	5.54	14.72
e (HC)	0.81	0.89	3.77

The emission rates E for all grid boxes for Munich Airport are calculated by inserting the above described parameters in eq. 1. The sum of aircraft emissions at Munich Aiport for an average day in 1993 in the 11 grid cells of the lowest layer up to 50 m height totals:

E(NOx) = 1349 kg/d

E(CO) = 2277 kg/d

E(HC) = 575 kg/d

Furthermore, the emissions of HC have been subdivided according to Spicer et al. (1992), with methane as the dominant species (78 % by weight), followed by ethene (12 %), acetylene (2.6 %), propene (2.5 %), benzene (1.4 %), toluene (1.2 %), ethane (1.1 %), propane (0.9 %), and additionally oxidized hydrocarbons like formaldehyde (15 %), acetaldehyde (5.3 %), and acetone (4 %).

Besides these primary emissions, we considered the production of secondary reaction products in the high-HO_x and high-temperature regime in the aircraft exhaust plumes at short time scales (until HO_x and T has lowered to ambient values). This was done by simulations with our own exhaust plume model, consisting of 173 reactions of the inorganic and the most important organic species. Time-dependent functions of decreasing temperature and increasing plume cross-section are also incorporated in this model. The most abundant secondary products in the plume are HNO_3 and HNO_2. But their concentration ratio to the emitted NO_x is only about 1 % and 3 %, respectively. Thus they are not very important to the emission inventory, especially when considering the low height level of our inventory, where the ambient concentrations of these two species are relatively high.

Motor vehicle emissions

The main motor vehicle traffic in the modelling domain occurs on highways A9, A92 and on the departure from the highway A92 to the airport. The emissions on these highways are estimated from the measured traffic density, a statistical engine distribution and mean emission factors for NO_x, CO and HC. The emissions of the total hydrocarbons are subdivided into the specific components after Bailey et al. (1990).

In table 2, the annual average emissions are summerized for the two categories. The emissions per gridcell are of the same order of magnitude. The CO and NO_x road traffic emissions are 3.2 and 1.4 times higher respectivly than the air traffic emissions. However the HC emissions from the aircraft are only slightly higher. The data for the total emission in the lowest layer show that the road traffic emissions predominate.

Table 2: Average air- and road-traffic emission per gridcell and summed over the model domain in the lowest layer (Unit: kg/d).

	air traffic per gridcell	road traffic per gridcell	air traffic total	road traffic total
CO	207	670	2278	46903
NO_x	122	168	1349	11817
HC	52	45	575	3167

RESULTS AND DISCUSSION

As stated above, the object of this paper is the simulation of an episode with the goal to calculate the effect of air traffic emissions on the resulting pollutant concentrations. We selected a period in March (12th to 15th March 1993), where the observed data show increasing temperature and pollutant concentrations with time. The NO_2-concentration reaches a maximum of 70 $\mu g/m^3$, and the mean concentration (50 $\mu g/m^3$) was about 50 % greater than the other days in March. The mean ozone-concentration also increased from 20 µg/m³ to 40 µg/m³. The ground measurements and radiosonde data indicated an easterly flow with 2 - 3 m/s winds near the surface, turning to south with height. The simulation was started on midnight at March 12th, and was initialized with the radiosonde data. Typical values for the atmospheric constitutents were used to initialize the chemical part of the model. At the inflow boundaries, these values were kept constant; at the outflow boundaries, a radiation condition was specified.

The simulated flow in the morning at 10 m above ground, shown in fig. 3, is represented in form of wind traces and indicates south-easterly flow with low wind speed. The influence of topography is very moderate due to a relativly flat terrain; the orography is rising from 400 m in the south of the model domain to 500 m in the north with a maximum slope of 30 m per km. At this time of the day, the mixed layer is well developed and thus the potential temperature profile is nearly adiabatic. Under these conditions a relativly undisturbed flow is to be expected.

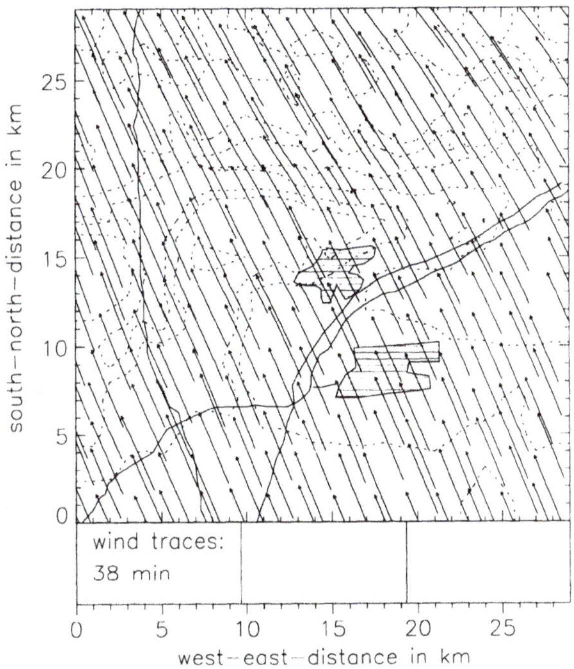

Fig. 3: Simulated flow filed at 10 m above ground in the morning of 13th March 1993. The arrows represent wind traces.

With this south-easterly flow, the emissions are transported to the north-west, and are chemically converted along the way. For example, consider the calculated NO-concentration presented in fig. 4. The highest values occur along highway A9. The surface aircraft emissions are transported 15 km north of the airport, and are influencing the immissions in the nearest town, Freising.

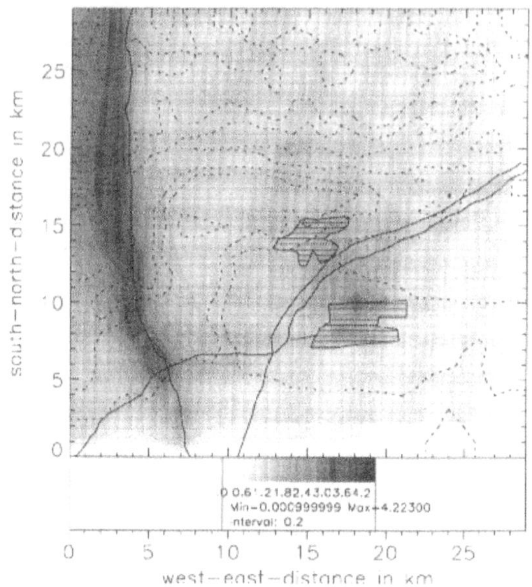

Fig. 4: NO-concentration ($\mu g/m^3$) at 10 m above ground in the morning of 13th March 1993.

Due to the prevailing south-easterly flow on this particular day, aircraft landings and take-offs where to the east. The immissions resulting from the aircraft operations are noticable up to a height of 1500 m above ground. In the early evening, when the air traffic has its second maximum (see fig. 1) and the oxidation of NO to NO_2 is effective, the NO_2-concentration reaches maximum values of 8 $\mu g/m^3$ in the flight tracks. This value is about 50 % greater than the prescribed initial background value. The oxidation of NO to NO_2 leads to a depletion of ozone. This chemical behaviour is also reproduced by the model as shown in fig. 5.

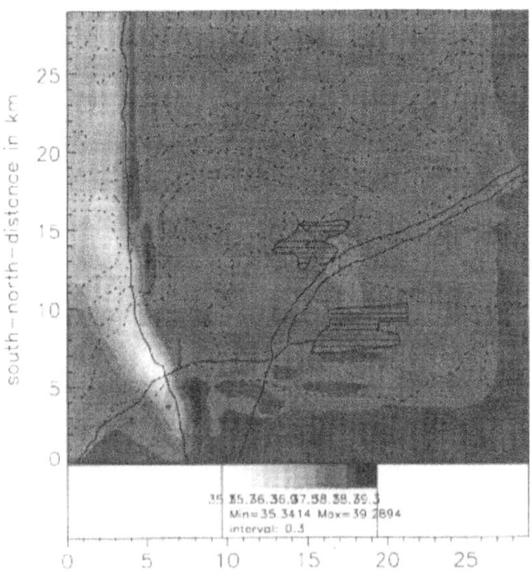

Fig. 5: Ozone-concentration ($\mu g/m^3$) at 10 m above ground in the early evening of 13th March 1993.

CONCLUSIONS

In this paper, initial air-pollution modelling results, using a combined dynamic-chemical model and considering emssions from aircraft and motor vehicles are presented. Air traffic emissions are small compared to road traffic emissions in the model domain. Nevertheless, the effect of aircraft on the immissions in the vicinity of Munich airport are seen in the results; especially on the simulated day, when the airport aera is not polluted from the highway emissions. However when averaged over a season or a year immissions resulting from aircraft emissions are expected to be small.

REFERENCES

Bailey, J.C., B. Schmidl, M.L. Williams, 1990: Speciated hydrocarbon emissions from vehicles operated over the normal speed range on the road, *Atmos. Environ. 24A, 17-29.*

Heimann, D., 1990: Three-dimensional modelling of synthetic cold fronts approaching the Alps, *Meteorl.Atmos.Phys. 42, 197-219.*

Hesstvedt, E., O. Hov, P. Speth, 1978: Quasi-steady-state approximations in air pollution modelling, *Comp. J. Chem. Kinetics X, 971-994.*

IACO, 1993: IACO engine exhaust emission data bank, *First edition, Defence Research Agency, Farnborough, Hampshire.*

Ruggaber A., 1994. Modellierung photolytischer Prozesse in der Troposphäre, *Dissertation an der Fakultät für Physik der Universität München.*

Schumann U., 1994: On the effect of emissions from aircraft engines on the state of the atmosphere, *Ann. Geophysicae 12, 365 -384.*

Spicer, C.W., M.W.Holdren, D.L.Smith, D.P. Hughes, M.D. Smith 1992: Chemical composition of exhaust from aircraft turbine engines, *J. Eng. for Gas Turbines and Power 114, 111-117.*

Stockwell, W.R., 1986: A homogeneous gas phase mechanism for use in a regional acid depostion model, *Atmos. Environ., 20, 1615-1632.*

DISCUSSION

B. YAMARTINO: Most of the HC and CO emissions from aircraft are emitted on the ground but NO_x is emitted aloft. Did you account for the vertical distribution of aircraft NO_x?

J. GRAF: The vertical distribution of all aircraft-emissions, most important for NO_x, was considered using a flight route model for the Munich airport and different emission factors for the respective engine modes.

MASS BALANCES AND INTERACTION OF BUDGET COMPONENTS OF CHEMICALLY REACTIVE AIR POLLUTANTS OVER THE FEDERAL STATE OF BADEN WÜRTTEMBERG (GERMANY)

Hans-Jürgen Panitz, Klaus Nester, and Franz Fiedler

Institut für Meteorologie und Klimaforschung
Forschungszentrum Karlsruhe/Universität Karlsruhe
Postfach 3640, D-76021 Karlsruhe, Germany

INTRODUCTION

The level of pollution of a region is not only dependent on the regional emissions from the industry, the traffic, and the private households, but also on the meteorological conditions, the long range transport of polluted air from outside into the area, the deposition, and on the chemical behaviour of emitted and advected pollutants.

The budget equation of a pollutant can be written as

$$\frac{\partial C}{\partial t} = P_{ADV} + P_E + P_{CH} - L_{ADV} - L_{DEP} - L_{CH}.$$

The left-hand side describes the temporal change of the concentration C of a species, the right-hand side represents the production P and the loss L due to advection (ADV), emission (E), deposition (DEP), and chemical transformation (CH). The integration over a volume gives a balance equation describing the temporal change of the mass of a pollutant. By calculating mass balances it is intended to quantify to what extent the different mass budget components - regional emission, deposition, chemical transformation, and transport - contribute to the pollution of a mesoscale area, and how they change with time. The paper focuses on the results achieved for Ozone (O_3) and two groups of ozone precursors, namely NO_x (= NO + NO_2), and those volatile organic compounds (VOC) which are emitted from the regional sources (e.g. Ethane and other higher Alkanes, Ethene and other Olefins, Isoprene, Toluene, Formaldehyde, Acetaldehyde, and Ketones).

SIMULATIONS

The non-hydrostatic mesoscale model KAMM [1] is used to calculate the space and time dependent distributions of the wind field, the temperature field, and the turbulence

Air Pollution Modeling and Its Application XI, Edited by
Gryning and Schiermeier, Plenum Press, New York, 1996

field, which are necessary input informations for the Eulerian grid diffusion model DRAIS [2]. The DRAIS model, which is coupled with the chemical transformation mechanism RADM2 [3], calculates the concentration and deposition distributions of reactive chemical species.

The simulations have been performed for September, 16th, 1992, the first day of the second intensive measuring phase of the TRACT-experiment [4]. The model domain comprises the south-west part of Germany including the Upper Rhine Valley and the Black Forest as well as parts of the Alsace (France) including the Vosges Mountains (Figure 1). The model area covers 260 km * 240 km with a horizontal resolution of 5 km. In the vertical the model domain is divided into 35 levels between the ground and 8000 m. The mass balance terms are calculated for the whole model volume which has a size of $5.0 \cdot 10^{14}$ m^3, approximately.

The KAMM model is forced by a large-scale basic state being derived from radiosonde profiles which have been performed every 3 hours during the intensive measuring phase at different locations in the TRACT area [5].

As far as available for the simulation day, aircraft measurements [6], and observations near the ground have been used to determine the initial concentration distributions for the different species being considered in the DRAIS model.

No information about the air pollution outside the model area has been yet available. In order to take into account the pollution of advected air, average emissions have been defined along the boundary of the model domain assuming that they reflect the emission conditions in the outer regions. This assumption probably leads to an underestimation of the importance of the advective flux term in the mass balance, especially, for secondary species like ozone. As soon as simulations are available it is intended to derive the initial and the boundary conditions for the DRAIS model from the results of the large scale EURAD model [7] which covers most parts of Europe and whose results can be nested into the mesoscale domain of the DRAIS model by an one-way nesting procedure [8]. This will allow a more realistic consideration of the long range transport.

The emission data which have been used comprise anthropogenic and biogenic emissions [9], [10]. The parameterization used in the DRAIS model [11] takes takes into account the biogenic emissions of isoprene, α-pinene, monoterpenes and other VOCs.

RESULTS

The Flow Field

Figure 1 shows the horizontal wind field in 10 m above the ground at 09:00 UTC. The channelling of the flow along the Rhine Valley is not as pronounced as in simulations which have been carried out for other meteorological situations [1], [8]. The flow field in the boundary layer is characterized by different local wind systems and a general change of the wind direction in the afternoon. The profiles of wind direction at the radiosonde station of Stuttgart reflect that, during the morning, the surface wind at this location is blowing from south-east (Figure 2 top). With increasing height the direction changes towards west and north-west. In the afternoon, the surface wind has a pronounced northerly component, whereas no siginificant change is observed in the free troposphere. The changing meteorological conditions can also be found in the observed and simulated diurnal variations of wind speed and wind direction at the the rural station Karlsdorf (20 km north of Karlsruhe, Figure 2 bottom).

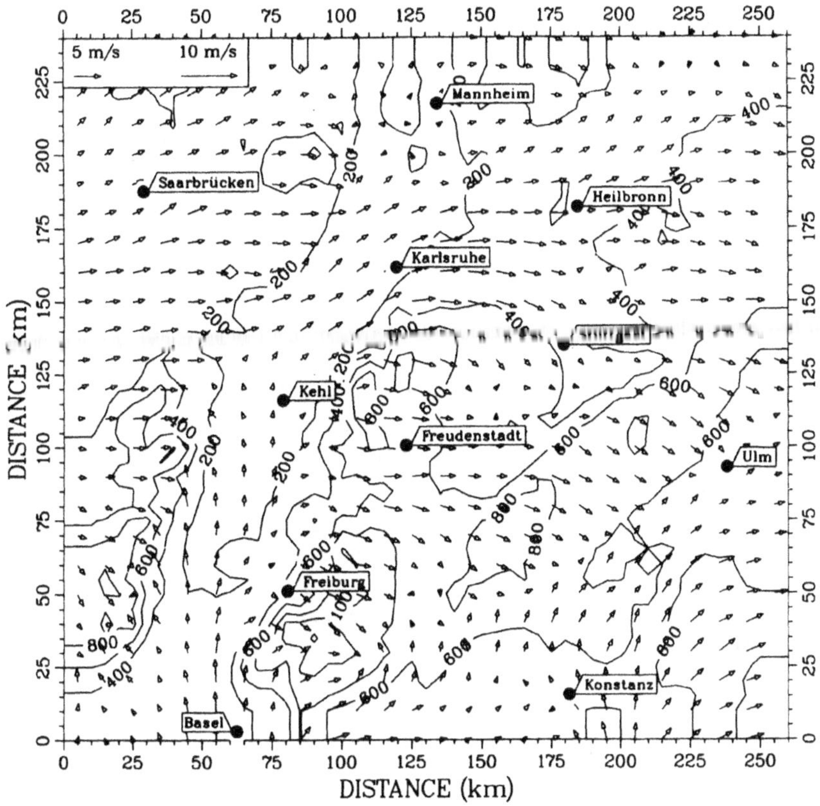

Figure 1. **Simulated wind field (m/sec) at 28 m above ground, 16. Sept. 1992, 09:00 UTC and topography (m)**

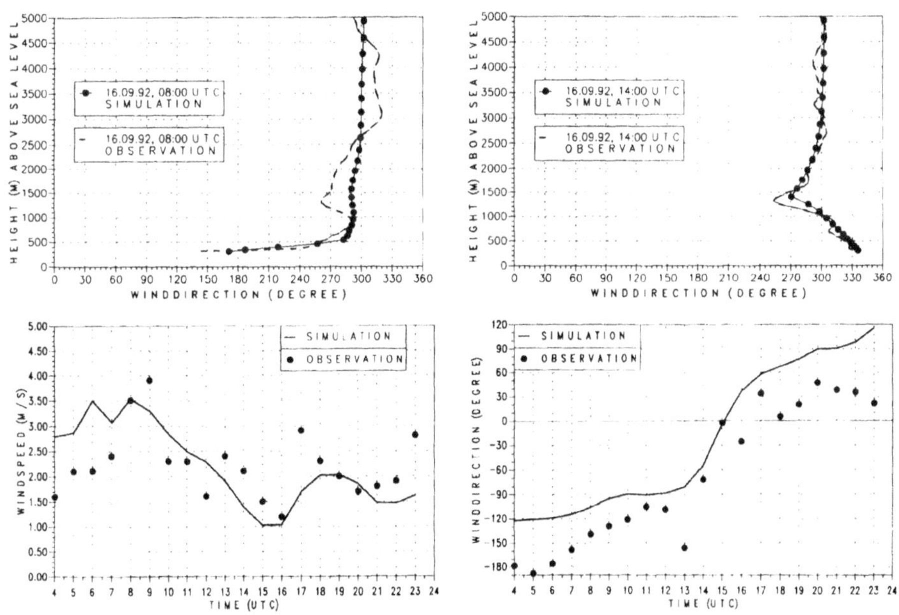

Figure 2. **Profiles of wind direction at Stuttgart (top) and time series of wind direction (negative from north to west) and speed at Karlsdorf (bottom)**

The Concentration Fields

As an example of the simulated concentration fields Figure 3 shows the Ozone distribution in 28 m above ground at 06:00 UTC. The field is influenced by the orography and the locations of NO sources. The highest concentrations between 30 and 35 ppb occur in the less industrialized regions of the Black Forest, the Vosges, and Swabian Mountains. Along the highways, and especially in the densely populated and highly industrialized areas around Stuttgart, Mannheim, Karlsruhe, and Heilbronn with high NO emissions, the O_3-concentrations are very low.

The diurnal variations of observed and simulated Ozone, NO, and NO_2 for the station Rastatt are shown in Figure 4. The station is located in the Rhine Valley on the outskirts of the city of Rastatt, close to a small road but only 800 m away from a highway with high traffic density. The high NO concentrations are underestimated by the model by more than a factor of three. High emissions occuring close to the observation point cannot be resolved in the model. The underestimation of NO explains also the difference between observed and measured Ozone in the morning. The high NO values at the station keep the Ozone concentration at a very low level because Ozone is destroyed by the chemical reaction with NO. The model simulates less NO. Therefore, less Ozone is converted into NO_2. During the evening the chemical interactions between Ozone and NO can be recognized as well in the measurements as in the simulation results. Due to an increased amount of NO, Ozone is destroyed which leads to an increase of the NO_2 concentration.

The Mass Balances

The mass balance components emission, advection, deposition, and chemical transformation being calculated for the whole model volume are illustrated in Figure 6 for Ozone, NO_x, and a lumped group of hydrocarbons. Since all contributions have been divided by the value of the integration volume, the figure presents the diurnal variations of hourly concentration changes.

During daylight chemical processes dominate the change of the Ozone concentration, whereas advection and deposition are less important. After noon, the increase of Ozone weakens and a loss of mass is observed between 4 p.m. and 6 p.m.. This is the time of the second NO emission maximum and the chemical reaction between NO and O_3 leads to negative changes of the O_3 concentrations. During the night hours positive changes of mass are caused by the transport of Ozone across the lateral boundaries of the volume. Integrated over the whole day, about 68% of the gain of mass of Ozone are due to chemical processes, advection contributes with an amount of 32% (Table 1). The most important sink of Ozone is its deposition with a contribution of 61% to the total loss. 39% of the loss are caused by chemical decomposition during the night hours.

Changes of NO_x are predominately caused by emissions of NO in the morning and by chemical transformations of NO and NO_2 with a midday maximum decomposition. The contributions of advection and of deposition can be neglected. More than 96% of the gain of NO_x are due to the emissions from the different sources inside the model area. Chemical processes are the most important sink of mass of NO_x (more than 90% of the net loss). At noon, they prevail the positive contribution of the emissions, so that the total mass is slightly drecreasing for a few hours. Towards the late afternoon, the total change is positive. To understand the maximum change at 5 p.m. it is necessary to consider the budgets of NO and NO_2 seperately. At that time, the chemical reaction between NO and O_3 causes a strong negative contribution to the NO budget and it leads to a pronounced

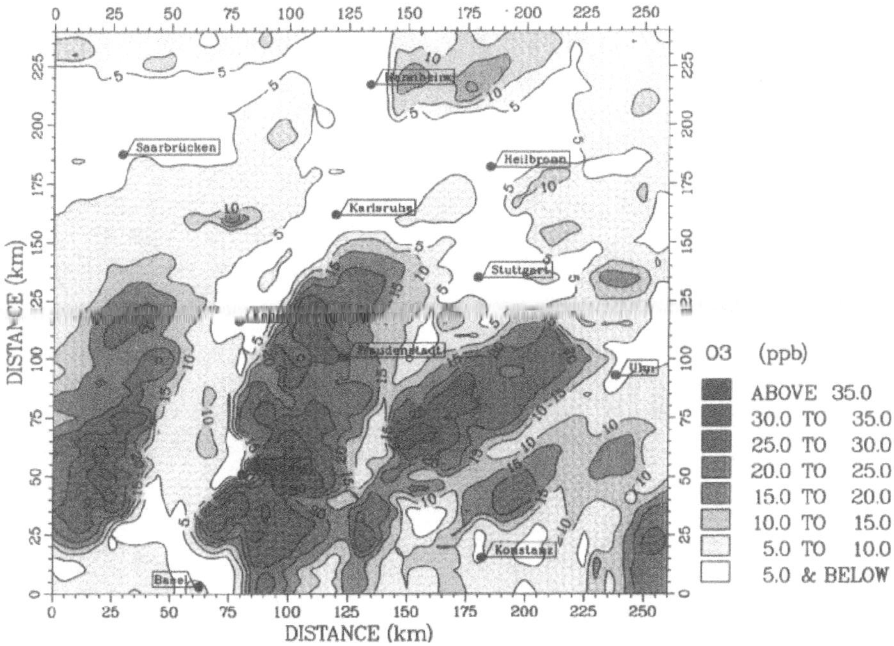

Figure 3. Simulated Ozone concentration at 28 m above ground, 16. Sept. 1992, 06:00 UTC

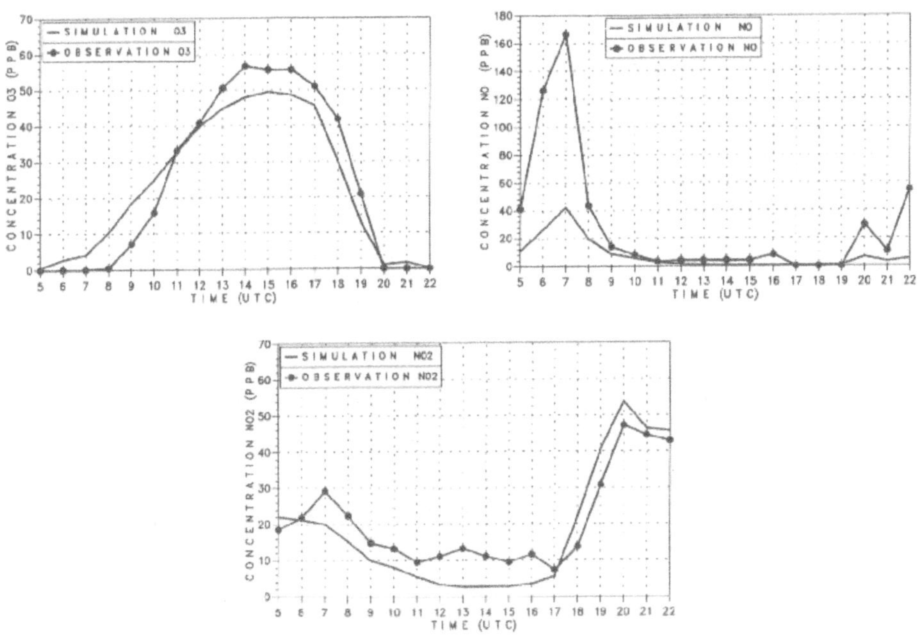

Figure 4. Time series of simulated and measured concentrations of O_3, NO, and NO_2, at Rastatt

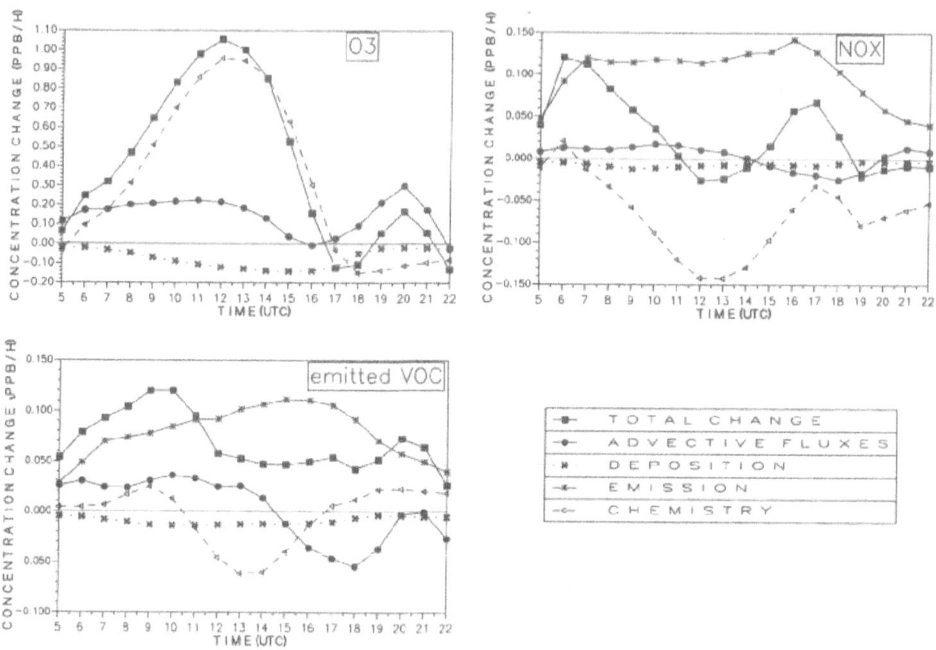

Figure 5. Hourly changes of mass budget components for O_3, NO_x, and emitted VOC

Table 1. Time integrated concentration budget for O_3, NO_x, and a lumped group of hydrocarbons being emitted

specie	initial concentration (ppb)	production (ppb)	loss (ppb)	contribution to production (%)			contribution to loss (%)		
				emission	chemistry	net flux	deposition	chemistry	net flux
O_3	53.8	9.3	2.2	-	68.4	31.6	61.1	38.9	-
NO_x	0.6	1.9	1.5	96.5	1.0	2.5	9.5	90.5	-
VOC	1.3	1.8	0.4	84.3	10.2	5.5	42.6	57.4	-

increase of the NO_2 concentration. Therefore, the net loss of NO_x due to chemical transformations approaches zero. Adding the positive contribution of NO emissions causes the net increase of NO_x. With decreasing emissions and due to chemical decompositions of NO and NO_2, the concentrations of NO_x are slightly decreasing during the evening and the night.

The budget of the VOCs represents a lumping of those hydrocarbons being emitted from anthropogenic and biogenic sources. The net budget of these species shows an accumulation of mass because the positive contributions of the emissions, the advection, and of the chemical processes dominate the deposition, the chemical decomposition in the middle of the day, and the negative contribution of the net fluxes over the lateral boundaries of the model volume. The positive net contribution of chemical processes in the night, the morning hours, and the evening is due to the net chemical production of Formaldehyde, Acetaldehyde, Acetic Acid, and Ketones. With the exeception of Acetaldehyde which experiences a net loss in the middle of the day due to a combination of photolysis and thermal reactions with OH and NO_3, the other three species show a 100% net gain of mass due to chemistry during the whole day. Their production is the result of a number of reactions between non-emitted hydrocarbons and OH, and between different kinds of peroxyradicals. It exceeds the decomposition of these species which is caused by photolysis, and reactions between OH and NO_3. All other hydrocarbons (Ethane and higher Alkanes, Ethene, Propene, Butene, Isoprene, and Toluene) which have been considered for the lumped budget are not at all produced by chemical processes. The only sources for these species are the emissions and the advective fluxes.

With respect to the overall budgets of the three specie groups considered, the advective flux of O_3 is the most important one (32% versus 2.5% for NO_x and 5.5% for the lumped VOCs). The time variations of the concentration changes due to the advective fluxes across the lateral boundaries of the model area reflect the change in the flow regime during the afternoon (see also Figure 2). In the early afternoon, the positive contributions of the advective fluxes to the mass budgets weakens, and in the late afternoon and the evening, the outflow even dominates the inflow for NO_x and the VOC group.

CONCLUSION

The contributions of different budget components to the net mass balance for Ozone, Nitrogen Oxide, and the lumped group of all emitted hydrocarbons have been analysed and discussed. The mass budget of O_3, which is a representative of the non-emitted chemical species, is dominated by the chemical transformations during daylight. The production of O_3 due to chemical processes inside the model volume is more than twice the positive contribution of the advective fluxes across the lateral boundaries of the model volume. But this result might be revised if the boundary values can be derived from the results of the large-scale model EURAD. The budget of the ozone precursor NO_x is dominated by the emissions of NO and the chemical transformations of NO and NO_2. Most of the time, the chemical decomposition of NO is more pronounced than the gain of mass due the emissions leading to a net decrease of NO. The changes of NO_2 are predominantely caused by chemical processes with positive net contributions to the NO_2 and NO_x budgets. The lumped group of VOCs comprises a number of different species which are emitted from anthropogenic and biogenic sources. The budget of this group shows an accumulation of mass over the whole day which is caused by the emissions and the advective fluxes. At the end of the day the content of the VOCs in the model volume is more than twice the initial value, whereas O_3 and NO_x increase by 13% and 67%, respectively.

The mass balance calculations have been carried out for the whole model volume. In the future it is intended to calculate the different budget components in smaller volumes which are part of the large one. Thereby it will be distinguished between urban and neighbouring rural areas in order to investigate, on the one hand, spatial variations of the budget components, and, on the other hand, the influence the urban plume on the rural region.

ACKNOWLEDGMENT

The research was funded under contract 2 93 002 with Projekt Europäisches Forschungszentrum für Maßnahmen zur Luftreinhaltung.

REFERENCES

[1] Adrian, G. and Fiedler, F., Simulation of Unstationary Wind and Temperature Fields over Complex Terrain and Comparison with Observations. *Contr. Atmos. Phys.*, 1991, **64**, 27-48.

[2] Nester, K. and Fiedler, F., Modelling of the Diurnal Variation of Air Pollutants in a Mesoscale Area. *Health and Ecological Effects. Proceedings of the 9th Clean Air Congress, Montreal, August 30-September 4, 1992, paper no. IU-16C.02* , Air & Waste Management Association, Pittsburgh, Pennsylvania.

[3] Stockwell, W.R., Middleton, P., Chang, J.S. and Tang, X., The Second Generation Regional Acid Deposition Model Chemical Mechanism for Regional Air Quality Modeling *J. Geophys. Res*, 1990, **95, No. D10**, 16,343 - 16,367.

[4] Fiedler, F., TRACT Field Measurements and Model Simulations: First Results - A Contribution to Subproject TRACT -. *in: Transport and Transformation of Pollutants in the Troposphere. Proceedings of EUROTRAC Symposium '94, Garmisch-Partenkirchen, 11. - 15. April 1994 (Ed.: P.M. Borell, P. Borell, T. Cvitas, and W. Seiler), pp. 614,*SPB Academic Publishing, Den Haag, 1994, ISBN 90-5103-095-9

[5] Binder, H.-J., Kalthoff, N, Fielder, F., Koßmann, M., Vögtlin, R., Corsmeier, U., and Schlager, H., The Boundary Layer Evolution during TRACT: A contribution to the subproject TRACT. *EUROTRAC Annual Report 1993,* 9 - 17, EUROTRAC International Scientific Secretariat, Garmisch-Partenkirchen, 1994

[6] Vögtlin, R., Koßmann, M., and Corsmeier, U., Aircraft Measurements in the Main Area During TRACT. *EUROTRAC Annual Report 1993,* 38 - 48, EUROTRAC International Scientific Secretariat, Garmisch-Partenkirchen, 1994

[7] Ebel, A., Neubauer, F.M., Raschke E. and Speth, P. (ed.), Das EURAD Model: Aufbau und erste Ergebnisse. *Mitteilungen aus dem Institut für Geophysik und Meteorologie der Universität zu Köln* , 1989.

[8] Nester, K., Panitz, H.-J. and Fiedler, F., Comparison of the DRAIS and EURAD Model Simulations of Air Pollution in a Mesoscale Area. *accepted for publication in Meteorology and Atmospheric Physics, 1995*

[9] Obermeier, A., Friedrich, R., John, C., Seier, J., Vogel, H., Fiedler, F., and Vogel,B., Photosmog - Möglichkeiten und Strategien zur Verminderung des bodennahen Ozons. *ECOMED Verlagsgesellschaft*, pp. 184, Landsberg, 1995, ISBN: 3-609-65320-5

[10] Vogel, B., Fiedler, F. and Vogel, H., The Influence of Topography and Biogenic VOC Emissions in the State of Baden-Württemberg on the Ozone Concentrations during Episodes of High Air Temperatures. *accepted for publication in J. Geophys. Res.*, 1995

[11] Pierce, T.E, Lamb, B.K., and Van Meter, A.R., Development of a Biogenic Emissions Inventory System for Regional Air Pollution Models. *paper 90-94.3, presented at the 83. Annual Meeting and Exhibition of the Air and Waste Management Association*, Pittsburgh, PA, June 24-29, 1990

DISCUSSION

J. GRAF: Why is the maximum chemical ozone production at noon, and not, like normally, in the afternoon between 5 and 7 p.m.?

H.-J. PANITZ: The chemical ozone production essentially follows the diurnal course of solar insolation which has its maximum at noon. Therefore, the maximum production occurs at the same time. Chemical production and concentration may not be mixed up. Of course, the ozone concentration has its maximum in the afternoon when the chemical mass budget component changes its sign from production to loss.

M. PIRINGER: Can you comment on the discrepancies between observed and predicted NO concentrations?

H.-J. PANITZ: The reason is that the high emissions close to the observation point lead to the high observed concentrations which are not resolved by the model. The emissions in the model are area averages over the size of a grid cell so that peak emission values are underestimated. The resulting concentrations are averaged over the volume of a grid box. This leads to an underestimation of the predicted concentrations of emitted species like NO.

G. KALLOS: What kind of lateral boundary conditions did you use?

H.-J. PANITZ: Along the boundaries of the model domain we used average emissions assuming that they reflect the emission conditions outside of the model area. These averaged values have been calculated from the area emissions in a 30 km wide strip extending along the boundaries inside the model domain. The assumption probably leads to an underestimation of the importance of the advective flux term in the mass balance, especially, for secondary species like ozone.

POSSIBLE MECHANISMS FOR LONG RANGE TRANSPORT IN THE EASTERN MEDITERRANEAN

G. Kallos, V. Kotroni, K. Lagouvardos, M. Varinou, and A. Papadopoulos

University of Athens, Department of Applied Physics, Meteorology Laboratory, Ippocratous 33, Athens 10680, Greece

INTRODUCTION

The Mediterranean Sea is closed from all sides and has approximately an elliptical shape. Its major axis of about 3600 km has a W-E orientation, while the N-S width is variable from 200-800 km. It is surrounded by high peninsulas and important mountain barriers. The most important are the Alps and the Balcan Peninsula to the N, the Iberian Peninsula to the W, the Atlas Mountains to the SW, and the Asia Minor to the NE. The Asia Minor Plateau separates the Mediterranean from the Black Sea. The gaps between these major mountainous regions act as channels for the air mass transport from/to the Mediterranean. This kind of transport is considered as very important for the air quality in the Mediterranean Region. These topographic features along with the significant variation of the physiographic characteristics are partially responsible for the development of various-scale atmospheric circulations. These locally produced atmospheric circulations are quite strong, especially during the warm period of the year. The most significant from these regional to mesoscale circulations in the area are described in the publication of the UK. Meteorological Office "Weather in The Mediterranean", (1962).

The Climate of the Eastern Mediterranean cannot easily be characterized as maritime or continental. The cold season (end of October - beginning of March) is the rainy period. The warm season (June - September) is the dry period with almost no rain. The rest of the year consists the transient seasons (spring and autumn) where the winter and summer-type weather patterns are interchanging. The main characteristics for these time periods can be summarized as follows:

i. During winter, cyclogenesis in the Mediterranean takes place in preferable locations such as Cyprus (El Fandy, 1946; Kallos and Metaxas, 1980), Southern Ionian Sea, and the Gulf of Genoa (Alpert et al., 1990). This cyclogenetic activity is associated with the low index circulation (Metaxas, 1978) and is a result of positive vorticity advection and invasion of cold air over relatively warm waters (Metaxas, 1978; Kallos and Metaxas, 1980). These cold air masses originate from Central or Eastern Europe and Scandinavia. Usually, the cold outbreak is associated either with the cyclogenesis in the preferable locations aformentioned or with the rejuvenation of dissipating lows moved in the area. The anticyclonic type of circulation during the winter is associated with a cold core anticyclone laying over the Central Europe or the Balkan. This high pressure system has a relatively long duration (two to four weeks) and is associated with light northerly flow in the Eastern Mediterranean.

ii. During the warm period of the year, the land of North Africa becomes very hot during the day-hours while the Mediterranean waters have a moderate temperature (23-27 ^{o}C). The land of South Europe becomes also hot during the day-hours but not as hot as the land of North Africa. Because of this differential heating, the Mediterranean region and South Europe are covered by an anticyclonic system which is relatively shallow. The general circulation is

Air Pollution Modeling and Its Application XI, Edited by Gryning and Schiermeier, Plenum Press, New York, 1996

moving northward (high index circulation). There is almost no rain in the region with the exception of some convective storms. Air masses from the Mediterranean are moving toward north Africa. These air masses are replaced by others from above (large-scale subsidence) or from the Central and Eastern Europe. Trade winds persist over areas like the Aegean Sea (Etesians). The Etesians is a regional-scale phenomenon with significant diurnal variation. The synoptic-scale pattern can be described as follows: A high pressure system covers the Eastern Mediterranean and Balkan area up to the Black Sea. Over the West and Central Mediterranean there is a ridge. A thermal low over the Anatolian Plateau is evident during the day-hours because of heating of the dry land (semi-desert). These systems are always in a critical balance. When the high pressure system weakens, the thermal low over the Anatolian Plateau extends towards the W and a strong pressure gradient over the Dardanelles and the Aegean is forming during the day which weakens at night. During these days, strong northerly winds persist across the Aegean and all the way to the North African coast. In the Middle East they become NW to W. When the high pressure system is strengthening, a weak northern flow prevails across the Aegean Sea. In these cases, the local thermal circulations dominate the region.

iii. During the transient seasons the weather type changes between summer and winter-type. This change occurs rapidly during spring and slower during autumn.

Because of these complicated flow interactions, the air pollutants released from various sources located in the surrounding areas can be transferred in long distances, in a complicated way. The insolation helps the chemical transformation of the released air pollutants. The transport of air masses from Europe toward the Mediterranean occurs at the lower levels of the atmosphere through the gaps between the mountains (channelling).

Early observations of the concentrations of particulates in the Eastern Mediterranean showed that the peak values are associated with the summer-type weather conditions, especially during the period of trade winds. There is limited information about these transport and transformation processes occurring in the Mediterranean Region, despite the fact that the impact of these processes on the ecosystems and climate, is very important. The transport and transformation mechanisms existing in the Eastern part of the Mediterranean Region was the subject of a three-year research project funded from the DG-XII of the European Union (AVICENNE project: Kallos et al., 1995). In this presentation, the possible mechanisms for transportation of air pollutants are investigated using the Regional Atmospheric Modeling System (RAMS) and Hybrid Particle Concentration and Transport (HYPACT) models.

MODEL DESCRIPTION

The model used for this study is the Regional Atmospheric Modelling System (RAMS), developed at Colorado State University. It is a merger of a nonhydrostatic cloud model (Tripoli and Cotton, 1982) and a hydrostatic mesoscale model (Mahrer and Pielke, 1977). A general description of the model and its capacities is given in Pielke et al (1992). However, some RAMS features are the two-way interactive nesting with any number of either telescoping or parallel fine nest grids, the terrain following coordinate surfaces with cartesian or polar stereographic horizontal coordinates, the cloud microphysics parameterization at various levels of complexity, the various turbulence parameterization schemes, the radiative transfer parameterizations (short and longwave) trough clear and cloudy atmospheres, the various options for upper and lateral boundary conditions and for finite operators, the various levels of complexity for surface-layer parameterization (soil model, vegetation etc.), the selection of either horizontally homogenous or variable initialization (isentropic analysis), and its portability (it runs in several types of computers).

For dispersion calculations, the Hybrid Particle and Concentration Transport Package (HYPACT) was used. It is a combination of a Lagrangian particle model and an Eulerian concentration transport model. It was developed at the Colorado State University and ASTeR, Inc (Tremback et al., 1993). The Lagrangian model attempts to deal with the subgrid-scale aspects of a particle release while the Eulerian model takes over when the contaminant 'cloud' becomes adequately resolved on the computational grid, thus reducing the number of particles needed and increasing the computational efficiency. With the velocity and turbulence fields simulated by RAMS, the Lagrangian model in HYPACT will advect the particles with mean and random turbulent wind components. HYPACT uses a level 2.5 turbulent closure scheme based on the prognostic turbulent kinetic energy from RAMS. The source configurations of

HYPACT are very flexible. Any number of sources may be specified anywhere in the domain and configured as a point, line, area, or volume source. The emissions from these sources can be instantaneous, intermittent, or continuous. The particles can be treated as gases or aerosols and a radioactive half-life or a settling velocity can be specified. Plume rise parameterizations and a dry deposition scheme have recently been added to HYPACT.

MODELS CONFIGURATION - DATA USED

In order to simulate the summer-type atmospheric circulations on an accurate way, RAMS and HYPACT were configured on different ways. For all the simulated cases an outer grid covering an area 4000 km x 3000 km, with 60 km horizontal grid increment centered 20 E and 40 N, was defined. Thirty levels following the topography were used in both grids, having a 80 m vertical spacing near the ground, stretching to 1 Km at about 6 Km altitude and remaining constant up to about 16 Km This large grid was selected in order to represent the entire Mediterranean, the Black Sea and the surrounding areas and therefore to avoid boundary value problems. The model was initialized using 1x1 degree ECMWF analysis files available every 6 hours at the standard levels. Simulations were performed for 3 days during June 1990, 10 day during September 1993, 25 days during June and July 1994 and 10 days during July 1995. For selected cases, two other grids were used, nested over the NE part of the Mediterranean in order to examine the role of some mesoscale circulations developed and their interaction with the regional-scale phenomena. The second grid has 15 km horizontal grid increment covering mainly the Balkan area, and partially Turkey and Italy. The third grid has 3.75 km horizontal grid increment and covers most of the eastern part of the Greek Peninsula. The horizontal grid increments selected were found to be adequate for describing all scales of thermal circulations developed in this region. The topographic data sets used are the NCAR 5 min longitude-latitude for the two outer grids. For the finer one, a data set retrieved from cartographic maps with 30 sec resolution was used. Mean-monthly Sea Surface Temperatures (SST) with 30 min resolution were also used. For vegetation and soil moisture, ECMWF data with coarse resolution were used. These data were interpolated at model grids because of lack of other better data sets.

Since emission data for the surrounding the Mediterranean Sea were not available, qualitative-type dispersion simulations were performed. For the various simulations performed, the sources under consideration are mainly large urban areas or areas known for their industrial activities. Area-type sources covering a surface of 10km x 10 km with a depth of 200 m were defined. This is considered as a typical urban area containing some industrial-type of releases. The areas from where releases were made are Athens, Thessaloniki, Istanbul, Constanza, Burgas, Milano, Messina, Catania and Barcelona.

MODEL RESULTS

From the simulated cases the results from two major synoptic groups will be presented: the first which favors the transport from Eastern Europe, across the Aegean Sea toward the Eastern Mediterranean and the second which favors a westerly-type of flow (from the western to the eastern part of the Mediterranean sea).

Transport across the Aegean Sea

The transportation through this route occurs during all seasons but it is more efficient during the warm period of the year, especially with the etesians. This is due to the fact that the washout mechanisms are stronger during winter than in summer. The etesians, is a thermal-type circulation at regional scale with some channelling characteristics between the Greek and the Asia Minor Peninsulas. A typical near-ground flow pattern at 4 July 1994, is shown in Fig. 1. As it is seen, there is a relatively strong flow from the Black Sea, across the Aegean towards Middle East. In general, the flow across the Aegean is stronger during the day and weaker during the night (Figs. 1a, b). This flow pattern is being modified from mesoscale circulations developed.

In order to study the dispersion at a regional scale of air pollutants from major air pollution sources (urban or industrial) simultaneous particle releases were made. Fig. 2 shows the particle positions for releases made from areas mentioned previously. It shows the plume positions after 51 hours of release (the release began at 0000 UTC, 2 July 1994). As it is seen, the plumes from the locations around the Black Sea are transported towards the Eastern part of the Mediterranean mainly across the Aegean Sea. According to the weather conditions in the Mediterranean, these plumes are directed either towards the Middle East or towards Egypt.

a.

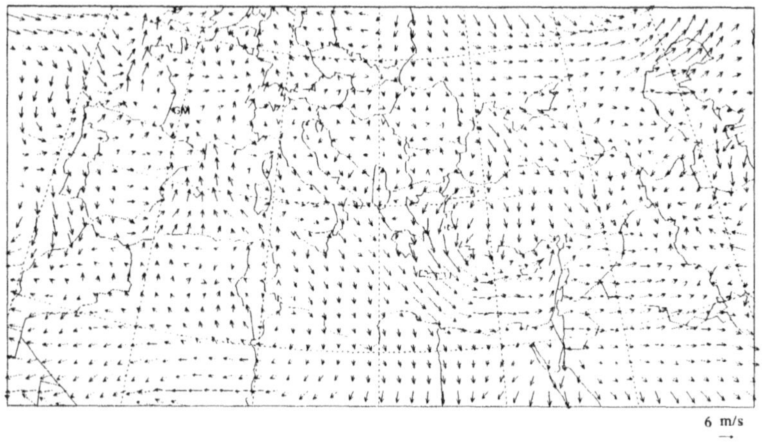

b.

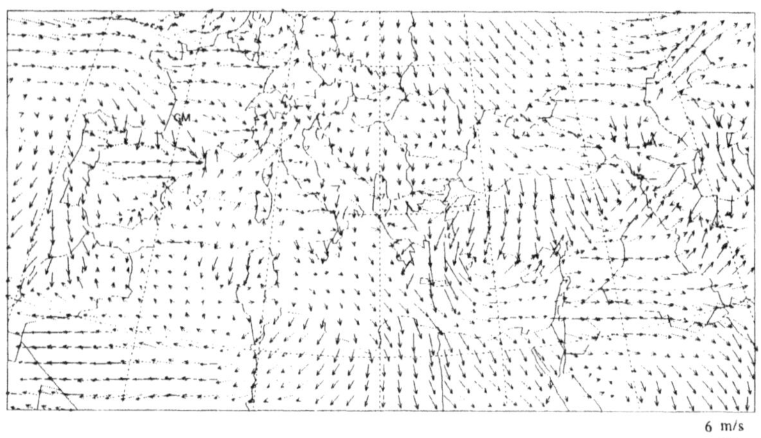

Figure 1. Wind fields at z=39 m at (a) 0000 UTC, (b) 1200 UTC, 4 July, 1994

Test simulations showed that these plumes will be transported deep south, mainly over Egypt, towards the equatorial zone within a very deep Planetary Boundary Layer (PBL). These plumes keep their characteristics for large distances from the source because they follow a path mainly over the sea, within a shallow PBL or above it where horizontal advection is strong. The time scales for the transport of a released plume in the area around the Black Sea until it reaches the Middle East is approximately 3-4 days. This is in good agreement with observations made in the coastal zone of Israel where the maximum concentration of sulfate particles was found to occur approximately 3 days after the initiation of the etesians (Kallos et al., 1995). In cases with relatively weak regional-scale forcing and therefore stronger mesoscale circulations this time scale is even longer because of recirculation or other kind of tidal movements. In all the cases (with or without strong regional-scale forcing) it was found that the urban plume of Athens might keep its main characteristics for several hundreds of kilometers, at least until the island of Crete. This was verified by aircraft flights during the last two years (Kallos et al., 1995).

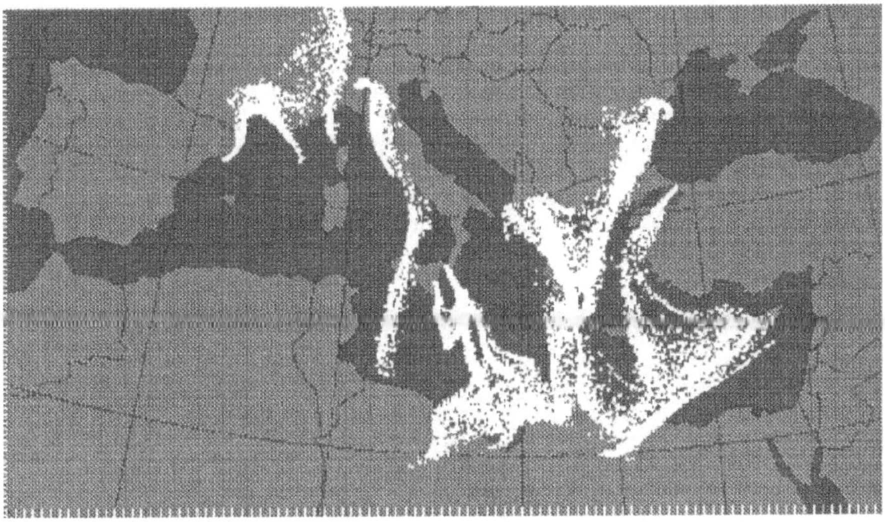

Figure 2. Particle projection at the first model level from the HYPACT dispersion model at 0300 UTC, 4 July, 1994.

A case study representative of relatively weak regional scale forcing is the one at 6-7 July 1994. During this period, a weak large-scale northerly flow prevails across the Aegean Sea and local thermal circulations dominate the region. These circulations are evident in simulations with finer nested grids. Figure 3 shows the resulting flow pattern within the two inner grids. As it is seen in the intermediate grid (Fig. 3a) the flow is diverted around Peloponnese and then directed towards the Saronic Gulf and Athens, enhancing the sea-breeze effect evident in the inner grid (Fig. 3b). When the same simulation was repeated with only the coarse grid, the simulated flow presented significant differences in the area around the Aegean Sea (not shown). This is due to the model capability to include a larger portion of the spectrum of the atmospheric motions in the case with the three nested grids. These mesoscale features developed in the region are considered as very important for some places like Athens with the well known air quality problems. As it was described in Kallos and Kassomenos (1993, 1994), and Kallos et al., (1995), the regional and mesoscale flow interactions in the area of the Aegean Sea are very important and define the air quality in the Greater Athens Area (GAA). Dispersion calculations were performed for the same period. Figure 4 presents the position of particles released from the region of Athens, after 39 hours of simulation (the release started at 0000 UTC 6 July 1994). The plume which is located over Athens consists of the particles released the same day (7 July 1994) while the particles found over the sea in the region south of the GAA have been released the previous day and they have been advected towards the north from the local circulations depicted in Fig. 3b.

Transport from the Western towards the Eastern Part of the Mediterranean

When in the western or central Mediterranean the synoptic flow is generally from west (e.g. when a relatively large cyclonic system is located over Europe) it is quite possible to have advection of polluted air masses located over the western or central part of the Mediterranean towards the Eastern part. Such a case is presented in Figure 5. In this case, only the coarse grid was used. From this pattern it is evident that after 39 hours of particle release the plumes from Messina are directed towards the Aegean Sea, through the gap between Peloponnese and Crete. This behaviour of the flow was described in Kallos and Kassomenos

(1993,1994) where it was shown the importance of this characteristic in the formation of the appropriate conditions for an air pollution episode in Athens.

When the westerly flow over the central part of the Mediterranean is stronger or from NW directions, the main path of the plumes passes from the south of Crete towards the eastern part of the Mediterranean. In some cases, due to the deflection of this flow towards Africa because of the strong sea breezes or because of the appearance of a low across north Africa the plumes might be found over the Gulf of Syrtis and the area around Libya and Tunis. In this case, the plumes exhibit a significant vertical extend due to the deep mixing as long as they move over the land. This is more pronounced during summer.

a.

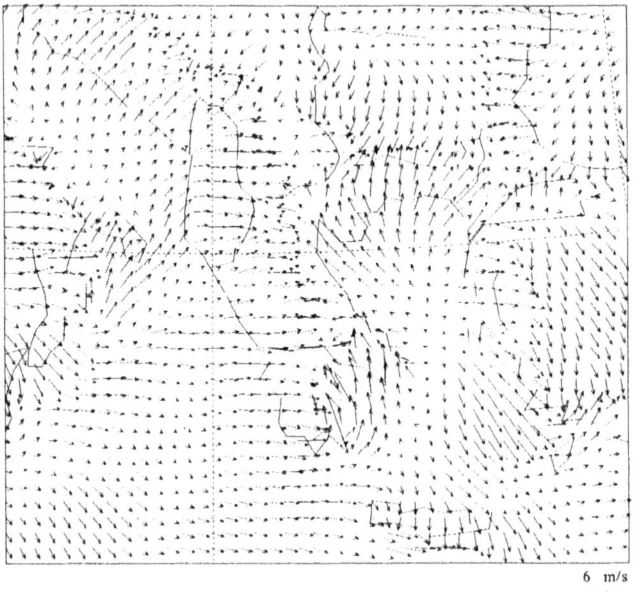

b.

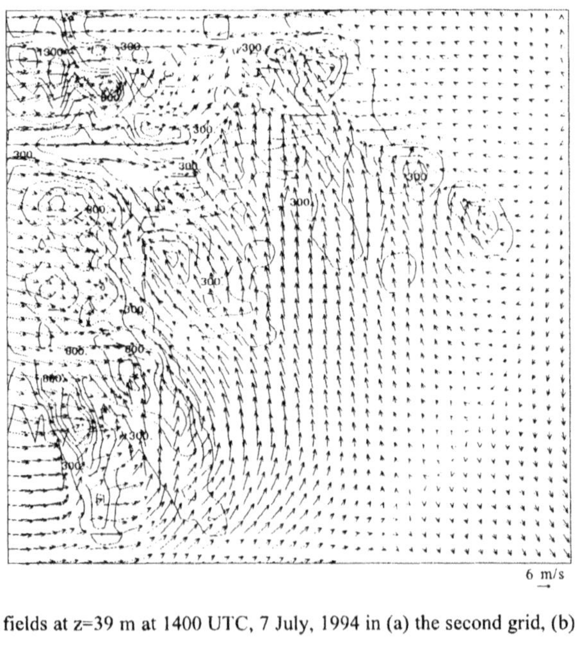

Figure 3. Wind fields at z=39 m at 1400 UTC, 7 July, 1994 in (a) the second grid, (b) the inner grid.

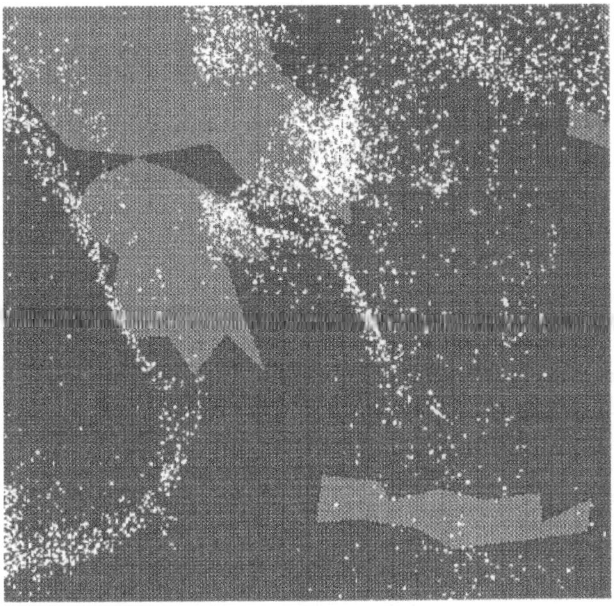

Figure 4. Particle projection at the first model level from the HYPACT dispersion model at 1500 UTC, 7 July, 1994, after 39 hours of particle release.

CONCLUSIONS

In the frame of this study, two main paths of the transport of air pollutants from Europe towards the eastern part of the Mediterranean have been identified and described: one path across the Aegean Sea and the other across the straights south of Italy and Greece. The first one seems to occur more frequently because of the synoptic conditions and the developed regional-scale circulations. It is the dominant mechanism during summer. The second one might occur either during summer or winter but mainly during spring.

Because of the stable conditions occurring over the Mediterranean waters the plumes, especially those released near coastal areas, might travel for several hundreds of kilometers keeping some of their characteristics.

Mesoscale local circulations of various types might play a significant role in the dispersion of such plumes and define air quality at local scales. This is very important in coastal urban areas around the Mediterranean Sea. The contribution from such kind of transport in the Mediterranean region might reach significant levels. For example, it was found that some aged plumes might transfer over cities ozone of the order of 60-100 ppbv (Kallos et al., 1995). The time scale for transport of air pollutants from sources located in Eastern Europe towards the Eastern Mediterranean is approximately two to three days, while for the transport from the western to the eastern part of the Mediterranean is in general longer.

Acknowledgements

This research was supported by the research projects South European Cycles of Air Pollution (SECAP) contract number EV5VCT910050 and the AVICENNE Initiative contract number EV*-CT92-0005 of the DG-XII of EU. We would like to thank Dr. Craig Tremback from ASTER Inc for his assistance of setting-up and running the models.

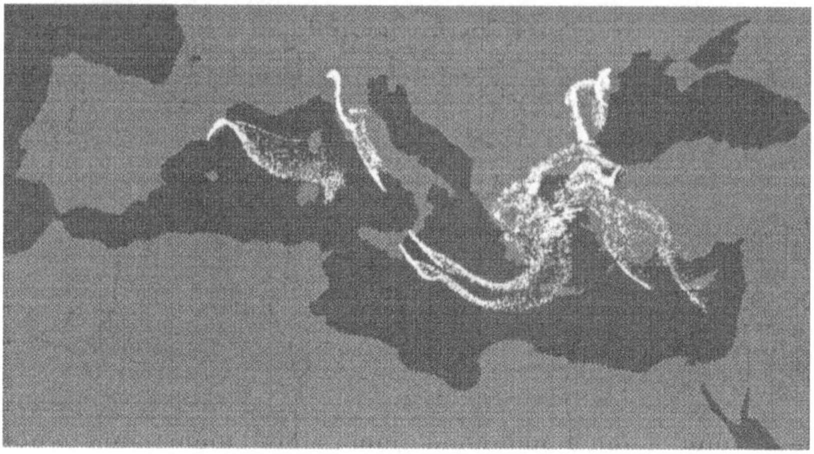

Figure 5. Particle projection at the first model level from the HYPACT dispersion model at 1500 UTC, 16 June, 1990, after 39 hours of particle release.

REFERENCES

Alpert, P., B.U. Neeman, and Y. Shay-El, 1990: Climatological analysis of Mediterranean cyclones using ECMWF data. Tellus, 42A, 65-77.

El Fandy, M.G., 1946: Barometric Lows in Cyprus. Quart. Roy. Met. Soc., 72, 291-306.

Kallos, G. and D.A. Metaxas, 1980: Synoptic processes for the formation of Cypus lows. Riv. Met. Aer., vol. XL, no 2-3. 121-138.

Kallos, G., P. Kassomenos, and R.A. Pielke, 1993: Synoptic and mesoscale weather conditions during air pollution episodes in Athens, Greece. Boundary-Layer Meteorol. 62, 163-184.

Kallos, G., and P. Kassomenos, 1994: Effects of the selected domain in mesoscale atmospheric simulations and dispersion calculations. Proc. of the 20th ITM of NATO/CCMS on Air Pollution Modeling and its Application. Editors: S. Gryning and M. Millan, Plenum Press, X, 35-44.

Kallos, G., Kotroni V., Lagouvardos K., and Varinou M., M. Luria, M. Peleg, G. Sharf, V. Matveev, D. Alper-SimanTov, A. Vanger, G Tuncel, S. Tuncel, N. Aras, G. Gullu, M. Idrees, F. Al-Momani, 1995: Transport and Transformation of Air Pollutants from Europe to the East Mediterranean Region. Environmental Research Program AVICENNE. Intermediate Technical Progress Report (for the DGXII of EU).

Mahrer, Y. and R.A. Pielke, 1977: A numerical study of the airflow over irregular terrain. Beitr. Phys. Atmos., 50, 98-113.

Metaxas, D.A., 1978: Strong cold outbreaks in the East Mediterranean. A synoptic study. Riv. Met. Aer., vol. XXXVIII, no 2, Rome. 95-105.

Meteorological Office, 1962: Weather in the Mediterranean. Vol. I, General Meteorology H.M. Stat. Office, London. Second Edition.

Pielke, R.A., W.R. Cotton, R.L. Walko, C.J. Tremback, W.A. Lyons, L.D. Grasso, M.E. Nicholls, M.D. Moran, D.A.Wesley,T.J. Lee, and J.H. Copeland, 1992: A comprehensive meteorological modelling system - RAMS. Meteorol. Atmos. Phys. 49, 69-91.

Tremback, C.J., W.A. Lyons, W.P. Thorson, and R.L. Walko, 1993: An emergency response and local weather forecasting software system. Preproceedings of the 20th ITM on Air Pollution and its Application, November 29 - December 3, 1993. Valencia, Spain. pp 8.

Tripoli, G. J., and W.R. Cotton, 1982: The Colorado Stae University three-dimensional cloud/mesocale model. Part I: general theoretical framework and sensitivity experiments. J. Rech. Atmos., 16, 185-220.

DISCUSSION

G. GRAZIANI: On the Mediterranean sea, which were the mesh sizes (in horizontal and vertical) used in RAMS? How were you able to smooth Sicilian topography included Etna volcano?

G. KALLOS: The horizontal grid increments used in these simulations is 60 km for the outer grid or the one-grid simulations, 15 km the intermediate and 3.7 the fine one. The two inner grids were used mostly over Greece and a part of Italy. The topography used is the 5 min topographic set widely available. In this set, Etna is represented on an acceptable way. In the RAMS simulation, Etna is also present with a smoothed peak. We are using silhouette averaging in topography preparation in order to keep the shapes of the mountains as much as we can. Now, we just obtained a fine topography set (30 arc sec lon-lat) which is considered as more appropriate for fine model resolution.

G. SCHAYES: The presentation shows the trajectories and plumes as seeing from a few selected cities. Considering the ensemble of sources one would obtain a much more fuzzy picture of air pollution concentrations where individual sources can no more be recognized.

G. KALLOS: These cities were selected more or less arbitrarily and for illustrative purposes on how the plumes can be transported over the stable Mediterranean boundary layer. Someone can select to present one, two or more sources. Simulating several urban plumes simultaneously, you are right, the picture is fuzzy, especially after the second day from the start of the release. These simulations were made in order to show the possible paths of the plumes during the next 2-4 days and how other places (mostly the coastal ones) should be affected from such a long transport.

MODEL-CALCULATIONS OF TRANSBOUNDARY EPISODES OF HIGH SO_2-CONCENTRATIONS DUE TO EMISSIONS IN THE KOLA PENINSULA

Trond Iversen[1], Anne K. Brenden[1], Marit H. Jensen[1],
Jan Erik Haugen[2], and Bjarne Sivertsen[3]

[1]Department of Geophysics, University of Oslo
P.O. Box 1022 Blindern, 0315 Oslo, Norway
[2]Norwegian Meteorological Institute,
P.O. Box 43 Blindern, 0313 Oslo, Norway
[3]The Norwegian Institute of Air Research,
P.O. Box 100, 2007 Kjeller, Norway

INTRODUCTION

The production of non-ferrous metals from ores containing up to 30% of sulphur, are causing large emissions of SO_2 close to the border of Norway at the Kola peninsula in north-west Russia. The total annual emissions released from smelters in the towns Nikel and Zapoljarnij, which are situated only 10-12 km from the Norwegian border, amounted to about 260 kilotonnes of SO_2 in 1992 (Sivertsen *et al.*, 1994). In comparison the total emissions from Norway for the same year was 37 kilotonnes of SO_2, and corresponding numbers for Finland and Sweden were 121 kilotonnes and 101 kilotonnes respectively (Barret and Seland, 1995). Another smelter complex is located about 400 km to the east of the Norwegian border in Montsegorsk with similar emission amounts (Kulmala, 1987). These very efficient SO_2-emitters will under favourable meteorological conditions cause transboundary environmental damage due to direct effects of SO_2. Hourly concentration levels exceeding recommended critical levels for natural vegetation (150 $\mu g(SO_2)m^{-3}$; IUFRO, 1978) and air quality guideline values for human health (350 $\mu g(SO_2)m^{-3}$; WHO, 1987) have been measured at sites on the Norwegian side of the border (Sivertsen *et al.*, 1992 and 1994). Extreme episodes of high hourly concentrations up to 2-3 $mg(SO_2)m^{-3}$ have been measured on some occasions.

Numerical modelling of the semi-annual averaged concentrations have been made with considerable skill (Sivertsen *et al.*, 1992). Modelling of hourly concentration variations during episodes have been attempted before with only limited degree of success (Sivertsen, 1995). Model calculations on small meso-scales have, however, been performed with time-resolved meteorological data in other places (Persson, *et al.* 1994; Lyons *et al.*, 1994). By going down in temporal and spatial scales, the stochastic nature of the flow field which determines the dispersion, increases in relative importance. Earlier investigations (e.g. Venkatram, 1981) have shown that averages over one hour is not sufficient to be fully equivalent with ensemble averages. For hourly averages one should expect stochastic fluctuations which amount to the hourly averages themselves (Venkatram, 1981).

The main purpose of the present study was to investigate the possibility for models to reproduce the very high concentrations measured episodically close to the border between Russia and Norway, as well as the meteorological conditions for them. We used the model code of the limited area numerical weather prediction model which was in routine use at the Norwegian Meteorological Institute up to 1995 (Grønås and Hellevik, 1982; Nordeng, 1986). A horizontal grid resolution of 5 km was applied, hence the model was named LAM5. Wind and turbulent exchange coefficients were taken from model runs and used in an extended version of a meso-scale dispersion model developed at the Norwegian Institute of Air Research (EPISODE) to calculate SO_2-concentrations (Grønskei *et al.*, 1993). The model had not been used under full heterogeneous and non-stationary conditions with fine resolution of the atmospheric boundary layer (ABL) before. Two specific periods were selected for investigation. One period included an episodic transport situation for which the emissions in Montsegorsk could be of specific importance, whilst another encompassed a more typical episode with shorter transport.

A SHORT DESCRIPTION OF THE MODEL TOOLS

The LAM5 uses the quasi-hydrostatic assumption and is thus limited to horizontal scales larger than about 5-10 km in general (e.g. Iversen and Nordeng, 1987). The model integrates in time the horizontal equation of motion, the mass continuity equation, the thermodynamic energy equation and conservation equations for the water substance. It employs terrain-following sigma-coordinates with 30 layers in the vertical. Vertical diffusion is parameterized according to Blackadar (1979). This parameterization most certainly underestimate vertical diffusion in the unstable ABL, but such conditions are less pronounced during pollution episodes over land. The model contains cloud and precipitation parameterization, and short- and long-wave radiation (Nordeng, 1986). We have chosen 15 layers in the lower 2500 m, thus a well resolved ABL was obtained. The model was nested into a model with 50 km grid resolution which produced analyzed meteorological data in a six-hourly intermittent data-assimilation cycle. The LAM5-results were therefore considered to be a small meso-scale interpretation of analyzed meteorological fields. Horizontal diffusion processes were not added, since numerical diffusion inherent in the approximations overestimates the physical effect.

The EPISODE model had earlier only been used with three layers in the vertical and with wind and static stability from interpolated measurements (Grønskei *et al.*, 1993). The main body of the model consists of an eulerian grid-point solution to the budget equation of a chemical component. Sub-grid diffusion from major point sources are treated with a gaussian plume module of the model. The centre axes of a plume is allowed to bend along streamlines. The plume size grows as a function of lagrangian time elapsed since emission release, until the standard deviation in either horizontal or vertical direction reaches 20% of the grid size in the eulerian model. When this occurs, the mass contained by the puff in the grid-box is included in the eulerian calculation and forgotten by the plume-model.

Table 1. Emission data used in the model.

Chimney	Lat.	Long.	Height	Temp.	Volume flux	SO_2-emission	
						April-92	June-92
			(m)	(K)	(m^3s^{-1})	($g(SO_2)s^{-1}$)	($g(SO_2)s^{-1}$)
Nikel 1	69°25'	30°15'	150	392	691	1162.4	830.3
Nikel 2	69°25'	30°15'	160	372	1154	3285.4	2348.7
Nikel 3	69°25'	30°15'	160	392	746	1123.8	802.7
Nikel 8	69°25'	30°15'	38	292	227	190.2	133.8
Nikel 9	69°25'	30°15'	35	292	1168	276.5	197.5
Nikel 75-79, 87-89	69°25'	30°15'	40	372	119	19.5	13.9
Nikel 277	69°25'	30°15'	90	372	151	43.3	30.9
Nikel sum1	69°25'	30°15'	35	292	122	446.3	318.8
Nikel sum2	69°25'	30°15'	10	292	44	139.3	99.5
Zapoljarnij 1	69°24'	30°45'	100	390	709	3092.5	2208.9
Zapoljarnij 2	69°24'	30°45'	80	453	429	183.8	131.3
Kirkenes	69°44'	30°03'	30	400	503	25.6	18.3
Montsegorsk	68°53'	32°58'	100	350	283	6342.0	----

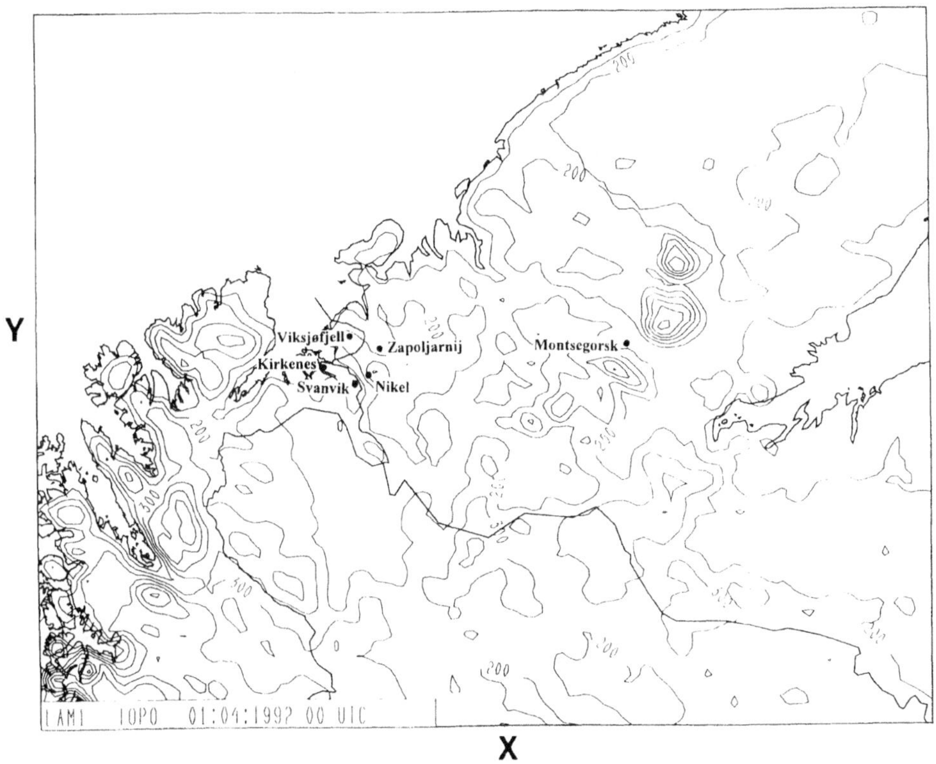

Figure 1. Map of integration domain. Nikel, Zapoljarnij and Montsegorsk are major emission points. Measurement sites are also denoted on the map. The model topography and the borders between Russia, Finland and Norway are also drawn.

Growth rates of the plume extensions as a function of lagrangian time are calculated by formulas used in a similar model at the US Environmental Protection Agency (Irwin, 1983). In our application of the model no chemistry was included. Over the ocean and over land in summer, the dry deposition velocity was assumed to be 8 mm/s. whilst over snow-covered land it was set to 2 mm/s. Wet deposition did not influence the pollution situation in our domain of interest during the chosen periods. In our case we grouped the different anthropogenic emissions in the smelter complexes into a few major sources in accordance with the spatial resolution of 5 km. Thus they were treated as 13 large point sources, including Montsegorsk and the Norwegian town Kirkenes. Table 1 gives emission data for each point source used in the model. Based on data from the emission stacks, Briggs' plume-rise formulas were used for efficient emission heights.

The 15 lowermost layers in LAM5 were used as calculation layers in EPISODE. The thickness of the layers increased gradually from 28 m in the lowermost to 223 m in the uppermost layer. The lowermost model level was about 14 m above the ground. The vertical turbulent exchange coefficient from LAM5 was used to estimate the standard deviation of the vertical wind fluctuation $\sigma_w = \sqrt{K/T_L}$ where K is the vertical exchange coefficient and T_L=600s is the assumed lagrangian timescale for the vertical component of the turbulence. Since LAM5 does not calculate any horizontal diffusion coefficient, the corresponding "horizontal turbulence" had to be heuristically estimated as $\sigma_v = 0.3\,|\vec{v}_H|$ where $\vec{v}_H$ is horizontal wind. All concentration calculations with EPISODE started out with zero concentrations, and no long-range transported part was included. Both LAM5 and EPISODE were run over an integration domain covering 131 times 101 gridpoints. Figure 1 shows the integration domain with the main emission points and measurement sites specified.

PERIOD I: APRIL 1. 1992 00 UT - APRIL 3. 1992 18 UT.

The large scale weather situation started out with a synoptic scale cyclone to the east of the integration domain. To the north and east of this cyclone a high pressure ridge was dominant. In our integration domain this resulted in quite strong winds parallel to the y-axes. As the cyclone moved towards the east and the ridge protruded south- and east-wards, the wind decreased and turned to become parallel with the x-axes (easterly to south-easterly). After the turning of the wind pollutants from Montsegorsk could possibly reach receptors in Norway with considerable concentrations. Figure 2 shows calculated and observed wind speed and direction at Viksjøfjell, which is at the peak of a smooth hill and is not particularly influenced by local topography. We see that the model reproduces the change in direction with very good accuracy. The underestimation in wind speed in the early part of the period is probably due to a too large roughness length (z_0) in the model.

Figure 3 shows the calculated ground-level concentration distribution well after the turning of the wind. On this time-level the influence of Montsegorsk on concentration levels on the Norwegian side of the border was estimated to be at its maximum. The Montsegorsk plume clearly provided a meso-scale background concentrations which added between 10 and 20 $\mu g(SO_2)m^{-3}$ to the peaks from the plumes originating from Nikel and Zapoljarnij. In this case about 350 $\mu g(SO_2)m^{-3}$ were calculated close to the border, but not at measurement sites. As an example of model-observation comparison we show results for Svanvik. At a first glance it may seem that the model did not perform very well. The observed concentration increase after hour 30 was delayed by 6 hours and the decrease between hours 40 and 47 by 8 hours, and the concentration peaks were overestimated by a factor 2. However, the model very clearly distinguish between the clean first part of the period and the period after the wind-turning. The phase delay of only a few hours in timing must be expected in view of the degree of randomness (Venkatram, 1981). Furthermore, Svanvik is situated in a local valley not resolved by the model topography, and this valley is close to being perpendicular to the main south-easterly winds during the polluted part of the period. The model data are strictly valid at 14 m above the model topography. At Viksjøfjell a similar comparison (not shown) reveals an underestimation by the model. Thus it is quite clear that both measurements and model calculations reflect generally polluted conditions at the Norwegian side of the border in the second half of the period, but that the plumes of the model does not hit the exact grid squares to the right time. Figure 4b) shows that most of the contribution to the peak concentrations are from sub-grid plumes. The remaining part predominantly stems from emissions in Montsegorsk.

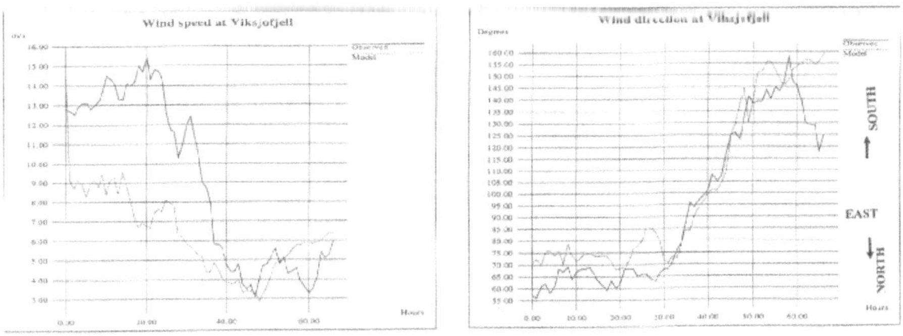

Figure 2. Wind speed (m/s, left) and direction (degrees, right) at Viksjøfjell calculated by LAM5 (dotted line) and observed (continuous line). The time 0 is April 1. 00 UT.

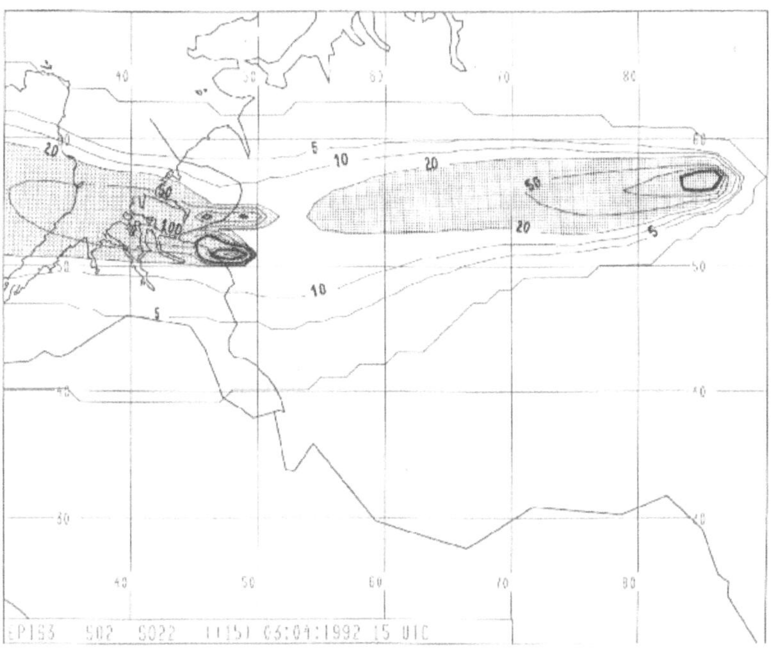

Figure 3: Model calculated ground-level concentrations of SO_2 in a section of the integration domain on April 3. 15 UT, (hour 63). Isolines in units of $\mu g(SO_2)m^{-3}$ are; 0.1, 5, 10, 20, 50, 100, 300, 500. Thick contours are contours of critical levels 150 and 350 $\mu g(SO_2)m^{-3}$. In hatched area: conc. > 20 $\mu g(SO_2)m^{-3}$. Every 10th grid-line is also shown.

PERIOD II: JUNE 26. 1992 12 UT - JUNE 29. 1992 00 UT

The episode occurring in this period is more classic; SO_2 accumulate in a thin ABL during stagnant and stable condition governed by widespread subsidence in an anticyclone. The situation gradually change in connection with an approaching cyclone from west and a slow propagation eastwards by the anticyclone. ABL air mass trajectories originating in Nikel or Zapoljarnij then approaches the Norwegian border, and a transboundary situation is initiated. Even though the pollution situation close to the emissions were much worse during the calm conditions than encountered in the border areas during the subsequent transport phase, this was a situation which caused measured hourly SO_2-concentrations to

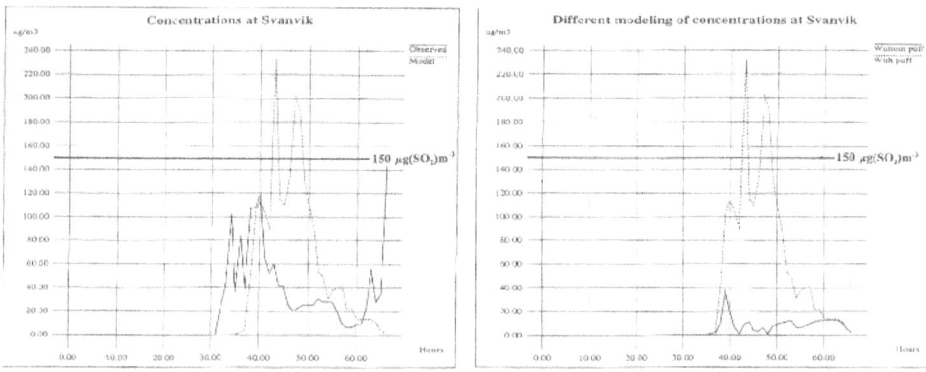

a) Dotted line: calculated
Continuous line: measured

b) Dotted line: total concentrations.
Continuous line: no sub-grid plume.

Figure 4. SO_2-concentration ($\mu g(SO_2)m^{-3}$) as a function of integration time in Svanvik. Time 0 is April 1. 00 UT. a) shows comparison between measurements and calculations, whilst b) shows the effect of not including the contribution from the sub-grid plume.

exceed air quality guidelines for human health (WHO, 1987) with more than 100 $\mu g(SO_2)m^{-3}$ at Svanvik. The episodic situation ends with a passage of the frontal system associated with the approaching cyclone from the west.

Figure 5 shows the calculated and observed wind at Viksjøfjell during the period. We notice a good match between the curves, and the change in wind-direction from northerlies to south-easterlies. The speed is all the time quite weak. Figure 6 shows the calculated and measured hourly SO_2-concentration at Svanvik and Viksjøfjell. In both places the model underestimated the observations, but the timing of the episodes was better than could be expected, in particular at Viksjøfjell. At Svanvik the peak was predicted four hours too early, and the concentration values were very good.

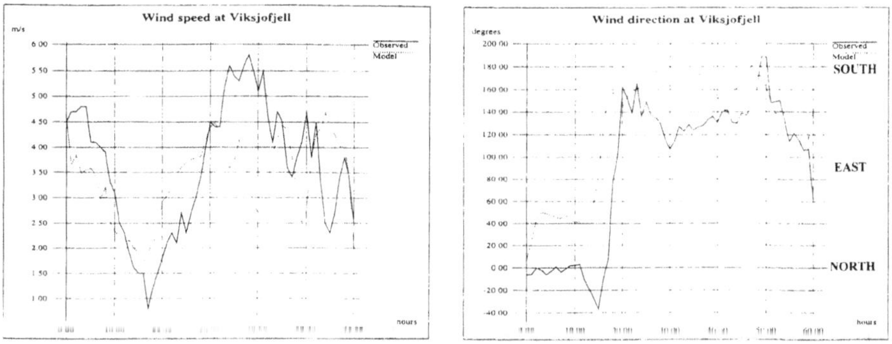

Figure 5. Wind speed (m/s, left) and direction (degrees, right) at Viksjøfjell calculated by LAM5 (dotted line) and observed (continuous line). The time 0 is June 26. 12 UT.

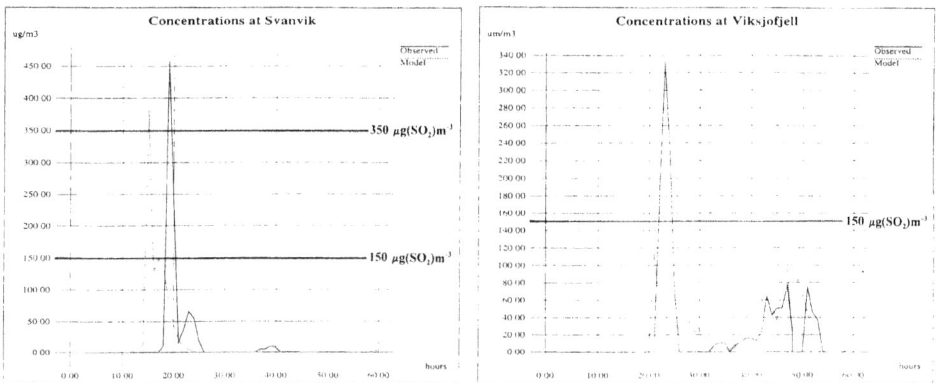

Figure 6. SO_2-concentration ($\mu g(SO_2)m^{-3}$) as a function of integration time at Svanvik (left) and Viksjøfjell (right). Time 0 is June 26. 12 UT. Dotted lines are calculations and continuous lines measurements.

CONCLUDING REMARKS

We have demonstrated the ability of a comprehensive meso-scale meteorological model to calculate hourly meteorological data with 5 km resolution and a detailed description of the ABL, which can be used for episodic transport calculations. We admit that we would like to see even better comparisons with measurements, but it is an open question how good the comparison ought to be, taking into account the stochastic nature of the flow field on the scales in question. It is clearly seen that the meteorological situation determines the episodic behaviour of the concentrations, and that the presently used model (LAM5) to a large extent was able to describe these governing conditions.

ACKNOWLEDGMENTS

The authors are indebted to Mr. Sam Erik Walker at the Norwegian institute of air research for his thorough help during the work with EPISODE. Mr. Anstein Foss at Norwegian Meteorological Institute is thanked for help in setting up the runs with LAM5.

REFERENCES

Barret, K. and Seland, Ø. 1995, European transboundary acidifying air pollution. EMEP/MSC-W Report 1/95. The Norwegian Meteorological Institute, Oslo, Norway.

Blackadar, A. K., 1979, High resolution models of the planetary boundary layer. *Advances in Environment and Scientific Engineering, 1*, Gordon and Branch, 276pp.

Grønskei, K. E., Walker, S. E. and Gram, F., 1993, Evaluation of a model for hourly spatial concentration distributions. *Atmos. Environ.*, **27B**, 105-120.

Grønås, S. and Hellevik, O. E., 1982, A limited area prediction model at the Norwegian Meteorological Institute. DNMI Tech. Rep. no. 61. The Norwegian Meteorological Institute, Oslo Norway.

Irwin, J. S., 1983, Estimating plume dispersion - a comparison of several sigma schemes. *J. Clim. Appl. Meteor.*, **22**, 92-114.

IUFRO, 1978, Air standards for sulphur dioxide. International Union of Forest Research Organization.

Iversen, T. and Nordeng, T. E., 1987, A numerical model suitable for the simulation of a broad class of circulation systems on the atmospheric mesoscale. NILU TR 2/87. The Norwegian institute for air research, Kjeller, Norway.

Kulmala, A., 1987, Sulphur dioxide emissions in the north-western region of USSR. Tech. Rep., Finnish Meteorological Institute, Helsinki, Finland.

Lyons, W. A., Pielke, R. A., Cotton, W. R., Tremback, C. J., Walko, R. L. and Uliasz, M., 1994, Recent applications of the RAMS meteorological and the HYPACT dispersion models. *Air Pollution Modelling and its Application X* S. E. Gryning and M. M. Millàn (Eds.). Plenum Press, New York. 19-26.

Nordeng, T. E., 1986, Parameterization of physical processes in a three-dimensional numerical weather prediction model. DNMI Tech. Rep. no. 65. The Norwegian Meteorological Institute, Oslo, Norway.

Persson, C., Langner, J. and Robertson, L., 1994, MATCH: a mesoscale atmospheric dispersion model and its application to air pollution assessments in Sweden. *Air Pollution Modelling and its Application X*, S.-E. Gryning and M. M. Millàn (Eds.). Plenum Press, New York. 9-18.

Sivertsen, B., 1995, Episodic impacts of air pollution in Norway from nickel smelters emissions in Russia. NILU F 9/95, The Norwegian institute for air research, Kjeller, Norway.

Sivertsen, B., Makarova, T., Hagen, L. O. and Baklanov, A. A., 1992, Air pollution in the border areas of Norway and Russia. NILU OR 8/92, The norwegian institute for air research, Kjeller, Norway.

Sivertsen, B., Baklanov, A. A., Hagen, L. O. and Makarova, T., 1994, Air pollution in the border areas of Norway and Russia. Summary report April 1991 - March 1993. NILU OR 56/94, The Norwegian institute for air research, Kjeller, Norway.

Venkatram, A., 1981, Model predictability with reference to concentrations associated with point sources. *Atmos. Environ.*, **15**, 1517-1522.

WHO, 1987, Air quality guidelines for Europe. WHO Regional Publications, European Series no. 23. World Health Organization, Copenhagen.

DISCUSSION

E. GENIKHOVICH: Did you take into account variations of the eddy diffusivity due to the terrain influence? How did you do it?

T. IVERSEN: The vertical eddy-diffusivity (K) is calculated in the LAM5-model by using the Blackadar (1979) - formulas which gives K as a function of the local Richardson's number (Ri). Horizontal diffusivity is not calculated explicitly by LAM5, since numerical diffusion is believed to overestimate the real in most cases, but σ_v is estimated for the EPISODE-model as proportional to the horizontal wind-speed ($|v_H|$).

Both Ri and $|v_H|$ is directly influenced by the topography which is resolved by the 5 km topography, thus it is taken into account in the diffusivity estimates. Subgrid-scale topography is, however, not accounted for explicitly. The incipient dispersion of emissions in narrow valleys under very stable conditions, as well as local fumigation effects, may be underestimated by this.

THE IMPACT OF THE EL NIÑO - SOUTHERN OSCILLATION ON THE AIR QUALITY OVER PORTUGAL : THE LISBON CASE

N. Barros[1], A. Rocha[2], and C. Borrego[1]

[1] Department of Environment and Planning
[2] Department of Physics
University of Aveiro
3800 Aveiro, Portugal

INTRODUCTION

The main purpose of this paper is to investigate the impact that global ocean surface temperature anomalies have on the atmospheric circulation over Portugal and its implications to air quality.

The Lisbon region, because of its industrial and urban importance, is an example of a region with high emissions levels and photochemical air quality problems. A preliminary analysis of ozone data over Lisbon shows that between 1989 and 1992 the Portuguese ozone standard has been several times exceeded in this city (Barros et al., 1995). The need to understand the behaviour of this airshed is most felt as a substantial increase of air pollutants emissions is foreseen for the future (Coutinho et al., 1992). On that perspective, this region will be used in this work as a case study.

The Lisbon coastal region is associated to a complex coastline with significant terrain features and sea/land breezes circulation. This results in a complex airflow pattern which cannot be represented by the available meteorological network.

Mesoscale meteorological models are the most appropriated tool to represent the sea-breeze circulation through the mathematical simulation of atmospheric physical processes (Borrego et al.,1994).

In this paper, the air quality evolution over Lisbon region, using ozone as an indicator, has been simulated by coupling a mesometeorological model, a version of the Colorado State University Mesoscale Model (CSUMM) (Barros, 1994), with a photochemical grid model, the Urban Airshed Model (UAM) (Ames et al., 1985). The anomalous synoptic atmospheric scenarios used as initial conditions by the CSUMM have been taken from a simulation performed with the Melbourne University General Circulation Model (MUGCM) which has been forced by anomalous sea surface temperatures (SSTs) associated to El-Niño / Southern Oscillation (ENSO) phenomenon.

THE MELBOURNE UNIVERSITY GENERAL CIRCULATION MODEL (MUGCM)

In the MUGCM the horizontal variance of the prognostic variables is represented in terms of spherical harmonics rhomboidally truncated at wave number 21. The model has nine vertical levels and originates from that described in detail by Bourke et al. (1977) and McAvaney et al. (1978). Subsequent changes are documented by Simmonds (1985) and Simmonds et al. (1988). Radiation is allowed to interact with carbon dioxide, ozone, water vapour and clouds, whose spatial distributions are prescribed to their January climatological values. SSTs are also prescribed in the model.

The climatology of the model has been assessed by Simmonds et al. (1988). The model generates credible simulations of the present climate, as evident also from an intercomparation of the present climate simulated by 14 GCMs (our being one of them) summarised by Boer et al. (1991).

THE MAR SYSTEM

A prognostic meteorological model which is a version of the Colorado State University Mesoscale Model (CSUMM), was used to generate the meteorological inputs need by the Urban Airshed Model (UAM). This integrated system, specially developed to be applied to coastal regions, has been named MAR (Modelo Atmosférico Regional) (Barros, 1994).

The meso-meteorological model used in this work is a version of the hydrostatic, incompressible, 3D primitive-equation model originally developed by Pielke (1974). The model includes a surface heat budget (Tremback et al., 1985) and parameterizations of the atmospheric surface and planetary boundary layers. The most thorough descriptions of the model formulation are found in Mahrer and Pielke (1977, 1978) and Arritt (1985).

The UAM permit the calculation of concentrations of both reactive and inert pollutants through simulation of the numerous physical and non-linear chemical mechanism that take place in the atmosphere. This Eulerian transport model is specially developed to aiming the calculation of ozone concentrations.

The CSUMM winds are vertically averaged within the UAM layers at each horizontal grid location and hour to generate the UAM wind fields, taking in account the mixing height predicted by the CSUMM as well. The UAM temperature field is generated by converting the CSUMM potential temperatures at the first level to actual temperatures.

CASE STUDY

Modelling region

A modelling domain of 120 km x 120 km was selected, centered in Lisbon. The horizontal grid spacing was fixed at 3 km. In the vertical direction the grid consists of 28 non-equidistant layers up to 8 km for the mesometeorological model and 4 layers up to 2 km for the photochemical dispersion model.

Lisbon is built in a very complex topographic region, dominated by a large estuary (320 km^2) and multiple hills reaching more than 400 m above sea level.

This region was selected taking into account the geographical distribution of the main industrial sources and of the most important urban centres of Portugal with about 3.5 million inhabitants (Figure 1).

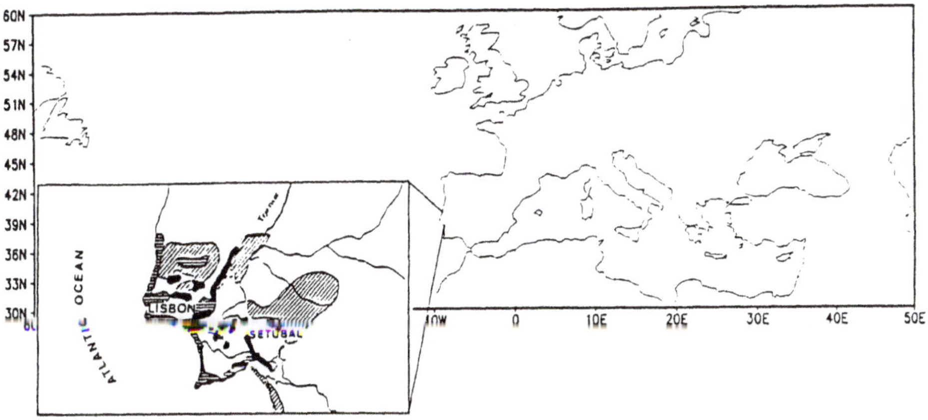

Figure 1. Lisbon region. The residential areas are squared, industrial areas are solid and forest and agricultural areas are marked with diagonals.

Mesoscale initial conditions

Ten simulations have been carried out with the MUGCM in perpetual mode for January where the model's atmosphere has been forced by SSTs for the 1982-83 ENSO event. SSTs anomalies with respect to the climatology are shown in Figure 2. Each simulation started from different initial conditions which were then averaged and compared with a 20 - year control simulation with climatological January SSTs.

The vertical profiles of the atmospheric variables needed as initial conditions by the mesoscale model have been obtained by the profile of the grid GCM nearest the Lisbon region. Two sets of initial conditions were obtained, one for the control simulation and the other for the perturbed simulation (ENSO conditions). As a response to ENSO the atmosphere circulation at the low levels near Lisbon strengthens and shifts from SSW to SSE. The climatological and the ENSO wind fields are displayed in Figure 3. and 4., respectively.

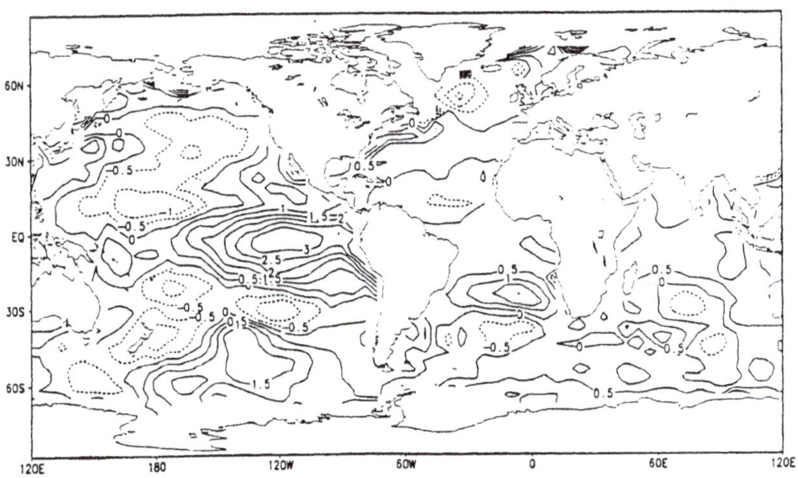

Figure 2. ENSO SSTs anomalies. The contour interval is .0.5 °C.

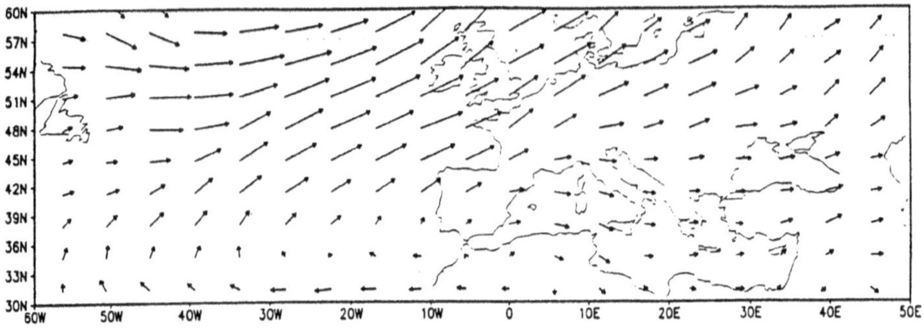

Figure 3. Wind field for the control simulation. The largest vector corresponds to 10 $m.s^{-1}$.

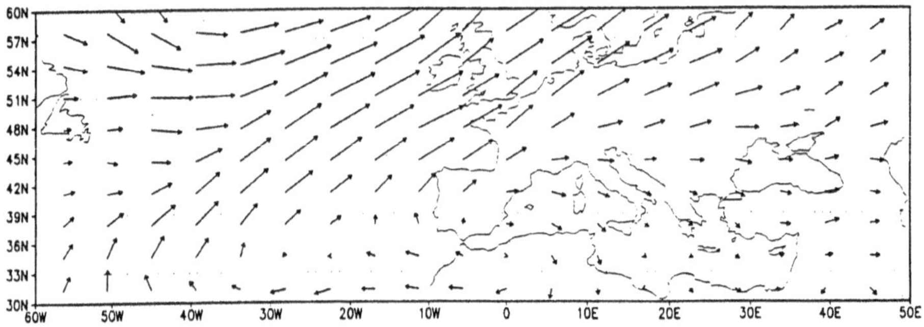

Figure 4. Wind field for the ENSO simulation. The largest vector corresponds to 10 $m.s^{-1}$.

Emissions and air quality input

Another important input in air quality models, besides the meteorological data, is the emission inventory. This problem is one of the most difficult to solve on the air quality modelling domain.

The CORINAIR is a emission inventory performed for each community country including Portugal. This emission inventory was performed under an European Commission contract and because of that, the base data of the inventory is not public. The inventory include area sources and point sources. The area emission data are aggregated in territory units of level III. That level of aggregation, with an area of several square kilometers associated, is not appropriate for an air quality modelling.

To solve this problem, some assumptions have been used to establish a relationship between the CORINAIR 90 NUT III emissions (Valadas et al., 1994) and different type of combustible consumption, population density, economical weight factors of each municipality. With this approximation, it was possible to performe a spacial discretization over the Portugal 1990 CORINAIR NUT III emissions inventory and generate a gridded data base at municipality level (Barros, 1994). Additionally to this area emission data base, a 21 point source data base for the power plants (with real data) and main industrial point sources were included.

In order to specify the boundary concentrations, a clean air approach was assumed (Borrego et al., 1990). The strong influence of the clean air advected from the Atlantic Ocean over the Lisbon airshed, as well as the results of field campaigns performed in rural and forest atmospheres were considered in the definition of typical trace constituents concentrations (Coutinho et al., 1991).

In this work, the emission data base and the boundary concentrations have been considered the same for both simulations, control and ENSO.

RESULTS AND DISCUSSION

Wind fields

The air flow simulated by the CSUMM within the modelling region at sun rise is characterised by downslope flows along the main mountain ridges for both simulations (control and ENSO). Over the Tejo's valley, in the case of the perturbed simulation, stagnations zones, characterised by very weak and not definite winds, are generated by the convergence of those circulations and the SSE background circulation. Opposed to that, on the control situation, a well-defined SW circulation is installed on that valley.

Early afternoon, the CSUMM simulates, for both situations, a strong onshore circulation along most of the coast line. At this time, the thermal potential between the sea and the land is not strong enough to develope a sea breeze. Mesoscale circulations show a small spatial extent and the airflow is dominated by the Atlantic circulation, particularly in for the control simulation.

During the evening, onshore flow is gradually replaced by nocturnal drainage flow. CSUMM develops a downslope flow on the main mountain ridges.

Photochemical dispersion results

The prognostic air flow combined with ozone field concentrations generated by MAR system at 10 m AGL are presented in Figures 5-8. The wind and concentration fields for 10.00 and 15.00 LST were select.

At sun rise, for the control situation, the precursors emitted in Lisbon earlly morning are transported onshore by the Tejo's valley circulation. In opposit, for the ENSO situation, those pollutants are parcialy trapped on the convergence region formed over the Tejo's valley allowing a local photochemical production. In Figure 5 and 6 is possible to verify that phenomena with higher ozone concentrations for the ENSO situation.

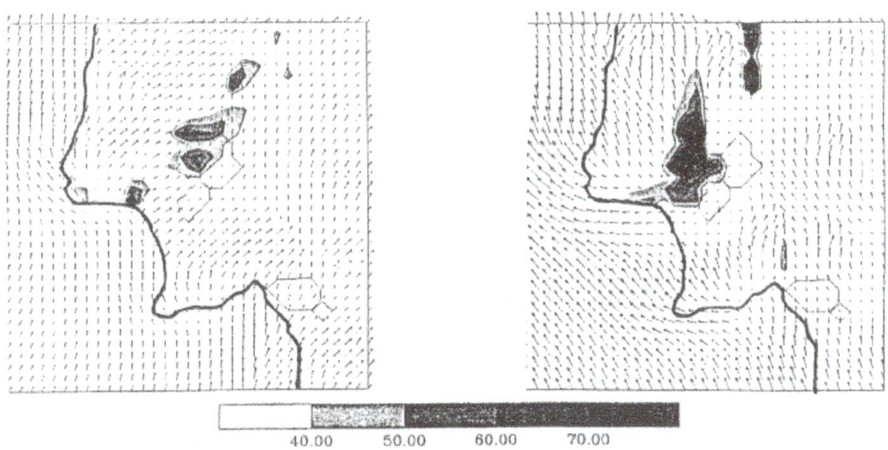

Figure 5. Ozone concentration field in ppb at 10.00 LST. Wind velocity is scaled at 8 m.s^{-1} per grid spacing (control simulation).

Figure 6. The same as in Figure 5, but for the ENSO simulation.

In the afternoon strong onshore circulation from SSW are simulated in both experiments (control and ENSO).This results in a transport of the ozone plume formed in the morning and dispersion of the precursors emitted in Lisbon region with the consequent decrease in ozone concentrations.

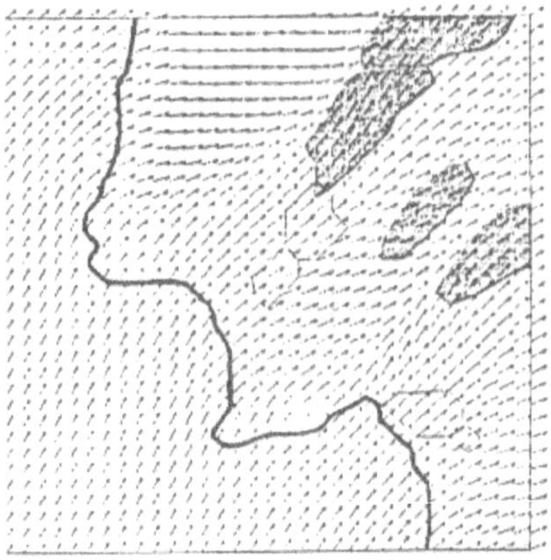

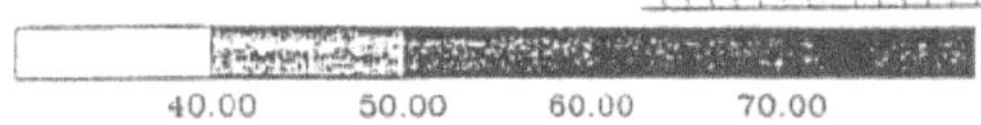

Figure 7. Ozone concentration field in ppb at 15.00 LST. Wind velocity is scaled at 8 m.s^{-1} per grid spacing (control simulation).

Figure 8. The same as in Figure 7. but for the ENSO simulation.

Figure 9 shows the percentual variation of ozone peak concentration during ENSO over the region of study.

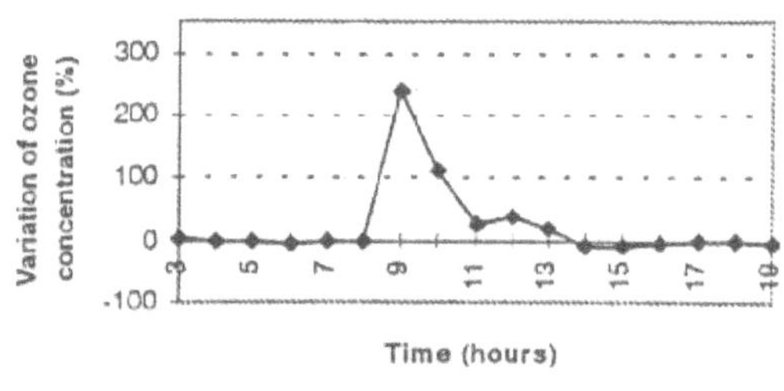

Figure 9. Variation of peak ozone concentration between the control simulation and the ENSO simulation.

CONCLUSIONS

Routine air quality observations are frequently insufficient to represent complex air pollutants transport patterns in coastal countries like Portugal, in particularly for ozone. The work performed over Lisbon region demonstrated that numerical models are an essential tool to study different scenarios of synoptical circulation and its impact on regional air quality. Even minor differences found in large scale forcing due to global changes can affect the mesoscale circulations.

The formation of stagnant regions, under certain large scale forcing, is particularly critical, representing areas where convergence of polluted air masses could occur causing an increase of the concentration of secondary pollutants.

This preliminary work seems to show that, in fact, a phenomenon like the ENSO can affect the Lisbon air quality.

The results show that this down-scaling approach can be a useful tool to assess the impact of anomalous large-scale atmospheric circulations on the air quality over Portugal.

ACKNOWLEDGEMENTS

The authors would like to thank the JNICT (PRAXIS XXI) and the University of Aveiro, Environment and Planning Department,for the Ph.D. grant of N. Barros.

REFERENCES

Ames J., Myers T., Reid L., Whitney D., Golding S., Hayes S., Reynolds S., 1985, SAI Airshed Model Operations Manuals, *EPA/600/8-85/007, U. S. Environmental Protection Agency.*

Arritt, R.W., 1985, Numerical Studies of Thermally and Mechanically Forced Circulations Over Complex Terrain. Cooperative Institute for Research in the Atmosphere, Colorado State University, Fort Collins, Colorado.

Barros, N., 1994, Modelação da Poluição Fotoquímica na Região de Lisboa. *Dissertation presented to Faculdade de Engenharia da Universidade do Porto to obtain the Master degree on Thermal Engineering.*

Barros, N. and Borrego, C., 1995, Influence of coastal breezes on the photochemical production over Lisbon region. To be publishing in *Air Pollution 95 Proc.*, Porto Carras, Grece, 26 - 28 September.

Boer, G. J., K. Arpe, M. Blackburn, M. Déqué, W.L. Gates, T.L. Hart, H. le Treut, E. Roeckner, D. A. Sheinin, I. Simmonds, R. N. B. Smith, T. Tokioka , R.T. Wetherald, D. Williamson, 1991, An intercomparation of the climates simulated by 14 atmospheric general circulation models. Word Meteorological Organization/International Council of Scientific Unions Word Climate Research Programme, *Report Nº 15, WMO/TD* - Nº 425, 37 pp plus maps.

Bourke, W. , B. McAvaney, K. Puri, and R. Thurling, 1977, Global modeling of atmospheric flow by spectral methods. *Methods in Comp. Phys. 17. General Circulation Models of Atmosphere*, ed. J. Chang, Academic Press, New York, 267-324.

Borrego, C., M. Coutinho and J. Rua, 1990, Eulerian modelling of atmsopheric dispersion over Portugal: background concentrations and emission data preparation. *Proc. of 18th ITM on Air Pollution Modelling and its Applications* , NATO-CCMS, Vancouver, 13-17 May.

Coutinho, M. and Borrego, C.,1991, Photochemical production on coastal areas of Portugal. *Proc. of 19th ITM on Air Pollution Modelling and its Applications , NATO-CCMS*, Ierapetra, Crete, Greece, 29 September - 4 October.

Coutinho, M., A.I. Miranda, C. Borrego and J. Rua. 1992, Emissions trends in Portugal: a global perspective. *Proc. of the 85th Annual Meeting of the Air & Waste Management Association*, AWMA, Kansas City.

Mahrer, Y. and R.A. Pielke. 1977: A numerical study of the air flow over irregular terrain. Contr. Atmos. Phys., **50**, 98-113.

Mahrer, Y. and R.A. Pielke. 1978: A test of an upstream spline interpolation technique for the advective terms in a numerical mesoscale model. *Mon. Wea. Rev.*, **106**, 818-830.

McAvaney, B. J., W. Bourke and K. Puri, 1978, A global spectral model for simulation of the general circulation. *J. Atmos. Sci.*, **35**, 1557-1583.

Pielke, R.A., 1974: A three-dimensional numerical model of the sea breeze over south Florida, *Mon. Wea. Rev.*, **102**, 115-139.

Simmonds, I., 1985, Analysis of the 'spinup' of a general circulation model. *J. Geophys. Res.*, **90**, 5637- 5660.

Simmonds, I., G. Trigg, R. Law, 1988, The climatology of the Melbourne University General Circulation Model. Publication Nº 31, Department of Meteorology, University of Melbourne, 67 pp. (NTIS PB 88 227491).

Tremback, C. and R.C. Kessler. 1985: A surface temperature and moisture parameterization for use in mesoscale numerical models. *Proc. 7^h Conferenceon Numerical Weather Prediction*, Montreal.

Valadas, B. and Gois, V., 1994, Inventário Nacional de Emissões Atmosféricas - CORINAIR 90 Portugal. Direcção Geral do Ambiente.

DISCUSSION

R. BORNSTEIN: Were you able to see any river breeze effects in your results ?

N. BARROS: In this particular application, the resolution of the grid, 3 km x 3 km, is not enough to see the river breeze effect in my results.

AIR POLLUTION ASSESSMENT STUDIES FOR SWEDEN BASED ON THE MATCH MODEL AND AIR POLLUTION MEASUREMENTS

Christer Persson, Joakim Langner, and Lennart Robertson

Swedish Meteorological and Hydrological Institute (SMHI)
S-601 76 Norrköping
Sweden

ABSTRACT

The MATCH (Mesoscale Atmospheric Transport and CHemistry) model has been developed as a tool for air pollution assessment studies on different geographical scales. Model versions covering Europe, Sweden and subregions of Sweden have been used as a basis for decision making concerning environmental protection in Sweden. MATCH is an Eulerian atmospheric dispersion model, including physical and chemical processes governing sources, atmospheric transport and sinks of oxidized sulfur and oxidized and reduced nitrogen. With the MATCH system, air pollution contributions from different source types like traffic, industry, shipping, farming etc. can be obtained. Using a combination of air and precipitation chemistry measurements and the MATCH model, the contribution of air pollution and deposition from long-range transport can be quantified in the model region. Also hourly mappings of ozone concentrations are obtained. The calculations for the year 1991 show that the Swedish import was about 4.5 times larger than the export for sulfur and about six times larger for reduced nitrogen, while the Swedish import of oxidized nitrogen only exceeded the export by 10%. Using the MATCH system we estimate the long-range transport in an independent way compared to EMEP. Comparison between the EMEP and MATCH calculations for 1991 show that the total deposition of oxidized nitrogen over Sweden is similar, while the EMEP-values for total deposition of oxidized sulfur and reduced nitrogen are 25% respectively 40% smaller than what is obtained from MATCH. The Swedish contributions to deposition over Sweden are also somewhat smaller (20-38%) in the EMEP model compared to MATCH. Results from MATCH for the year 1994 show a decrease in sulfur deposition over most of Sweden, both from Swedish sources and long-range transport, except for south-east Sweden where a pronounced increase in long-range transport contribution was obtained.

1. INTRODUCTION

Deposition of acidifying pollutants is a very important environmental problem in Sweden as in many other countries. Considerable efforts have been made on both national and international scales in order to quantify and understand the phenomenon. The work on the European scale has been coordinated within EMEP (Co-operative programme for monitoring and evaluation of the long range transmission of air pollutants in Europe, EMEP, 1980). EMEP runs a model covering Europe which is capable of allocating the deposition of oxidized sulfur and oxidized and reduced nitrogen compounds on a 150x150 km horizontal grid to the emitting countries on an annual basis (Eliassen and Saltbones, 1983).

In addition to EMEP most countries run national acidification programmes. In Sweden monitoring now includes measurements of sulfur and nitrogen compounds at several background locations as well as a dense network of through-fall measurements in the southern part of Sweden. Modeling work aimed at utilizing these data to provide additional information and generalizations has so far been limited in Sweden. To optimize efforts such as liming and national emission controls to limit the effects of acid deposition, information about deposition of acidifying pollutants with higher resolution than currently available from EMEP is desirable. In describing the effects of acidification on the ecosystem level, information should be available at least on the size of the ecosystems. The work presented here is an attempt to meet these requirements by combining model calculations for national emissions with careful analyses of atmospheric chemistry measurements linked with high resolution meteorological data and land-use information.

Another important environmental problem is the occurrence of high ozone concentrations in the air near ground. A detailed geographical coverage of information about ozone concentrations is desired. Also for that purpose attempts have been made to combine model calculations for national emissions with a careful analysis of background ozone measurements and high resolution meteorological data.

2. MODELING SYSTEM

The MATCH modeling system consists of three parts: A regional atmospheric dispersion model including modules for emission, chemistry and deposition of sulfur and nitrogen compounds. An objective analysis system for air- and precipitation chemistry data, and an objective analysis system for meteorological data.

2.1. Dispersion Model

The MATCH (Mesoscale Atmospheric Transport and CHemistry) model (Persson and Robertson, 1991; Persson et al., 1994) is a three dimensional Eulerian atmospheric dispersion model. The model is a so called "off-line" model, meaning that it requires meteorological data from an external archive at regular time intervals (usually three hours) in order to calculate transport, chemistry and deposition. The model is designed to be flexible with regard to horizontal and vertical resolution. It is used for several different applications in dispersion modelling at SMHI, with horizontal grid resolutions ranging from 5 to 100 km. In this paper a description is given of the model version, covering Sweden or subregions of Sweden, applied to air pollution assessment studies in Sweden.

The model version used here has three layers in the vertical. The first layer has a fixed height of 75 m. The top of the second layer is taken to be the same as the mixing height, and the top of the third layer is fixed at a certain level (~1.5 km in winter, ~2.5 km in summer). It is easy to add additional layers if necessary, but for calculations over areas of the size of Sweden or smaller three layers was judged to be sufficient. The horizontal resolution for the calculations over Sweden is 20x20 km while a 5x5 km resolution has been used for subregions in Sweden.

Horizontal advection is calculated using a fourth order flux correction scheme (Bott 1989a, 1989b). Vertical advection is calculated using an upstream scheme. Vertical transport is also induced by turbulent vertical diffusion and the spatial and temporal variations of the mixing height. For the numerical solution of the combined horizontal and vertical transport, chemistry and deposition an operator split time integration scheme is used.

2.1.1. Emissions. Emissions can be specified both as area and point sources. Surface area sources are introduced into the lowest layer of the model. The initial dispersion from point sources is described with a Gaussian puff model including plume rise calculations. The puffs are advected until they have reached the size of the horizontal grid when they are merged into the large scale concentration field. The emission totals used for 1991 are given in Table 1.

2.1.2. Chemistry and deposition. The chemistry in the model deals with sulfur oxides and oxidized and reduced nitrogen compounds and is almost identical to that used in the EMEP model (Iversen et al., 1989). The main difference compared with the EMEP model is in the specification of ozone (O_3) concentrations. Here analysed O_3 distributions with one hourly time resolution are generated from background observations (c.f. section 2.2). A local adjustment of the O_3 concentration with regard to local NO- and NO_2-concentration, solar radiation and dry deposition is then done.

Wet scavenging of the different species is proportional to the precipitation rate and a species specific scavenging coefficient. Dry deposition is proportional to the concentration and a species specific dry deposition velocity at 1 m height. Since the lowest model layer has a height of 75 m, the dry deposition flux calculation is transformed to the middle of that layer using standard similarity theory for the atmospheric surface layer. Dry deposition velocities are specified as a function of the surface characteristics (fraction forest, field etc). Scavenging coefficients and most deposition velocities have values close to those used in the EMEP calculations. For some compounds (mainly NO_2 and particles) the deposition velocities to forest have been somewhat larger than used by EMEP.

2.2. Analysis Of Air And Precipitation Chemistry Data

The dispersion model described above, combined with national emission estimates and meteorological data provides daily estimates of concentrations in air and precipitation as well as dry and wet deposition of the simulated sulfur and nitrogen compounds. These results refer to contributions from sources within the model area (in this case Sweden). To derive distributions of the contribution from sources outside Sweden the following method is employed: Model calculated daily contributions from Swedish sources are deducted from observed daily values of concentration in air and precipitation at background locations on a point by point basis (about 50 stations in and around Sweden are used). The residual is termed long-range transport contribution. These residuals are analysed using an optimum interpolation method, where differences in observation quality can be accounted for, to give distributions of long-range transport contributions of concentrations in air and precipitation over the whole modeling domain. The basic idea behind this method is that the long-range transport contributions to the concentrations can be expected to vary more smoothly in space than the total concentrations which are affected to some extent by local sources, and should therefore be more suitable for interpolation. Long-range transport wet deposition is then calculated by multiplying with the observed precipitation field (c.f. section 2.3). Long-range transport dry deposition is calculated by running the long-range transport air concentrations through the dry deposition module of the dispersion model. Considerable efforts have been spent on quality control of both input chemical observation data and resulting analyzed concentration distributions. The objective analysis scheme is a very useful tool for identifying different kinds of errors in the observations. A similar method using total concentrations has been developed by van Pul et al. (1994). The difference here is the possibility to separate the national and long-range transport contributions. The method assumes however, that the model calculated national contribution is correct. This is not necessarily the case. In order to gain confidence in the method comparisons have been made with independent air chemistry measurements. One such comparison taken from a study over the Swedish west coast region using exactly the same method as described here is given in Figure 1. It shows a time series of the concentration of NO_2 in central Gothenburg for the second half of 1991. Both the overall concentration level and episodes are captured favorably. Additional comparisons have been performed and indicate that the model calculations are reasonable.

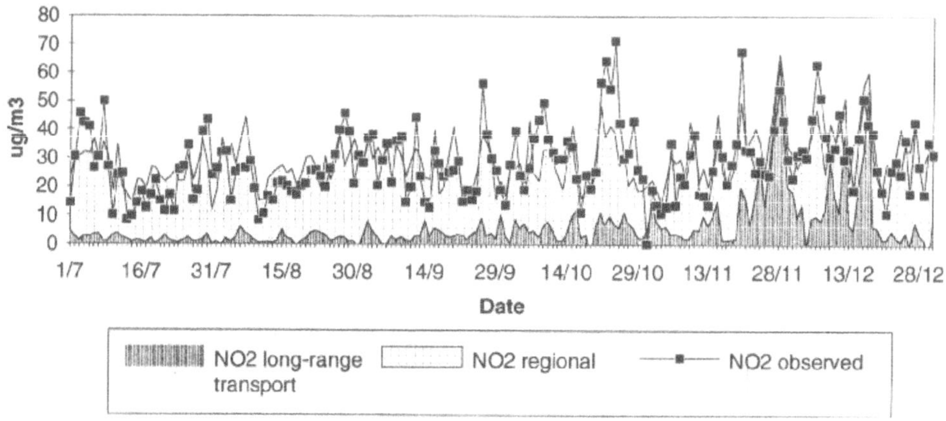

Figure 1. Comparison, for a measuring station in central Gothenburg, between long-range transport contribution (dark shaded) plus regional contribution (light shaded) from the Swedish west-coast area and observed (black squares) total NO_2 concentrations in air for the second half of 1991 (μg N/m^3).

2.3. Meteorological Analysis

The dispersion model requires meteorological data to calculate transport, chemistry and deposition processes. For studies over Sweden an objective meteorological analysis system has been developed. The system makes use of routine meteorological observations to derive the wind, turbulence and precipitation fields required by the dispersion model. The analyses are performed at three hour intervals. A high resolution data base for topography and land use (fraction of forest, field, water, urban) is used in the analysis.

The precipitation analysis is given special attention: About 800 stations over Sweden measuring daily precipitation is combined with precipitation and weather information from synoptic stations to give precipitation fields with three hourly time resolution and high horizontal resolution. Corrections for sampling losses and topography are also applied.

3. RESULTS AND DISCUSSION

Examples of results from the MATCH-Sweden preliminary calculations for 1994 are shown in Figures 2 and 3. Figure 2 shows the annual average contributions to the NO_2-concentration from emissions in Sweden, long-range transport and also the total NO_2-

concentration. As can be seen Swedish contributions dominate in southern Sweden and in the populated areas along the east coast of central and northern Sweden.

Figure 3 shows similar results for total (dry + wet) deposition of oxidized sulfur for 1994. In this case the long-range transport dominates over most of Sweden. Note the maximum in deposition over the southern part of the Swedish west-coast area which is due to a combination of high concentrations of sulfate in air and precipitation and a maximum in precipitation on the edge of the highlands in southern Sweden. This maximum is normally present every year. The sulfur deposition has decreased in Sweden during the last 10-15 years with as much as about one third and the decrease continued during 1994 for most of Sweden. In south-east Sweden, however, there was a pronounced increase in the long-range transport contribution compared to earlier years. It can be seen as a secondary maximum in south east and was mainly caused by meteorological conditions. Large precipitation amounts occurred in connection to atmospheric transport from source areas in eastern Europe. The largest sulfur deposition values are obtained on the coast of Lithuania, in the south-east corner of the model area

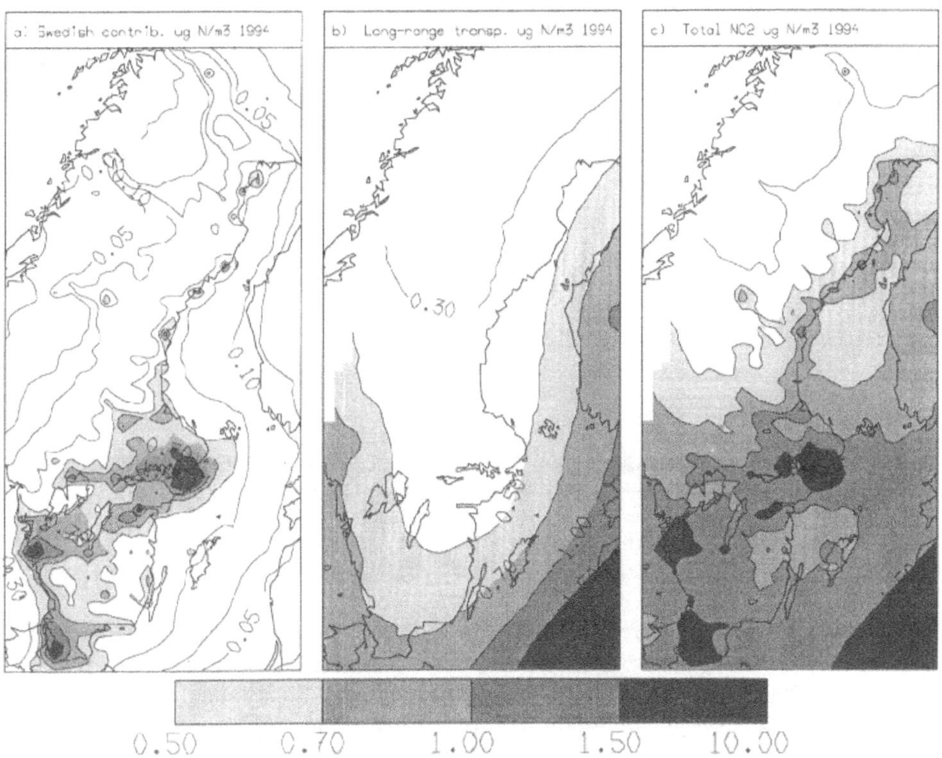

Figure 2. Calculated preliminary annual average (year 1994) NO_2-concentration: a) contribution from Swedish sources, b) contribution from long range transport, c) total concentration ($\mu g\ N/m^3$).

3.1. Comparison with EMEP

Using the MATCH system, we estimate the long-range transport in an independent way compared to EMEP (Tuovinen et al., 1994). Comparison between the EMEP and MATCH calculations for 1991 (Table I) show that the total deposition of oxidized nitrogen over Sweden is similar, while the EMEP-values for total deposition of oxidized sulfur and reduced nitrogen are 25 and 40% smaller respectively than what is obtained from MATCH. The Swedish contributions to deposition over Sweden are also somewhat smaller (20-38%)

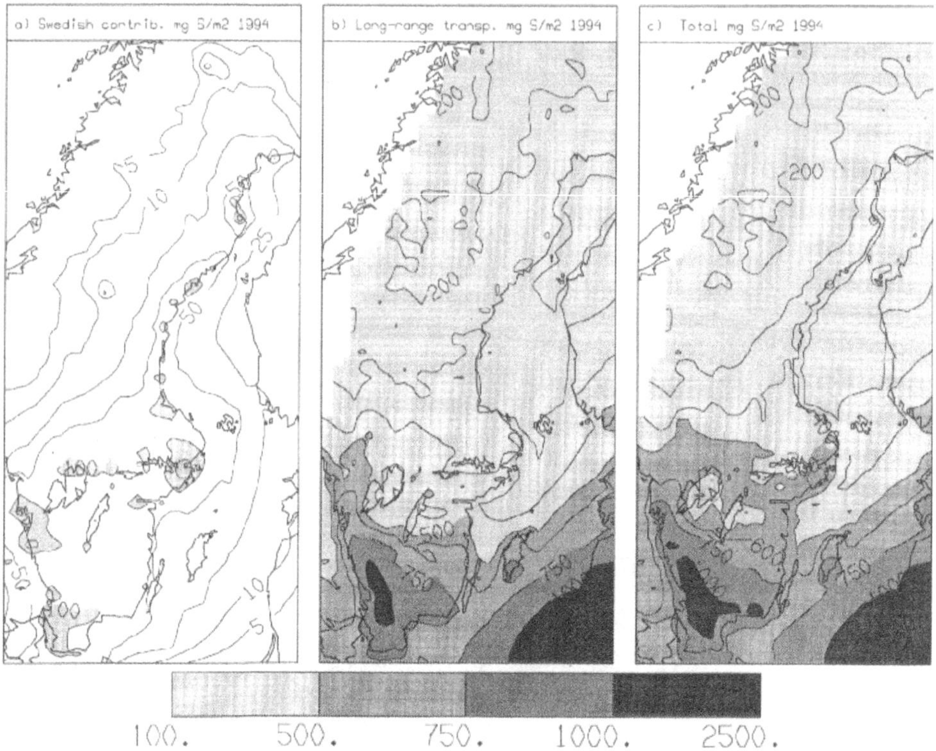

Figure 3. Calculated preliminary annual (year 1994) dry + wet deposition of oxidized sulfur: a) contribution from Swedish sources, b) contribution from long range transport, c) total (mg S/m^2).

in the EMEP model compared to MATCH. When looking closer at different regions in Sweden even larger differences can be noted. The coarser resolution of the emission data in the EMEP model here obviously plays an important role. Some small changes, both in emission estimates and model parameters were made in the version of the EMEP model used by Tuovinen et al. compared to previous versions. Comparing the MATCH calculations with the previous version of the EMEP model (Sandnes, 1993) gives a closer agreement both for total and Swedish contributions to deposition over Sweden

Table 1.

Comparison between EMEP and MATCH estimates of the Swedish and total contribution to deposition of oxidized sulfur and oxidized and reduced nitrogen over Sweden for 1991 (100 ton/year). The assumed Swedish emissions in the different models are also given.

	Sulfur	NO_x- nitrogen	NH_x - nitrogen
Swedish emissions (incl. national shipping):			
EMEP	550	1184	420
MATCH	700	1184	440
Swedish deposition over Sweden:			
EMEP	152	148	220
MATCH	243	207	276
Total deposition over Sweden:			
EMEP	1740	1200	720
MATCH	2300	1300	1260

4. CONCLUSION

A modeling system capable of providing high resolution estimates of national and long-range transport contributions to concentrations in air and to dry and wet deposition of sulfur and nitrogen compounds over Sweden has been developed. Comparisons with the EMEP calculations indicate substantial differences for total deposition of oxidized sulfur and reduced nitrogen With lower estimates derived from the MATCH system. Calculations for 1991 and 1994 (not yet completed) have been performed and the intention is to continue and expand the work to include other chemical components in the future.

ACKNOWLEDGMENTS

This work has received financial support from the Swedish Environmental Protection Agency. Air and precipitation chemistry data for background stations in Sweden, Norway, Finland, Denmark and Lithuania were provided by Swedish Environmental Research Institute (IVL), Norwegian Institute for Air Research, Finnish Meteorological Institute, National Environmental Research Institute in Denmark and Institute of Applied Environmental Research in Sweden respectively. IVL has also provided some Swedish emission data. Traffic emission data has been obtained from Swedish Road and Transport Research Institute.

REFERENCES

Bott, A., 1989a, *Mon. Wea. Rew.* 117, 1006 - 1015.

Bott, A., 1989b, *Mon. Wea. Rew.* 117, 2633 - 2626.

Eliassen, A., and Saltbones, J., 1983, *Atmos. Environ.*, 17, 1457-1473.

EMEP, 1980, *Summary report of the Western Meteorological Synthesizing Centre for the first phase of EMEP*, EMEP/MSC-W, Norwegian Meteorological Institute, Oslo, Norway.

Iversen, T., Saltbones, J., Sandnes, H., Eliassen, A., and Hov Ø., 1989, *Airborne Transboundary Transport of Sulphur and Nitrogen over Europe - Model Description and Calculations*, EMEP/MSC-W Report 2/89, DNMI, Oslo, Norway.

Persson, C., and Robertson, L., 1991, In H. van Dop and D. G. Steyn (eds.), *Air Pollution Modeling and Its Application VIII*, Plenum Press, New York, pp 649-650.

Persson, C., Langner, J., and Robertson, L., 1994, In S-E. Gryning and M. M. Millán (eds.), *Air Pollution Modelin and Its Application X,* Plenum Press, New York, pp 9-18.
van Pul, W. A. J., Erisman, J. W., van Jaarsveld, J. A., and de Leeuw, F. A. A. M., 1994, In S-E. Gryning and M. M. Millán (eds.), *Air Pollution Modelin and Its Application X,* Plenum Press, New York, pp 625-627.
Sandnes, H., 1993, *Calculated budgets for airborne acidifying components in Europe: Calculated fields and budgets 1985-93,* EMEP/MSC-W Report 1/93, DNMI, Oslo, Norway.
Touvinen, J-H., Barret, K., and Styve, H., 1994, *Transboundary acidifying pollution in Europe: Calculated fields and budgets 1985-93,* EMEP/MSC-W Report 1/94, DNMI, Oslo, Norway.

DISCUSSION

G. COSEMANS: Are the differences in deposition over Sweden according to EMEP and MATCH due to differences in dry deposition, wet deposition or both?

C. PERSSON: The Swedish contribution to dry as well as wet deposition over Sweden is larger according to the MATCH model compared to EMEP. The total wet deposition is also larger in MATCH, while the total dry deposition is roughly of the same size (some differences between compounds) in MATCH and EMEP.

T. IVERSEN: Firstly, please accept my compliments for your very good work which can serve as a guide for other national centres in Europe when they wish to use EMEP-data as a background for detailed studies on the national level. My question: Do you have a detailed, land-use dependent dry deposition module in MATCH, and does this module give different dry deposition velocity than EMEP-models?

C. PERSSON: We calculate the dry deposition with regard to land-use information (sea, open field, forest, urban). For the time being we use a rather simple dry deposition module which is similar to what EMEP used in Report 1/94 and earlier, except for that we in MATCH use somewhat larger dry deposition velocities to forest. This adjustment have been introduced due to large observed throughfall values in Sweden.

PRELIMINARY RESULTS: MODELING FINE PARTICULATE MASS FOR THE EASTERN UNITED STATES USING THE EPA REGIONAL PARTICULATE MODEL

Jason Ching*, Francis S. Binkowski*, and Jonathan E. Pleim*

Atmospheric Sciences Modeling Division
Air Resources Laboratory
National Oceanic and Atmospheric Admisistration
Research Triangle Park, NC 27711

*On assignment to the National Exposure Research Laboratory
U.S. Environmental Protection Agency

INTRODUCTION

PM-10, Particulate Matter less than 10 μm, is one of several criteria air pollutants for which National Ambient Air Quality Standards (NAAQS) are established by the United States Environmental Protection Agency (EPA) to protect human health and welfare. Recently, attention has been focussed on finer particles, those less than about 2.5μm. Typically, particles in this size range arise as products of atmospheric reactions of sulfur, nitrogen and organic pollutants. These pollutants are also primary natural and anthropogenic emissions from urban areas and from point sources in rural areas. The atmospheric lifetimes of aerosol particles can be several days. Strategies designed to reduce emissions to achieve NAAQS for particulate matter must incorporate inventories for both primary particulate emissions and primary gaseous precursor emissions. Assessments of current loadings from such emissions and development and analysis of control strategies to achieve compliance with NAAQS require regional-to-urban scale photochemical and aerosol models.

This paper briefly describes the Regional Particulate Model (RPM), and illustrates some of its capabilities with preliminary results. The model was designed for addressing both the assessments and emissions control strategy options required for the NAAQS review. This model provides a basis for understanding the complex temporal and spatial distributions of air pollution on scales ranging from urban-suburban to regional (sub-continental) scales.

The model predicts the chemical composition and size distribution of airborne aerosol particles. RPM includes the main processes of aerosol dynamics such as growth, coagulation, and new particle production as well as a parameterization of size-dependent dry deposition. RPM is based on the EPA's Regional Acid Deposition Model (RADM) which has been run for assessment strategies to determine the benefits to be achieved from the Clean Air Act Amendments of 1990 (CAAA). Aqueous production of sulfate aerosol mass in RPM is thus based upon the corresponding parameterization used in RADM. Preliminary model results of atmospheric concentrations of fine particles for current emissions levels presented here illustrate the potential for addressing candidate future emissions reductions scenarios.

Air Pollution Modeling and Its Application XI, Edited by
Gryning and Schiermeier, Plenum Press, New York, 1996

DESCRIPTION OF THE MODEL

As noted, RPM is based upon the paradigm of RADM (Chang et al., 1990), that is, RPM is an Eulerian model which uses input fields of meteorology and emissions. The meteorological fields are provided from the Fourth Generation Penn State-NCAR Mesoscale Meteorological model with Four Dimensional Data Assimilation (Stauffer and Seaman, 1990; Stauffer et al.,1991) on an hourly basis.The horizontal extent of the model domain is as shown in the figures, the vertical domain is from the surface to 100 mb (about 16 km) in 15 unequally spaced layers. Gridded emissions data account for both anthropogenic sources (area, point, and mobile sources) as well as biogenic sources and represent the source strengths for the year 1990. Using these data, the model generates gridded outputs of gas and aerosol particle concentration fields and deposition values.

A unique feature of the RPM is the use of a bi-modal aerosol size distribution represented by the superposition of two interacting lognormal sub-distributions called modes (Binkowski and Shankar, 1995), an idea first proposed by Whitby (1978). The smaller or Aitkin mode contains newly produced particles, and the larger or accumulation mode contains aged particles and particles produced by cloud processes. New particles are produced in the model by emission of primary sulfates and by a parameterization of the nucleation process (Kerminen and Wexler, 1994). This parameterization will be briefly described below.

The current version of RPM has been formulated to make hourly predictions of gridded fields of sulfate, nitrate and organic aerosols. Particles in RPM are internally mixed, and can grow or shrink with respect to water content. The particles are assumed to be aqueous solutions when the relative humidity exceeds the humidity of crystallization. That is, the particles are assumed to be liquid even though they are below the relative humidity of deliquescence. There is evidence (Rood et al., 1989) that this metastable state is common in the atmosphere. Both model results and measurements indicate that in the eastern half of North America, sulfate aerosols are the dominant secondary aerosols species, and that the aerosol particles are frequently acidic. This acidity is neutralized by the presence of ammonia which is emitted by both biogenic and anthropogenic sources. Unfortunately, ammonia emissions inventories have a high degree of uncertainty at present.

Sulfate aerosol is formed by two mechanisms. The first is through hydroxyl radical attack on sulfur dioxide which is directly emitted from combustion of fossil fuels. The second path is through cloud water processes and is discussed below. The hydroxyl radical reaction produces sulfuric acid vapor in the presence of water vapor. This material will either condense upon existing particles or nucleate to form new particles. The criterion for whether condensation or nucleation occurs is given by the Kerminen and Wexler parameterization. In this parameterization the product of the time scale for condensation on the existing aerosol and the chemical production rate due to hydroxyl radical attack forms a concentration which can be viewed as a steady state value over some time interval. This concentration is compared with a critical concentration value derived from thermodynamical theory. If the steady state value exceeds the critical value, then new particles are formed from the excess and added to the Aitkin mode, the rest being condensed upon the existing particles in both modes. If the critical value is not exceeded, then all of the newly produced sulfate is condensed upon the existing particles. The water content of this aqueous solution is determined by the relative humidity and the ammonium to sulfate ratio in the solution. The amount of nitrate in the aerosol is a strong function of the acidity also a function of the ammonium content, thus increasing the relative importance of ammonia. The equilibration of nitric acid vapor, ammonia, and the sulfuric acid solution is done with a modification of the method given by Saxena et al. (1986). This modification uses the multicomponent method of Pitzer and Mayorga (1974) for the activity coefficients. The water content of the aerosol is very important because particles grow or shrink as they take on or lose water in response to the relative humidity, thus shifting the size distribution to larger or smaller values. This size shift, combined with the change in refractive index with changing composition, strongly influences the role of the aerosols in radiative transfer, especially visibility. As noted, the nitrate content of the aerosol particles is a strong function of the ammonia levels in the atmosphere. The simulations done here show that, as expected, the strong acidity of the aerosols inhibits the presence of nitrates. Sensitivity analyses (Alapaty and Binkowski, 1995) show that suggested levels of uncertainty in ammonia emissions yield different nitrate content for the particles. Dennis (1995) has noted that the ammonia emissions source strengths are very likely significantly underestimated, which would lead to a bias to underestimate the nitrate content of aerosol particles. Sulfate is also produced through oxidation of dissolved sulfur in cloud water. The main oxidants are hydrogen peroxide, ozone and oxygen catalyzed by the presence of iron and manganese. The cloud algorithm assigns the sulfate produced in cloud water to the accumulation mode upon evaporation of the cloud water as is observed in the atmosphere. Except for the oxidation by hydrogen peroxide, the aqueous reactions are strongly inhibited by acidity.

Over eastern North America organic aerosols are next in importance to sulfates. A major part of this organic material is of secondary origin. Secondary organic aerosols (SOA) are produced in the atmosphere from gas-phase precursors by chemical reactions. RPM uses the approach of Pandis et al. (1992) which provides aerosol yields for oxidation of various reactive organic gases (ROG) by hydroxyl radicals for the lumped gas-phase species of the RADM 2 mechanism (Stockwell et al., 1990),

$$\text{Production rate} = [OH]^* \text{ sum } \{C_n k_n [ROG]_n\}$$

where C_n and k_n are empirical constants and reactivities specific to n classes. This version of RPM models five ROGs. These are HC8 (alkanes), OLI (internal alkenes, including monoterpenes), and three aromatics TOL (toluene and less reactive aromatics), CSL (cresol and other hydroxy substituted aromatics) and XYL (xylene and more reactive aromatics). The production rate of SOA is then used to determine the rate of condensation of the organic carbon on the existing particles. No nucleation of new particles by SOA is permitted in the model. Thus RPM incorporates either by directly modeling or by parameterization all important contributing atmospheric processes that affect the transport and changes in the atmospheric aerosol composition, size distribution and concentration.

3.0 RESULTS OF SIMULATIONS

RPM results from a 72 hour simulation beginning 00 GMT, July 17, 1985 are shown in Figures 1-5 which display examples of key model output fields. The initial weather for this study was characterized by a synoptic scale high pressure system centered in northern Michigan and flanked by two cold fronts, one along the east coast and another oriented SW-NE over the Dakotas. With a southeastward translation, the center of the high moved to central Pennsylvania by the end of the simulation period. Dry conditions characterized this period, with the exception of rain associated with the passage of the front in the western part of the modeling domain. Other modeling information including emissions are found in Binkowski and Shankar (1995).

Using emissions from the 1990 base case and results of a RADM simulation, the aerosol chemical and physical features were computed. Results for 1800Z on the third day of this simulation are shown in the following set of figures to illustrate a few of the outputs of the model. Figure 1 shows the sulfate ($\mu g/m^3$) concentration field in the first model layer ($\Delta z \sim 80m$). The maximum concentration is located over the major sulfur source area along the Ohio River. The aerosol water content field (Figure 2) is highly correlated with the sulfate field. This is expected because the sulfate, the major chemical constituent, (other than the water) is acidic and thus strongly hygroscopic. Figure 3 illustrates the organic portion of the aerosols. The local maxima of organic aerosols associated with major urban areas are primarily of anthropogenic origin. The broad regions of organics in the southeast and upper midwest are mainly attributable to biogenic sources of monoterpenes. The aerosol surface area (Figure 4) is well correlated with the maxima in sulfate and water fields. The same is true for the visual index (Figure 5).

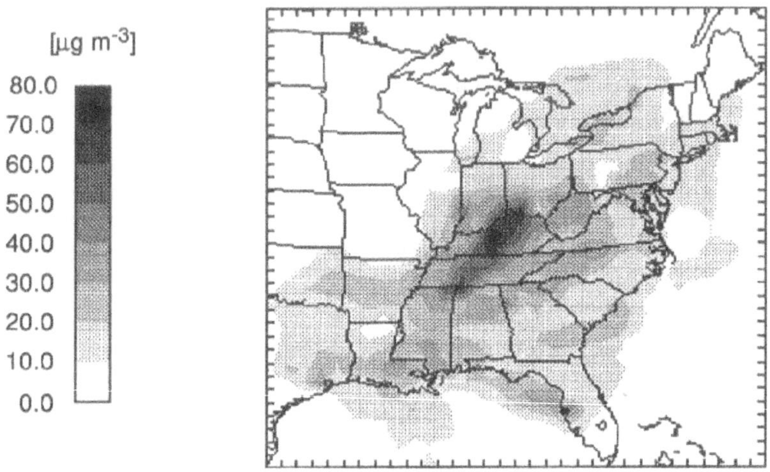

Figure 1: Simulation of aerosol sulfate [μg m^{-3}] in the first layer [Δz~80m] for 1800Z, July 19, 1985.

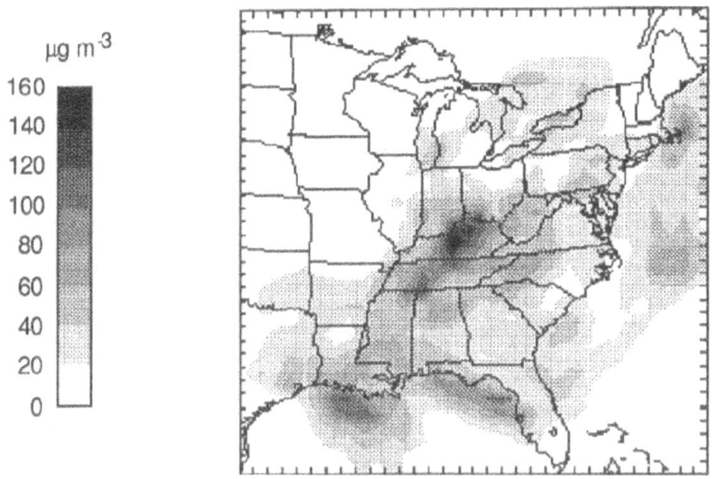

Figure 2: Simultion of aerosol water content [μg m^{-3}] in the first layer [Δz~80m] for 1800Z, July 19, 1985.

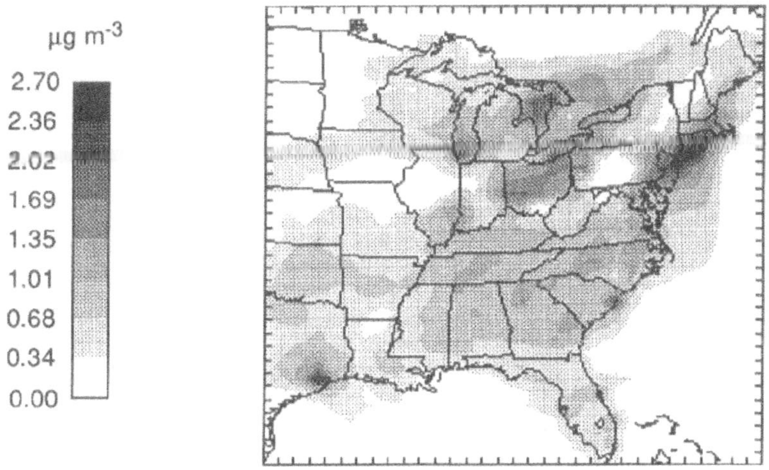

Figure 3: Simulation of organic component of aerosol [µg m^{-3}] in the first layer [Δz~80m] for 1800Z, July 19, 1985.

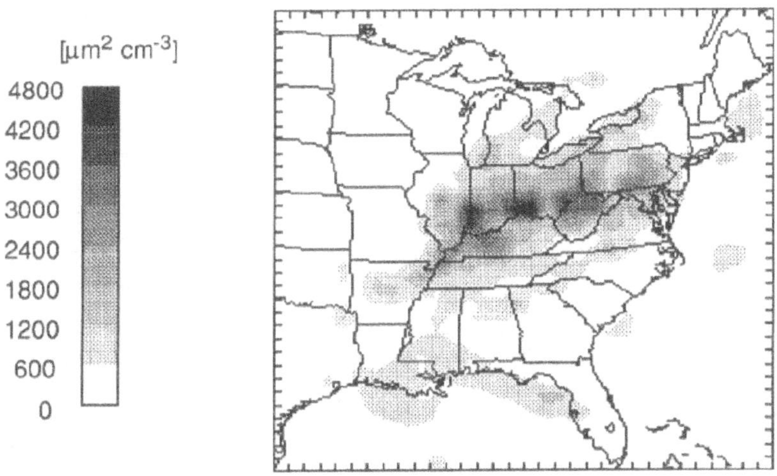

Figure 4: Simulation of total aerosol surface area [μm^2 cm^{-3}] in the first layer [Δz~80m] for 1800Z, July 19, 1985.

DISCUSSION

The modeling system reported here is an interim version of a state-of-science model for secondary aerosols. It is an extension of the original sulfate-ammonium-water species model and now includes the nitrate and organic contributions. At this juncture in the development of RPM, several opportunities for applications as well as model evaluation and refinements are discussed. One of the major applications of RPM will be to investigate and provide regional particulate distributions for conducting assessments of current and future emissions projection scenarios in support of the NAAQS standard for PM-10 and a potential standard for fine particulate matter, PM-2.5. Given that RPM operates in an episodic mode, providing gridded particulate fields on an hourly basis, seasonal and annual fields of particulate matter concentration and deposition are computed using aggregation techniques.

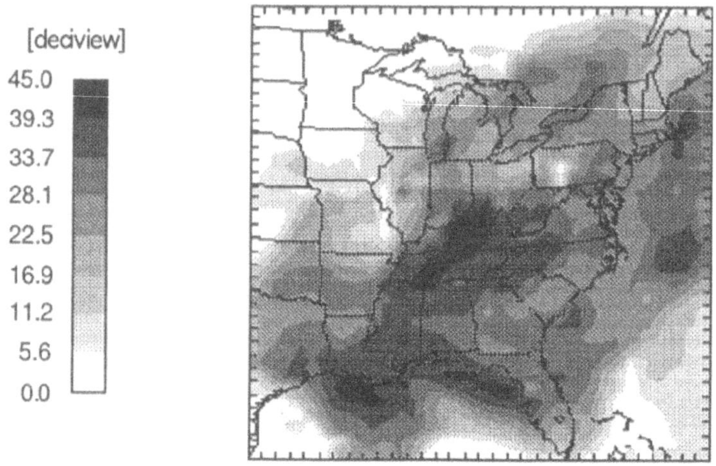

Figure 5: Simulated visual index in deciview units [deciview = 10 $\ln(B_{ext} / 0.01)$] in the first layer [$\Delta z \sim 80$m] for 1800Z, July 19, 1985, where B_{ext} is the aerosol extinction coefficient [km^{-1}].

With the composition, size distribution, and concentration levels known, optical parameters can be computed. A measure of visibility is the deciview = 10 $\ln[B_{ext}/0.01]$ as discussed by Pitchford, and Malm (1994). Figure 5 shows results of visibility in deciview units with values of zero (extremely good visibility) to 40 (very poor visibility) indicated. The typical range of median summer visibility in the eastern US is 20-32 (Pitchford and Malm, 1994). Another application of RPM (described in Ching, et al., 1995) is the transport and wet and dry deposition of toxic airborne semi-volatile organic compounds (SVOCs) applicable on a regional scale. In principle, these compounds cycle between aerosol and gas phases depending primarily on their vapor pressure, ambient temperature and particle surface area. The particles provide sites for the condensation or volatilization of SVOCs. They show that the RPM toxics approach can handle the full range of volatility of various pollutants. Their demonstration analysis was for relatively lower volatility organoclorines and relatively higher volatility persistent aromatic hydrocarbon pollutants. In their formulation, they used gas-particle partition functions based on formulations of Pankow (1993) in conjunction with predicted aerosol surface area (Figure 4).

The CAAA mandates controls for various pollutant categories and issues, including photochemical oxidants, acid deposition, and toxics to meet pollutant targets for NAAQS. To meet this mandate RPM will be integrated into the EPA Models-3 framework (Byun et al., 1995). The objective of Models-3 is a flexible and general modeling system to address issues in a comprehensive manner. Models-3 is designed to support computational scalability for multipollutant and multiscale air quality simulation while taking advantage of the enhanced computational capabilities provided by high performance computing and communication (HPCC) architectures. The inclusion of particulates in air quality simulation models will allow the capability for modeling heterogeneous processes. The various processes inclusive of transport and deposition as well as the chemistry is therefore much more adequately and credibly simulated.

RPM currently features the particulate matter in the size range below about 2.5μm. To capture other components of particulate matter that are transportable on urban to regional scales, the addition of a coarse mode for larger particles is envisioned. Major tasks to be undertaken are performance and diagnostic evaluations. This will be difficult because data bases from special field study programs that have performed aerometric sampling which include size and composition information are very limited in number and scope.

DISCLAIMER

This paper has been reviewed in accordance with the U.S. Environmental Protection Agency's peer and administrative review policies and is approved for presentation and publication. Mention of trade names or commercial products does not constitute endorsement or recommendation for use.

REFERENCES

Alapaty, K., and F.S. Binkowski, 1993, A windowed model for applications in air quality modeling, (in press) Proceedings of AWMA Specialty Conference on Regional Photochemical Measurements and Modeling Studies, November 7-12, 1993, San Diego, CA.

Binkowski, F.S. and U. Shankar, 1995, The regional particulate matter model: Part I, model description and preliminary results, (in press), *J. Geophys. Res.*

Byun, D.W., A.H. Hanna, C.J. Coats, and D. Hwang, 1995, Models-3 air quality model prototype science concept development. (in press) Proceedings of the AWMA Specialty Conference on Regional Photochemical Measurement and Modeling Studies, San Diego, CA, November 7-12, 1993.

Chang, J. S., F. S. Binkowski, N. L. Seaman, D. W. Byun, J. N. McHenry, P. J. Samson, W. R. Stockwell, C. J. Walcek, S. Madronich, P. B. Middleton, J. E. Pleim, and H. L. Landsford, 1990, The regional acid deposition model and engineering model, NAPAP SOS/T Report 4, in *National Acid Precipitation Assessment Program, Acidic Deposition: State of Science and Technology, Volume I,* Washington, D.C.

Ching, J. K.S., F.S. Binkowski, O.R. Bullock,Jr., 1995, Deposition of semi-volatile air toxic pollutants to the Great Lakes: A regional modeling approach. Accepted for publication as book chapter in *Atmospheric Deposition of Contaminants and Nutrients to the Great Lakes and Coastal Waters.* by Society of Environmental Toxicology and Chemistry.

Dennis, R.L., 1995, Personal communication.

Kerminin, V.-M., and A. S. Wexler, 1994, Post-fog nucleation of H_2SO_4-H_2O particles in smog. *Atmos. Environ.* 28: 2399-2406.

Pankow, J. F., 1993, A simple box model for the annual cycle of partitioning of semi-volatile organic compounds between the atmosphere and the earth's surface. *Atmos. Environ.* 27A: 1139-1152

Pitchford, M.L., and W.C. Malm, 1994, Development and applications of a standard visual index. *Atmos. Environ.* 28: 1049-1054.

Pitzer, K.S. and G. Mayorga, 1974, Thermodynamics of electrolytes, III. Activity and osmotic coefficients for 2-2 electrolytes. *J. Solution Chem.* 3: 539-546.

Pandis, S.N., R.A. Harley, G.R. Cass, and J.H. Seinfeld, 1992, Secondary organic aerosol formation and transport, *Atmos. Environ.* 26A: 2269-2282.

Rood, M.J., M.A. Shaw, T.V. Larson, and D. S. Covert, 1989, Ubiquitous nature of ambient metastable aerosol, *Nature* 337: 537-539.

Saxena, P., A. B. Hudischewskyj, C. Seigneur, and J. H. Seinfeld, 1986, A comparative study of equilibrium approaches to the chemical characterization of secondary aerosols, *Atmos. Environ.* 20: 1471-1483.

Stauffer, D.R. and N.L. Seaman, 1990, Use of four-dimensional data assimilation in a limited area mesoscale model. Part I: Experiments with synoptic scale data., *Mon. Wea. Rev.* 118: 1250-1277.
Stauffer, D.R., N.L. Seaman, and F.S. Binkowski, 1991, Use of four-dimensional data assimilation in a limited area mesoscale model. Part II: Effects of data assimilation within the planetary boundary layer. *Mon. Wea. Rev.* 119: 734-754.
Stockwell, W.R., P. Middleton, and J.S. Chang, 1990, The second generation regional acid deposition model chemical mechanism for regional air quality modeling. *J. Geophys. Res.*, 95: 16,343-16,367.
Whitby, K.T., 1978, The physical characteristics of sulfur aerosols, *Atmos. Environ,* 12: 135-159.

DISCUSSION

D. STEYN: How well does computed visibility ranges compare with observations?

J. CHING: Both the magnitude and the range of modeled deciview values agree with the analyses of seasonal distribution of deciview values based on observations for the eastern north America region published by Pitchford and Malm (Atm. Env. 1994). We plan a program of further model comparisons against observations on an episodic basis, including composition and size distribution measurements from aircraft measurements as well as optical data from surface measurements.

E. GENIKHOVICH: How could your results be changed when you add coarse particles and wind erosion/resuspension processes to your model?

J. CHING: The size distribution of this model is currently limited to particles whose diameter is less than 2.5 μm. Larger particles can be modeled in the current framework by the superposition of a coarse mode which is expected to be significant under high wind speed, for exposed land surfaces and for areas with high fugitive emissions such as urban-industrial areas. We expect such particles to be rapidly deposited by gravitational settling. Thus the addition of a coarse mode will be to add additional variability to predicted concentrations. The addition of a coarse mode within RPM is a future development activity.

A. HANNA: Comment on question concerning comparison between deciview and observed visibility: A comparison has been made between observed visibility and deciview (calculated using the Particulate Model). The results show reasonable agreement during 19 July 1985 haze episode that dominated most of the model domain.

J. CHING: I appreciate the comments by Dr. Hanna who is very actively engaged in visibility modeling research. His results provide much encouragement since the model comparison was performed against data on an episode basis.

R. DENNIS: Comment to D. Steyn's Question: Marc Pitchford has computed DeciViews, based on measurements, that are in the low 40's for poor visibility conditions in the eastern U.S. This compares fairly well to RADM/RPM results in terms of magnitude of visibility degradation (± 25 to ± 50 %).

H. van JAARSVELD: What is the role of primary emitted aerosols in relation to visibility? What is in your model the approximate ratio of primary and secondary aerosol concentration?

J. CHING: The primary particle emission in RPM is sulfate obtained from the models emissions inventory. In results by Binkowski and Shankar (in press, JGR), the ratio of emission rate for primary sulfate to secondary sulfate production rate is typically very small. They show that the ratio vary between one and two orders of magnitude throughout the modeling domain, with the smallest ratios found near the sulfur source regions. This ratio also varies with time of day.

UNCERTAINTIES ASSOCIATED WITH SOURCE-RECEPTOR RELATIONSHIPS FOR VISIBILITY REDUCING SPECIES DERIVED FROM A GRID MODEL

Prakash Karamchandani and Prasad Pai

ENSR Consulting and Engineering
1220 Avenida Acaso
Camarillo, CA 93012

ABSTRACT

A non-linear three-dimensional grid model was used to derive source-receptor relationships, referred to as transfer coefficients, for visibility reducing species (such as sulfate) for 78 sources and 16 Class I receptors in the western United States for the calendar year 1992. The transfer coefficients will be used to assess the impact of changes in emissions on visibility at sensitive receptors, such as the Grand Canyon National Park. Uncertainties associated with model inputs and model formulation and assumptions lead to uncertainties in the derived transfer coefficients. This paper examines the uncertainties in the source-receptor relationships that are associated with grid resolution as well as the effect of assuming that levels of SO_2 oxidants, such as H_2O_2, do not change when calculating transfer coefficients.

INTRODUCTION

The reduction of visibility in the Grand Canyon region has been the subject of several studies conducted over the past 15 years (NPS, 1989; NRC, 1990). The 1990 amendments to the Clean Air Act have been the driving force for several studies that are currently being conducted to determine the contributions of various sources and source regions to visibility degradation in the Grand Canyon and other Class I locations. The sources of interest include both nearby point sources and urban regions, as well as sources hundreds of kilometers away (e.g., Los Angeles, San Francisco, Mexico) from the Class I receptors. Source-receptor relationships between precursor emissions at the source and particulate concentrations at the receptor are being used to perform this assessment of source contributions. A variety of approaches are being employed by various groups to derive the necessary source-receptor relationships, which are also referred to as transfer coefficients. The **V**isibility **A**ssessment of **R**egional **E**mission **D**istributions (VARED) study, sponsored by the Electric Power Research Institute (EPRI) and a consortium of Western utilities, is using a non-linear three-dimensional grid model, called the **Vis**ibility and **H**aze in the **W**estern **A**tmosphere (VISHWA) model, to calculate transfer coefficients. The development of the model, and its evaluation with data from the 1992 Project MOHAVE (**M**easurements **o**f **H**aze and **V**isual **E**ffects) field study, are described in Venkatram et al. (1995). To facilitate the computation of transfer coefficients for a large number of sources for an entire year, certain assumptions were made in the formulation and application of the model. This paper examines the effect of some of these assumptions on the derived transfer coefficients for secondary sulfate.

DEFINITION OF TRANSFER COEFFICIENTS

A transfer coefficient (TC) for a given visibility reducing species (e.g., sulfate) is defined as the concentration of that species at the receptor due to unit emissions of the precursor species (e.g., SO_2) at the source. In other words, a transfer coefficient can be considered as the concentration at the receptor normalized by emissions at the source. A transfer matrix for the species consists of an array of transfer coefficients — the dimensions of the array are the number of sources times the number of receptors.

The product of the transfer matrix with the vector representing the actual emissions of the various sources results in a vector which contains the actual concentration at the receptor due to all the sources included in the analysis. Mathematically, if **T** is the (m x n) sulfate transfer matrix, where m is the number of receptors, and n is the number of sources, and **Q** is the vector of SO_2 emissions of length n, then the matrix multiplication of **T** and **Q** yields a sulfate concentration vector **C** of length m:

$$\mathbf{C} = \mathbf{TQ} \tag{1}$$

The elements of the concentration vector represent the total sulfate concentration at the receptors due to all the sources considered in the analysis. For example, the sulfate concentration at receptor "r", c_r, is given by:

$$c_r = \sum_{s=1}^{n} t_{sr} q_s \tag{2}$$

where t_{sr} is the sulfate transfer matrix coefficient for source "s" and receptor "r", and q_s is the SO_2 emission rate for source "s".

CALCULATION OF TRANSFER COEFFICIENTS

A three-dimensional Eulerian grid model was used to calculate the transfer coefficients (TCs) for visibility-reducing species such as sulfate, nitrate, organic carbon, and elemental carbon. The advantage of using the Eulerian approach is that is possible to explicitly treat subsidence and shear effects, as well as to account for multi-layer transport. These features are important for simulating regional transport in a realistic manner.

The grid model used in our study is called the **V**isibility and **H**aze in the **W**estern **A**tmosphere (VISHWA) model. It is based on a comprehensive air quality model, the **A**cid **D**eposition and **O**xidant **M**odel (ADOM), which has been developed, tested, and improved over the last decade (Venkatram et al., 1992). For the Grand Canyon visibility application, some of the detailed chemistry modules in ADOM were replaced by simpler modules to enhance the computational efficiency of the model. These modifications, and the testing of the VISHWA model against data from Project MOHAVE, are described in Venkatram et al. (1995).

The transfer coefficient calculations were initiated after the testing of the VISHWA model was completed. Because the transfer coefficients were required for an entire year for a large number of sources, we used a two-step approach for computational efficiency. In the first step (base case simulation), we ran the full VISHWA model with all sources in the emission inventory (without distinguishing among different sources) for the year of interest (1992) to characterize the "base case oxidation state" of the region. The oxidation state parameters were saved for use in the next step, when the model was exercised separately for each source.

The oxidation state, for the purposes of our study, is determined by the concentrations of the OH radical (which is responsible for the gas phase oxidation of SO_2 to sulfate and NO_x to inorganic nitrate, as well as the formation of secondary organic aerosols), the organic radical pool species, RP (which oxidizes NO_x to organic nitrate), and the ratio of base case H_2O_2 to base case SO_2 (which determines the extent of oxidation of SO_2 to sulfate in the liquid phase). The relevant chemistry is described in Venkatram et al. (1995).

In the second step, the model was applied separately for each source of interest. Figure 1 shows the locations of the 45 point sources and 33 area sources that were explicitly simulated. These sources were selected to provide a good geographical distribution as well as to include most of the major emissions influencing air quality and visibility on the Colorado Plateau. The selected sources consist of both existing and hypothetical sources. Unit emissions of precursors were released at each source and the precursors and secondary aerosol products (sulfate, nitrate, and secondary organic carbon) were tracked separately for each source. The chemical transformation rates were calculated using the oxidant concentrations stored in the base case simulations. Transfer coefficients were extracted for the 16 Class I receptors shown in Figure 2.

SOURCES OF UNCERTAINTY IN TRANSFER COEFFICIENTS

In addition to the usual uncertainties associated with numerical modeling of air pollution transport (such as uncertainties in meteorology, emissions, dispersion, chemistry), two other sources of uncertainty are likely to affect the calculated transfer coefficients. The first is related to the resolution of the grid model used in the transfer coefficient calculations. The second is related to the decoupling of the chemistry governing oxidant formation from the chemistry governing the oxidation of the precursor species. As discussed below, these uncertainties can be important for certain situations.

Effect of Grid Resolution

As described in Venkatram et al. (1995), a horizontal grid resolution of 50 km was used for the base case and transfer coefficient calculations. Because the model was being applied to a large domain (approximately 3000 km by 3000 km), much finer resolution than 50 km would be prohibitively expensive for the transfer coefficient calculations. However, for large point sources that are close to the receptors of interest (e.g., within a few grid cells), the relatively coarse resolution of 50 km can introduce a bias, since the pollutant emitted from the point source is instantaneously mixed into the 50 km by 50 km grid cell in which the release occurs (determined by stack location, stack height, and plume rise). This artificial mixing creates a broad plume which is likely to impact nearby receptors more frequently than the actual plume from the point source which has to travel several grid cells before its dimensions approach those of the model resolved plume. This can overemphasize the effect of the source on a receptor which is near the source, but is usually upwind of the source, for both the emitted species as well as species formed by the chemical transformation of the primary species. For nearby downwind receptors, the effect of grid resolution on primary species concentrations is expected to be less important because the higher frequency of impact calculated by the grid model is offset by the artificial dilution of the plume in the model. However, as discussed below, this is not necessarily true for secondary species.

The grid resolution can introduce an additional error for secondary species such as sulfate, whose formation is controlled by the availability of oxidants such as hydrogen peroxide. H_2O_2 concentrations are typically orders of magnitude lower than SO_2 concentrations in the vicinity of a major SO_2 source. This is likely to be true even when the plume is artificially diluted in the grid model. Then, sulfate concentrations in both the actual plume and the grid plume in the vicinity of the source will be comparable and equal to the H_2O_2 concentration. Since the model plume impacts nearby receptors more frequently than

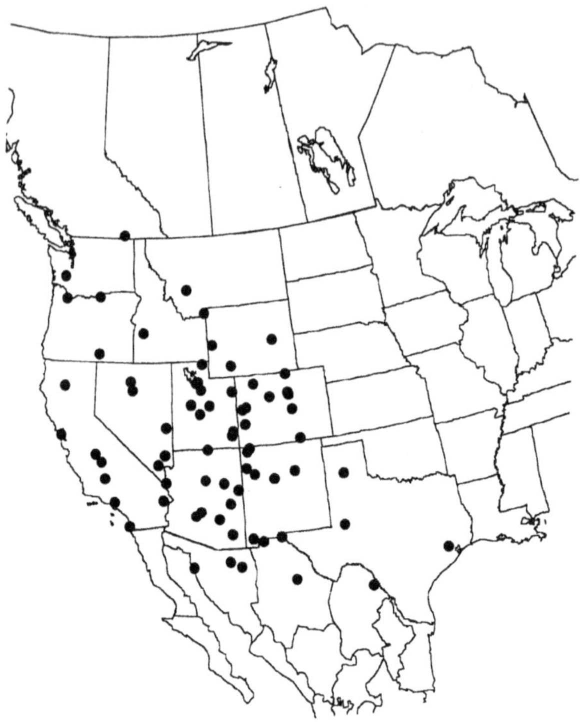

Figure 1. Source locations for transfer coefficient computations.

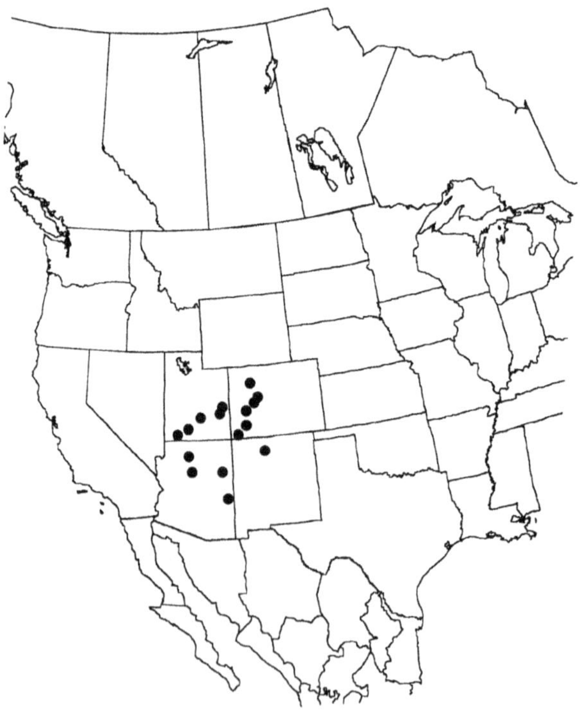

Figure 2. Receptor locations for transfer coefficient computations.

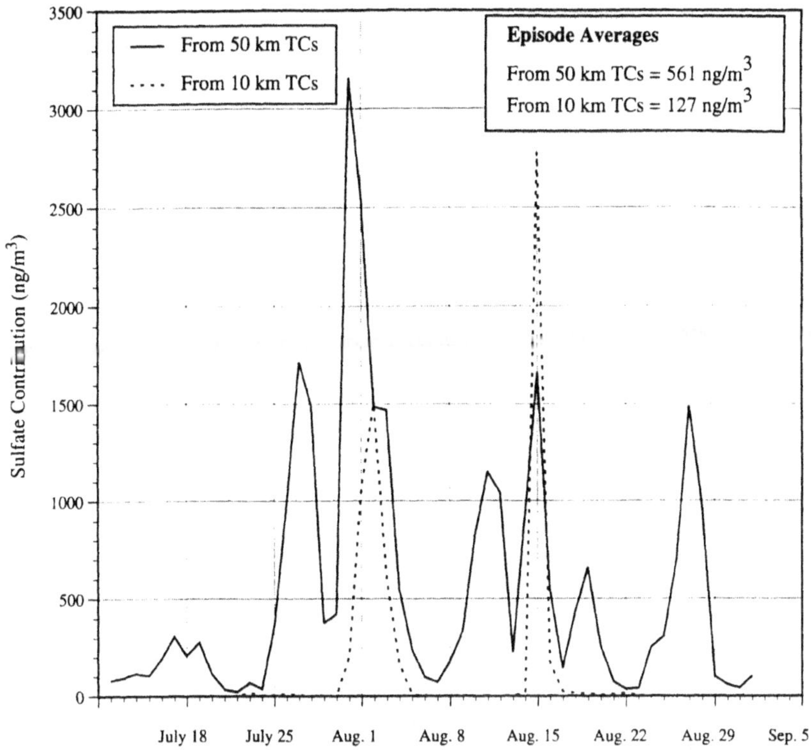

Figure 3. Effect of grid resolution on contribution of Source 1 to 24-hour average sulfate concentrations at Receptor 1 for summer intensive period, July 12 to September 1, 1992.

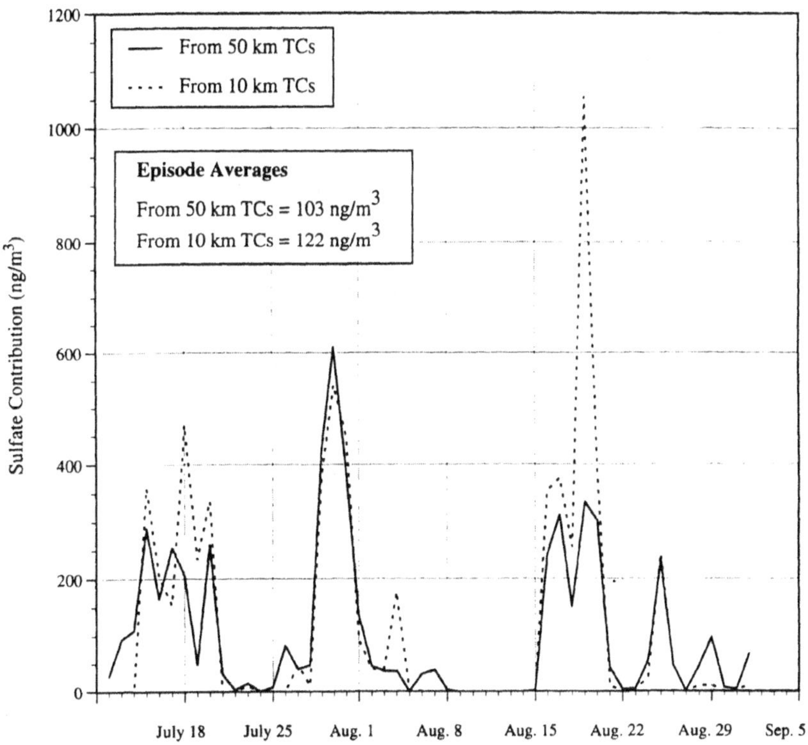

Figure 4. Effect of grid resolution on contribution of Source 1 to 24-hour average sulfate concentrations at Receptor 1 for summer intensive period, July 12 to September 1, 1992.

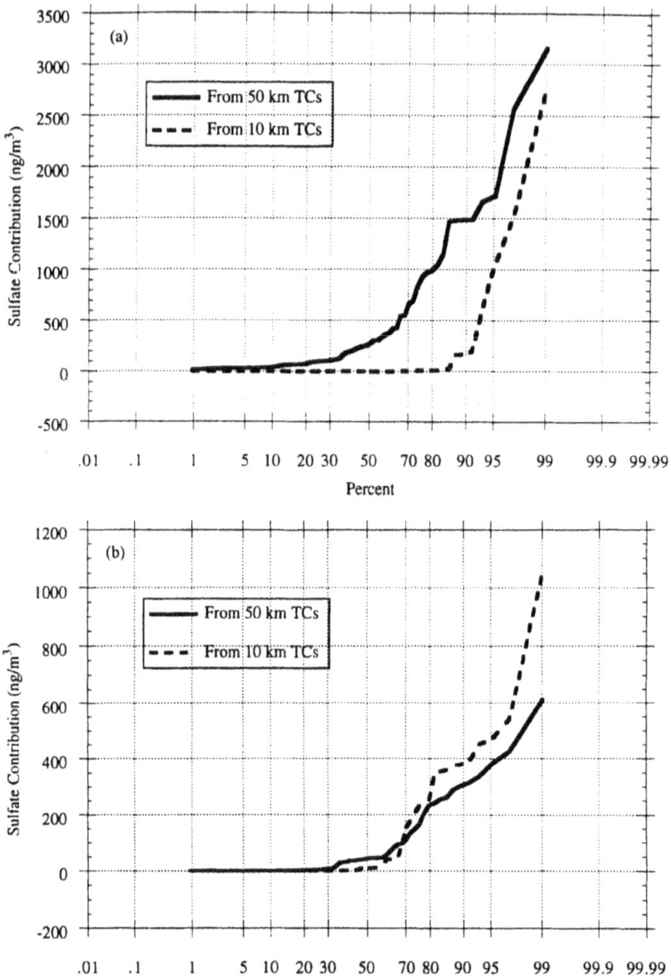

Figure 5. Frequency distributions of (a) Source 1 and (b) Source 2 contributions to 24-hour average sulfate concentrations at Receptor 1 for the summer intensive period.

the actual plume, this is likely to result in an overestimation of sulfate concentrations at receptors near the source.

To investigate the effects of grid resolution on the point source transfer coefficients, we recalculated transfer coefficients for the summer period of July 12, 1992 to September 1, 1992 for a smaller domain (800 km x 800 km) around the Grand Canyon region, with a grid resolution of 10 km. The 10 km simulations were performed for a subset of the 45 point sources considered in the 50 km simulations.

Figure 3 shows the 24-hour average 50 km and 10 km sulfate transfer coefficients for a source-receptor pair (Source 1 and Receptor 1, respectively) that are separated by a distance of 120 km (i.e., within two grid cells apart for the 50 km grid and 12 grid cells for the 10 km grid). The orientation of this source-receptor pair is such that the wind rarely blows from Source 1 to the receptor during summer conditions (i.e., Source 1 is usually upwind of the receptor). Thus, we expect this source to impact the receptor infrequently. As seen in the figure, however, with the 50 km grid resolution Source 1 contributes sulfate to the receptor throughout the episode, with an episode averaged contribution of 561 ng/m^3. This concentration is about 50% of the episode averaged sulfate concentration measured at this receptor. With the 10 km grid spacing, Source 1 has a significant impact on the receptor on only two occasions: August 2nd and August 15th. The episode averaged contribution with the 10 km grid resolution is 127 ng/m^3, nearly 5 times lower than with the 50 km grid spacing.

Figure 4 shows the effect of grid resolution for Source 2, which is separated from Receptor 1 by approximately 240 km or about 5 grid cells for the 50 km grid resolution. Furthermore, the receptor is often downwind of Source 2, and is expected to be impacted by it more frequently than by Source 1. Thus, the artificial mixing introduced by the 50 km grid is expected to have a smaller impact for this source than for Source 1. As shown in Figure 4, both the 50 km and 10 km grid transfer coefficients are fairly well correlated (the correlation coefficient is 0.8) and show similar impact frequencies, and the episode averaged sulfate contributions for the two grid resolutions are comparable.

An alternative way of understanding the dramatically different effects of grid resolution for Sources 1 and 2 on sulfate concentrations at Receptor 1 is by examining the frequency distributions of the 24-hour average concentrations. The frequency distributions are shown in Figures 5a and 5b for Sources 1 and 2, respectively. As shown in Figure 5a, with the 50 km grid resolution, Source 1 impacts the receptor over 95% of the time. On the other hand, with the 10 km grid resolution, the frequency of impact is reduced to 15%. For Source 2, shown in Figure 5b, the differences between the 50 km and 10 km grid resolution results are not as large.

EFFECT OF DECOUPLING CHEMISTRY

The approach used to calculate source contributions, or transfer coefficients, has been discussed above. Briefly, unit emissions of precursors are released from the source of interest, with zero emissions from all other sources in the domain, and zero background concentrations. The chemical transformations are performed using results from a base case simulation which includes all the sources in the domain. An alternative approach to calculate the contribution of a source is to perform a full simulation with all sources except the source of interest. Then, the source contribution is the difference between results from this simulation and the "base case" simulation. The second approach is expected to be more detailed and accurate than the first approach, since it accounts for variations in the oxidant fields due to the perturbation of the source. However, this detailed approach is also more time consuming and resource-intensive than the adopted approach, and would not be practical for the large number of sources that were considered in our study.

To examine the effect of using the computationally efficient approach that we adopted versus the more detailed approach, we conducted a full simulation of the summer intensive period using the 10 km grid resolution, with and without a major SO_2 source (Source 1, discussed earlier in this paper).

Figure 6 shows the difference between the detailed approach and the transfer coefficient approach for 24-hour averaged sulfate contributions of Source 1 to Receptor 1, which, as discussed earlier, is about 120 km from Source 1 and is frequently upwind of Source 1. The

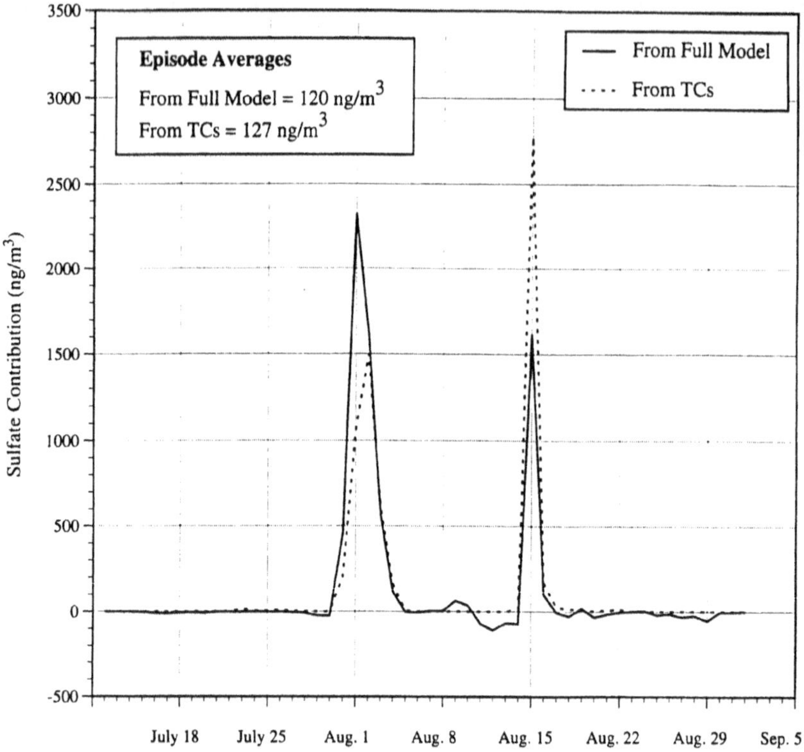

Figure 6. Effect of decoupling chemistry on contribution of Source 1 to 24-hour average sulfate concentrations at Receptor 1 for the summer intensive period, July 12 to September 1, 1992.

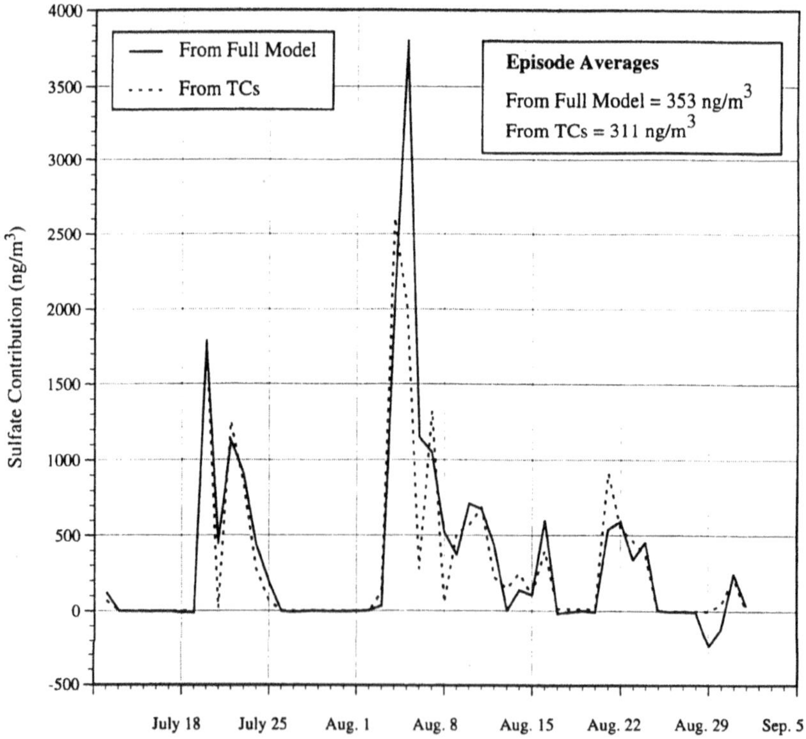

Figure 7. Effect of decoupling chemistry on contribution of Source 1 to 24-hour average sulfate concentrations at Receptor 1 for the summer intensive period, July 12 to September 1, 1992.

estimated contributions from the two approaches are well correlated (with a correlation coefficient of 0.85), although there are differences in the peak contributions. The detailed approach also estimates a few negative contributions, i.e., removing Source 1 sometimes results in higher sulfate concentrations at Receptor 1. This is probably due to changes in the H_2O_2 concentration fields when Source 1 is removed from the system. This effect cannot be reproduced in the transfer coefficient approach, since it assumes that oxidant concentrations are unchanged. Note that both approaches yield very similar event-averaged contributions, with a slightly higher (6%) average contribution predicted by the transfer coefficient approach than the detailed approach.

Figure 7 shows the effect of the transfer coefficient approximation on a receptor (Receptor 2) that is about 220 km downwind of Source 1. Again, the results from the two approaches are highly correlated (r = 0.88). There are some differences in the peak contributions, but the temporal patterns are similar. As in the case of Receptor 1, there are some instances of negative contributions calculated by the detailed approach. The event-averaged contributions are also similar, with the transfer coefficient approach estimating a 11% lower contribution than the detailed approach.

CONCLUSIONS

This paper has examined the role of grid resolution and "pseudo-linearization" of the chemistry in assessing point source contributions to sulfate concentrations at sensitive receptor locations using a grid model. The results show that estimates of contributions from sources very close to the receptor can be significantly overestimated with a coarse grid resolution. The effect of grid resolution diminishes as the distance between the source and receptor increases. The assumption that oxidant levels are unchanged when SO_2 emissions from a single source are perturbed does not appear to introduce much error in the calculation of transfer coefficients. However, the uncertainties associated with this assumption are likely to be larger when emissions from a large number of sources are altered simultaneously.

ACKNOWLEDGMENTS

The work described in this paper has been supported by the Electric Power Research Institute (EPRI) and a consortium of Western utilities. We thank Dr. Robert Goldstein (EPRI), Mr. Vince Mirabella (Southern California Edison), Dr. Ivar Tombach (ENSR), and Dr. Akula Venkatram (University of California, Riverside) for their guidance on the study.

REFERENCES

NPS (National Park Service) 1989. National Park Service Report on the Winter Haze Intensive Tracer Experiment (WHITEX). Final Report, Air Quality Division, National Park Service, Fort Collins, CO.

NRC (National Research Council) 1990. Haze in the Grand Canyon: An Evaluation of the Winter Haze Intensive Tracer Experiment. National Academy Press, Washington, D.C.

Venkatram, A., P. Karamchandani, G. Kuntasal, P.K. Misra, and D.L. Davies 1992. The development of the Acid Deposition and Oxidant Model (ADOM). Environ. Pollut., 75: 189–198.

Venkatram, A., P. Karamchandani, P. Pai, C. Sloane, P. Saxena and R. Goldstein 1995. The development of a model to examine source-receptor relationships for visibility on the Colorado Plateau. Accepted for publication in the Journal of Air and Waste Management.

DISCUSSION

C. NAPPO: Can you list the order of importance of model uncertainties?

P. KARAMCHANDANI: The largest uncertainties are probably in emissions from Mexico, which are not well documented, as well as emissions associated with wild fires and controlled fires. Other important sources of uncertainty include the specification of clouds, chemistry parameterization, and the subject of this paper, namely grid resolution for large point sources near the receptors of interest.

G. BRUSASCA: Do you consider the plume rise for major sources?

P. KARAMCHANDANI: Yes, plume rise is calculated for all major point sources. The emissions are released in the appropriate layer.

J. CHING: What role does the presence (or lack) of clouds have on the overall uncertainty of the model source-receptor relationships? For the sensitivity studies, how important was aqueous chemistry in contributing to differences between the fine and coarse grid results?

P. KARAMCHANDANI: As stated in the answer to the first question, the specification of clouds is an important source of uncertainty. Sensitivity studies with different cloud fields showed a significant effect of clouds. Because of oxidant limitation, aqueous-phase chemistry is probably one of the largest sources of the differences between the fine and coarse grid results.

EFFECTS OF EMISSION CONTROLS AND STRATOSPHERIC OZONE LOSS ON TROPOSPHERIC OZONE PRODUCTION

Kevin C. Crist[1], Gregory R. Carmichael[2], Brian T. Hotchkiss[2], and Kuruvilla John[3]

[1]Department of Health Science, Ohio University, Athens Ohio 45701
[2]Department of Chemical and Biochemical Engineering and Center for Global and Regional Environmental Research University of Iowa, Iowa City, Iowa 50613
[3]Department of Environmental Engineering, Teaxa A&M-Kingsville, Kingsville Texas 78363

INTRODUCTION

In the past, air pollution generated from human activity was viewed as a local phenomenon. However, it is becoming well recognized that political boundaries do not contain air pollutants. These regional or even global consequences of anthropogenic emissions have challenged atmospheric scientists and policy decision makers. The complexity of regional scale air pollution problems is especially true for the eastern United States. The eastern United States represents one of three major areas on the globe with a high density urban/agricultural corridor (Chameides, et al., 1994). This mix of economic vitality have lead to intensified energy consumption resulting in a major source area of tropospheric pollutants.

One of the major species associated with tropospheric pollution is ozone. Within the urbanized corridors of the eastern United States, ozone is formed at high concentrations often exceeding the National Ambient Air Quality Standard (NAAQS) set by EPA of 120ppbv (National Research Council, 1991). At high concentrations tropospheric ozone affects both human and biotic systems.

For ozone, tropospheric increases have not been the sole environmental concern. Over the last several decades there has been a growing public concern that the protective shield (stratospheric ozone layer) surrounding the planet has been thinning due to anthropogenic emissions. It is this layer of ozone which protects life from the harmful segment of the solar spectrum. Anthropogenic emissions of chloroflurocarbons (CFCs) have been the major contributor to destruction of stratospheric ozone documented over the last decade. Stratospheric ozone depletion was originally associated with the polar regions of the southern hemisphere remote from heavily populated areas. However, recent analysis of data obtained from the Nimbus 7 satellite has shown a decline in the total ozone column over a majority of the globe (Herman et al., 1991; Stolarski et al., 1991). This analysis has shown a significant downward trend (0.4%-0.8% a year) over the last decade for much of the northern hemisphere including the heavily populated areas of northern United States, Canada, Scandinavia, Europe, and the former Soviet Union.

The loss of stratospheric ozone and the resulting increase in ultraviolet radiation (UVR) not only has significant environmental health implications (UNEP,1994) but also has a direct impact on the tropospheric photo-oxidant cycle. UVR is the driving force behind tropospheric chemistry. For example, in the troposphere the formation of ozone results from the photolysis of NO_2:

$$NO_2 + hv \rightarrow NO + O \quad (1)$$

$$O + O_2 + M \rightarrow O_3 + M \quad (2)$$

Ozone is also directly destroyed by the absorption of UVR via

$$O_3 + hv \rightarrow O(^1D) + O_2 \quad (3)$$

This reaction in turn produces OH (the major oxidizing agent in the troposphere)

$$O(^1D) + H_2O \rightarrow 2OH \quad (4)$$

which propagates reactions leading to HO_2 formation, and subsequent reactions involving ozone destruction and production.

$$HO_2 + O_3 \rightarrow OH + 2O_2 \quad (5)$$

$$HO_2 + NO \rightarrow NO_2 + OH \quad (6)$$

$$NO + O_3 \rightarrow NO_2 + O_2 \quad (7)$$

The production/destruction of ozone also depends upon a complex set of reactions involving VOCs which are targeted precursor emissions under the Clean Air Act Legislation in controlling ozone production. VOCs become involved in the ozone production cycle via organic peroxy radicals (RO_2) generated by the reactions of VOCs and OH. These proxy radicals are then available to convert NO to NO_2 without consuming ozone.

$$RO_2 + NO \rightarrow RO + NO_2 \quad (8)$$

An increase in UVR to the troposphere will alter the ozone distribution. Ozone and its precursors photochemically react by the absorption of UVR. Depending on the ambient concentrations of precursor species, an increase or a decrease in tropospheric ozone levels could occur. The current understanding of the interaction of UVR and the tropospheric ozone/production cycle is limited to simple box model studies. Liu and Trainer (1988) conducted a series of box model simulations analyzing the sensitivity of tropospheric ozone production/destruction to an increase in UVR as a function of NO_x levels. These modeling studies (Liu and Trainer, 1988) indicated that in areas with elevated NO_x an increase in ozone would be expected due to the increase of odd-hydrogen radicals which are available to oxidize NO to NO_2 again without consuming ozone. However, when NO_x is in short supply, HO_2 can react with ozone resulting in a reduction of tropospheric ozone levels.

This has significant implications for the States in the eastern U.S. and their State Implementation Plans (SIPs) for addressing control strategies to bring ozone non-attainment areas into compliance. The urban corridors along the eastern seaboard of the United States are typically NO_x rich environments (National Research Council, 1991). This would suggest that a probable increase in tropospheric ozone levels could occur under current trends of stratospheric ozone loss. Since most of the eastern United States is NO_x limited (National Research Council, 1991), a potential reduction in surface ozone concentrations can be expected.

At present most chemical models used to evaluate the regional scale air pollution problems calculate photolysis rates using off line techniques involving look-up tables or empirical equations that are functions of solar zenith angle, surface albedo, height, and cloud cover. These models are not equipped to accurately simulate the photolytic changes that occur as the result of UVR increases under a stratospheric ozone loss. Under conditions of stratospheric ozone loss there is a change in quantity and quality of the radiative flux to the troposphere. Detailed spectral trends calculated by Crist (et al., 1994) illustrated a significant shift in the quality of the solar spectrum reaching the troposphere as the result of a stratospheric ozone loss. The loss of stratospheric ozone results in an increase in the radiative flux to the surface (quantity) and in the wavelength distribution of the incoming solar radiation with a shift toward the lower wavelengths (quality). Under these conditions a disproportional increase in photo-sensitive reaction rates can occur when the radiatively important species is a stronger absorber in the shorter wavelengths. This non-linear behavior in photolytic response for radiatively important trace gas species can have alternate effects on ambient ozone levels depending on the local chemical regime as reviewed previously.

RESULTS AND DISCUSSION

A detailed delta two stream radiative transfer code (Zdunkowski et al.,1980) was coupled with a three dimensional regional scale photochemical model (STEM-II) (Carmichael, et al., 1991) to investigate the effects of increased UV radiation as the result of stratospheric ozone loss on tropospheric ozone distribution. This study was conducted using the 1985 National Acid Precipitation Assessment Program (NAPAP) emission inventories and total ozone measurements and trends obtained from the Nimbus 7 satellite for the eastern United States.

Photolysis Rate Calculations

The radiative transfer code was used to analyze the effects of stratospheric ozone loss on the reaction rates of tropospheric photo-active trace gas species. A series of sensitivity studies were conducted on the reaction rates of these photosensitive species based on a 16 percent reduction in stratospheric ozone. This loss of stratospheric ozone corresponds to potential conditions in the year 2004 based on the analysis of TOMS data by Stolarski (1991) showing an 8 percent per decade reduction in total ozone over the northern latitudes including the eastern United States. However, the increase in UVR flux is not constant for each wavelength. This change in the quality of the radiation is further illustrated by trend analysis of the solar wavelength distribution. Detailed spectral trends were calculated based on the 1984 and 2004 simulations and reported by Crist et al. (1994). As can be seen from Figure 1 the percent change in UVR increases sharply in the shorter wavelengths of the UV spectrum.

For photolysis rates of tropospheric trace gas species the non-linear spectral changes in UVR will have profound effects. Figure 2 shows the calculated changes (increases) in the tropospheric photo reaction rates for Columbus Ohio at noon 7 June as a result of a 16 percent loss of total ozone. As can be seen from Figure 2, the reaction rates do not respond uniformly to overhead changes in total ozone. Enhanced photo-dissociation responses were experienced for ozone (39 percent) and HNO_3 (23 percent) to the 16 percent change in total ozone. On the other hand, the reaction rate for the primary production pathway of tropospheric ozone, the photolysis of NO_2, is relatively insensitive to overhead changes in total ozone.

Loss of Stratospheric Ozone: Effects on Ozone Production

A series of three-dimensional studies of the perturbations to ozone over the eastern United States based on 16 percent loss of total ozone were perfomed. The impacts to tropospheric ozone production were studied with three-dimensional STEM-II model simulations. The radiative transfer code was coupled to STEM-II off line via the calculations for the photolysis rates for the photo-active trace gas species. The rate constants for the photolytic dependent reactions were computed on the same grid system established for STEM-II which covered 6776 grid locations over the eastern United States.

The impacts of changes in UVR were studied by performing a series of simulations for an episodic event during 7-11 June, 1984. This period is typical of the stagnant high pressure systems which frequently occur in the eastern United States, and which lead to severe episodes of elevated ozone (Shin, 1990; Shin et al., 1993). The horizontal grid spacing is 80 km and the vertical resolution is 500m extending to 5 km. A scenario 20 years into the future from 1984, with a decrease in stratospheric ozone outlined above was also run.

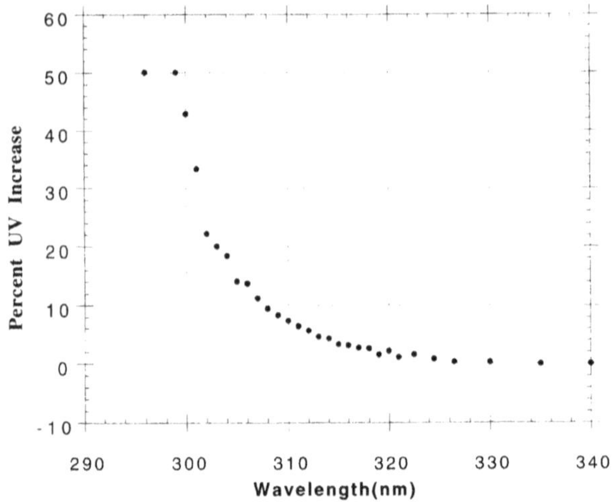

Figure 1. Wavelength dependent trends in ground level radiation calculated from simulations for a 16% decrease in stratospheric ozone.

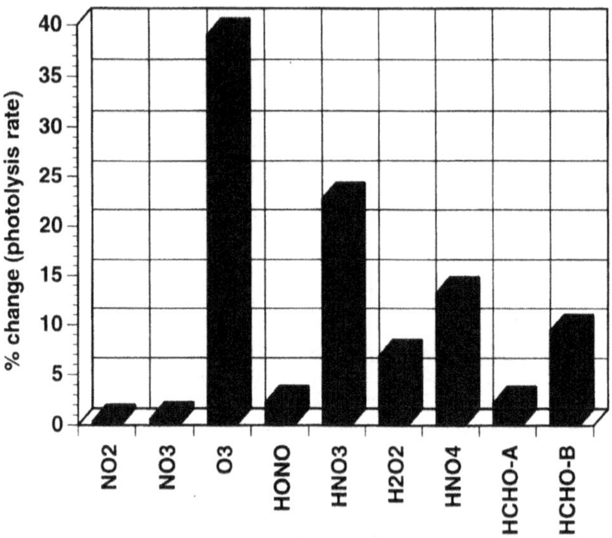

Figure 2. Calculated changes in photolysis rates of tropospheric trace gas species for a 16% decrease in stratospheric ozone.

Figure 3 demonstrates the sensitivity of tropospheric ozone to UVR perturbations. Reductions in daily averaged surface ozone values of four to six percent were calculated for Maine through New York, down into the western sections of Pennsylvania, and into West

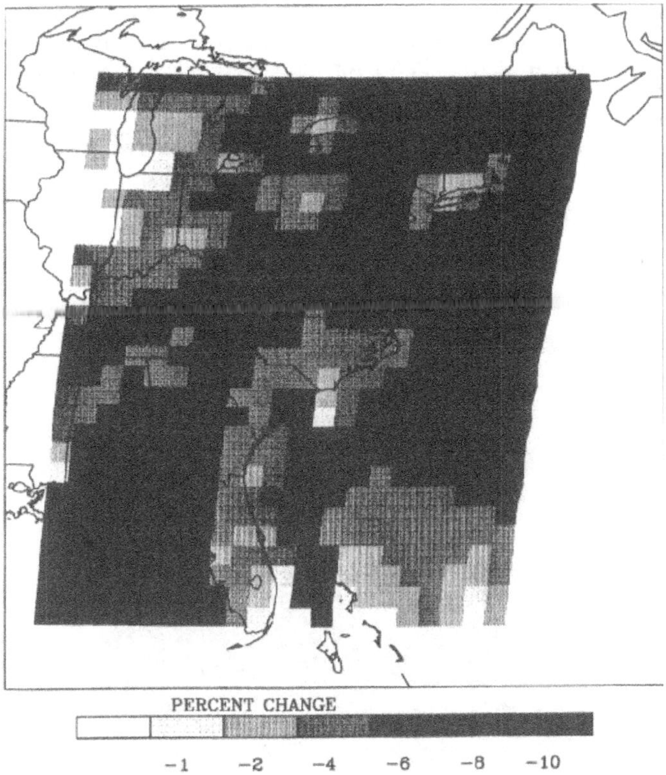

Figure 3. Simulated changes in surface level ozone for noon 7 June based on a 16% reduction in stratospheric ozone.

Virginia. Two to four percent decreases were experienced through Florida, and along the east coast of Georgia, South Carolina, North Carolina, and Virginia.

The unique features of these results is the loss of ground level ozone over the continental grid cells as a result of the reduction in stratospheric ozone. Previous box model calculations (Liu and Trainer,1988) have shown that changes in ozone production exceeded the changes in ozone loss, as the result of a reduction in stratospheric ozone for NO_X rich environments. These NO_X rich environments were at NOx levels in excess of 1ppb which occurs over most of the eastern United States (National Research Council, 1991). The study by Liu and Trainer (1988) indicated an NO_X trigger, where increases in UVR would reduce tropospheric ozone with NO_X levels less then 1ppb. Conversely, an increase in ozone would be expected with NO_X levels greater than 1ppb. Simulations conducted with the radiative code/STEM II coupled system also indicated this NO_X trigger (Crist, 1994). However surface level ozone reductions were shown to be the result of vertical transport of ozone depleted air. The NO_X decreased drastically with height dropping to below 1ppb at 500m where significant ozone reductions occurred. This ozone depleted air was mixed with the surface level ozone changes resulting in a net surface ozone loss (Crist, 1994).

The efficacy of NOx control strategies, is dependent upon local and regional concentrations of ozone precursors. For example, ozone formation is relatively insensitive to changes in NOx emissions in areas where the ratio of VOCs to NOx is low. STEM II simulations conducted by Crist (1994) for the eastern United States utilizing a 33% reduction in NOx emissions from the June 1984 eposidic event experienced the largest reductions in ozone (10 to greater than 15 percent) in the southeast from Florida through Virginia. This area also experiences high VOC/NO_X. Across the northern sections of the domain from Chicago to New York, where VOC/NO_X is lower, the ozone concentrations were less sensitive (0-5 percent reduction in ground level ozone) to the reduction in NO_X(Crist, 1994).

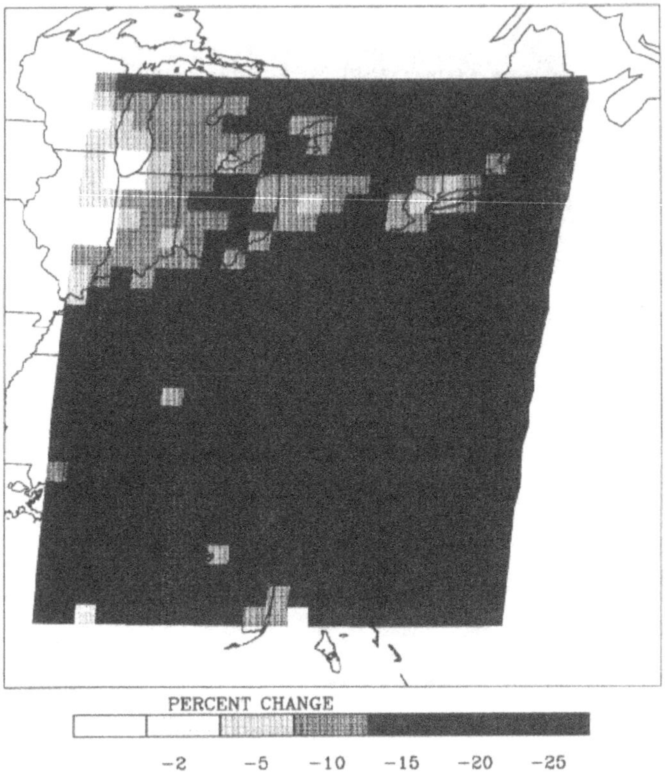

Figure 4. Percent change in surface ozone from base case: Reduced NO_X emission (33%) with a 16% stratospheric ozone reduction.

Figure 4 shows the result of NO_x emission reductions overlaying a 16% reduction in stratospheric ozone loss. As expected, the reductions in ozone were enhanced in the high VOC/NO_x regions which experienced ozone reductions from 15 to 30 percent (southeastern states). In the northern sections of the domain, in areas of low VOC/NO_x, ozone was reduced by 5-10 percent.

Under a "status quo" policy, wherein a 33 percent increase in NO_x emissions from the 1984 base case was assumed, similar effects on ozone concentrations were experienced

(Crist, 1994). In a serious of STEM-II simulations conducted by Crist (1994) under the status quo policy ozone concentrations in areas with low VOC/NOx, experienced throughout much of the north east, resulted in less than a 2 percent change in ground level ozone. However, in the southeast which maintains larger VOC/NOx ratios an increase of 8 to 12 percent in ground level ozone was predicted(Crist, 1994).

For the simulation with increased NO_X emission and a decline in stratospheric ozone a very different pattern of ground level ozone develops. Figure 5 shows the percent change in ground level ozone for a 33 percent increase in NO_X emissions, and a 16 percent loss in stratospheric ozone, as compared to the base case. A striking contrast between the northern, and the southern regions can be seen. Predicted ozone concentrations from southern Illinois northeast through upstate New York, were reduced from 2 to greater than 6 percent. While from Florida through Virginia, ground level ozone increased from 2 to 6 percent.

CONCLUSION

With the loss of stratospheric ozone the troposphere will become more reactive significantly impacting regional ozone distributions. Based on a 16 percent loss of stratospheric ozone, photolysis rates were calculated for all radiatively important trace gas species across the eastern United States. The increases were non-linear with the photolysisof ozone increasing by approximately 40 percent, while photolysis rates for NO_2 were virtually unchanged.

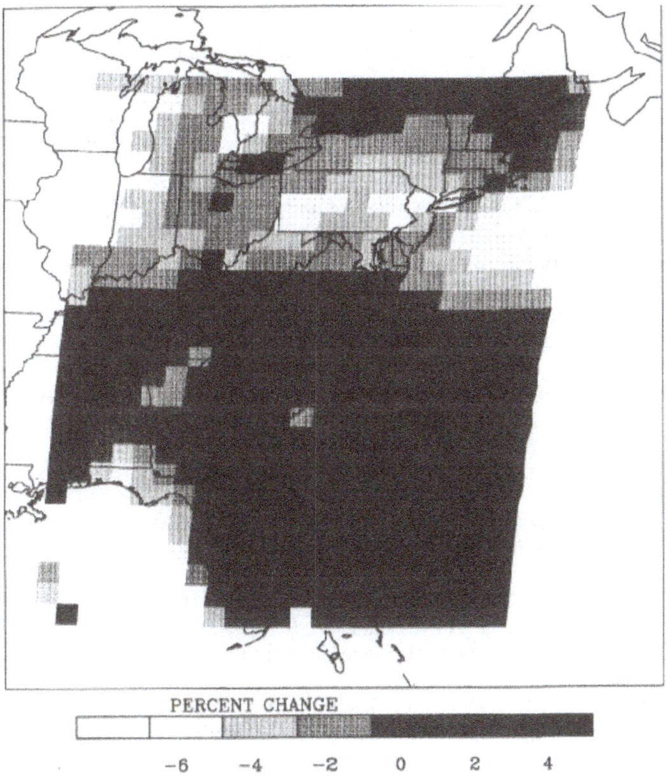

Figure 5. Percent change in surface ozone concentrations from base case: Increased No_X emission(33%) - 16% stratospheric ozone reduction.

Current scientific understanding suggests the enhanced photolysis rates of radiatively important trace gas species as the result of UVR increases for the urban corridors will increase surface ozone levels (UNEP, 1994; Liu and Trainer, 1988). In these preliminary findings a decrease in surface level ozone with an increase in UVR was experienced. This implies a large gap in the state of science. Recent studies suggest that region wide NO_X coupled with urban specific VOC controls provide an effective method for ozone attainment (Rao and Sistla, 1993; John et al. 1994). The efficiencies of these control strategies will be impacted under increasing UVR. These uncertainties need to be quantified for a better determination of the efficacy of emission control strategies in light of the present negative trends in stratospheric ozone levels.

REFERENCES

Carmichael G. R., L. K. Peters, and R. D. Saylor, The STEM-II regional scale acid deposition and photochemical oxidant Model: I. An overview of model development and applications, *Atmos. Envir.*,25A, 173-188, 1991.

Chameides W. L., P. S. Kasibhatla, J. Yienger, and H. Levi. II, The growth of continental-scale metro-agro-plexes,regional ozone pollution, and world food production, submitted to *Science*, 1994.

Crist, K.C., G.R. Carmichael, and K. John, UV-B exposure and atmospheric ozone: evaluation of radiative flux to changes in ambient ozone levels, J. Haz. Mat., 37, 527-538, 1994

Crist K.C., Comprehensive regional scale air pollution/radiative modeling: An analysis of the relationships between changes in radiative transfer and atmospheric trace gas species, Ph.D. thesis, University of Iowa, Iowa City, Iowa, 1994

Herman J. R., R. Hudson, R. McPeters, R. Stolarski, Z. Ahmad, X. Y. Gu, S. Taylor, and C. Wellemeyer, A new self-calibration method applied to TOMS and SBUV backscattering ultraviolet data to determine long-term global ozone change, *J. Geophys. Res.*,96,7531-7545, 1991.

John K., S.T. Rao, G. Sistla, N. Zhou, W. Hao, K.L. Schere, S. Roselle, N. Possiel, and R. Scheffe, Examination of the efficacy of VOC and NOx emissions reductions on ozone improvement in the New York Metropolitan Area, in Gryning S.E. and Millan, M. (ed), Air Pollution Modeling and Its Application X, Plenum Publishing Corportation, New York, NY. 1994

Liu S. C., and M. Trainer, Responses of tropospheric ozone and odd hydrogen radicals to column ozone change, *J. Atmos. Chem.*,6,221-221, 1988.

National Research Council, *Rethinking the ozone problem in urban and regional air pollution*, National Academy Press, Washington, D.C., 1991.

Rao S.T., and Sistla G., Efficacy of nitrogen oxides and hydrocarbon emissions controls in ozone attainment strategies as predicted by the Urban Airshed model, *Water, Air and Soil Pollution,* 67:95, 1993

Shin S. G., K. John, and G. R. Carmichael, The regional distribution of oxidants in the eastern Unites States and its sensitivity to reductions in anthropogenic emissions, submitted to *Atmos. Envir.*, 1993.

Shin W. C. Comprehensive air pollution modeling on multiprocessing environments: Application to regional scale problems, Ph.D thesis, University of Iowa, Iowa City, 1990.

Stolarski R. S., P. Bloomfield, R. D. McPeters, and J. R. Herman, Total ozone trends deduced from Nimbus 7 TOMS data, *Geophys. Res. L.*,18,1015-1018, 1991. UNEP, Environmental effect of ozone depletion: 1994 assessment, UNEP, Nairobi, Kena, 1994

Zdunkowski W. G., R. M. Welch, and G. Korb, An investigation of the structure of t ypical two stream methods for the calculation of solar fluxes and heating rates in clouds, *Atmos. Phys.*,53,147-167, 1980.

DISCUSSION

A. HANNA: Did you look at the effect of changing surface albedo on the photolysis rates?

K. CRIST: No we were conducting sensitivity studies as related to total column ozone.

R. DENNIS: How does the 16 % change studies compare to the temporal variability (and spatial) of the ozone column within an ozone sensor?

K. CRIST: Daily fluctuations of plus or minus 10 % are experienced with even larger seasonal fluctuations. One of our research objectives is to analyse the sensitivity of tropospheric ozone production to this temporal variability in stratospheric ozone concentrations.

P. BUILTJES: Did you take into account in your model calculations the reduction of stratospheric influx due to the reduction of stratospheric ozone and if so what would be the influence on your model and scenario results?

K. CRIST: No we did not. Stratospheric ozone influx would not change the total integrated column ozone. There maybe a slight effect on surface level radiation since tropospheric ozone appears to be a more efficient filter of UV. However, I do not anticipate that this would have a significant impact on the tropospheric photolytic rates.

R. YAMARTINO: What is the mechanism that enables a 16 % reduction of stratopsheric ozone to creat a 45 % increase of radiation at 290 nm. Is this non-linearity due to physics or chemistry?

K. CRIST: Absorption of radiation when broken down into its individual constituents and wavelengths is quantified by the Beer-Lambert law which holds that the decrease of the radiant intensity, passing through a homogenous medium follows an exponential (not a linear) relationship related to the absorption capabilities of the radiative species, its concentration. and the pathlength.

H. KAMBEZIDIS: Have you calculated the UVR increase due to stratospheric ozone depletion alone.

K. CRIST: Yes, please refer to two publications:

Crist K. C., G. R. Carmichael, and K. John, UV-B exposure and atmospheric ozone: evaluation of radiative flux to changes in ambient ozone levels, J. Hazardous Mat.,37 (1994), 527-538

Ping-Chi Li, G.C. Carmicael, and K.C. Crist, The trends of biological active ultraviolet radiation exposure in Tiapei, Taiwan, TAO,6 (1995), 461-471

GLOBAL AND LONG-RANGE TRANSPORT

chairman: J. Kretzschmar

rapporteur: J. Ching

SULFUR DEPOSITION IN ASIA: AN ASSESSMENT OF THE PRESENT SITUATION AND FUTURE SCENARIOS

Gregory R. Carmichael and Richard L. Arndt

Department of Chemical and Biochemical Engineering
Center for Global and Regional Environmental Research
University of Iowa, Iowa City IA 52242 USA

INTRODUCTION

The management of the atmospheric environment is complicated by the fact that pollutants can be transported over long distances and across political boundaries. Through the mechanisms of long range transport chemical species emitted into the atmosphere from urban and industrial centers can be distributed over large regions and as a result such widespread pollution problems as regional elevated oxidant levels, acid deposition, and visibility impairment can occur. While the long range transport of air pollutants has been widely studied in the past decade, most of these studies have focused on the industrial areas of the western hemisphere.

One of the fastest growing regions of the world is Asia. The potential for large-scale pollution problems is great because of high emissions and the close geographical proximity of many of the major industrial and urban centers (e.g., Tokyo, Seoul, Taipei, Shanghai, Hong Kong, etc.). Through long range transport processes the pollutants emitted from the various countries can be expected to contribute in adverse ways to the air quality of Asia.

The subject of long range transport of pollutants is really just in the beginning stages in Asia (Kotamarthi and Carmichael, 1990; Kurita et al., 1985; Merrill et al., 1985). However, the future air quality of the region is at risk because of Asia's large and rapidly increasing population, its growing economy, and the associated systems of energy consumption and production. Asia currently accounts for more than 55% of the world's population and is projected to reach four billion people by the year 2010. In addition, the economic growth rate for Asia is higher than any other region of the world. These factors, combined with the use of indigenous coal as the region's primary fuel source, will result in increased emissions of pollutants into the environment.

In July 1992, researchers in Europe, the USA, and Asia began a collaborative project to study the effects of sulfur deposition in Asia. The broad goals of this international collaborative project on Acid Rain and Emission Reduction in Asia are to:

* develop approaches that assist in formulating national and regional policies to address the rapidly emerging problems of atmospheric emissions in Asia;
* provide preliminary assessments needed to catalyze the process of inter-governmental policy dialogue on acid rain precursor emission control in Asia;
* strengthen Asian research capacity for the support of national and regional policies.

Air Pollution Modeling and Its Application XI, Edited by
Gryning and Schiermeier, Plenum Press, New York, 1996

Based on the framework of an acid rain integrated assessment for Europe (RAINS) (Hordijk, 1991), data and models were brought together on energy use, emissions of sulfur dioxide, abatement of emissions, long-range transport and deposition, and environmental effects. This RAINS-ASIA project has produced the first estimates of the regional impacts of the expanding energy use in Asia (Bhatti et al., 1992). The project was funded by using multi-national trust funds through The World Bank and The Asian Development Bank.

In this paper the results of the analysis of the emissions and deposition of sulfur oxides in Asia are presented and discussed. The analysis includes estimates of the emissions of SO_2 arising from anthropogenic activities with a 1° by 1° resolution. Results representative of 1990 and those for various future scenarios of energy and associated sulfur emission growth, are presented and discussed. The implications for the ecosystem impacts are also presented.

OVERVIEW OF RAINS-ASIA MODEL & MAJOR FINDINGS

RAINS-ASIA is a policy oriented model which provides a framework for integrated assessment of acid deposition. The basic structure of RAINS-ASIA is shown in Figure 1. The model consists of three sub-models (modules): the Energy and Emissions module; the Acid Deposition module; and the Ecosystems Impact module. These modules are combined together into a microcomputer based policy analysis model. The model allows users to describe emissions in current and future years on regional, country or sub-country basis, follow these emission through the acid deposition processes, and assess the potential of those emissions to impact critical ecosystems.

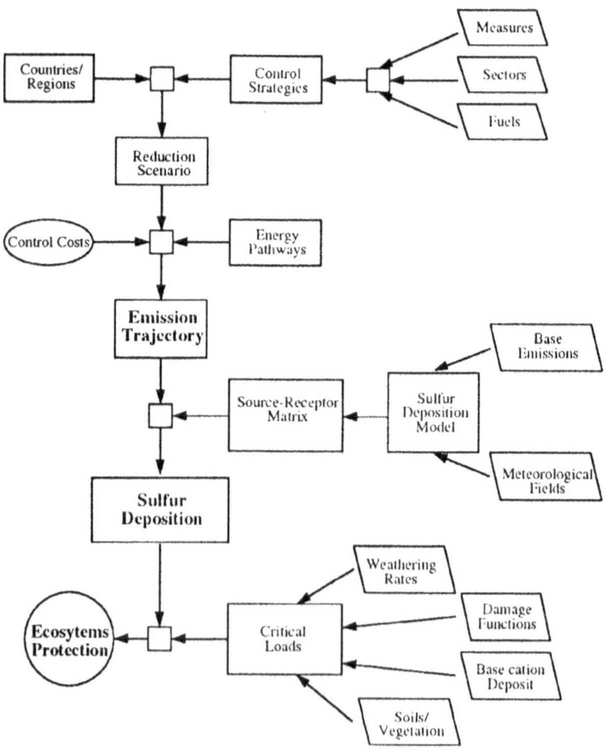

Figure 1. Schematic diagram of the modules in the RAINS-Asia model.

The region of study is shown in Figure 2 and includes 23 countries which are further divided into 95 sub-national regions for which energy scenarios are developed and emissions estimated. Of these 95 regions, approximately 30 are large metropolitan areas (mega-cities).

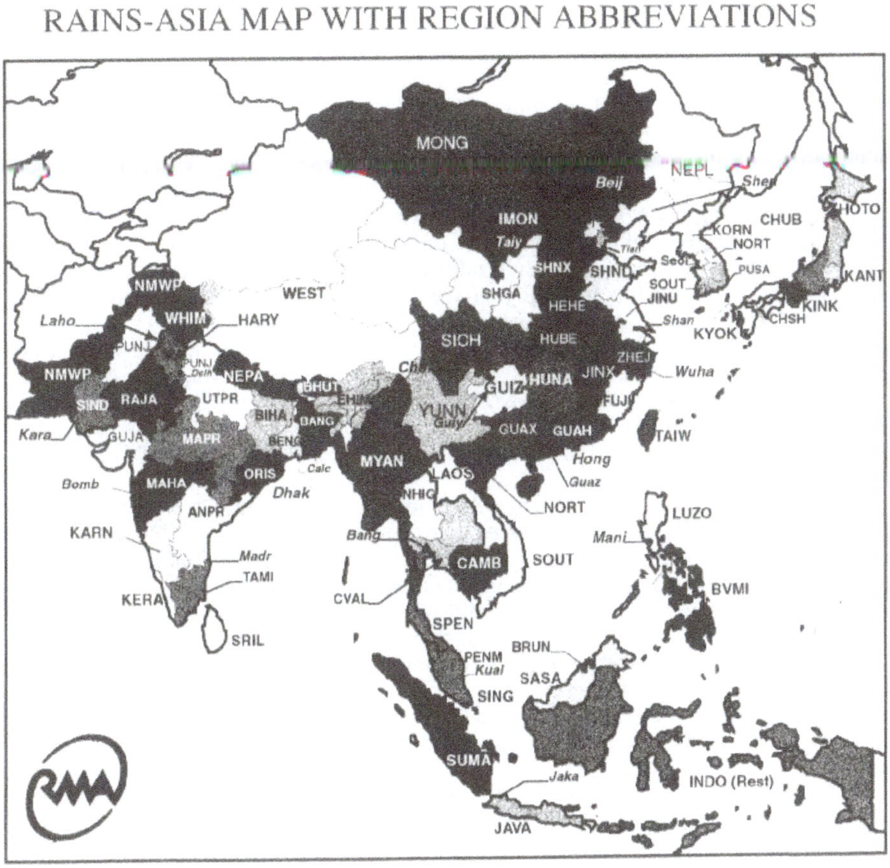

Figure 2. Countries and regions used in the study. Map prepared by Resource Management Associates, Madison, WI, U.S.A.

Emissions

Developing an accurate and detailed picture of sulfur dioxide emissions in Asia is a crucial component of integrated assessment and serves multiple purposes. It is an output of the characterization of the energy and industrial situation that links the economic development of the region with the damage it causes to the environment. By studying the use of fossil fuels for provision of energy services and production levels in industrial manufacturing, it is possible to estimate the releases of sulfur dioxide to the atmosphere. These emissions then serve to drive the atmospheric transport, transformation, and deposition model which calculates the temporal and spatial fate of the emissions. Subsequently, this information is used to estimate the vulnerability of ecosystems to these patterns of deposition. By projecting future changes in energy and industrial development, future emissions can be calculated and future threats to the environment estimated.

The picture of current emissions from different types of facilities also serves as a reference point for estimating the potential for reducing emissions. The cost and performance of emission control technologies and related techniques are provided within the RAINS-ASIA model and used to develop scenarios of reduced emissions and incremental costs. By transforming the patterns of reduced emissions through the atmospheric deposition and impacts modules, a view of the consequent reduced environmental damage can be obtained in physical terms. If these physical measures could be converted to economic benefits, a full picture of the cost-effectiveness of different emission control options could be obtained.

Emissions-related activities were focused on two areas: (1) estimation of base year 1990 SO_2 emissions; and (2) development of energy and emissions scenarios over the period 1990 to 2020. The base-year emissions inventory represents a snapshot in time of current sulfur dioxide emissions in Asia. The base-year inventory was compiled on a $1^\circ \times 1^\circ$ grid for the entire Asian region and subsequently aggregated to 95 regions and 23 countries. All anthropogenic sources of emissions were included, both land-based and from international shipping. In contrast to some other inventories, sulfur dioxide emissions from the burning of agricultural wastes, dried animal wastes, and fuel wood were specifically included for many of the countries. Steady-state (non-eruptive) emissions from volcanoes were also included.

The inventory is a combination of two data sources. For Southeast Asia (10 countries), the Indian subcontinent (6 countries), and Mongolia, emissions were calculated specifically for the RAINS-ASIA project using the methodology described below. For the remaining six countries of East Asia (China, Hong Kong, Japan, North Korea, South Korea, and Taiwan), the emission estimates of Akimoto and Narita (1994) were used.

For Southeast Asia, the Indian sub-continent, and Mongolia, information was initially compiled for major point sources in the region (Bhatti, et al. 1992). Fossil-fuel-fired power plants and industrial process sources that emit sulfur dioxide (e.g., copper smelters) were included. In all, 252 point sources were included for these 17 countries, 94 of them being power generation facilities. Each point source was assigned to its appropriate grid cell. The contribution of these point sources to total energy use and industrial production was then subtracted from national totals. The remaining energy use and industrial activity was then converted to sulfur dioxide emissions using standard emission factors and shared out to grid cells . The emission estimates also contain specific information on some 400 existing and planned large point sources (LPS) in the region.

Energy and emissions scenarios are derived in the ENEM module which deals with sectoral energy consumption, calculated SO_2 emissions and estimates of cost of abatement. ENEM contains an energy and emissions database and an algorithm to derive emissions from energy use. The core of ENEM is a sectoral- and fuel-specific end-use energy model, applied to the sub-regions to calculate energy consumption and production for each of the sub-regions and countries in the study domain (Foell, et al. 1995). Consumption is dis-aggregated according to the industrial, transportation, residential, commercial, and "other" sectors. The energy supply and transformation system is disaggregated according to the major sources of emissions including electricity generation, oil refining, and other energy sector operations which result in combustion of fossil fuel.

This framework is used to calculate emissions for the sub-regions for a range of energy scenarios based on a variety of socio-economic (such as population and Gross Domestic Product (GDP)) and technical assumptions. From these energy scenarios sulfur emissions from fossil-fuel combustion are calculated, based on fuel characteristics, combustion technology, and emission control assumptions. Technological abatement options for limiting sulfur emissions can be applied employing assumptions regarding boiler size, capacity utilization, and sulfur content of fuel.

There are also a number of control options available to reduce emissions of sulfur including: energy conservation; fuel switching; fuel desulfurization; and flue gas desulfurization. The ENEM module contains the flexibility to develop and analyze various energy scenarios. For scenario analysis, users can select one of several preset energy-use scenarios or can create their own, based on expectations of fuel use. Base year and future year emissions are resolved as both low level area and high level point sources. The user can then select a consistent package of control measures to develop an integrated abatement strategy for each country.

Sample results of the emissions estimates are shown in Table 1. Shown are country level total SO_2 emissions for the base year 1990, along with estimates for the year 2020 under three different energy/emissions pathways: no further controls (business as usual assumptions); the same energy pathway but with the application of basic control technologies; and an energy efficiency pathway along with a local application of technology. The energy efficiency pathway assumes significant energy efficiency improvements and fuel substitutions. The local and basic control technologies scenarios are based on the use of relatively inexpensive control measures, including add-on controls along with fuel switching. These scenarios indicate the range of the growth in emissions in Asia. The 1990 emissions for Asia are estimated to be 33.6 million tonnes of SO_2 per year. Under the no further control case, emissions are seen to rise by nearly a factor of 4, while the energy efficiency pathway results in a 33% increase in emissions.

Table 1. National emissions' estimates for several energy scenarios. SO_2 emissions at the country level are shown for the base year, 1990, and three different energy/emissions pathways in the year 2020: no further controls beyond current regulations, the same energy pathway but with the application of basic control technologies in China, India, and Pakistan, and an energy efficiency pathway along with local application of technology. Units: 1000 tonnes SO_2/yr

EMISSIONS SCENARIO:	BASE YEAR EMISSIONS	NO FURTHER CONTROLS	BASIC CONTROL TECHNOLOGIES	LOCAL APPLICATION OF TECHNOLOGY
COUNTRY				
BANGLADESH	118.0	524.6	258.3	394.7
BHUTAN	1.5	11.9	4.2	10.1
BRUNEI	6.3	18.4	17.0	13.2
CAMBODIA	22.2	147.5	69.4	78.6
CHINA	21908.3	60687.7	38124.7	28557.9
HONG KONG	139.6	247.0	68.6	54.3
INDIA	4471.7	18549.2	13054.0	9462.7
INDONESIA	630.0	3162.3	785.0	574.8
JAPAN	835.5	1119.7	1047.8	724.8
NORTH KOREA	343.1	1345.4	707.4	792.7
SOUTH KOREA	1640.3	5537.4	1469.6	1089.8
LAOS	3.4	11.7	6.9	11.0
MALAYSIA	205.7	409.8	245.6	183.5
MONGOLIA	77.7	167.7	81.3	127.3
MYANMAR	18.1	39.6	36.9	38.9
NEPAL	122.3	247.4	229.5	195.3
PAKISTAN	614.1	7527.6	3609.4	4893.8
PHILIPPINES	390.7	2036.9	439.9	318.6
SEA LANES	243.3	511.9	307.2	511.9
SINGAPORE	190.9	1033.2	220.8	161.1
SRI LANKA	41.9	239.0	52.5	121.4
TAIWAN	499.5	1478.7	754.7	748.1
THAILAND	1037.6	4637.7	813.8	568.9
VIETNAM	113.2	654.7	345.6	556.4
SUM	**33674.7**	**110346.8**	**62750.1**	**50189.9**

TRANSPORT AND DEPOSITION

The sulfur deposition patterns resulting from the estimated emissions are calculated using a regional-scale trajectory model (called ATMOS). The model is a modified version of NOAA's Air Resources Laboratory Branching Atmospheric Trajectory model (Heffter, 1984). The ATMOS model is a three-dimensional, multiple-layer Lagrangian model which includes the calculation of SO_2 and sulfate surface concentrations, and wet and dry deposition amounts. It was also modified to include the capability for modeling both elevated and surface emission sources.

The ATMOS model provides a one-by-one degree resolution of the concentrations and deposition of SO_2 and sulfate. Within the modeling domain SO_2 emission plumes are modeled as puffs released every three hours from the emission source location. Each puff is assigned a mass proportional to the source strength, and is assumed to mix uniformly in the vertical throughout an assigned layer, and to diffuse with a Gaussian distribution in the horizontal. Area emissions are modeled as surface sources (released at the center of the grid) while large point sources and volcanoes are treated as elevated sources. Individual emission puffs are followed throughout their transport and deposition "lifetimes". Each puff's transport is followed for up to five days (or until the mass falls below a cut-off value). Puffs which are transported beyond the modeling domain are no longer tracked. As the puffs are transported, SO_2 is chemically converted to sulfate, and SO_2 and sulfate are deposited to the surface. The deposition of the two pollutants is separated into wet and dry components. Running these trajectories for an entire year allows the calculation of annual deposition throughout the region resulting from a particular source. Repeating the calculation for all sources allows the total deposition from the regional emissions to be calculated.

The ATMOS model requires as inputs the emission rates of SO_2, meteorological winds and temperature, precipitation rates, and estimates of the dispersion coefficients, dry deposition velocities, and wet scavenging coefficients. The ATMOS model calculates the chemical- conversion rates based on simple first-order rate constants, the dry deposition rate as a function of the surface concentrations and the dry deposition velocity, and the wet scavenging coefficient as a simple function of precipitation rate (Arndt et al., 1995).

The ATMOS model is run for each elevated and surface source estimated by ENEM model. Sulfur deposition is calculated on the 1° by 1° grid, and the contribution of each source to the grid is calculated. The results are aggregated to provide source/receptor information for each of the 95 sub-regions to the acid deposition patterns in the study region, and further aggregated to provide country-to-country source/receptor information.

The model calculated annual total deposition in $g\text{-}S/m^2\text{-}yr$ for the base year 1990 is presented in Figure 3. Shown are the contributions from all anthropogenic sources included in the model (LPS and area sources) and volcanoes. There are very few regions in Asia which are not impacted by sulfur deposition. The high sulfur deposition regions follow closely the spatial distribution and the density of the emissions . For example, the dense emission regions in eastern and southern China, S. Korea, northern Thailand, and eastern India all show elevated sulfur deposition. The highest annual deposition (~10.5 $g\text{-}S/m^2\text{-}yr$) occurs around the city of Chongqing in Sichuan province. The strong continental outflow from east Asia is also clearly depicted. Sulfur emissions in the latitude band 20° to 40° N result in high sulfur deposition virtually throughout the western Pacific Ocean at these latitudes. Transport and deposition off the eastern coast of India over the Bay of Bengal is shown as is the high sulfur deposition around Malaysia, Singapore and the western parts of Indonesia. The total deposition patterns are also heavily influenced by the annual precipitation patterns. High annual precipitation amounts are found to occur in northern India, Nepal, southeastern China, and Southeast Asia.

Elevated sulfur deposition also occurs along the major shipping lanes. The contribution of ships to sulfur deposition exceeds 10% for vast regions of the study area below 20° N latitude. In many oceanic regions the deposition due to emissions from ships is the dominant source. Ships also contribute to the on-shore deposition throughout Malaysia, Indonesia, Vietnam and the Philippines.

Further insight into the sulfur deposition patterns are found by looking at the sulfur deposition arising from LPS. Most of the deposition from LPS occurs in eastern China, central India and northern Thailand. The maximum amount of deposition due to point sources is ~70%, whereas on average LPS account for ~10% of the sulpher deposition.

This is consistent with the LPS emissions comprising about 10% of the total anthropogenic sulfur emissions in Asia. Deposition due to volcanic sources in Asia have also been studied. Sulfur deposition from this natural source contributes significantly to sulfur deposition in Japan, Indonesia and the Philippines. Volcanoes account for ~30% of the total deposition in Japan, ~50% in Indonesia, and ~20% in the Philippines.

The deposition of sulfur due to dry and wet removal processes was also calculated. For the case of SO_2, the domain-integrated total deposition due to dry and wet removal mechanisms are of similar importance. In contrast, the wet deposition of sulfate dominates the sulfate deposition process. In most locations wet deposition of sulfate accounts for > 90% of the sulfate deposition. Wet removal processes account for 60% of the total sulfate deposition.

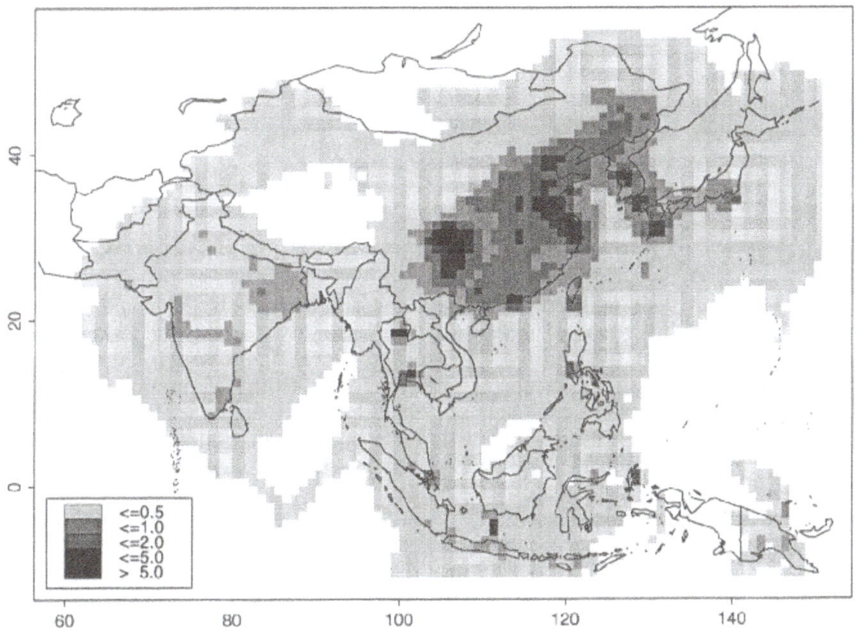

Figure 3. Calculated annual sulfur deposition in 1990. Units in g- $S/m^2/yr$.

The ATMOS model calculates the deposition from each source directly and this information can be used to analyze a variety of policy-related questions. For example, the deposition from a specific LPS, region, or country can be viewed separately. The contribution of sulfur deposition from volcanoes and ships previously presented illustrate this capability. This information can also be used to identify which sources contribute to the deposition at a specified receptor. This information can be aggregated to provide source-receptor information at a country-to-country or region-to-region level. (Please note that the source-receptor information discussed is based on only one year of meteorology and must be considered preliminary.) Table 2 demonstrates how such data can be interpreted to quantify the impact of a specific country's deposition on the region. For example, 83% and 14% of China's emissions which are deposited in the study region fall on China and the region's oceans, respectively. The remaining 3% falls on other nations (i.e. 0.8% on N. Korea and 0.5% on Japan). Column three presents the fraction of the receptor country's total deposition resulting from China emissions. For example, 96% of the sulfur deposited in China is from Chinese sources while 35% and 22% of the sulfur deposited in N. Korea and Vietnam, respectively, is due to Chinese emissions. Table 2 provides an interesting perspective on the region's deposition. Although 97% of China's emissions deposited in the region fall either within China or on the region's oceans, the remaining 3% can account for significant percentages of the neighboring countries' total deposition.

Table 2. China's contribution to sulfur deposition in the region, expressed both as a percentage of China's total deposition in the region and as a percentage of the receptor-country's total deposition.

COUNTRY-TO-COUNTRY SOURCE-RECEPTOR RELATIONSHIP

RECEPTOR	% OF CHINA'S DEPOSITION IN ASIA	% OF RECEPTOR'S TOTAL DEPOSITION
China	83	96
Oceans	14	36
North Korea	0.8	35
South Korea	0.4	13
Japan	0.5	10
Vietnam	0.4	22

ENVIRONMENTAL IMPLICATIONS OF SULFUR DEPOSITION

The environmental implications of sulfur deposition cannot be evaluated simply by examining the sulfur deposition amounts at a specific location. Rather, deposition values must be compared with the ability of the receptor locations to assimilate the sulfur deposited. The environmental impacts of sulfur deposition in Asia are assessed in RAINS-ASIA through use of estimates of critical loads (Hettelingh, 1991). A critical load is the maximum level of pollutant that can be deposited on a specific location without environmental damage. Critical loads for Asia are calculated as part of the in the IMPACT module (Hettelingh et al., 1995).

The IMPACT module consists of dynamic models describing processes in key ecosystems, e.g., soils, surface waters, and vegetation systems. These models include the computation of the depletion of the acid buffer capacity of ecosystems under the influence of precipitation, evaporation, water flows, and budgets of chemical ecosystem constituents (Hettelingh et al., 1991).

The critical loads are compared with the estimates of sulfur deposition to determine which ecosystems may be at risk under various emission scenarios. Presented in Figure 4 are the calculated sulfur exceedances (i.e., the difference between sulfur deposition and the sulfur 20%-levels, i.e., the critical loads which represent sulfur deposition amounts that protect 80% of the ecosystems) for the base year 1990. To account for uncertainties in the estimate of critical loads, 20%-levels were chosen, which are more lenient then the 5%-levels used in Europe All areas with positive values indicate those regions where sulfur deposition exceeds the critical load, and thus those areas where ecosystems are predicted to be at risk. Vast regions of Asia are predicted to be in excess of the critical load. These areas include most of Korea, Thailand, Taiwan, the Malaccan Straits, eastern China, and parts of India and Bangladesh.

These results indicate that there are regions in Asia which are presently at risk to damage due to acid deposition. It is also important to explore how this situation may change in the future. The rapid increase in energy consumption in Asia, will certainly result in a large growth in sulfur emissions. Without the introduction of additional emissions controls to counter this growth, elevated pollutant levels can be anticipated. Shown in Figure 5 are the exceedances for the year 2020 under the no further control scenario. This scenario reflects how energy, emissions and deposition would respond to a situation where economic growth in the region continues along present trends with the continued increase in energy consumption without further sulphur emissions controls or modifications to energy production methods (e.g., replacement of coal burning with natural gas usage). As presented in Table 1, the emissions are predicted to grow by a factor of 4 in the region under these assumptions. Now most of eastern China, the Korean Peninsula, Japan, Thailand, and large regions of India, Nepal, Bangladesh, The

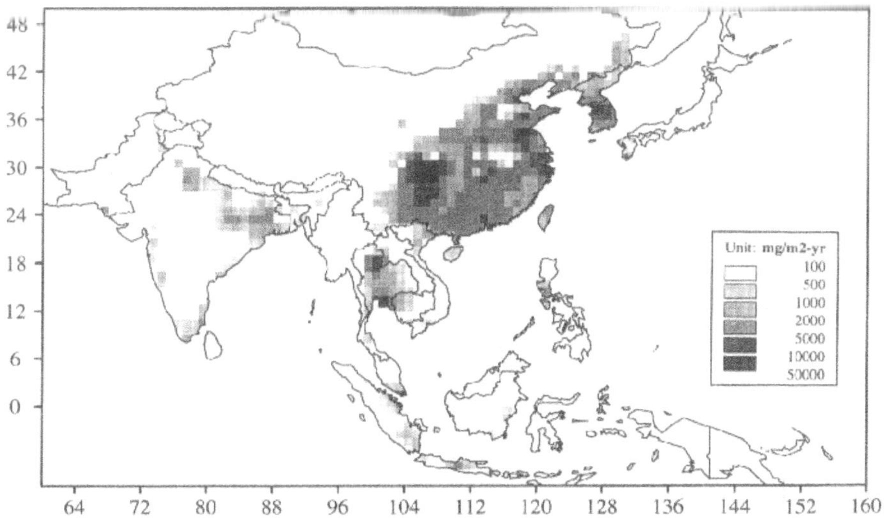

Figure 4. Sulfur exceedances for the base year 1990. Exceedances are expressed in g-S/m^2/yr. and are based on the 20% critical loads.

Philippines, Malaysia, Indonesia, and sections of Vietnam have deposition in excess of the critical loads. This situation represents a probable worst-case scenario. As shown in Table 1, lower energy pathways and/or the use of control technologies can reduce the growth rates in emissions and the associated environmental impacts. The estimated annual costs to achieve these emission pathways in the year 2020 are each ~$30 billion USD.

CLOSURE

RAINS-ASIA is a tool which provides a framework for integrated assessment of acid deposition in Asia. The model combines energy use, emissions, abatement approaches, long-range transport and deposition, and regional environmental impacts. The results are being made available to national and international bodies and research institutions to facilitate discussions on actions to alleviate current and future environmental degradation.

Results from the first integrated assessment have been presented. The present situation in Asia is that sulfur deposition at levels sufficient to cause ecosystem damage (as estimated by critical loads for Asia) are calculated for vast regions in southern China, S. Korea, Taiwan, Thailand and parts of India, Indonesia, Bangladesh, and Vietnam.

This situation most likely will change dramatically as a consequence of the very high economic and population growth rates of the region. The expansion of fossil fuel energy systems, combined with a major fuel shift to indigenous coal, will undoubtedly result in a significant increase in atmospheric emissions for the Asian countries. Substantial portions of these emissions will be transported by winds hundreds of kilometers from their source.

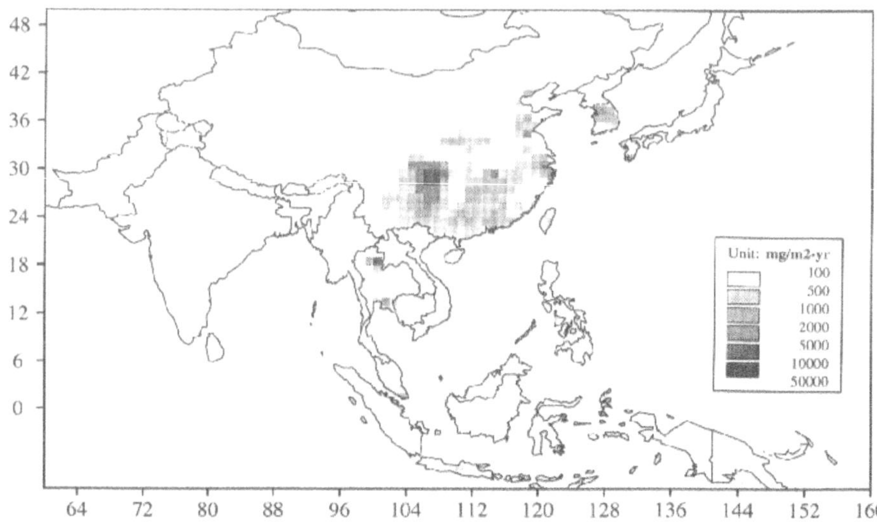

Figure 5. Sulfur exceedances for the year 2020 under the no-further control emission scenario. Exceedances are expressed in g-S/m^2/yr. and are based on the 20% critical loads.

To help quantify and anticipate environmental impacts associated with these emissions it is imperative that we develop a greater understanding of the mechanisms of long range transport of pollutants in Asia. Transboundary transport issues appear to be of potential importance in east Asia (involving eastern China, Korea and Japan), southeast Asia, and among those countries bordering India. Acid deposition problems throughout Asia are anticipated as a result of the rapid expansion in energy consumption in the region. By developing an understanding of the relationship between sulfur emissions and their resulting deposition, we can better predict and assess the impact of Asia's growing energy demands on the environment. Clearly much more work is required to better improve the scientific inputs into these studies, to improve the quality of the input data, and to evaluate the model against observational data. Work in all these areas is ongoing, and it is hoped that this integrated assessment activity will help to facilitate further research in this area.

Acknowledgments

This research was supported in part by funds from The World Bank and The Asian Development Bank as part of the RAINS-ASIA project. Special thanks to the collaborators on the RAINS-ASIA Phase-1 project. They include: M. Amann, J. Cofala, W. Foell, C. Green, L. Hordijk, J. Shah, W. Schopp, D. Streets, M. Chadwick, J. Kuylenstiering, S. Cinderby, D. Zhao, M. Mohan, and many others.

REFERENCES

Akimoto, H. and H. Narita, 1994, Distribution of SO_2, NO_x, and CO_2 Emissions from Fuel Combustion and Industrial Activities in Asia with 1° x 1° Resolution, *Atmospheric Environment*, 28 (2), 213-225.

Arndt, R., G. Carmichael, D. Streets, and N. Bhatti, 1995, Sulfur Dioxide Emissions and Sectorial Contributions to Sulfur Deposition in Asia, *Atmospheric Environment*, submitted.

Bhatti, N., D.G. Streets, and W.K. Foell, *Environmental Management, 16*, 541, 1992.

Foell, W., C. Green, M. Amann, D. Streets, G. Carmichael, and J. Hettelingh, 1995, Energy Use, Emissions and Air Pollution Reduction Strategies in Asia, *Journal for Water, Air and Soil Pollution*, in press.

Heffter, J.L., 1983, Branching Atmospheric Trajectory (BAT) Model, NOAA Technical Memorandum, ERL ARL-121.

Hettelingh, J.P., H. Sverdrup, and D. Zhao, 1995, Deriving Critical Loads for Asia, *Journal for Water, Air and Soil Pollution*, in press.

Hettelingh, J-P. 'Guidelines for Design of Acid Rain Policy for Asia', pp. 123-145, Proceedings of the Second Annual Workshop on Acid Rain in Asia. AIT Bangkok, 1991

Hordijk, L. 1991. Use of the RAINS Models in Acid Rain Negotiations in Europe. *Environmental Science and Technology*, Vol. 25, No., 4, p. 596-603.

Kotamarthi, V. and Carmichael. G. The Long Range Transport of Pollutants in the Pacific Rim Region. *Atmospheric Environment*, Vol. 24A, pp. 1521-1534, 1990

Kurita, H., Sasaki, K., Murago, H., Ueda, H. and Wakamatshu, S. Long-range Transport of Air Pollutants under Light Gradient Winds. *Journal of Climate and Applied Meteorology*. Vol. 24, pp. 425-434, 1985

Merrill, J. R. Bleck and Avita, L. Modeling Atmospheric Transport to Marshall Island. *Journal of Geophysical Research*, Vol. 90, pp. 12927-12936, 1985

DISCUSSION

T. IVERSEN: Your area of interest covers both tropical and extratropical domains. How well does the Lagrangian model describe especially the complex tropical circulations? And are there not special problems connected with the quite fast vertical exchange in those areas?

G. CARMICHAEL: We don't know the answer to this. We know that convective transport is very important. That is why we used a 3-layer model. More work is needed before we can assess the limitations of this approach.

P. PAI: I am assuming you are using the ATAD model to generate your trajectories. ATAD uses mixed layer average winds that are calculated from the NWS meteorological stations. Can you tell us about the existing meteorological network in your study domain and are there any plans to enhance/augment the network to obtain better/more data.

G. CARMICHAEL: The network consists of approximately 200 stations. Because of the large ocean areas, we plan to go to a grided meteorological data sets in the future.

P. KARAMCHANDANI: You mentioned that the model did not capture the low SO_2 concentrations observed in Bangladesh during the monsoon season. Can that be attributed to the quality of cloud and precipitation data during those months.

G. CARMICHAEL: At many sites we do capture the monsoon variation, but at Bangladesh we don't. This could be due to precipitation/cloud data, but could also be due to other factors (e. g. source distribution).

R. BORNSTEIN: What is the contribution of China to the deposition over Japan?

G. CARMICHAEL: We estimate that Chinese sources contribute approximately 10% of Japan's deposition. This is highly uncertain and based on one year of meteorology.

C. NAPPO: Did your calculations and production estimates include considerations of alternate energy productions?

G. CARMICHAEL: Yes, we do include alternatives. They are estimated to be rather small. Energy efficiency was also considered and found to be very important with big savings.

C. NAPPO: Are these regions considering increased nuclear production?

G. CARMICHAEL: No, not at this time. Nuclear is not "popular". The immediate plan is to use coal and then see what happens.

THE EUROPEAN LONG RANGE TRACER EXPERIMENT (ETEX): PRELIMINARY EVALUATION OF MODEL INTERCOMPARISON EXERCISE

Graeme Archer[1], Franco Girardi[1], Giovanni Graziani[1], Werner Klug[2], Sonia Mosca[1], and Katrin Nodop[1]

[1]Environment Institute
JRC of the EC
21020 Ispra (VA), Italy
[2]Institut für Meteorologie
Hochschulestrasse 1
Darmstadt 6100, Germany

ABSTRACT

This report presents the preliminary results of a model intercomparison exercise based on the ground level concentration evolution consequent to the 23 October 1994 ETEX tracer release performed in North-West France.

The three-hour-average observed values, measured at over one hundred stations for a period of sixty hours, have been compared with predictions made by real-time long-range dispersion models at the same points in space and time. Twenty-eight models from twenty-four separate organisations, mainly national meteorological services and nuclear safety institutions, participated in the exercise. The models used various information concerning the forecasted circulation patterns over Europe in the two-and-a-half days following release, in order to run their long range dispersion models.

The analysis includes the comparison of simple results (times of arrival, maximum values) and a more complex statistical evaluation, based on the bias, normalized mean square error, correlation coefficients and Figure-of-Merit in time and space.

1 INTRODUCTION

The need for a controlled tracer release experiment aimed at validating long range dispersion models became evident following the conclusions to the ATMES study (Klug *et al.*, 1989). The main drawbacks to this earlier exercise were:

1. The release characteristics of the radioactive material at Chernobyl remained uncertain.

2. The evaluation of the model results took place several years after the event.

3. The measurement data base was very inhomogeneous with regard to sampling methods and intervals.

To address these issues, it was decided to launch in Europe an inert tracer experiment in which the evolution of the concentration field would be forecast by modelers at prescribed locations and time intervals, these forecasts being made in *real-time*. Thus **ETEX** (the **E**uropean **T**racer **EX**periment) was born. A large effort was made to organise the operation of sampling stations, overcoming budgetary, language and bureaucratic difficulties and establishing communications with all the stations. It was envisaged that the easiest weather situation for an experiment would be one with westerly flow over middle Europe (most common in Autumn-Winter period) which suggested immediately that the source location should be in north-western France.

The preparation phase of the exercise took almost two years, during which time three dry runs were executed to test communications between the various centres where decisions were made, measurements taken and calculations executed. Finally, two actual tracer releases were planned and executed. This paper deals only with the results of the first one, since the chemical analysis of air samplers for the second release is still being undertaken.

2 PARTICIPANTS AND REAL TIME MODELS

In the modelling phase of ETEX, participants had to forecast both the flow and diffusion field to calculate concentration evolution in real time. Therefore discrepancies between concentration forecasts of different models are due to the initialisation of flow models, to the flow model characteristics and to the dispersion model used. There was a larger number of participants in the exercise than had been expected. Table 1 lists the participants, with the main characteristics of the flow and dispersion models employed.

3 DESCRIPTION OF THE WEATHER SITUATION DURING FIRST RELEASE

During the alert period (15 October - 15 December 1994), Sunday 23 October was selected for the first release since the main desired weather criteria were matched:

- a rather strong west to southwesterly flow, advecting the tracer in the correct direction;
- no centre of high-pressure or low-pressure, no extending ridges or troughs were thought likely to pass, or be very close to the release site;
- no frontal system should pass the release site shortly before or after the release.

On that day, the synoptical situation was characterised by a cold front far east between 10 and 15 degrees easterly longitude.

A deep Low, 975 hPa east of Scotland was slowly moving north, maintaining a strong southwesterly flow over the release site (Rennes). The advected air was very unstable, with showers, some accompanied by thunder and squall-lines (see Figure 1). The synopsis for 24 October continues to show an unstable flow over the release site and the advection area. However, the wind in the advection area has decreased, because of the northerly movement of the ground Low over the North Sea.

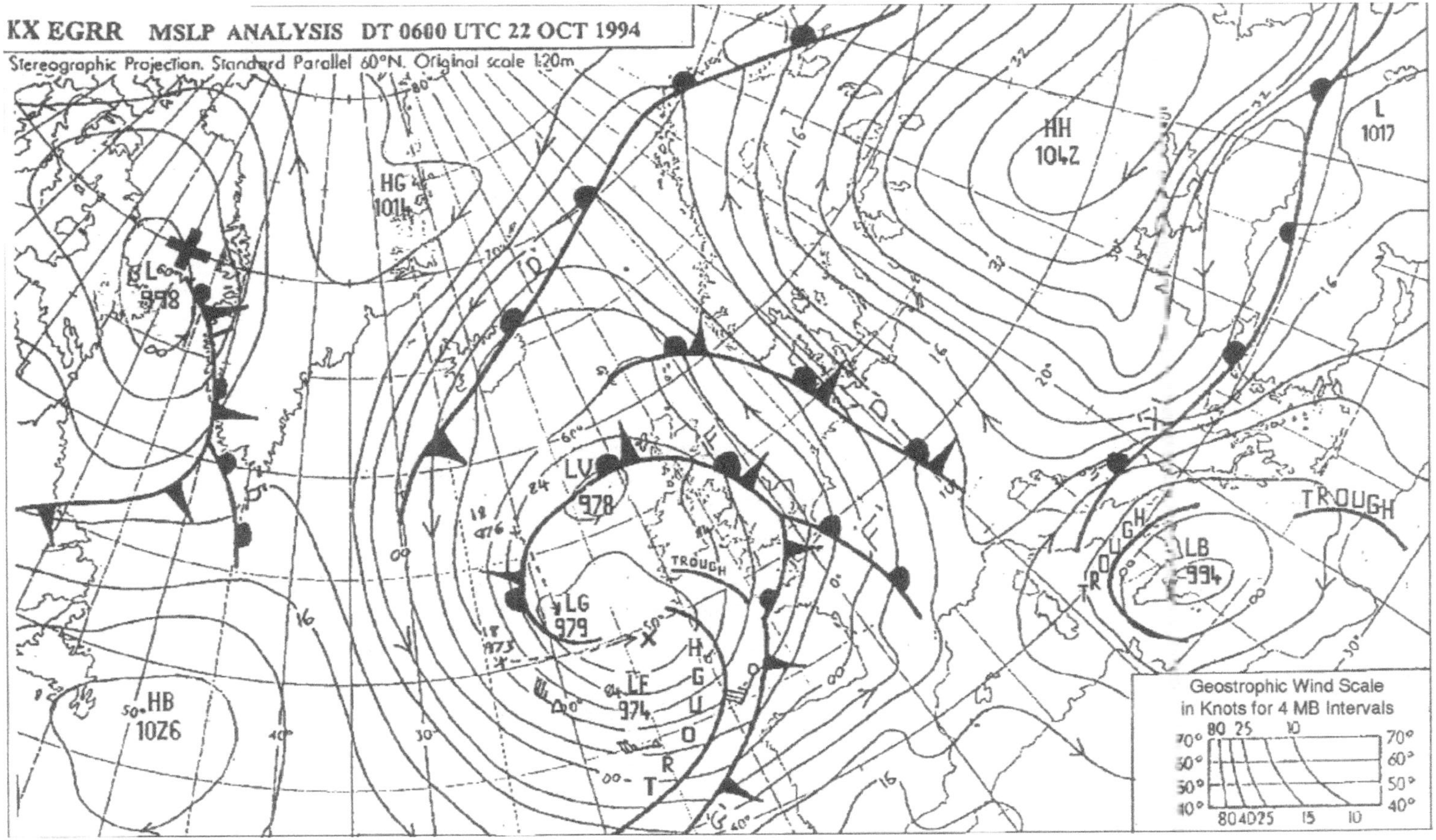

Figure 1: Synoptical situation on 22 October 1995.

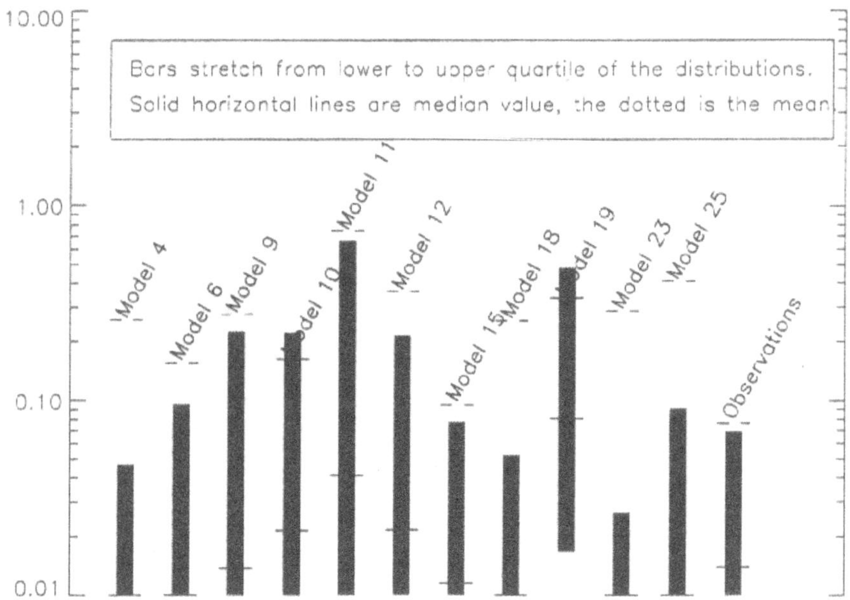

Figure 2: Global boxplots of fully forecasted predictions from eleven models, and the observed values.

4 STATISTICAL INDICATORS USED FOR THE COMPARISON

Statistical comparison was performed in three ways: (i) globally, pairing model predicted values with observations over all the time-points and spatial locations; (ii) time-dependent analysis, where the data were examined at a number of selected sites; (iii) space-dependent analysis, for given time intervals.

Table 1. Participant names and main model characteristics.

	IN	FLOW MODEL	GLOB	DISPERSION MODEL	EUL	LAGRANGIAN		OTHER
						PUFF	PART.	
Czech R.	CHI	LOTR1	N	LOTR1		X		
Sweden	SMHI	ECMWF	N	MATCH	X			
Netherlands	RIVM	ECMWF	N	knmi PUFF		X		
France	METEO1	ARPEGE	Y	ARPEGE1	X			
France	METEO2	ARPEGE	N	ARPEGE2	X			
United Kingdom	UKMO	NAME II	Y	NAME-II			X	
Finland	FMI	HIRLAM	N	TRADOS			X	
Germany	DWD1	EUROPA	N	LPDM			X	
Germany	DWD2	EUROPA	N	ETM	X			
Israel	IIBR	HIRLAM	N	LORA		X		
Japan	JAERI	WSPEEDI	Y	WSPEEDI			X	
Switzerland	SMI	ECMWF	N	LORAN		X		
Bulgaria	BMI	NAME II	Y	LED		X		
United States	NOAA	AUN	Y	HY-SPLIT				MULTIPUFF
United States	ARAC	NOGAPS	Y	ADPIC				EUL/LAG
Italy	ANPA	ECMWF	N	APOLLO			X	
Netherlands	KNMI	ECMWF/ HIRLAM	N	PUFF		X		
France	IPSN	ARPEGE	Y	SIROCCO		X		
Denmark	DMI	HIRLAM	N	DERMA		X		
Canada	CMC	CANERM	Y	CANERM	X			
Unites States	SRS	RAMS	N	LPDM			X	
Russia	TYPHOON		N	STADIUM			X	
Rumania	RMI		N		X			
Norway	NMI	HIRLAM	N	SNAP			X	
Slovak R.	SHMU1	EGRR	N	SHMI-I		X		
Slovak R.	SHMU2	EGRR	N	SHMI-II		X		
Belgium	KMI1	ECMWF	N	PUFF		X		
Belgium	KMI2	ECMWF	N				X	

Listed here are all the statistical measures used to examine how well model predictions matched the observed values. They are well known in the literature and are similar to those used for ATMES (Klug *et al.*, *ibid.*). There is space to display only a sample of (the most useful of) these. The full set of measures consisted of:

- global boxplots displaying upper and lower quartile, mean and median value of the predictions and the observations;
- global scatter diagram, with FA2 and FA5 (no. of points within factor 2 and 5);
- normalised mean square error, a measure of the dispersion of the results;
- bias, the difference between the mean value of computed and observed tracer level and used to examine a model's tendency for over - or under - prediction;
- Pearson's correlation coefficient, a measure of linear dependence between two random variables;
- fractional bias, fractional standard deviation, suggested as possible model evaluation techniques for transport models by Hanna (1988);
- Figure of Merit in Time (FMT), which is the total percentage agreement between observed and predicted values over the twenty time points at any spatial location;
- Figure of Merit in Space (FMS), which is evaluated for a fixed time. It is defined as the percentage of common area between two spatial distributions. The **FMS** can be useful not only to evaluate the percentage of agreement between two sets of data, but also for a graphical estimate of the geometrical shape and extension of the two areas.

5 RESULTS

Participants were asked to provide, in real-time, forecasted data and analysed data (that is, predictions calculated on the basis of forecast meteorological information, and with the weather situation fully known, respectively), as well as all the intermediate solutions. The full forecasted results will be discussed here, since the observations are as yet preliminary and incomplete. Only eighty-six stations have been examined, mainly in those European countries hit by the cloud. In the following, attention is focussed on the results of a few participants, selected from those with global box plots similar to the distribution of observations (Figure 2). The figure shows the 25th and 75th percentiles, together with the median and the mean values.[1] As one can see, only eleven models out of twenty-eight have been selected. For them, FA2 varies from 10% to 30%, whilst FA5 is within 24% - 57% interval. Among these models, five underestimate and six overestimate the concentrations.

These excellent results, however, hide very different simulations of the cloud evolution, as shown by Figures 3a and 3b. Here, the 0.1 ng/m^3 cloud contour is shown at the time T_0+48 (hours), as derived from observations and calculated by two different models. One can see that, even if both participants were able to forecast correctly the main cloud trajectory, horizontal dispersion is different in the two simulations. The FMS values 48 hours after the release (22% and 49%) represent the lower and upper limit for the group of models selected. Another four models among the eleven have an FMS greater than 40% at T_0+48. This result changes if the significant level is increased.

The good performances of this group of models is confirmed by the analysis of time series in eleven chosen locations, belonging to two arcs of points approximately equidistant from the source and covering the north-south cloud extension. The first is located easterly to the cloud position at about T_0+24, the other westerly to cloud position at T_0+48. Figures

---

[1]To maintain anonymity, the identification numbers of participants reported in Figure 2 and elsewhere have no relations with the participant list of Table 1.

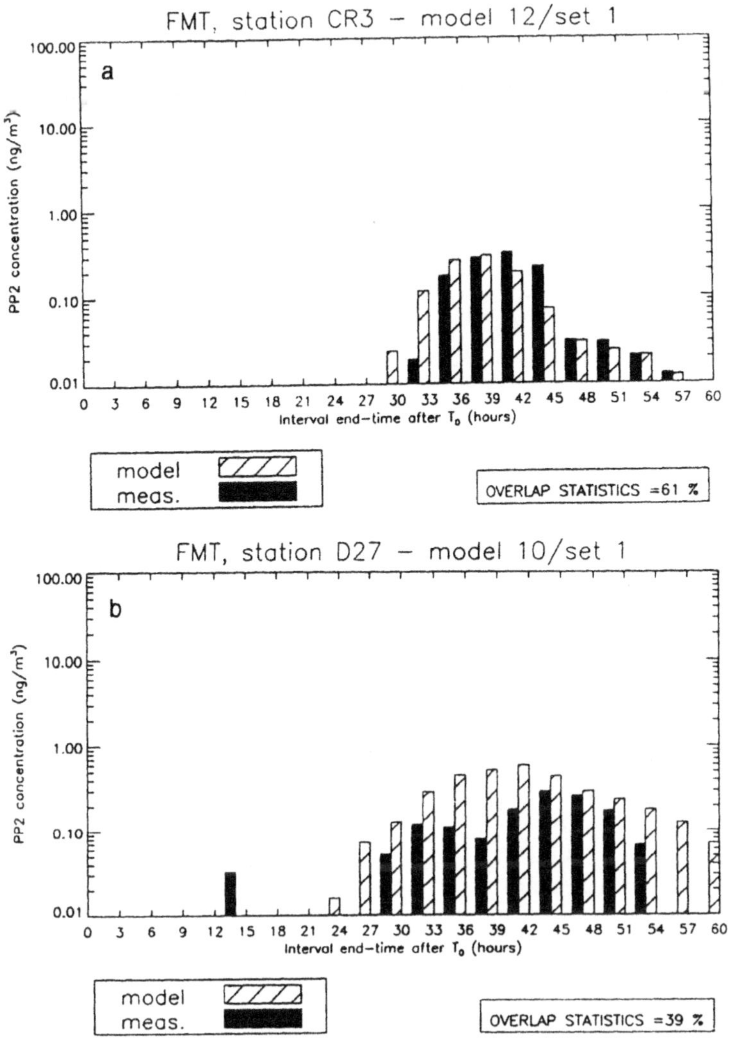
FMT, station CR3 – model 12/set 1
a
PP2 concentration (ng/m³)
100.00
10.00
1.00
0.10
0.01
0 3 6 9 12 15 18 21 24 27 30 33 36 39 42 45 48 51 54 57 60
Interval end–time after T₀ (hours)
model
meas.
OVERLAP STATISTICS =61 %
FMT, station D27 – model 10/set 1
b
PP2 concentration (ng/m³)
100.00
10.00
1.00
0.10
0.01
0 3 6 9 12 15 18 21 24 27 30 33 36 39 42 45 48 51 54 57 60
Interval end–time after T₀ (hours)
model
meas.
OVERLAP STATISTICS =39 %

Figure 3

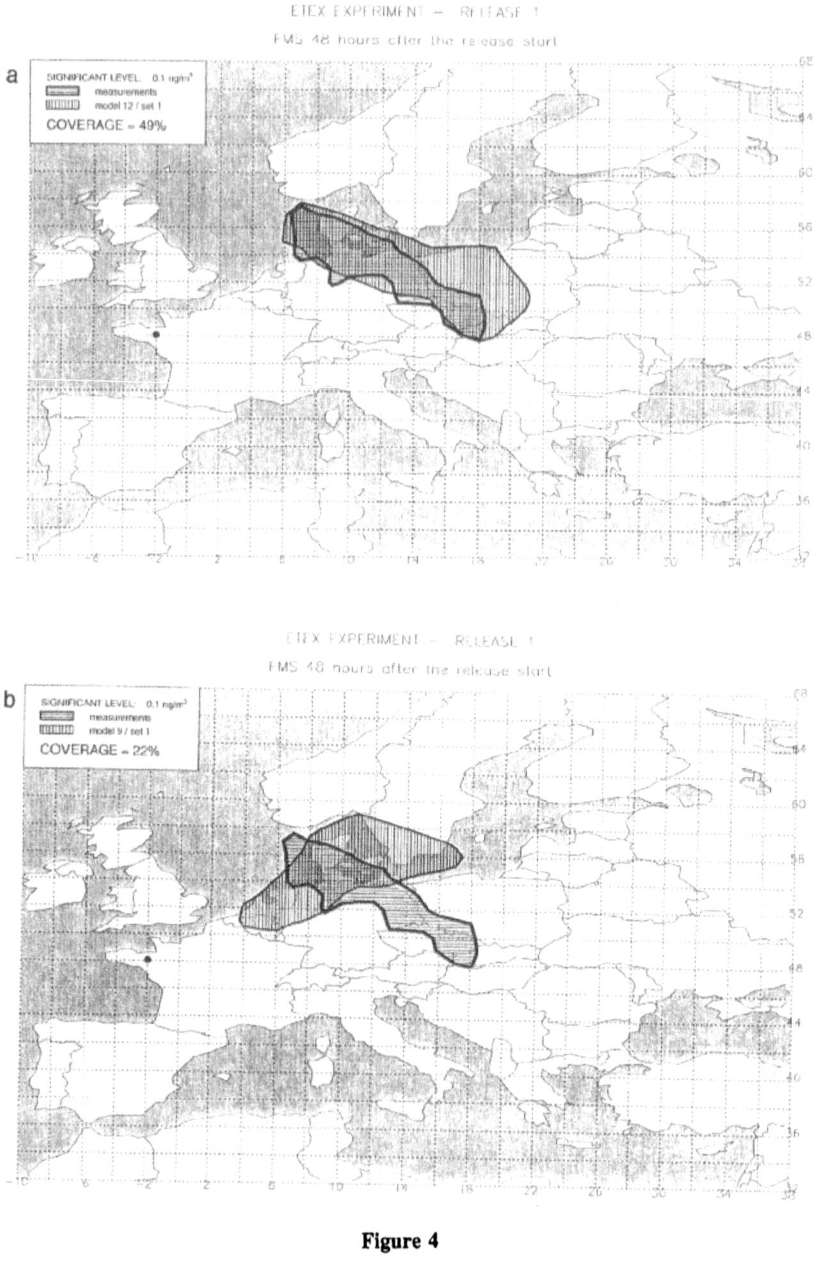
ETEX EXPERIMENT – RELEASE 1
FMS 48 hours after the release start
a
SIGNIFICANT LEVEL: 0.1 ng/m³
measurements
model 12 / set 1
COVERAGE = 49%
ETEX EXPERIMENT – RELEASE 1
FMS 48 hours after the release start
b
SIGNIFICANT LEVEL: 0.1 ng/m³
measurements
model 9 / set 1
COVERAGE = 22%

Figure 4

4a, 4b show the concentration time evolution (observed and calculated by one participant) in two of these locations for the second arc, when the cloud is larger and therefore the peak is better defined.

The performances of the selected models are listed in Table 2 using FMT, since this indicator summarises the discrepancies in time-of-arrival, duration and peak concentration.

Table 2. FMT of eleven good models at eleven sampling stations.

Mod/ Loc	B5	F15	NL1	NL5	CR1	CR3	D19	D27	D8	DK5	H2
4	0	44	21	0	66	18	0	18	8	0	0
6	18	25	27	45	22	68	31	30	28	3	35
9	3	13	9	12	0	4	10	13	25	50	0
10	6	4	14	25	0	0	17	39	26	45	0
11	7	20	7	38	58	10	7	10	12	24	58
12	6	35	11	56	56	61	26	20	28	51	47
15	4	39	13	0	61	35	28	35	1	0	15
18	11	52	19	0	46	26	13	14	13	33	39
19	11	15	21	30	32	38	15	10	25	79	18
23	6	18	11	8	21	15	14	28	28	2	27
25	0	6	43	0	28	10	4	20	14	38	25

6 CONCLUSIONS

Comparison of model results is still under way. However, it can already be concluded that atmospheric long range models have improved considerably since the Chernobyl accident. A consistent group of participants were able to reproduce the cloud position and concentration levels at different time intervals in a correct way. Nonetheless, it should be remembered that the synoptic situation was rather simple and could be more complex in the case of a real nuclear accident. Further, large discrepancies have been found in the concentration time evolution at single sites. In particular, time-of-arrival, tracer cloud duration above a given threshold and peak value of the concentration were often not well described, even by the selected "top" eleven models. This indicates that research should continue to aim at the improvement of model performance, in order to reduce these kind of uncertainties.

REFERENCES

Klug, W., Graziani, G., Grippa, G., Pierce, D. & C. Tassone, 1992, *Evaluation of Long Range Atmospheric Models using Environmental Radioactivity Data from the Chernobyl Accident: the ATMES Report*, Elsevier Science Publishers, Barking, England.

Hanna, S.R., 1988, Air Quality Model Evaluation and Uncertainty, *JAPCA*, **38**, 406-412.

DISCUSSION

T. IVERSEN: What are the levels of uncertainty of the measured tracer concentrations as the time increases (compared to the discrepancies between models)?

G. GRAZIANI (reply): Final evaluation of the observed concentrations is still under way. At the end of this process we hope to get an overall uncertainty in the measurements in the order of 25 - 30%.

LONG RANGE TRANSPORT OF NO_x, SO_x AND O_3 OVER EAST ASIA AND THE NORTHERN PACIFIC OCEAN CAUSED BY TYPHOONS

Toshihiro Kitada[1], Seiji Isogawa[1], and Yutaka Kondo[2]

[1]Dept. of Ecological Engineering, Toyohashi University of Technology
Toyohashi 441, Japan
[2]Solar-Terrestrial Environment Laboratory, Nagoya University
Toyokawa 442, Japan

INTRODUCTION

The PEM-WEST (A), Pacific Exploratory Mission West experiment was conducted during September and October 1991. The experiment was a part of a large region data acquisition campaign launched to investigate the characteristics of concentration-fields of various trace chemical species in the troposphere over the northern Pacific called the NASA Global Troposphere Experiment. In this paper, reported are a 3-D Eulerian model for transport/chemistry/deposition of trace chemical species in semi-global scale, and the model simulation results for the PEM-WEST (A) campaign, particularly for Mission 8 and 9, which were carried out using Yokota as their base on 25 and 27 September 1991, respectively, when the Typhoon Mireille (Typhoon 9119) passed over Japan area. The model reproduced quite well measured concentrations of O_3, HNO_3, PAN and H_2O_2, but not those of NO, NO_2, C_2H_4 etc. The measurement was made for the whole depth of the troposphere from boundary layer to lower stratosphere. In the paper, the model performance will be evaluated, and features of O_3 transport by typhoon will also be discussed.

TRANSPORT/TRANSFORMATION/DEPOSITION MODEL

The model is a σ-coordinate global model consisting of a set of equations that account for unsteady state, advection, diffusion, chemical reactions and dry/wet depositions. This model is then applied to the northern Pacific. To solve the equations, a solution procedure which is a combination of time-splitting technique and finite element method has been adopted. For example, the original governing equation for a chemical species is first split into three one-dimensional transport equations for ϕ, θ, and σ directions, and one chemical reaction equation. Then they are solved in turn to advance the simulation by one time step, where the

Air Pollution Modeling and Its Application XI, Edited by
Gryning and Schiermeier, Plenum Press, New York, 1996

finite element method for the transport equation and a semi-analytical method for the chemical reaction rate equation are used[1]. The governing equations, in which the variable transformation into the σ -coordinate is applied with some simplification on the diffusion terms, are as follows.

$$C\frac{\partial X_i}{\partial t} + CU\frac{\partial X_i}{\partial x} + CV\frac{\partial X_i}{\partial y} + C\dot{\sigma}\frac{\partial X_i}{\partial \sigma} = \frac{\partial}{\partial x}(CE_\phi\frac{\partial X_i}{\partial x})$$

$$+ \frac{1}{\cos\theta}\frac{\partial}{\partial y}(C\cos\theta E_\theta\frac{\partial X_i}{\partial y}) + \frac{\rho g^2}{\pi^2 r^2}\frac{\partial}{\partial \sigma}(C\rho r^2 E_\sigma\frac{\partial X_i}{\partial \sigma}) + R_i \quad (1)$$

$$\dot{\sigma} = \{ (\frac{\partial z}{\partial t})_\sigma + \vec{V}\cdot\nabla_\sigma z - W \}\frac{\rho g}{\pi} \quad (2)$$

where $dx = r\cos\theta\, d\phi$, $dy = r\, d\theta$, $\sigma = (P - P_T)/\Pi$, $\Pi = P_S - P_T$, Xi is the non-dimensional concentration of the i th species; C is the air density in kmol m^{-3}; ρ is the air density in kg m^{-3}; θ is the latitude; ϕ is the longitude; r is the distance from the center of the Earth; P_S is the atmospheric pressure at ground level; P_T is the atmospheric pressure at the top boundary which is set at 55 hPa; z is the altitude of the σ-surface; U and V are the horizontal flow velocities; W is the vertical flow velocity; g is the gravitational acceleration; E_ϕ and E_θ are horizontal eddy diffusivities in ϕ and θ directions, respectively; and E_σ is the vertical eddy diffusivity and R_i is the chemical reaction rate for the i th species. For better description of the planetary boundary layer activity, outputs with a k-ϵ turbulence model[2] are adopted for E_σ. The model has 22 advected species of NO, NO_2, HNO_3, PAN, $RONO_2$, HNO_2, CO, SO_2, SO_4^{2-}, O_3, H_2O_2, CH_4, C_2H_6, C_3H_8, C_2H_4, ALKENEs, HCHO, ALDEHYDEs, HCOOH, ACETIC ACID, ROOH and Lumped KETONEs, and 15 steady-state assumption applied species such as OH, HO_2, NO_3 and N_2O_5,. These species are tied by a system of 86 chemical reactions. The system is an adaptation from that by Lurmann et al. [3] Dry deposition was treated using conventional deposition-velocity concept for species such as SO_2, SO_4^{2-}, NO_2, HNO_3, H_2O_2, O_3, PAN, HCHO and ALDHYDEs. For example, typical velocity values for SO_2 were 0.5 cm s^{-1} for land surface and 0.1 cm s^{-1} for sea surface, while those for SO_4^{2-} were 0.1 cm s^{-1} for both land and sea surfaces. For wet removal processes, semi-empirical equations were used for SO_2, SO_4^{2-}, HNO_3, H_2O_2, O_3, the equations were derived on the basis of capture of aerosol particle by cloud and rain drops and gas absorption into those drops[4]. Meteorological parameters used in the wet deposition equations were heights of cloud-top and -base, and precipitation intensity.

FLOW FIELD, EMISSION SOURCES AND SIMULATION SETTING

Input meteorological data for simulations, such as air flow, temperature, humidity and pressure, were outputs of the global analysis model by the Japanese Meteorological Agency. The resolutions of the data were 1.875 degrees in both longitudinal and latitudinal directions, and 12 hours with respect to time. In addition, fields of precipitation rate, cloud-top and -base heights were available every three hours for a limited area over Japan and its vicinity. These precipitation-related data were the outputs of the JSM, i.e. the Japan Spectral Model, of JMA.

Emission sources were assumed for NO, NO_2, SO_2, CH_4, other volatile organic compounds (C_2H_6, C_3H_8, C_2H_4, and other alkenes) and CO. The source strength of NO_x, SO_2 and CH_4 was determined using independent information, while that of VOCs and CO was estimated so as to be proportional to the NO_x source. For example, SO_2 annual emission in east Asian countries in 1986 are as follows: 18×10^6 ton from China, the Korean Peninsula, and Taiwan; 11×10^5 ton from Japan; and 7×10^5 ton from volcanic sources in Japan.

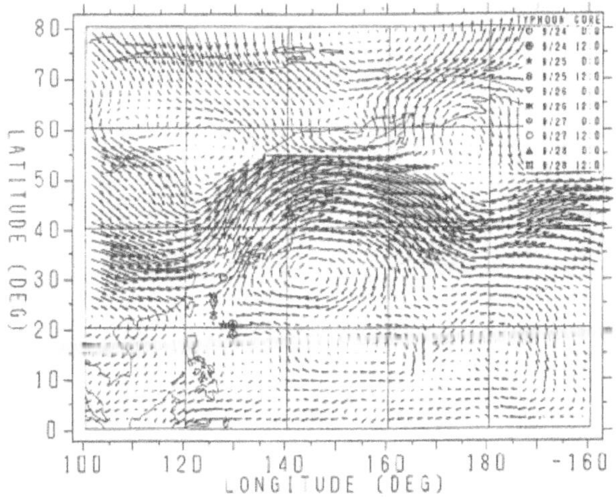

Figure 1. Calculation domain and flow field at 10 km high at 06GMT 27 Sept. The location of the Typhoon Mirelle is plotted every 12 hours from 00GMT 24 Sept. to 00GMT 28 Sept.

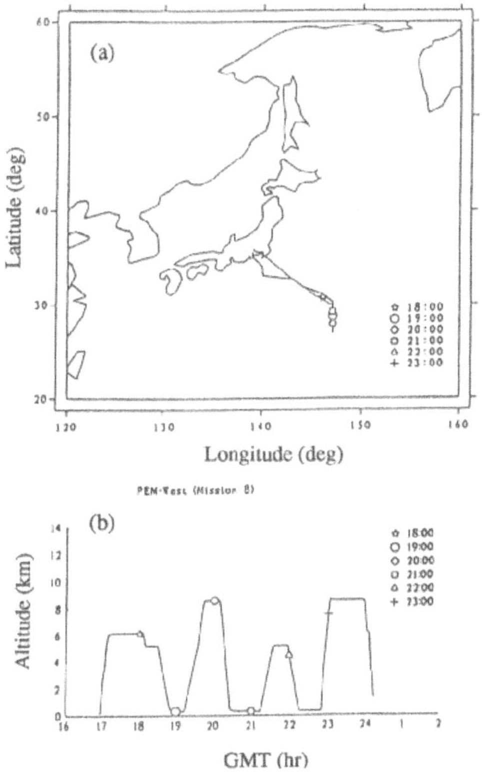

Figure 2. Flight route of Mission 8 on 25 Sept. 1991: (a) horizontal route, and (b) flight's height.

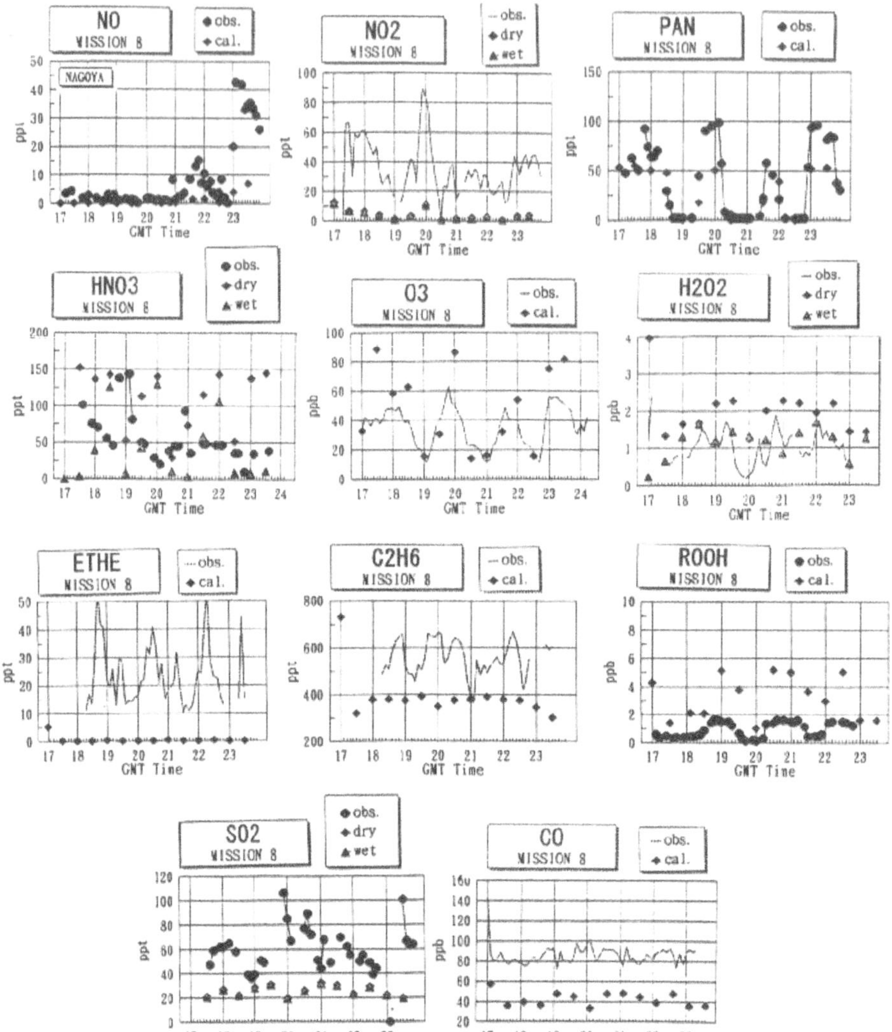

Figure 3. Comparison between calculated and observed concentrations for Mission 8.

The simulation domain was 99.375 °E to 159.375 °W in longitude and 0°N to 80.625°N in latitude, with grid size of 1.875 degree for both directions. In the vertical, 16 variable grids were used. Time step was 30 minutes. Initial concentration fields were estimated using both climatological data and the data from the PEM-WEST(A) campaign itself. Under these conditions, two simulations with- and without-wet deposition processes have been carried out for the period from 00GMT 20 September to 00GMT 28 September. Figure 1 shows the calculation domain, and, as an example, the input flow field at 10 km high at 06GMT on 27 Sept. In the figure, the locations of the Typhoon Mireille are also plotted every 12 hours during 00GMT 24 Sept. to 00GMT 28 Sept.

COMPARISON OF CALCULATED AND OBSERVED CONCENTRATIONS

Mission 8

The model were tested against the observation of Mission 8 on 25 Sept., 1991, the flight route for which is shown in Figure 2: (a) the course mapped in horizontal plane, and (b) the flight height. Since the typhoon was far from the Mission 8 flight route, this observation probably represents undisturbed troposphere over the west Pacific ocean near Japan. In Figure 3, calculated concentrations are compared with those measured during the flight. Figure 3 indicates that (1)the model could reproduce quite well observed variations of O_3, PAN, and H_2O_2, (2)for H_2O_2, the simulation case in which wet deposition process was included shows better agreement with observation, (3)the calculated CO and SO_2 are somewhat lower than those observed, (4)the model predicted much lower NO, NO_2, and C_2H_4 (ETHE) concentrations than observation. With these comparisons we judge the model performance to be acceptable. However, the comparison also suggests necessity of additional NO_x sources such as lightning and aircraft emission in the middle and upper troposphere, probably together with larger surface source of C_2H_4.

Mission 9

Figure 4 shows flight route for Mission 9 on 27 Sept., 1991. That mission is characterized by the flight over the typhoon's eye area between 5 and 6 GMT near Nagasaki, Japan. Figure 5 presents the model calculated and observed concentrations along the route. These figures generally indicate that difference between calculated and observed concentrations has tendency similar to that in Figure 3. Particularly, higher concentrations in some species such as C_2H_6 and CO, and lower concentrations in O_3 and PAN found in observation over "eye" area at 5 and 6 GMT are not so clear in calculation. This may be due to lack of sub-grid scale convective motion in this model and also coarse time and spatial resolutions of the meteorological input data. These coarse resolutions are not probably appropriate for the analysis of the phenomena, such as mass-tranport over typhoon eye area, where sharp gradients usually exist in various parameters. The symbol (X) in figures of O_3, HNO_3, and PAN shows these species concentrations re-evaluated by taking slight delay in simulation of the typhoon eye's passing time near the flight route into consideration. The delay was caused by the coarse temporal resolution of 12 hours in original flow data, although input flow fields to the simulation were provided every 30 minutes with time interpolation.

CALCULATED O_3 FIELDS

Figure 6 shows simulated O_3 concentration fields at heights of (a) 1 and (b) 10 km at 21GMT on 25 Sept., 1991. At this stage, a big typhoon (i.e., tropical cyclone in east Asia), called Mireille in the USA and T9119 in Japan, was located at about 126°E and 24°N. Figure

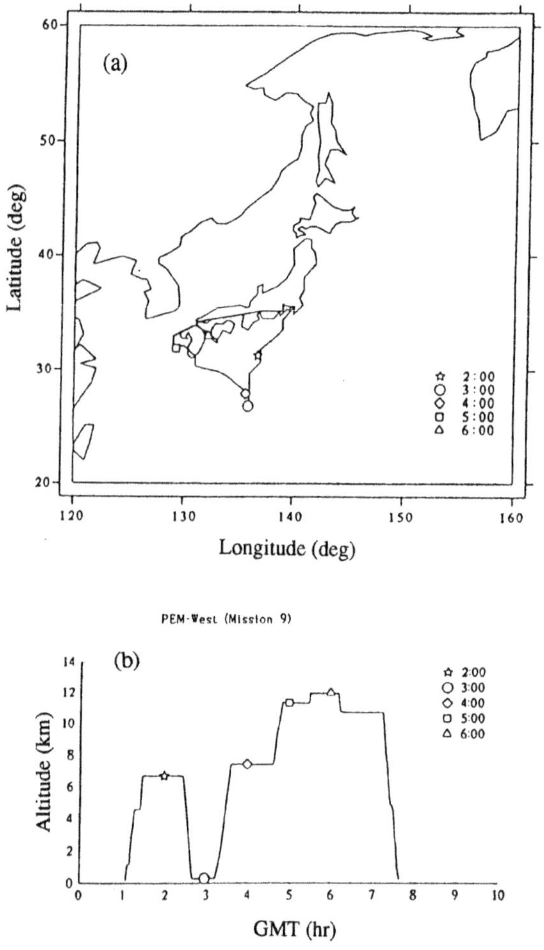

Figure 4. Flight route of Mission 9 on 27 Sept. 1991: (a) horizontal route, and (b) flight's height.

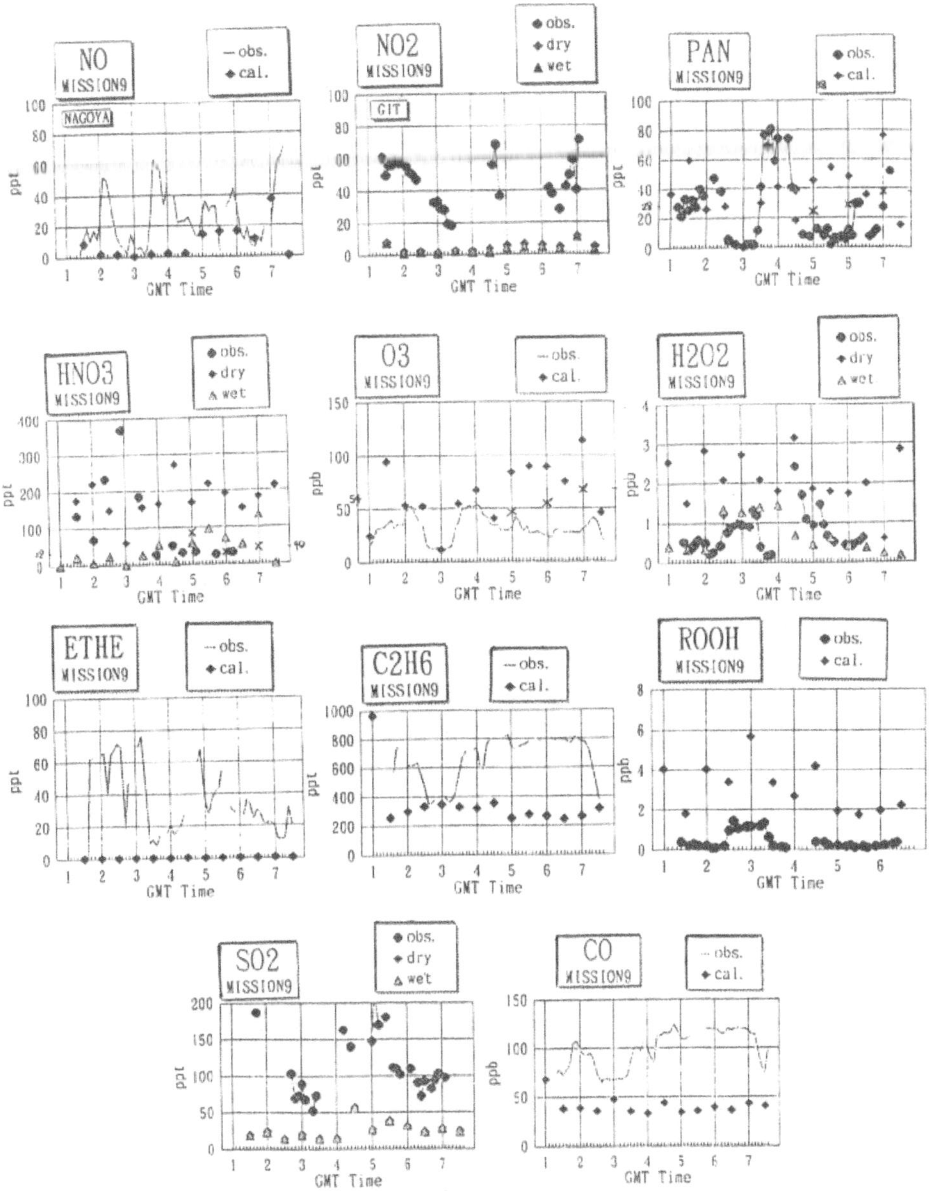

Figure 5. Same as in Fig. 3, but for Mission 9.

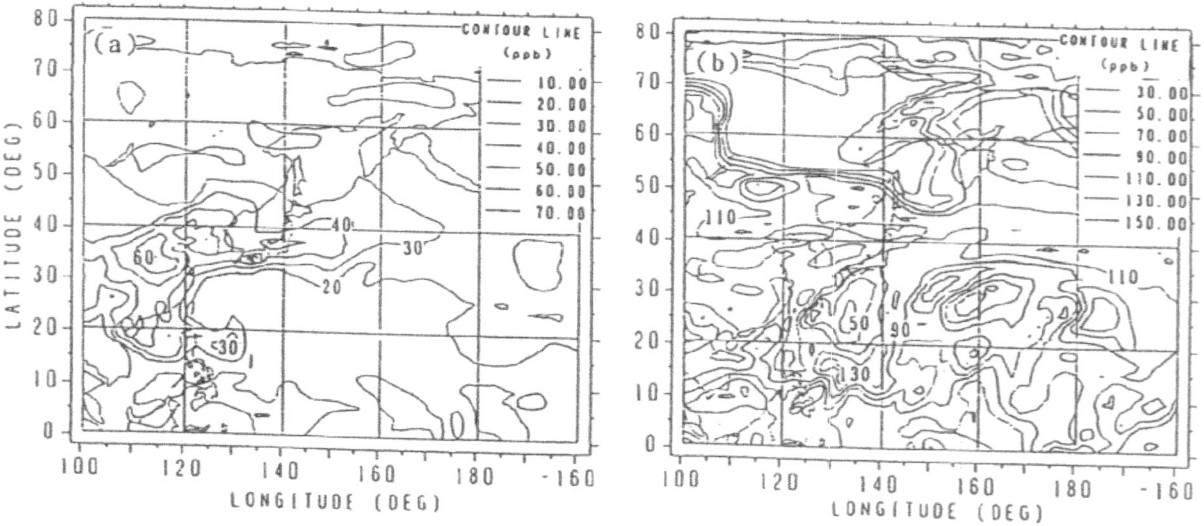

Figure 6. Calculated O_3 concentration field at heights of (a) 1, and (b) 10 km at 21GMT on 25 Sept. 1991 for Mission 8.

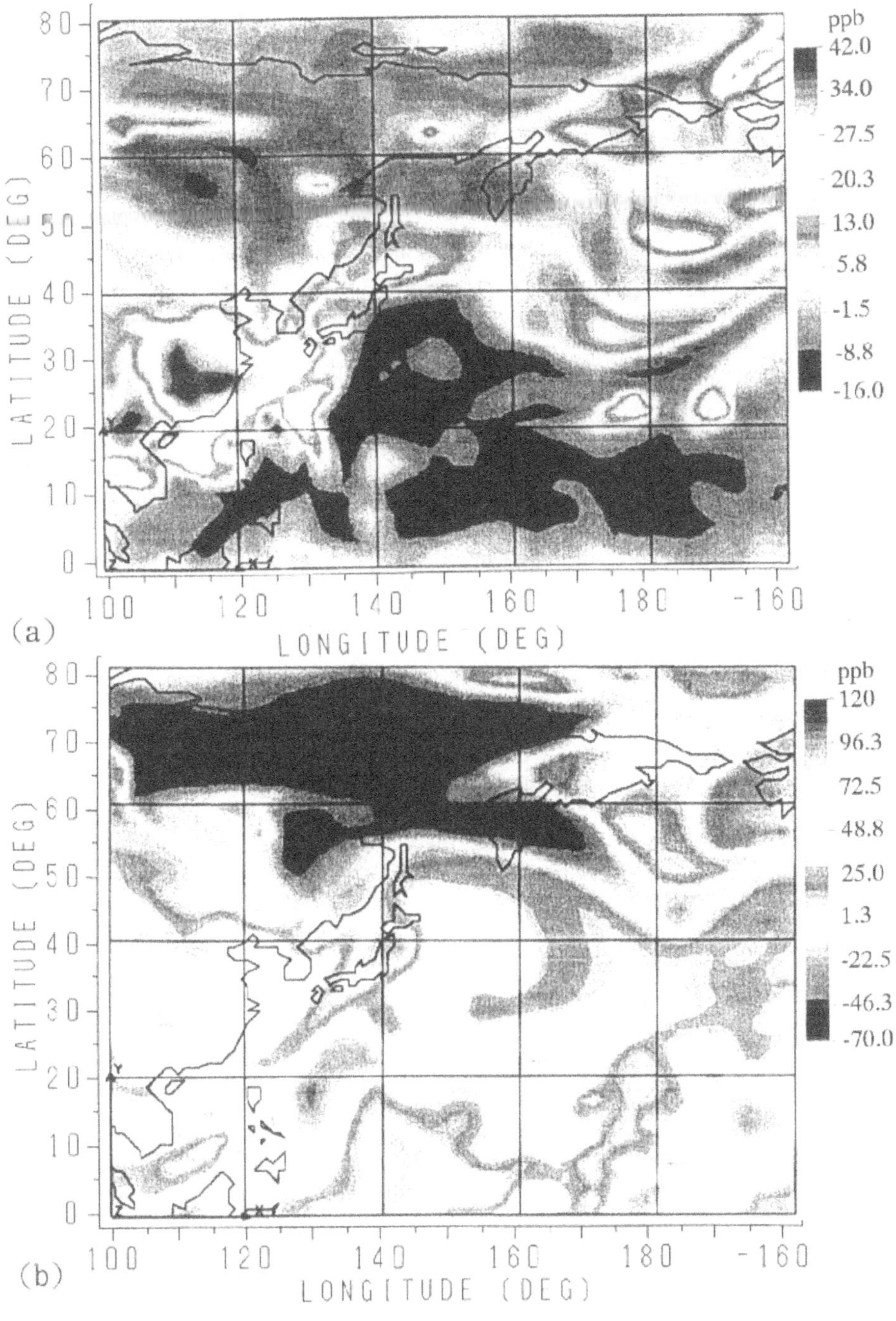

Figure 7. Net gain and loss of O_3 in ppb after 8 days simulation, from 00GMT 20 Sept. to 00GMT 28 Sept. 1991, at heights of (a) 1.7, and (b) 10 km.

6(a) demonstrates how the strong vortical motion associated with the typhoon affects horizontal mass transport in upper boundary layer. The typhoon brings O_3-poor tropical marine air to the coast of east Asia between 20°N and 30°N. At the same time it brings continental polluted air mass over the south Pacific ocean south of 20°N. The high O_3 spot found at 130°E and 17°N may be due to stratospheric O_3, which was brought down by the downward motion associated with the typhoon, although we do not have observational evidence for this. Another remark to be made on Figure 6(a) is, since typical natural O_3 concentration at 1 km high in mid-latitude is around 35 ppb, high O_3 over 60 ppb covering northern China, the Korean peninsula and Japan is obviously due to photochemical production from anthropogenic precursors of NO_x and VOCs.

High O_3 concentrations found in Figure 6(b) should be stratospheric origin. Band of high O_3 along 40°N line well corresponds to the zone of strong westerly of the jet stream. Effects of the typhoon Mireille at this 10 km high can be found at three locations. Namely, two spots of low O_3 at 125°E & 132°E and 25°N are caused by upward air mass movement over near typhoon center and the place east to the center. In contrast to these spots, high O_3 found at 135°E and 15°N is due to downward air motion on the south of the typhoon center, and O_3 transport by this downward motion resulted in high O_3 at altitude of 1km, which was described above and shown in Figure 6(a).

Figure 7 shows net gain and loss of O_3 in ppb after 8 days simulation, from 00GMT20 Sept to 00GMT28 Sept, at heights of (a) 1.7 and (b) 10 km. At lower troposphere, as shown in Fig. 7(a), photochemical O_3 production over China is remarkable. Its highest value was 43 ppb for 8 days. At this 1.7 km high, O_3 increase is also found around 130°E and 25°N, but this time mainly due to intrusion from stratosphere. Its strength was about 27 ppb for 8 days. This downward O_3 flux was caused by the subsidence associated with the typhoon. The subsidence was much stronger at 10 km high, and has resulted in the strong O_3 gain of 100 ppb for 8days at 130°E and 20°N, as shown in Figure 7(b).

SUMMARY

The global scale transport/chemistry/deposition model for trace chemical species reported here, quite well simulated concentrations of key chemical species which were collected during the PEM-WEST(A) campaign (see Figures 3 and 5), though the model did not explicitly include sub-grid scale process of deep convective motion. Characteristics and importance of O_3 transport in horizontal and vertical directions by the typhoon were indicated in Figures 6 and 7. For further improvement of the present model, inclusion of NOx sources such as lightning and aircraft emission, better description for VOC sources, and sub-grid scale parameterization for convective transport are definitely necessary. These procedures are now under way.

REFERENCES

1. For example, G.R. Carmichael, L.K. Peters, and T. Kitada, A second generation model for regional scale transport/deposition/chemistry, *Atmos. Environ.* 20:173(1986).
2. T. Kitada, Turbulence structure of sea breeze front and its implication in air pollution transport -Application of k-ϵ turbulence model, *Bound.-Layer Meteor.* 41: 217(1987).
3. F.W. Lurmann, A.C. Lloyd, and R. Atkinson, A chemical mechanism for use in long-range transport/acid deposition computer modeling, *J. Geophys. Res.* 91:10905(1986).
4. T. Kitada, Transport, transformation and deposition model for acidic species, *Meteorological Research Notes*, Meteor. Soc. of Japan, No.182:95(1994) (in Japanese).

DISCUSSION

G. CARMICHAEL: Lightning can be an important upper troposphere source of NO_x. Did you include it in the analysis?

T. KITADA: In this simulation, the NO_x source by lightning was not included. I feel this was one of the most serious deficits of the simulation as well as lack of NO_x source by aircraft.

DEPOSITION OF PERSISTENT ORGANIC POLLUTANTS OVER EUROPE

W.A.J. van Pul[1], J.A. van Jaarsveld[1], and C.M.J. Jacobs[2]

[1]National Institute of Public Health and the Environment (RIVM), Bilthoven, The Netherlands
[2]Agricultural University Wageningen, Department of Meteorology, Wageningen, The Netherlands

INTRODUCTION

Persistent Organic Pollutants (POPs) are compounds that are resistant to photolytic, biological, and chemical degradation. Due to this persistence these pollutants accumulate in the environment and food chains of humans and animals, where they can induce toxic effects (UN-ECE, 1994). POPs originate to a large extent from anthropogenic activities. Examples of POPs are: pesticides (DDT, hexachlorocyclohexane HCH), and industrial (by-) products (PCB, PAH, Dioxins). Many POPs are semi-volatile at atmospheric conditions. Because of these characteristics POPs have an atmospheric lifetime of several weeks or more and are subject to long-range atmospheric transport. During this transport POPs are deposited but for the more volatile POPs a re-emission from soil and water surfaces might occur. For this reason the description of the deposition process generally used for components only depositing, such as most acidifying components and heavy metals, cannot be applied. Hence, the deposition should be considered as the sum of the deposition and re-emission fluxes, or in other words, as a net-deposition flux. In terms of effects, this net deposition of POP in the environment is a more relevant quantity.

The availability of measurements of POP is not yet sufficient to assess the environmental impact of POP from measurements alone. An emission-based modeling approach is followed because of this lack of measurements and to gain insight in the cause-effect chain and to identify source categories and areas. The main goal of this approach is to establish whether it is possible to simulate measured concentrations in air and rain with the existing emission data and knowledge on the exchange processes of POP at the earth's surface.

This study presents the transport and deposition of γ-HCH (lindane) simulated with the EUROS model. The net-deposition flux or accumulation of lindane on a European scale was calculated using European databases on emissions and soil properties. The model was extended with a description of the exchange process of POPs at the soil and water surfaces. The calculations are preliminary, and the results should therefore be considered as indicative only. Lindane was chosen as an example because of the emission data available for this pesticide. Under atmospheric conditions lindane is mainly present in the gas phase (van Jaarsveld et al., 1994). All POPs present in the gas phase can be modeled using this approach.

GENERAL MODEL DESCRIPTION

Transport Model

EUROS is an eulerian air pollution transport model with which the dispersion and transport of a component in the lower troposphere can be calculated (van Rheineck Leyssius et al., 1990).

The troposphere in EUROS is represented by means of three layers: the atmospheric boundary layer (ABL), a residual layer (RL) and an upper layer (UL). The ABL height is parameterized to have a diurnal variation. The total height of the three layers is set to a fixed value of 3000 m. The concentration in the upper two layers is assumed to be mixed homogeneously with height. In the upper 90% of the ABL the concentration is also constant with height. In the lower 10% of the ABL, the concentration profile is modified to allow the influence of dry deposition or (re-)emission. The lindane emissions are considered as area sources at the surface.

Advection in each layer is solved numerically with the second-moment method (Egan and Mahoney, 1971) with a time step of 30 min. Advection in the ABL is described with the 1000 hPa wind field, and in the RL and UL with the 850 hPa wind field. These wind fields, as well as observations of temperature, precipitation and cloud cover, were obtained from ECMWF. The modeled area covers a great part of Europe, with a 0.5° x 0.5° longitude-latitude resolution on shifted pole coordinates, i.e., approximately 60 x 60 km (see also Figure 2).

Exchange with Soil and Water

The gaseous exchange of POP at soil and sea surfaces is parameterized using a resistance analogy in the atmosphere. The flux of a POP, F_d, at the surface is driven by the difference between the POP concentration at a reference level, C_{zr} and the concentration in the gas phase in the top layer of the soil or just above the sea, $C_{soil,sea}$, and is controlled by a set of resistances assumed to operate in series:

$$F_d = \frac{C_{zr} - C_{soil,sea}}{r_a + r_b + r_s} \qquad [1]$$

Here, r_a is the aerodynamic resistance between the reference level and the aerodynamic roughness length, z_0, r_b the boundary layer resistance that applies between z_0 and the average sink or source height of the substance, z_{0c}, and r_s, the surface resistance. The aerodynamic and quasi-laminar boundary layer resistances (r_a and r_b) are calculated in the usual way (e.g., Hicks et al., 1987).

In the case of soil, a formulation for r_s and C_{soil} can be derived that is consistent with both [1] and the theory of transport and storage of POPs in the soil. In the description of the transport and storage of POPs in the soil we followed Jury et al. (1983; 1991). It is assumed that concentrations of POP present in the solid, liquid and gas phases in the soil are in equilibrium. The total concentration, C_T, is partitioned over the solid phase (C_s), the liquid phase (C_l) and the gas phase (C_g) as follows:

$$C_T = \rho_s C_s + \theta C_l + a C_g \qquad [2]$$

where ρ_s is the density of the solid material per cubic meter of soil volume, θ the volumetric water content of the soil and *a* the volumetric air content of the soil. The pollutant is assumed to obey Henry's law, so that:

$$C_g = K_H C_l \qquad [3]$$

where K_H is the dimensionless Henry coefficient. This coefficient depends on temperature; here, we use the parameterization as suggested in GESAMP (1989). The ratio of C_s to C_l is assumed to be given by:

$$C_s = f_{oc} K_{oc} C_l \qquad [4]$$

where K_{oc} is the organic carbon distribution coefficient (for lindane: 1.3 $m^3\ g^{-1}$) and f_{oc} the fraction of organic carbon in soil. Using [2]-[4], partition coefficients for the liquid phase, the adsorbed phase and the gas phase (R_l, R_s and R_g, respectively) can be defined such that

$$C_T = R_l C_l = R_s C_s = R_g C_g \qquad [5]$$

The flux of POP in the soil is given by:

$$J_T = - D_E \frac{\partial C_T}{\partial z} \qquad [6]$$

where D_E is the so-called effective gas-liquid diffusion coefficient:

$$D_E = \frac{\xi_g D_g}{R_g} + \frac{\xi_l D_l}{R_l} \qquad [7]$$

Here, D_g and D_l are the molecular diffusion coefficients of the POP in gas and water, respectively. Furthermore, ξ_g and ξ_l are the so-called gas and liquid tortuosity factors, by which the effect of the reduced flow area and the longer diffusion path in the soil is accounted for.

The gaseous exchange is assumed to occur between the atmosphere and a thin top layer, having a thickness of Δz_1 and a mid-layer concentration $C_{T,1}$. Furthermore, we assume a constant flux between the reference level and mid-way the top layer. Then, it can be shown that r_s and C_{soil} are given by:

$$r_s = \frac{1}{R_g \left(\frac{2D_E}{\Delta z_1} \right)} , \quad C_{soil} = \frac{C_{T,1}}{R_g} \qquad [8]$$

In the case of lindane, r_s may vary between about 500 s m^{-1} for cool, wet soils and 10,000 s m^{-1} or more in the case of warm soils with moderate water content. For completely dry soils, $r_s \approx 1300$ s m^{-1}. The value of r_s is independent of f_{oc}, whilst C_{soil} decreases strongly with increasing f_{oc} due to the increased adsorption of material to the solid phase [4]. Note that the latter effect can never be captured in the classical approach, where $C_{soil} \equiv 0$. Soil organic matter data in the topsoil over Europe, given by Fraters et al.(1993) were used to model this C_{soil}. In large parts of Europe the typical percentage of organic matter is 2%, in Scandinavia and Russia this can be up to 10%.

For water surfaces a similar approach can be applied to define r_s and C_{sea}. Here, we followed the theory presented by Liss and Slater (1974). Surface waters are assumed to be well-mixed, having a concentration, C_w. For the depth of these surface waters, a typical value of 25 m is used. At the top of the water surface, a thin water layer directly adjacent to the atmosphere is present where transport is dominated by molecular diffusion. The r_s and C_{sea} are defined by:

$$r_s = K_H r_w \quad ; \qquad C_{sea} = K_H C_w \qquad [9]$$

where r_w is the resistance to POP transport in this thin water layer given by:

$$r_w = \frac{\Delta Z_{top}}{D_l} \qquad [10]$$

Here, ΔZ_{top} is the depth of the molecular layer. Liss and Slater (1974) found $\Delta Z_{top} \approx 4\text{x}10^{-5}$ m for oxygen, which we assume to be valid for POPs as well. For lindane, the r_s for water is small as compared to the other resistances (< 10 s m^{-1}). This means that the exchange of lindane at the water surface is primarily controlled by atmospheric transport.

Degradation

POPs are defined as components that are rather persistent to degradation, i.e., the degradation time scale is much larger than the transport time scale. Nevertheless, they are subject to some photolytic, chemical, and biological degradation in the environment. For long-term scenario runs and mass balance studies of these components, the degradation must be taken into account. Degradation of POPs is described as a first-order decay. For lindane, the effective degradation constant, k_d, is taken as $3\text{x}10^{-8}$ s^{-1} in soil (Jury et al., 1983; Strand and Hov, 1994) and $2.5\text{x}10^{-7}$ s^{-1} in air (Warmenhoven et al., 1989). The degradation in water bodies is much slower (dozens of years) and is ignored here.

Implementation of the exchange module in EUROS

The theory described above was implemented in a numerical model, and coupled to EUROS. In the soil, five layers are defined with depths of 0.5, 0.5, 1, 2, and 11 cm. Small layers at the top of the soil are defined to be able to simulate the saturation of POP in the soil. As a first approximation, the soil is assumed to be homogeneous, so that each layer is equal with respect to soil characteristics, temperature (taken to be the 2-m air temperature), and soil moisture content (θ = 0.3 in the simulations presented here). Concentrations are defined mid-way each layer and at fluxes at the interfaces between layers. As the boundary conditions at the top of the soil model domain, the sum of the average dry deposition or re-emission flux (F_d) and the wet deposition flux (see next section) are taken. It is assumed that no pollutant escapes from the model domain, i.e., $J_T(15cm)=0$. A simple explicit numerical scheme proved to be sufficiently accurate. The module is operated with a time step of six hours.

Wet Deposition

Wet deposition is computed by assuming equilibrium partitioning between the gas-phase concentration in the atmosphere, $C_{g,a}$ and the water-phase concentration, $C_{w,a}$ so that

$$C_{g,a} = K_H C_{w,a} \qquad [11]$$

Now, the concentration change due to wet deposition in an air layer of depth H_a is assumed to be given by:

$$\frac{\partial C}{\partial t} = MAX\left[\frac{R_i W_{sc}(C - C_u)}{H_a}, 0\right] \qquad [12]$$

where R_i denotes rain intensity in m s^{-1}, W_{sc} the scavenging ratio defined by $C_{w,a}/C_{g,a}$ and taken equal to $1/K_H$, and C_u, the concentration in a layer just above the layer being considered. For each time step, the analytical solution to [12] is applied. Note that the diffusion towards the water phase is modeled to stop if the water is saturated with pollutant. Finally, the wet deposition flux, F_w, is given by:

$$F_w = R_i C_{w,a} \qquad [13]$$

Emissions of Lindane

Emissions of lindane into air over Europe were taken from an inventory by TNO, based on application data for 1990 as a part of the so-called ESQUAD project. These data cover the model domain on a 1° x 0.5° longitude-latitude resolution (Berdowski and Veldt, 1994). The quantity of lindane applied to soil and vegetation is considered as a source to the topsoil (0.5 cm) and is twice the emission to air. The emissions, only present in agricultural areas, are considered as area sources. Information on the agricultural area per grid cell was obtained from a European land-cover database (van de Velde et al., 1994).

The uncertainty in the emission values is large and the accuracy is estimated to be a factor 2-5. Lindane is used as a pesticide on crops as well as in soil. It is assumed that vegetation does not accumulate lindane to a large extent. POPs taken up by vegetation are expelled later and will finally reach the soil surface (van Jaarsveld et al., 1994). The application of lindane is concentrated in spring and early summer (i.e. 10, 15, 25, 25, 25% of lindane is emitted in February through June).

Background

The influence of emissions outside the model area are accounted for by adding an influx of air containing a background concentration at the boundaries of the model area. The background concentration of lindane in air was taken to be 0.015 ng m^{-3} (van Jaarsveld et al., 1994).

RESULTS AND DISCUSSION

The transport and deposition of lindane were calculated with EUROS for a simulation period of 10 years. Due to a lack of data, these simulations were made by running the model ten times with the meteorological conditions of 1989, taking the last values for lindane concentrations in all compartments of the previous year as input for the following year run.

Results shown below are for non-agricultural areas. From the model's calculations it was concluded that agricultural areas where lindane was applied were emitting lindane throughout the year, even in the period when lindane was not applied. The results presented are preliminary and should therefore be considered as indicative only.

Above land the atmospheric concentration clearly increases to a maximum in June (Figure 1 a). When the emissions stop (July), a strong decrease occurs because of the dry and wet deposition of lindane and the transport out of the model area. Above the sea, lindane concentrations are much lower due to a larger distance from the emission areas and the larger dry deposition. Wet deposition is also much less, since it is highly dependent on the air concentration.

Dry deposition above the sea shows a yearly course similar to the air concentrations (Figure 1 b). However, after year 5 the accumulation of lindane has reached a certain level at which re-emission takes place when the air concentration drops in fall and winter. Recently, Bidleman et al. (1995) showed that lindane is re-emitted from some northern seas due to a drop in the air concentrations. The results shown here will be somewhat exaggerated because water current is not taken into account. Water advects and mixes lindane horizontally, leading to lower water concentrations and a reduction in the re-emissions.

At land surfaces, lindane is generally dry deposited during the emission period caused by the high air concentrations (Figure 1 c). When the emissions stop and the air concentrations of lindane drop, lindane is re-volatilized from the soil. The concentration of lindane in the topsoil is high due to the large input of lindane by wet deposition (particularly in the months 4 and 6). During successive years, re-emission of lindane starts earlier every year because the concentration in the topsoil is slowly increasing. After year 5 the dry deposition above land is more-or-less stationary. At this point the dry and wet deposition and degradation of lindane are more-or-less in equilibrium. Due to the re-volatilization of lindane the concentration in air increases above land and sea, and also reaches a constant value after a few years.

Generally, for non-agricultural areas wet deposition is the most important input pathway of lindane; for the sea this is the dry deposition. However, for land surfaces, a considerable part of the lindane deposited will re-volatilize, leading to a much lower net deposition than calculated from the wet and dry deposition alone. This means that wet deposition data, as such, cannot be used to estimate the net input to the surface. A deposition flux balance made for several grid cells showed that the re-emission of lindane can be 50% of the wet deposition. This percentage depends on the distance from emission areas, surface properties and meteorological conditions.

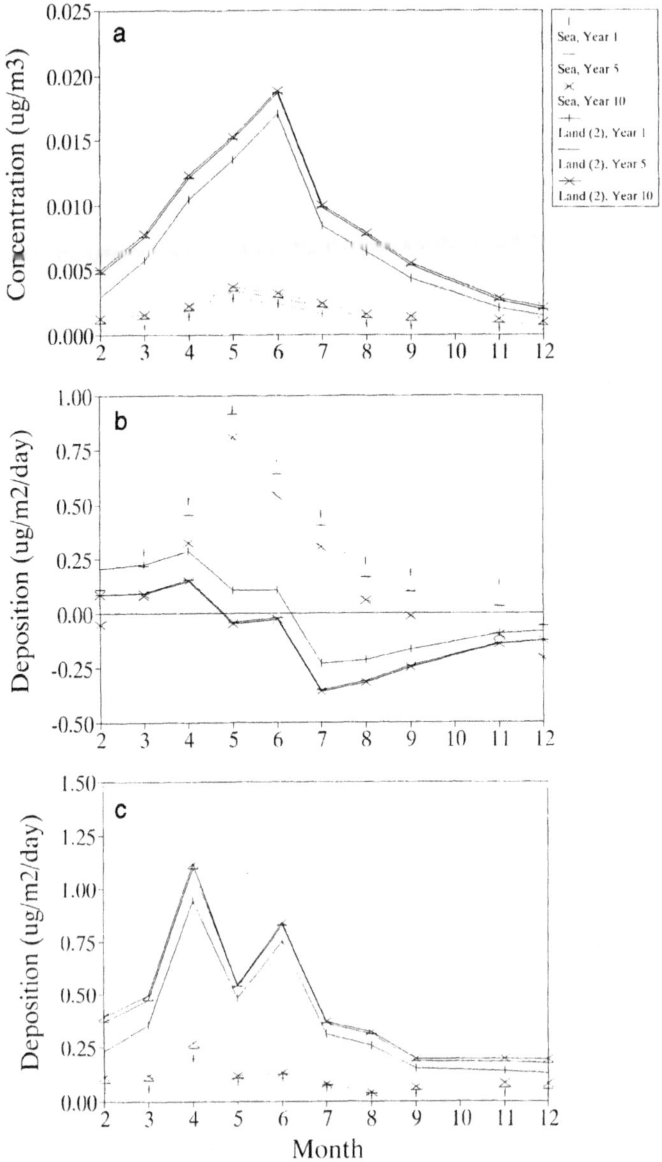

Figure 1. (a) Monthly average lindane concentration in the atmospheric surface layer for the modeled area above sea and land after 1, 2, and 5 years. (b) As in Fig. 1(a), but for dry deposition of lindane on non-agricultural areas. (c) As in Fig. 1(a), but for wet deposition.

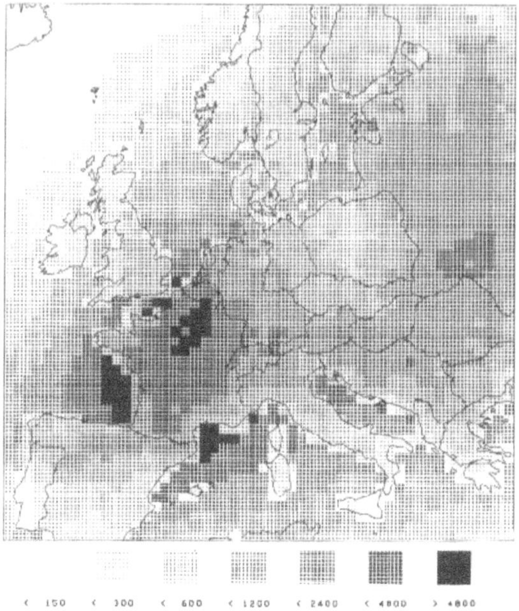

Figure 2. Net deposition (0.01 g ha^{-1} y^{-1}) of lindane on non-agricultural areas over Europe after 10 years of simulation.

The net deposition of lindane on non-agricultural areas varies strongly over Europe (Figure 2). The net deposition on land is the highest in Western and Eastern Europe due to large emissions in those areas. Also, the deposition on the sea is higher close to these emission areas. In Southern Europe the net deposition is lower because of the lower rainfall and higher surface temperatures. A higher temperature leads to a larger Henry coefficient and subsequently to a higher soil-gas concentration and re-volatilization rate. In Northern Europe the reverse effect is noticeable: the net-deposition is relatively high due to higher rainfall and lower surface temperatures. The latter is known as the "cold condensation" effect of semi-volatile POPs and indicates that those pollutants migrate towards colder regions of the earth (MacKay and Wania, 1995).

Water bodies such as the North Sea and Mediterranean Sea act as important sinks for lindane, which accumulates in these water bodies because of the very small degradation in water. A mass balance for the model area showed that after 10 years about 12% of the emitted lindane is present in the model compartments (8%, 4% and 0.2% in water, soil and air), 47% had been degraded (42% and 5% in soil and air) and 41% had been transported out of the model area. However, these figures are strongly dependent on the size of the model area and the sea to land distribution within this area.

A comparison between modeled concentrations in precipitation and observed concentrations (OSPARCOM, 1994) showed a moderate correlation (linear regression through origin: c.c.=0.56), but the observations are overestimated by a factor of 3. Part of this discrepancy lies in the uncertainty in the emissions, i.e., application amount and emission factors such as application depth. Also, the transport of POP in the soil by leaching and evaporation of water is not accounted for in the model. Leaching, however, is an effective mechanism that transports lindane to greater depths, leading to lower topsoil concentration of lindane. Particularly in emission areas this will give rise to a decrease in re-emission of the applied lindane. Consequently, wet deposition of lindane will decrease as well because of the lower air concentrations. Besides, the measurements of lindane in precipitation have a limited accuracy.

CONCLUSIONS

Model calculations of the net deposition of lindane over Europe were made with the EUROS model coupled to a POP-surface exchange module. For the modeled non-agricultural land area, lindane was deposited mainly during the application period. At the same surfaces in the application-free period, clearly, a re-volatilization of lindane took place. Wet deposition is the major input pathway for non-agricultural areas, particularly at land surfaces. Up to 50% of this wet deposition can be re-volatilized, depending on the distance from emission areas, surface properties, and meteorological conditions. This means that input estimates of lindane at land surfaces cannot be based on precipitation measurements alone.

At sea, dry deposition of lindane is the most important input pathway. Lindane accumulates in water bodies because of the low degradation in water.

ACKNOWLEDGMENTS

We thank the ECMWF and Royal Netherlands Meteorological Institute (KNMI) for making available the ECMWF wind fields and the WMO synops data. Frank de Leeuw is thanked for his useful comments and Ruth de Wijs for editing the paper.

REFERENCES

Berdowski, J.J.M., and Veldt, C., Emission of cadmium, copper,lead, benzo(a)pyrene and lindane to air in Europe in 1990. Background report of the ESQUAD project, TNO Delft, the Netherlands, TNO- report (In preparation).

Bidleman, T.F., L.M.Jantunen, R.L.Falconer and L.A.Barrie, 1995, Decline of hexachlorocyclohexane in the arctic atmosphere and reversal of air-sea gas exchange. Geophysical Research Letters, Vol.22, No.3:219-222.

Egan, B.A., and Mahoney, J.R., 1971, Numerical modelling of advection and diffusion of urban area sources pollutants. *J.Appl.Meteor.* 11: 312-322.

Fraters, D., Bouwman, A.F., and Thewessen, T.J.M., 1993, Soil organic matter of Europe. Estimates of soil organic matter content of the topsoil of FAO-Unesco soil units, RIVM, Bilthoven, The Netherlands. Report no. 481505004.

GESAMP-IMO/FAO/UNESCO/WMO/WHO/IAEA/UN/ UNEP Joint group of experts on the scientific aspects of marine pollution, 1989, The atmospheric input of trace species to the world oceans, WMO, Geneva. Report and studies no.38.

Hicks B.B., Baldocchi, D.D., Meyers, T.P., Hosker R.P. Jr., and Matt, D.R., 1987, A preliminary multiple resistance routine for deriving dry deposition velocities from measured quantities. *Water Air Soil Pollut.* 36: 311-330.

Jaarsveld van, J.A., Pul W.A.J. van, and Leeuw, F.A.A.M. de, 1994, Modelling the long-range transport and deposition of persistent organic pollutants over Europe and its surrounding marine areas, in: *Air Pollution and its Application.* S-E. Gryning and M. Milan, eds., Plenum Press, New York.

Jury, W.A., Spencer, W.F., and Farmer, W.J., 1983, Behavior assessment model for trace organics in soil: I Model description, *J. Environ. Qual.*12: 558-564.

Jury, W.A., Gardner, W.R., and Gardner, W.H., 1991, *Soil Physics*, 5th ed., John Wiley & Sons, Inc., New York.

Liss, P.S., and Slater, P.G., 1974, Fluxes of gases across the air-sea interface, *Nature* 247: 181.

MacKay, D. and F.Wania, 1995, Transport of contaminants to the Arctic: partitioning, processes and models. The Science of the Total Environment 160/161: 25-38.

OSPARCOM, 1994, Calculation of atmospheric input to the North Sea 1987-1992. Oslo and Paris Commisions,

Rheineck Leyssius H.J. van, de Leeuw, F.A.A.M. and H.Kesseboom, 1990, A regional scale model for the calculation of episodic concentrations and depositions of acidifying components. *Water Air Soil Pollut.* 51: 327-344.

Strand, A., and Hov, O., A model strategy for the simulation of chlorinated hydrocarbon distributions in the global environment (Submitted for publication)

UN-ECE, 1994, State of Knowledge Report of the UN ECE Task force on Persistent Organic Pollutants for theconvention on Long-range Transboundary Air Pollution.

Velde, van de R.J., W.S. Faber, V.F. van Katwijk, J.C.I. Kuylenstierna, H.J. Scholten, T.J.M. Thewessen, M.Verspuij and M. Zevenbergen, 1994, The preparation of a European land use database. RIVM report 712401001.

Warmenhoven, J.P., Duiser, J.A., Leu L.T.H. de, and Veldt, C., 1989, The contribution of the input from the atmosphere to the contaminant of the North Sea and the Dutch Wadden Sea, TNO, Delft. Report R89/349A.

BUDGETS OF OZONE ON AN EUROPEAN AND GLOBAL SCALE

Michiel Roemer, Ger Boersen, Peter Builtjes, and Paul Esser

TNO Institute of Environmental Sciences
P.O. Box 6011
2600 JA Delft
The Netherlands

INTRODUCTION

The budget of ozone in the troposphere consists of four terms: (net) influx of ozone from the stratosphere, chemical production, chemical destruction and dry deposition. Integrated over the entire troposphere and over one year these terms nearly balance, the rest being responsible for a trend, if any. However, if regional budgets are considered large surplusses or deficits may occur which are then compensated by interregional transport of ozone.

A global tropospheric 2D-model and a regional 3D-model for Europe were used to calculate the contribution of each of the aforementioned terms to the ozone budgets on a global as well as regional scale.

THE OZONE BUDGET ON A GLOBAL SCALE

The Model

A detailed description of the formulation of the basic model is given in several publications (Isaksen and Rodhe, 1978; Isaksen *et al.*, 1985; Isaksen and Hov, 1987). The model extends horizontally from 85°N to 85°S with a resolution of 10° and vertically from ground level up to 16.75 km with a resolution of 0.5 km for the lowest six layers, 0.75 km for the seventh and 1.0 km for the remaining layers. A no cross boundary flux condition is set at 85° N and 85°S.

Air Pollution Modeling and Its Application XI, Edited by
Gryning and Schiermeier, Plenum Press, New York, 1996

The flux at the upper boundary is a prescribed function of latitude and season. The annual average stratospheric influx of ozone is 7.4E10 molecules/cm/s which is equal to 930 Tg/yr. Influx from the stratosphere of NO and HNO_3 are 0.5 Tg N/yr and 2.0 Tg/yr respectively. The flux at the lower boundary is determined by dry deposition and emissions.

With respect to the original version the main modifications applied to the TNO version are on chemistry and emissions (Roemer and van den Hout, 1992 and references therein). The chemistry now describes the oxidation chain of fourteen precursor non-methane hydrocarbons (NMHCs): ethane, ethene, acetylene, propane, propene, normal butane, isobutane, normal pentane, isopentane, hexane, benzene, toluene, xylene(s) and isoprene.

Ozone

The calculated distribution of ozone for the month of June is shown in Fig. 1. The two-dimensional distribution of ozone over the troposphere shows the well known characteristics that are apparent in other model's results. Mixing ratios of summer time ozone are markedly higher in the lower troposphere of the northern midlatitudes than in the southern midlatitudes. In the middle troposphere (2-8 km) the lowest ozone mixing ratios are calculated for the tropics. The downward fluxes of ozone at the tropics are smaller than at midlatitudes. Moreover the tropical middle troposphere acts as an important chemical sink to ozone thereby effectively reducing ozone mixing ratios.

Compared with ozone profiles from ballonsondes (Logan, 1981: Logan, 1994) the agreement in the lower and middle troposphere is acceptable. Differences occur in the upper troposphere due to the rather simplified parametrisation of the cross tropopause fluxes.

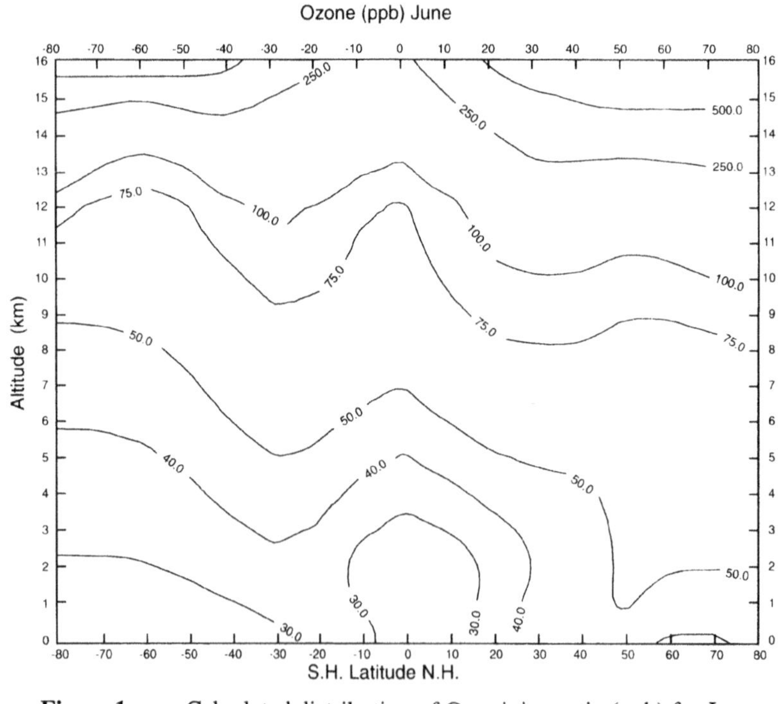

Figure 1 Calculated distribution of O3 mixing ratio (ppb) for June.

Budgets

The budget of ozone in the troposphere consists of four terms: (net) influx of ozone from the stratosphere, chemical production, chemical destruction and dry deposition. Averaged over the entire troposphere these four terms nearly balance, the small difference being responsible for the trend, if any. Fig. 2 shows the two-dimensional distribution of the contribution of stratospheric influx, net chemistry and dry deposition to the ozone budget for the month of June.

Altitude	S.H. Latitude N.H.																
(km)	-80	-70	-60	-50	-40	-30	-20	-10	0	10	20	30	40	50	60	70	80
16.25	+++	+++	++++	++++	++++	++++	+++	+++	+++	+++	+++	++++	++++	++++	++++	++++	++++
15.75	0	0	0	0	0	0	0	0	0	0	0	0	0	0	0	0	0
15.25	0	0	0	0	0	0	0	0	0	0	0	0	0	0	0	0	0
14.75	0	0	0	0	0	0	0	+	+	+	0	0	+	0	0	0	0
14.25	0	0	0	0	0	0	0	+	+	+	+	+	+	0	0	0	0
13.75	0	0	0	0	0	0	+	+	+	+	+	+	+	+	0	0	0
13.25	0	0	0	0	0	0	+	+	+	+	+	+	+	+	0	+	+
12.75	0	0	0	0	0	+	+	+	++	++	+	+	+	+	+	+	+
12.25	0	0	0	0	0	+	+	++	++	++	++	+	+	+	+	+	+
11.75	0	0	0	0	0	+	+	+	++	++	++	+	+	+	+	+	+
11.25	0	0	0	0	0	+	+	+	++	++	++	+	+	+	+	+	+
10.75	0	0	0	0	0	+	+	+	++	++	+	+	+	+	+	+	+
10.25	0	0	0	0	0	+	+	+	+	+	+	+	+	+	+	+	+
9.75	0	0	0	0	0	+	+	+	+	+	+	+	+	+	+	+	+
9.25	0	0	0	0	0	+	+	+	+	0	0	+	+	+	+	+	+
8.75	0	0	0	0	0	+	0	0	0	0	0	0	+	+	+	+	+
8.25	0	0	0	0	0	+	0	0	0	0	•	0	0	0	+	+	+
7.75	0	0	0	0	0	0	0	0	•	•	•	•	0	0	0	0	+
7.25	0	0	0	0	0	0	0	•	•	•	•	•	0	0	0	0	0
6.75	0	0	0	0	0	0	0	•	•	•	••	•	0	0	0	0	0
6.25	0	0	0	0	0	0	•	•	••	••	••	••	0	0	0	0	0
5.75	0	0	0	0	0	0	•	•	••	••	••	••	•	0	0	0	0
5.25	0	0	0	0	0	0	•	••	••	••	••	••	•	0	0	0	0
4.75	0	0	0	0	0	0	•	••	••	••	••	••	•	0	0	0	0
4.25	0	0	0	0	0	0	•	••	••	••	••	••	•	0	0	0	0
3.75	0	0	0	0	0	0	•	••	••	••	••	••	•	0	0	0	0
3.25	0	0	0	0	0	0	••	••	••	••	••	••	•	0	0	+	0
2.75	0	0	0	0	0	0	••	••	••	••	••	••	•	0	+	0	+
2.25	0	0	0	0	0	0	••	••	••	••	••	•	0	+	+	+	+
1.75	0	0	0	0	0	0	•	••	••	••	••	•	+	++	++	+	+
1.25	0	0	0	0	0	0	•	•	•	••	••	+	++	+++	++	++	+
0.75	0	0	0	0	+	++	++	++	+	0	+	+++	+++	+++	+++	++	++
0.25	••	••	•••	•••	•••	••	•	0	0	••	••	++	•••	•••	••••	•••	••

++++ :	> 10	ppb/day	0 :	> -0.3	ppb/day
+++ :	> 3	ppb/day	• :	> -1	ppb/day
++ :	> 1	ppb/day	•• :	> -3	ppb/day
+ :	> 0.3	ppb/day	••• :	> -10	ppb/day
			•••• :	> -30	ppb/day

Figure 2 Calculated distribution of the ozone budget (ppb/day) for June.

Regionally large surplusses or deficits occur. The middle troposphere of the tropics acts as a large chemical sink of ozone. Net chemical production of ozone occurs at northern midlatitudes, especially at the lowest 2-3 km where most of the anthropogenic emissions takes place, and more moderately in the upper troposphere. The ozone budget in the upper troposphere is dominated by the influx of ozone from the stratosphere. Averaged over the entire troposphere chemistry is a net source of ozone. The net ozone produced by chemistry almost equals the amount of ozone transported from the stratosphere (Tab. 1).

Table 1 Budgets of ozone (Tg/month) for June and December.
(+: production or net import, –: destruction, or net export.)

	85-25 NH 0-16.75 km	25 NH-25 SH 0-16.75 km	25-85 SH 0-16.75 km	GLOBAL 0-16.75 km	55-35 NH 0-2 km
June					
• stratospheric influx	+48.9	+7.4	+16.5	+72.8	0
• dry deposition	–76.1	–49.0	–19.5	–144.6	–39.6
• chemical production	+203.0	+242.8	+31.7	+477.5	–70.8
$HO_2 + NO$ (%)	64	50	62	57	63
$CH_3O_2 + NO$ (%)	18	24	24	21	15
• chemical destruction	-121.5	–250.3	–22.8	–394.6	-25.6
$H_2O + O(^1D)$ (%)	34	52	25	45	26
$O_3 + HO_2$ (%)	41	34	49	37	37
$O_3 + OH$ (%)	14	11	14	12	9
$O_3 + NO_2$ (%)	6	1	8	3	12
• net production	+81.5	–7.5	+8.9	+82.9	+45.2
• interregional transport	–50.4	+54.5	–4.1	0	-5.6
December					
• stratospheric influx	+19.8	+7.4	+47.5	+74.5	0
• dry deposition	–54.6	–50.3	–19.3	–124.2	–28.9
• chemical production	+104.2	+244.4	+64.7	+413.3	+35.4
$HO_2 + NO$ (%)	58	50	63	54	56
$CH_3O_2 + NO$ (%)	14	22	27	21	10
• chemical destruction	–52.4	–249.0	–71.4	–372.8	–14.2
$H_2O + O(^1D)$ (%)	15	51	39	44	2
$O_3 + HO_2$ (%)	39	35	41	37	21
$O_3 + OH$ (%)	8	10	17	11	3
$O_3 + NO_2$ (%)	23	2	2	8	39
• net production	+51.8	–4.6	–6.7	+40.5	+21.2
• interregional transport	–25.5	+49.3	–23.8	0	+7.7

In the equatorial box (25°NH-25°SH; 0-16.75 km) the chemical production almost equals the chemical destruction (Tab. 1). The main loss process of ozone is through photolysis of ozone into the $O(^1D)$ radical and the subsequent reaction with water vapour which takes one ozone molecule out of the O_3-$O(^3P)$-$O(^1D)$ cycle. The second largest ozone destructive reaction (on a global scale) is the reaction of ozone with the hydroperoxyde radical. In the extratropical areas this is the most important ozone destructive reaction especially in the winter. Smaller contributions to the destruction of ozone come from the reactions of ozone with the hydroxy radical and with nitrogen dioxyde respectively. In NO_x rich areas such as the mixed layers of the northern midlatitudes the reaction of O_3 with NO_2 becomes in the winter the main chemical process leading to ozone destruction.

In the lowest 2 km of the northern midlatitudes (35°NH-55°NH; 0-2 km) the net ozone production is in June about twice as large as in December. Table 1 shows that the net ozone production in this region contributes considerably to the global ozone budget. The net ozone production, 45 Tg/month in the summer and 21 Tg/month in the winter, equals the amount of ozone imported from the stratosphere into the Northern Hemisphere. In the budget however much of the produced ozone is balanced by dry deposition. In the summer this area acts as a net exporter of ozone, exporting at a rate of 5.6 Tg/month. In the winter there is a deficit in the ozone balance and ozone is imported from elsewhere (mainly the free troposphere) to compensate.

THE OZONE BUDGET OVER EUROPE

The Model

LOTOS (LOng Term Ozone Simulation) is a three dimensional Eulerian model that calculates the distribution of ozone and related species in the lowest 2.5 km of the troposphere over Europe (Builtjes, 1992; Esser, 1993; Roemer *et al.*, 1993). The horizontal resolution is 0.5° latitude x 1.0° longitude which is about 60x60 km^2. In vertical direction three prognostic layers are used. The lowest layer follows the diurnal variation of the mixed layer, the other two layers are so-called reservoir layers. The description of the chemistry is based on the CBM-IV mechanism. The current LOTOS model uses 17 longlived (transportable) species and 7 shortlived (radical) species reacting with each other in 52 reactions.

Boundary and initial conditions are set by the calculated concentrations from the two-dimensional global, described in section 2. The boundary concentrations vary as a function of height, latitude and month.

Budgets

Budgets for all species were calculated as a function of time during the July-August 1990 period. Four regions were distinguished. Northern Europe (Norway, Sweden, Finland, Denmark and the Baltic States), Western Europe (Germany, Benelux, Northern France and England), Southern Europe (Europe south of the Alps from Portugal to Greece) and Europe as a whole. The main results for O_3, NO_x and CO are summarised in Tab. 2.

Calculations with the LOTOS model for the year 1990 indicate a net chemical production of ozone over the entire European area (Iceland excluded) and in the lowest 2.5 km of the troposphere of about 21 Tg ozone per month in the summer (July-August), see Tab. 2. Most of the ozone is lost by dry deposition (17 Tg/month). Net transport out of Europe accounts for about 4 Tg/month.

Jacob *et al.* (1993) argue that the true export of ozone from a continent like the US (or Europe) is much higher due to the fact that much of the ozone deposited to the continent is imported from outside (exogenous). By extracting this exogenous background in a simulation with no chemistry (ozone regulated solely by transport and deposition) they calculated that about 70% of the ozone regionally produced by a continent is exported into the global troposphere. Such a distinction of ozone as where it comes from is not made in this study. The comparison of net ozone transport rates between the different models is just based on how much ozone is net imported or exported according to the transport modules of the models irrespective the origin of the ozone.

Table 2. Budgets in Europe of O_3, NO_x and CO; NO_x in NO_2 eq.

	North 54.5-69N; 05E-26E 0-2.5 km	West 47-54.5N; 04W-14E 0-2.5 km	South 36-47N; 09W-26E 0-2.5 km	Europe 35.5-69N; 09W-58E; 0-2.5 km
Area (10^6 km^2)	2.6	1.0	3.6	16.7
O_3 (Tg/month)				
chemistry	2.2	2.7	5.5	21.3
dry deposition	-1.9	-1.6	-3.4	-16.8
wet deposition	-0.0	-0.0	-0.0	-0.0
net transport	-0.3	-1.1	-2.1	-4.2
NO_x (Gg/month)				
emissions	114	523	400	1742
chemistry	-124	-377	-356	-1526
dry deposition	- 19	- 62	- 56	- 235
wet deposition	- 0	- 0	- 0	- 0
net transport	12	- 79	- 6	- 81
CO (Tg/month)				
emissions	0.5	2.2	1.6	7.8
chemistry	-0.4	-0.4	-0.9	-4.0
dry deposition	-0.0	-0.0	-0.0	-0.0
wet deposition	-0.0	-0.0	-0.0	-0.0
net transport	0.1	-1.7	-0.6	-3.0

The intensity of the (net) chemical production of ozone is highest in Western Europe. Covering only 6% of the European area it is responsible for almost 13% of the net chemical production of ozone. Consequently, also the contribution of dry deposition and net transport out of the domain is disproportionally higher. If scaled per unit area the intensity of the chemical ozone production in Northern and Southern Europe is about 30% and 60% respectively of that in Western Europe. In all regions loss by dry deposition is the most important sink to ozone.

The large differences in the ozone production intensity among the European regions is not surprising seen in the light of distribution of precursors sources. About 30% of the European emissions of NO_x, CO and anthropogenic VOC is emitted in Western Europe. Due to its short lifetime most of the NO_x emitted is chemically transformed in other nitrogen species within the region itself. (This of course is not entirely true. The chemistry applies on ambient NO_x concentrations which are determined by not only emissions but also by import from other regions.) It seems that in all regions production by emissions and loss by chemistry are in balance with each other within 15%, the remainder accounted for by dry deposition. Only in Western Europe net export is an important term in the budget of NO_x. About 15% of the NO_x emitted is advected out of the region.

Due to its much larger lifetime transport is an important term in the budget of CO, whereas chemical decay is less important than it is for NO_x. About 40% of the CO emitted in Europe is transported out of the European domain.

The ozone production in Europe calculated with the LOTOS model for the 1990 summer is more than two times as large as the calculated ozone production of 1.7 Tg/month for the 1985 summer in the US (Jacob *et al.*, 1993). The area of the American model domain is with 14.10^6 km^2 somewhat smaller than the LOTOS model domain, but the land area is considerably smaller. So, if scaled on land area the intensity of ozone production per unit land area in Europe is about 50% larger than in the US. Given the differences between both regional models an exact comparison of ozone production rates for both continents is not possible. Nevertheless, it indicates that the mixed layers of Europe and the US contribute in comparable quantities to the global ozone budget in terms of local ozone production.

Compared with the calculated ozone production rates in the zonal belt (35-55 NH; 0-2 km) it means that in the summer the mixed layers of Europe and the US together export almost about as much ozone (4.2 + 1.7 Tg/month) as the entire zonal midlatitudinal belt. Given the discrepancies in geographical location and volumes the comparison of ozone production rates is only indicative. Within this midlatitudinal belt there are however more ozone producing areas such as East Asia. Taking this into account the comparison suggests that the sum of ozone exported directly from the three major source areas (US, Europe, East Asia) exceeds the net zonal export of ozone. As a consequence, the north midlatitudinal belt also includes areas which act, even in the summer, as a net ozone importer.

CONCLUSIONS

Calculations with a global 2D-model show that the net ozone production in the lowest 2 kilometers of the northern midlatitudes contributes considerably to the global ozone budget. Due to dry deposition the net transport of ozone to or from the free troposphere is much smaller. In the summer this area acts as a net exporter of ozone, in the winter when much less ozone is produced, the area acts as a net importer of ozone.
Comparison with ozone budgets derived from continental models for Europe and the US indicates that in the summer Europe and the US together export as much ozone as the entire lowest 2 km of the northern midlatitudes. Including other important source areas such as East-Asia it suggests that within the northern midlatitudes there are also areas where there is, even in the summer, a net import of ozone into the mixed layer.

Acknowledgements

This study was financially supported by grants from the National Research Programme on Global Air Pollution and Global Change (NOP) under contract nr. 852072 and by additional grants from the Ministery of Public Housing, Spatial Planning and the Environment under contract nr. DS.2.C.1.

REFERENCES

Builtjes, P.J.H.,1992, The LOTOS - LOng Term Ozone Simulation project; Summary Report. TNO report R92/240, Delft, The Netherlands.

Esser, P.,1993, An evaluation of the LOTOS model for the growing season of the year 1985. TNO report R94/008, Delft, The Netherlands.

Isaksen I.S.A. and Rodhe H.,1978, A two-dimensional model for the global distribution of gases and aerosol particles in the troposphere. Report AC-47. Department of Meteorology, University of Stockholm, Stockholm, Sweden.

Isaksen I.S.A. and Hov O.,1987, Calculation of trends in the tropospheric concentrations of O_3, OH, CO, CH_4 and NO_X. Tellus, 39B:271.

Isaksen I.S.A., Hov O., Penkett S.A. and Semb A.,1985, Model analysis of the measured concentration of organic gases in the Norwegian Arctic. J.atmos.Chem.3:3.

Jacob, D., J. Logan, G. Gardner, R. Yevich, C. Spivakovsky, S. Wofsy, S. Sillman and M. Prather,1993, Factors regulating ozone over the United States and its export to the global atmosphere. J.Geophys. Res.98:14817.

Logan J.A., M.J. Prather, S.C. Wofsy and M.B. McElroy,1981, Tropospheric chemistry: A global perspective. J. geophys. Res.86:7210.

Logan J.A.,1994, Trends in the vertical distribution of ozone: an analysis of ozonesonde data. J.geophys. Res.99:25553.

Roemer M.G.M. and van den Hout K.D. (1992) Emissions of NMHC and NO_X and global ozone production. in: Proceedings of the nineteenth NATO/CCMS ITM on Air Pollution Modelling and Its Application (eds.: van Dop and Kallos), September 29-October 4, 1991 Crete, Greece. Plenum Press, New York.

Roemer, M.G.M., P.J.H. Builtjes , P.J. Esser and G. Boersen ,1993, Oxidant modelling over Europe-results of the LOTOS model system. in: Proceedings of the EMEP workshop on the control of photochemical oxidants in Europe (eds.: Antilla, P.), April 20-22, 1993, Porvoo, Finland

DISCUSSION

R. KUNZ: The input data for the coarse grid dispersion calculations were derived from LOTOS simulations results.

P. BUILTJES: Thank you for this comment. LOTOS model results, as they represent O_3 and other concentrations over Europe on a grid of 0.5 deg times 1 deg over the whole year on a 3 hourly basis have been used as boundary conditions for more local studies regularly. The LOTOS results are available on request.

THE EFFECT OF HISTORICAL EMISSION CHANGES AND METEOROLOGICAL VARIABILITY ON SULPHUR LOADS IN EUROPE

Sophia Mylona

Norwegian Meteorological Institute
P.O. Box 43 Blindern
N-313 Oslo, Norway

INTRODUCTION

Acidification of terrestrial and aquatic ecosystems in Europe has a history of several decades. In order to evaluate the role of sulphur compounds in the environmental degradation over the years it is necessary to study the historical evolution of the air concentration and deposition of oxidised sulphur. Recognising that atmospheric sulphur loading has been determined by the combined action of emissions, meteorological conditions and the presence of oxidants, it follows that investigation should focus on the long-term effect of emission changes, meteorological variability and the changing oxidising capacity of the atmosphere.

The present study is a part of a project which aims at showing the development of the concentration and deposition patterns of oxidised sulphur in Europe during the historical period 1880-1991. Results may shed some light on problems related to the evolution of acidification and the causes of the observed environmental effects. They may also support research on climate change since sulphate aerosol particles seem to affect the radiative budget of the Earth.

The purpose of this paper is to outline completed work on the effect of emission changes, and present in brief the method and first results from the ongoing study into the effect of meteorological variability.

APPROACH

Calculations are performed with an earlier version of the Lagrangian, one-layer, trajectory acid deposition model (Sandnes and Styve, 1992) operating at the Norwegian Meteorological Institute (DNMI) which acts as the Meteorological Synthesising Centre-West (MSC-W) of the European Monitoring and Evaluation Programme (EMEP). The model has a spatial resolution of 150x150 km^2 and uses as input 6-hourly meteorological data from a

Air Pollution Modeling and Its Application XI, Edited by
Gryning and Schiermeier, Plenum Press, New York, 1996

Numerical Weather Prediction model operating at DNMI, annual emission data and prescribed oxidant and background concentrations. Using a system of linear differential equations, the sulphur module calculates gridded air concentrations of sulphur dioxide (SO_2) and particulate sulphate plus dry and wet deposition of oxidised sulphur.

Work is divided into two parts. Investigation in the first part is concentrated on the evolution of sulphur fields as a result of changes in the spatial and temporal distribution of anthropogenic emissions of sulphur dioxide. It includes the compilation of an historical emission inventory for SO_2 in Europe and its implementation into the EMEP/MSC-W model assuming meteorological conditions as occurred in 1991. Variations in the sulphur releases are considered to play the most important role in the long-term perspective, so such an approach is expected to constitute a good first approximation to the problem.

The second part of this work investigates the long-term influence of the interannual meteorological variability alone assuming 1991 level emissions. Although such an assessment has been performed for recent years deterministically (Mylona, 1991; Iversen, 1993; Barrett et al. 1995), direct use of pollution models historically becomes impossible due to the paucity of detailed meteorological information. Compromises as to the methods to apply to find relationships between the available meteorological variables and sulphur amounts become inevitable with statistical methods on regression being often the most attractive option. Such an approach has been followed in the present study.

Although results from the first part are now available for each EMEP grid square in Europe, work on the second part is still in process. For the sake of consistency the effect of emission changes and meteorological variability will be presented only for one grid element over the western coast of Ireland.

THE EFFECT OF HISTORICAL EMISSION CHANGES

The compilation of the historical emission inventory along with model estimates is documented in detail in Mylona (1995). In brief, the applied approach is as follows.

Anthropogenic SO_2 releases are calculated from energy and industry statistics (Mitchell, 1978, 1982; UN statistics; OECD statistics; National statistics) for each European country since 1880 with a time step of 5 years. The processes considered are combustion of solid fuels and liquid fuels, production of non-ferrous metals, sulphuric acid, wood pulp and cement, production and consumption of coke in the iron and steel making and oil refining. National emission estimates are further disaggregated spatially to the EMEP grid on the basis of historical information about the distribution of sources.

Long-term emission estimates reveal that in most countries combustion of coal was the dominant emission source before World War II and combustion of liquids thereafter, with releases from industrial processes being relatively small. The highest SO_2 releases occurred mainly in the 1960s and 1970s, while emission control measures resulted in gradual reductions in most countries in the 1980s. In Europe as a whole, coal combustion remained the major emission source throughout the century. Releases peaked in the 1970s at about 55 million tonnes SO_2.

Figure 1 illustrates the evolution of the emission pattern in Europe as employed in this work. It is clear that the location of the European emission sources has shown over the years a progressive detachment from the coalfields towards a widespread distribution. In the last decades considerable reductions took place over north-western and parts of central Europe, while in the south and south-east emissions increased substantially.

Calculations with the dispersion model reflect to a great extent the emission pattern with maximum air concentrations and depositions over parts of north-western and central Europe. Accumulated depositions over the period 1880-1991 in these areas reach 600 g (S) m^{-2}.

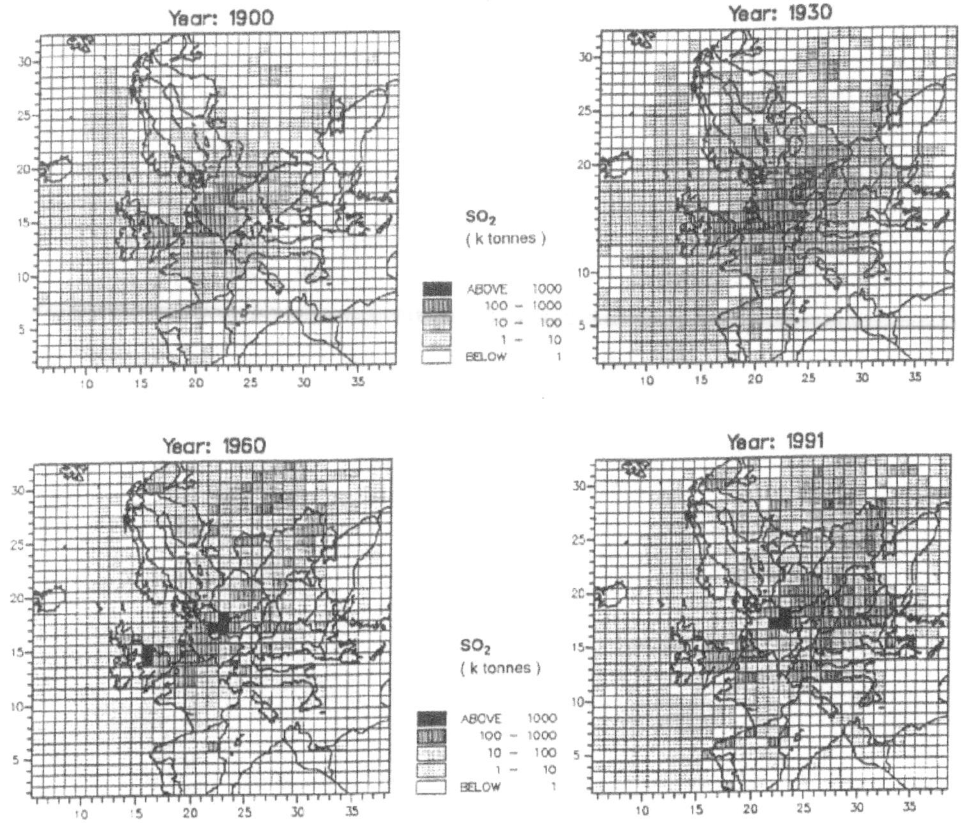

Figure 1. The spatial disaggregation of the European emissions of SO_2 over the EMEP grid. For 1900 and 1930 the emission distribution is calculated. For 1960 and 1991 the distribution is as reported by the countries in the late 1980s. The numbers on the axes of the maps denote the EMEP grid squares constituting the modelled domain.

In low emission areas comparable concentrations of SO_2 and particulate sulphate reveal the importance of long range transport in determining the atmospheric composition far away from the sources.

The historical development of the air concentration and deposition of sulphur in a region of western Ireland (EMEP grid square (13,13)) is shown in Figure 2, featuring a progressive increase from 1880 to peak values in the 1970s and considerable reductions in the 1980s.

THE EFFECT OF METEOROLOGICAL VARIABILITY

When links between sulphur concentrations and meteorological variables are sought, it is necessary to remove as far as possible the effect of emissions variance from the sulphur data before they undergo regression. This is not an easy task when sulphur data are measurements because relationships are bound to be obscured by the influence of emissions especially in high source areas. Model calculations, however, can handle this problem easily when the atmospheric behaviour of the component in question is limited to linear processes.

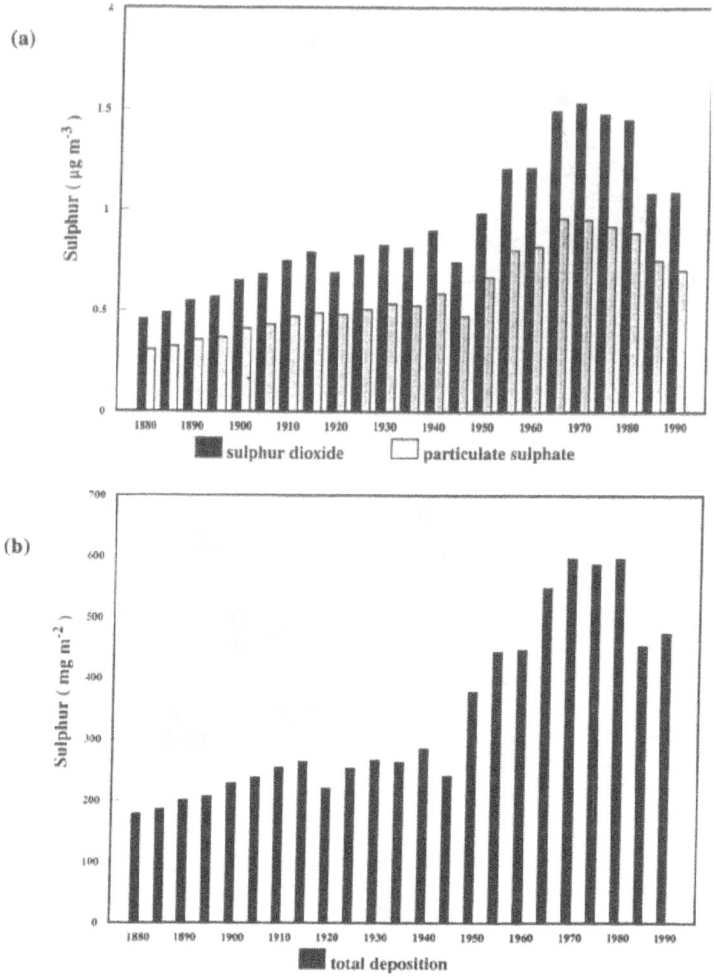

Figure 2. Modelled historical trends of oxidised sulphur at one EMEP grid square in western Ireland. (a) Air concentrations of SO_2 and particulate sulphate. (b) Total (dry plus wet) annually accumulated deposition of oxidised sulphur. Calculations are based on the meteorological conditions of 1991.

In the present study both the model output and the meteorological variables are preprocessed in a simple way described in the following.

Data preprocessing

Modelled sulphur data. Calculations with the dispersion model consist of monthly mean concentrations of SO_2 and particulate sulphate plus dry and wet deposition of oxidised sulphur in the 10-year period 1985-1994. Removal of the variance due to emissions is achieved by running the model for all the above years with the corresponding actual meteorological conditions but with emissions as released and distributed in 1991 - hereafter referred to as *the 1991 adjusted calculations*. The actual 1991 calculations (derived on the basis of the 1991 meteorology and 1991 emissions) are then subtracted from the 1991 adjusted data. These differences are referred to as *the processed data* in the following and represent the part in the concentrations or depositions attributed to the difference between the meteorological conditions of a particular year and those of 1991 i.e. the effect of meteorological variability. Thus, the effect of emission changes does not enter the processed data explicitly.

Meteorological data. Historical meteorological information is based on Mean Sea-Level Pressure (MSLP) data provided by the UK Meteorological Office and the European Centre for Medium-Range Forecasts (ECMWF). These data have a spatial resolution of 10^0 Longitude x 5^0 Latitude and are available over Europe on a daily basis since 1881. For the purposes of this work they are interpolated to the EMEP grid and are expressed in geographical coordinates. On the basis of these data the following variables are calculated for each EMEP grid square:

$$ZI = p_S - p_N$$

$$MI = p_E - p_W$$

$$\zeta \propto \frac{\partial^2 p}{\partial x^2} + \frac{\partial^2 p}{\partial y^2}$$

$$\frac{\partial \zeta}{\partial t} \text{ and } \frac{\partial p}{\partial t}$$

where p_N, p_S, p_W and p_E are respectively the MSLP in the EMEP grid square north, south, west and east of the grid element in question, ZI and MI are termed the zonal and meridional indices respectively, x and y are the actual geographical coordinates of the grid cell under consideration and ζ is an approximation for relative vorticity assuming geostrophic winds derived from the MSLP patterns. The zonal and meridional indices provide information about the strength and direction of surface winds. The approximated vorticity imforms as to the cyclonic or anticyclonic character of the wind fields. The rate of change of pressure and vorticity can contribute with some information on the intensity of the cyclonic activity at the region of interest, but the poor spatial resolution does not encourage high expectations.

Monthly means are calculated for all the above variables and MSLP from 1881 to 1994. The data are subsequently preprocessed in the fashion described above i.e. for each of the six variables monthly values are subtracted from the corresponding monthly means of 1991.

The applied statistical method

The multivariate technique used in this application is called Partial Least Squares Regression (PLS-R) and is based on Principal Component Analysis (PCA) theory (Esbensen et. al., 1994). PCA is a data analytical method which decomposes a given data set into a number of independent variables, and finds which ones are really important for the total variation of the data set. This information is then extracted and constitutes a PC model. Thus, the original data set is reduced into the fewest number of orthogonal components which explain a large percentage of the total variability. When data sets are viewed as points and point swarms in the data space, a PC model may be seen as a *transformation* where a coordinate system of many dimensions is transformed into another coordinate system with fewer dimensions. The directions of the new coordinate axes (*principal components*) are chosen to describe the largest variations. The localisation of the samples in the new system i.e. their coordinates related to the principal components are called *scores*. The corresponding relationship between the original variables and the new principal components are called *loadings*. The differences between the coordinates of the samples in the new and old system, lost information due to projection onto fewer dimensions, can be regarded as the modelling error or lack of fit.

Multivariate regression problems involve mathematical relations between two data sets **X** (input variables) and **Y** (output variables). On the basis of a representative set of known corresponding samples of both X- and Y- variables (*the training set*), a so-called *calibration model* is established ($Y = f(X)$). This model is used to predict unknown Y-values from the measurements of new X-variables. PLS is based upon this principle and is known to give optimal prediction ability in the strict statistical sense of minimum prediction error. In particular, it performs a simultaneous and *interdependent* PCA decomposition in both **X**- and **Y**- matrices in such a way that the correlated information in the **Y**-matrix is used directly as a guide for optimal decomposition of the **X**-matrix and then performs regression of **Y**.

In this application, the training data set refers to the period 1985-1993 and consists of a 96 (months) x 6 (meteorological variables) matrix **X** and a 96 x 1 matrix **Y** of monthly mean SO_2 concentrations. All matrix elements are *processed* data and refer to the Irish EMEP grid square. First, a PLS calibration model is constructed. On the basis of this model and the available historical meteorological information, SO_2 concentrations are predicted for the period 1950 onwards. The period 1880-1949 has not been considered because it involves more preprocessing work - not yet completed - due to the two different emission distributions used in the model runs for the periods 1880-1919 and 1920-1949, respectively.

Data for 1994 are not used in the training set but are kept aside as an independent set for validation of the predictability of the method.

Results

PLS indicates that for the grid cell under consideration the first PLS component explains almost all the variability in the **Y**-data set. The meteorological variable contributing strongest to this factor is the zonal index. This is of course expected as Ireland is clearly exposed to the westerlies. The meridional index is much less important, whilst the contribution from all other variables is insignificant.

Figure 3 illustrates how the projections of the data on the first PLS component (scores) of the **X**-matrix (t_1) are related to the corresponding projections of the **Y**-matrix (u_1). The correlation coefficient is as high as 0.88, meaning that 77% of the variance in the SO_2 concentrations can be explained by considering primarily the zonal component of the surface wind over the western coast of Ireland.

The predictive ability of the PLS method is tested by applying the constructed calibration model on the monthly estimates of the meteorological variables for 1994. Figure 4 shows the predicted values versus the *processed* SO_2 concentrations as calculated by the EMEP/MSC-W model. The high correlation coefficient (0.88) shows a very good agreement.

The calibration model can now be used to calculate the *1991 adjusted* concentrations for the period 1950 onwards. Figure 5 presents both the effect of emission changes and meteorological variability on the SO_2 concentrations in the Irish EMEP grid square since 1950. The horizontal line shows the calculated 1991 concentrations. The thick solid line exhibits the effect of emission changes on a 5-year basis by showing concentrations calculated on the basis of historical emissions and 1991 meteorology. Concentrations computed considering varying meteorological conditions and 1991-level emissions present the effect of meteorological variability and are shown in both annual means (thin solid line) and 5-yearly averages (dashed line). Differences between the curves and the horizontal line indicate the importance of the respective effect on the SO_2 concentrations when 1991 is used

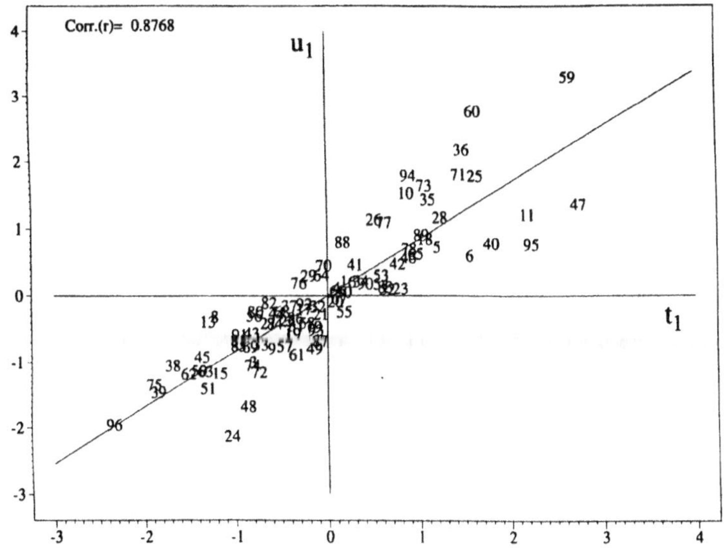

Figure 3. Inter-score plot (t_1-u_1) for the first PLS component. *Processed* SO_2 concentrations ($\mu g\ m^{-3}$ as sulphur) are indicated by numbers referring to the months in the period 1985-1993. Number of outliers removed: 4.

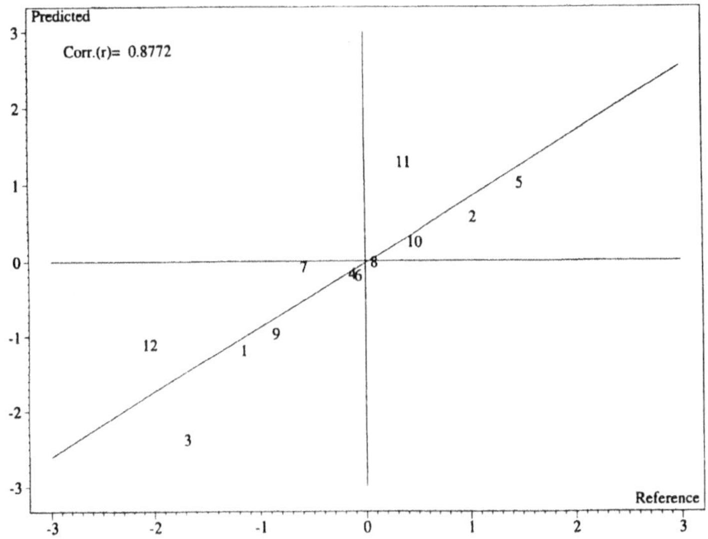

Figure 4. Validation plot. *Processed* SO_2 concentrations ($\mu g\ m^{-3}$ as sulphur) as predicted by PLS (y-axis) versus calculations from the EMEP model (x-axis) for 1994.

as the base year for the calculations.

It is clear that the region under consideration is exposed to strong year-to-year meteorological variations. Sulphur concentrations are affected correspondingly and the influence appears often to be comparable and even larger than that of emission changes. The comparability of the two effects in magnitude and the fact that they run either in parallel or in opposition imply that assessments on the historical evolution of sulphur based solely on emission changes are justified only for regions experiencing small meteorological variations.

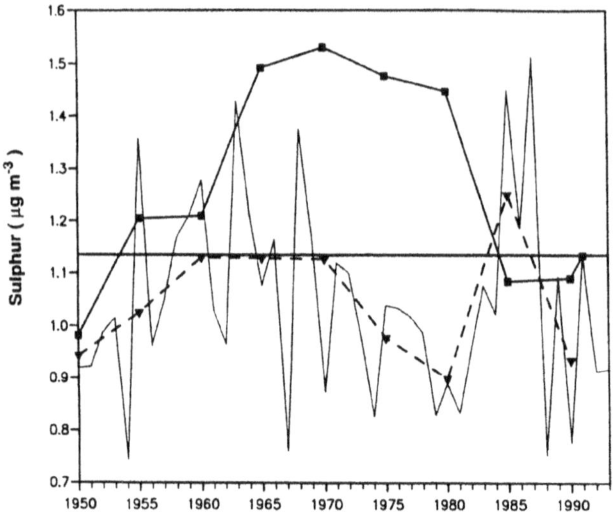

Figure 5. The effect of emission changes and meteorological variability on the SO_2 concentrations at the EMEP grid square (13,13) during the period 1950-1993. The horizontal line shows the calculated 1991 concentrations. The thick solid line shows concentrations derived on the basis of historical emissions and 1991 meteorology. The thin solid line indicates concentrations based on 1991 emissions and varying meteorology. The dashed line displays 5-year averages of the latter.

CONCLUDING REMARKS

Historical emissions of sulphur dioxide seem to have affected substantially the regional concentrations and depositions of oxidised sulphur in Europe. The effect of meteorological variability is studied by means of a multivariate technique applied on a limited set of available meteorological and sulphur data. For a low emission region influenced strongly by the westerlies the method performs very well. First results show that the meteorological influence can be very important and should by no means be disregarded when the evolution of the air quality is sought. Further work will examine the performance of the method on other sulphur components and in regions exposed to high emission loads and/or more complex meteorological conditions.

REFERENCES

Barrett, K., Foss, A., Mylona, S., Sandnes, H., Styve, H., and Tarrasón, L., 1995, European transboundary acidifying air pollution-Ten years calculated fields and budgets to the end of the first sulphur protocol, EMEP/MSC-W Report 1/95, Norwegian Meteorological Institute, Oslo.

Esbensen, K., Schönkopf, S., and Midtgaard T., *Multivariate Analysis in Practice*, Camo AS, Trondheim (1994).

Iversen, T., 1993, Modelled and measured transboundary acidifying pollution in Europe-Verification and trends, *Atmospheric Environment* **27A**: 889-920.

Mitchell, B.R., 1978, *European Historical Statistics 1750-1970*, The Macmillan Press LTD, London and Basingstoke.

Mitchell, B.R., 1982, *International Historical Statistics-Africa and Asia*, The Macmillan Press LTD, London and Basingstoke.

Mylona, S., Detection of sulphur emission reductions in Europe during the period 1979-1986, in: *Air Pollution Modelling and its Application, VII*, H. van Dop, ed., Plenum Press, New York (1991).

Mylona, S., 1995, Sulphur dioxide emissions in Europe 1880-1991 and their effect on sulphur concentrations and depositions, *Tellus* (accepted for publication).

Sandnes, H., and Styve, H., 1992, Calculated budgets for airborne acidifying components in Europe, 1985, 1987, 1988, 1989, 1990 and 1991, EMEP/MSC-W Report 1/92, Norwegian Meteorological Institute, Oslo.

DISCUSSION

D. STEYN: Are you able to use proxy data (ice cores, lake cores, tree rings) to investigate the reality of your modelled sulphur loadings?

S. MYLONA: Understanding the influence of both meteorology and emission changes upon modelled sulphur loadings is a necessary precursor to proper comparison with historical proxy data. Initial inquiries into the availability and validity of such data are being made. This will be pursued once the model validity and the proxy data validity is understood.

G. COSEMANS: In the EMEP model 15% of the SO_2 emission in a grid-square of 150x150 km^2 is deposited onto that square without any consideration of meteorology. How would this figure of 15% change if meteorological variability is taken into account?

S. MYLONA: The recently revised routine EMEP/MSC-W model accounts for both meteorology and emission height in its estimates of the so-called local deposition factor. In areas where only meteorological variation exists, the revised value lies between 4 and 15%. Elsewhere, the combined emission height/meteorological dependent value lies between these values. Due to this modification the relative change in the concentration and deposition sulphur fields varies spatially mostly within the range $\pm$ 15%.

OZONE MODELING OVER A LARGE CITY BY USING A MESOSCALE EULERIAN METEOROLOGICAL AND TRANSPORT MODEL: MADRID CASE STUDY

Roberto San José, Luis M. Marcelo, Belen Moreno and Arturo Ramírez-Montesinos.

Group of Environmental Software and Modelling.
Computer Science School - Technical University of Madrid
Boadilla del Monte- 28660, Madrid (Spain)

INTRODUCTION

Air Quality models are essential tools in the understanding of pollutant dynamics in the atmosphere. In recent years, our understanding of the scientific foundations of the chemical and physical phenomena occurring in the atmosphere has continued to expand. We are able to construct comprehensive models that describe the dynamics of the air pollution. The inherent complexity and nonlinearity of the governing equations has made air quality modelling a computational "Grand Challenge". Because grid models cannot explicitly simulate processes which occur on spatial scales smaller than the grid cells, the importance of such processes must be assessed. If a particular subgrid processes is found to be significant to the objective of the simulation, either the grid resolution of the model should be increased or a subgrid parameterization should be developed to account for the effect of the subgrid process. This is an important and common factor to all mesoscale and regional atmospheric chemistry models, such as the Acid Deposition and Oxidant Model (ADOM[1]) and Regional Acid Deposition Model (RADM[2]) and the Sulfur Transport Eulerian Model (STEM[3]). Because of this limiting factor and essentially for the final objectives of these simulations, comparison with monitoring data can only be made taking into account that the data has very different nature and limiting processes in the simulation and measuring sides are present. With this in mind, we present in this contribution the results of a photochemical numerical mesoscale Eulerian simulation over the Madrid (Spain) urban and suburban area by using the data from a meteorological sounding at 0h00 and 12h00 in the Madrid International Airport (Barajas) as input meteorological information and Madrid Municipality Pollution Monitoring Network at two stations into the Urban Area. Because of enormous computing requirements we have restricted the investigation to one day (August, 15th, 1991) and also because of the data availability. Further investigations on speeding-up the code and parallelization should be performed.

The NUFOMO model is composed on several modules which have been previously developed by other authors and by ourselves. The MEMO (Mesoscale Meteorological Model) model[4] is a non-hydrostatic mesoscale model which is able to prognostic and diagnostic the wind components, temperature and humidity at every three dimensional cell. This module has been integrated into the NUFOMO system adapting the surface boundary conditions. The MEMO module includes an Eulerian transport module which allows to integrate the transport equation by using the meteorological fields and fluxes from the MEMO module. We have adapted this module to extend the transport to different pollutants and groups (in the case of organic volatile compounds). Figure

1 shows a scheme of the different codes developed to generate the interface for connecting to the chemical module which will be discussed further.

The NUFOMO integrated system contains an Emission Module which is able to describe the emissions from the different grid cells in the domain. These emissions are formed by the point sources, line sources, area sources and volume sources and include the biogenic and anthropogenic sources. The system includes a deposition module (DEPO)[14] based on the resistance approach[5] which has been slightly adapted to the South European ecosystems[15]. The canopy resistance depends on the land use type, on the pollutant, solar radiation and reactivity. The photochemical module is based on the scheme proposed by Jacobson and Turco[6] which is a Gear-type code that efficiently solves ordinary differential equations in large grid domains.

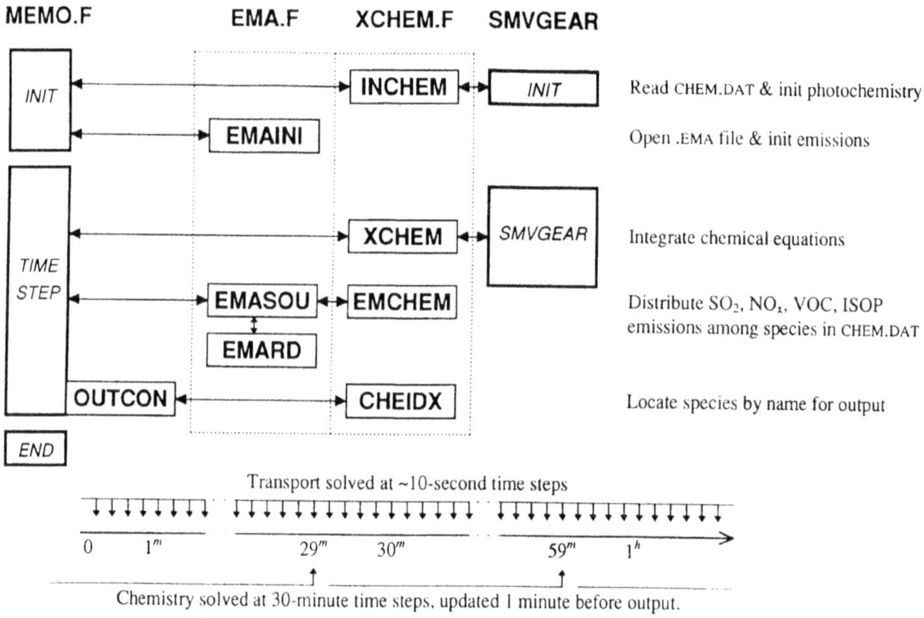

Figure 1. NUFOMO Modelling system.

In this contribution, we have used 45 species, 92 chemical reactions and 15 photodissociation processes. This code -which is able to be run for any number of chemical reactions, species and photochemical processes- is much faster than the classical Gear technique and it is as efficient as that one. In spite of that, the CPU time for running the NUFOMO system continue to be uncomfortable for medium workstations. The system is able to predict 24 hours of the SMVGEAR test reaction set in 88 hours, CPU time. We have used a RISC/6000/380 workstation. Parallelization and vectorizing the code for new generation of massively parallel computers seems to be the further step in order to reduce drastically the simulation time.

The domain topography is shown in Figure 2. The selected day is a holiday and is made during a period with low industrial and traffic emissions. The Meteorological conditions show a thermic low over the Spanish Peninsula which is expected to produce typical mesoscale clockwise turns in wind flows which is reproduced by the model.

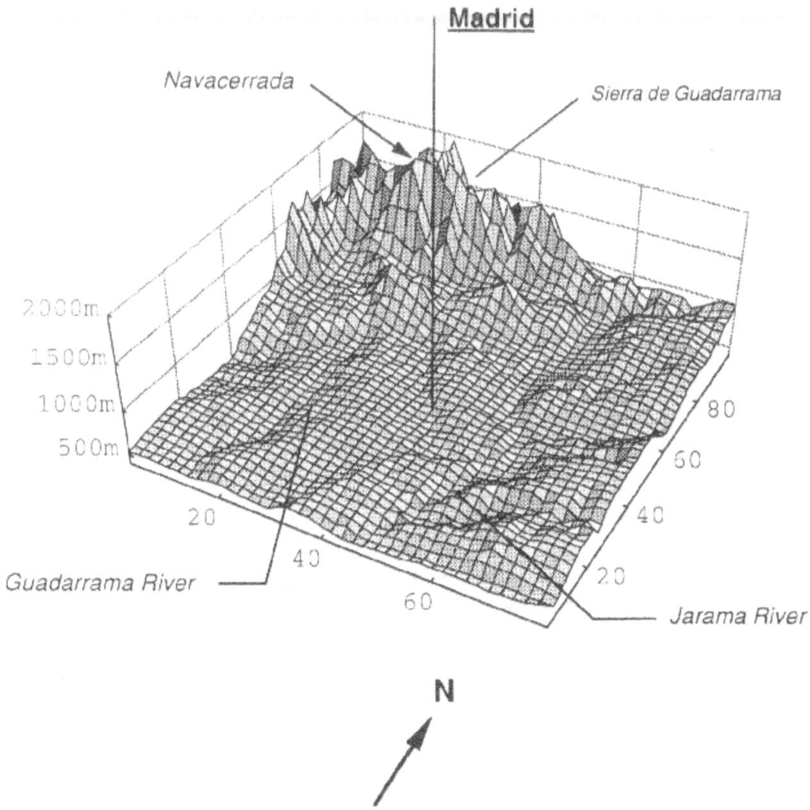

Figure 2. Domain Topography. Madrid, Spain (80 x 100 km^2).

EMISSION DATA FOR THE EPISODE

The emission database (EMIMA 2.0, Emission database for Madrid Area) for this study was prepared based on the CORINAIR[7] and EPA[8] methodologies and includes emission factors for SO_2, NO_x and VOC's (anthropogenic and biogenic). The VOC's biogenic emissions were based on a land-use classification and includes isoprene and monoterpene emissions. Point sources include the data provided by the INYPSA[9] report. This includes the most important industries in the Madrid Area. Line sources are mainly modelled by the urban and highway traffic. This traffic is modelled by using the mean velocity on each driving mode and calculate the emission factor with this velocity and calculate the oil consume. This is made for the main seven National roads crossing in Madrid and for the two Madrid rings (M-30 and M-40). The area sources are parameterized by considering the tertiary and residential activities which involve the production of hot water and heat.

Biogenic and hydrocarbon emissions depend strongly on the intensity of the light and the temperature. These effects are critical on the emission rate factors. Isoprene and monoterpene emissions are specifically estimated by using Andreani-Aksoyoglu and Keller[10] and Lamb et al.[11]. Emissions are input as kilograms per second per grid cell for the five species/groups SO_2, NO_x, anthropogenic VOCs, (biogenic) isoprene and (biogenic) terpenes. These emissions are then properly distributed among the species contained in each reaction set (e.g. 90% of NO_x emitted as NO, 10% as NO_2) and added to the cells at the bottom layer of the model grid at every time step

PHOTOCHEMICAL MODULE

The transport model is an extension of that used by MEMO 5.1 which has been extensively validated for a number of different locations[12]. This model reads initial meteorological soundings and local topography and land use and calculates non-hydrostatic wind, temperature, pressure and humidity fields. The transport model is incorporated to the Navier-Stokes wind and temperature partial differential equation system defined by MEMO, the transported quantity in this case being the mass of each chemical species. Terms included are advection, turbulent diffusion (both horizontal and vertical) and surface deposition. The deposition flux is incorporated by using the Wesely and Hicks[13] formulation which accounts for the resistance of the air to have deposition of a species. The deposition model is the DEPO 3.0[15] code in this version.

The kinetic equations for the chemical reactions are solved using the Sparse-Matrix Vectorized Gear code (SMVGEAR) developed by Jacobson and Turco[6]. This highly optimized code greatly contributes to reduce the computational requirements of our simulation by enabling us to accurately solve the kinetic equations using time steps as large as several minutes to half an hour. This means that SMVGEAR solver needs only to be called once for every many 10-to-20-second MEMO integration steps (Fig. 1). Another interesting feature of SMVGEAR is that the chemical reaction set is not hardcoded, but rather is read from an external file, so different reaction schemes may be tested with relative ease. Figure 3 shows the different modules of the chemical code with the temperature dependence and photochemical rates. This scheme is based on those adopted by Jacobson and Turco[6] in the SMVGEAR code as a test set.

RESULTS AND DISCUSSION

The ozone is the only element which we are going to use to compare with results. However, we will show data from SO_2 and NO_x. The meteorological model for this day is tested by using the meteorological sounding at Madrid International Airport. Results present a complex interaction surface-atmosphere. The wind flow on surface is North-South early in the morning and reverse clockwise during the strong convective conditions during the day because south area generates much higher values of sensible heat flux. This process is expected to send most of the ozone to the west side of the domain.

Figure 4 shows that at 18h00 most of ozone high concentrations is located in that side. Figure 5 shows the agreement between observed and simulated (photostationary -only NO_x and complete SMVGEAR module). The agreement at Puerta de Toledo station is excellent although the evening reduction is simulated early and stepped than in the measured values at the monitoring station. Maximum is located slightly early than observed which could be adjusted by tunning the urban emissions on NO_x and $VOC's$.

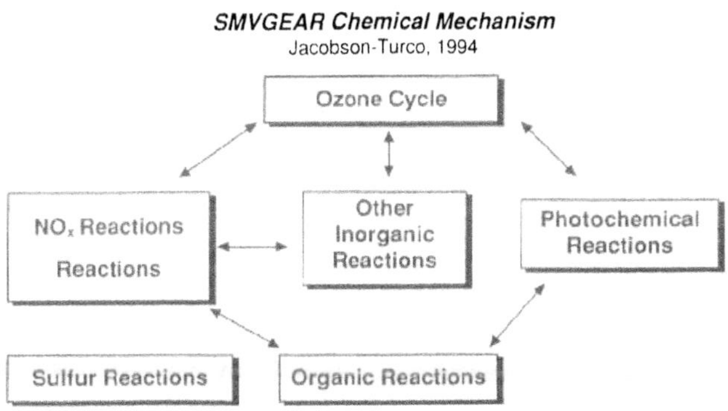

Figure 3. CBM-IV (SMVGEAR based) reaction scheme.

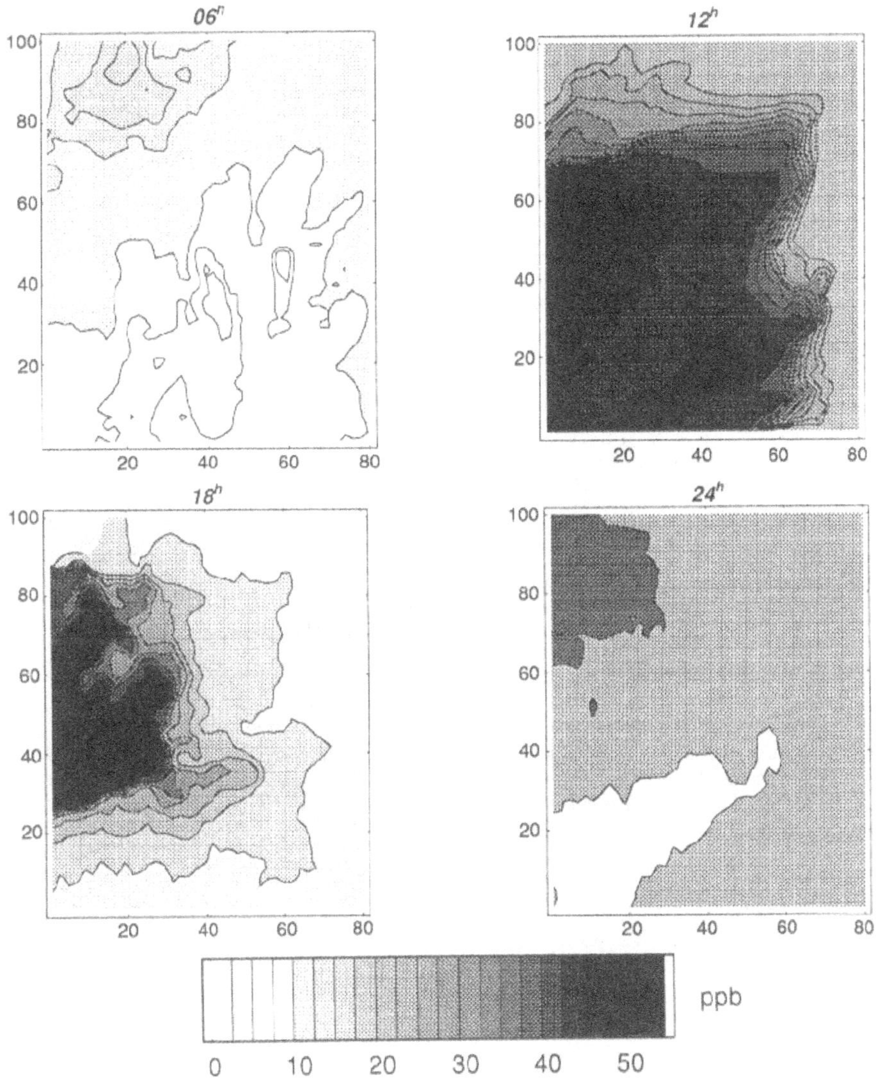

Fig. 4 Predicted ozone concentrations at surface level. CBM-IV-based chemistry, August, 15, 1991.

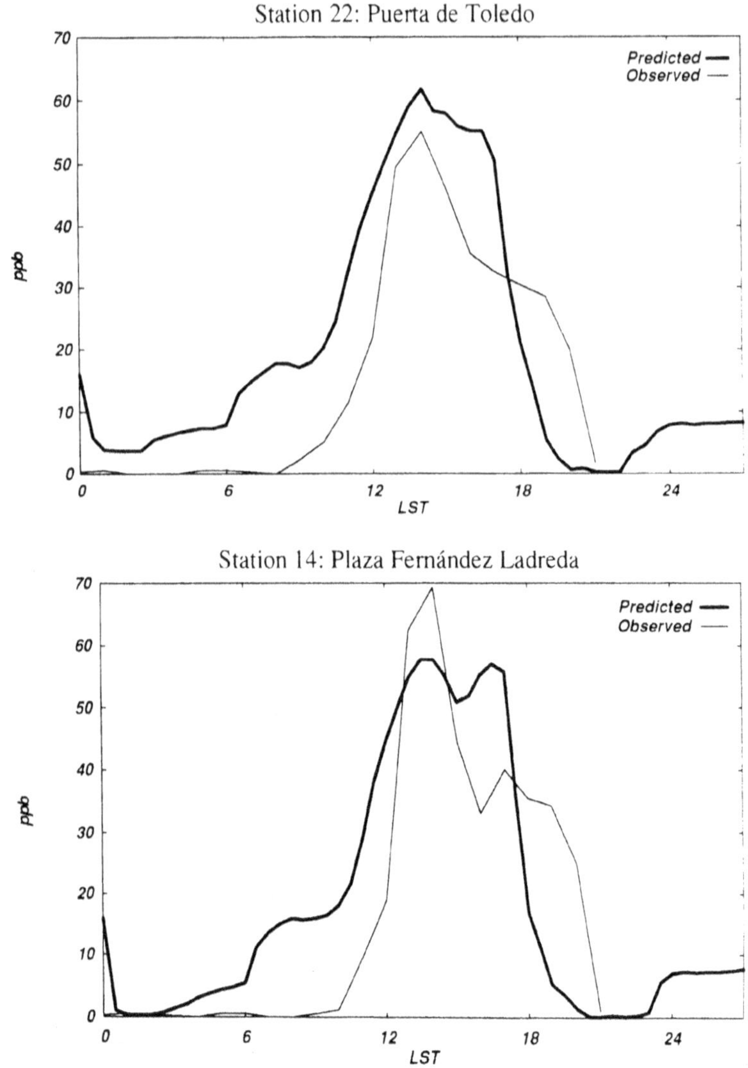

Fig. 5 Predicted and observed surface ozone concentration. August, 15th, 1991.

Further simulations should be performed on different days. Particularly, simulations during working days at different seasons (winter, spring and autumn are also needed). The emission is intended to be improved by adding the vaporative emissions. Longer simulations are also needed and they will be performed in further work. The preliminary results showed in this contribution encourages us for future deeper and longer simulations.

ACKNOWLEDGEMENTS

We wish to thank to Prof.-Dr. N. Moussiopoulos and Dr. T. Flassak for the MEMO model. Dr. Jacobson for the SMVGEAR code. The Madrid Municipality for the monitoring data. The

Spanish Meteorological Institute for the Madrid International Airport meteorological soundings. Pro.-Dr. Rosa González Barras from the Complutense University of Madrid for her help on the interpretation of the meteorological fields of the MEMO model.

REFERENCES

1.- Venkatram A., Karamchandani P. and Misra P. (1988) "Testing a comprehensive acid deposition model", Atmospheric Environment, 22, 737-747, 1988.

2.- Chang J.S., Brost R.A., Isaksen I.S.A., Madronich S., Middleton P., Stockwell W.R. and Walcek C.J. "A three dimensional Eulerian acid deposition model: physical concepts and formulation", J. Geophys. Res., 92, 14081-14700, 1987.

3.- Carmichael G.R., Peters L.K. and Kitada T. "A second-generation model for regional-scale transport/chemistry/deposition", Atmospheric Environment, 20, 173-188, 1986.

4.- Flassak T. "Ein nicht-hydrostatisches mesoskaliges Modell zur Beschreibung del Dynamik der Planetaren Grenzschicht", Ph. D., Fakultät für Maschinenbau der Universität Karlsruhe (Germany), 1989.

5.- Wesely M.L. "Parameterization of surface resistances to gaseous dry deposition in regional-scale numerical models", Atmospheric Environment, 23, 1293-1304, 1989.

6.- Jacobson M.Z. and Turco R.P. "SMVGEAR: A sparse-matrix vectorized gear code for atmospheric models", Atmospheric Environment, 28, 273-284, 1994.

7.- Bouscaren R., Veldt C. and Zierock K. H. "CORINE: Emission Inventory Project, Final Report", European Union, 1986.

8.- E.P.A. "Compilation of Air Pollutant Emission Factors", 3rd Edition, Report AP-42, 1977.

9.- INYPSA. "Estudio Inventario de emisiones contaminantes a la atmósfera de Madrid y su entorno industrial" 1986.

10.- Andreani-Aksoyoglu S. and Keller J. "Estimates of monterpene and isoprene emissions from the forests in Switzerland" (personal communication), 1994.

11.- Lamb B., Gay D. and Westberg H. "A biogenic hydrocarbon emission inventory for the U.S.A. using a simple forest canopy model", 1987.

12.- Moussiopoulos N.(Ed.) "The EUMAC Zooming Model. Model structure and applications", EUROTRAC, International Scientific Secretariat Garmisch-Partenkirchen, March, 1994.

13.- Wesely, M.L. and Hicks B.B "Some factors that affect the decomposition rates of sulfur dioxide and similar gases on vegetation", APCA J. 27, 1110-1126, (1977).

14.- San José R., Bilbao J., de Miguel A., Cachorro V.E., Viloria R.F., Tricio V., García A., Cancillo M.L., Casado H., Encinas D. and Palacios P. "An experimental and modelling study of SO_2 and O_3 dry deposition over Castilla-León (Spain)", Air Pollution Research Report 47, Ed. J. Slanina, G. Angeletti and S. Beilke, pp.133-138, (1993).

15.- Gery M.W., Whitten G.Z., Killus J.P. and Dodge M.C. "A photochemical kinetics mechanism for urban and regional scale computer modeling", Journal of Geophysical Research, 94, D10, 12925-12956, (1989).

DISCUSSION

I. ZIOMAS: Since the results of your model for ozone shows maximum ozone concentrations of about 60 ppb is your model practically reproducing the diurnal variations of background ozone? If so there is no effect of local NO_x emissions?

R. SAN JOSE: The model is run for a summer day in the middle of the holiday and tourist season. Background concentrations are considerably lower and the results are shown only for showing the performance of the complete tool. Highest ozone values are produced on June and July were the traffic is much more heavier.

X. LIN: How does the modelling system have a right representation of emission regionality? In other words, how do you set up the boundary conditions for advective chemical pollutants?

R. SAN JOSE: The chemical boundary conditions are set by using information from regional stations (outside of the area domain). However, the intention is to nest the actual system to a model over the spanish or even european domain in the near future.

H.-J. PANITZ: Did you only consider biogenic isoprene emissions or also hydrocarbons like alpha-pinene, monoterpenes and other?

R. SAN JOSE: We have considered the alpha-pinene as the most abundant monoterpene emitted. The model computes isoprene and monoterpene separately but no split is made for different types of monoterpenes.

THE ROLE OF MINERAL AEROSOLS IN TROPOSPHERIC CHEMISTRY

Yang Zhang[1], Li-Ling Chen,[2] Gregory R. Carmichael,[2] and Frank Dentener[3]

[1]Environmental and Energy Sciences Division
Battelle, Pacific Northwest Laboratories
Richland, WA 99352
[2]Department of Chemical & Biochemical Engineering
Center for Global & Regional Environmental Research
The University of Iowa
Iowa City, Iowa 52240
[3]Department of Air Quality
Wageningen University
The Netherlands

INTRODUCTION

Mineral aerosols are a significant component in the Earth and atmosphere system. These aerosols are mainly emitted from arid and semi-arid areas, which cover about one third of the earth's land surface. Estimate of the global source of mineral aerosol is of ~2000 Tg yr^{-1} (d'Almeida et al., 1991). In comparison, the global estimate of the secondary aerosols (e.g. carbonaceous substances, organics, sulfate, and nitrate) mainly form fossil fuel burning is ~400 Tg yr^{-1} (Preining, 1991).

Mineral aerosols play an important role in the global climate system and in many biogeochemical cycles. For example, they can affect the concentration and lifetime of many important atmospheric species by providing reaction sites and serving as carriers and/or sinks. Zhang et al. (1994) have shown that the presence of mineral dust can reduce the daytime NO_3, N_2O_5 and HNO_3 levels by up to 60% , the concentrations of OH, HO_2, and H_2O_2 by 25-98%. The decrease in these precursor concentrations consequently results in a decrease in O_3 production by up to 25% under typical dust storm conditions. Hirai et al. (1991) proposed that sulfate and nitrate are absorbed and/or formed on the surfaces of wetted dust particles, and displace the carbonates during the long range transport processes. Analyses of precipitation sampled during Asian dust events in Korea and Japan are consistent with this hypothesis and show that up to 75% of the carbonate is displaced by nitrate and sulfate by the time the particles reach Korea and Japan. Additional evidence shows that 70% or more of SO_4^{2-} and NO_3^- are formed on the dust surface relative to those on the surface of non-mineral aerosols (Nishikawa et al., 1991).

The preexisting deserts and the ongoing large scale desertification have caused the constant seasonal cycle of severe dust storms over many regions including the north Pacific, the mid-Atlantic, the Mediterranean, east Asia, Europe, Africa, and South America (Schutz and Jaenicke, 1974; Prospero et al., 1981; Nishikawa et al., 1991). For example, long range transport of dust is a common occurrence in east Asia in the springtime. It is estimated that 6-12 x 10^6 tons of Asian dust are transported annually to the central North Pacific (Uematsu et al., 1983). During dust storms the particle concentrations can exceed 600 $\mu g\ m^{-3}$ (Kang and Sang, 1991) with particle diameter ranging from 0.01 to 40 μm (Nishikawa et al., 1991). Recent observational data have also shown that elevated nitrate,

Air Pollution Modeling and Its Application XI, Edited by
Gryning and Schiermeier, Plenum Press, New York, 1996

sulfate and ozone concentrations are found on dust particles which are transported long distances.

In this paper the role of the mineral aerosols on the atmospheric chemical processes are investigated using a regional 3-D model (STEM-II) and a global 3-D transport model (MUGUNTIA). The models combine the heterogeneous chemistry on mineral aerosols with the gas-phase chemistry to evaluate the effects of the mineral dust on the photochemical oxidant cycle and the formation of nitrate and sulfate. The importance of dust surface reactions on the regional and the global scales are discussed.

MODEL DESCRIPTIONS

The Regional 3-D STEM-II model

The STEM-II (Sulfur Transport Eulerian Model-II) is an Eulerian model that accounts for the transport, chemical transformation and deposition of atmospheric pollutants in gas, cloud, rain, snow, and aerosol phases. The chemical mechanism is based on the Lurmann et al.'s condensed chemical mechanism (Lurmann et al., 1986), and is modified to include low NO_x conditions and explicit treatment of isoprene (Carmichael, 1986 and 1991). In this work, mineral aerosols are modeled as a transported species with a multi-component in composition and a multi-layer in micro-structure. A log-normal distribution is employed for dust particles with a diameter ranging from 0.1- 40 μm. The simulated aerosol dynamic processes include nucleation, condensation/evaporation, coagulation, absorption and deposition. Heterogeneous reactions of sulfur (including SO_2 and DMS), nitrogen (including NO_3, N_2O_5 and HNO_3), and odd hydrogen species (including OH, HO_2, H_2O_2) are simulated on the wetted dust surface. The aerosol nitrate is formed through the gas to particle conversions of NO_3, N_2O_5 and HNO_3. The sulfate formation mechanism includes the gas to particle conversion of SO_2 and nucleation of sulfuric acid and water vapor.

The 3-D simulations are performed for a dust storm period of May 1-14, 1987 for east Asia. This region covers about 63 degrees in longitude from 99° to 162° E, 36 degrees in latitude from 19.626° to 55.514° N and 0-8 km in altitude. The dimension of each grid box is 1.125°, 1.122° and 400 m (total 21 vertical layers including the surface level) in the zonal, meridional and vertical directions. The wind field data are obtained from the European Center for Medium Range Weather Forecasting (ECMWF) database. The emission inventories of NO_x and SO_2 are based on Fujita (1992) and Hameed and Dignon (1988). The anthropogenic emissions of CO, CH_4 and NMHCs are interpolated from the work of Saylor et al. (1991) and Piccot et al. (1992). A dust emission rate of ~2 x 10^4 g grid^{-1} s^{-1} is obtained using Gillette and Passi's method (1988). This rate results a daily dust emission of 4.3 x 10^4 to 5.5 x 10^5 tons during this dust period. Different initial values for the major gaseous and aerosol species are given over land and over sea. For example, the initial O_3 and dust concentrations are set to be 50 ppb and 50 μg m^{-3} over land and 20 ppb and 10 μg m^{-3} over sea, respectively.

To study the dust effects on tropospheric chemistry, a seperate simulation is performed to include the dust heterogeneous chemistry for the period of May 4-9, 1987. The results are compared against those obtained when the dust chemitry is not included. Results from these runs are also analyzed and compared against the observational data collected in east Asia.

The Global 3-D MUGUNTIA Model

The global 3-D transport MUGUNTIA Model is a coarse resolution model which can be used to simulate the global sources, transport and removal of mineral aerosols. The results of the dust simulations are stored and used to provide aerosol surface and mass for heterogeneous reactions of SO_2, HO_x, NO_y and O_3. The MUGUNTIA model used in this work has a horizontal resolution of 10° x 10° and 10 vertical layers of 100 hPa thickness (Zimmermann, 1988). Transport is described by monthly average winds and eddy diffusion profiles based on the standard deviations of those winds. In addition to the monthly mean transport, deep cumulus convection is paramaterized. Convective transport, however, is only applied to relatively insoluble trace gases, whereas all aerosol components including mineral aerosols are assumed to be effectively scavenged in the cumuli. The model includes background C_3-C_2-CH_4-CO-NO_x-HO_x photochemistry, coupled with reduced nitrogen (NH_x) and sulfur (DMS, SO_2, and SO_4) chemistry. Details of the chemistry calculations are presented by Langner and Rodhe (1991) and Dentener and Crutzen (1993 and 1994).

In the MUGUNTIA model no instantaneous winds are available so a statistical approach is chosen to describe the dust sources. Desert areal coverage for arid regions in Africa, Asia, Australia, South America and North America is calculated from Pye (1987). The monthly amount of dust storm days is calculated from Pye (1987), Schutz (1979) and Uematsu (1983). In areas for which we do not have detailed information on the seasonal distribution of dust storm days, the yearly amount of dust storm days are monthly distributed using the model winds. Aerosol size distributions for the dust storm are prescribed based on the measurements by d'Almeida and Schutz (1983) in Matam, Senegal. Matam is situated in an alluvial flood plain, with a relatively large fraction of small particles, which can be subjected to long-range transport after being uplifted into the atmosphere. Mineral aerosols are transported in 10 discrete size intervals.

MODEL RESULTS AND DISCUSSIONS

Regional Distributions of Mineral Dust

The concentrations of mineral dust exhibit highly varied spatial and temporal distributions during the dust storm period. The travel time for dust from the source areas to east Asia and the western North Pacific is on the order of 3-5 days. Figure 1 shows the dust concentrations in east Asia at 1.2 km at 3:00 p.m. for May 1 and May 4, 1987. The dust was emitted and uplifted into mid troposphere from the Gobi and the loess areas of Huang River by strong winds under a low pressure system on May 1. The maximum dust concentrations exceeded 200 $\mu g\ m^{-3}$ at and near the source regions after 6-hour emission. The dust front traveled eastward and southeastward under the westerly and northwesterly winds and covered almost the entire Pacific Rim region by May 4. The ground level dust concentrations ranged from 100-600 $\mu g\ m^{-3}$ during the dust storm in the region. After May 10, the ambient dust concentration was significantly reduced (< 50 $\mu g\ m^{-3}$) due to deposition into the North Pacific. These values are consistent with the measurement data.

Table 1 and Table 2 show the dust concentrations and the available surface areas at various altitudes for four locations: Beijing, Cheju, Happo, and Tokyo after 3-day and 7-day transport, respectively. After 3-day transport the maxima dust concentrations occurred between 0.8-2 km in altitude, with concentrations ranging from 79 to 221 $\mu g\ m^{-3}$. The dust concentrations at and near the surface were relatively low due to a large surface deposition. After 7-day transport, 25-70% of the dust deposited onto the surface or into the low layers. However, a high dust loading layer still persisted with dust concentrations ranging from 59.5 to 183 $\mu g\ m^{-3}$ at these locations. The available dust surface areas ranged from 0.1-2.64 $cm^2\ m^{-3}$. These surface areas are of great importance in providing the reaction sites for various absorbed species, and thus in affecting the photochemical oxidant cycle and the formation of sulfate and nitrate in the troposphere.

Table 1. Dust concentrations and surface areas at selected locations after 3-day transport.

	Beijing		Cheju		Happo		Tokyo	
Altitude km	Conc. $\mu g\ m^{-3}$	Area $cm^2 m^{-3}$	Conc. $\mu g\ m^{-3}$	Area $cm^2 m^{-3}$	Conc. $\mu g\ m^{-3}$	Area $cm^2 m^{-3}$	Conc. $\mu g\ m^{-3}$	Area $cm^2 m^{-3}$
0	38.5	0.5	16.2	1.9	14.3	0.2	10.5	0.1
0.4	91.8	1.1	83.7	1.0	55.0	0.7	35.7	0.4
0.8	137.0	1.6	124.0	1.5	98.0	1.2	79.0	0.9
2	159.0	1.9	221.0	2.64	129.0	1.5	123.0	1.5
4	83.3	1.0	122.0	1.5	75.2	0.9	71.8	0.9
6	47.8	5.7	79.6	1.0	33.6	0.4	45.6	0.5

Table 2. Dust concentrations and surface areas at selected locations after 7-day transport.

	Beijing		Cheju		Happo		Tokyo	
Altitude km	Conc. $\mu g\ m^{-3}$	Area cm^2m^{-3}	Conc. $\mu g\ m^{-3}$	Area cm^2m^{-3}	Conc. $\mu g\ m^{-3}$	Area cm^2m^{-3}	Conc. $\mu g\ m^{-3}$	Area cm^2m^{-3}
0.0	30.5	0.4	6.5	0.1	14	0.2	8.38	0.1
0.4	94.6	1.1	31.1	0.4	47.8	0.6	29.1	0.3
0.8	183.0	2.2	51.3	0.6	57.9	0.7	52.5	0.6
2.0	106.0	1.3	67.2	0.8	60.8	0.7	56.9	0.7
4.0	85.4	1.0	82.5	1.0	56.5	0.7	59.5	0.7
6.0	59.5	0.7	33.7	0.4	29.8	0.4	37.9	0.5

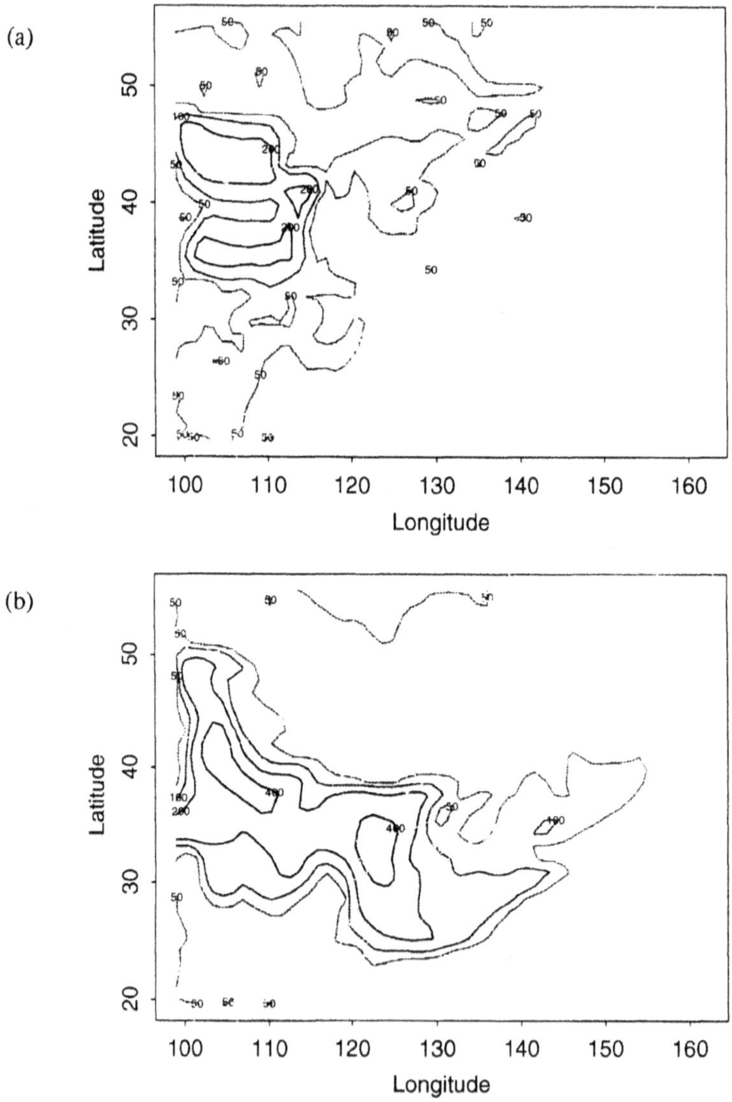

Figure 1. Dust concentration contours in $\mu g\ m^{-3}$ at 1.2 km at 3:00 p.m. local time on (a) May 1, and (b) May 4, 1987.

Chemical Interactions of Dust and Trace Gas Species

Coincident measurements of aerosols and O_3 at Happo, a high altitude (~2 km) surface site in Japan show that during dust storms the O_3 concentrations typically increase along with the dust loading. On hourly time scales, however, the concentrations of O_3 and dust appear to be anti-correlated (i.e., the O_3 concentration minima occurred with the dust maxima). Similar anti-correlations between O_3-n.s.s. (non-sea-salt) SO_4^{2-} and O_3-NO_3^- are also observed in the free troposphere over the eastern North Atlantic (Prospero and Savoie, 1989).

The reason for the complicated relationships between O_3 and dust is not clear. The elevated concentrations could result from the long range transport of O_3 along with the dust storm, the downward transport of O_3 from the stratosphere, the local photochemical formation, or a combination of these processes. The anti-correlations between dust and O_3 could also arise from a variety of processes. For example, a decrease in O_3 under a high dust loading could be due to a decrease in the production of O_3 in the presence of dust or the aerosol-rich air may have lower O_3-producing precursors.

Our results show that O_3 and dust were similar in their regional distributions, transport trends and plume dimensions, i.e., high O_3 concentrations occurred at regions where there was a high dust loading, and low O_3 values coincided with the low dust loading. This similarity is consistent with the observational elevated O_3 levels during the dust storms. While dust has a surface source, high O_3 air comes from the upper troposphere. Transport processes mix these species together into the middle troposphere and cause these spatial and temporal correlations.

Strong anti-correlations between dust and O_3 were also predicted when the dust heterogeneous processes were included in our simulation. On the average, dust caused a reduction of up to 40 ppb (~50% change!) in O_3 concentrations. Figure 2 shows the time series concentrations of mineral dust, ozone without dust effects, and ozone when dust-photochemical interactions were included at 0.8 km for Tokyo and Happo, Japan. In the absence of dust, the elevated ozone concentrations were found during the period of May 1-10, with the peak values of 90 and 70 ppb for Tokyo and Happo, respectively. The ozone concentrations were reduced 8-30 ppb in the presence of 20-150 $\mu g\ m^{-3}$ of dust. For example, a decrease of 5-22 ppb in O_3 concentrations was associated with to a variation of 20 to 80 $\mu g\ m^{-3}$ in dust concentrations at Happo. For comparison, the observed O_3 at Happo fluctuated by 5 to 25 ppb when the observed dust concentrations varied from 10 to 70 $\mu g\ m^{-3}$. The stronger anti-correlations between dust and O_3 concentrations were predicted at Happo because the decrease of O_3 concentrations due to dust heterogeneous chemistry was predominant in the relatively clean atmosphere. On the other hand, the variation in O_3 concentrations was not simply caused by the change in dust loading because a stronger photochemical production or destruction of O_3 could occur as a result of high O_3 precursor levels in the local air at Tokyo. The effects of dust were also propagated into the mid troposphere in this region.

Mineral dust is shown to provide important surface for n.s.s. sulfate and nitrate formation. The regional distributions of n.s.s sulfate and nitrate were very similar to those of the dust. High concentrations of sulfate and nitrate were usually found at those locations where high dust concentrations occurred. The predicted nitrate concentrations ranged from 0.007 to 0.33 $\mu g\ m^{-3}$. In comparison, the nitrate concentrations measured in the central Pacific by Prospero et al. (1989) were on the order of 0.3 -0.4 $\mu g\ m^{-3}$. The predicted n.s.s. sulfate ranged from 0.03-1.7 $\mu g\ m^{-3}$. A ratio of n.s.s. sulfate to nitrate was found to be about 4-5, which matched with our observational data.

Global Effects of Mineral Dust

The global aspects of reactions of trace gases on mineral aerosols is also investigated using the MUGUNTIA model. Figure 3 (a) and (b) show the annual averaged ratios at the earth's surface of sulfate present on mineral aerosols to total sulfate, and nitrate present on mineral aerosols to total nitrate, respectively. The fractions of sulfate associated with mineral aerosols in and near the dust source regions is very large, ranging from 60-80%. The regions where over 20% of the sulfate is associated with the mineral aerosols extend from east Asia out into the central Pacific Ocean and the southern hemisphere region of the Indian Ocean, span from central Africa to the northeastern portions of South America, and cover vast regions of the western USA, southern South America, and Australia. The calculated ratio of dust-nitrate to total nitrate is almost unity in Northern Africa, the Middle East and Asia, indicating that the buffering capacity by mineral dust is sufficient to neutralize the strong acidity by HNO_3. The regions where at least 40% of the total nitrate is found on the mineral aerosol cover vast regions of the Northern and Southern Hemispheres. The consequence of HNO_3 being present on mineral dust and not as much in the gas phase or in accumulation range aerosols may be significant. For example, HNO_3 reacting with mineral aerosols in or close to the dust source regions may be faster removed than the gas-phase HNO_3 by the sedimentation of the coarse mode aerosols. On the other hand, HNO_3 reacting with the fraction of the mineral aerosols subject to long range transport, may substantially enhance the transport distance of nitrate, and thus provide a means for HNO_3 to be transported to remote atmospheric regions, where it can participate in photochemistry after photodissociation of HNO_3.

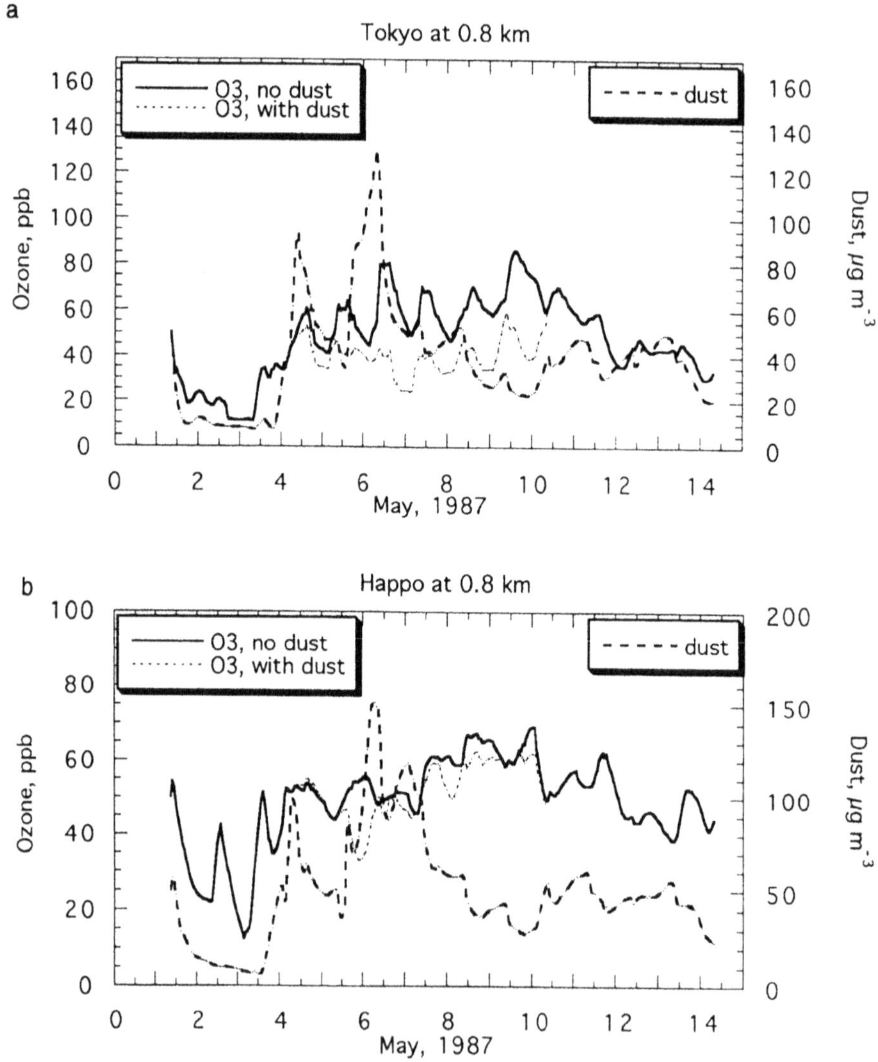

Figure 2. Time series concentrations of mineral dust, ozone without dust effects and ozone with dust effects at 0.8 km for (a) Tokyo, and (b) Happo Japan. The starting simulation time is 0 Z (zero Zulu) on May 1, i.e., midnight Greenwich Mean Time (GMT). This time corresponds to 9 a.m. Japan Standard Time (JST).

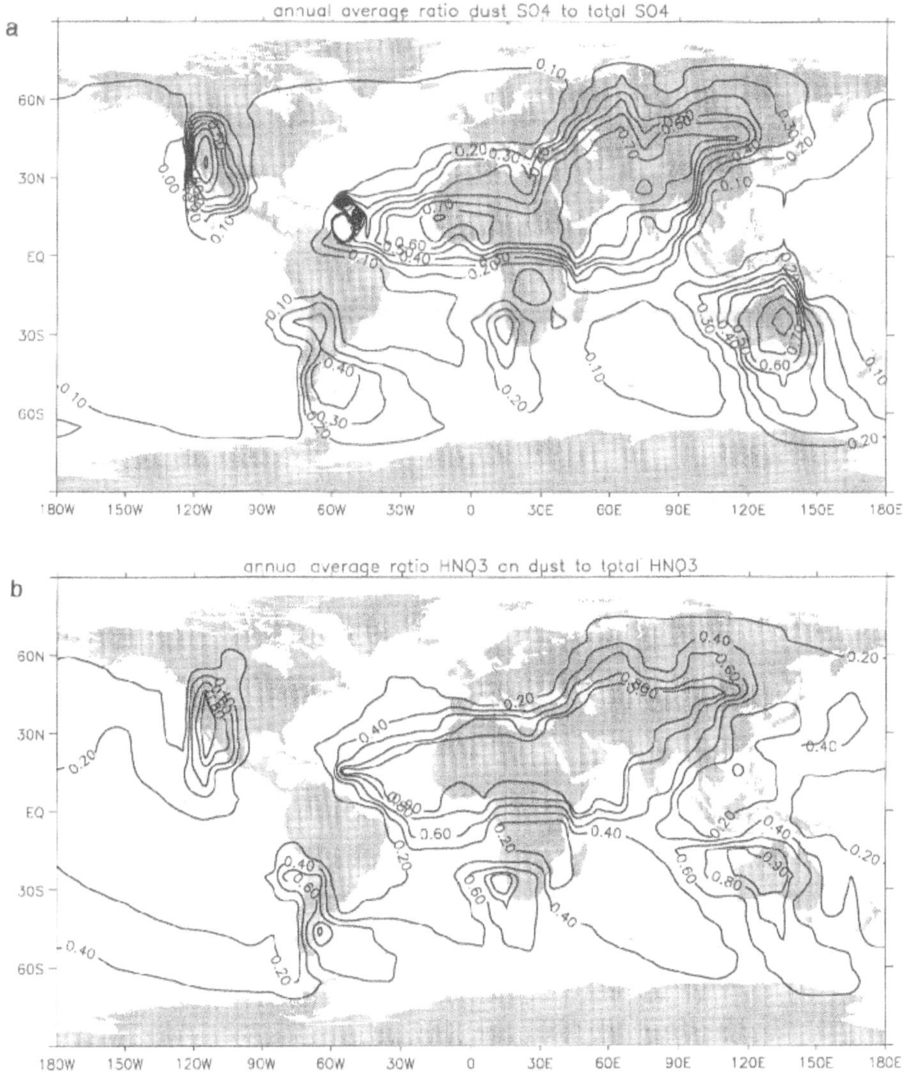

Figure 3. The annual averaged ratios at the earth's surface of (a) sulfate present on mineral dust to total sulfate, and (b) nitrate present on mineral aerosol to total nitrate.

Mineral aerosols may also affect the photochemical oxidant cycle by providing additional reaction pathways for species such as N_2O_5 and HO_2, which in turn can influence the ozone production and destruction rates. In addition, the direct destruction of ozone on the dust surface may also take place. The effects of the reactions of N_2O_5, HO_2 and O_3 on dust particles on ozone levels are evaluated. Ozone is reduced by more than 10% by including these reactions in the dust source regions during the high dust season and up to 8% on the yearly average. Outside the high dust regions ozone decreases are found to be very small. The HO_2 radical concentrations are calculated to decrease by about 10% in the dust source regions. However, the contribution of this reduced HO_2 concentrations to O_3 decrease is negligible, as the regions of highest dust concentrations generally have low

NO_x concentrations, and the photochemical production of O_3 by the reaction of NO with HO_2 is not important. Most of the calculated decrease of O_3 (2-6 %) is caused by the assumed direct O_3 destruction reaction on aerosols. The destruction rates of O_3 on mineral aerosols are extremely uncertain, with lower values making destruction negligible and higher values having some importance. Destruction of N_2O_5 on mineral aerosols could by itself be effective in removing NO_x from the atmosphere. However, in our calculations (following Dentener and Crutzen (1993)), we use, in addition to mineral aerosols, an accumulation range 'background' sulfate concentration, on which N_2O_5 is also removed. As a consequence of the SO_2 reactions on dust, this 'background' sulfate decreases, and the role of sulfate in removing N_2O_5 is taken over by dust. The removal of N_2O_5 on dust therefore has only a small "additional" effect compared to our reference simulation, which ignores the effects of dust.

CONCLUSIONS

The role of mineral aerosols in tropospheric chemistry has been studied using a regional 3-D model and a global 3-D model. A series of 3-D simulations using observed meteorological information has been conducted for the May 1-14, 1987 dust episode for east Asia to understand the observed chemical interactions of dust with tropospheric ozone, sulfate and nitrate. Results show that the heterogeneous chemistry of nitrogen and radical species on dust surface play am important role in affecting the total O_3 concentrations and regional distributions. This appears to be the major mechanism causing the anti-correlations observed in the hourly O_3 and dust measurement data. The presence of dust loading of 100-1000 $\mu g\ m^{-3}$ can cause a maximum of 40 ppb decrease (~50% !) in O_3 concentrations. On the global yearly average ozone and HO_2 are reduced by 8% and 10% when the dust heterogeneous reactions are included. The direct O_3 reaction on aerosols could be an important pathway for ozone destruction. In addition, mineral aerosols can provide important surface for sulfate and nitrate formation. At least 20 % of the predicted sulfate and nitrate are associated with mineral aerosol in vast regions of the Northern and Southern Hemispheres. The inclusion of dust transport and chemistry helps characterize tropospheric O_3 and provides insights for many aspects of tropospheric chemistry.

ACKNOWLEDGMENTS. We sincerely thank Mr. Mark Maclenna and Ms. Ching-Yi Chang for their assistance in 3-D model simulations and data analyses. This work was supported in part by NASA grant # *NAGW*-2428.

REFERENCES

Carmichael, G.R., Peters, L.K., and Kitada, T., 1986, A second generation model for regional-scale transport/chemistry/deposition, *Atmos. Environ.* 20(1): 173-188.

Carmichael, G.R., Peters, L.K., and Saylor, R.D., 1991, The Stem-II regional scale acid deposition and photochemical oxidation model-I. an overview of model development and applications, *Atmos. Environ.* 25A(10):2077-2090.

d'Almeida, G.A., Koepke, P., and Shettle, E.P., 1991, *Atmospheric Aerosols: Global Climatology and Radiative Characteristics*, A. DEEPAK Publishing, Hampton, Virginia.

d'Almeida, G.A., and Schutz, L., 1983, Number, mass and volume distributions of mineral aerosol and soils of the Sahara, *J. Clim. Appl. Meteor.* 22:233-243.

Dentener, F.J., and Crutzen, P.J., 1993, Reaction of N_2O_5 on tropospheric aerosols: impact on the global distributions of NO_x, O_3, and OH, *J. Geophys. Res.* 98:7149-7163.

Dentener, F.J., and Crutzen, P.J., 1994, A three-dimensional model of the global ammonia cycle, *J. Atmos. Chem.* 19:331-369.

Fujita, S., 1992, Acid deposition in Japan, in: *Report of the Central Research Institute of Electric Power Industry*, Japan.

Gillette, D. A., and Passi, R., 1988, Modeling dust emission caused by wind erosion, *J. Geophys. Res.* 93(D11):14233-14242.

Hameed, S., and Dignon, J., 1988, Changes in geographical distribution of global emission of NO_x and SO_x from fossil fuel combustion between 1966 and 1980, *Atmos. Environ.* 22: 441-450.

Hirai, E., Miyazaki, M., Chohji, T., Lee, M., Kitamura, M., and Yamaguchi, K., 1991, Effect of Kosa aerosol on inorganic components in rainwater collected from circum-pan-Japan-Sea area, in: *Proceedings of the 2nd IUAPPA Regional Conference on Air Pollution*, 2:27-34, *published by Korea Air Pollution Research Association, Seoul, Korea*,

Kang, K.H., and Sang, S.E., 1991, Influence of yellow sand on TSP in Seoul, in: *Proceedings of the 2nd IUAPPA Regional Conference on Air Pollution*, 2:1-7, *published by Korea Air Pollution Research Association, Seoul, Korea.*

Langner, J., and Rodhe, H., 1991, A global three-dimensional model of the tropospheric sulfur cycle, *J. Atmos. Chem.* 13225-13265.

Lurmann, F., Lloyd, A., and Atkinson, A., 1986, A chemical mechanism for use in long range transport/acid deposition computer modeling, *J. of Geophys. Res.* 91(10):10905-10936.

Nishikawa, M., Kanamori, S., Kanamori, N., and Mizoguchi, T., 1991, Kosa aerosol as eolian carrier of anthropogenic material, *The Science of the Total Environment* 107:13-27.

Piccot, S.D., Watson, J.J., and Jones, J.W., 1992, A global inventory of volatile organic compound emissions from anthropogenic sources, *J. of Geophys. Res.* 97:9897-9912.

Prospero, J.M.,, and Savoie, D.L., 1989, Nitrate concentrations over the Pacific: oceanic background and continental impacts, *Nature* 339: 687-689.

Preining, O., 1991, Aerosol and climate: an overview, *Atmos. Environ.* 25 A(11):2443-2444.

Prospero, J. M., Claccum, R. A., and Nees, R. T., 1981, Atmospheric transport of soil dust from Africa to South America, *Nature* 289:570-572.

Pye, K., 1987, *Aeolian Dust and Dust Deposits*, Academic, San Diego, California.

Saylor, R.D., and Peters, L.K., 1991, The global numerical simulation of the distribution of CO in the troposphere, in : *Air Pollution Modeling and Its Application VIII*, 485-495, H.V. Dop and D.G. Steyn, ed., Plenum Press, New York.

Schutz, L., 1979, Sahara dust transport over the North Atlantic ocean- model calculations and measurements, in: *Saharan Dust: Mobilization and Transport*, 267-277, C. Morales, ed., John Wiley, New York.

Schutz, L. W., and Jaenicke, R., 1974, Particle number and mass distributions above 10^{-4} cm radius in sand and aerosol of the Sahara desert, *J. of Applied Meteorology* 13(8):863-870.

Uematsu, M., Duce, R.A., Prospero, J.M., Chen, L., Merrill, J.T., and McDonald, R.L., 1983, Transport of mineral aerosol from Asia over the North Pacific ocean, *J. of Geophys. Res.* 88(C9):5343-5352.

Zhang, Y., Sunwoo, Y., Carmichael, G.R., and Kotamarthi, V., 1994, Photochemical oxidant processes in the presence of dust: an evaluation of the impact of dust on particulate nitrate and ozone formation, *J. of Applied Meteorology* 33(7): 813-824.

Zimmermann, P.H., 1988, Moguntia: A handy global tracer model, in: *Air Pollution Modeling and Its Applications VI*, H.V. Dop, ed., Plenum Press, New York.

DISCUSSION

J. CHING: Due to the large (relatively) size of the aerosols settling is active. The spatial distribution is likely strongly dependent on the wind speed and synoptic episodicity. Please discuss further.

G. CARMICHAEL: Yes indeed. Only particles > 20 μm are lifted into the free troposphere and then made available for long range transport. These particles do settle out, but can be transported as far as the Central Pacific.

B. HICKS: How sensitive are your conclusions to assumptions about ammonia and sea salt?

G. CARMICHAEL: We did include NH_3-sulfate aerosol simultaneously with the dust, so these effects are considered. Sea salt interactions we are just now looking into.

R. BORNSTEIN: Do your data provide any information on the impact of biomass burning in Brazil?

G. CARMICHAEL: No, I don't know the problems associated with smoke.

THE DANISH EULERIAN HEMISPHERIC MODEL

Jesper Christensen

National Environmental Research Institute
Frederiksborgvej 399
DK-4000 Roskilde
Denmark

ABSTRACT

Three dimensional Eulerian hemispheric air pollution model is in process of development at the National Environmental Research Institute (NERI). The model has been used to study long-range transport of pollution on the Northern Hemisphere. The present version of the model includes the long-range transport of sulphur di-oxide (SO_2), particulate sulphate (SO_4), humidity and liquid cloud water. The chemistry in the model is described by a simple linear oxidation of SO_2 to SO_4, and the wet deposition of SO_2 and SO_4 is estimated by using the amount of precipitation, which is calculated from the contents of liquid cloud water (see Christensen, 1994, 1995a, 1995b).

The model has been used to study the air pollution in the Arctic. Results for 3½ years run and an intensive analysis of the results will be presented: Analysis of the episodes and the meteorological conditions during the episodes are visualised using animation technique and will be presented by using video.

DESCRIPTION OF THE MODEL

The model is based on a set of coupled continuity equations (see Pudykiewicz, 1991)

$$\frac{\partial q_i}{\partial t} = -u\frac{\partial q_i}{\partial x} - v\frac{\partial q_i}{\partial y} - \dot{\sigma}\frac{\partial q_i}{\partial \sigma} + \frac{\partial\left(K_x\frac{\partial q_i}{\partial x}\right)}{\partial x} + \frac{\partial\left(K_y\frac{\partial q_i}{\partial y}\right)}{\partial y} + \frac{\partial\left(\Gamma^2 K_z\frac{\partial q_i}{\partial \sigma}\right)}{\partial \sigma} \tag{1}$$

$$- Q_i(c_1,c_2,\cdots,c_{nq})q_i + P_i(c_1,c_2,\cdots,c_{nq}) \qquad , \quad i=1,nq$$

where x and y are the horizontal coordinates in the east and north direction respectively, and u and v are wind velocities for the respective directions. The vertical coordinate σ is a terrain following coordinate and is defined by $\sigma = p/p_s$, where p is the pressure and p_s is the surface pressure, and $\dot{\sigma}$ is the vertical velocity. K_x and K_y are the horizontal diffusion coefficient and K_z is the vertical diffusion; $\Gamma = g\sigma/RT$, where g is the acceleration due to the gravity, R is the gas constant and T is the temperature. P_i and Q_i are the production term and loss term for specie i. nq is the number of species.

In the present version nq=4, where the four species are SO_2, $SO_4^{=}$, humidity and cloud water. The equations are coupled through chemistry, condensation and evaporation of cloud water, and the release of precipitation.

NUMERICAL SOLUTIONS OF THE TRANSPORT EQUATIONS

The Northern Hemisphere was transformed to a plane, by using the polar stereographic projection system true at 60° North (see, Christensen, 1995a). The horizontal space of the model is defined on a regular 96x96 grid that covers most of the Northern Hemisphere with a grid-distant of 150 km at 60°N, and the vertical space is defined on an irregular grid with 12 grid-points, which cover most of the whole troposphere up to σ level 0.4 ($\approx$ 7km). The time integration is done by splitting (1) into four sub-models:

1. 3-d Advection
2. Horizontal diffusion
3. Vertical diffusion (with dry deposition and surface emissions)
4. Upper level emissions, chemistry, condensation-evaporation-precipitation and wet deposition

The three dimensional advection equation is solved by using a pseudospectral algorithm (see, e.g., Christensen, 1993, 1995a) for the horizontal space derivatives and by using an one-dimensional finite element method for the vertical space derivatives (see Pepper et al., 1979, and Christensen, 1993, 1995a). The time integration of the advection equation is performed by using a predictor-corrector method with several correctors (see Zlatev, 1984; Christensen, 1993, 1995a). A horizontal Forester filter is applied in order to reduce spurious oscillations after each timestep (see Forester, 1976). This combination of the pseudospectral method and a Forester filter gives a very accurate numerical solution to the horizontal transport equation (see, e.g., Chock and Winkler, 1994).

The horizontal diffusion equation is solved by using a pseudospectral method (see Christensen, 1995a), and the vertical diffusion equation is solved by applying a finite element algorithm (see Pepper et al., 1979; Christensen, 1995a).

The pseudospectral algorithm demands periodical boundary conditions, which are imposed by some artificial sinks on the horizontal boundaries (Zlatev et al., 1992; Christensen, 1993). The boundary conditions at the ground are defined by that the fluxes out or in of the model domain are only due to the dry deposition and the surface emission. Free boundary conditions are applied for the upper boundary.

PHYSICAL PARAMETERS

In this section the main physical processes will be described. A much more detailed description is given in Christensen (1995a,1995b).

Meteorological parameters are from the European Centre for Medium-range Weather Forecasts (ECMWF), ECMWF/TOGA Basic Level III data sets and the supplementary sets on a 2.5°x2.5° grid. All the meteorological data are interpolated to the grid, used in the model. The data are given with a time resolution of 6 hour for the cloud cover, wind stress and the heat flux (the last two are accumulated fields) and 12 hour for the other, which are 3-d wind, temperature, humidity and geopotential for the upper levels (1000hPa, 850hPa, 700hPa, 500hPa, 400hPa, 300hPa, 250hPa, 200hPa, 150hPa, 100hPa, 70hPa, 50hPa, 30hPa, 10hPa, after 1. January 1992 also 925 hPa are available) and 10 meter wind, 2 m temperature, 2 m dewpoint, surface pressure, surface temperature and surface geopotential for the surface level.

The advection is calculated using the 3-d wind fields from the meteorological input. Horizontal diffusion is assumed to be constant: $K_x=K_y=2\cdot10^5$ m²/s.

The vertical diffusion is parameterized by using a K_z profile based on the Monin-Obukhov similarity theory coupled to either observations or computationally generated data (see e.g. Seinfeld, 1986) for the surface layer, and this K_z profile for the surface layer is extended to the whole boundary layer by using a simple extrapolation, which ensures that the value of K_z is low (=0.1 m²/s) above the boundary layer (see Christensen, 1994, 1995a, 1995b).

The height of the boundary layer in the model is calculated by a simple parameterization. This parameterization is based on a simplified energy balance equation for the internal boundary-layer, where the change of the advected height of the boundary layer depends on the mechanically and convectively generated energy and the dissipation of the energy (see Christensen, 1994, 1995a, 1995b).

Two parts of the condensation scheme described in Sundqvist et al. (1989), the stratiform precipitation and a simple parameterization of convective precipitation, are included in the hemispheric model. This will give a more physically correct description of precipitation, where transport of humidity and cloud water, lifetime of different types of clouds, evaporation of cloud water or rain drops, are all taken into account (see Christensen, 1994, 1995a, 1995b).

The emissions of anthropogenic sulphur are based on the global GEIA inventory of sulphur emissions, version 1A.1, for 1985 on a $1^o \times 1^o$ grid (see Benkowitz et al., 1995) . These emissions are redistributed to the grid used in the model. The EMEP emissions for 1990 (see Sandnes and Styve, 1992) are used for the part of grid, which is equal the EMEP grid. The emissions are distributed evenly up to 800 m above the surface. There are no seasonally variations of the emissions.

Natural emissions of dimethyl sulphide (DMS) from the open oceans and shelf waters are also included in the model, and they depend on the concentrations of DMS in the oceans and the windspeed in 10m (see, e.g. Tarrason et al., 1995, Christensen, 1995b). In the model it is assumed, that 44% of the DMS flux are going to SO_2 and 4% are going to SO_4 (see Hertel et al., 1993).

The chemistry is simple linear, where the oxidation rate of SO_2 to SO_4 depends on the height of the sun at midday ($=\theta(\text{lat},\tau)$, where lat is the latitude and τ is the time in the year). This spatial and temporal variation of the oxidation rate simulates the variations of OH in the gas phase and H_2O_2 and O_3 in the liquid phase, which all are depending on the solar radiation. The maximal oxidation rate is about $5.6 \cdot 10^{-6}\ s^{-1}$, where the sun is in zenith at midday, and the minimal rate is $0.3 \cdot 10^{-6}\ s^{-1}$ in the Arctic night. The actual expression for the oxidation rate is

$$ox_{SO_2} = \left[4.7 \cdot 10^{-6} \cdot \max\left(0.,\frac{\sin(\theta(lat,\tau))}{\sin(\theta(50^o N, 23 june))}\right) + 0.3 \cdot 10^{-6}\right] s^{-1} \qquad (2)$$

This oxidation rate ensures that the rate is close to the rate in the EMEP-model (Sandnes and Styve, 1992) at 50°N, and in the Arctic night the rate is close to the rate in the hemispheric model by Iversen (Iversen, 1989; Tarrason and Iversen, 1992).

u_* for land is calculated by using the similarity profile for the windspeed and the roughness, where the roughness is given in Voldner et al. (1986) for 7 different land surface types. The land use data are based on the work by Wilson and Henderson-Sellers (1985). The roughness and u_* for open water is computed using the 10 m windspeed from the dependence between roughness and u_* (see, e.g., Voldner et al., 1986).

The dry deposition velocity of SO_2 is based on the resistance method as given in Voldner et al. (1986):

$$v_{d_{SO_2}} = (r_a + r_b + r_c)^{-1} \qquad (3)$$

where r_a is the aerodynamic resistance, r_b is the laminar resistance and r_c is the surface resistance (see Voldner et al., 1986; see also Christensen, 1995b). For snow or ice covered surface there is added an extra surface resistance, which ensures that the deposition is very low, if the surface temperature is below -2°C (see Wesely, 1989).

The dry deposition velocity for $SO_4^=$ over land is (see Walcek et al., 1986; Seland et al., 1995)

$$v_{d_{SO_4}} = u_* / a \qquad (4)$$

where a is 500 except for a forest with leaves, where a is 100. For unstable conditions there is a correction to deposition velocity (see Walcek et al., 1986). The dry deposition for $SO_4^=$ over open water is based on the work by Slinn and Slinn (1980).

The wet deposition is parameterized by using a simple scavenging ratio formulation with different in-cloud and below-cloud scavenging ratios (see Tarrasón and Iversen, 1992; Christensen, 1994, 1995a, 1995b; Jakobsen et al., 1995). For SO_2 the in- and below-cloud scavenging ratios are $3 \cdot 10^5$ and $1.5 \cdot 10^5$ respectively, while for $SO_4^=$ they are $7 \cdot 10^5$ and $1 \cdot 10^5$.

NUMERICAL RESULTS

The model has been run from October 1990 to May 1994. The calculated mixing heights and precipitation have been compared for a part of the period with data given by the EMEP Meteorological Synthesizing Centre - West, Norwegian Meteorological Institute. These comparisons show a reasonably agreement (see Christensen, 1994, 1995a, 1995b).

The calculated concentrations of SO_2 and $SO_4^=$ have been compared with the European measurements of EMEP. These comparisons show that there is as good agreement between the calculated and measured concentrations, both on seasonally, monthly and even daily basis for many EMEP stations. (see Christensen 1995a, 1995b); for the wet depositions and precipitation the agreement is not so good as for the air concentrations, mainly because it is very difficult to calculate the right amount of precipitation at

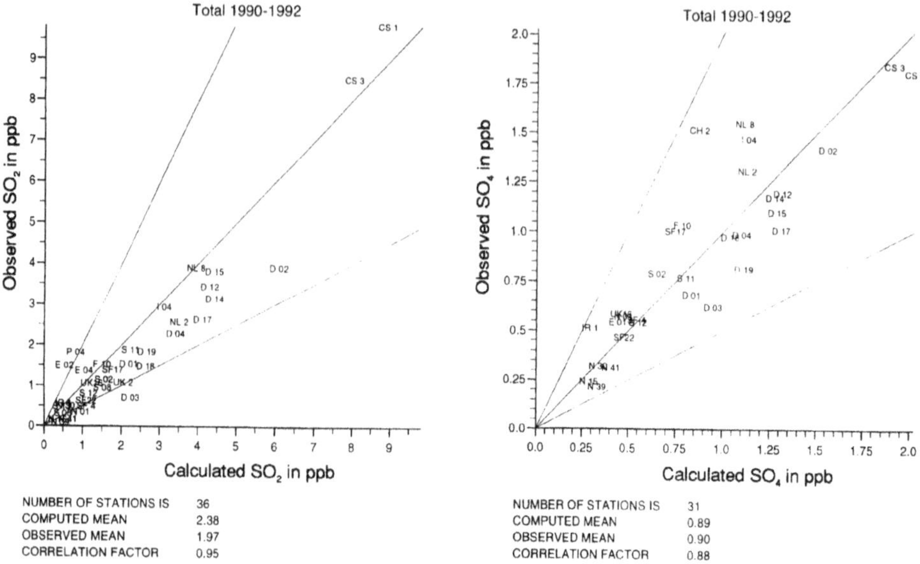

Figure 1. Scatterplots for the comparison between the average concentrations from October 1990 to December 1992 of SO_2 and $SO_4^=$ as monitored by EMEP and calculated by DEHM.

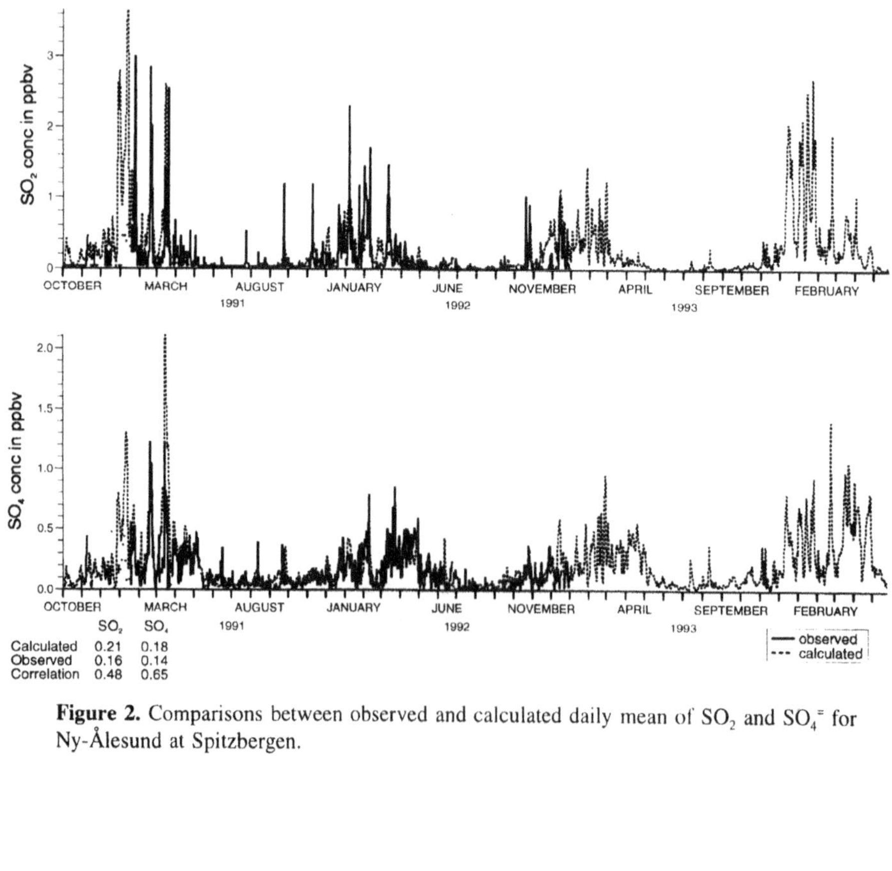

Figure 2. Comparisons between observed and calculated daily mean of SO_2 and $SO_4^=$ for Ny-Ålesund at Spitzbergen.

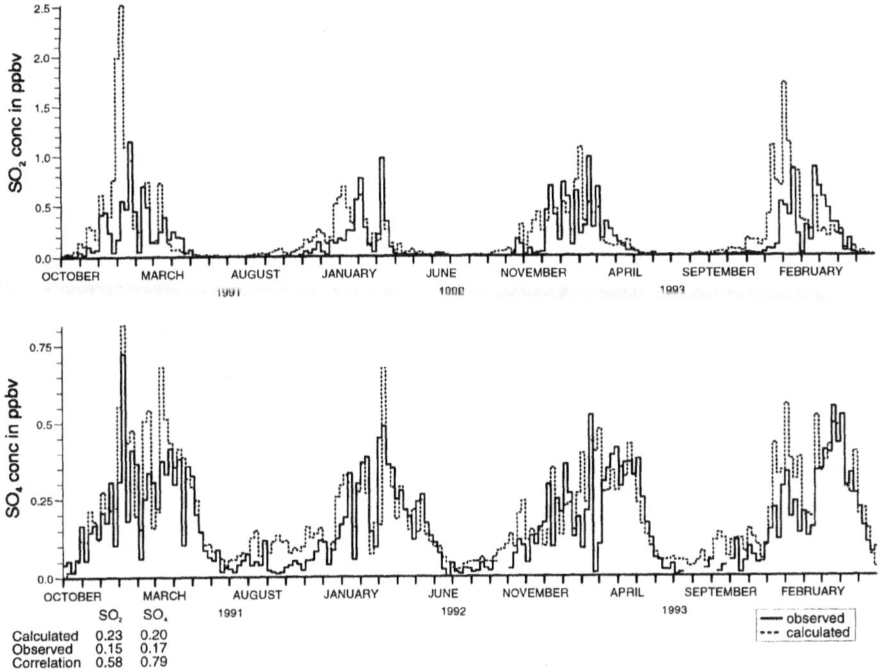

Figure 3. Comparisons between observed and calculated weekly mean of SO_2 and $SO_4^=$ for Station Nord in North-Eastern Greenland.

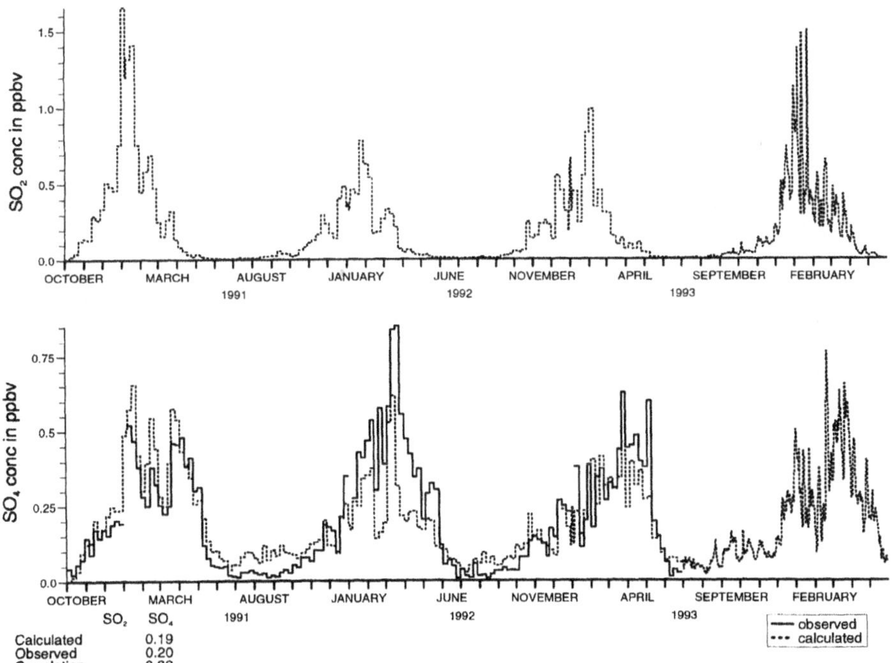

Figure 4. The calculated SO_2 and the comparisons between observed and calculated weekly mean of $SO_4^=$ for Alert, Northwest Territories, Canada.

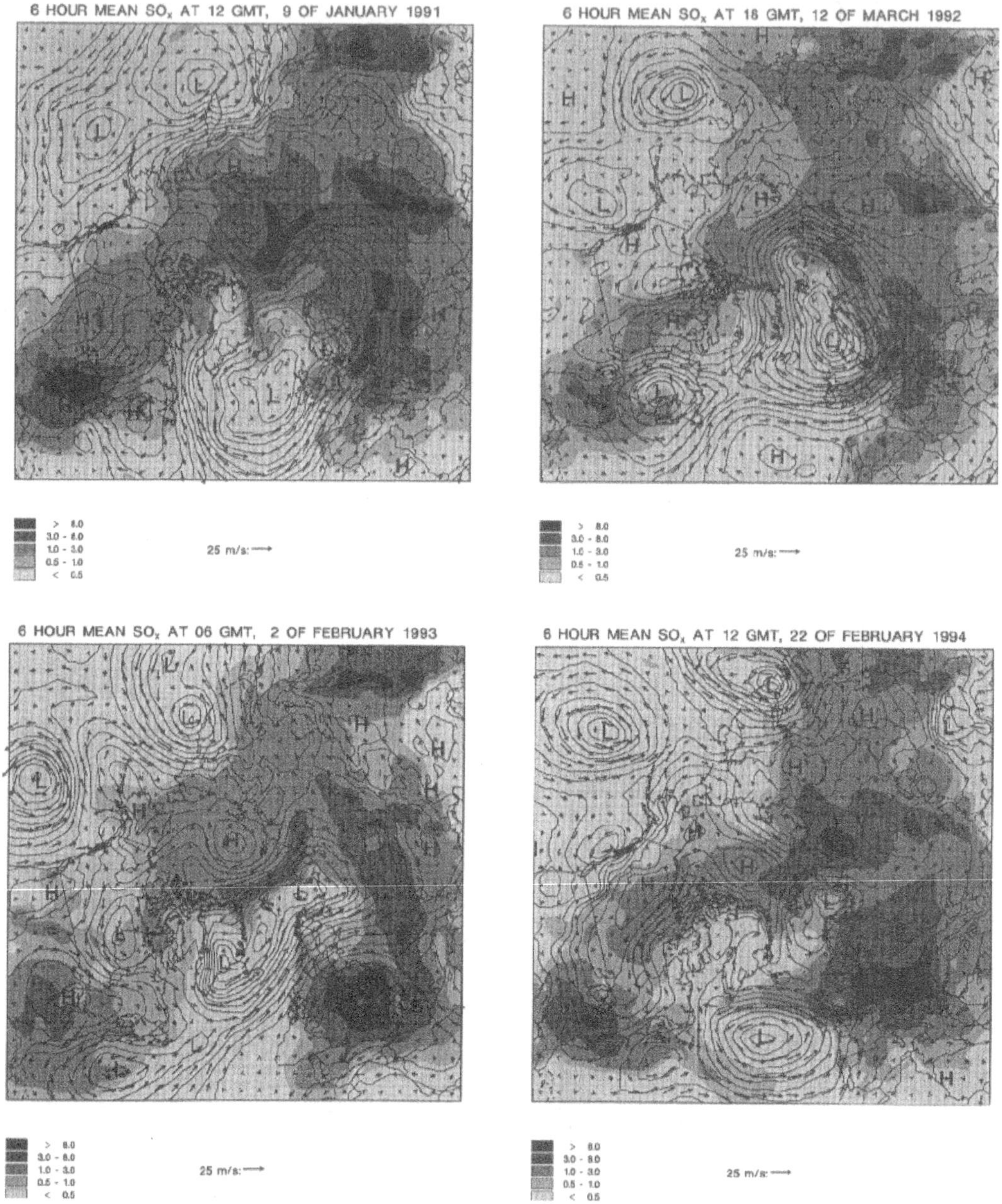

Figure 5. Four examples of transport of SO_x to the Arctic. The concentrations are for the surface level in ppbv, the contour lines are the MSL pressure, and the windfield is for σ level 0.935.

the right time. Fig. 1 shows two scatter plots with the total calculated and observed mean concentrations of SO_2 and $SO_4^=$ for the period from October 1990 to December 1992 for these stations, which have measured continuously for the whole period. This figure shows that there is good correlation between observed and calculated concentration of SO_2 and $SO_4^=$.

For the Arctic areas the model has been compared with measurements from three sites: Station Nord in northeastern Greenland (81° 36' N, 16° 40' W) for the whole period, see Wåhlin (1993), Ny-Ålesund at Spitzbergen (78° 54' N, 11° 53' E) from October 1990 to December 1992, see Schaug et al. (1992, 1993, 1994), and Alert in Northwest Territories, Canada, (82° 50' N, 62° 30' W) for the period from October 1990 to June 1993 and only of $SO_4^=$, see Li and Barrie (1993). For all three stations there is a rather good agreement between the calculated and observed concentrations of SO_2 and especially for $SO_4^=$, see Fig. 2, Fig. 3 and Fig. 4.

A video animation of the model calculations is available. Such an animation is a strong tool to visualize and analyze the model calculations, because it shows the temporal variations of both the meteorology data, which drives the model, and the calculated concentrations. In Fig. 5 there is four still pictures from the animation, which show examples of transport episodes in the Arctic, one for each winter period. Shown are the surface concentrations of SO_x ($=SO_2+SO_4^=$), the windfield at σ level 0.935, the contour lines for the Mean-Sea-Level pressure, and the centres for the low and high pressure systems at the figure.

CONCLUDING REMARKS AND PLANS FOR THE FUTURE

A Eulerian long-range model, covering most of the Northern Hemisphere, was presented here. The chemistry is a simple linear SO_2-$SO_4^=$ chemistry. The model has a condensation scheme to calculate the rate of the release of precipitation, which is very important for the deposition of SO_2 and $SO_4^=$.

The model has been run for 3½ years from October 1990 to May 1994. The model calculations of sulphur species have been compared with measurements in Europe (EMEP stations) and with three monitoring stations in the Arctic. The model reproduced very well the measured concentrations both in Europe and in the Arctic.

A video with an animation of the calculated concentrations is made.

Plans for the future are improvement of the parameterization of physical processes. It is planned to extend the model for other species, which are especially important for the Arctic: Persistence Organic Pollutants, Heavy Metals, and finally to improve the chemistry, both the gas-phase and the aqueous-phase.

The model development is an important contribution to the Danish part of the International Arctic Monitoring and Assessment Programme, AMAP.

ACKNOWLEDGEMENTS

The computing expencies were financed by the Danish Science Research Council. The ECMWF is acknowledged for the meteorological data: "ECMWF 1994. The Description of the ECMWF/WCRP Level III-A Global Atmospheric Data Archive". Len Barrie at Atmospheric Environment Service in Canada has provided the measurements at Alert. Jan Schaug at NILU, Norway, provided the EMEP measurements, especially from Spitzbergen. The staff at NERI and especially the field staff at Station Nord is acknowledged for the measurements from Station Nord.

REFERENCES

Benkowitz, C., M., J. Dignon, J. Pacyna, T. Scholtz, L. Tarrasón, E. Voldner and T. E. Graedel (1995): Global Inventories of Anthropogenic Emissions of SO_2 and NO_x. (in preparation).

Chock, D. P. and S. L. Winkler (1994): A Comparison of Advection Algorithms Coupled with Chemistry. Atm-Env.,28, 2659-2675.

Christensen, J. (1993): Testing Advection schemes in a Three Dimensional Air Pollution Model. Mathl. Comput. Modelling, 18, 75-88.

Christensen (1994): A Three Dimensional Hemispheric Air Pollution Model Used for the Arctic. In: "Air Pollution Modelling and its Applications X". (S.-E. Gryning and M. M. Millán, eds), Plenum Press, New York 119-127.

Christensen, J. (1995a): Transport of Air Pollution in the Troposphere to the Arctic. PhD thesis. National Environmental Research Institute, DK-4000 Roskilde, Denmark, 377 pp.

Christensen, J. (1995b): The Danish Eulerian Hemispheric Model - Three Dimensional Air Pollution Model Used for the Arctic. In preparation.

Forester, C. K. (1976): Higher Order Monotonic Convective Difference Schemes. J. Comput. Phys., 23, 1-22.

Hertel, O., J. Christensen and Ø. Hov (1994): Modelling of the Endproducts of the Chemical Decomposition of DMS in the Marine Boundary Layer. Atm. Env., 28, 2431-2449.

Iversen, T (1989): Numerical Modelling of the Long Range Atmospheric Transport of Sulphur Dioxide and Particulate Sulphate to the Arctic. Atmos. Env., 23, 2571-2595.

Jakobsen, H. A., E. Berge, T. Iversen and R. Skålin (1995): Status of the Development of the Multilayer Eulerian Model. Note No 3/95. EMEP Meteorological Synthesizing Centre - West, Norwegian Meteorological Institute, PO Box 43, Blindern, N-0313 Oslo 3, Norway.

Li, S.-M. and L. A. Barrie (1993): Eleven Years of Methane Sulphonic Acid Observations in the High Arctic. In: "Proceedings of the 5th International Symposium on Arctic Air Chemistry, Copenhagen, Denmark, September 8-10, 1992". (N. Z. Heidam, ed). NERI Technical Report NO. 70, National Environmental Research Institute, P. O. Box 358, Frederiksborgvej 399, DK-4000 Roskilde, Denmark.

Pepper, D. W., C. D. Kern and P. E. Long, Jr, (1979): Modelling The Dispersion of Atmospheric Pollution Using Cubic Spline and Chapeau Functions. Atmos. Env., 13, 223-237.

Pudykiewicz, J. (1991): Environmental Prediction Systems: Design, Implementation Aspects and Operational Experience with Application to Accidental releases. In: "Air Pollution Modelling and its Applications VIII". (H. van Dop and D. G. Stein, eds), Plenum Press, New York 145-159.

Sandnes, H. and H. Styve (1992): Calculated Budgets for Airborne Acidifying Components in Europe, 1985, 1987, 1988, 1989, 1990 and 1991, Report No 1/92. EMEP Meteorological Synthesizing Centre - West, Norwegian Meteorological Institute, PO Box 43, Blindern, N-0313 Oslo 3, Norway.

Schaug, J., U. Pedersen, J. E. Skjelmoen and I Kvalvågnes (1992): Data Report 1990. Part 2: Monthly and Seasonal Summaries, EMEP/CCC-Report 3/92. Norwegian Institute For Air Research, PO Box 64, N-2001 Lillestrøm, Norway.

Schaug, J., U. Pedersen, J. E. Skjelmoen and I Kvalvågnes (1993): Data Report 1991. Part 2: Monthly and Seasonal Summaries, EMEP/CCC-Report 5/93. Norwegian Institute For Air Research, PO Box 64, N-2001 Lillestrøm, Norway.

Schaug, J., U. Pedersen, J. E. Skjelmoen and K. Arnesen (1994): Data Report 1992. Part 2: Monthly and Seasonal Summaries, EMEP/CCC-Report 5/94. Norwegian Institute For Air Research, PO Box 100, N-2007 Kjeller, Norway.

Seinfeld, J. H. (1986): Atmospheric Chemistry and Physics of Air Pollution. John Wiley & Sons, New-York.

Seland, Ø., A. Pul, A. Sorteberg and J.-P. Tuovinen (1995): Implementation of a Resistance Dry Deposition Module and a Variable Local Correction Factor in the Lagrangian EMEP Model. Report No 3/95. EMEP Meteorological Synthesizing Centre - West, Norwegian Meteorological Institute, PO Box 43, Blindern, N-0313 Oslo 3, Norway.

Slinn, A. A. and W. G. N. Slinn (1980): Predictions for Particle Deposition on Natural Waters. Atm. Env., 14, 1013-1016.

Sundqvist, H., E. Berge and J. E. Kristjansson (1989): Condensation and Cloud Parametrization Studies with a Mesoscale Numerical Weather Prediction Model. Mon. Wea. Rev., 117, 1641-1657.

Tarrasón, L. and T. Iversen (1992): The Influence of North American Anthropogenic Sulphur Emissions over Western Europe. Tellus, 44B, 114-132.

Tarrasón, L., S. Turner and I. Fløisand (1995): Estimation of Seasonal Dimethyl Sulphide Fluxes over the North Atlantic Ocean and their Contribution to European Pollution Levels. J. Geophys. Res., 100, 11623-11639.

Voldner, E. C., L. A. Barrie and A. Sirois (1986): A Literature Review of Dry Deposition of Oxides of Sulphur and Nitrogen with Emphasis on Long-Range Transport Modelling in North America. Atm. Env., 20, 2101-2123.

Walcek, C. J., R. A. Brost, J. S. Chang and M. L. Wesely (1986): SO_2, Sulfate and HNO_3 Deposition Velocities Computed using regional Landuse and Meteorological Data. Atmos. Env., 20, 949-964.

Wesely, M. L. (1989): Parameterization of Surface Resistances to Gaseous Dry Deposition in Regional-Scale Numerical Models. Atm. Env., 23, 1293-1304.

Wilson, M. F. and A. Henderson-Sellers (1985): A Global Archive of Land Cover and Soils Data for Use in General Circulations Models. J. Clim., 5, 119-143.

Wåhlin, P. (1993): A Receptor Model Applied to Aerosol Data from Northeastern Greenland. In: "Proceedings of the 5th International Symposium on Arctic Air Chemistry, Copenhagen, Denmark, September 8-10, 1992". (N. Z. Heidam, ed). NERI Technical Report NO. 70, National Environmental Research Institute, P. O. Box 358, Frederiksborgvej 399, DK-4000 Roskilde, Denmark.

Zlatev, Z. (1984): Application of Predictor-Corrector Schemes with Several Correctors in Solving Air Pollution Problems. BIT, 24, 700-715.

Zlatev, Z., J. Christensen and Ø. Hov (1992): A Eulerian Model for Europe. J. Atmos. Chem., 15, 1-37.

NEW DEVELOPMENTS

chairmen:	H. Erbrink R. San Jose
rapporteurs:	R.J. Barthelmie H.-J. Panitz

MESOSCALE MODELING OF CLOUDS AND AEROSOL PARTICLES

Andrea I. Flossmann

Laboratoire de Météorologie Physique
Université Blaise Pascal - C.N.R.S
24, avenue des Landais
63177 Aubière Cedex
France

INTRODUCTION

Aerosol particles are present everywhere in the atmosphere. They are emitted naturally or anthropogenically into the atmosphere, or are produced by gas to particle conversion. They grow through coagulation and adsorb various trace gases, among those water vapor. On the particle surfaces a variety of chemical reactions can occur.

Recently, aerosol particles have received renewed attention as they not only participate in the atmospheric pollution but are also suspected to influence our climate through their direct and indirect radiative properties.

The loading of the atmosphere with aerosol particles is determined on the one hand side by the rate at which they are introduced into the atmosphere and on the other hand by the rate at which they are removed by dry deposition and wet deposition involving clouds and precipitation.

In the following, we will study the interaction of clouds and aerosol particles. These interactions are numerous. In the first place, aerosol particles serve as cloud condensation nuclei forming the cloud drops. Thus, these aerosol particles are scavenged from the atmosphere and when the cloud precipitates this contributes to a cleaning of the air. Also, by determining the number of cloud drops formed the aerosol particles determine the optical properties of the cloud (indirect radiative effect or Twomey effect; Twomey et al, 1984), i.e. cloud albedo. And when the cloud evaporates, the aerosol particles released back into the atmosphere will have changed due to in-cloud processes. We will investigate these processes in the troposphere with the help of high resolution detailed cloud models. Here, we will distinguish between pure liquid phase clouds (warm clouds) and mixed clouds (cold clouds), disregarding pure ice clouds (cirrus). We will shortly review the scavenging processes occurring in these clouds and then study their effect and their contributions in the framework of our detailed dynamical and microphysical model DESCAM. We will look at drop size dependent effects as well as the total numbers of material processed by convective clouds.

Air Pollution Modeling and Its Application XI, Edited by
Gryning and Schiermeier, Plenum Press, New York, 1996

SCAVENGING PROCESSES IN LIQUID PHASE CLOUDS

Nucleation scavenging

Aerosol particles normally contain at least a small fraction of soluble material, i.e. salts. Consequently, they take up water vapor from the ambient air. In sub-saturated air their size represents an equilibrium value which is determined by the mass of the dry particle, its chemical composition and the relative humidity and is described by the Köhler equation (see e.g. Pruppacher and Klett, 1978):

$$\frac{e_a}{e_{sat,w}} = \exp\left[\frac{2M_w\sigma_{s,a}}{RTa\rho_w} - \frac{\upsilon\Phi_s\varepsilon_v M_w\rho_s r_N^3}{M_s(a^3 - r_N^3)\rho_w}\right]$$

Here, e_a is the water vapor pressure over the curved moist aerosol particle and $e_{sat,w}$ is the saturation vapor pressure over a flat and pure water surface. M_w and M_s are the molecular weight of water and the salt of the aerosol particle, $\sigma_{s,a}$ is the surface tension over the solute droplet, R is the universal gas constant, T is the temperature, ν is the number of ions into which the salt dissociates, Φs is the osmotic coefficient of the solution, ε_v is the volume fraction of the soluble material, ρ_w and ρ_s are the densities of the water and the salt, a and r_N are the radii of the moist and the dry aerosol particle, resp. This equation is schematically displayed in Fig.1.

The equation remains valid even in slightly supersaturated air, depending on the radius of the particle. Eventually, however, the size of the aerosol particles passes the critical point given by the maximum of its Köhler curve. Then, it doesn't need a further increase in humidity to continue its growth. The solution droplet will grow even though the supersaturation might drop. Consequently, we will call such a droplet *activated*. The more soluble and the larger an aerosol particle is, the easier it can activated, i.e. the lower the necessary supersaturation.

The number of aerosol particles activated, i.e. the number of cloud condensation nuclei, is a function of the aerosol particle loading of the atmosphere. In marine air masses ~100 particles per cm^3 activate while in highly polluted areas as much as several 1000 drops per cm^3 will form which might represent 10 - 90% of the initial aerosol number concentration. Among those are preferably the largest aerosol particles which represent the main particle pollution mass. Consequently, about 80 - 99% of the initial aerosol particle mass concentration are scavenged into the drops during the process of nucleation.

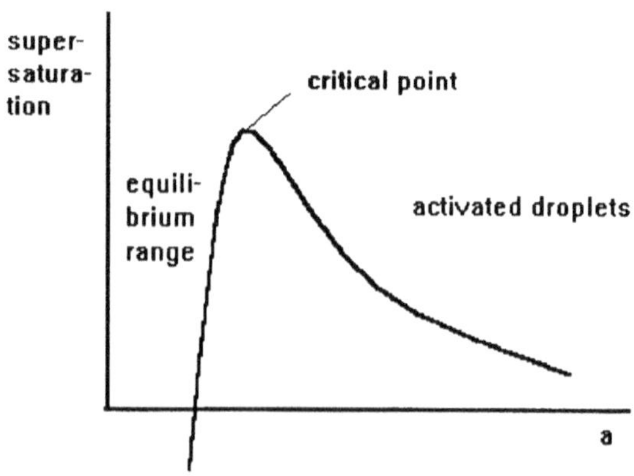

Fig.1: Schematic display of the Köhler curve.

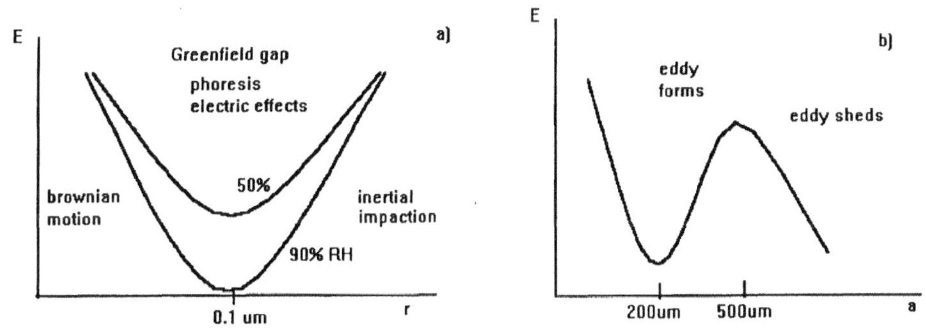

Fig.2: Schematic display of the impaction scavenging efficiency E as a function of aerosol particle radius r (a) and drop radius a (b).

Impaction scavenging

Once formed the drops grow through condensation or evaporate. They collide and coalesce which in warm clouds is the process of rain formation and when they become too big they will break up. During their entire lifetime drops will capture unactivated aerosol particles in the air due to the processes of brownian motion, inertial impaction, phoretic and electric effects. This process is called impaction scavenging. It is negligible in the cloud as the small particles which were left unactivated between the drops as the so-called interstitial aerosol only would contribute little mass even if completely impacted. However, this is the dominant scavenging process below cloud base in the rainfall area.

Impaction scavenging is described by the so-called impaction scavenging efficiency E. The qualitative behavior of E with respect to aerosol particle and drop radius is given in Fig.2 after Grover et al (1977), Wang and Pruppacher (1977), and Carstens and Martin (1982), among others. In Fig.2a we see the dependency on the particle radius. When the particle radius is small (radius below 10^{-2} μm), the efficiency of the impaction process is high as the particles participate in the brownian motion of the air and collide easily with the drop. The efficiency decreases as the particle size increases. Once the particles are large (radius larger 1μm), the efficiency again is large as the particle has a sufficient inertia to resist the flow that wants to carry it around the drop. In between, there is a minimum called the Greenfield gap, located around 0.1μm radius. Here, only the phoretic and electric processes are active. They tend to fill this gap when the relative humidity decreases or the charges on the particle or the atmosphere increase. The relative behavior of the efficiency with drop radius is displayed in Fig.2b. We see a decrease with increasing particle radius due to the increase in flow field strength which tends to carry the particles around the drops. The minimum is located around the drop radius where the rear drop eddy forms. As the eddy gains in strength, more and more particles are captured at the rear of the drop. The maximum is located around 500 μm drop radius where the eddy begins to shed.

MODEL DESCRIPTION

In order to investigate the interactions of aerosol particles and clouds we have developed the DESCAM (DEtailed SCAvenging and Microphysics) model. This model follows in a size dependent way the distributions of the aerosol particles in the air and in the drops, as well as the drop size distribution itself (Flossmann et al, 1985).

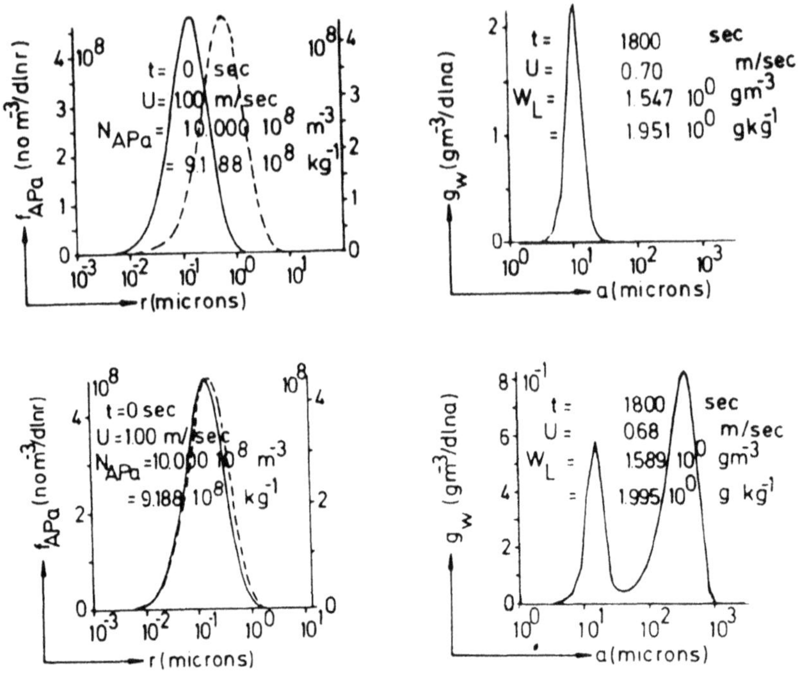

Fig.3: Aerosol number density distribution function at t=0s (a and c) and water mass density distribution function after 30min of model time (b and d) in an ascending and entraining air parcel. Case 1 (a and b): continental aerosol consisting of $(NH_4)_2SO_4$ particles. Case 2 (c and d): same size distribution but particles only 1% of $(NH_4)_2SO_4$.

The model can be driven by a dynamic framework of various complexity ranging from a simple ascending and entraining air parcel model (Flossmann et al, 1985) all the way to a two or three dimensional dynamics (Clark 1977, 1979; Clark and Gall, 1982; Clark and Farley, 1984; Hall, 1980; Flossmann and Pruppacher, 1988; Flossmann, 1991). In addition, modules of liquid phase chemistry (Flossmann et al, 1987; Flossmann, 1994) or ice microphysics (Alheit et al, 1990; Respondek et al, 1995) can be attached.

INFLUENCE OF AEROSOL PARTICLE COMPOSITION ON CLOUD FORMATION AND EVOLUTION

Internal mixtures and their role in cloud formation

As was already discussed in the previous sections, the number of activated particles depends significantly on the characteristics of the atmospheric aerosol. This is illustrated in Fig.3 for an internally mixed aerosol. These simulations where done with the DESCAM model coupled to an ascending and entraining air parcel dynamics (Flossmann et al, 1985). Displayed with a solid line in Fig.3a and c is a dry aerosol particle spectrum of a continental total number concentration, i.e. 1000 particles per cm^3. In a first case, illustrated in Fig.3a, we assumed that the particles entirely consist of $(NH_4)_2SO_4$ and in a second case in Fig.3c it is assumed that only 1% is soluble $(NH_4)_2SO_4$ and the rest is insoluble.

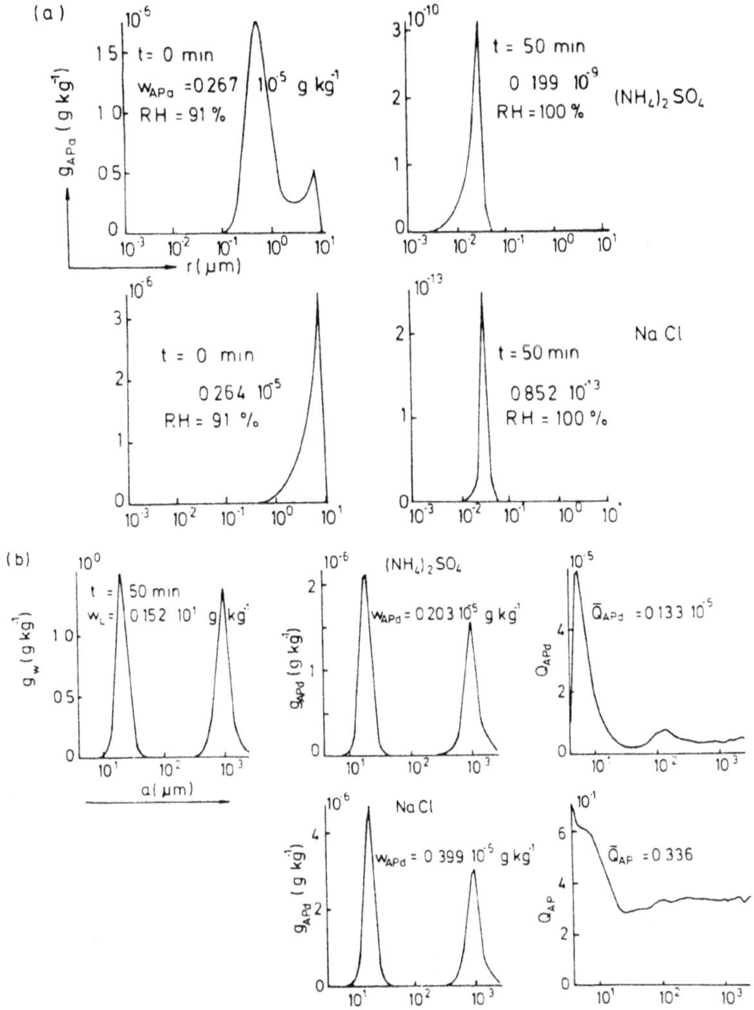

Fig.4. (a) mass density distribution function $g_{APa}(m_{AP})$ for $(NH_4)_2SO_4$ and NaCl particles in air at model times t=0s and t=50min and (b) water mass density distribution function $g_w(m)$, mass density distribution function for $(NH_4)_2SO_4$ and NaCl in the drops, $g_{APd}(m)$, mixing ratio of aerosol mass in the drops Q_{APd} (m) and the fraction of $(NH_4)_2SO_4$ compared to the total scavenged aerosol material Q_{AP} at model time t=50min at x=13.4km and z=1.5km.

The resulting moist aerosol particle spectrum being in equilibrium with an ambient relative humidity of 99% is displayed by the dashed curve which belongs to the right ordinate.

We see that highly insoluble particles generally grow far less than soluble particles. The liquid water spectrum that evolved after 30 min of cloud time is totally different for the two cases considered. In the first case, for which the liquid water spectrum is displayed in Fig.3b, about 800 drops/cm^3 formed directly at cloud base. These were too numerous to grow by condensation to the extend that collision and coalescence of drops would become effective. Consequently, no precipitation sized drops formed as can be seen from the narrow water mass spectrum centered around 10μm in radius. Totally different is the case where only 1% of the total material of the particle was soluble. There, only 500 drops/cm^3 were formed (Fig.3d) and consequently, rain drops of larger sizes could be formed as can be seen in the second maximum of the cloud liquid water spectrum.

This does not only result in a totally different cloud microphysics and dynamics but also has consequences on the radiative properties of the cloud. Furthermore, this determines

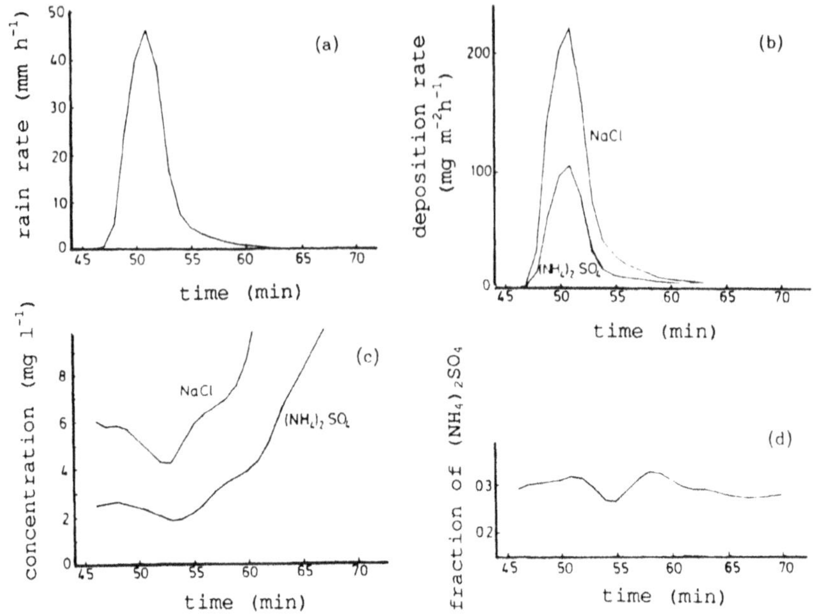

Fig.5: Evolution of the rainfall rate (a), the amount of deposited material by the rain (b), the concentration of scavenged material in the rain (c), and the % of $(NH_4)_2SO_4$ in the total deposited material (d) at 13.4km averaged over 1 min intervals.

the total scavenged pollutant mass by particles and through that the wet phase chemistry and the wet deposition. This has to be attributed not only to the fact that the scavenged particle mass often dominates masswise the total pollutant mass scavenged but also to the fact that the resulting ions and acidity inside the drops as well as the size of the drops influence the uptake of other gaseous constituents.

In the above presented case the aerosol particles represented an internal mixture, i.e. all particles were mixed particles with the same volume fractions of chemical species. The other extreme of aerosol particle composition would be an external mixture. Here, the individual particles would be homogeneous but have a different composition than the particles next to them.

External mixtures and their role in cloud formation and evolution

We will use a marine aerosol particle spectrum which is known to represent such an external mixture (Fitzgerald, 1991) and study the distribution of the different chemical aerosol species in the air and in the drops.

For this study we used our DESCAM model embedded in the 2-D dynamic model developed by Clark and coworkers for the simulation of a warm precipitating convective cloud at day 261 of the GATE campaign (Flossmann, 1991).

The initial aerosol particle spectrum was assumed to be of maritime type consisting of a superposition of three log-normal distributions. The small modes were assumed to consist of $(NH_4)_2SO_4$ particles and the large model was set to hold only NaC! particles. The particle spectra were assumed to decrease with height considering that practically no NaCl exists above 2.5km. The two different chemical species of aerosol particles were followed in the air, inside the drops in the cloud, and inside the drops which have arrived as rain on the ground. The dynamical and microphysical results compared quite well with the observations made during that day. The aerosol scavenging results regarding pollution concentrations in the rainwater agreed well with typical marine situations.

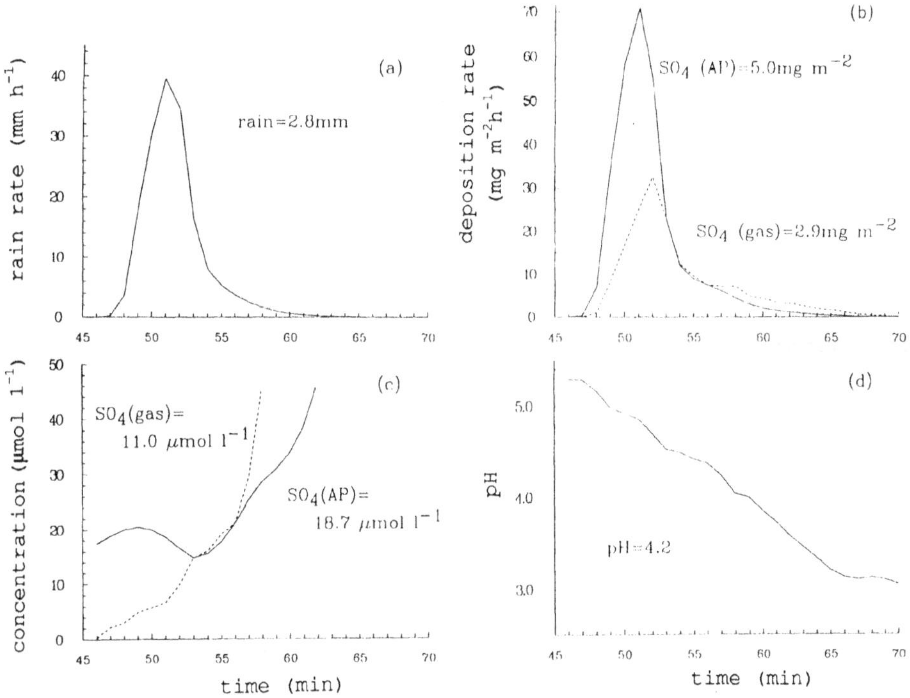

Fig.6: Evolution of the rainfall rate (a), the amount of deposited sulfate from the particulate and gaseous phase by the rain (b), the concentration of the scavenged sulfate in the rain (c) and the pH of the rain (d) at 13.4km averaged over 1min intervals.

In Fig.4, we have displayed for a chosen grid point of the 2-D model domain (20km horizontal, 10km vertical) the evolution of the spectra of the interstitial aerosol particles and the spectra of the drops and the scavenged aerosol particle mass in the cloud for a location near cloud base. In Fig.4a the two mass density distribution function for $(NH_4)_2SO_4$ and NaCl in the air are displayed. In the first column we see the spectra at t=0s in equilibrium with the environmental relative humidity before cloud formation as prescribed by the Köhler equation. The second column displays the interstitial aerosol particle spectra at 50min of model time at the same grid point. The two spectra show a cutoff at the larger end due to nucleation scavenging at the same moist aerosol particle radius irrespective of the chemical composition of the particles. This is due to the fact that at the critical point the aerosol particles have already taken up so much water that the "solute effect" looses its importance and the activation radius is predominantly determined by the supersaturation (Ahr et al, 1990). Comparing the numbers we note that the total aerosol particle mass is reduced by more than 99% as already indicated above.

In Fig. 4b the distributions inside the cloud water are displayed. We note that at model time 50min which corresponds to a cloud life time of 24min a bimodel liquid water spectrum has evolved. The mode at the small drop sizes is due to new drops formed by nucleation while the large mode is associated with drops which have been formed by collision and coalescence aloft and are falling towards the ground. The aerosol particle mass scavenged in the drops closely follows this behavior as we can see in the adjacent two spectra for the $(NH_4)_2SO_4$ and NaCl mass is the drops. The other two spectra displayed represent the total mixing ratio of material in the drops Q_{APd} and the fraction of $(NH_4)_2SO_4$ in the total scavenged aerosol particle mass. In the mixing ratio Q_{APd} the history of evolution of the drop spectrum is reflected. It displays a maximum at the small drop sizes whenever the supersaturation and the entrainment of particles allows the formation of new drops. Then, drops with water masses small as compared to the nucleus mass are introduced into the spectrum.

As these drops grow to larger sizes the maximum in the mixing ratio moves with the spectrum until it eventually disappears due to further condensation and collision and coalescence. This, we can see in Fig.4b. The maximum in concentration associated to the initial nucleation of cloud drops at cloud base during cloud formation is only barely visible and has migrated to drops sizes around 100μm radius.However, freshly entrained particles are continuously nucleated resulting in the sharp increase in concentration below 10μm. Also displayed in Fig.4b is the fraction of $(NH_4)_2SO_4$ in the aerosol mass scavenged by the drops. We note here that the small drops have a larger fraction of $(NH_4)_2SO_4$ which corresponds to the fact that also the small aerosol particles mainly consist of $(NH_4)_2SO_4$. The larger aerosol particles (>1μm), however, consist mainly of NaCl which grow most and are thus in the larger drops which activated early. When collision and coalescence starts the aerosol particle mass becomes well mixed in the precipitation sized drops. However, the $(NH_4)_2SO_4$ mass never reaches the value of the NaCl mass, resulting in the fact that the aerosol mass in the precipitation sized drops consists of about 70% NaCl and only 30% $(NH_4)_2SO_4$. The aerosol material scavenged into the drops at this point in the cloud is almost exclusively determined by nucleation scavenging as already pointed out above. Impaction scavenging only becomes important when the drops leave cloud base and fall towards the ground in the rain.

The evolution of the rainfall at a given location on the ground is displayed in Fig.5. This corresponds to the situation of a precipitating cloud moving overhead of the measuring point. The maximum of the rainfall intensity is reached around 50min of model time. Comparing this to Fig.5c we notice that the concentration of aerosol material in the drops decreases till the maximum of the rainfall rate passed overhead. Then, the concentration increases again. This is caused by a stronger evaporation in time of low precipitation resulting in a higher concentration of pollutant mass is the drops. Fig.5b gives the deposition of material in $mg/m^2/h$. We see here two things: first, the aerosol deposition curves follow again the evolution of the water mass. This is a general finding, resulting from the fact that the scavenged pollutant mass is transported along attached to the water mass in the cloud. The second point is that less $(NH_4)_2SO_4$ is deposited than NaCl. Calculating the fraction of $(NH_4)_2SO_4$ as compared to the total aerosol mass fraction in the rain (Fig.5d) we see that again 30% of the mass is $(NH_4)_2SO_4$ and 70% is NaCl, as was already found in Fig.4b for the spectra inside the cloud.

Considering the overall results, we found that the aerosol scavenging efficiency of the cloud (~40%; 34% for $(NH_4)_2SO_4$ and 43% for NaCl) closely corresponds to its precipitation efficiency (~46%). The major amount of pollutant mass (87%, 93% for $(NH_4)_2SO_4$ and 85% for NaCl) were incorporated into the cloud through nucleation scavenging. However, 32% of the overall aerosol material in the rain at the ground (22% for $(NH_4)_2SO_4$ and 36% for NaCl) were scavenged below cloud base due to impaction scavenging.

This seems large as compared to the 87% of nucleation scavenging. We have to keep in mind, however, that only 46% of the water condensed actually precipitated. Consequently, a lot the cloud water remains in the air along with the scavenged particle mass. The percentage that actually precipitates scavenges additional mass through impaction and thus, this material contributes about 30% to the mass in the rain.

These investigations were aimed at understanding the role of internal and external mixtures of aerosol particles on the formation and the evolution of the cloud.

However, aerosol material is not exclusively taken up due to nucleation and impaction scavenging. It can also be formed due the uptake of gases and chemical processes in the wet phase. In order to investigate the importance of this process for the overall evolution of the particle mass associated with the cloud process, we have added to the previously shown simulation the uptake of gaseous SO_2, H_2O_2 and O_3.

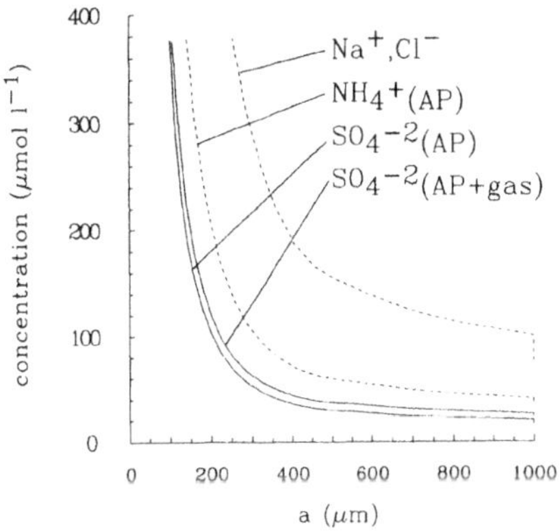

Fig.7: Concentration of ions in the raindrop spectrum arriving at the ground after 50min of model times at x=13.4km and z=0.1km.

Influence of in-cloud sulfate production on the cloud and the sulfate scavenging

Fig.6 shows the evolution of the rainfall characteristics at the surface with the precipitating cloud moving overhead similar to Fig.5. Displayed again is the rainfall rate with a total rain of 3.2mm. In Fig.6b we give the sulfate content in the rain due to gas and particle uptake in terms of the wet deposition in $mg/m^2/h$. The sulfate from the particulate phase is the same as in Fig.5 and amounts to a total of $5mg/m^2$. Now in addition to that sulfate is formed due to uptake of SO_2 gas into the drop and due to an oxidation of S(4) to S(6) inside the liquid phase. For remote marine conditions with an initial SO_2 concentration of 0.5ppb(v) (H_2O_2=05.ppb(v), O_3=30ppb(v)) the gas phase contributes another $2.9mg/m^2$. Not considered thereby is the S(4) in the raindrops which could add a sulfate mass of 0.09 mg/m^2 if all S(4) mass would immediately be oxidized.

We notice that the scavenging of sulfate from the particles is almost twice that from the gas which, of course, is a result of the initial conditions. Fig.6c displays the mixing ratio of sulfate. We see here that the curve for the particulate sulfate shows roughly an inverse correlation with the rainfall rate which was already found in Fig.5. The average particulate sulfate concentration is 18.7μmol/l. The curve for the gaseous sulfate, however, increases continuously. Here, the evaporation process is coupled with the time factor of the oxidation process. At the beginning of the rainfall the oxidation completely masks the evaporation while at the end the evaporation causes an even steeper increase in the sulfate concentration. This results in an average gaseous SO_4 concentration of 11μmol/l. This time factor also becomes obvious for the time evolution of the pH in Fig.6d which continuously decreases from 5.3 at the beginning all the way down to 3.1 in the last 5 min of precipitation where the rainfall rate is very small and consequently evaporation is the dominant process. Consequently, the average pH is 4.2 derived from the average concentration of the H^+ions. During the time of the main rainfall between 49 and 54 min, however, the mean pH is around 4.7 which lies within the range observed.

The evaporation process below cloud base also becomes obvious from Fig.7 which displays the evolution of the concentration across the raindrop spectrum during the time of the strongest precipitation. We see here the higher concentration of the ions predicted by the model in the small drops.

In general, we can conclude that the sulfate produced by the gas phase is always smaller than the particulate sulfate. From the total sulfate processes, i.e. uptaken or produced by the cloud within 75min of model time 87% was scavenged by nucleation of drops on sulfate containing aerosol particles. 6% of sulfate entered the drop through impaction

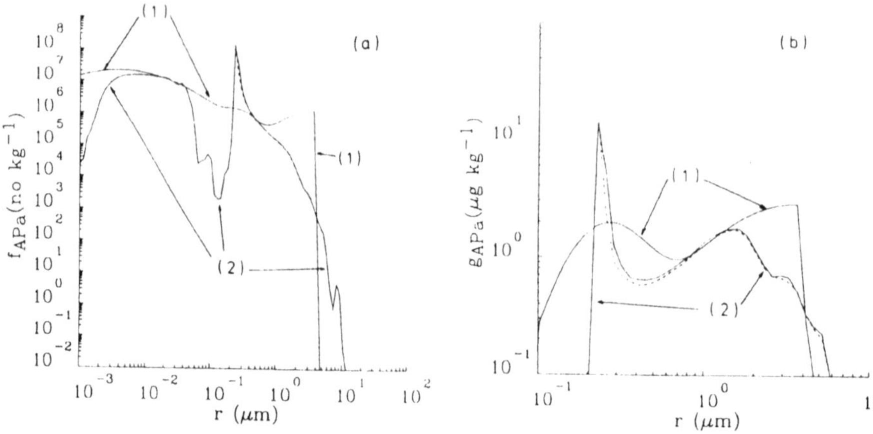

Fig.8: Aerosol particle number and mass density distribution function f_{APa} (a) and g_{APa} (b) in the air at x=15.4km and z=1.5km (1) initial spectrum, (2) spectrum after 70min of model time (corresponds to 44 min after the first cloud formed) as a superposition of the interstitial aerosol particle spectrum and the spectrum resulting from immediate drying-off of the drops (The dashed curve disregards the contribution of gas scavenging).

scavenging below cloud base. Almost 7% was produced by oxidation by H_2O_2 and only 0.2% by oxidation of O_3. Evidently, this low value for the influence of ozone oxidation is a result of the low pH values of the drops.

Averaging over space and time the model calculated a pH of 4.4 in the rain and about 2 mg/l in the rain stem from particulate sulfate and roughly 1 mg/l from gaseous sulfate.

Influence of in-cloud sulfate production on the processed aerosol particle spectrum

We also investigated the effect of in-cloud processes and the subsequent evaporation of the cloud on the aerosol population. Herefore, we assumed that at the end of our computational run the cloud would evaporate instantaneously and thus unload into the atmosphere all the pollutant mass captured during the cloud cycle. We assumed in doing so that each drop would produce only one aerosol particle composed of all the aerosol mass captured by the drop and all gaseous mass scavenged and subsequently converted to non-desorbing species. The results of this investigation is shown in Fig.8 where also the original aerosol particle spectrum at the beginning of the simulation is given. We notice that the final aerosol particle spectrum consists of two parts: one from the interstitial aerosol particle spectrum (<1μm) which remained unactivated in between the drops during the entire cloud life time and which became depleted by impaction scavenging at the lower end (compare the efficiencies given in Fig.2). The other part results from the dried-off droplets. The peak between 0.2 and 0.3μm is an artifact of the model resulting from the fact that the smallest drop bin was 4μm in radius. We see, however, that the aerosol particle spectrum now consists generally of larger particles than before. This is a result of the collision and coalescence still going on in the remaining cloud. The gas scavenging only caused a small shift in the particle spectrum which is illustrated by the difference between the solid and the dashed curve (compare results from Fig.6 and 7).

In Fig. 9 the corresponding CCN spectra are displayed. The solid line gives the original CCN spectrum while the dashed curve displays the dried-off particle spectrum of Fig.8. We see that this aerosol particle spectrum which was processes by the cloud now results in an enhanced number of CCN in the low supersaturation range. Consequently, these particles

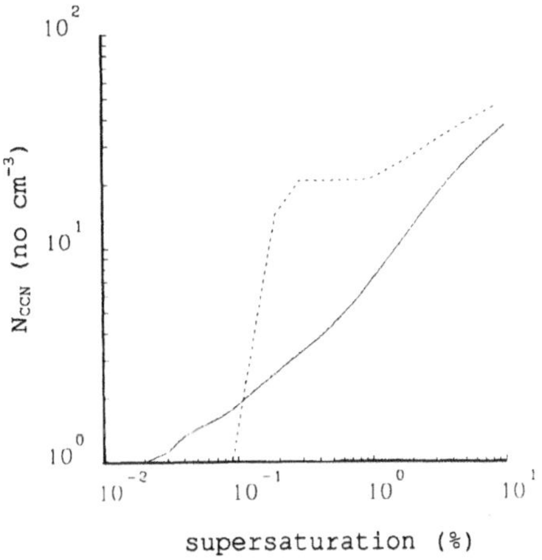

Fig.9: CCN spectrum at x=15.4km and z=1.5km; solid line: initial CCN spectrum, dashed line: CCN spectrum after 70min of model time resulting from the aerosol particle spectra displayed in Fig.8.

are better nuclei in a subsequent cloud cycle. These findings, however, are dependent on the initial conditions and the life history of the cloud.

SCAVENGING PROCESSES IN CLOUDS CONTAINING AN ICE PHASE

Nucleation scavenging

Ice initiation above -40°C requires the presence of so-called ice nuclei (IN) which are aerosol particles preferably with a crystalline structure similar to the one of ice. These ice nuclei are activated depending on the supercooling of the cloud and the relative humidity. Currently, we distinguish three different mechanisms for the IN to produce ice crystals: the deposition mode (i.e. water deposition on IN), the contact mode (i.e. collision of IN with drops), and the freezing mode (i.e. activation of IN inside existing drops) (Pruppacher and Klett, 1978). IN have a typical radius of 1μm as observed by Rosinski et al (1980) and Stein and Georgii (1983) and e.g. a concentration of $1\ l^{-1}$ (at -20°C) as suggested by Fletcher (1962). From these facts we can conclude that an additional nucleation scavenging due to the formation of ice in the mixed clouds considered is negligible.

Impaction scavenging

Similar to drops also ice crystals collide with the interstitial aerosol particles and scavenge them. However, an additional complexity enters here as the impaction scavenging efficiency is a function of the crystal shape (compare e.g. Martin et al, 1980, and Wang, 1985) and of temperature.

Riming scavenging

For the ice phase a third scavenging process exists which is not a direct uptake from the ambient air but an indirect via the liquid phase. Here, the possibility exists that drops which have scavenged aerosol particle mass through nucleation and impaction become ice. This happens on the one hand side through a freezing of the drop via the activation of an IN (see above) or via the process of riming. Here, the drop collides with an already existing ice

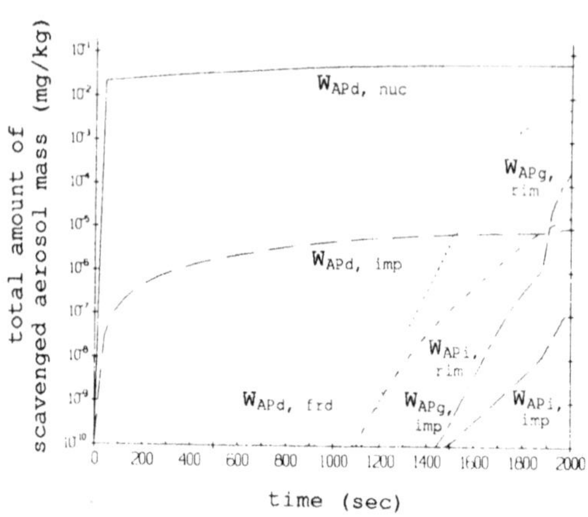

Fig.10: Time evolution of the total amount of aerosol mass scavenged by drops via nucleation scavenging of drops ($w_{APd,nuc}$), by graupel via riming ($w_{APg,rim}$), by ice crystals via riming ($w_{APi,rim}$), by ice particles via drop freezing ($w_{APd,frd}$), by drops via impaction scavenging ($w_{APd,imp}$), by graupel via impaction scavenging ($w_{APd,imp}$), and by ice crystals via impaction scavenging ($w_{APi,imp}$)

particle and freezes forming a new aggregate with the ice crystal. This is known to be a very effective processes in converting a liquid cloud into ice and forming precipitating graupels.

PATHWAYS OF AEROSOL PARTICLE SCAVENGING IN CLOUDS CONTAINING AN ICE PHASE

Distribution of the aerosol particle mass through different scavenging processes

We have investigated with the DESCAM model and the attached ice module in the framework of the ascending and entraining air parcel model the relative contribution of the different scavenging mechanisms of aerosol particles (Alheit et al, 1990). This model follows planar type ice crystals which are predominantly formed in the temperature range between -10 and -20 °C. These ice crystals rime and eventually become graupel particles. Graupel is also formed by the direct freezing of large drops. The aerosol particle mass contained in the various size classes of these two ice particle types are followed in the model together with the processes in the liquid phase. The results are displayed in Fig.10.

We notice that over the whole model time nucleation scavenging by drops ($w_{APd,nuc}$) dominates impaction scavenging of drops ($w_{APd,imp}$) by drops as well as all other scavenging modes. In the presented simulation ice initiation took place via drop freezing. Then, the uptake and incorporation of aerosol particles into the ice phase proceeds via five different mechanisms: (i) by impaction of aerosol particles on the formed ice crystals ($w_{APi,imp}$), (ii) by impaction of aerosol particles on the formed graupel ($_{wAPg,imp}$), (iii) by freezing of drops containing previously captured aerosol mass ($w_{APd,frd}$), (iv) by riming of ice crystals ($w_{APi,rim}$) and (v) by riming of graupel ($w_{APg,rim}$). We find here, that inside the cloud, impaction scavenging by ice particles ($w_{APi,imp}$, $w_{APg,imp}$) starts fairly late and remains inefficient until the cloud air becomes subsaturated (~1900 model seconds). Transfer of aerosol mass from the liquid phase into the ice phase by drop freezing ($w_{APd,frd}$), and scavenging of aerosol mass by riming of graupel ($w_{APg,rim}$) dominates the scavenging of aerosol mass due to the riming of ice crystals ($w_{APi,rim}$) as well as due to the impaction on ice particles ($w_{APi,imp}$, $w_{APg,imp}$). The effici-

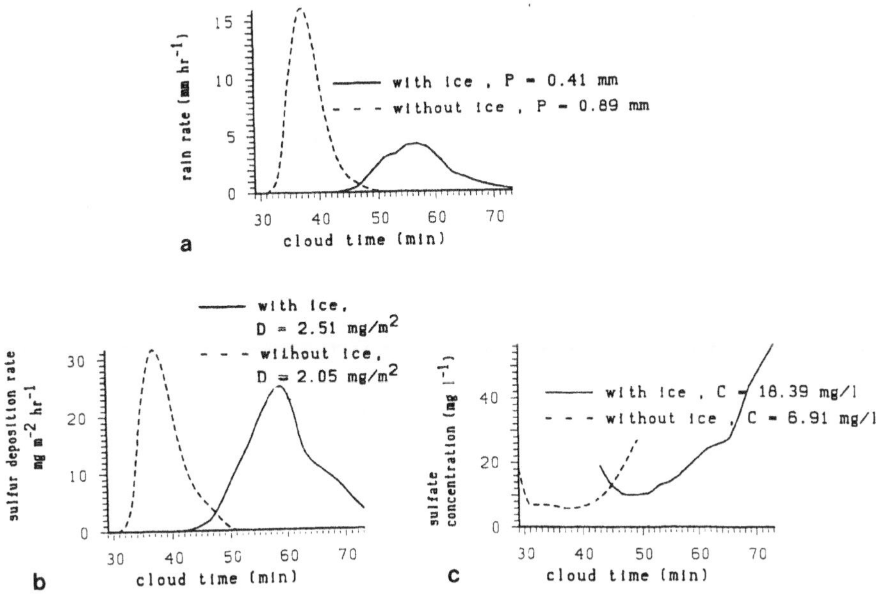

Fig.11 (a) Rain rate (mm h^{-1}), (b) sulfur deposition rate ($mg/m^2/h$), and (c) sulfate concentration (mg/l) averaged over cloud domain for model run with ice (solid lines) and without ice (dashed lines); P=cumulative precipitation (mm), D=cumulative sulfur deposition (mg/m^2), and C=precipitation-weighted mean sulfate concentration (mg/l).

ent uptake of aerosol mass by riming of graupel is due to the fact that the graupel acquire drops which themselves have previously, very efficiently scavenged aerosol particles by nucleation scavenging ($w_{APd,nuc}$). So we can conclude that scavenging of aerosol particles by riming of graupel is an extremely efficient process due to the large efficiency of aerosol particle scavenging by nucleation of drops and the efficient uptake of drops by graupel so that the transfer of aerosol mass into the ice phase by riming of graupel and freezing of drops dominates all other transfer mechanisms.

Importance of ice phase for dynamic, microphysical and scavenging evolution of the cloud

In numerical models, the ice phase is often neglected due to the complexity it introduces via the different ice crystal shapes occurring and their different dynamical, microphysical and scavenging behavior. In a last example, we will study the effect of this neglect on the rainfall and the sulfate deposition resulting exclusively from the scavenging of aerosol particles. Herefore, we have applied our DESCAM model with the ice module coupled to the 2-D model to the CCOPE experiment on 19 July 1981. The model is able to represent quite realistically the dynamical and microphysical features observed during that day as is described in Respondek et al (1995). Then, in an attempt to investigate the impact of the ice phase on the cloud evolution we purposely inhibited the formation of the ice phase. The results are displayed in Fig.11 and show that not only the total amount or precipitation differs significantly but also the onset and duration of precipitation. In the presented case, the warm rain processes could proceed without interruption and produce rain fast while the ice case froze the drops while still being relatively small. Consequently, a time lag occurred for the ice particles to grow to a precipitation size. Thus, the case with ice precipitates much later and with a smaller intensity. These features are reflected in the curves for the sulfur deposition and the sulfate concentration in the rain drops. In other clouds, the neglect of the ice phase might produce different results but we can, nevertheless, speculate that they will be significantly different from a simulation with ice.

REFERENCES

Ahr, M., A.I. Flossmann, and H.R. Pruppacher, 1990: On the effect of the chemical composition of atmospheric aerosol particles on nucleation scavenging and the formation of a cloud interstitial aerosol. *J. Atmos. Chem.*, **9**, 465-478.

Alheit, R.R., A.I. Flossmann, and H.R. Pruppacher, 1990: A theoretical study of the wet removal of atmospheric pollutants. Part IV. *J. Atmos. Sci.*, **47**, 870-887.

Carstens, J.C., and J.J. Martin, 1982: In-cloud scavenging by thermophoresis, diffusiophoresis and Brownian motion. *J. Atmos. Sci.*, **39**, 1124-1129.

Clark, T.L., 1977: A small scale dynamic model using terrain-following coordinate transformation. *J. Comput. Phys.*, **24**, 186-215.

Clark, T.L., 1979: Numerical simulations with a three dimensional cloud model. *J. Atmos. Sci.*, **36**, 2191-2215.

Clark T.L., and Gall, R., 1982: Three dimensional numerical model simulations of air flow over mountainous terrain: A comparison with observation. *Mon. Weather Rev.*, **110**, 766-791.

Clark, T.L and Farley, R.D., 1984: Severe downslope windstorm calculations in two and three spatial dimensions using anelastic interactive grid nesting. *J. Atmos. Sci.*, **41**, 329-350.

Fitzgerald, J.W., 1991: Marine aerosols: A review, *Atmos. Environ.*, **25A**, 533-545.

Fletcher, N.H., 1962: Physics of Rain Clouds. Cambridge University Press, London.

Flossmann, A.I., 1991: The scavenging of two different types of marine aerosol particles using a two-dimensional detailed cloud model. *Tellus*, **43B**, 301-321.

Flossmann, A.I., 1993: The effect of the impaction scavenging efficiency on the wet deposition by a convective warm cloud. *Tellus*, **45B**, 34-39.

Flossmann, A.I., 1994: A 2-D spectral model simulation of the scavenging of gaseous and particulate sulfate by a warm marine cloud. *J. Atmos. Res.*, **32**, 255-268.

Flossmann, A.I., W.D. Hall, and H.R. Pruppacher, 1985: A theoretical study of the wet removal of atmospheric pollutants. Part I. *J. Atmos. Sci.*, **42**, 582-606.

Flossmann, A.I., H.R. Pruppacher, J.H. Topalian, 1987: A theoretical study of the wet removal of atmospheric pollutants. Part II. *J. Atmos. Sci.*, **44**, 2912-2923.

Flossmann, A.I., and H.R. Pruppacher, 1988: A theoretical study of the wet removal of atmospheric pollutants. Part III. *J. Atmos. Sci.*, **45**, 1857-1871.

Grover, S.M., H.R. Pruppacher, and A.E. Hamielec, 1977: A numerical determination of the efficiency with which spherical aerosol particles collide with spherical water drops due to inertial impaction and phoretic and electric forces. J. Atmos. Sci., 34, 1655-1663.

Hall, W.D., 1980: A detailed microphysical model within a two-dimensional dynamic framework: Model description and preliminary results. *J. Atmos. Res*, **37**, 2486-2507.

Martin, J.J., P.K. Wang, and H.R. Pruppacher, 1980: A theoretical determination of the efficiency with which aerosol particles are collected by simple ice crystal plates. *J. Atmos. Sci.*, **38**, 2462-2469.

Pruppacher, H.R., and Klett, J.D., 1978: Microphysics of Clouds and Precipitation. D. Reidel, 714pp.

Respondek, P.S., A.I. Flossmann, R.R. Alheit, and H.R. Pruppacher, 1995: A theoretical study of the wet removal of atmospheric pollutants. Part V. *J. Atmos. Sci.*, **52**, 2121-2132..

Rosinski J., G. Morgan, C.T. Nagamoto, G. Langer, G. Yamate and F. Parungo, 1980: Size distribution of inorganic and organic ice forming nuclei present in downdrafts of convective storms. *Meteor. Rundschau*, **33**, 97-106.

Stein, D., and H.W. Georgii, 1983: Messungen der Groessenverteilung atmosphaerischer Gefrierkerne unter Beruecksichtigung der Feuchte. *Meteo. Rundschau*, **35**, 1735-1743.

Twomey, S.A., M. Piepgrass, and T.L. Wolfe, 1984: An assessment of the impact of pollution on global cloud albedo. Tellus, 36B, 356-366.

Wang, P.K., 1985: A convective diffusion model for the scavenging of submicron aerosol particles by snow crystals of arbitrary shapes. *J. Rech. Atmos.*, **19**, 185-191.

Wang, P.K., and H.R. Pruppacher, 1977: An experimental determination of the efficiency with which aerosol particles are collected by water drops in subsaturated air. *J. Atmos. Sci.*, **34**, 1664-1669.

DISCUSSION

J. CHING: Please discuss, why the contribution of gas (SO_2) contributing to cloud droplet sulfate to the size distribution of evaporated cloud relative to that of the aerosol contribution was so small (with about 1/3 of sulfate produced by SO_2 to SO_4 in cloud it was surprising to see so little difference in the contribution to the spectra in the evaporated cloud by gas vs aerosol). Can the result be generalized for precipitating and non-precipitating clouds?

A. FLOSSMANN: There are basically three reasons why the SO_4 from the gas contributed so little to an effective growth of the dried-off aerosol particles:

a) the relation between mass and radius: even if, e.g., a $1\mu m$ sulfate aerosol particle would double its mass by sulfate from the uptake of SO_2 and subsequent oxidation to SO_4, this would only result in a new aerosol radius of $1.26\mu m$.

b) the low concentration of atmospheric sulfur dioxide in this marine case: about a factor of 3 lower than the total input of particulate sulfate

c) the life time of individual cloud drops is never much longer than about 10min; consequently, the time for gas uptake and subsequent oxidation is limited.

While reason b) applies only to this specific run, reason a) and c) apply generally. Consequently, we expect this result also to apply to other similar scenarios, irrespective of the fact that clouds precipitate or not.

B. KARGIN: What kind of the clouds did you have in mind? It is well known that the processes of nucleation and coagulation vary strongly depending on the thermodynamical situation. So, the question is, how did you take into account the thermodynamical and hydrodynamical aspects of the problem?

A. FLOSSMANN: The calculations were done for a small convective warm cloud. We used for the dynamical calculations a 2-D version of the 3-D model of Clark and co-workers. Attached was our DESCAM model, calculating microphysics and chemistry. The coupling is fully interactive, e.g., the vertical velocity fields interfere in the precipitation and the release of latent heat is considered for the dynamics. The specific results presented, however, depend on the initial conditions, i.e. for continental clouds the evolution of the cloud, the precipitation and the scavenging will be different.

A NEW MC2/ADOM MODELING SYSTEM AND ITS APPLICATION TO THE STUDY OF OXIDANTS IN THE CANADIAN SOUTHERN ATLANTIC REGION

Xiude Lin[1], Wanmin Gong[2], and Sylvain Menard[3]

[1] Ontario Hydro Technologies, Toronto, Ontario, Canada
[2] Atmospheric Environment Service, Downsview, Ontario, Canada
[3] 185, 10th Avenue, Ile Perrot, Quebec, Canada

INTRODUCTION

ADOM, a regional, comprehensive Eulerian Acid Deposition and Oxidant Model (Venkatram et al, 1988), was developed in the early and mid 80s primarily for acid rain related modelling needs. With the emergence of oxidant related air quality issues in recent years, there have been attempts to apply ADOM to oxidant modelling since the chemistry mechanism also includes regional oxidant chemistry. ADOM is included in an oxidant model evaluation study coordinated by CIRAC (the Canadian Institute for Research in Atmospheric Chemistry) for the EMEFS-1 intensive measurement campaign (Hansen et al., 1991) period. The study shows that although in general ADOM reproduces the diurnal cycle of surface ozone reasonably well, it significantly underpredicts the peak values for a number of EMEFS' stations. There are obvious weaknesses in the model that may have contributed to its poor performance. Firstly, the original ADOM has a horizontal grid spacing of 127 km (at 60°N), and this relatively coarse resolution can give rise to a number of problems. For example, it will overpredict dispersion of a plume, especially close to the sources. It will also introduce artificial dilution of the emissions since all pollutants emitted are mixed instantly in their residing grid which will lead to lower concentration of these species as well as reactions amongst them while in reality they may not meet at all in that particular grid. Secondly, the ADOM input meteorological fields are derived from a combination of the CMC (Canadian Meteorological Centre) large scale spectral weather prediction model, used as a dynamic interpolator to obtain hourly meteorological fields from the objective analysis of the observed data, and a 1-D boundary layer model developed by the MEP company in Toronto (Scholtz et al., 1986). Although this ensures that the ADOM meteorological fields are compatible with the available observations, on a synoptical scale, the disadvantages of the original ADOM meteorological driver include 1) the possible inconsistency in the meteorological fields due to the combination of observations and model calculations, 2) the inconsistency between the lower and upper flow fields due to the combination of the spectral model and the 1-D boundary layer model which does not account for horizontal advection, 3) the inability (due to the limitation of the spectral model) of representing mesoscale flow, which could be important in the nonlinear processes of the transport and transformation of oxidants, and 4) the cumbersome procedures.

As one of the three ozone non-attainment regions in Canada, the Canadian Southern Atlantic Region (SAR) is recognised as a region where oxidant problems are strongly influenced by long range transport. Due to the coastal setting, the local circulations can be extremely important in this region. The modelling of SAR oxidants therefore requires, among other things, 1) sufficiently high resolution to properly resolve the coastal situation and terrain features, and 2) a meteorological driver capable of simulating the marine-coastal boundary layer and its evolution. These modelling needs were the initial driving force for the work that are presented here.

In the present study, we replaced the original ADOM meteorological driver with a mesoscale model, MC2 (mesoscale compressible community model). The new MC2/ADOM system is more flexible and has a nesting capability to allow higher resolution modelling focusing on the area of interest. In this paper, we will present some preliminary results from a simulation of photo-oxidants in SAR during the EMEFS-1 intensive measurement period using the new modelling system. We will compare the model simulation with the observations as well as with the original ADOM results for the same period, and will discuss some of the aspects, e.g. meteorology and resolution of oxidant modelling.

THE MC2/ADOM MODELING SYSTEM AND THE SAR EMEFS-1 SIMULATION

The formulation and structure of the ADOM is given in Venkatram et al (1988). Here we will only give a brief description to the new meteorological driver and the ADOM nesting scheme.

MC2 was developed (and is still evolving) during the past few years by a group of scientists under the coordination of the Canadian Centre for Research in Mesometeorology (CCRM). The main feature of this model is that it employs a semi-Lagrangian and semi-implicit numerical scheme, based on the work by Tanguay et al (1990), to achieve efficient integration of the fully elastic equations. The model can, in principle, be applied to all scale of flows with appropriate physics (Bergeron et al, 1994). The model is formulated in cartesian coordinates on a polar stereographic projection of a sphere in the horizontal and a modified Gal-Chen terrain-following coordinate with stretched spacing in the vertical. The dynamic nesting for all the prognostic variables is done through sponge zones, in both the horizontal and vertical, where the modelled fields are gradually blended with the boundary conditions. The physics to date includes boundary-layer and surface processes, turbulence closure of order of one and a half (solving the prognostic equation for turbulent kinetic energy), orographic gravity wave drag, shallow and deep convection, cloud microphysics and radiation.

The EMEFS-1 intensive field campaign was mostly during the month of August 1988, when rather extensive chemistry measurements were made at a large number of sites in the Eastern United States and Canada. This field study provided a valuable set of data for model evaluations. The present simulation period is from the 30th of July to the 5th of August in 1988. Both the meteorological model, MC2, and the oxidant model, ADOM, were run on two grids, a coarse grid with 127 km resolution in horizontal and a nested fine grid with 21.17 km resolution. The coarse ADOM domain consisted of 33 by 33 horizontal cells (the original ADOM grids) covering most of eastern north America (as shown in figure 1), while the coarse MC2 domain consisted of 70 x 60 horizontal cells to cover most of the North American continent. The nested ADOM fine mesh grids, consisting of 108 by 60 horizontal grid cells, were designed to mostly focus on SAR, with its left boundary extending west of Ohio (see figure 2) to include some of the major source regions in the north-eastern U.S., the Detroit-Windsor Quebec City Corridor and most of the industrial Ohio Valley. The nested MC2 fine mesh run had a slightly larger domain than that of ADOM with 137x89 grids so that the MC2 sponge (nesting) zones were outside the ADOM domain. There were 12 vertical layers in ADOM between 1 and 10000 m with 8 layers located below 2000 m, while for the MC2 model runs there were 25 vertical levels between 0 and 25 km with 9 levels below 2000 m in the coarse case and 31 levels between 0 and 15 km with 14 levels below 2000 m for the fine case.

The CMC global objective analysis provided the initial and temporal boundary conditions for the MC2 coarse resolution run, which in turn provided the initial and boundary conditions for the fine resolution run. For this 7-day period, the mesoscale meteorological model was run in an "assimilation" mode, i.e. the model was re-initialised every 24 and 18 hours for the coarse and fine case respectively to avoid significant departure from the objective analysis. Discontinuity was minimized by overlapping (staggering) the runs. As for the oxidant modelling, all the meteorological fields for the ADOM runs (coarse and fine mesh) were derived (through interface programmes) from the corresponding MC2 runs. The hourly output of concentrations of 29 advected species from the ADOM coarse mesh run was used to construct the time-dependant boundary conditions for the nested ADOM fine mesh runs. In the ADOM nested fine mesh run, the original 30-minute time-step for advection and diffusion calculations was reduced to 5 minutes to ensure that the Courant number remained less than one.

RESULTS AND DISCUSSION

In additional to comparing the present simulation with the observations, comparisons were also made with the "original" ADOM simulation (with the original meteorological driver), as well as between the coarse and the fine mesh runs. For the sake of simplicity, we will refer to the original ADOM run as ADOM-MEP, the MC2/ADOM coarse grid run as ADOM-MC2 and the MC2/ADOM fine mesh run as ADOM-NEST.

Generally speaking, the two ADOM coarse grid runs, ADOM-MEP and ADOM-MC2, are comparable, both predicting a regional surface ozone episode with maximum values exceeding 82 ppbv during the first week of August, 1988. The modelled ozone episode, developed in the north-eastern U.S. and extending into southern Ontario, is in general agreement with observations (Trainer et al., 1995). Figure 1 shows the geographical distributions of the modelled surface daily maximum ozone for August 2, 1988 from the two coarse grid runs. It is seen that high ozone levels reside over a broad area extending from Ohio to New York and new England states while the ocean and most of the Canadian territory are in a low ozone regime.

Significant improvements were achieved with the fine mesh nested simulation. For comparison, we plotted the modelled surface ozone at three daytime hours on the nested domain from the ADOM-MC2 and the ADOM-NEST runs in Figures 2. Clearly, the ADOM-NEST run produced more structure in the ozone field in the major emissions areas than the coarse grid runs; the plumes were narrower and more concentrated. Interestingly, the results from the ADOM-NEST run exhibited two distinct high ozone systems, one developed in the Ohio Valley-Detroit area to southern Ontario and the other in the north-eastern U.S.. The two systems were naturally divided by the Appalachian Mountains. In contrast, the coarse grid simulation smoothed out the spacial variations in ozone concentration and did not give clear indication of the separation of the two systems.

The comparison shows that the model prediction of the high ozone extremes as well as the diurnal variations are significantly enhanced by reducing the model grid spacing (from 127 km to 21 km in the present case). This is, as briefly discussed in the introduction, partly due to better resolution of the emissions, smaller numerical dispersion of the plumes and therefore better representation of the chemistry. However, more detailed investigation is needed to further quantify the impact of spatial resolution on the roles of chemical and dynamic representations in model prediction of oxidants in this highly nonlinear system.

It is interesting to see from the present study the role of meteorology in the ozone prediction, particularly in regions where transport dominates over the in-situ photochemistry in determining the local ozone variations. This is especially the case in the SAR, where local emissions of both NOx and volatile hydrocarbons are minimal. We will attempt to illustrate this through comparisons among the model results and the observations at a rural site, Kejimkujik in Nova Scotia.

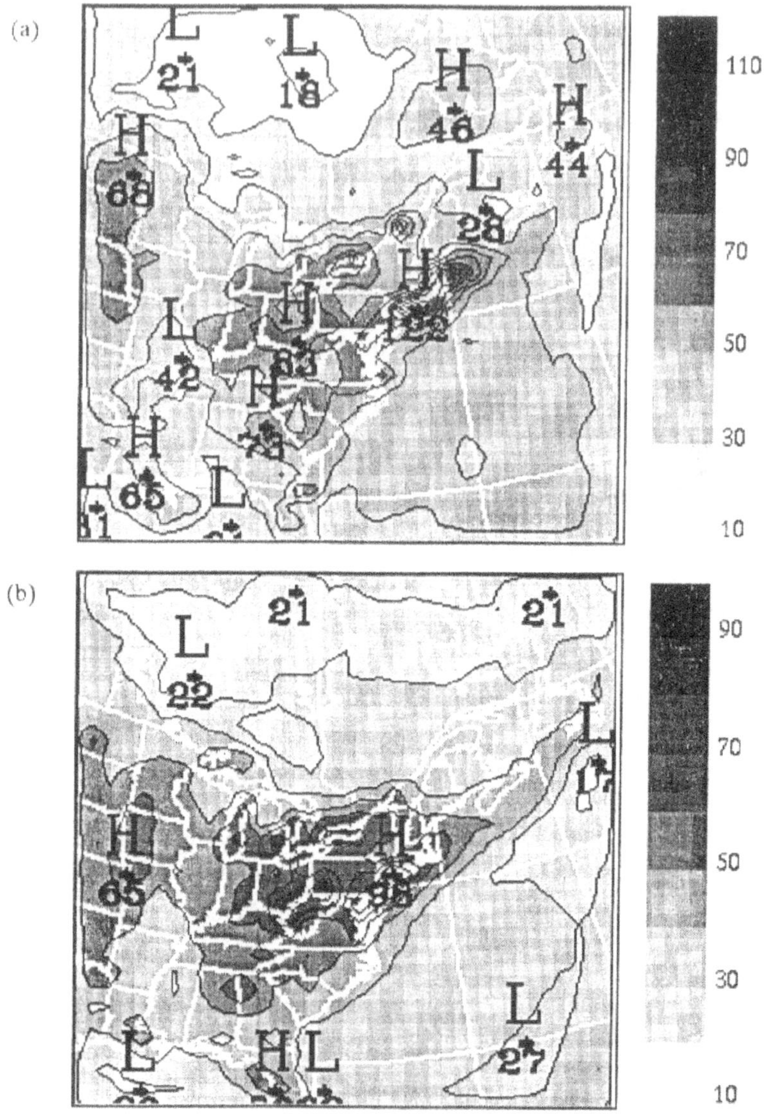

Figure 1. Modelled maximum surface ozone (first model layer) for August 2, 1988. (a). ADOM-MEP, (b). ADOM-MC2.

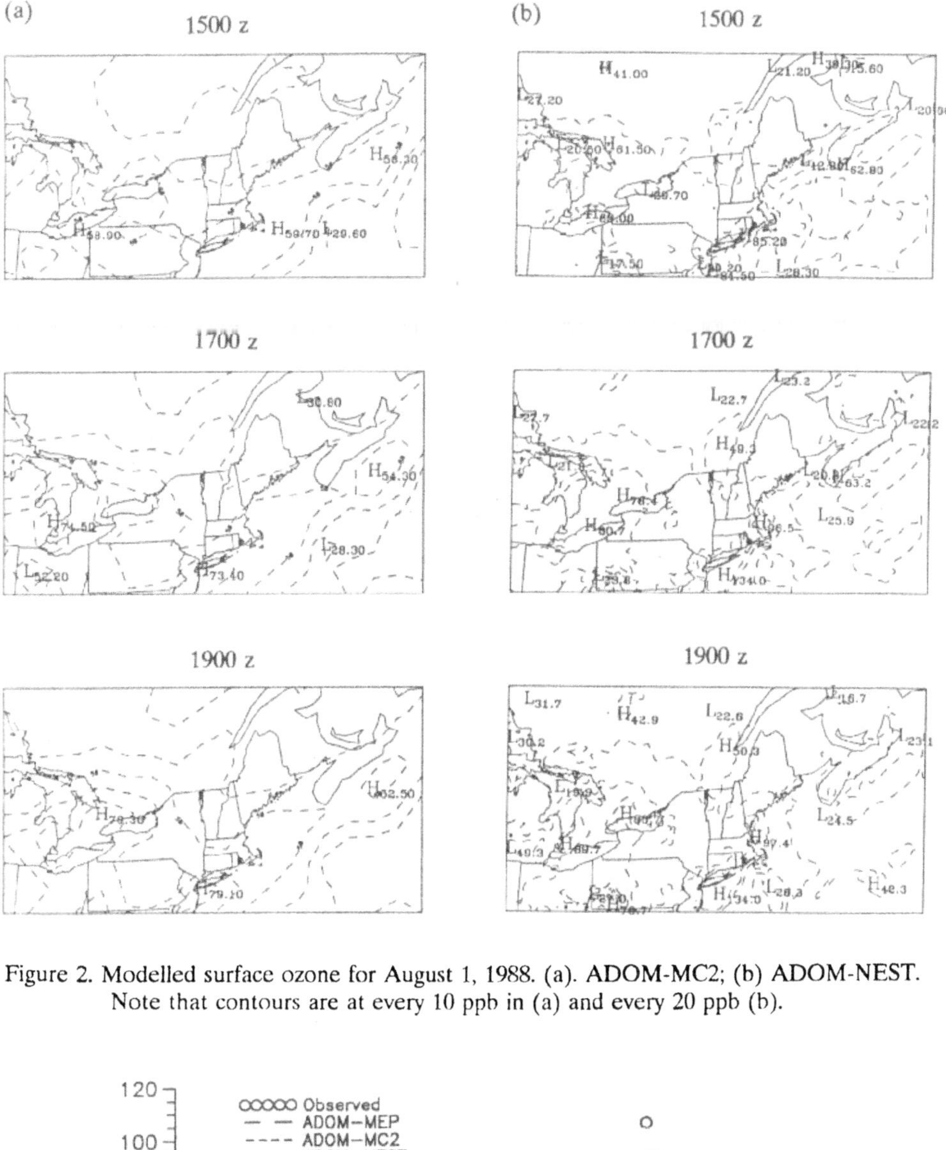

Figure 2. Modelled surface ozone for August 1, 1988. (a). ADOM-MC2; (b) ADOM-NEST. Note that contours are at every 10 ppb in (a) and every 20 ppb (b).

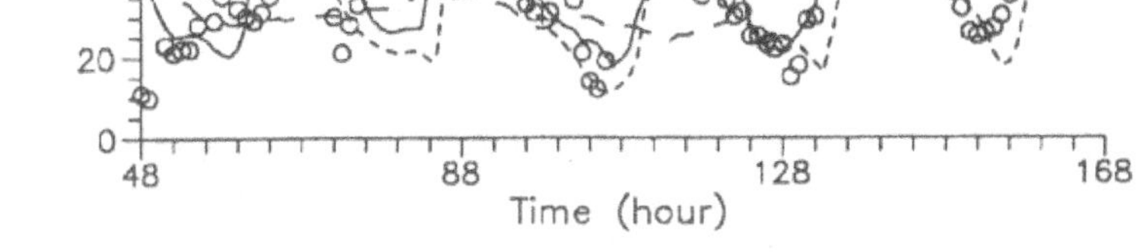

Figure 3. Time series of the observed and the modelled surface ozone at the Kejimkujik site from August 1 to August 5, 1988, (hour 48 = 00 Z, August 1, 1988).

The Kejimkujik site (44.43°N and 65.21°W) is situated in a rural environment in the central Nova Scotia peninsula. It usually takes about one day for an air mass travelling from the major emissions sources of ozone precursors in the north-eastern U.S. across the sea to arrive at the site. Figure 3 shows the ozone monitoring data compared with various model predictions of the first model layer (1 to 56 m). As seen, the ozone measurements at the site displayed clear diurnal variations during the period of August 1 to 5, 1988 when the region was under a general south-westerly flow. The daily minimum values occurred in the early morning while the daily maxima were reached around noon to afternoon with the highest value of 105 ppbv appearing on August 4. It is evident that the ADOM-MEP did not reproduce the basic features of the measured variations. The amplitudes of diurnal variations and the maximum levels were underpredicted. In particular, the modelled ozone variations were clearly out of phase with the observation after August 2. In comparison, ADOM-MC2 appeared to generate surface ozone variations in reasonably good agreement with the observations (in phase) though the amplitude of variations were still underpredicted.

Table 1. Surface ozone and its hourly budget due to various processes at Kejimkujik, for a 24-hour period from Eastern Standard Time (EST) 7 p.m., August 3 to 6 p.m., August 4, 1988, from the ADOM-MEP run.

EST (hour)	O_3	$\delta(O_3)$ chemistry	$\delta(O_3)$ horizontal transport	$\delta(O_3)$ vertical transport	$\delta(O_3)$ vertical diffusion	$\delta(O_3)$ vertical advection	$\delta(O_3)$ dry deposition
19:00	28.09	-0.24	4.20	-2.61	2.89	-1.73	-3.77
20:00	28.60	-0.22	4.90	-4.16	1.87	-2.17	-3.86
21:00	32.72	-0.20	5.41	-1.08	1.83	-0.98	-1.94
22:00	36.82	-0.17	4.89	-0.59	4.32	-1.17	-3.74
23:00	41.82	-0.15	3.08	2.10	8.36	0.41	-6.67
0:00	45.29	-0.14	2.27	1.33	8.06	2.48	-9.21
1:00	49.88	-0.12	0.12	4.60	10.82	4.91	-11.12
2:00	53.05	-0.13	0.31	2.98	11.44	5.38	-13.84
3:00	57.43	-0.13	-0.13	4.59	13.28	7.98	-16.67
4:00	59.81	-0.13	0.02	2.46	9.28	5.50	-12.32
5:00	60.99	-0.11	-0.74	2.03	6.25	4.16	-8.39
6:00	60.31	-0.06	-1.00	0.38	9.16	5.22	-14.00
7:00	57.22	0.06	1.55	-4.76	8.95	0.32	-14.03
8:00	54.82	0.15	0.82	-3.43	9.18	0.64	-13.25
9:00	51.34	0.20	4.15	-7.91	8.91	-2.95	-13.87
10:00	49.79	0.14	2.25	-3.97	7.88	-0.71	-11.13
11:00	47.61	0.14	1.84	-4.36	8.35	-0.82	-11.89
12:00	44.86	0.18	4.54	-7.41	8.66	-3.42	-12.65
13:00	42.45	0.53	2.06	-5.07	10.69	-0.94	-14.82
14:00	42.70	0.53	0.23	-0.54	10.11	0.60	-11.25
15:00	44.32	0.43	-0.11	1.18	7.34	0.97	-7.13
16:00	45.50	0.22	3.02	-2.02	4.65	-0.74	-5.92
17:00	50.05	-0.01	1.34	3.23	4.16	0.64	-1.57
18:00	53.50	-0.15	2.85	0.77	3.96	0.04	-3.23

Units are in ppbv.

To investigate further, contributions of various physical/chemical processes to hourly ozone budgets at this site were calculated for the two ADOM coarse grid runs and are presented in Tables 1 and 2. The vertical component is further broken down into three processes: vertical diffusion, vertical advection and dry deposition. One may draw the following conclusions from these two tables:

1). In both the cases, horizontal advection generally makes a positive contribution to the ozone changes reflecting the long-range transport of ozone on the regional scale.

2). In the ADOM-MEP case, substantial downward diffusion persists during the night and reaches a maximum around 3 a.m.. Its positive contribution to the ozone change coupled with the contribution from horizontal advection overwhelms the loss due to dry deposition and leads to an ozone increase during the night. In the daytime, ozone production is minimal in accord with a very low NOx level (0.2-0.3 ppbv) calculated. With the level of the downward diffusion about the same as during the night, the enhanced dry deposition loss dominates and results in an ozone decrease.

Table 2 Same as Table 1 but the values are generated from the ADOM-MC2 run.

EST (hour)	O_3	$\delta(O_3)$ chemistry	$\delta(O_3)$ horizontal transport	$\delta(O_3)$ vertical transport	$\delta(O_3)$ vertical diffusion	$\delta(O_3)$ vertical advection	$\delta(O_3)$ dry deposition
19:00	35.80	-0.18	5.00	-8.74	0.00	-3.17	-5.57
20:00	32.59	-0.18	4.37	-7.26	0.00	-2.10	-5.16
21:00	30.29	-0.17	4.16	-6.22	0.00	-1.35	-4.86
22:00	28.57	-0.17	3.82	-5.33	0.00	-0.68	-4.64
23:00	27.39	-0.16	3.67	-4.68	0.00	-0.20	-4.48
0:00	26.48	-0.14	3.22	-3.98	0.00	0.41	-4.39
1:00	25.83	-0.14	2.69	-3.20	0.00	1.12	-4.32
2:00	25.28	-0.15	2.13	-2.52	0.00	1.74	-4.26
3:00	24.59	-0.16	1.39	-1.93	0.00	2.29	-4.22
4:00	23.76	-0.16	0.94	-1.61	0.00	2.51	-4.12
5:00	22.80	0.03	0.64	-1.68	0.00	2.31	-3.98
6:00	20.85	0.50	0.96	-3.51	0.00	1.45	-4.96
7:00	18.36	0.97	1.89	-5.39	0.00	-0.44	-4.95
8:00	16.23	1.41	1.94	-5.54	0.00	-0.60	-4.94
9:00	25.23	1.23	1.81	5.90	20.31	-1.66	-12.75
10:00	35.57	0.72	0.81	8.72	32.93	-2.23	-21.98
11:00	45.04	0.41	0.46	8.47	30.73	-3.98	-18.28
12:00	47.23	0.46	1.96	-0.33	17.40	-2.18	-15.55
13:00	52.33	0.34	3.78	0.89	24.49	-3.00	-20.60
14:00	52.70	0.41	4.48	-4.56	25.52	-4.33	-25.74
15:00	50.13	0.41	3.47	-6.45	16.80	-3.05	-20.21
16:00	50.34	0.25	3.46	-3.47	12.98	-2.60	-13.85
17:00	51.70	-0.03	3.25	-1.79	9.70	-2.23	-9.26
18:00	51.58	-0.19	2.80	-2.65	9.24	-2.52	-9.37

3). In the ADOM-MC2 case, the downward diffusion from aloft vanishes at night, and the dominance of dry deposition causes a continuous drop of ozone level. During the day the rapid increase in downward diffusion, in response to daytime heating and the steep vertical gradient of ozone formed through the night, overcomes the loss through dry deposition and prompts an ozone build up at ground level. Again, the daytime ozone chemical production is negligible owing to low NOx levels (0.2-0.3 ppbv).

It can be seen that the difference in the ozone predictions at this site between the two coarse ADOM runs is mainly caused by the difference in the vertical diffusion. In the case of ADOM-MEP, there is no significant diurnal variation in the eddy diffusivity at the site (resembling marine situations), with generally low values (4 m^2/s maximum). In contrast, there is a marked diurnal variation in the diffusivity in the case of ADOM-MC2 with a daytime maximum reaching

12 m^2/s. Clearly, the local meteorological parameters are of critical importance in model predictions of ozone at this site.

As shown in Figure 3, the ADOM-NEST run gives a significant improvement in the overall simulation of the surface ozone at the Kejimkujik site in comparison with the coarse ADOM runs especially the daytime maximum, though the phase is somewhat behind the observation on the August 4. A closer look at the meteorological fields revealed that higher values of vertical diffusivity at Kejimkujik were generated by the MC2 fine mesh run than by the MC2 coarse grid run. For example, the maximum diffusivity on August 4 was 38 m^2/s in the ADOM-NEST case whereas 12 m^2/s in the ADOM-MC2 case. Obviously, the high diffusivity was responsible for the elevated daytime ozone level. The ozone level was further enhanced by photochemistry - the daytime NOx level was doubled to ~ 0.4 to 0.6 ppbv in the ADOM-NEST case. Such a NOx level is able to produce a significant amount of ozone (Liu et al., 1987). The surface ozone budgets show a substantial increase in the positive contribution from vertical transport later in the evening on the 4th, which is primarily responsible for the modelled high ozone at that time.

SUMMARY

In this study, we have established a MC2/ADOM modelling system with a nesting capability. The new system was utilised in a high resolution simulation of the SAR oxidants during the EMEFS-1 intensive measurement campaign. Comparisons were made with the observations as well as between different versions of model simulation. Preliminary results show marked improvement in ozone prediction with the new modelling system at higher resolution compared with the previous ADOM simulation. A more detailed analysis of the model predictions of ozone at Kejimkujik revealed the importance of local meteorological parameters in model prediction of ozone at this rural site.

REFERENCES

Bergeron, G., Laprise, R. Caya, D., Robert, A., Giguère, M, Benoit, R., and Chartier, Y., 1994, Formulation of the mesoscale compressible community model, *CCRM*.

Hansen, D.A., Puckett, K.J., Jansen, J.J., Lusis, M., and Vickery, J.S., 1991, The eulerian model evaluation field study (EMEFS), *Proc. 7th Joint Conference on Applications of Air Pollution Meteorology with AWMA*, AMS, pp 58-62.

Liu, S.C., Trainer, M., Fehsenfeld, F.C., Perrish, D.D., Williams, E.J.,Fahey, D.W., Hubler, G., and Murphy, P.C., 1987, Ozone production in the rural troposphere and the implications for regional and global ozone distributions, *J. Geophy. Res.* Vol. 92, pp 4191-4207.

Scholtz, M.T., Weisman, B., Mahrt, L., and Christie, A.D., 1986, Generation of meteorological data fields for the ADOM eulerian regional model, *5th Joint AMS/APCA Conference on Applications of Air Pollution Meteorology*, Chapel Hill, NC, pp 145-150.

Tanguay, M., Robert, A., and Laprise, R., 1990, A semi-implicit semi-Lagrangian fully compressible regional forecast model, *Monthly Wea. Rev.*, Vol. 118, pp 1970-1980.

Trainer, M., Perrish, D.D., Buhr, M.P., Norton, R.B., Fehsenfeld, F.C., Anlauf, K.G., Bottemheim, J.W., Tang, Y.Z., Weibe, H.A., Roberts, J.M., Tanner, R.L., Newman, L., Bowersox, V.C., Meagher, J.F., Olszyna, K.J., Rodgers, M.O., Wang, T., Berresheim, H., Demerjian, K.L., Roychowdhury, U.K., 1993, Correlation of ozone with NOy in photochemically aged air, *J. Geophys. Res.*, Vol.98, pp 2917-2925.

Venkatram, A., Karamchandani, P.K., and Misra, P.K., 1988, Testing a comprehensive acid deposition model, *Atmos. Environ.*, Vol. 22, no. 4, pp 737-747.

DISCUSSION

J. CHING: Please explain the reason for the large differences in vertical diffusivity between the MC2 nested version versus the previous model of meteorology usen in ADOM.

X. LIN: The large difference in vertical diffusivity at the Kejimkujik site is attributable to two factors. The MC2 calculates the boundary layer parameters in a 3-D mode while the previous vertical diffusivity was generated by a 1-D PBL model. More significantly, the nested MC2 run in the present study produced the vertical diffusivity on a horizontal resolution of 21 km. In contrast, the vertical diffusivity generated by the 1-D PBL model corresponds to a 127 km by 127 km grid. Since the grid of the Kejimkujik site contains a large portion of ocean, a strong marine signal is included in the calculated boundary layer parameters.

TEST OF A NEW CONCENTRATION FLUCTUATION MODEL FOR DECISION-MAKERS

Per Løfstrøm[1], Hans Jørgensen[2], Erik Lyck[1] and Torben Mikkelsen[2]

[1]National Environmental Research Institute
Department of Atmospheric Environment
P.O. Box 358
DK-4000 Roskilde, Denmark
[2]Risø National Laboratory
Department of Meteorology and Wind Energy
P.O. Box 49
DK-4000 Roskilde, Denmark

INTRODUCTION

It is well known that changes in meteorological conditions also change air pollution levels. But even under apparently constant meteorological conditions still fast and random concentration fluctuations exist due to the turbulent motion of the atmosphere. Dispersion models of today can estimate reasonable accurate one-hour average concentrations for operational use. Operational models describing concentrations on shorter averaging times are needed for assessing odour problems, nonlinear toxic effects, risks of infections, risks of explosions, and fast atmospheric chemistry.

The goal, which this study is a part of, is to establish a new formula based concentration fluctuation model suitable for implementation in one-hour-mean plume models already containing information on dispersion parameters σ_y and σ_z, Monin-Obukhov length etc.. The model should be fast, operational and designed for decision-makers often making calculation for a complete year of hourly meteorological data, when they assess odour annoyance originating from low point sources at short distances. The new model is based on an existing semi-empirical Gaussian concentration fluctuation model developed on wind tunnel data by Wilson et al. (1985,1982a,1982b). The model describes the spatial distribution of concentration fluctuation intensities (rms/mean) from a point source. This distribution is closely connected to the mean plume dispersion parameters σ_y and σ_z. In the new model the so-called variable 'variance source' is described in a new way by the dispersion parameters σ_y and σ_z, the turbulent length scales and source diameter.The main point in the model is to predict the downwind variation of the variance source strength.

The new model is developed and tested on two different series of full scale atmospheric dispersion experiments conducted in Denmark, where joint SF_6-tracer and Lidar measurements were performed on non-buoyant elevated and ground-level releases.

Firstly, the model is verified by use of data from the experiments conducted during the summer of 1992, BOREX'92. Secondly, the new parameterization is tested against data collected in the summer of 1994, BOREX'94.

The concentration profiles are analyzed regarding crosswind (y,z) distribution of fluctuation intensities. At several selected positions in the plume the cumulative distribution functions (cdf) are examined, and they are fitted well by clipped normal cdf's.

With the model for concentration fluctuation intensity, the measured form of the cdf's, and with one-hour mean values calculated by the Danish Gaussian plume model OML (Olesen et al., 1991), a new model will be established, which describes the statistical concentration fluctuations within an hour.

EXPERIMENTS

The two dispersion experiments were conducted during the summer of 1992 and 1994. Smoke and SF_6 were released simultaneously either at the top of a meteorological mast (22 m and 24 m, respectively) or at the ground (0.5 m, only in 1994) on the flat Borris Moor in Western Jutland. The design of the smoke generator resulted in an initial plume diameter of about 1 m.

A backscatter mini-Lidar system (Jørgensen and Mikkelsen, 1993) was operated just above one crosswind SF_6 line to obtain instantaneous crosswind concentration profiles of the smoke plume with high spatial (1.5 m) and temporal (1/3 Hz) resolution. The Lidar data was used for calculating plume statistics.

Ground-level (0.5 m) half-hour average concentrations of SF_6 were measured (Lyck, 1990) along three crosswind lines at distances between 100 and 700 m from the point of release. The SF_6 data was used for validating the Lidar measurements and during horizontal Lidar measurements also for estimating σ_z from the crosswind integrated SF_6 concentrations utilizing a reflected Gaussian plume assumption. In experiments with simultaneous lidar and SF_6 measurements the lidar measured either horizontally just above the SF_6 samplers, or the Lidar was scanning the smoke plume above the SF_6 samplers.

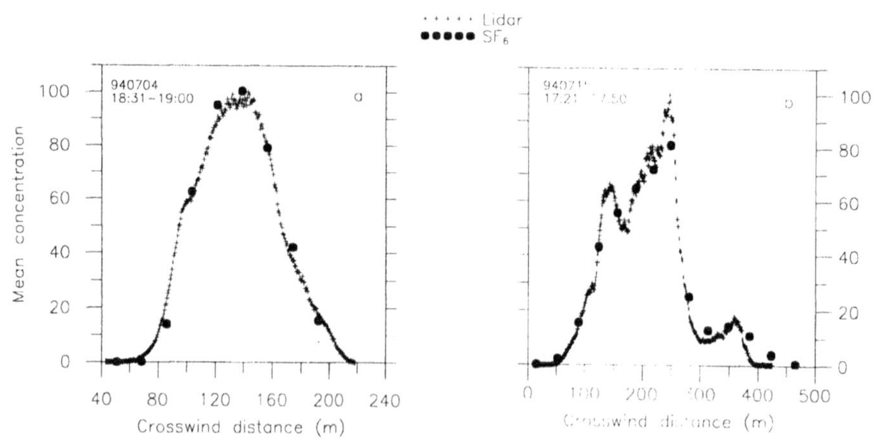

Figure 1. Examples of half-hour mean concentrations (relative units) measured by SF_6 (0.7m) and Lidar (2.0m) a) 200 m and b) 400 m downwind of surface release during unstable conditions.

Surface-layer scaling parameters were obtained using sonic at a height of 7 m and vertical profile measurements of wind speed and direction from the release mast. Balloon soundings were performed, but for many experiments in 1992 only routine soundings were available. Most of the dispersion experiments were conducted during moderate and strong unstable atmospheric conditions and a few during moderate stable conditions. Each individual experiment lasted about one half to one hour.

Two examples of the high quality of the two different measurement techniques are shown in Figure 1.

THE MODEL

The semi-empirical Gaussian plume model is founded on the theoretical considerations by Wilson et al. (1985,1982a,1982b). The model is based on wind tunnel measurements of the different terms in the variance-budget equation expressing a balance between production, advection, dissipation and diffusion. In the model the variance is not constrained by conservation of variance, and a variable variance source strength q is used to take this into account. They found that there was an increased dissipation of σ_c^2 near the ground and accounted for that by introducing an image sink with strength αq. By assuming that the vertical and crosswind spread of σ_c^2 occurs with the same turbulent diffusion processes and eddy diffusivities as the mean field, the final expression for the variance is given by

$$\sigma_c^2 = \left(\frac{q}{2\pi U\sigma_y\sigma_z}\right)^2 \exp\left(-\frac{1}{2}\left(\frac{y}{\sigma_y}\right)^2\right) \left[\exp\left(-\frac{1}{2}\left(\frac{z-h_v}{\sigma_z}\right)^2\right) - \alpha \exp\left(-\frac{1}{2}\left(\frac{z+h_v}{\sigma_z}\right)^2\right)\right] \qquad (1)$$

where σ_y and σ_z are the usual dispersion parameters for the mean concentration, U is the wind speed, $h_v=[h^2+(\phi_z\sigma_z)^2]^{0.5}$ accounts for the position of the effective variance source and h is the release height. Wilson et al. estimated α and ϕ_z from the wind tunnel data and found values of 0.6 and 0.7, respectively.

The mean concentration <c> is modelled by a reflected Gaussian plume, and the fluctuation intensity $i=\sigma_c/<c>$ then becomes

$$i(y,z) = \left(\frac{q}{Q}\right) \exp\left(\frac{1}{4}\left(\frac{y}{\sigma_y}\right)^2\right) \exp\left(\frac{1}{2}\left(\frac{z-h}{\sigma_z}\right)^2 - \frac{1}{4}\left(\frac{z-h_v}{\sigma_z}\right)^2\right) \left[1-\alpha \exp\left(-2\frac{h_v z}{\sigma_z^2}\right)\right]^{0.5} \left[1+\exp\left(-2\frac{hz}{\sigma_z^2}\right)\right]^{-1} \qquad (2)$$

where Q is the emission rate.

Based on wind tunnel observations of approximately balance between dissipation and advection of variance, because production and diffusion generally have opposite sign and are smaller, they developed a relation for the normalized variance source

$$\frac{q}{Q} = \frac{k_1 \lambda}{\lambda^2 + \lambda_0^2} \tag{3}$$

where

$$\lambda_0 = \frac{k_2}{k_3 \Lambda / d + 1} \tag{4}$$

and λ is the normalized downwind distance

$$\lambda = \sqrt{\sigma_y \, \sigma_z} \; / \; H \tag{5}$$

here, H is the boundary layer height, d is the source diameter, Λ is the average integral length scale for the turbulence, which is approximated by $0.6*h*exp((-\sigma_z/h)^2)$. The empirical constants k_1, k_2 and k_3 for the wind tunnel data were *0.34*, *0.12* and *0.055*, respectively. The model has proven to fit the wind tunnel data well for $h/H=0.192$ and for a wide range of d within the experimental range $4<\Lambda/d<400$ and $0.02<\lambda<0.23$. Applying the model outside this ranges should be done with caution, but some extrapolation may be justified due to the physical arguments used (Wilson et al., 1982b).

However, the limitations of the wind tunnel experiments, i.e. only neutral conditions and a relative limited range of meandering, calls for the model to be tested against measurements obtained in the atmospheric surface layer.

DISTRIBUTION OF INTENSITY

To validate the spatial distribution of the intensity as predicted by (2) the following procedure was used starting with the cross-wind analyzes. For a downwind position, where σ_y and σ_z are known, the formula (2) predicts the shape of the horizontal distribution of intensity: $i(y) \approx exp(1/4(y/\sigma_y)^2)$. The shape of $i(y)$ has been examined for the experiments with horizontal lidar measurements and compared with the theoretical predictions using the observed σ_y. Figure 2 shows an example of the measured intensity from BOREX'94 and the shape predicted by (2), when the level of the intensity is fitted. All the analyzed experiments during both BOREX'92 (Løfstrøm et al., 1994) and BOREX'94 fits well with the theoretical shape of $i(y)$ as others also have found (Jørgensen, 1994). However, there is a tendency that the u-shape of measured intensities opens in stable conditions. This tendency was of course not revealed in the earlier mentioned wind tunnel experiments.

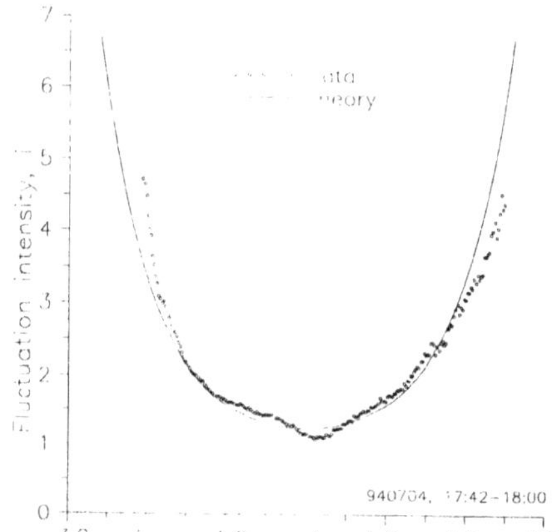

Figure 2. An example of the cross-wind distribution of concentration intensity at surface (2m) 200m downwind of surface release in unstable conditions during BOREX'94.

To analyze how the intensity function depends on the vertical position z, the intensity profiles $i(y,z)$ from Lidar scans have been normalized with $exp(1/4(y/\sigma_y)^2)$, assuming the shape of the cross-wind profile follows the theory as just shown. This means that at a certain height z the off-centerline intensities have been transformed to their theoretical value at the centerline. The corresponding theoretical function $i_n(z)=i(y,z)/exp(1/4(y/\sigma_y)^2)$

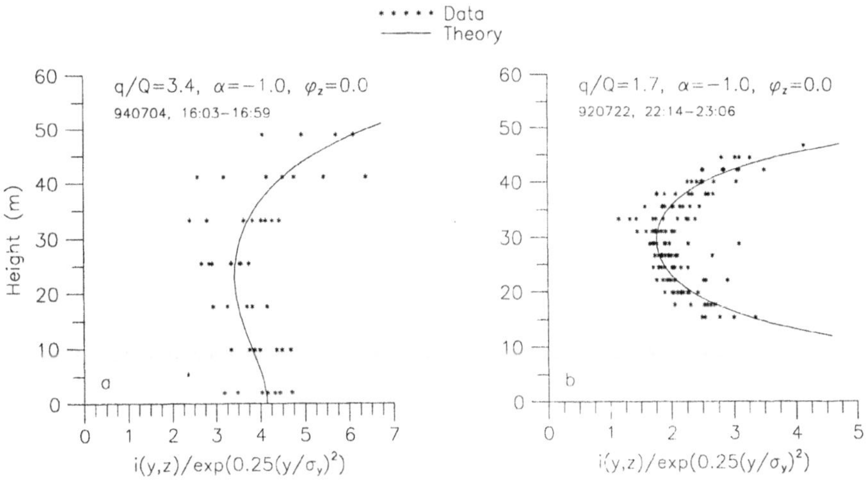

Figure 3. Two examples of the measured normalized vertical distribution of intensity. a) Very unstable conditions 200 m downwind, and b) moderate stable conditions 650 m downwind.

has then been fitted to the measured normalized data to obtain an estimate of α, ϕ_z and q/Q. Only data where $<c> > 20\% <c>_{max}$ have been used due to great uncertainty of the Lidar statistics at the plume edges. Two examples from BOREX'92 and BOREX'94 are shown in Figure 3. Examination of all the scanned plumes in both sets of experiments shows that the best fit in most cases was obtained with $\alpha = -1$ and $\phi_z = 0$.

Although the lowest Lidar measurements were performed at 2 m above the surface, the value of α indicates a full reflection of the variance at the ground, opposite to the findings of Wilson et al. (1985). Additionally, this means that the 'artificial' h_v can be substituted by h in the equation for the intensity $i(y,z)$.

NEW NORMALIZED VARIANCE SOURCE STRENGTH

The downwind variance of the normalized variance source strength q/Q has been examined in an earlier study (Løfstrøm et al., 1994) using the expression of λ in (5) and an expression of the integral scale Λ that differs from that proposed by Wilson et al. (1985). Wilson et al. mentioned that for large values of Λ/d there is a large difference between the distance to the downwind location of maximum intensity predicted by (3) as being proportional to $dH\Lambda^{-1}$ and the proportionality of $d^{2/3}\Lambda^{1/3}$ predicted by wind tunnel and a simple model by Sykes et al. (1984). Because of this and the large uncertainty in operational determination of H used in (5), a new non-dimensional expression for λ is used in this study

$$\lambda = \sqrt{\sigma_y \sigma_z} \; / \; \Lambda \tag{6}$$

The downwind distance to the maximum intensity is in this case proportional to d. (Using the expression $\lambda = (\sigma_y\sigma_z)^{1/2}/(\Lambda^4/d)^{1/3}$ ensures the same proportionality of the downwind location of the maximum intensity as predicted by Sykes et al., but preliminary analyzes using this expression do not seem to improve the model.)

In this study the integral scale $\Lambda(z)$ at height z is approximated by the length scale at which the eddies are carrying the maximum energy in the energy spectra of the vertical velocity field as parametrized in the surface-layer by Panofsky et al. (1984) using z and L. Above the surface-layer, $z>0.1H$, it is assumed that $\Lambda(z)=\Lambda(0.1H)$. Involved here are only two stable experiments where $z>0.1H$. For these stable experiments H were estimated by $H=2400\,u_*^{1.5}$ (Venkatram, 1980). Introducing the horizontal length scale, as proposed by Wilson et al. (1985), spoiled the analyses, and was not included.

When $\sigma_z<h$ the height z used for calculating $\Lambda(z)$ is h. But when $\sigma_z>>h$, which particularly is the case during surface release experiments and also for several elevated releases, z should depend also on σ_z. The height z used for calculating $\Lambda(z)$ should be related to the vertical position of the plume from release to the position in concern. Therefore, z is calculated as the time-weighted mean height of $max(h, 0.675\sigma_z)$, where $0.675\sigma_z$ is the half-mass-height for a reflected Gaussian plume for a surface release. When σ_z is assumed to increase linearly in time, this results in

$$z = 0.5(0.675\sigma_z + h^2/0.675\sigma_z) \qquad , \textit{for } 0.675\sigma_z>h \tag{7}$$

For $\sigma_z>>h$, z approximates $0.338\sigma_z$.

For experiments with Lidar scans q/Q was extracted from the data as described in the last section in connection with determining α and ϕ_z.

For the experiments where the lidar measured at a fixed horizontal level, q/Q was determined indirectly. Firstly, σ_z was estimated using a reflected Gaussian plume formulation, the observed crosswind integrated SF_6 concentration, the SF_6 emission rate, h and U. Then q/Q was estimated from (2) using measured values of σ_y, $i(0,z=2)$, and the estimated σ_z together with the values of $\alpha=-1.0$ and $\phi_z=0.0$ determined from the scanning experiments. In this way 17 estimates of $q/Q(\lambda)$ from BOREX'92 were obtained for $0.2<\lambda<2.1$ and $7<\Lambda/d<191$.

The fit on BOREX'92 data of the constants k_i in (3) and (4) was obtained using the observed values of σ_y and σ_z, the estimated values of Λ together with the found constant values of α, and ϕ_z. The fit resulted in the following values $k_1=3.483$, $k_2=1.874$ and $k_3=0.0289$. Figure 4a shows the observed and the calculated values of q/Q. In the figure q/Q is closely related to the centerline intensity. From (3), using the full scale experimental findings: $h_v=h$ and $\alpha=-1$, it is seen that at the plume centerline for $h>>\sigma_z$: $i=q/Q$ and for $\sigma_z>>h$: $i=q/(Q2^{1/2})$. The reasonably good fit in Figure 4a might not be surprising using 17 data points and 3 fitted constants. Therefore the empirical constants have been used unchanged on the analyzed data from BOREX'94 with surface releases and with a somewhat higher release height for the elevated releases. The comparisons between 20

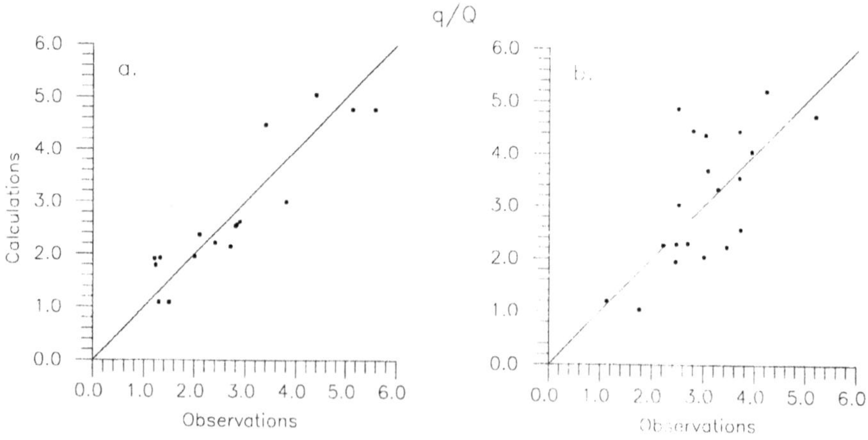

Figure 4. Comparison between predicted and observed normalized variance source strength. a) BOREX'92, empirical constants dependent on data, b) BOREX'94, constants from BOREX'92.

measured and calculated values of q/Q ($0.1<\lambda<2.4$ and $7<\Lambda/d<154$) are shown in Figure 4b, indicating as for 4a a good correlation between observation and the model predictions. In Figure 4b the use of (7) is crucial for the performance of the surface releases.

CUMULATIVE DISTRIBUTION FUNCTION

To describe within e.g. an hour the total time of the exceedance of instantaneous concentrations above a certain level, the cumulative distribution function (cdf) is useful. Accordingly, the concentrations at several selected positions in the plumes have been analyzed regarding the cumulative clipped normal distribution function. The clipped normal cdf is chosen because it has been confirmed for surface releases for neutral and slightly unstable (Mylne and Mason, 1991) and stable conditions (Mylne, 1992), for elevated releases (Lewellen and Sykes, 1986), and because of the relative simple and limited numbers of input parameters σ_0 and μ_0, which are uniquely related to σ_c^2 and $<c>$ (Lewellen and Sykes, 1986). The clipped normal cdf is given by

$$cdf(C) = \int_{-\infty}^{C} \frac{1}{\sqrt{2\pi}\sigma_0} \exp\left(-\frac{(c-\mu_0)^2}{2\sigma_0^2}\right) dc \qquad , C \geq 0$$

An illustration of the measured cdf for normalized concentrations, $c/<c>$, is shown in Figure 5. The measurements are taken in the plume centerline 650 m downwind in stable atmospheric conditions during BOREX'92. A clipped normal cdf will appear as a straight line. The full straight line corresponds to the best fit within the chosen concentrations range and the stippled line is extrapolation. The deviation from the clipped normal cdf for large values of $c/<c>$ is probably due to statistical noise, because of the limited amount of observations in this range. For the whole data set there is a tendency for the measured intensity to be a little less than the intensity for the best clipped normal fit, but the measured data are represented rather well by a clipped normal cdf, disregarding the low concentration values.

CONCLUSIONS

The comparison between the measurements and the modified semi-empirical model of Wilson et al. (1985) has shown to describe well the vertical and crosswind distribution

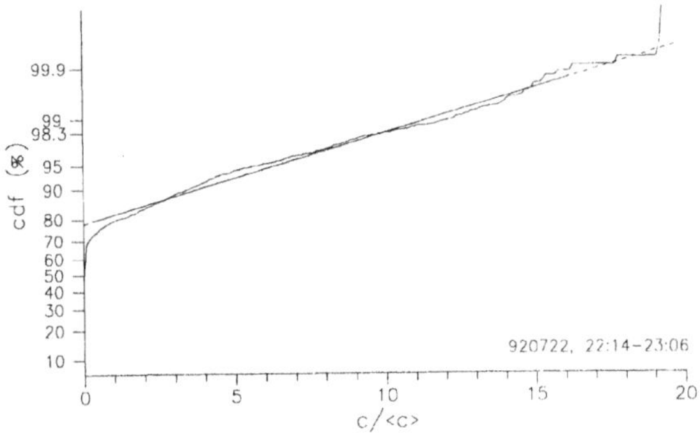

Figure 5. An example of a measured cumulative distribution function cdf. Elevated centerline, stable conditions, 650 m downwind. The straight line correspond to a clipped normal cdf.

of the concentration fluctuation intensities and the downwind variation of the variance source strength for the present two sets of full-scale atmospheric data. The surface reflection of fluctuations, as described by the value of α, has been found to be very different as compared to the wind tunnel studies. Atmospheric data indicates a full reflection of the variance at the ground, opposite to the findings in the wind tunnel, where the major part of the variance is dissipated.

The combination of the model and the clipped normal cdf's creates the possibility to make a simple formula based model for concentration fluctuations to be included in an fast operational model for decision-makers, where often a complete year of hourly meterological data are processed.

The model are so far restricted to a certain range of atmospheric conditions due to the full scale atmospheric data mostly limited to unstable conditions. Experimental data from the summer 1995 are going to be analyzed for further improvement of the model.

ACKNOWLEDGMENTS

This work was sponsored by the Danish Environmental Research Programme and the Danish Environmental Protection Agency.

REFERENCES

Jørgensen, H.E. and Mikkelsen T., 1993, Lidar measurements of plume statistics, *Bound. lay. Met.*, vol 62, 361:378.

Jørgensen, H.E., 1994, Studies of concentration fluctuations in the atmospheric boundary surface layer, *Risø-R-729(EN)*.

Lewellen, W.S. and Sykes, R.I., 1986, Analysis of concentration fluctuations from lidar observations of atmospheric plumes, *J. Climate Appl. Meteor.*, 25, 1145:1154.

Lyck, E., 1990, A tracer technique for full-scale atmospheric dispersion experiments, in: Preprints of the Ninth Symposium on Turbulence and Diffusion, American Meteorological Society, Boston.

Løfstrøm, P., Jørgensen, H., Lyck, E. and Mikkelsen, T., 1994, A concentration fluctuation model for decision-makers based on joint tracer and lidar measurements from a non-buoyant elevated plume, in: *Air Pollution II Volume 1: Computer Simulation*, Eds. J.M. Baldasano, C.A. Brebbia, H. Power and P. Zannetti, Computer Mechanics, Publications, Southampton Boston.

Mylne, K.R. and Mason, P.J., 1991, Concentration fluctuation measurements in a dispersing plume at a range up to 1000 m, *Q. J. R. Meteorol. Soc.*, 117, 177:206.

Mylne, K.R., 1992, Concentration fluctuation measurements in a plume dispersing in a stable surface layer', *Q. J. R. Meteorol. Soc.*, 60, 15:48.

Olesen, H.R., Løfstrøm, P., Berkowicz, R. and Jensen, A.B., 1991, 'An improved dispersion model for regulatory use - the OML model, in: *Air Pollution Modeling and its Applications IX*. Eds. H. van Doop and G. Kallos, NATO-CCMS, vol.17, Plenum Press, New York.

Panofsky, H.A. and Dutton, J.A., 1984, *Atmospheric Turbulence*, John Wiley & Sons, New York, p.189.

Sykes, R.I., Lewellen, W.S. and Parker, S.F., 1984, A turbulent transport model for concentration fluctuations and fluxes, *J. Fluid Mech.*, 139, 193:218.

Venkatram, A., 1980, Estimating the monin-obukhov length in the stable boundary layer for dispersion calculations, *Boundary-Layer Met.*, 19, 481:485.

Wilson, D.J., Robins, A.G. and Fackrell, J.E., 1985, Intermittency and conditionally-averaged concentration fluctuation statistics in plumes, *Atmospheric Environment*, 19, 1053:1064.

Wilson, D.J., Robins, A.G. and Fackrell, J.E., 1982a, Predicting the spatial distribution of concentration fluctuations from a ground-level source, *Atmospheric Environment*, 16, 497:504.

Wilson, D.J., Fackrell, J.E. and Robins, A.G., 1982b, Concentration fluctuations in an elevated plume: a diffusion-dissipation approximation, *Atmospheric Environment*, 16, 2581:2589.

DISCUSSION

S. ZILITINKEVICH: How is the horizontal velocity variance parameterized in the model?

P. LØFSTRØM: It is not parameterized directly, because an attempt to include a measure of the horizontal integral length scale (analogous to the method for the vertical scale) by using the length scale at the spectral peak of the horizontal wind field, spoiled the analyses. This may be due to the parameterized vertical scale being dominated by the much larger horizontal scale dependent on z/L and mixing height, respectively, and the mixing heights were only roughly estimated from routine soundings 200 km away. It is assumed that the high-frequency part of the energy spectra of the horizontal velocity variance is proportional to the vertical spectra, and in this way the horizontal variance is indirectly accounted for in the length scales for the vertical wind field.

S. ZILITINKEVICH: Does the model reflect the effect of large-scale structures on the transport properties of convective boundary layers?

P. LØFSTRØM: Due to the good capability of the model to predict the variance source strength, one could say - at least for the convective surface layer, where most of the data are collected - that the large-scale structures are contained in the empirical constants and the lambda parameter, which includes the mean horizontal dispersion parameter.

E. GENIKHOVICH: What could you say about possible dependence of the coefficient 'alfa' on the atmospheric stability and surface characteristics (roughness, wet or dry surface, and so on)?

P. LØFSTRØM: Wet and dry surfaces will only have effect on deposition and I cannot comment on that. Increasing roughness or decreasing stability will increase turbulence intensity. I think the effect on 'alfa' will be small, because as the experiments show, chance in turbulence intensity due to stability does not influence 'alfa', and I do not think that changes in roughness will have an effect on alfa either.

MODEL OF INTERNAL BOUNDARY LAYER FORMATION

OVER INHOMOGENEOUS TERRAIN

Ekaterina Batchvarova[1,2] and Sven-Erik Gryning[1]

[1]Risø National Laboratory, DK-4000 Roskilde, Denmark
[2]National Institute of Meteorology and Hydrology, Sofia 1784, Bulgaria

INTRODUCTION

The transformation of an air mass passing from the cooler sea surface over the highly heated land in summer is connected with internal boundary layer formation which is of great importance for the dispersion of pollutants in coastal areas. Thus the delightful breezes can be connected with air pollution episodes instead of freshening of the air. A distinct example of such a phenomenon occurs in the area of Athens, where the complex orography and typical weather conditions lead to the formation of shallow internal boundary layers and accumulation of pollutants over the city (Pilinis et al., 1993; Ziomas et al., 1994; Lalas, 1983).

In this study the internal boundary layer height model of Gryning and Batchvarova (1990) is extended to account for an irregular coastline and variations of the wind field both in time and space and is applied to a region in the central part of Athens. The model is based on parameterizations of the energy balance equation for the boundary layer and the potential temperature jump at the top of it. At the initial stage the internal boundary layer is predicted to grow proportionally with the friction velocity. As the layer grows the effect of mechanical turbulence diminishes and convective turbulence becomes more important. The input for the model consist of wind speed and direction, friction velocity and kinematic heat flux in time and space for the area, and the potential temperature gradient and the mean vertical air motion above the internal boundary layer (Batchvarova and Gryning, 1994).

MODEL

The framework is the zero-order mixed-layer model for a fully convective atmosphere with the extension to near-neutral and unstable atmospheric conditions suggested by Gryning and Batchvarova (1990) and Batchvarova and Gryning (1991). The model incorporates a parameterization of the turbulent kinetic energy budget within the mixed-layer. Furthermore the temperature jump at the mixed-layer top is assumed to be in equilibrium with the actual meteorological conditions. The effect of subsidence warming on the temperature jump is not accounted for.

Advection is explicitly introduced by inserting the Euler relation

$$\frac{dh}{dt} = \frac{\partial h}{\partial t} + u\frac{\partial h}{\partial x} + v\frac{\partial h}{\partial y} \tag{1}$$

of the material derivative dh/dt, into the expression for the growth of the mixed-layer. This results in the following Equation for the height h of the internal boundary layer:

$$\left\{\frac{h^2}{(1+2A)h - 2B\kappa L} + \frac{Cu_*^2 T}{\gamma h((1+A)h - B\kappa L)}\right\}\left(\frac{\partial h}{\partial t} + u\frac{\partial h}{\partial x} + v\frac{\partial h}{\partial y} - w_S\right) = \frac{\left(\overline{w'\theta'}\right)_S}{\gamma} \tag{2}$$

where u and v are the horizontal components of the mean wind speed in the mixed-layer in the x- and y- directions, t is time, L the Monin-Obukhov length, κ- the von Karman constant, A, B and C are parameterization constants. Commonly accepted values are A=0.2, B=5, and C=8. Knowing the mean vertical motion of air, w_s (negative of the subsidence velocity) and the potential temperature gradient γ, above the internal boundary layer, the vertical kinematic heat flux at the surface $(\overline{w'\theta'})_s$, the friction velocity u_*, the buoyancy parameter g/T, and the wind field over the area, Equation (2) can be solved numerically for the height of the internal boundary layer as function of time and space.

The mean vertical motion of the air at the top of the internal boundary layer can be estimated when the horizontal divergence of the large-scale flow field is known as a function of height. For the special case in which the horizontal divergence, div, is constant with height, w_s can be expressed as

$$w_S = -\mathrm{div}\, h\,. \tag{3}$$

It can be also determined from the warming rate of the free atmosphere above the boundary layer (Steyn, 1990). The parameter is mainly important for large boundary layer heights.

DESCRIPTION OF THE AREA

The densely populated area of Athens is surrounded by mountains of elevation 450 m to the west and more than 1000 m to the north and to the east. The valley is open towards Saronikos Bay to the south-west, Figure 1. Several gaps between the mountains channel the air from the north-east and west into the city area. The existence of the semi-persistent circulation system during the warm period of the year is a peculiarity of the eastern Mediterranean and is connected with north-easterly winds known in Athens as Etesians. The Etesians give maximum ventilation of the city, but the cases when these northerly synoptic winds become weaker are favorable for sea breeze development and internal boundary layer formation over Athens - meteorological conditions typical for air pollution episodes during warm seasons in the area.

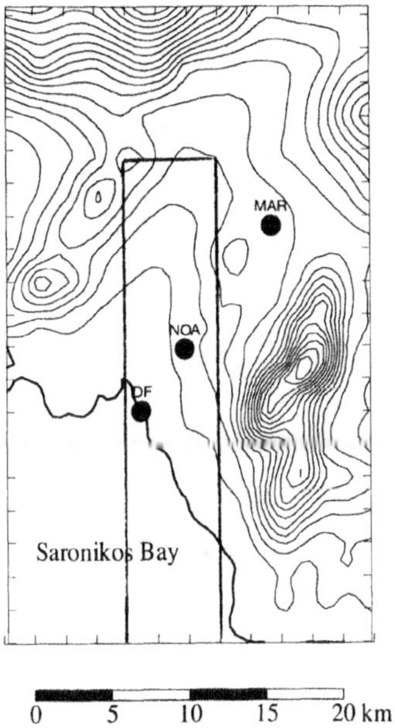

Figure 1. A map of the Greater Athens Area with observation sites: DF - Delta Faliro (shoreline meteorological mast), NOA (turbulence measurements), MAR - Marousi site (turbulence measurements). Contour lines are drawn in altitude steps of 50 m. The box shows the modeling domain.

EXPERIMENTAL DATA

The measurements used for the present application of the internal boundary layer growth model were carried out in August-September 1994 by Risø National Laboratory, Denmark, as a part of the MEDCAPHOT-TRACE Project within the EU Environmental Research Program (Progress Report, 1994).

A 9 m high meteorological mast for wind speed, wind direction and air temperature measurements situated at Delta Faliro 2 meters from the shoreline was a part of the measuring program, Figure 1. These records reflect the conditions near the coastline and provide estimates of the onset of the sea breeze over the Athens area.

Kaijo-Denki DAT/TR-61B sonic anemometers were used for turbulence measurements at two positions within the city - one near the National Observatory of Athens (NOA) about 4 km inland from the shoreline mast and the second, so-called Marousi site, close to Kifissias street at the Agricultural Research Station of Athens, about 13 km inland. Both the sonic anemometers were installed at 15 m above the ground. The records of the wind speed, wind direction, friction velocity and turbulent kinematic heat flux from NOA are used as input for the internal boundary layer modeling, presented in this paper.

MODEL SIMULATIONS

September 14 is chosen for the present simulations because it is the start of a typical air pollution episode in Athens.

Figure 2 illustrates the strong northerly winds on September 12. The wind speed decreases on the 13th and a weak non-stationary sea breeze is formed. Further weakening

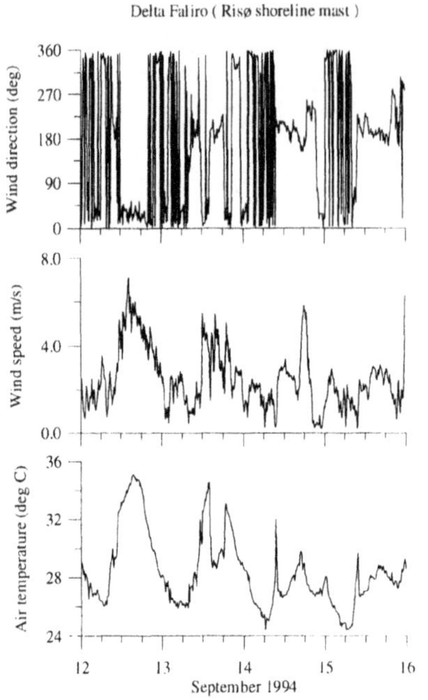

Figure 2. Wind speed, wind direction and air temperature at 9 m height at Delta Faliro site (shoreline mast) for the period 12-15 September 1994.

of the northern flow occurs on the 14th and 15th and a sea breeze (south-westerly flow) is observed starting from 10 a.m. and lasting until 9 p.m. on both days.

The night aerological sounding on 14 September at the Airport of Athens shows a strong northerly synoptic flow. The following noon sounding shows a layer with a westerly flow near the ground and northerly flow higher up. The potential temperature gradient in the layer above the internal boundary layer used for the simulations is estimated to be 0.005 K/m from the temperature profile at 00 a.m. on 14 September.

Figure 3 shows the data from the sonic-anemometer measurements of the wind direction, wind speed, friction velocity, air temperature and turbulent kinematic heat flux at NOA on September 14. The rapid change in wind direction, speed and temperature that occurs shortly before 10 a.m. indicates the start of the sea breeze. The wind direction remains relatively constant and from the south-west until 6 p.m. For the following four hours the wind is westerly. The north-easterly winds appear after 10 p.m. The comparison with shoreline mast data, Figure 2, shows the same flow pattern - almost the same time for the onset of the sea breeze and the same wind direction shift in the late afternoon of September 14.

The measurements on 14 September at the Marousi site, Figure 4, do not show the characteristics of a sea breeze. The wind direction through the day changes gradually from north over west to south and becomes easterly and stronger between 4 and 8 p.m. The wind speed and kinematic heat flux are lower compared to NOA. The air temperature and friction velocity are higher which reflects the influence of the surrounding urban area.

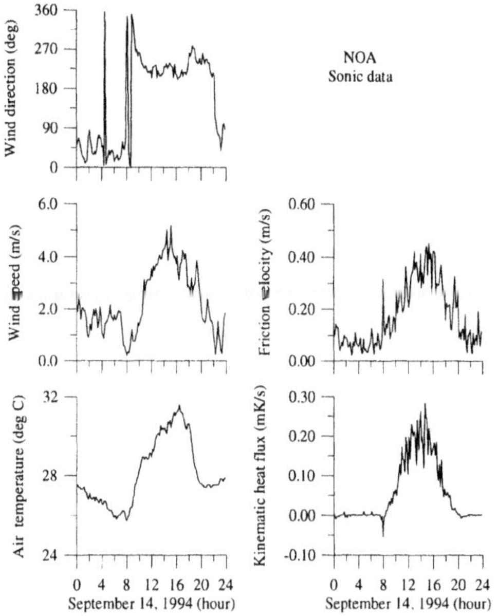

Figure 3. Sonic-anemometer measurements at NOA on 14 September 1994.

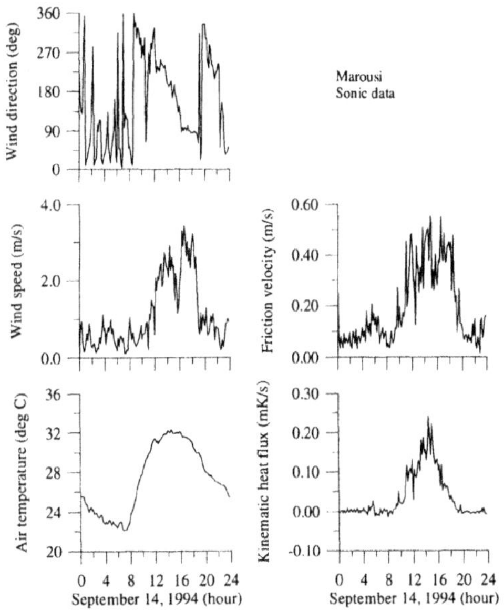

Figure 4. Sonic-anemometer measurements at Marousi on 14 September 1994.

The simulations are performed over the area of Athens basin, chosen to exclude the mountains around the city. The modeling domain is 6 km in the west-east direction and 29 km in the south-north direction. The grid-size is 250 m and the time step is 15 s. The topography is not accounted for. The shoreline is irregular, but the surface roughness, friction velocity and kinematic heat flux are assumed to differ only from water to land.

The calculations start at 10 a.m. with an initial internal boundary layer height of 10 m over the water. The growth of the internal boundary layer over the area can be followed on Figure 5.

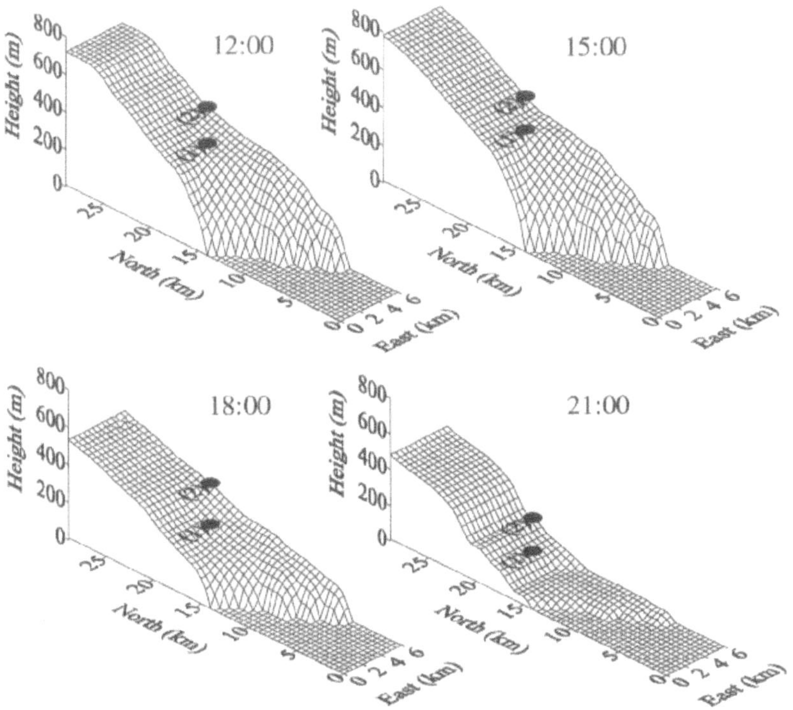

Figure 5. Simulation of the internal boundary layer height over Athens on 14 September 1994. The positions (1) and (2) correspond to the grid points where the changes of the internal boundary layer height with time are shown on Figure 6.

Two points along the flow direction are chosen for the illustration of the internal boundary layer development with time throughout the day, Figure 6. The first corresponds to NOA (1) and is 4 km inland from the shoreline mast towards north-east and the second (2) is 3 km further inland - the region north of Likavitos hill. The height of the internal boundary layer is relatively constant until 4 p.m. when it starts to decrease. An explanation for the secondary growth after 6 p.m. can be the westerly wind which prevails for about one hour and reflects land conditions. Later the internal boundary layer decreases again as the wind direction turns to south-west bringing air from the sea.

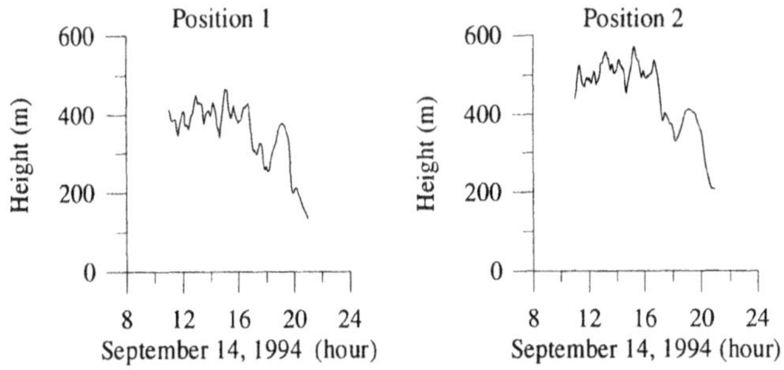

Figure 6. Evolution of the internal boundary layer on 14 September 1994. Position (1) corresponds to a grid point close to NOA and position (2) corresponds to a grid point north of Likavitos hill.

DISCUSSION

The model for the evolution of the internal boundary layer used for the simulations requires as input the friction velocity, kinematic heat flux, wind speed and direction as functions of time and space over the area, the potential temperature gradient above the internal boundary layer and an estimate of the subsidence velocity if available.

The topography of the greater Athens area does not allow application of such a model for the whole domain shown in Figure 1 but the simulations for the area of the city give consistent results although based on measurements at only one location (which then is assumed to be representative for the whole land area). With its relative simplicity and fast computations on a 486 type PC, the model presented here is an useful applied tool for air pollution studies even in regions like Athens. A more physically justified approach will be to use detailed input information from a meteorological measuring network or calculated by meteorological models that reflect the inhomogeneous nature of the area.

Acknowledgments

The study is a part of the Risø National Laboratory and National Institute of Meteorology and Hydrology participation in MEDCAPHOT-TRACE Project and was supported by the Danish Environmental Research Program 1992-1996.

REFERENCES

Batchvarova, E., and Gryning, S.-E., 1991, Applied model for the growth of the daytime mixed layer, *Boundary- Layer Meteorol.,* **56**, 261-274.
Batchvarova, E., and Gryning, S.-E., 1994, Applied model for the height of the daytime mixed layer and the entrainment zone, *Boundary-Layer Meteorol.,* **71**, 311-323.
Gryning, S.-E., and Batchvarova, E., 1990, Analytical model for the growth of the coastal internal boundary layer during onshore flow, *Quart. J. R. Met. Soc.,* **116**, 187-203.
Lalas, D.P., Assimakopoulos, D.N., Deligiorgi, D.G., and Helmis, C.G., 1983, Sea-breeze circulation and photochemical pollution in Athens, Greece, *Atmos. Environ.,* **17**, 1621-1632.
Pilinis, C., Kassomenos, P., and Kallos, G., 1993, Modelling of photochemical pollution in Athens, Greece. Applications of the RAMS-CALGRID modelling system, *Atmos. Environ.,* **27B**, 353-370.
Progress Report, 1994, Mediterranean campaign of photochemical tracers - transport and chemical evolution (MEDCAPHOT-TRACE). LAB-AUTH, Aristotle University of Thessaloniki, Greece.
Steyn, D., 1990, An advective mixed-layer model for heat and moisture incorporating an analytical expression for moisture entrainment, *Boundary-Layer Meteorol.,* **53**, 21-31.
Ziomas, I. C., Suppan, P., Rappengluck, B., Balis, D., Tzoumaka, P., Melas, D., Papayannis, A., Fabian, P., and Zerefos, C.Z., 1994, A contribution to the study of photochemical smog in the Greater Athens Area, *Beitr. Phys. Atmosph.,* in press.

DISCUSSION

P. MARTANO: How many wind data points did you use for the computations?

E. BATCHVAROVA: Only one measuring position is used, situated at NOA and equiped with sonic anemometer. The time discritization for this meteorological input was 10 minutes.

P. MARTANO: Which initialization was used as boundary condition above the sea?

E. BATCHVAROVA: Kinematic heat flux was assumed to be 0 mK/s and the IBL height was set to 10 m. Wind speed and direction were taken from the sonic anemometer at NOA.

M. PIRINGER: Did you compare the calculated IBL heights with measured data?

E. BATCHVAROVA: Some measurements of the IBL height were performed during the experiment using tethered balloon soundings at 2 positions. Our results are in good agreement with those data but the comparisson was not presented here. The model is validated on data sets for less complex terrain.

M. PIRINGER: How large would you judge to be the influence of local topography on the development of IBL height? As I understand, topography is not included explicitely in your model.

E. BATCHVAROVA: It is right that topography is included in the model implicitely through wind field and heat fluxes measurements or calculations. We have chosen the modelling domain to cover only the central part of Athens basin relatively far from the mountains. The purpose of our study is to provide fast estimation of the IBL height which is needed for air pollution assesments and is important when there is no time or funding to perform the complex mesoscale models.

S. ZILITINKEVICH: How did you determine the potential temperature flux on the right-hand-side of your equation for the IBL depth?

E. BATCHVAROVA: The potential temperature heat flux is measured with the sonic anemometer at NOA and the time discretization of these data is 10 minutes.

HIGH RESOLUTION DISPERSION MODELING NEAR TUNNEL EXITS

Roland Ries

Institute of Atmospheric Physics
Johannes Gutenberg University
Mainz, F.R.G.
and
Lahmeyer International GmbH
Frankfurt/Main, F.R.G.

1. INTRODUCTION

This contribution is concerned with a critical problem which usually occurs at motor-traffic tunnels within towns. All exhaust gases accumulating along the entire tunnel distance readily leave the tunnel exits in a very concentrated form. Since modern urban traffic management affords more and more tunnel constructions within highly built-up areas, a reliable calculation of the resulting immission distribution in the neighbourhood of tunnel exits is of great importance.

2. PRESENTLY AVAILABLE INVESTIGATION METHODS

From the fluid mechanics point of view the problem has to be described by a three-dimensional turbulent wall-jet penetrating into a cross flow which itself is characterized by turbulent shear.

In the literature about the theory of turbulent jets[1,2] there are no derivations of formulas taking into account the total number of influencing quantities simultaneously. In most cases the results are given by considering the influence of one single quantity ignoring the effect of interacting influences. Therefore it is not practicable to calculate the dispersion of tunnel exhaust air by the use of simple analytical models.

Since presently there are no calculation schemes available which are directly based on the underlying physics, in practice mostly semiempirical models are used with wind-tunnel measurements as foundation.

To give an example the TOP (Tunnel Outlet Pollution) model[3] describes the concentration decrease along the jet axis as an exponential function whose two empirical parameters have been estimated from field experiments.

The dispersion model from Iversen[4] goes beyond this by separating the dispersion process into a linear free jet with a concentration decrease according to the TOP model followed by a Gaussian plume model with a certain initial lateral extent propagating in the direction of the outside mean flow. Both models are not able to give particular information about the range of the pollution load distribution.

Much work has been done by Haerter and Steinemann[5,6,7] in their large-scale tests. They found out that the first stage of the free jet is most important for the dilution of the exhaust gases and additionally they show that at low ambient temperatures the free jet can lift off from the ground. For practical applications general planning proposals are given such as leading the exhaust gases along the road axis (e.g. by means of walls or cuttings) as long as possible to favour vertical mixing. Concerning immission calculations with numerical dispersion models near tunnel exits in built-up terrain they take into account the dilution during the free jet stage by spreading the tunnel emission source over a distance L along the jet axis.

The main difficulty rests with the determination of the length L which is a function of the geometric structure of the tunnel and the surrounding buildings, the traffic amount, the ventilation conditions in the tunnel tube and the meteorological conditions (wind direction and velocity, stability). Up to now there are not enough data available to give systematic guidelines covering the whole spectrum of appearing cases.

3. THE NEW CONCEPT

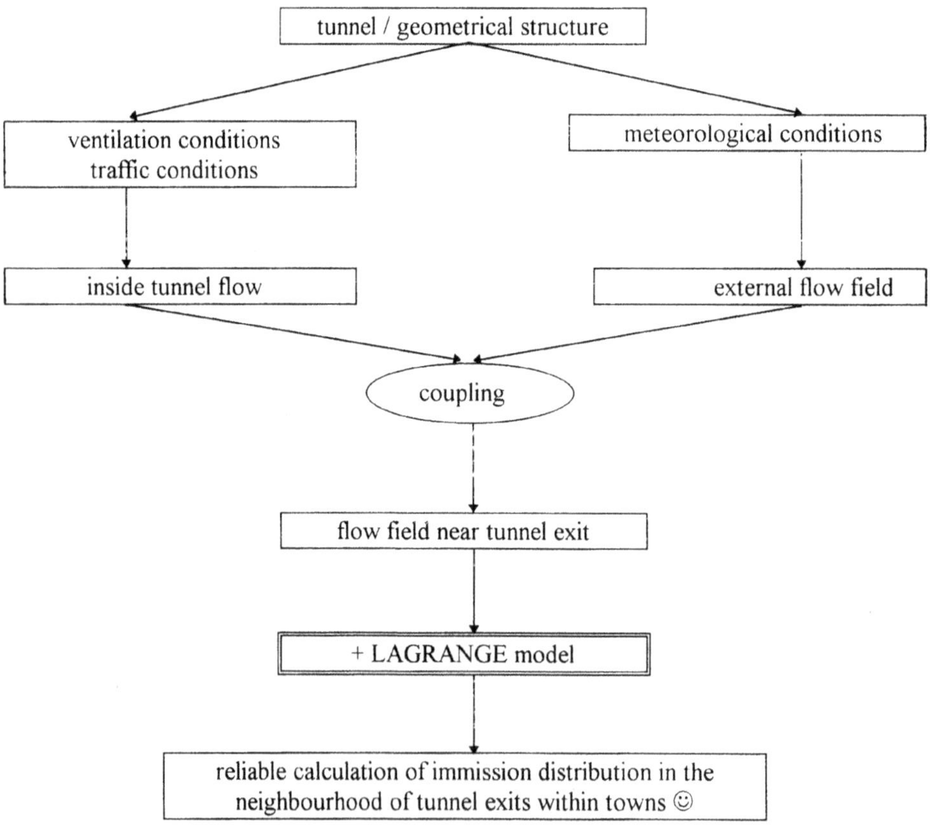

Figure 1. General arrangement of information flow

Nevertheless reliable planning instruments are needed since the conception of urban tunnels represents a mutual interaction process between the ventilation layout and the optimization of the air-hygienic situation outside the tunnel on the other hand.

In cooperation with the Lahmeyer International Company the three-dimensional flow and dispersion model MISCAM[8], **Micro Scale Air Pollution Model**, (see section 4.) has been extended to handle the flow and pollution fields near tunnel exits by coupling the inside tunnel flow with the external flow field and by inclusion of a Lagrangian particle dispersion module.

The work has been designed to provide a validated planning tool which can consider the characteristics of any particular situation.

4. THE BASIC MODELS

The numerical model MISCAM has been developed at the University of Mainz/Germany to evaluate special flow- and dispersion conditions within highly built-up urban areas. The model consists of a three-dimensional non-hydrostatic flow-model, combined with a set of dispersion models for different applications (Eulerian model, Lagrangian model). Up to now MISCAM is used by many institutions in Germany and has been applied to several planning projects. In addition, comparison of model results with wind tunnel experiments as well as dispersion measurements have been performed with success[9]. Further validation of the model is documented in another part of this volume.

The microscale model MISCAM can be applied to the microscale α-domain[10] with typical horizontal grid sizes of 2 - 20 m. It can handle particular urban areas such as street canyons or groups of buildings which may be of special interest. In this model buildings are explicitly taken into account in form of three-dimensional rectangular flow obstacles.

For use of reference the flow equations will be listed next:

$$\frac{\partial u_i}{\partial t} + \frac{\partial u_j u_i}{\partial x_j} = -\frac{1}{\rho}\frac{\partial p}{\partial x_i} + \frac{\partial}{\partial x_j}\left[K_m\left\{\frac{\partial u_i}{\partial x_j} + \frac{\partial u_j}{\partial x_i}\right\}\right] \tag{1}$$

$$\frac{\partial u_j}{\partial x_j} = 0 \tag{2}$$

Here K_m is calculated as a function of the three-dimensional velocity deformation. A similar procedure is used by Schuhmann et al.[11]:

$$K_m = \begin{cases} l^2\sqrt{\frac{1}{2}(\nabla v + v\nabla)^2}(1 - Ri) & Ri < 1 \\ 0 & Ri \geq 1 \end{cases} \tag{3}$$

The mixing length l is taken at each grid point as the minimum value of the distances to the next obstacles in every spatial direction.

The local Richardson-Number Ri is used in the form:

$$Ri = \frac{g}{\Theta}\frac{\partial\Theta / \partial x_3}{\frac{1}{2}(\nabla v + v\nabla)^2} \tag{4}$$

To simulate the flow sources due to ventilation and vehicle motion inside the tunnel tube a method was adopted using an adequate profile of the interior longitudinal velocity as constant boundary value.

Near tunnel exits the dispersion of pollutants is handled by use of a Lagrangian particle-diffusion model. In contrast to the Eulerian approach, which describes the behavior of the pollutant relative to a fixed coordinate system, the Lagrangian consideration describes the pollutants behavior relative to the moving fluid. This type of model is superior to the Eulerian K-diffusion model, especially in the given case of high numerical resolution to handle strong inhomogenity. Furthermore turbulence physics can be accounted for more completely while numerical diffusion is suppressed.

The conception of the Lagrangian model is based on the extended Langevin equation for inhomogeneous conditions, e.g. for the u-component:

$$\frac{du}{dt} = -\alpha u + \lambda \xi(t) + F_u \tag{5}$$

The first two forces contain the contributions of dynamic friction and intermediate collisions, whereas F_u describes the contributions due to inhomogeneous distributions of the velocity variances.

The solution leads to the following simulation equation for the discrete Markov sequence:

$$u'(t) = u'(t-\Delta t)\exp\left(-\frac{\Delta t}{T_{Lu}}\right) + \sqrt{1-\exp\left(-\frac{2\Delta t}{T_{Lu}}\right)}\,\sigma_u \xi(t) + F_u T_{Lu}\left(1-\exp\left(-\frac{\Delta t}{T_{Lu}}\right)\right) \tag{6}$$

For simplicity, $F_u = \frac{\partial \sigma_u^2}{\partial x}$ is assumed according to the drift correction of Legg and Raupach[12].

The turbulence parameters σ_u, σ_v, σ_w as well as T_{Lu}, T_{Lv}, T_{Lw} are computed from the turbulent kinetic energy as proposed by Groß et al.[13].

Since presently the prognostic equations for the turbulent kinetic energy and for the dissipation rate (see the contribution of Eichhorn „Validation of a Microscale Pollution Dispersal Model") have not yet been applied to the inside tunnel flow, the turbulent kinetic energy E is obtained indirectly from the Prandtl-Kolmogorov relation (see e.g. McComb[14]).

$$K = l\sqrt{cE} \tag{7}$$

In this relation c is an empirical constant, which is chosen as c = 0.2 according to Hanna and Strimaitis[15].

5. VERIFICATION OF THE MODEL

One important measure for the verification is the model's capability to describe the penetration of the free jet into the external flow field. Many measurements for individual meteorological situations are available from Haerter[6] and Klein[16].

To compare model results with wind tunnel experiments a cross wind situation without neighbouring flow obstacles was chosen with different ratios of wind speeds inside (v_{Tun}) and outside (v_{Win}) the tunnel. Using a combined flow and particle dispersion model calculations have been performed with two different atmospheric wind conditions and for different tunnel flow velocities. The results of the calculations have been combined to a nomogram (see Figure 2) which covers a range of v_{Tun}/v_{Win} from 0,5 .. 5. E.g. for a ratio of $v_{Tun}/v_{Win} = 2$ the measurements give an offset of the tunnel jet of about 40m up to 45m [6,15] which is in good agreement with the numerical simulation results.

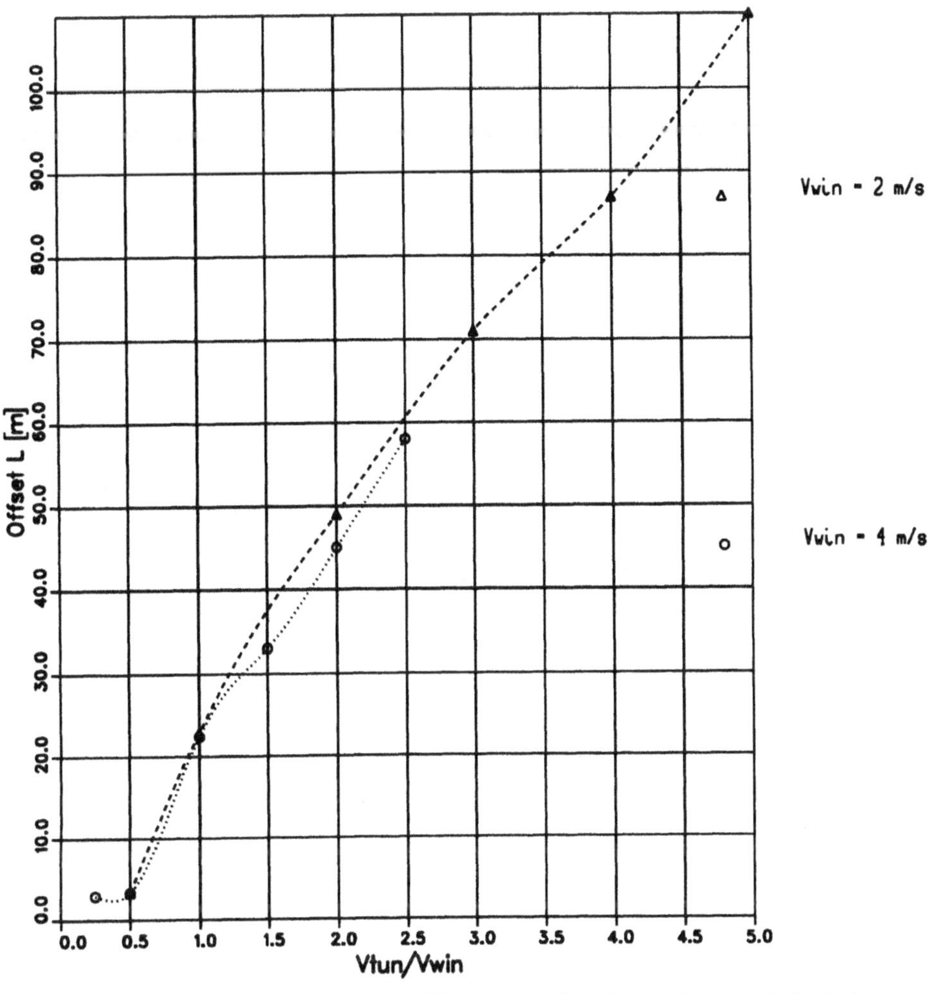

Figure 2.Offset of tunnel exhaust air for different ratios of v_{Tun}/v_{Win} (cross wind situation).

6. APPLICATIONS

Even if some model improvements are still in the testing phase, the questions resulting from daily problems lead to first applications. E.g. at double pipe tunnels the common recirculation effect may occur depending on meteorological and ventilation conditions (see Figure 3).

Normally this effect should be avoided by separating walls because the shortening of the ventilation circuit prevents effective ventilation. But on the other hand there are special situations (e.g. very close buildings) where the reverse flow of the exhaust air is desired to protect residents from critical immission concentrations.

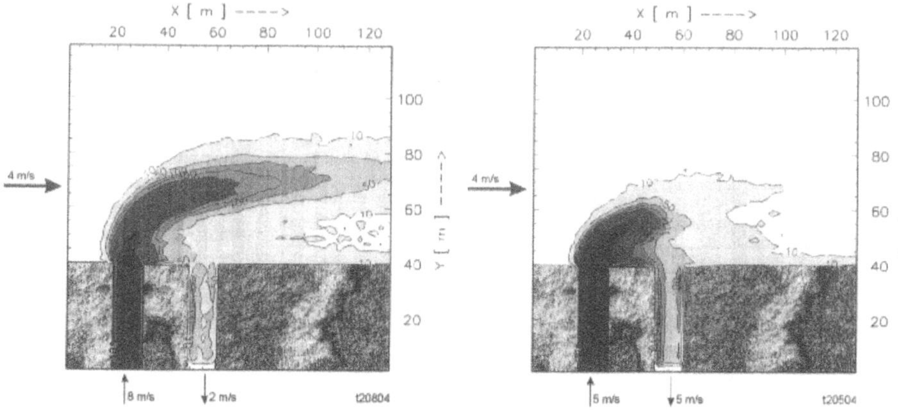

Figure 3. Computed distribution of a double pipe tunnels exhaust air in a cross flow for different ventilation conditions. Concentrations are in [$\mu g/m^3$] with source strength of 200 mg/s (only in the left tube).

7. CONCLUSIONS

The three-dimensional flow and transport model MISCAM has been extended to examine configurations with flow through obstacles, e.g. tunnel exits.

By inclusion of a Lagrangian particle model the dispersal of pollutants within highly complex configurations can be treated adequately. An examination of first model calculations shows promissing results. Principal characteristics of the coupled flow field can be reproduced and the depth of influence of the free stream in the external flow field is simulated within an acceptable error range which may be reduced by model improvements (e.g. inclusion of the prognostic turbulence scheme of the basic model). This suggests that MISCAM may also be used to optimate planning of urban tunnels.

ACKNOWLEDGEMENTS

The author wishes to express his gratitude to Dr. J. Eichhorn, Institute of Atmospheric Physics, University of Mainz, and to Dr. O. Heil and H. Rottmann, Lahmeyer International GmbH, Frankfurt/Main, for their assistance and for the many valuable discussions.

REFERENCES

[1] G. N. Abramovich. *The Theory of Turbulent Jets*, MIT Press, 1963.

[2] N. Rajaratnam. *Turbulent Jets, Developments in Water Science*; 5, American Elsevier Publishing Company, 1976

[3] N. Ukeguchi, H. Okamoto, Y. Ide. *Prediction of vehicular emission pollution around a tunnel mouth*, 4th International Clean Air Congress, Proceedings Paper II-18, Tokyo, 16-20.5.1977.

[4] T. Iversen. NILU, Oppdragsrapport Nr. 27/82, August 1982.

[5] W. Meyeroltmanns. *Abluftemissionen aus Tunnelportalen*, s+t, Band 36, 1982.

[6] A. Haerter. *Tunnellüftungssysteme mit geringer Abluftimmission*, Forschung + Praxis, Heft 23.

[7] A. Haerter. *Impact of Tunnel Air out of a Portal or Chimney on the Environment and Corresponding Tunnel Ventilation Systems*, OECD seminar on road tunnel management, Reports, Lugano-Switzerland, 26-29.11.1990.

[8] J. Eichhorn. *Entwicklung und Anwendung eines dreidimensionalen mikroskaligen Klimamodells*, ph.D. Thesis, University of Mainz, Germany, 1989.

[9] J. Eichhorn, N. Beltz, W. Jaeschke, R. Ries, O. Heil. *Naturmessungen verkehrsbedingter Immissionen als Grundlage zur Validierung des mikroskaligen Ausbreitungsmodells MISKAM*, VDI-Tagung: Emissionen des Straßenverkehrs; Immissionen in Ballungsgebieten, Berlin, 7.-8.9.1995.

[10] Orlanski. *A Rational Subdivision of Scales for Atmospheric Processes*, Bull.Amer.Meteor.Soc., 56, 527-530, 1975.

[11] U. Schuhmann, T. Hauf, H. Höller, H. Schmidt, H. Volkert. *A Mesoscale Model for the Simulation of Turbulence, Clouds and Flow over Mountains: Formulation and Validation Examples*, Beitr. Phys. Atmosph. **60**, 413-446, 1987.

[12] B.J. Legg, M.R. Raupach. *Markov-Chain Simulation of Particle Dispersion in Inhomogeneous Flows: The Mean Drift Velocity Induces by a Gradient in Eulerian Velocity Variance*, Boundary-Layer Meteorology **24**, 3-13, 1982.

[13] G. Groß, H. Vogel, F. Wippermann. *Dispersion Over and Around a Steep Obstacle for Varying Termal Stratification - Numerical Simulations*, Atmos. Environ. 21, 483-490, 1987.

[14] W.D. McComb. *The Physics of Fluid Turbulence* (1st edition), 126-128- Oxford Engineering Science Series **25**, Clarendon Press, Oxford, 1990.

[15] S.R. Hanna, D.G. Strimaitis. *Rugged Terrain Effects on Diffusion*, Meteorological Monographs **23**, Atmospheric Processes over Complex Terrain, Chapt. 6, 114-121, American Meteorological Society, Boston, 1990.

[16] P. Klein. *Untersuchung von Ausbreitungsvorgangen in einer turbulenten Windkanalgrenzschicht. Immissionsbelastung im Bereich von Tunnelportalen*, Thesis, University of Karlsruhe, Germany, 1993.

DISCUSSION

A. FLOSSMANN: MISCAM uses upstream-differencing for the advection. This scheme is known to produce unacceptable numerical dispersion. How much of the results presented is physical and how much is determined by this numerical scheme?

R. RIES: In fact the upstream advection scheme produces high numerical dispersion. For this reason we use this scheme only for the momentum advection. We have carried out model runs with the McCormack advection scheme in comparison. It was found that a significant change of the results only occurs in the case of single isolated flow obstacles. Within highly built-up areas, and that is MISCAM is concepted for, the changes for the simulation of the flow field can be rather neglected. For the calculation of pollutant dispersion in MISCAM the advection scheme of Smolarkiewicz has been applied to the Eulerian dispersion model. The Lagrangian particle model, which has been used for these calculations, naturally is free of numerical dispersion.

E. GENIKHOVICH: How did you take into account effects of the fluctuations of the horizontal component of the wind direction? These effects are very important in the description of the structure of the wind flow and dispersion of pollutants near bluff bodies (like buildings) in field conditions.

R. RIES: The flow and dispersion simulations with MISCAM are always carried out to stationarity. For the results presented turbulence was still modelled by use of a simple K-closure. The turbulent fluctuations of the velocity components are parameterized according to Ogawa and Oikawa (1982) on the basis of turbulent kinetic energy, which for these calculations has been still obtained indirectly from the Prandtl-Kolmogorov relation. But in near future the new prognostic E-ϵ turbulence closure of the basic model MISCAM will be implemented for the application of the tunnel flow calculation as well. From this improvement we expect a more realistic turbulence modelling near the buildings.

Lit.: Ogawa, Y. and S. Oikawa, 1982: A Field Investigation of the Flow and Diffusion around a Model Cube, Atmospheric Environment, 16, p207-222.

PERFORMANCE OF A FLUX CONSERVING AND A SEMI-LAGRANGIAN ADVECTION SCHEME IN SIMULATING A PHOTOCHEMICAL EPISODE

Prasad Pai,[1,2] Prakash Karamchandani,[1,2] and Akula Venkatram[3]

[1]ENSR Consulting & Engineering
1220 Avenida Acaso, Camarillo, CA 93012
[2]Presently at Atmospheric and Environmental Research, Inc.
2682 Bishop Drive, Suite 120, San Ramon, CA 94583
[3]University of California, Riverside, CA 92521

ABSTRACT

This paper examines the performance of two advection schemes in simulating a multiday ozone episode in the San Joaquin Valley of California. The advection schemes, viz. the flux conserving Smolarkiewicz scheme and a semi-Lagrangian scheme, were tested within the framework of a three-dimensional Eulerian comprehensive photochemical model. The modeling framework allowed us to test the performance of the advection schemes for the following two realistic test cases: (a) transport of inert species under the realistic flow conditions occurring during the ozone episode, and (b) transport of ozone precursors and formation of ozone during the four day episode. Our results indicate that the semi-Lagrangian scheme is not sensitive to deviations from mass consistency of the wind field, which is especially important in photochemical models that are driven by diagnostic meteorological models, such as the model used in the present study. Furthermore, we show that the semi-Lagrangian scheme preserves concentration peaks better than the more diffusive flux-conserving Smolarkiewicz scheme, used currently in a few operational photochemical models.

INTRODUCTION

Most urban and regional photochemical models (Carmichael et al., 1986; Chang et al., 1987; Venkatram et al., 1992; Scheffe and Morris, 1993;) solve a numerical approximation of the mass conservation equation. The mass conservation equation represents the combined problem of transport and source and sink terms such as emissions, chemistry, and removal at the surface. Accurate numerical treatment of advection, i.e., transport of the pollutant by the mean or bulk motion of air, is crucial in the numerical modeling of atmospheric pollution. The advection process is generally the first process to be numerically solved in the method of fractional steps (Yanenko, 1971) that is employed by most photochemical models. Therefore, inaccuracy in the numerical solution of advection terms can propagate in the numerical solvers of other processes, i.e., diffusion, chemistry and scavenging, resulting in erroneous results. Both the development (Brooks and Hughes, 1982; Bott, 1989) and testing (Long and Pepper, 1981; Schere, 1983; Chock, 1985) of advection algorithms has been the subject of research for the last two decades.

The work described here evaluates the performance of two advection schemes within the framework of a photochemical grid model. The impetus for this study developed during the analysis of the modeling database of the San Joaquin Valley Ozone Modeling Study also known as SARMAP. During this study a diagnostic meteorological wind field model (Pai et al., 1993) was used as a meteorological data analysis tool and a photochemical model (Karamchandani et al., 1990) was used as an air quality data analysis tool. While analyzing the various databases, it was found that the Smolarkiewicz (1983) scheme that was originally implemented in the photochemical model generated spurious ozone concentration peaks during the August 3-6, 1990 episode. The Smolarkiewicz scheme was replaced by an alternative advection solver based on the semi-Lagrangian scheme (Rancic and Sindjic, 1989; Seibert and Morariu, 1991; Smolarkiewicz and Pudykiewicz, 1992; Lee, 1993). The semi-Lagrangian solver was found to perform quite well for the complex flow conditions occurring during the episode. This paper presents the results from the application of the two advection solvers to simulate inert and reactive species transport for the episode.

SOLUTION OF THE MASS CONSERVATION EQUATION

The basis for most urban and regional photochemical models is the mass conservation equation, which expresses all the atmospheric processes in mathematical form as follows :

$$\frac{\partial C_i}{\partial t} = -\frac{\partial (uC_i)}{\partial x} - \frac{\partial (vC_i)}{\partial y} - \frac{\partial (wC_i)}{\partial z} + \frac{\partial}{\partial x}\left(K_H \frac{\partial C_i}{\partial x}\right) + \frac{\partial}{\partial y}\left(K_H \frac{\partial C_i}{\partial y}\right) + \frac{\partial}{\partial z}\left(K_v \frac{\partial C_i}{\partial z}\right) + R_i + S_i - L_i \qquad (1)$$

where C_i represents the three-dimensional concentration of pollutant "i", expressed in units of mixing ratio (e.g., ppb or ppm); u and v are the x- and y-direction components of the horizontal wind field, respectively; w is the vertical wind speed; K_H and K_v are the horizontal and vertical turbulent diffusion coefficients, respectively; R_i is the net rate of production of pollutant "i" by chemical reactions; S_i is the emission rate of pollutant "i"; and L_i is the net rate of removal of pollutant "i" by surface uptake processes.

Equation (1) with appropriate initial and boundary conditions, is numerically solved on a three-dimensional grid with lateral grid distances, Dx and Dy, on the order of 10 km and vertical grid distance, Dz, on the order of 100 m. The first three terms on the right-hand side of Equation (1) are known as advection terms and represent the transport of the pollutant by the mean or bulk motion of air. The mean motion of air is provided as hourly and three-dimensional varying winds at every point in the three-dimensional grid. The next three terms on the right-hand side of Equation (1) represent the turbulent diffusion terms (i.e., the transport of pollutants by fluctuations in the wind field that are not resolved by the grid). The last three terms on the right-hand side of Equation (1) represent the various sources and sinks of the pollutant.

Most grid models, employ some kind of operator splitting technique to solve Equation (1). In the operator splitting technique, the horizontal, vertical, and chemistry terms of the mass-conservation equation for a pollutant are marched in time by the method of fractional steps (Yanenko, 1971). The operator splitting technique first solves the horizontal transport, which consists of the x-direction advection and diffusion operator followed by the y-direction advection and diffusion operator. The solution of the horizontal transport is followed by the solution of the vertical advection, diffusion, and emissions injection operators. Finally, the chemistry is integrated. The advection time step is determined by the Courant number.

The primary difficulty in the numerical solution of the transport terms [first six terms on the right-hand side of Equation (1)] comes from the interaction between the advection and diffusion terms. The numerical solution of the advection equation typically leads to first order, quasi-linear or non-linear terms whereas the diffusion process leads to second-order linear terms. Also, the advection process is inherently non-symmetric whereas the diffusion process is symmetric. These terms drastically change the mathematical and physical behavior of the transport equation. The dominance of advection makes the equations hyperbolic, while diffusion makes the equations elliptic in nature. The boundary conditions, and the numerical methods required for satisfactory representation of these terms depend upon the nature of the equation and the relative importance of the physical processes involved.

As discussed above, the operator splitting technique divides the complexity of numerically solving the mass conservation equation into subproblems. One such subproblem is the solution of the advection terms, i.e., the first three terms on the right-hand side of Equation (1). Since our present study deals with the uncertainties in the numerical solution of the advection terms, we will restrict our discussion in the remainder of this section to the advection equation.

THE ADVECTION EQUATION

The transport of pollutants by mean motion of the air is represented by the advection terms in the mass conservation equation. The mean motion of the air is provided as input to most Eulerian grid models in the form of hourly and three-dimensional varying winds at every grid point. Typically, the input wind field is provided by a diagnostic or prognostic wind model.

The two-dimensional advection equation for a scalar quantity C is

$$\frac{\partial C}{\partial t} + \frac{\partial (uC)}{\partial x} + \frac{\partial (vC)}{\partial y} = 0 \qquad (2)$$

where C denotes the concentration of the scalar expressed as mass per unit volume (e.g., mg/m^3); u and v are the x- and y-direction components of the velocity field, respectively.

This equation is the flux form of the two-dimensional advection equation. Equation (2) can also be written as:

$$\frac{\partial C}{\partial t} + u\frac{\partial C}{\partial x} + v\frac{\partial C}{\partial y} + C\left(\frac{\partial u}{\partial x} + \frac{\partial v}{\partial y}\right) = 0 \qquad (3)$$

If the flow is assumed to be divergence free, i.e., if

$$\frac{\partial u}{\partial x} + \frac{\partial v}{\partial y} = 0 \qquad (4)$$

then Equation (2) becomes :

$$\frac{\partial C}{\partial t} + u\frac{\partial C}{\partial x} + v\frac{\partial C}{\partial y} = 0 \qquad (5)$$

Equation (5) is the total derivative form of the two-dimensional advection equation. Note that Equation (4) is implicitly assumed in most air quality models. However, interpolation or transformation procedures that are used to convert the outputs of a wind model to inputs for an air quality model can introduce divergences which might violate the divergence-free assumption of Equation (4).

The numerical solution of the advection equation can result in two undesirable effects: numerical diffusion and numerical dispersion. Numerical diffusion refers to the lowering or smoothing of initially peaked concentrations and is introduced when low-order numerical approximations are used. Numerical dispersion, on the other hand, is manifested as unphysical oscillations around the true solution. Numerical dispersion is caused due to the lagging between the true solution and the Fourier components in the solution. Since each wave number component travels with a different speed, the lagging causes the different Fourier components to spread apart, or disperse, as the numerical solution proceeds. The myriad of numerical techniques available represent attempts by various researchers to reduce these two errors. The variety of numerical schemes to solve Equation (2) used currently in air quality models suggests that there is little consensus in the modeling community on the most suitable scheme. However, there seems to be some consensus on the desirable features of a numerical scheme for advection in an air quality model. These are:

- the numerical solution should be stable;
- the numerical solution should preserve peak values in the concentration field;
- the numerical solution should not contain negative values;
- the numerical solution should be monotonic (non oscillatory), and
- the numerical solution should have reasonable computing time and computer storage requirements.

For the present study we have evaluated the suitability of the Smolarkiewicz (1983) scheme and the semi-Lagrangian scheme (Seibert and Morariu, 1991) to simulate a multi-day oxidant episode in the San Joaquin Valley of California. The Smolarkiewicz (1983) scheme is a positive-definite scheme that uses first-order upwind differencing together with an antidiffusive velocity to reduce numerical diffusion. This scheme solves the flux form of the advection equation [Equation (2)]. We used the one-dimensional version of this scheme in a time-splitting framework. The semi-Lagrangian scheme is based on the total derivative form [Equation (5)] of the advection equation which states that the scalar C does not change along a trajectory. Thus, the scalar C at the next time step can be determined by: (a) calculation of a backward trajectory to determine the starting point, and (b) estimating the scalar at this point, since it will generally not coincide with a grid point.

EVALUATING THE PERFORMANCE OF ADVECTION SCHEMES

The evaluation of numerical schemes for advection in air pollution models has been the subject of research for the last two decades. Table 1 lists some tests that are commonly used to evaluate the performance of advection schemes.

The following criteria are used to evaluate the accuracy of advection schemes :

- Maximum and minimum of the advected field; C_{max}, C_{min}
- Mass conservation ratio $\sum C(t) / \sum C(0)$
- Mass distribution ratio $\sum C^2(t) / \sum C^2(0)$
- Average absolute error $\sum | C(t) - C_e(t) | / N^2$; N is the number of grid points
- Maximum absolute error $\text{Max} | C(t) - C_e(t) |$

In addition to the above accuracy indicators, the numerical schemes are also evaluated in terms of execution time, storage requirements and ease of implementation in the grid model.

Table 1. Some commonly used problems for testing advection schemes

Test	Concentration Distribution	Velocity Field
Square Pulse (McRae et al., 1982)	$C(x) = 1 \quad \lvert x \rvert \leq \theta/2$ $= 0 \quad \lvert x \rvert > \theta/2$	u = constant
Triangle Pulse (McRae et al., 1982)	$C(x) = 1 - \lvert x \rvert/\theta \quad \lvert x \rvert \leq \theta$ $= 0 \quad \lvert x \rvert > \theta$	u = constant
Gaussian Pulse (McRae et al., 1982)	$C(x) = \exp\left[\Pi\left(\frac{x}{\Theta}\right)^2\right]$	u = constant
Rotating Cone Long and Pepper (1981)	$C(x,y) = 50\left[1 + \cos\left(\frac{\pi R}{4}\right)\right] \quad R \leq 4$ $= 0 \quad R > 4$ $R^2 = (x - x_0)^2 + (y - y_0)^2$ $(x_0 y_0)$ = coordinates of the center of the cone	u = -wy; v = wx, w = angular velocity

While the tests listed in Table 1 give good insight into the amplitude and phase errors produced by the numerical schemes, they are not very rigorous for realistic flow situations since u does not depend on x and v does not depend on y. Therefore in a time-splitting framework the tests listed in Table 1 subject one-dimensional schemes to constant velocity in the direction of advection in this test. Thus, by design these tests are free of splitting errors and therefore perform quite well.

Seibert and Morariu (1991) introduced another problem that provides a more stringent test in the time-splitting framework because the velocities are not constant along the direction of advection. In this problem, an initial square block of mass is subject to a purely deformational flow defined by u = -w x and v = w y where w is the angular velocity. The flow is periodically reversed after moving the block of mass every half cycle. The analytical solution is a pure squeezing and stretching about the center of the mesh without a change in the initial shape. The semi-Lagrangian scheme used in the present study was extensively tested by Seibert and Morariu (1991) and Lee (1993) using standard tests described above. Therefore, we will not repeat the tests here except to state that the semi-Lagrangian scheme scored high in the evaluation criteria stated above.

Our experience with advection schemes indicates that testing a numerical scheme with realistic wind fields often provides valuable information regarding the accuracy of the scheme. A typical wind field pattern in the study domain is shown in Figure 1. The major wind flow into the modeling domain is near the Sacramento River Delta where a split in the flow occurs. At the split, some of the winds diverge north to the Sacramento Valley, and the remaining flow goes south to the San Joaquin Valley. We tested the semi-Lagrangian and the Smolarkiewicz scheme by exercising only the advection module of the air quality model. The air quality model is driven by the three-dimensional wind fields generated by a diagnostic meteorological model (Pai et al., 1993) for the August 3-6, 1990 episode and consists of a constant concentration field as its initial condition. For a divergence-free flow, one would expect the concentration to remain constant at its initial value at the end of each time step. However, if the individual divergences are large, then the concentration will not

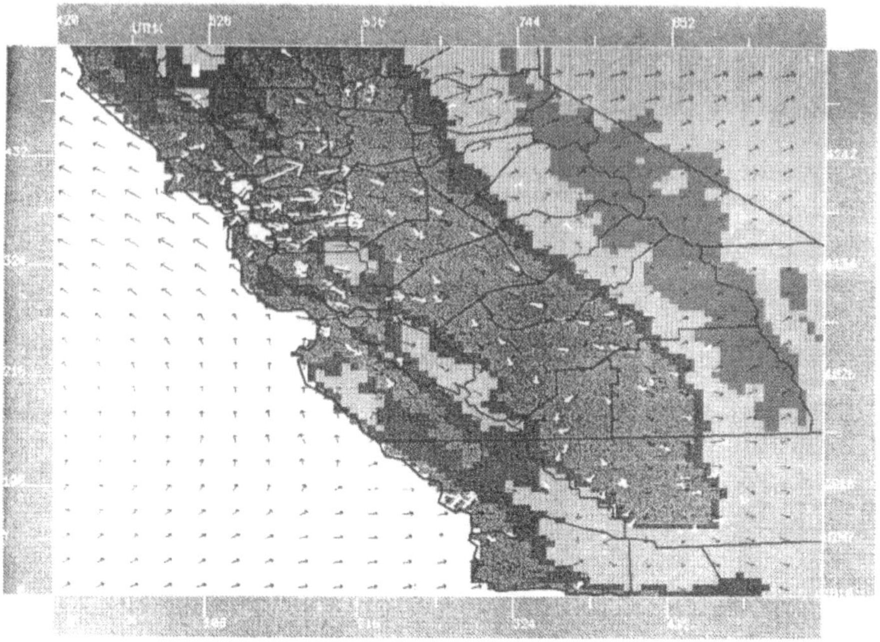

Figure 1. Surface layer winds predicted by the diagnostic meteorological model at hour 01 on August 3, 1990. The observed winds are shown as highlighted arrows and the wind fields are superimposed on the terrain (shaded) features of the modeling domain.

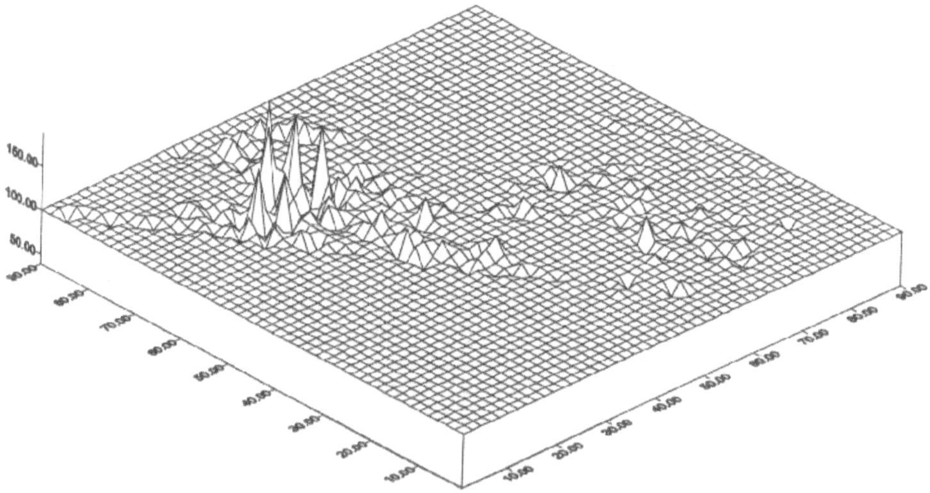

Figure 2. Concentration field after one hour of advection (advection time step = 900 seconds) with the Smolarkiewicz (1983) horizontal advection solver. The wind fields for hour 01 on August 3, 1990 shown in Figure 1 were used to advect an initial field of constant concentration.

remain constant after each one-dimensional advection and the error will continue to grow at every advection step, resulting in a non-constant concentration field at the end of the advection time step. Previous studies by Smolarkiewicz and Clark (1986) and Easter (1991) discuss ways to handle large one-dimensional divergences in flux-conserving schemes.

The semi-Lagrangian scheme solves the total derivative form (Equation (5)) and therefore an initial concentration field was undisturbed after an advection time step. On the other hand the Smolarkiewicz scheme changed the concentration field quite dramatically as shown in Figure 2. The reason for this is that even though the diagnostic wind field generated a near zero total divergence wind field, the non-zero components of the divergence (du/dx, dv/dy) during each one-dimensional advection step appear as source (if du/dx>0) or sink (if du/dx<0) terms in the advection equation (Petschek and Libersky, 1975). Thus, one step of the time splitting solution will decrease the concentration in a grid cell (if du/dx>0) and the next step will increase the concentration in a comparative manner, but the two changes do not exactly balance. The individual horizontal divergences are shown in Figures 3 and 4. Note that the Smolarkiewicz scheme shows the maximum deviations from the initial concentration field at grid cell (32,61) as shown in Figure 2. The grid cell is in the vicinity of the Sacramento River Delta region (see Figure 1) which represents the infow into the modeling domain and therefore the wind field exhibits significant variability. Thus the individual divergences in the vicinity of grid cell (32,61) are almost equal in magnitude but opposite in direction (peaks versus valleys) as shown in Figures 3 and 4.

The results shown in Figure 2 were obtained using an advection time step of 900 seconds. We repeated the above test with varying time steps to study the sensitivity of the scheme. Table 2 shows the maximum and minimum in the concentration field at the end of one hour of advection for different time steps. The initial concentration field, constant at 100, can increase and decrease by an order of magnitude at a time step of 1800 seconds. As smaller time steps are used, the performance of the one-dimensional Smolarkiewicz scheme improves and the concentration field remains relatively uniform at its initial value. However, the improvement in performance is obtained at the expense of computational efficiency.

Table 2. Effect of time step on the concentration field with the Smolarkiewicz scheme

Time step	1800	900	360	180	120	90
Maximum	1502	320	142	115	114	113
Minimum	1	3	38	70	78	83

EFFECT OF ADVECTION SCHEMES ON OZONE ESTIMATES

In this section, we present results from our studies to assess the sensitivity of ozone estimates from a photochemical grid model to the advection scheme. It is our contention that in photochemical grid modeling, where several species are transported and react with each other, it is important to test transport solvers within a framework that allows (a) simultaneous transport of more than one species, and (b) interaction of the transport solver with the chemistry solver. We believe that model estimates of a secondary pollutant, such as ozone, can be very sensitive to errors in the transport of its precursor species, i.e., ROG and NOx, and that these transport errors can then propagate through the chemistry solver leading to erroneous results. Previous studies (Hov et al., 1989; Tran and Mirabella, 1992; Chock and Winkler, 1994) have indicated the importance of testing advection schemes with numerical solvers of other processes such as chemistry in photochemical grid models.

We tested the ability of the two advection schemes to simulate the August 3-6, 1990 ozone episode in the San Joaquin Valley of California. For this simulation we used a time step of 90 seconds for the Smolarkiewicz scheme and a time step of 900 seconds for the

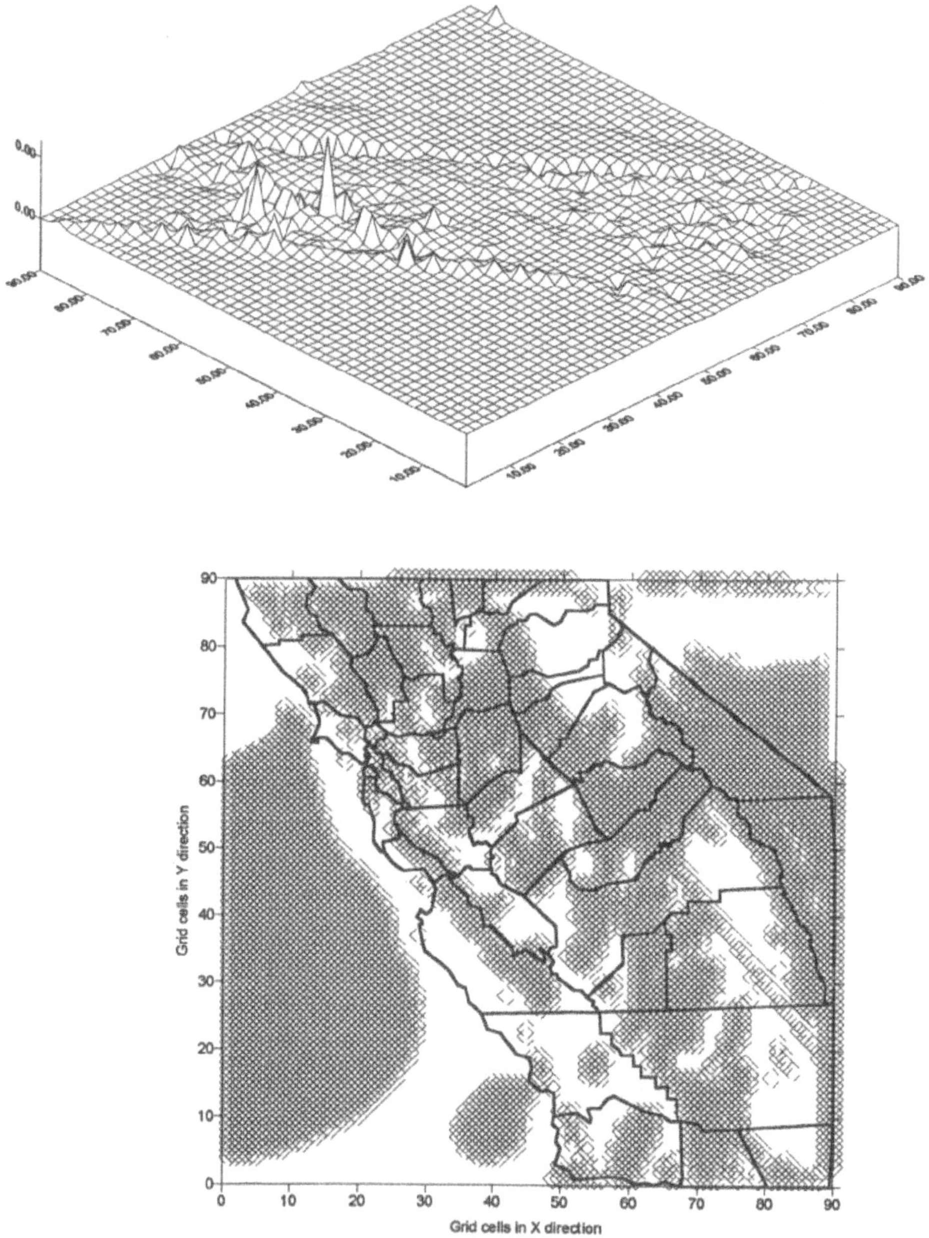

Figure 3. The horizontal x-direction divergence shown as (a) surface plot with the mountains and valleys representing the positive and negative values of the divergence, and (b) contour plots with the filled patterns representing the negative values of the horizontal divergence. The range of values for (du/dx) is (-0.0012, +0.0015) s^{-1}.

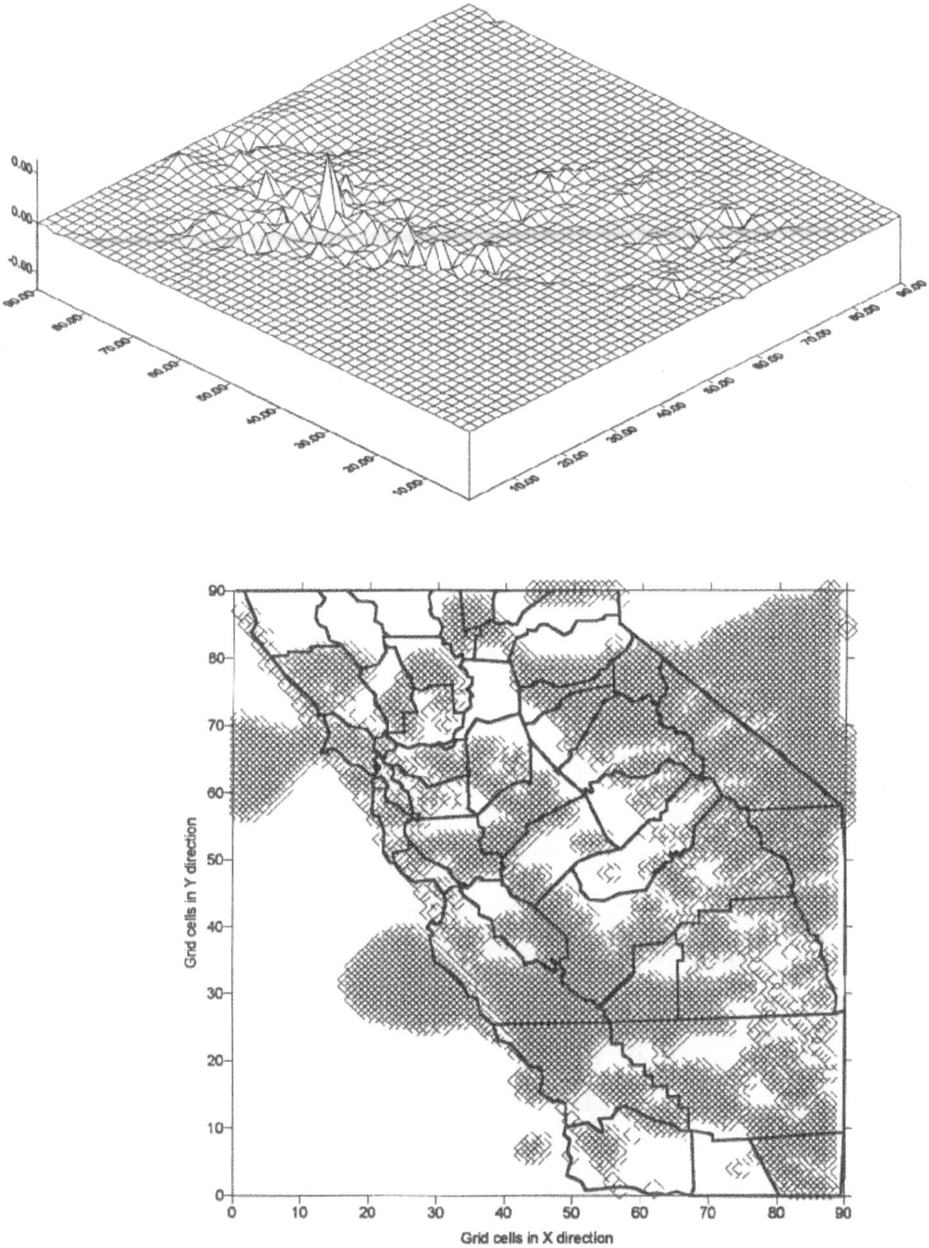

Figure 4. The horizontal y-direction divergence shown as (a) surface plot with the mountains and valleys representing the positive and negative values of the divergence, and (b) contour plots with the filled patterns representing the negative values of the horizontal divergence. The range of values for (dv/dy) is (-0.0018, +0.0014) s^{-1}.

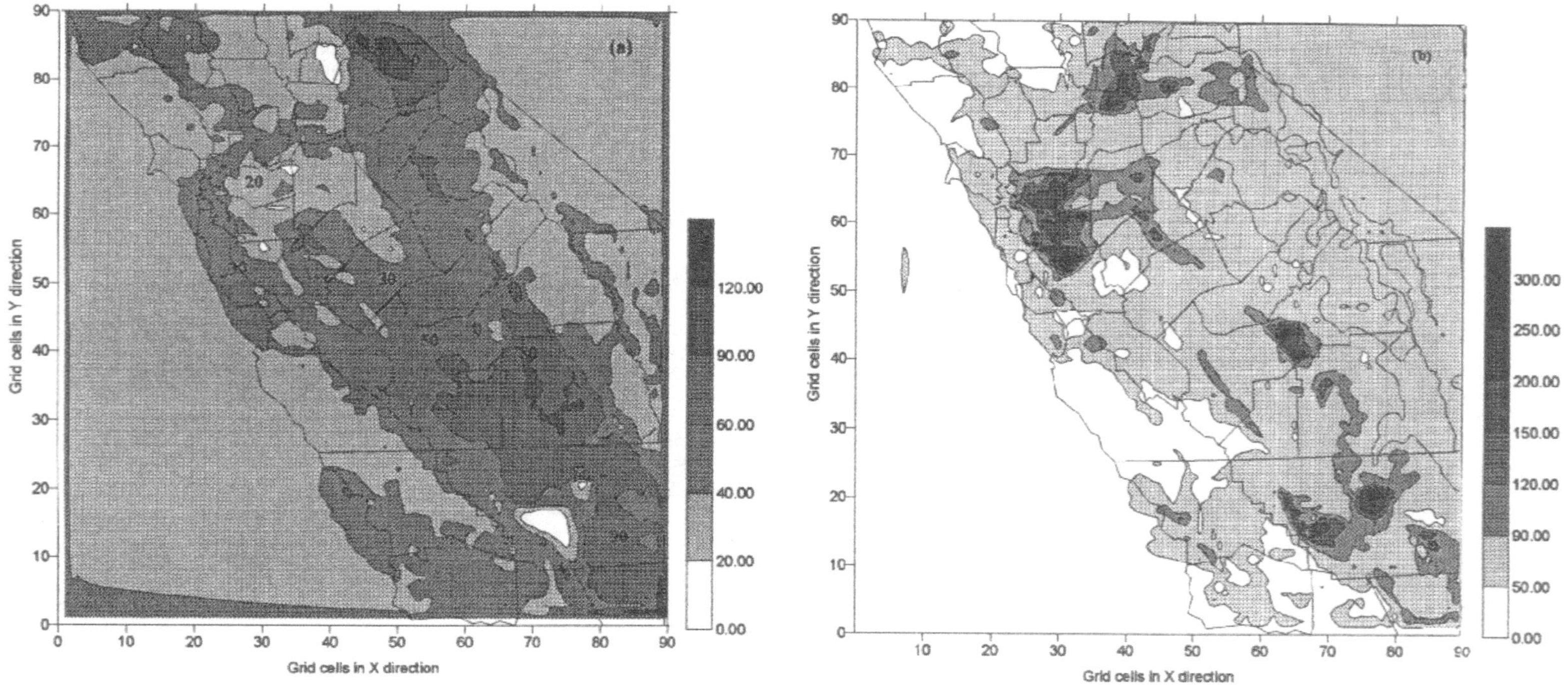

Figure 5. Ozone concentration (in ppb) estimated by the model on August 3, 1990, at 11:00 AM with (a) semi-Lagrangian scheme (maximum value = 129ppb) and (b) Smolarkiewicz scheme (maximum value=312ppb). The observed ozone concentration at various locations in the domain are shown in the figure.

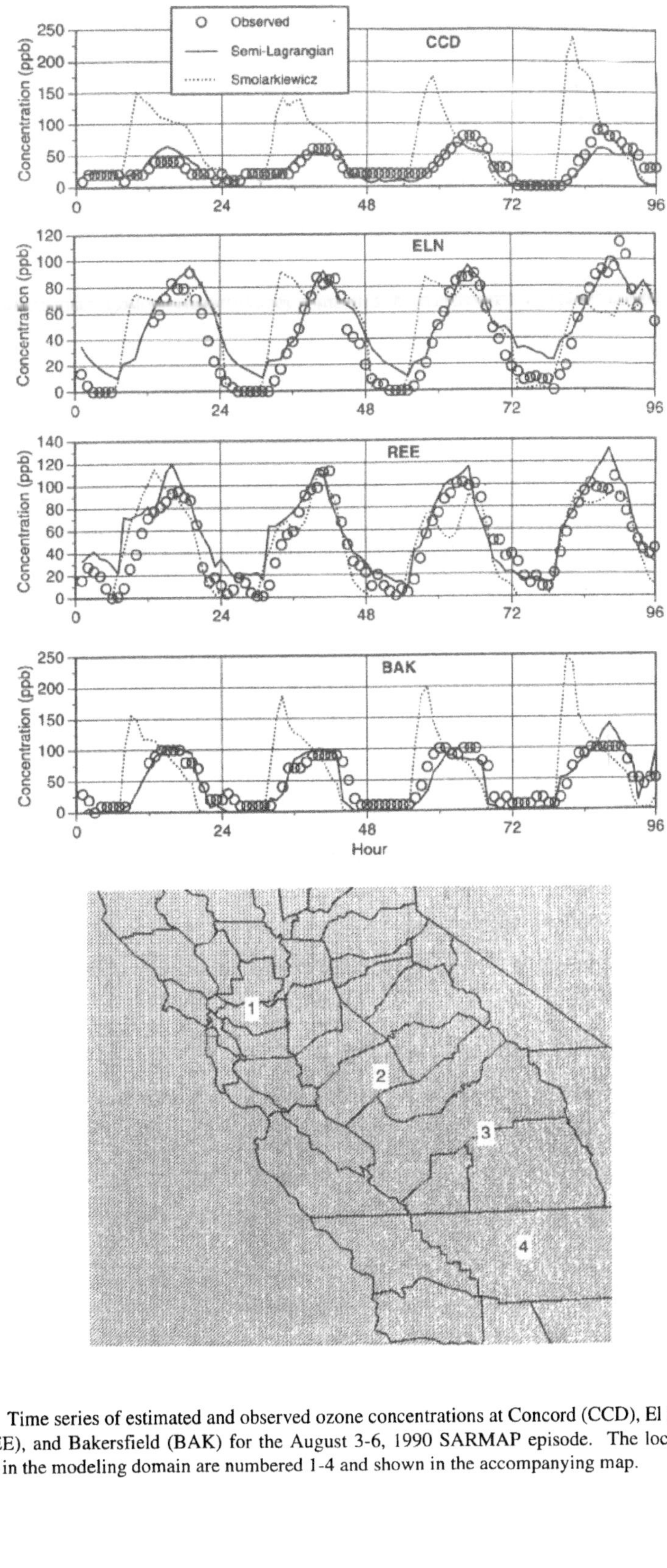

Figure 6. Time series of estimated and observed ozone concentrations at Concord (CCD), El Nido (ELN), Reedley (REE), and Bakersfield (BAK) for the August 3-6, 1990 SARMAP episode. The locations of the four stations in the modeling domain are numbered 1-4 and shown in the accompanying map.

semi-Lagrangian scheme. The photochemical model used for this simulation is described in Karamchandani et al., 1990 and Pai et al., 1993. During the August 3-6, 1990 episode, ozone concentrations exceeded state and federal standard of 90 ppb and 120 ppb respectively, at several times and in several locations. Figure 5 show the ozone estimates from the model with the two advection schemes on August 3 at 11 am. The observed concentrations at this hour are also shown in the figure. We see that the Smolarkiewicz (1983) advection scheme produces an artificial peak in the ozone concentration in the northern part of the domain which corresponds to the area that was shown to have large individual divergences in the earlier section. When the semi-Lagrangian scheme is used, this artificial peak disappears, and model estimates of ozone concentrations are in good agreement with observations throughout the domain.

Figure 6 shows the time series of model estimates and observations of ozone on all four days of the episode at four stations in the modeling domain. The four stations, Concord (CCD), El Nido (ELN), Reedley (REE), and Bakersfield (BAK), were chosen to be representative of the northern, north-central, central and southern regions of the San Joaquin Valley, respectively. The southern part of the valley experienced high concentrations of ozone on all four days of the August 1990 episode. As seen in Figure 6, when the semi-Lagrangian scheme is used, the model is able to capture the peaks and diurnal patterns in observed ozone concentrations at stations CCD, ELN, REE, and BAK very well. On the other hand, model estimates with the Smolarkiewicz scheme show significant disagreement with ozone observations at the northern (CCD) and southern (BAK) sites, with better performance at the north-central (ELN) and central (REE) sites. At CCD and BAK, the model with the Smolarkiewicz scheme consistently overestimates the peak ozone concentrations and there is a shift in the observed and simulated diurnal ozone concentration patterns.

Recall from our earlier discussion that the diagnostic model wind fields used in the above simulations had one-dimensional divergences that are equal in magnitude and opposite in direction in the northern and southern regions of the domain (see Figures 3 and 4). These divergences introduced errors in the time-split implementation of the Smolarkiewicz method, resulting in the poor performance of the model at these locations. Since the semi-Lagrangian scheme is not sensitive to these divergences in the wind field, the accuracy of the scheme was not compromised. In terms of computer execution times, the Smolarkiewicz scheme takes 224 hours and the semi-Lagrangian scheme takes 48 hours with time steps of 90 seconds and 900 seconds respectively, to simulate the August 3-6, 1990 episode.

CONCLUSIONS

The results presented in this paper show that the estimates from photochemical grid models can be sensitive to the choice of the advection scheme in the model. While standard advection tests are important to understand the fidelity of numerical advection solvers, it is even more important that these solvers be tested with realistic wind fields within the framework of the photochemical model. The results from simulating the August 3-6, 1990 episode showed that advection schemes based on the semi-Lagrangian method show great promise for air quality models in terms of accuracy and flexibility. This is because the semi-Lagrangian schemes preserves concentration peaks quite well and are not sensitive to deviations from mass consistency of the wind field. This is especially important for air quality models that depend on diagnostic meteorological models for the wind field.

REFERENCES

Bott, A. 1989. A positive definite advection scheme obtained by nonlinear renormalization of the advective fluxes. Mon. Wea. Rev., 117: 1006-1015.

Brooks, A.N. and T.J.R. Hughes 1982. Streamline upwind/Petrov-Galerkin formulations for convection dominated flows with particular emphasis on the incompressible Navier-Stokes equations. Comput. Meth. Appl. Mech. Engng, 32: 199-259.

Carmichael G.R., L.K. Peters, and T. Kitada 1986. A second generation model for regional-scale transport/chemistry/deposition. Atmos. Environ, 20: 173-188.

Chang, J.S., R.A. Brost, I. Isaksen, S. Madronich, P. Middleton, W.R. Stockwell, and C.J. Walcek 1987. A three-dimensional Eulerian acid deposition model: physical concepts and formulation. J. Geophys. Res, 920: 14681-14700.

Chock, D.P. 1985. A comparison of numerical methods for solving the advection equation - II. Atmos. Environ, 19(4): 571-586.

Chock, D. and S. Winkler. 1994. A Comparison of Advection Algorithms coupled with Chemistry. Atmos. Environ, 28(16): 2659-2675.

Easter, R.C. 1991. Two modified versions of Bott's positive definite numerical advection scheme submitted to Mon. Wea. Rev.

Hov, O., Z. Zlatev, R. Berkowicz, A. Eliassen, and L. Prahm. 1989. Comparison of numerical techniques for use in air pollution models with non-linear chemical reactions. Atmos. Environ, 23: 967-983.

Karamchandani, P., A. Venkatram, K. Kashanian, M. Tubino, and P. Ulricksen. 1990. Development and testing of an urban grid model for the Santiago, Chile region. In Proceedings of NATO/CCMS 18th ITM on Air Pollution Modeling and its Applications. Vancouver, Canada.

Lee, H.N. 1993. A semi-Lagrangian transport scheme with spectral interpolation. JAM, 32: 1908-1918.

Long, P.E. and D.W. Pepper. 1981. An examination of some simple numerical schemes for calculating scalar advection. JAM, 20: 146-156.

McRae, G.J., W.R. Goodin, and J.M, Seinfeld. 1982. Numerical solution of the atmospheric diffusion equation for chemically reacting flows. J. Comput. Phys, 45: 1-42.

Pai, P., A. Venkatram, and P. Karamchandani. 1993. Application of a diagnostic modeling approach to understand the August 3-6, 1990 oxidant episode in the San Joaquin Valley of California. In Proceedings of AWMA International Specialty Conference on Regional Photochemical Measurements and Modeling Studies. San Diego, CA.

Petschek, A.G. and L.D. Libersky. 1975. Stability, accuracy, and improvement of Crowley advection schemes. Mon. Wea. Rev, 103: 1104-1109.

Rancic, M. and G. Sindjic. 1989. Noninterpolating semi-Lagrangian advection scheme with minimized dissipation and dispersion errors. Mon. Wea. Rev, 117: 1906-1911.

Schere, K. 1983. An evaluation of several numerical advection schemes. Atmos. Environ., 17(10): 1897-1907.

Scheffe, R.D. and R.E. Morris 1993. A review of the development and application of the Urban Airshed Model. Atmos. Environ, 27B(1): 23-39.

Seibert, P. and B. Morariu. 1991. Improvements of upstream, semi-Lagrangian numerical advection schemes. JAM, 30: 117-125.

Smolarkiewicz, P.K. 1983. A simple positive definite advection scheme with small implicit diffusion. Mon. Wea. Rev, 111: 479-486.

Smolarkiewicz, P.K. and T.L. Clark. 1986. The multidimensional positive definite advection transport algorithm: further development and applications. J. Comput. Phys, 67: 396-438.

Smolarkiewicz, P.K. and J.A. Pudykiewicz. 1992. A class of semi-Lagrangian approximation for fluids. J. Atm. Sc, 49(22): 2082-2096.

Tran, K.T. and V. Mirabella. 1991. A comparison of advection schemes in existing photochemical grid models. AWMA 91-66.2, 84th Annual Meeting, Vancouver, British Columbia.

Venkatram, A., P. Karamchandani, G. Kuntasal, P.K. Misra, and D.L. Davies. 1992. The development of the Acid Deposition and Oxidant Model (ADOM). Environ. Pollut., 75: 189-198.

Yanenko, N.N. 1971. The method of fractional steps. Springer, New York.

DISCUSSION

J. PLEIM: What is the relative performance of the two advection schemes in terms of CPU time.

P. PAI: In terms of computer execution times, the Smolarkiewics scheme takes 224 hours and the semi-Lagrangian scheme takes 48 hours with time steps of 90 seconds and 900 seconds respectively, to simulate the August 3-6, 1990 episode. It was necessary to use the smaller (90 seconds) time step for the Smolarkiewicz scheme to get meaningful results.

B. KARGIN: Which quantity did you accept for Lagrangian scale of turbulence and what kind of interpolation did you use in your test calculations.

P. PAI: Both linear and cubic splines were used in the interpolation scheme of the present study. The interpolation scheme does not explicitly calculate a Lagrangian scale of turbulence.

T. IVERSEN: Do you have curved or straight line trajectories over the 900 second time step? If curved what is the time-step along the trajectories?

P. PAI: Trajectories (actual) are calculated from the gridded wind field for a time step of 900 seconds.

SO_2 DRY DEPOSITION PARAMETERIZATION IN A CHEMISTRY-GENERAL CIRCULATION MODEL: MODEL DESCRIPTION AND DEVELOPMENT

Laurens Ganzeveld, Jos Lelieveld, and Geert-Jan Roelofs

Department of Air Quality
Wageningen University
P.O. Box 8129, 6700 EV
Wageningen, The Netherlands

INTRODUCTION

The dry deposition flux of trace gases is often parameterized in models as the concentration of the trace gas at a specific height multiplied by a deposition velocity (V_d) which depends on atmospheric parameters as well as specific surface parameters. This deposition velocity is usually expressed in terms of an aerodynamic resistance, which is a function of the physical state of the atmosphere and a surface resistance, which is a function of the chemical, physical and biological properties of the surface (Chameides, 1987).

A dry deposition model, similar to that presented by Hicks et al. (1987), has been incorporated into the chemistry-general circulation model ECHAM (Roelofs and Lelieveld, 1995) to improve the description of the removal of chemically reactive trace gases. Different descriptions of the dry deposition process on a global scale have been used in previous studies. Penner et al. (1991) use a constant deposition velocity for each trace gas while Dentener and Crutzen (1993), Levy and Moxim (1989) and Kasibhatla et al. (1993) use a constant surface uptake rate, derived from observed deposition velocities, and a parameterization of turbulent transfer calculated from the drag coefficient. An inventory of the global distribution of emission and dry deposition velocities of trace gases by Müller (1992) takes into account a surface uptake rate dependent on surface characteristics. However, Müller (1992) calculates deposition velocities from assumed surface uptake rates and constant turbulent transport rates. Our scheme calculates deposition velocities according to the 'big-leaf' concept (Hicks et al., 1987) from the turbulent transfer and vegetation activity computed by the ECHAM model, supplemented with representative uptake rates for soil, water and snow/ice on a global scale. The main purpose of the development of this more comprehensive dry deposition scheme for ECHAM is to improve the description of trace gas exchange between the atmosphere and the surface, consistent with temporal and spatial dependencies of the model physics and chemistry. Moreover, the degree of detail of the new dry deposition parameterization must be compatible with that of other process descriptions in ECHAM. The scheme uses all the relevant ECHAM calculated parameters. It is not expected

Air Pollution Modeling and Its Application XI, Edited by
Gryning and Schiermeier, Plenum Press, New York, 1996

to reproduce local observed deposition velocities since differences between the spatial and temporal scales of ECHAM and micrometeorological processes may be large. However, the scheme is implemented to reproduce the specific global scale differences between distinct receptor surfaces exposed to very different meteorological conditions, e.g., equatorial tropical forest, desert, high latitude tundra, etc.

THE GCM ECHAM AND DRY DEPOSITION PARAMETERIZATION

The GCM ECHAM evolved from the numerical weather prediction model developed at the European Centre for Medium Range Weather Forecasts (ECMWF) (Roeckner et al., 1992). For the development of the basic dry deposition parameterization for this GCM we used the T21 horizontal resolution corresponding with grids squares of about 5.6 degrees and a time step of 40 minutes. The model has 19 vertical layers in a hybrid σ-p coordinate system. Prognostic variables are vorticity, divergence, temperature, surface pressure, humidity and cloud water. The model contains parameterizations of radiation, cloud formation and precipitation, convection and horizontal and vertical diffusion. The seasonal cycle of the sea surface temperature is prescribed as a boundary condition. Land surface processes are described by a 5-layer heat conductivity soil model and by a hydrological model to determine evaporation and runoff. Over land, each grid square is subdivided into 4 fractions to distinguish between snow coverage, bare soil, water in the skin reservoir (water stored within the canopy and on bare soil) and vegetation. Permanent ice cover over land is prescribed by a glacier mask. The vegetation fraction of each grid square is representative for the biological state of the vegetation type assigned to each grid square according to Wilson and Henderson-Sellers (1985). Their classification system discerns 6 major ecotype classes. Over land the roughness length is geographically prescribed while over ice-free sea it is calculated following Charnock (1955) (DKRZ, 1992, references therein). Transport of water vapour and trace gases is described by a semi-Lagrangian advection scheme. ECHAM is coupled to a chemistry scheme developed by Roelofs and Lelieveld (1995). The scheme calculates the NO_y (NO, NO_2, HNO_3, HNO_4, NO_3 and N_2O_5), OH and O_3 concentrations based on the background CH_4-CO-NO_x-HO_x photochemistry, taking into account wet deposition of HNO_3 and H_2O_2 and the role of chemical reactions of N_2O_5 on aerosol surfaces and clouds, and the resulting loss of NO_x (Dentener and Crutzen, 1993). Recently, the model has been extended with a sulfur cycle. Oxidation of SO_2 and DMS by OH and in-cloud sulfate formation are included explicitly. Emissions of DMS, SO_2 (Spiro et al., 1992), NO_x, CO and CH_4 are considered and the wet deposition calculations use the ECHAM parameterization schemes for large scale and convective clouds (Roelofs and Lelieveld, 1995).

The deposition velocity is related to specific characteristics of surfaces and the atmospheric conditions above these surfaces:

$$V_d = \frac{1}{R_a + R_b + R_{surf}} \tag{1}$$

where R_a is the aerodynamic resistance, R_b is the quasi-laminar boundary layer resistance (QBR) and R_{surf} is the combined resistance of all transfer pathways which play a role in the uptake of trace gases by the surface. The aerodynamic resistance is given by

$$R_a = \frac{1}{u_* k} [\ln(\frac{z-d}{z_0}) - \Psi(\frac{z-d}{L})] \tag{2}$$

where u_* is the friction velocity (m s^{-1}), k is the *Von Karman's* constant ($\approx$ 0.4), z is the reference height which is half the average height of the model's lowest layer (about 60 m, so $z \approx 30$ m) and Ψ is a dimensionless stability correction term which is a function of the height and the *Monin-Obukhov* length L (m). In our scheme the large scale displacement height d is assumed to be zero since d is already incorporated in the model's surface level. For the assumed to be zero since d is already incorporated in the model's surface level. For the roughness length z_0, ECHAM's surface roughness for momentum z_{0m} (DKRZ, 1992) is used. The stability correction term is calculated from the model's stability in the lowest model layer based on the *Dyer and Hicks* flux-profile relationships for heat (Brutsaert, 1973; references therein).

The aerodynamic resistance for a specific trace gas X (R_{aX}) can be expressed as the sum of the aerodynamic resistance, and an additional quasi-laminar boundary layer resistance R_{bX};

$$R_{aX} = R_a + \ln\left(\frac{z_{0m}}{z_{0X}}\right)\frac{1}{ku_*}(Sc/Pr)^{2/3} \tag{3}$$

where the second term on the right hand side represents R_{bX}. *Sc* is the Schmidt number and *Pr* is the Prandtl number (Hicks et al., 1987). For vegetated areas a value of 2 has been adopted for ln (z_{0m}/z_{0X}) (Garrat and Hicks, 1973) while for snow, ice, water and bare soil, a value of 1 has been adopted (Brutsaert, 1973).

In ECHAM, each land grid square is divided into 4 subgrids, defined by a snow (ice) covered fraction, one with water in the skin reservoir, one with bare soil and one with vegetation (DKRZ, 1992), while over sea the ice covered fraction is defined. The surface resistance of compounds with respect to the sea and wet skin reservoir, snow, ice and bare soil is defined by

$$R_{surf} = r_{wat/snow/ice/soil} \tag{4}$$

and that over vegetation by

$$R_{surf} = \frac{1}{LAI/r_{leaf} + 1/r_{soil}} \tag{5}$$

where LAI is the single-side total area of leaves/needles per area surface, r_{leaf} is the leaf/needle resistance which is the resultant resistance of the serial mesophyll and stomatal resistance, r_{mes} and r_{stom}, and a parallel cuticular resistance, r_{cut}. In ECHAM, the r_{stom} of the canopy is calculated as a function of the photosynthetically active radiation (PAR) and the available water in the root zone according to Sellers et al. (1986). The state of the canopy in the used ECHAM version is not expressed by the LAI, which has a constant value of 4 for all vegetation types, independent of time and location, but by a seasonally dependent vegetation area fraction, representative for each of the six vegetation classes assigned to each grid square according to Wilson and Henderson-Sellers (1985). The grid-average deposition velocity is computed as the area weighted average of the deposition velocities for each subgrid fraction from R_{aX} and the R_{surf}.

Table 1. Selected soil, cuticle, mesophyll, water, snow/ice resistance (s m^{-1}) for SO_2.

r_{soil}	250
r_{cut}	10^5
r_{mes}	0
$r_{wat/ws}$	10/100
$r_{snow/ice}$	max(10,f(T_{surf}))

The dry deposition parameterization in ECHAM has initially been developed for the trace gases O_3, HNO_3 NO_x (NO and NO_2). In this study we present the preliminary results of the extension of the dry deposition scheme to SO_2. Table 1 shows the adopted values for the r_{soil}, r_{cut}, r_{mes}, r_{wat} and $r_{snow/ice}$ for SO_2. A distinction has been made between the water resistance and the wet skin resistance, r_{ws}, since SO_2 deposition to wetted surfaces over land is controlled by both the aerodynamic and the surface resistance, in contrast to SO_2 deposition to the oceans which is solely controlled by turbulent transfer. The surface resistance of oceans is very small due to the high SO_2 solubility for the of ocean water pH of about 7 (Wesely, 1989). The SO_2 snow/ice resistance is calculated from ECHAM's surface temperature (Johansson and Granat, 1986; Valdez, 1987; Wesely, 1989).

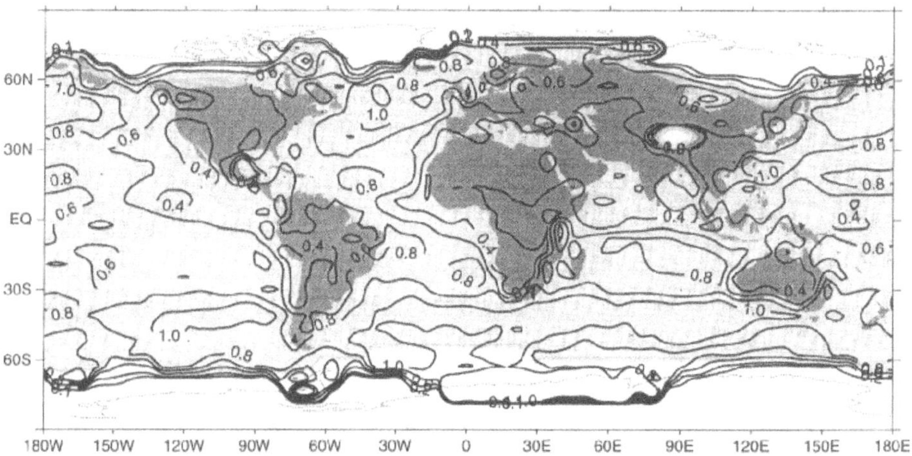

Figure 1. Average V_{dSO2} [cm s^{-1}], October. The isolines are 0.1, 0.2, 0.4, 0.6, 0.8 and 1.0 cm s^{-1}.

The soil resistance of SO_2 is a function of the pH of the soil and the relative humidity (Baldocchi, 1993). However, in this study a value of 250 s m^{-1} is assumed to be a representative average for a wide range of soil types, soil pH and relative humidity (Baldocchi, 1993; Erisman and Van Pul, 1994; Wesely, 1989; Payrissat and Beilke, 1975; Judeikis and Stewart, 1976). In the future more sophisticated parameterizations for the soil and wet skin uptake processes will be incorporated.

RESULTS

For more information concerning the evaluation of the diurnal and seasonal cycles of the calculated and observed O_3 deposition velocities and comparison of the results of our deposition scheme for O_3, HNO_3 and NO/NO_2 with those of a scheme in which the deposition velocities were kept constant, we refer to (Ganzeveld and Lelieveld, 1995).

Figure 1 shows the average SO_2 deposition velocity for October, indicating a distinct spatial distribution over the continents, which is mostly related to variations in surface characteristics. V_{dSO2} is about 0.5 cm s^{-1} over vegetated areas with values up to 0.8 cm s^{-1} over wetted surfaces, e.g., in north-western Europe, Scandinavia and south east Asia. In regions with small vegetation cover, e.g., north Africa, the Middle East, western North America, and Australia, the relatively large contribution of the soil uptake yields SO_2 deposition velocities of about 0.3 cm s^{-1}. The large V_{dSO2} over the Himalaya and a small region of Antarctica is the result of a very small snow resistance due to a surface temperature close to 273 K. Over sea, V_{dSO2} ranges between 0.4 cm s^{-1} in subsidence areas and more than 1.0 cm s^{-1} in areas with high windspeeds.

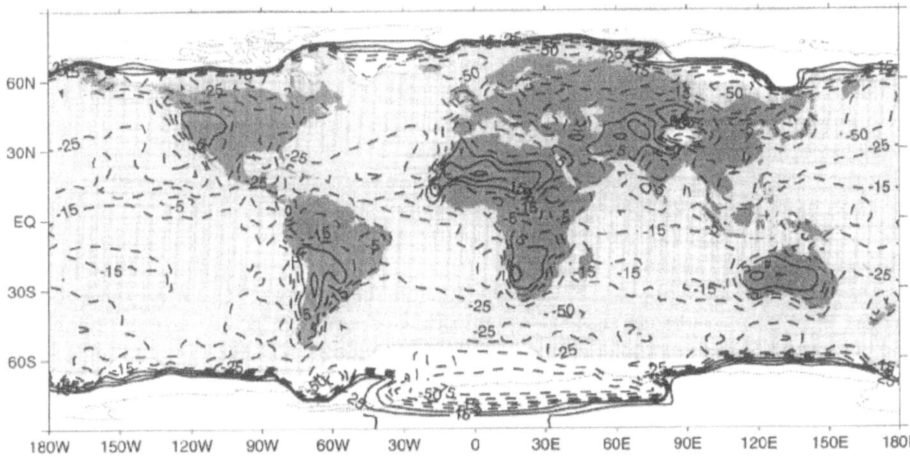

Figure 2. Relative difference (%) (relative differences are calculated as (new scheme minus "constant V_{dSO2}" scheme) / ("constant V_{dSO2}" scheme)) of SO_2 surface layer concentration between the new scheme and the "constant V_{dSO2}" scheme, October. Dashed and solid lines indicate a decrease and increase, respectively, of concentration calculated by the new scheme compared to the "constant V_{dSO2}" scheme. The isolines are: -75, -50, -25, -15, -5, 5, 15, 25.

Figure 2 shows the relative differences between the October average SO_2 concentrations in the model's surface layer with the new scheme and a "constant V_{dSO2}" scheme (V_{dSO2} is 0.6 cm s^{-1} over land without snow/ice cover, 0.1 cm s^{-1} over snow/ice surfaces and 0.8 cm s^{-1} over sea (Feichter et al., 1995)). Over vegetated areas and wetted surfaces, the SO_2 surface layer concentrations calculated with the new scheme are generally smaller compared with those calculated by the "constant V_{dSO2}" scheme with relative decreases up to 50%. In the regions where deposition is largely controlled by soil uptake, relative increases are as large as 20%. Over sea, the SO_2 concentrations calculated with the new scheme are smaller with relative decreases compared to the "constant V_{dSO2}" scheme up to 75% in the regions with high windspeeds. However, except for some coastal regions in the NH, marine SO_2 concentrations and thus deposition, are small compared to continental concentrations. The performance of the scheme will be investigated more thoroughly in future by studying the differences between the seasons and the impact of the incorporation of more sophisticated parameterizations of uptake processes, e.g., by soil, wetted foliage, snow and ice.

DISCUSSION

The new model routine improves the calculation of deposition velocities at different locations with various coverages. However, there are still shortcomings which will be addressed in future development of the dry deposition parameterization for ECHAM. For a thorough discussion of the model, e.g., violation of the constant flux approach due to reactions, the use of the large scale surface roughness, the reference height and the crudely defined canopy structure, we refer to Ganzeveld and Lelieveld (1995). Here we will discuss briefly the uptake process of SO_2. Application of dry deposition models on a global scale requires a sophisticated representation of uptake processes by soils since large areas have small vegetation cover, e.g., desert regions, tundras, grass and shrub. Most of the continents have a fraction covered with bare soil larger then 0.25 up to about 1 for North Africa and the Middle East, which implies that soil uptake plays an important role in the deposition process over the continents. For these locations more sophisticated parameterizations of the soil uptake process of SO_2 are required as a function of relative humidity and the pH of the soil (Baldocchi, 1993; Erisman and Van Pul, 1994; Payrissat and Beilke, 1975). In future versions of the model a soil resistance as a function of ECHAM's relative humidity and a soil pH derived from global soil databases, e.g., Wilson and Henderson-Sellers (1985), will be incorporated in the model.

The effect of foliage wetness on the deposition process has been investigated in several dry deposition studies (Baldocchi, 1993; Chameides, 1987; Wesely, 1989). Recent models make a distinction between foliage wetness caused by rain and by dew to account for their different chemical compositions (Wesely, 1989). In our scheme, the effect of foliage wetness on SO_2 dry deposition due to rain or dew has not been treated separately since foliage wetness in ECHAM is the net result of both processes. A study of the distribution of wet skin fraction on a global scale shows that the monthly average fraction exceeds 0.25 for a small area indicating that uptake by this fraction does not contribute significantly in the global deposition budget. However, local and transient uptake processes might be affected significantly by foliage wetness. In future, the parameterization by Chameides (1987) may be incorporated.

Calculated SO_2 deposition to snow and ice covered surfaces is about zero except of some regions with surface temperatures close to 273 K. A value of r_{snow} of 500 s m^{-1} for surface temperatures below 272 K is used by Erisman and Van Pul (1994), yielding SO_2 deposition velocities of about 0.1 - 0.15 cm s^{-1}. An interesting case to investigate will be the impact on the calculated deposition by these two parameterizations to a large area of North America, Canada, Europe and Russia, covered with snow in the NH winter. Despite the relatively small differences in V_{dSO2}, differences in total deposition might be significant since the SO_2 concentrations are large as a result of large emissions of SO_2, a low efficiency of the oxidation of SO_2 (Feichter, 1995), and suppressed vertical exchange in the boundary layer as a result of stable stratification.

Besides the incorporation of improved parameterizations of surface uptake processes, another important subject of future study will be the representation of the canopy structure, expressed by the LAI, a local displacement height and local surface roughness. With the incorporation of more sophisticated parameterizations of surface uptake processes, the deposition process might become sensitive for transport through the atmosphere, into and within the canopy. Especially over surfaces with a small SO_2 surface resistance, e.g., wet soils with a large pH, a dew wetted canopy and snow covered surfaces with temperatures close to 273 K, the SO_2 deposition velocity will be largely controlled by R_a. A major tool for the improvement of the representation of surface characteristics will be the Olson vegetation data set (Olson et al., 1985) in which the distribution of 46 global ecosystems and their characteristics are defined.

CONCLUSIONS

Even though considerable uncertainties remain, the dry deposition scheme appears to calculate realistic deposition velocities of SO_2 over most locations for different meteorological conditions, consistent with surface characteristics that control dry deposition. Incorporation of the scheme in the chemistry-general circulation model ECHAM yielded significant changes in the deposition fluxes and concentrations in the lower troposphere compared to a scheme using constant deposition velocities. For example, the new scheme calculates up to about 50% lower SO_2 concentrations in the surface layer in the continental NH. Dry deposition of SO_2 over sea is to a large extent controlled by the aerodynamic resistance R_a, while over land it is determined mostly by R_{surf}, i.e., r_{soil}, r_{veg}, r_{ws} and $r_{snow/ice}$. Future studies will focus on the development and incorporation of more sophisticated parameterizations of uptake processes for O_3, HNO_3, NO_x, SO_2 and other trace gases, e.g, H_2O_2, NO_3, N_2O_5, CH_3O_2H and aerosol particles. However, R_a is also strongly influenced by surface characteristics, which emphasizes the great importance of realistic representations of these characteristics in future versions of the dry deposition scheme.

Acknowledgements. This work was performed within the SINDICATE project (Study on the Indirect and Direct Climate Influences of Anthropogenic Trace Gas Emissions). We thank the Deutsches Klimarechenzentrum / Max Planck Institut für Meteorologie in Hamburg, Germany, for the use of computer facilities and support.

REFERENCES

Baldocchi, D. D., 1993, Deposition of gaseous sulfur compounds to vegetation, in *Sulfur nutrition and assimilation and higher plants*, edited by L. J. De Kok et al., 271-293, SPB Academic Publishing bv, The Hague, The Netherlands.

Brutsaert, W., 1973, *Evaporation into the Atmosphere*, Kluwer, Dordrecht, The Netherlands.

Chameides, W. L., 1987, Acid dew and the role of chemistry in the dry deposition of reactive gases to wetted surfaces, *J. Geophys. Res.* , 92:11,895.

Dentener, F. J., and Crutzen, P. J., 1993, Reaction of N_2O_5 on tropospheric aerosols: Impact on the global distributions of NO_x, O_3, and OH, *J. Geophys. Res.*, 98:7149.

DKRZ, Deutsches Klimazechenzentrum, Modellbetreuungsgruppe, 1992, Technical Report No. 6, *The ECHAM3 Atmospheric General Circulation Model*, Hamburg.

Erisman, J. W., and Van Pul, A., 1994, Parameterization of surace resistance for the quantification of atmospheric deposition of acidifying pollutants and ozone, *Atmos. Environ.*, 28:2595.

Feichter, J, Kjellström, E., Rodhe, H., Dentener, F., Lelieveld, J., and Roelofs, G. J., 1995, Simulation of the troposheric sulfur cycle in a global climate model, *Atmos. Environ.*, in press.

Ganzeveld, L.N., and Lelieveld, J., 1995, Dry deposition parameterization in a chemistry general circulation model and its influence on the distribution of reactive trace gases, *J. Geophys. Res.*, 100:20,999.

Garratt, J. R., and Hicks, B. B., 1973, Momentum, heat and water vapour transfer to and from natural and artificial surfaces, *Quart. J. R. Met. Soc.*, 99:680.

Hicks, B. B., D. D. Baldocchi, D. D., Meyers, T. P., Hosker, R. P., Jr., and Matt, D. R., 1987, A preliminary multiple resistance routine for deriving dry deposition velocities from measured quantities, *Water, Air, and Soil Pollution*, 36: 311-330.

Johansson, C., and Granat, L., 1986, An experimental study of the dry deposition of gaseous nitric acid to snow, *Atmos. Environ.*, 20:1165.

Judeikis, H. S., and Stewart, T. B., 1976, Laboratory measurements of SO_2 deposition velocities on selected building materials and soils, *Atmos. Environ.*, 10:769.

Kasibhatla, P. S., Levy, H., II, and Moxim, W. J., 1993, Global NO_x, HNO_3, PAN, and NO_y distributions from fossil fuel combustion emissions: A model study, *J. Geophys. Res.*, 98:7165.

Levy, H, II, and Moxim, W. J., 1989, Simulated global distribution and deposition of reactive nitrogen emitted by fossil fuel combustion, *Tellus*, 41B:256.

Müller, J., 1992, Geographical distribution and seasonal variation of surface emissions and deposition velocities of atmospheric trace gases, *J. Geophys. Res.*, 97:3787.

Olson, J., Watts, J. A., and Allison, L. J., 1985, *Major World Ecosystem Complexes Ranked by Carbon in Live Vegetation: A Database*, Report NDP-017, Oak Ridge National Laboratory, Oak Ridge, Tennessee, 164 pp.

Payrissat, M., and Beilke, S., 1975, Laboratory measurements of the uptake of sulfur dioxide by different European soils, *Atmos. Environ.*, 9:211.

Penner, J. E., Atherton, C. S., Dignon, J., Ghan, S. J., and Walton, J. J., 1991, Tropospheric nitrogen: A three-dimensional study of sources, distributions and deposition, *J. Geophys. Res*, 96:959.

Roeckner, E., Arpe, K., Bengtsson, L., Brinkop, S., Dümenil, L., Esch, M., Kirk, E., Lunkeit, F., Ponater, M., Rockel, B., Sausen, R., Schlese, U., Schubert, S., and Windelband, M., 1992, *Simulations of the Present-day Climate with the ECHAM Model: Impact of Model Physics and Resolution*, Report No. 93, Max-Planck-Institut für Meteorologie, Hamburg.

Roelofs, G.-J., and Lelieveld, J., 1995, Distribution and budget of O_3 in the troposphere calculated with a chemistry-general circulation model, *J. Geophys. Res.*, in press.

Sellers, P. J., Mintz, Y, Sud, Y. C., and Dalcher, A., 1986, A simple biosphere model (SiB) for use within general circulation models, *J. Atmos. Sci.*, 43:505.

Spiro, P. A., Jacob, D. J., and Logan, J. A., 1992, Global inventory of sulfur emissions with 1^0x 1^0 resolution, *J. Geophys. Res.*, 97:6023.

Valdez, M. P., Bales, R. C., Stanley, D. A., and Dawson, G. A., 1987, Gaseous deposition to snow. I. experimental study of SO_2 and NO_2 deposition, *J. Geophys. Res.*, 92:9779.

Wesely , M. L., 1989, Parameterization of surface resistances to gaseous dry deposition in regional-scale numerical models, *Atmos. Environ.*, 23:1293.

Wilson, M. F., and Henderson-Sellers, A. A., 1985, A global archive of land cover and soils data for use in general circulation climate models, *J. Climatology.*, 5:119.

DISCUSSION

R. SAN JOSE: What kind of method has been used to measure the dry deposition velocities you compare with your modelling data?

L. GANZEVELD: The model derived SO_2 deposition velocities have not yet been compared with observations. However, this will be one of the major issues in improving and validating the model in future work. You mentioned the uncertainy in observations of SO_2 deposition velocities but we feel that this uncertainty is still significantly smaller compared to the difference between the previously used constant SO_2 deposition velocities and the ones calculated by the new model. It is also very likely that differences due to scale differences between model derived and observed deposition velocities and the neglect or rough representation of controlling processes, e.g. transport through the canopy and uptake by soils, are much more significant. However, validation of the model through comparison with observations over very different surfaces and for very different seasons can certainly give an indication about the performance of the model.

THE DYNAMIC EXCHANGE OF POLLUTANTS AT THE AIR-SOIL INTERFACE AND ITS IMPACT ON LONG RANGE TRANSPORT

J.A. van Jaarsveld

National Institute of Public Health and the Environment (RIVM)
P.O. Box 1
3720 BA Bilthoven
The Netherlands

INTRODUCTION

Emission and deposition rates of volatile substances applied in the soil or at the soil surface are strongly affected by physico-chemical properties of the substance and soil properties. Some of these depend strongly on such meteorological conditions as wind speed, temperature, and precipitation. This category of substances contains pesticides applied to soils for agricultural purposes, but also, for example, ammonia resulting from the spreading of manure.

The DEPASS model (Dynamic Exchange of Pollutants between Air and Soil Surface) has been developed to describe the vertical transport and diffusion in both soil and the atmosphere, and the exchange of pollutants between the compartments dependent on actual meteorological conditions (Figure 1). This one-dimensional numerical model was developed with the intentions:

- to study the exchange process and its dynamics
- to derive effective emission rates, depending on application modes
- to derive effective deposition rates
- to derive diurnal and seasonal emission patterns
- to derive parameterizations for inclusion in LRT models

The model is briefly described in this paper, with some typical applications given using lindane as being representative for a group of widely used pesticides. Some preliminary results of a comparison with field data are also given.

MODEL DESCRIPTION

Soil compartment

The following one-dimensional time-dependent equations describe the transport and diffusion of contaminants and moisture in an unsaturated soil:

contaminant in gas phase:

$$\frac{\partial C_{SG}}{\partial t} = \frac{\partial}{\partial z}\left[D_G e(\theta)\frac{\partial C_{SG}}{\partial z} - J_{SG}C_{SG}\right] \tag{1}$$

contaminant in liquid phase:

$$\frac{\partial C_{SL}}{\partial t} = \frac{\partial}{\partial z}\left[D_L e(\theta)\frac{\partial C_{SL}}{\partial z} - J_{SW}C_{SL}\right] \tag{2}$$

contaminant in solid phase:

$$\frac{\partial C_{SS}}{\partial t} = -\mu_S C_{SS} \tag{3}$$

soil moisture:

$$\frac{\partial \theta}{\partial t} = \frac{\partial}{\partial z}\left[D_H(\theta)\frac{\partial \theta}{\partial z}\right] + \frac{\partial k_H(\theta)}{\partial z} \tag{4}$$

where:

C_{SG}	=	contaminant concentration in soil gas [g m^{-3}]
C_{SL}	=	contaminant concentration in soil moisture [g m^{-3}]
C_{SS}	=	contaminant concentration in soil solids [g m^{-3}]
$D_Ge(\theta)$	=	effective diffusivity of the contaminant in the interstitial air [m^2 s^{-1}]
$D_Le(\theta)$	=	effective diffusivity of the contaminant in the interstitial water [m^2 s^{-1}]
θ	=	moisture filled soil fraction
J_{SG}	=	air flux in the soil [m^3 m^{-2} s^{-1}]
J_{SW}	=	water flux in the soil [m^3 m^{-2} s^{-1}]
μ_S	=	first-order degradation rate constant [s^{-1}]
T_S	=	soil temperature [K]
z	=	soil depth below or atmospheric height above the surface [m]
t	=	time [s]
$D_H(\theta)$	=	moisture-dependent hydraulic diffusivity [m^2 s^{-1}]
$k_H(\theta)$	=	moisture-dependent hydraulic conductivity [m s^{-1}]

A three-phase local equilibrium is assumed throughout the soil medium. The adsorbed-liquid partitioning is expressed through a linear relationship:

$$C_{SS} = K_D \rho_b C_{SL} \tag{5}$$

where ρ_b is the soil bulk density [g m^{-3}] and K_D the slope of the adsorption isotherm [m^3 g^{-1}]. Jury *et al.* (1983) relate K_D to the organic carbon fraction f_{oc} of the soil by defining an organic carbon partition coefficient K_{oc} [m^3 g^{-1}] where $K_D = K_{oc}\, f_{oc}$.
Liquid-vapor partitioning is presented through Henry's Law:

$$C_{SG} = K_H C_{SL} \tag{6}$$

where K_H is the (dimensionless) Henry's Law constant of the contaminant ($K_H = K_H'/RT$; *R* is the gas constant (8.3 Pa m^3 mol^{-1} K^{-1}), *T* the temperature (K) and K_H' the Henry coefficient (Pa m^3 mol^{-1})).

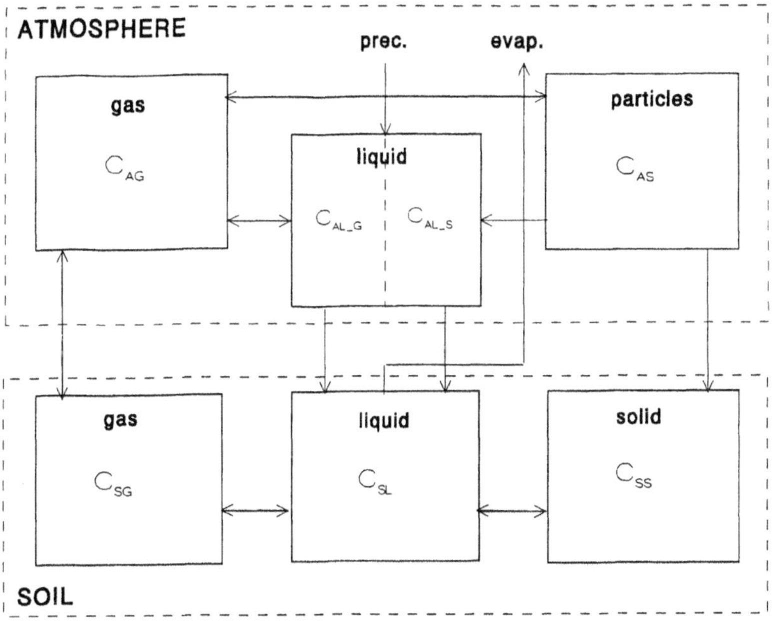

Figure 1. Schematic overview of the DEPASS model.

Atmospheric compartment

The change in the atmospheric concentration of a contaminant due to vertical transport, diffusion, wet removal, and chemical degradation is, for gas-phase contaminants, given by:

$$\frac{\partial C_{AG}}{\partial t} = \frac{\partial}{\partial z}\left[K_z(z)\,\frac{\partial C_{AG}}{\partial z}\right] - k_{WG}(C_{AG} - C_{AL}K_H) - k_{CHG}C_{AG} \tag{7}$$

where C_{AG} is the gas-phase concentration [g m^{-3}], $K_z(z)$ the height-dependent vertical diffusivity [m^2 s^{-1}], k_{WG} and k_{WP} the wet removal rates [s^{-1}], and k_{CHG} the chemical degradation rate [s^{-1}]. C_{AL} is the concentration of the dissolved gas in rain drops [g m^{-3}]. Note that z is here defined as the height above the surface. A similar expression as in Eq. 7 can be written for the particle-phase concentration C_{AP}, assuming that all the contaminant is attached to small particles, of which the behavior in a turbulent atmosphere may be approached by that of gases. Since the uptake of particles in rain drops is determined by physical processes such as Brownian diffusion, interception, and inertial impaction, the concentration in the rain drops is of no importance.

The (equilibrium) partition between the phases is described using Junge's model (1977):

$$\frac{C_{AP}}{C_{AG}} = \frac{c\,\varphi}{p^0L + c\,\varphi} \tag{8}$$

where φ is the available aerosol surface area (m^2 m^{-3} air), p^0_L the (subcooled) liquid-phase vapor pressure of the contaminant (Pa) and c a constant that depends on heat of condensation and molecular weight (Pankow, 1987). Junge assumed $c \sim 0.17$ Pa m for high molecular-weight organics. Whitby (1978) reports average aerosol total surface area in the range of 4.2 x 10^{-5} (clean continental background) to 1.1 x 10^{-3} (urban), with an average background of 1.5 x 10^{-4} m^2 m^{-3} air. The latter value is adopted here.

Considering the range of atmospheric environments, organics having $p^0_L > 10^{-2}$ Pa will be present almost entirely in the gas phase, while those having $p^0_L < 10^{-6}$ Pa almost entirely in the particle phase. Due to the strong temperature dependency of p^0_L, the absorbed fraction will show a distinct diurnal and seasonal pattern.

Fluxes across the air-soil interface

Dry flux. The exchange of gas-phase contaminant between atmosphere and soil is driven by the concentration gradient over a so-called laminar sublayer in the atmospheric compartment

$$\frac{\partial C_{AGI}}{\partial t} = - \frac{\partial C_{SGI}}{\partial t} = \frac{\partial}{\partial z}\left[D_G \frac{\partial C_{A\_S}}{\partial z}\right] \tag{9}$$

where D_G is the diffusivity of the gas in air [m^2 s^{-1}] and C_{A_S} the concentration in the laminar surface layer. C_{AGI} and C_{SGI} are the concentrations in the layers immediately adjacent to the surface layer in the atmosphere and soil, respectively. In flux terms, Eq. 9 can be written as:

$$F_{A\_S} = D_G \frac{\partial C_{A\_S}}{\partial z} \tag{10}$$

Here, F_{A_S} is defined as the atmosphere to soil flux [g m^{-2} s^{-1}]. Assuming that the laminar surface layer has a small thickness δ_s, Eq. 10 can be written in a discrete form:

$$F_{A\_S} = \frac{D_G}{\delta_s}(C_{AGI} - C_{SGI}) \tag{11}$$

where δ_s is a function of atmospheric parameters and the roughness of the surface. Equation 11 is frequently used in emission models (e.g. Jury *et al.*, 1983), often assuming that C_{AGI} is zero. Since no direct observations of δ_s are available, parameterizations have to be used. In deposition models the flux to soil or vegetation is often described using an electrical resistance analogy:

$$F = V_d (C_{AG} - C_{SG}) \qquad \text{with} \quad V_d = (r_a + r_b + r_c)^{-1} \tag{12}$$

where r_a is the aerodynamic resistance of the lower part of the boundary layer (e.g. over 1 m), r_c the resistance against uptake of the contaminant by soil elements and/or vegetation and r_b the resistance of the laminar surface layer. In deposition models, C_{SG} is often assumed to be zero. Using the resistance analogy in the present model, it can be concluded that r_a is already included in the diffusion process in the atmospheric layers, and r_c by the diffusion and transport processes in the soil. The remaining resistance, r_b, can be considered as equivalent to $\delta s/D_G$ given in (11). An expression for r_b is, for example, as in Hicks *et al.*(1987):

$$r_b = \frac{1}{\kappa u_*} \ln(\frac{z_0}{z_{0c}}) \tag{13}$$

where z_0 and z_{0c} are roughness length scales [m] for momentum and contaminant, respectively, κ the von Kármàn constant (~0.4) and u_* the friction velocity [m s^{-1}]. This expression basically quantifies the way pollution transfer differs from momentum transfer in the immediate vicinity of the surface. This expression still contains an unknown length scale, z_{0c}. Wesely and Hicks (1977) approximated r_b by:

$$r_b = \frac{2}{\kappa u_*} \left[\frac{N_{Sc}}{N_{Pr}} \right]^{\frac{2}{3}} \tag{14}$$

where N_{Sc} and N_{Pr} are the respective Schmidt and Prandtl numbers. $N_{Sc} = \nu/D_G$, where ν is the kinematic viscosity of air (~0.15 x 10 $^{-4}$ m^2 s^{-1}); N_{Pr} is 0.72.

Wet flux. The change of the wet deposition flux Fw_{AG} with height is equal to the flux of the gaseous contaminant to individual droplets integrated over the entire droplet-size spectrum, or:

$$\frac{dFw_{AG}}{dz} = k_{WG}(C_{AG} - C_{AL} K_H) = \int Fw'(D)\, dD \tag{15}$$

where $Fw'(D)$ is the flux of contaminant to drops with size D, defined as:

$$Fw'(D) = k'_G \pi D^2 (C_{AG} - C_{AL} K_H) \frac{dN}{dD}. \tag{16}$$

Here, dN/dD represents the drop-size distribution of a cloud or rain event, expressed in this situation as the number of drops, m^{-3} m^{-1}. k'_G is the convective-diffusive mass transfer coefficient, which can be estimated by the semi-empirical Frössling equation:

$$k'_G = \frac{D_G}{D} \left[2 + 0.6 \left[\frac{D v_t}{\nu} \right]^{1/2} \left[\frac{\nu}{D_G} \right]^{1/3} \right] \tag{17}$$

where v_t is the terminal velocity of the drop [m s^{-1}]. k_{WG} can be calculated from (15), (16), and (17). Janssen and Ten Brink (1985) parameterized k_{WG} as a function of the precipitation intensity R_i [m s^{-1}] on the basis of a drop-size distribution given by Best (1950) as follows:

$$k_{WG} = c_1 D_G^{c_2} R_i^{c_3} \tag{18}$$

For a lower drop-size limit of 0.000125 m, c_1 = 4167 m^{-3} s, c_2 = 0.744 m^{-3} s and c_3 = 0.628 m^{-3} s.

The expressions (15) to (17) actually describe the situation for rain drops falling through layers of contaminated air (below-cloud scavenging). They indicate that the concentration of the contaminant in rain drops is not necessarily in equilibrium with the concentration in the gas phase, especially not when K_H is very low or when there are strong vertical gradients in gas-phase concentrations. In a real situation the residence time of gas-phase contaminants in clouds is much higher than the below-cloud residence time of droplets. For this reason it can be assumed that liquid and gas-phase concentrations are in equilibrium within clouds. It is further assumed that precipitation-forming takes place in the upper layer of the one-dimensional model and that the drop-size distribution below this layer is independent of height.

Similar expressions can be derived for the (below-cloud) wet scavenging of particles. In this case the uptake is not limited by the concentration in the droplets. The total scavenging is, however, dominated by in-cloud processes, which are difficult to describe. Considering the formation of particles as a secondary process, which takes place throughout the atmosphere, the wet scavenging is described independent of height using a scavenging ratio, W:

$$k_{WP} = \frac{W\ R_i}{z_i} \qquad (19)$$

where z_i is the mixing height of the pollutant. The scavenging ratio, W, used in the model is derived from simultaneously measured Pb concentrations in precipitation and air.

Model implementation

Soil compartment. The soil compartment contains 10 non-equidistant layers where each layer has its gas, liquid, and solid fraction. The model assumes linear equilibrium between vapor, liquid, and absorbed chemical phases, net first-order degradation, and exchange of pollutant to the atmosphere through the gas phase. Parameterizations of effective diffusion in both soil gas and water, and descriptions of phase partitioning are taken from the model from Jury *et al.* (1983). Vertical (re-)distribution of moist is modeled as a diffusion mechanism using a moisture-content-dependent hydraulic diffusivity. In the present model, a mechanism to describe a time-dependent vertical temperature distribution has not yet been implemented. For the time being, it is assumed that the soil temperature is equal to the ambient air temperature.

Atmospheric compartment. This compartment also contains 10 non-equidistant layers. Vertical diffusion is described using the K theory with parameterizations from Brost and Wyngaard (1978). Windspeed profiles are described using the Monin-Obukhov similarity theory with parameterizations from Van Ulden and Holtslag (1985). An expression for horizontal advection is included to account for the limited horizontal dimensions of actual application plots. Moist exchange between atmosphere and soil takes place through precipitation and evaporation. Evaporation rates are estimated according to a parameterization from Van Ulden and Holtslag (1985).

Meteorological parameters. The model is driven by observed meteorological parameters such as wind speed, temperature, global radiation, and precipitation. Secondary meteorological parameters such as Monin-Obukhov length and friction velocity are calculated using routines from Beljaars and Holtslag (1990).

RESULTS

Lindane (γ-HCH) was selected as a representative for a series of relatively good water-soluble pesticides, which are predominantly in the gas phase. Henry's constant (in Pa m^{-3} mol^{-1}) was calculated from: log K_H = -2703/T + 8.7 (Kucklick *et al.*, 1991). Other lindane-specific parameters used were: K_{oc}= 1.3 x 10^{-3} m^3 g^{-1}; μ_S= 3 x 10^{-8} s^{-1} (Jury *et al.*, 1983) and k_{CHG} = 2.5 x 10^{-7} s^{-1} (Warmenhoven *et al.*, 1989). Selected soil parameters (unless otherwise noted) were: ρ_b = 1.3 x 10^6 g m^{-3}; f_{oc}= 0.72% and a soil porosity of 0.4.

Figure 2 gives time series of gas-phase concentrations in the atmosphere (Cg air) and in the soil air (Cg soil) over four days after a momentary atmospheric release of γ-HCH at 00.00 h, April 10, 1994. Dry and wet vertical fluxes are also given. It turns out that the topsoil layer becomes nearly saturated within 24 h, by which the deposition flux decreases dramatically. At noon on day 2 (36 h after release) soil-gas concentrations are even higher than the gas' concentration in air, leading to a soil to air flux (emission). The strong diurnal variations are due to variations in temperature (indirectly, also in evaporation) which causes changes in gas-liquid partitioning.

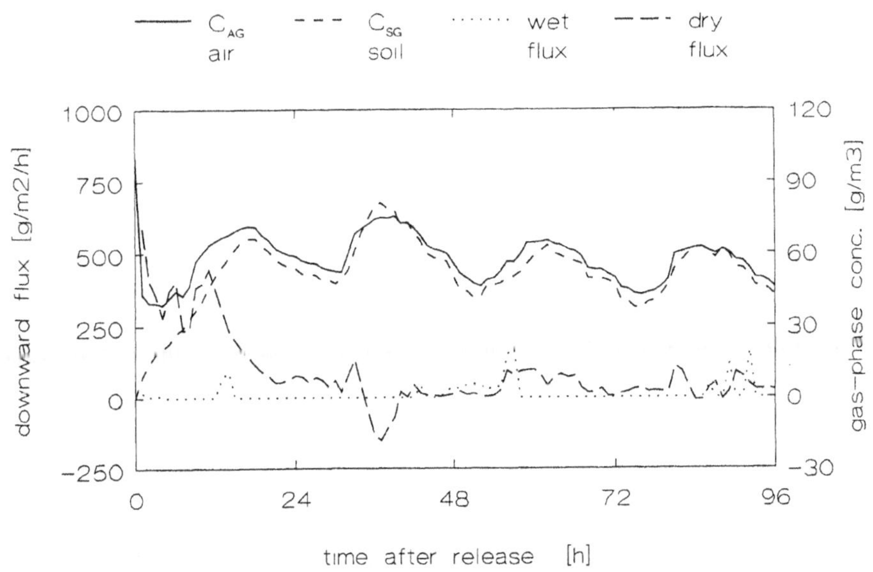

Figure 2. Development of concentrations of γ-HCH in air and soil, and the corresponding exchange fluxes after a momentary release of γ-HCH in air.

Figure 3 gives the volatilization flux of lindane incorporated in the top 10 cm of the soil on June 1, 1992. The flux varies from day to night, covering almost two orders of magnitude. The cumulative volatilized fraction is given in the same figure. Due to the low degradation rate in the soil (half-life of 292 days), soil emission is taking place even months later. Depending on the organic carbon fraction of the soil, 10-30% of the incorporated lindane will evaporate within a year.

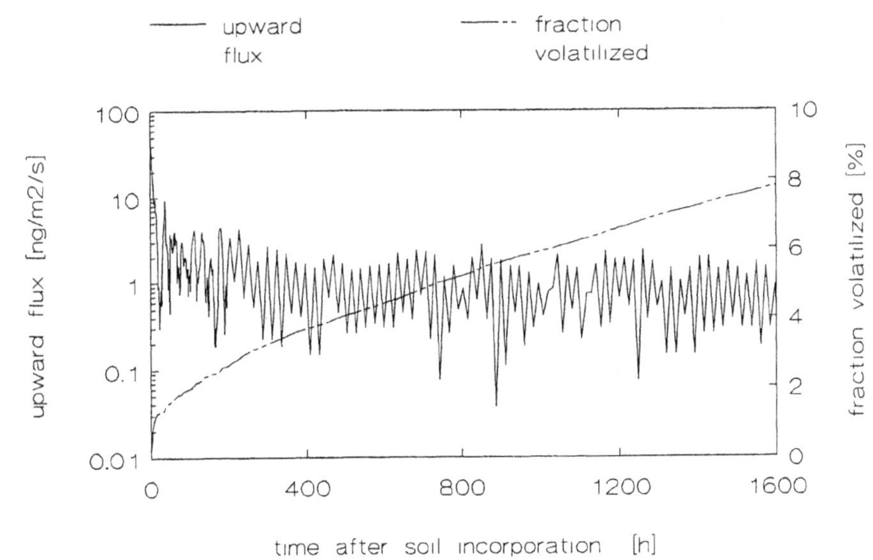

Figure 3. Volatilization flux and cumulative volatilized fraction of lindane incorporated in the top 10cm layer of the soil.

In Figure 4 some typical diurnal exchange flux profiles are given for a situation where the atmospheric concentration is kept constant. For this example, a simulation period of two years was used. The driving force behind the diurnal variations is mainly the temperature variation, which influences the partitioning between gas and liquid phase in the topsoil. But also the moist evaporation, which is highest during daytime, contributes significantly to the variation shown. The deposition-emission process may be expected to vary with the different seasons. To investigate this, the upward (emission) and downward (dry, wet) fluxes were separately accounted for on a monthly basis during the same two-year period, starting with zero lindane concentrations in the soil medium. The results are given in Figure 5. The first year of the simulation may be considered as a building-up period for the soil. The second year is therefore more representative for an actual situation. It turns out that in the winter period the flux is mainly downward and in the summer months mainly upward, leading to one conclusion: that measured wet deposition fluxes in the summer months in no way represent actual deposition fluxes. Another conclusion is that application of a substance such as lindane in summer or winter has a large effect on the long-range transport: the buffer capacity of a soil in summertime is much less than in wintertime.

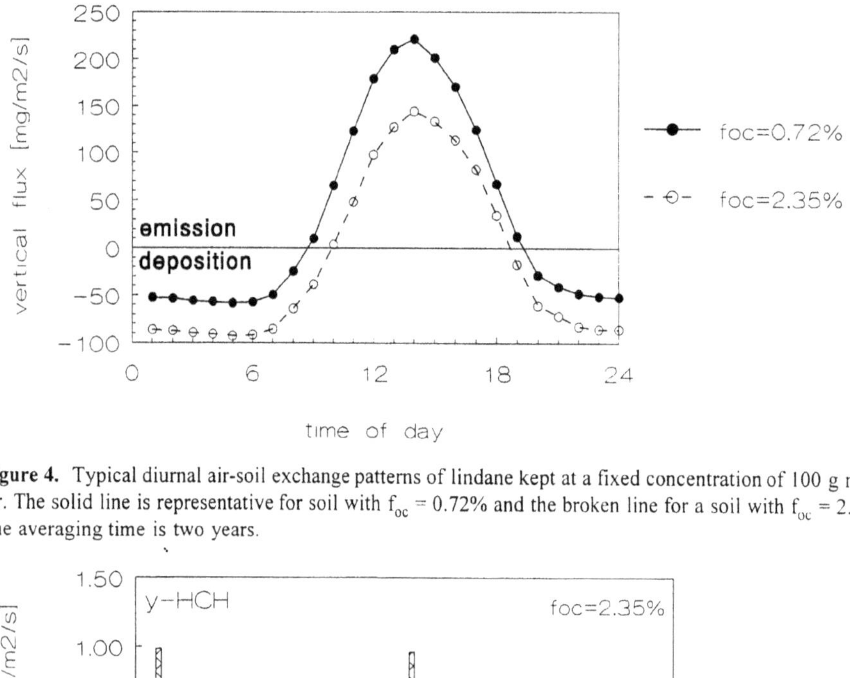

Figure 4. Typical diurnal air-soil exchange patterns of lindane kept at a fixed concentration of 100 g m^{-3} in air. The solid line is representative for soil with f_{oc} = 0.72% and the broken line for a soil with f_{oc} = 2.35%. The averaging time is two years.

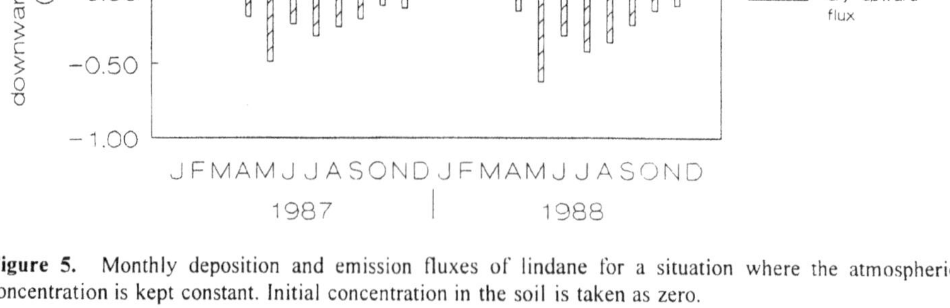

Figure 5. Monthly deposition and emission fluxes of lindane for a situation where the atmospheric concentration is kept constant. Initial concentration in the soil is taken as zero.

The DEPASS model is used to calculate effective dry deposition velocities for a series of POPs (Persistent Organic Pollutants). These dry deposition velocities have been used in a transport model to calculate net deposition loads to the North Sea and other parts of Europe (Baart *et al.*, 1995). The effective dry deposition velocity is calculated from:

$$V_d eff = \frac{(\overline{F_{net}} - \overline{Fw_{AG}})}{\overline{C_{AG}}} \tag{20}$$

where F_{net} is equal to the sum of the downward deposition fluxes minus the upward (emission) flux averaged over two years. The effective dry deposition velocity may even become negative if much of the wet deposited substance evaporates afterwards. According to the calculations with DEPASS, this is the case when the organic carbon fraction of the soil is low or if the organic carbon partition coefficient is low. For example, in the case of γ-HCH and an assumed organic carbon fraction of 0.7 % in the topsoil, a net emission flux is calculated and $v_d eff$ amounts to -8×10^{-4} m s^{-1}. If this fraction is taken as 2.3 %, then $v_d eff$ increases to 2.6×10^{-4} m s^{-1}. Some calculated $v_d eff$ values are given in Table 1, all on the basis of an assumed organic carbon fraction of 2.3 %

Table 1. Calculated effective deposition velocities and Henry's constants for some persistent organic pollutants predominantly in gas-phase

Pollutant	v_d eff soil	$1/K_H$
	cm/s	dimensionless
1,1,1-trichlorethane	0.0004	2
2,4-D	-0.005	9.8×10^3
bentazon	0.1	1.1×10^{12}
dichlorvos	0.043	2.8×10^4
diuron	0.52	4.4×10^6
endosulfan	0.042	1.7×10^4
MCPP	0.85	7.0×10^7
mevinfos	0.97	1.6×10^{10}
lindane	0.026	4.2×10^4
parathion-ethyl	0.51	4.4×10^5
pentachlorophenol	0.49	1.2×10^5
trifluralin	-0.0036	1.3×10^3
fluoranthene	0.11	5200
PCB52	-0.0017	113
anthracene	0.016	1400
benzo(a)anthracene	0.41	9.4×10^3
benzo(b)fluoranthene	-0.000015	1.0×10^5
chrysene	-0.28	8.4×10^4
phenantrene	0.024	1700
naphtalene	0.0006	125

COMPARISON WITH OBSERVATIONS

A first comparison with measurements was carried out using observations of ammonia emission fluxes after application to the surface of bare soil (Van der Molen *et al.*, 1990a). Specific phase-partitioning parameters for ammonia and ammonium were taken from Van *der Molen *et al.*(1990b). Meteorological data was taken from the nearest station of the Royal Netherlands Meteorological Institute (KNMI). Initial results are given in Figure 6.

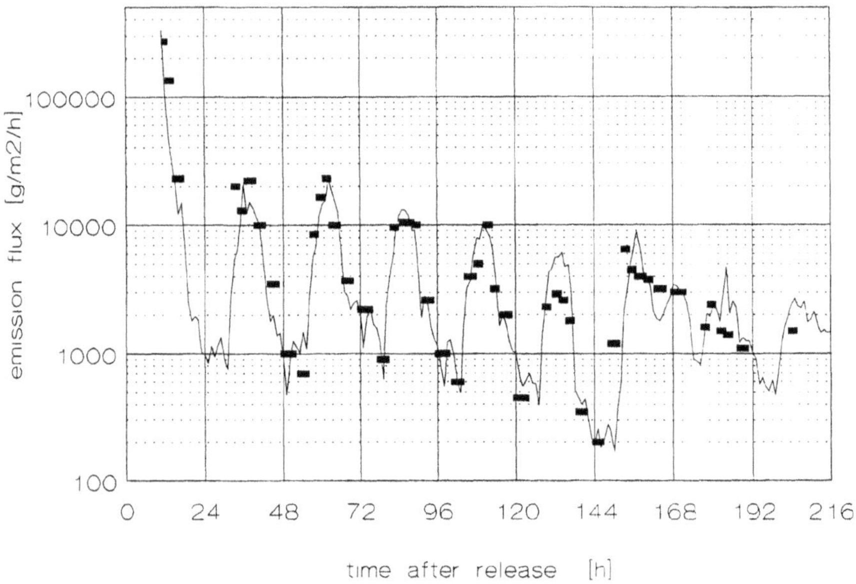

Figure 6. Comparison of calculated NH_3 emission fluxes with results of the field experiment carried out by Van der Molen *et al.* (1990b). Solid line: model predictions on an hourly basis. Squares: observations.

In this experiment it has been shown that more than 30% of the surface-applied ammonia has evaporated within 24 h and that the emission flux has shown a distinct diurnal pattern. The model follows the diurnal changes fairly well. Part of the good correlation between model and measurements ($r = 0.91$) is the result, however, of the variations in wind speed, which directly affect the emission fluxes in both the measured and modeled results.

CONCLUSIONS

The one-dimensional model described in this paper integrates the soil and atmospheric compartments with their specific vertical diffusion and transport mechanisms. Under the assumption that the absorption of contaminants on organic carbon is reversible, the dynamics of the vertical air-soil exchange of the contaminant depends strongly on meteorological conditions. Soil temperature, water evaporation and atmospheric turbulent diffusion explain many of the variations in the exchange fluxes. Physico-chemical properties of the contaminants (vapor pressure, Henry's constant and degradation rates) and the organic carbon fraction of the soil determine the magnitude of the average fluxes. Distinct diurnal and seasonal variations in exchange fluxes are predicted by the model. It is clear that the temporal change in buffer capacity of the soil has a large impact on the long-range transport of POPs. Models used for the simulation of the long-range transport of POPs should therefore include the soil compartment with its basic processes. A model such as described in this paper can be helpful in parameterizing these processes.

REFERENCES

Baart, A.C, Berdowski, J.J.M., and Van Jaarsveld, J.A., 1995, Calculation of atmospheric deposition of contaminants on the North Sea. Report, TNO-MW-R95/138, TNO, Delft.

Beljaars, A.C.M., and Holtslag, A.A.M., 1990, A software library for the calculation of surface fluxes over land and sea, *Environ. Software* 5:60-68.

Best A.C., 1950, The size distribution of raindrops, *Q. J. R. Meteorol. Soc.* 76:16-36.

Brost, R.A., and Wyngaard, J.C., 1978, A model study of the stably stratified boundary layer, *J. Atmos. Sci.* 35:1427 1440.

Janssen, A.J., and Ten Brink H.M., 1985, Plume wash-out: modelling, calculation and validation, Report ECN-170, Netherlands Energy Research Foundation, Petten, The Netherlands (in Dutch with English summary).

Junge, C.E., 1977, Basic considerations about trace constituents in the atmosphere as related to the fate of global pollutants, in: *Fate of Pollutants in the Air and Water Environment*. Part I, I.H. Suffet, ed., (Advances in environmental science and technology, Vol. 8), Wiley-Interscience, New York.

Jury, W.A., Spencer, W.F., and Farmer, W.J, 1983, Behaviour assessment model for trace organics in soil: I. Model description, *J.Environ. Qual.* 12:558-564.

Hicks, B.B., Baldocchi, D.D., Meyers, T.P., Hosker, R.P. Jr., and Matt, D.R., 1987, A preliminary multiple resistance routine for deriving dry deposition velocities from measured quantities, *Water, Air Soil Pollut.* 36:311-330.

Kucklick, J.R., Hinckley, D.A., and Bidleman, T.F., 1991, Determination of Henry's Law constants for hexachlorocyclohexanes in distilled water and artificial seawater as a function of temperature, *Marine Chem.* 34:197-209.

Pankow, J.F., 1987, Review and comparitative analysis of the theories on partitioning between the gas and aerosol particulate phases in the atmosphere, *Atmospheric Environment* 22:2275-2283.

Van der Molen, J., Van Faassen, H.G., Leclerc, M.Y., Vriesema, R., and Chardon, W.J., 1990a, Ammonia volatilization from arable land after application of cattle slurry. 1. Field estimates, *Neth. J. Agric. Sci.* 38:145-158.

Van der Molen, J., Beljaars, A.C.M., Chardon, W.J., Jury, W.A., and Van Faassen, H.G., 1990b, Ammonia volatilization from arable land after application of cattle slurry. 2. Derivation of a transfer model, *Neth. J. Agric. Sci.* 38:239-254.

Van Ulden, A.P., and Holtslag, A.A.M., 1985, Estimation of atmospheric boundary layer parameters for diffusion applications, *J. Clim. Appl. Meteor.* 24:1196-1207.

Warmenhoven, J.P., Duiser, J.A., de Leu, L.Th., and Veldt, C., 1989, The contribution of the input from the atmosphere to the contamination of the North Sea and the Dutch Wadden Sea, TNO Report R 89/349A, Delft.

Wesely, M.L., and Hicks, B.B., 1977, Some factors that affect the deposition rates of sulphur dioxide and similar gases on vegetation, *J. Air Pollut. Contr. Assoc.* 27:1110-1116.

Whitby, K.T., 1978, The physical characteristics of sulphur aerosols, *Atmospheric Environment* 12:135-159.

DISCUSSION

D. STEYN: If you replace lindane (for example) by water, can your model simulate the evaporation? If it can, you will have solved a problem (that of closing the surface energy budget) that has vexed boundary & mesoscale modellers for more than 20 years.

J.A. van JAARSVELD: The model assumes that the concentration of the contaminant in all phases is low (in the order of parts per million or less). Therefore the energy involved in evaporating or condensating the contaminant will not influence the surface energy budget. Such an assumption is of course not valid when water is taken as the contaminant because of the much higher concentrations. At present the surface energy budget in the DEPASS model is parameterized according to Van Ulden and Holtslag (1985). This or similar methods are applied in most mesoscale models. Although the evaporation of water plays an important role in the evaporation of its contaminants, I do not expect that, for example, effective deposition velocities will be much different when the surface energy budget is modelled in a more elementary way. Nevertheless, the model offers a framework in which the energy budget can be described in much more detail then is done in the present version.

MODELING OF CLOUD EFFECTS ON OZONE OVER EUROPE

Jan Matthijsen[1], Peter J.H. Builtjes, and Ernst W. Meyer[2]

TNO Institute of Environmental Sciences
P.O. Box 6011, 2600 JA Delft, the Netherlands
[1]Present affiliation: Laboratoire d'Aerologie O.M.P. 14,
Av. E. Belin, 31400 Toulouse, France
[2]Present affiliation: Institute of Public Health and Environment
P.O. Box 1, 3720 BA Bilthoven, the Netherlands

INTRODUCTION

The effect of anthropogenic and biogenic emissions on ozone concentrations is still an important issue. Over Europe emissions of the dominant ozone precursors NO_x and non-methane hydrocarbons (nmhc) are among the highest in the world. On the European continent in summer time under stagnant meteorological conditions O_3 concentrations frequently surpass human health standards in so called photochemical smog episodes. Also the growing season ozone limit value of 40 ppb is surpassed nearly everywhere in Europe. Although numerous studies have been carried out on the mechanisms which drive the O_3 formation, not all processes are well understood, such as the effect of clouds. The effect of clouds on O_3 and O_3 precursors has been the subject of several studies over the last decade (Schwartz, 1984; Jacob, 1986; Lelieveld and Crutzen, 1990; Jonson and Isaksen, 1993; Dentener, 1993; Walcek *et al.*, 1993; Jacob and Liang, 1995). All these studies indicate that O_3 concentrations can be significantly affected due to clouds. Clouds affect photochemistry in the troposphere by altering radiation transfer and vertical transport. Furthermore, clouds affect photochemistry through aqueous-phase reactions in cloud droplets and wet removal.

So far the overall effect of clouds on O_3 has been elaborated for global and hemispheric scales using or implicitly assuming a relatively coarse grid and climatological data. The effect of actual cloudy conditions on actual O_3 concentrations on a regional scale has not got much attention yet.

In this study we have investigated the effect of clouds on actual O_3 formation and concentrations over Europe with a regional scale photochemical dispersion model, LOTOS (long term ozone simulation), Builtjes, 1992. Model runs have been performed for a cloudy period of 10 days in August 1990. The LOTOS-model uses actual meteorologic input data and emission data based on a European emission inventory. To account for cloud effects we have incorporated parameterizations for altered radiative transfer, wet deposition, and aqueous-phase chemistry with and without transition metal chemistry of

Air Pollution Modeling and Its Application XI, Edited by
Gryning and Schiermeier, Plenum Press, New York, 1996

iron (Fe) and copper (Cu). The effect of deep cloud convection, which is a phenomenon with a low frequency of occurrence during photochemical episodes, has not been taken into account yet.

Approach

We have estimated the impact of clouds on the actual O_3 concentration and formation over Europe with five model runs for a cloudy period from 14 august to 23 august 1990. A clear sky base run (A) is defined by excluding all parameterizations for cloud effects. The second run (B) incorporates a description for wet deposition of a number of water soluble species. The third run (C) carries a description for the altered radiative transfer due to clouds. The fourth and fifth run are similar to the third, but in addition the effect of aqueous-phase chemistry is included without (D) and with (E) Fe and Cu reactions. The aqueous-phase mechanism takes in its current formulation a considerable amount of computation time. Although the LOTOS-model is essentially able to predict the effects of aqueous-phase chemistry, for computation time reasons we have estimated the impact of aqueous-phase chemistry on O_3 with and without Fe and Cu aqueous-phase reactions using a box-model description of the LOTOS gas- and aqueous-phase mechanism only in the cloud.

Radiative transfer

To model the effect of clouds on the radiation transfer and henceforth the photodissociation rate constants we have used an analytic three-layer cloud parametrization developed by Van Weele and Duynkerke (1993). This parameterization calculates a multiplication factor which determines the photodissociation rate constants in the presence of clouds given the clear sky values. The multiplication factor is calculated as a function of height, wavelength, cloud-optical thickness, and height of the cloud-top and the cloud-bottom. Furthermore, the parameterization accounts for changes in the multiplication factor above the cloud due to wavelength dependent diffusion of the radiation, and it describes the in-cloud multiplication factor related to an assumed linear profile of the in-cloud optical thickness. The cloud radiation parametrization is extended to model the radiative effect of two separate cloud layers (low and middle clouds) with fractional cloud cover.

Wet deposition

The parameterization of wet deposition used in run B of gaseous species in the LOTOS-model is obtained by equilibrium and mass balance between the aqueous-phase and gas-phase components of the species. A first order scavenging rate is calculated for a species with a grid-cell averaged concentration from the concentration in rain droplets assuming Henry's law equilibrium, the rainfall rate, and the height of the region top .

Aqueous-phase chemistry

The aqueous-phase chemical mechanism has been adapted to CBM-IV the chemical gas-phase scheme used in LOTOS. Although CBM-IV implicitly describes the degradation of methane it does not account for oxidation products needed by our aqueous-phase chemical mechanism; the methyl peroxide radical (CH_3O_2), methyl hydro peroxide (CH_3OOH) and formic acid ($HCOOH$). These methane oxidation products can not be simply derived from the oxidation rate of methane, since box-model runs for averaged continental and polluted continental photochemical regimes indicate that less than 60% and less than 20%, respectively, of the methane-adduct production originates from the oxidation of methane.

The cloud droplets are assumed to be monodisperse and to instantaneously reach the average droplet radius of 5 μm or 10 μm for droplets from continental or marine origine, respectively. Initial cloud water concentrations for Fe, Cu, H^+, SO_4^{2-}, NH_4^+ and NO_3^- are defined for marine and averaged continental and polluted continental conditions. Marine, averaged continental and polluted continental air masses are distinguished based on the NO_x and SO_2 concentration in the below-cloud layer .

The main mechanism of O_3 reduction due to aqueous-phase chemistry is:
O_3 depletion in the aqueous phase is due to reaction with the O_2^- radical, which is the dissociated form of HO_2 ($HO_2 \rightarrow O_2^- + H^+$: $pK_h \approx 4.8$), the rate of this reaction is, therefore, quite sensitive to the pH of cloud water, which ranges roughly from 3 to 6. Furthermore, HO_2 is depleted in the aqueous phase in the self-reaction with O_2^-, which leads in the gas phase to a reduction of the O_3 production rate through reaction $HO_2 + NO \rightarrow NO_2 + OH$, and a reduction of water soluble O_3 precursors like formaldehyde due to depletion in the aqueous phase.

When the O_3 relevant reactions with Fe and Cu are included the depletion of HO_2 in the aqueous phase substantially increases compared to when Fe and Cu reactions are not included, which leads in the gas phase to an enhanced reduction of the O_3 production rate through reaction $HO_2 + NO \rightarrow NO_2 + OH$. Moreover, aqueous-phase OH concentrations increase, which leads to enhanced depletion of for instance formaldehyde losing most of its O_3 formation potential. Furthermore, the aqueous-phase reaction of O_3 with Fe is important for Fe concentrations typical for continental cloud water (Fe $\geq$ 0.1 M).

The selected 10 day cloud period from 14 August to 23 August 1990 is characterized by an average precipitation amount of 25 mm and a median value of 5 and a 90 percentile of 80 was calculated for the cloud optical thickness. The cloud optical thickness is a measure for the radiative effect of a cloud. The mean observed cloud cover for all clouds besides cirrus during the cloudy period is about 20%. Although we speak here of a cloudy period, the values for the total amount of precipitation and the cloud optical thickness are not extreme. The average cloud optical thickness is calculated from the average lwp. The average cloud thickness was found to be approximately 500 m. Over the cloudy period approximately 5% of the model sub-volume was occupied by clouds. For more detailed information on the LOTOS-model and the cloud parameterizations we refer to Builtjes (1992), Matthijsen (1995), Matthijsen et al. (1995) and Matthijsen et al. (1996).

Results

Model results are shown in the form of changes of the 99 percentile O_3 concentration relative to the clear-sky values for run A, run B and run C. Furthermore, we show for all runs the domain average effort of clouds on the O_3 formation and degradation rates relative to the clear-sky rates and in absolute numbers. The results are presented for only a part of the total LOTOS-area (35°-70°N, 10°W-60°E). The sub domain reaches from 40° to 65°N and from 8°W to 30°E.

The results of the clear sky base run (A) for the 99-percentile is given in Figure 1.
Figure 2 and 3 show deviations of the 99 percentile of the O_3 concentration in the pbl calculated from run B (wet deposition taken into account) and C (radiative cloud effect taken into account) in respect to the 99 percentiles of the clear sky simulation, run A (figure 1).

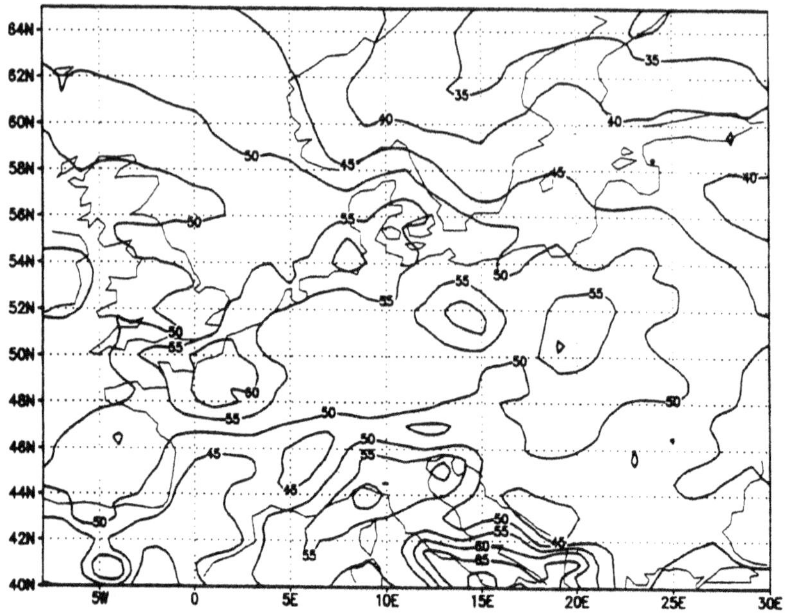

Figure 1. 99 percentile (ppb) of O_3 (run A) in the pbl for the sub-domain.

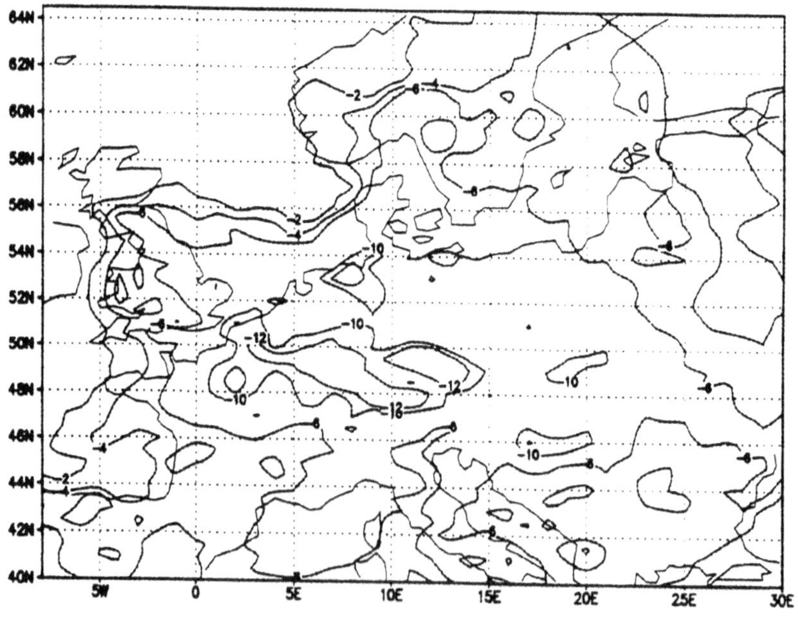

Figure 2. Deviations (%) of the 99 percentile of O_3 (run B) from clear sky 99 percentiles (Fig. 1).

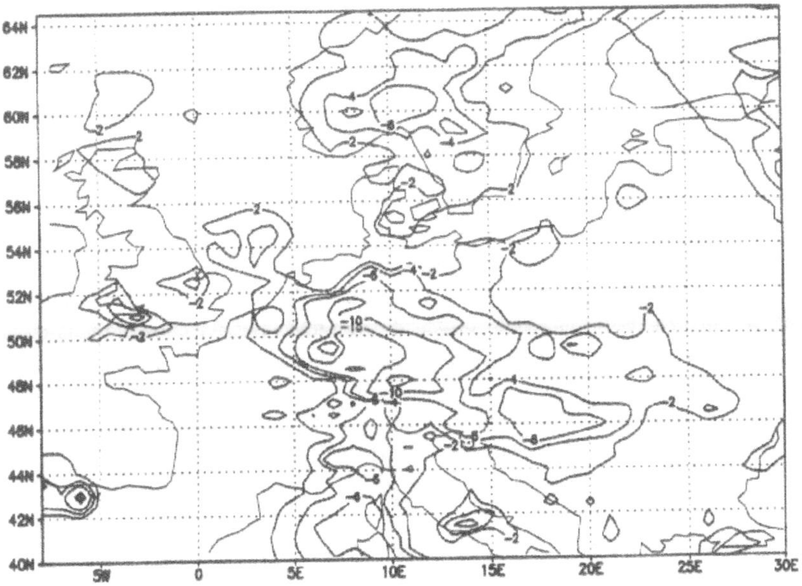

Figure 3. Deviations (%) of the 99 percentile of O_3 (run C) from clear sky 99 percentiles (Fig. 1).

For both processes, the O_3 reductions have values of upto 10%. The radiative cloud effect on the highest O_3 concentrations is more a local effect than the wet deposition effect and is mainly bound to continental areas.

Figure 4 shows the sub-domain average O_3 concentration deviation in the pbl due to the cloud effects in run B and C compared to the clear sky simulation, run A. On average the cloud effects are relatively small, we found an average reduction of O_3 concentrations over the period and area considered of approximately 5% due to either wet deposition or the radiative cloud effect.

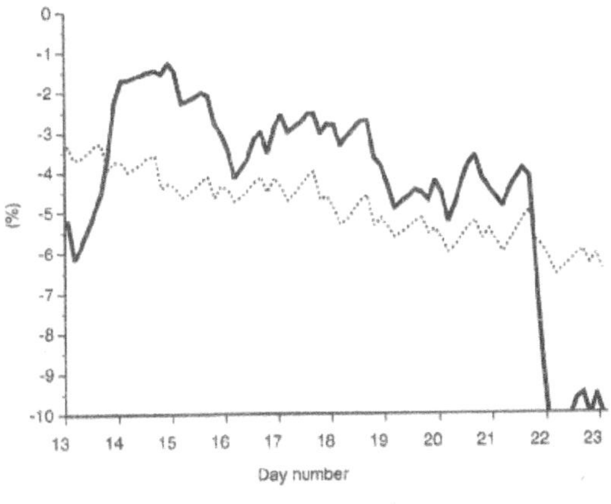

Figure 4. O_3 concentration deviation (%) in the pbl due to wet deposition (dashed line) and due to radiative cloud effect (solid line); averaged over the sub-domain and the period from 14 August to 23 August, 1990.

The in-cloud O_3 concentrations are on average reduced with respect to run C by approximately 65% when aqueous-phase chemistry is included (run D) and by an extra 25% up to 90% when Fe and Cu reactions are added (run E). Given the average cloud volume of 5%, this would lead to an extra O_3 concentration reduction compared to run C of 3% and 5% respectively.

However, estimates of the changes in the net O_3 formation rates when aqueous-phase chemistry is included indicate that O_3 concentrations will be less affected. This is shown in the following paragraph.

The effect of aqueous phase chemistry is presented by considering O_3 production, loss and formation rates for all runs.
The estimated overall effect of aqueous-phase chemistry on the net O_3 formation rates is an extra reduction of 3% (run D) and 4% (run E) (table 1). These reductions are expected to lower O_3 concentrations by 1% or less in analogy with the average reduction of net O_3 formation and concentration found for run B and C. Above we estimated that O_3 concentrations were reduced by as much as 5% due to aqueous-phase chemistry, based on in-cloud O_3 concentration reductions and cloud volume. The difference indicates the inaccuracy caused by extrapolating the results of run D and E to an overall effect. Including aqueous-phase reactions leads to an important reduction of the in-cloud O_3 concentration and net formation, but the overall effect is relatively small. The reduction by 3% of O_3 concentrations due to aqueous-phase chemistry seems therefore a rather good estimate as also calculated from a simple analysis by Jacob and Liang (1995). However, it should be noted that results of earlier studies show a rather wide range of cloud effects on ozone. This indicates that rather large uncertainties still exist in the determination of cloud effects.

Table 1. Production (P), loss (L) and net formation (N) rates of O_3 (105 molecules.cm^{-3}.s^{-1}) time and volume averaged. Between brackets the relative changes (%) compared to the clear sky case (run A) are given.

	P	L	N
run A	33.4	14.8	18.6
run B	28.6 (-14)	13.3 (-10)	15.3 (-18)
run C	28.7 (-14)	12.8 (-14)	15.9 (-15)
run D	27.9 (-16)	12.6 (-15)	15.3 (-18)
run E	27.6 (-17)	12.6 (-15)	15.0 (-19)

The in-cloud net O_3 formation rates of run C are on average reduced from 15 to 3.2 10^5 molecules.cm^{-3}.s^{-1} by approximately 80%. When Fe and Cu reactions are included (run E) the net O_3 formation rates of run C are on average reduced from 15 to -1.5 10^5 molecules.cm^{-3}.s^{-1} with an extra 30% up to 110%.

CONCLUSIONS

The influence of clouds on O_3 over Europe has been studied using the LOTOS-model, a regional scale photochemical dispersion model. Included in the model are aqueous-phase chemistry, wet deposition and the effect of clouds on radiation. In addition O_3 relevant reactions with iron and copper are incorporated into the aqueous-phase chemical mechanism. The effect of wet deposition and radiative changes leads, in a case study with a domain average cloud cover of 20%, to a combined reduction of ozone of roughly 10% in the planetary boundary layer. The effect of aqueous-phase chemistry on ozone was estimated to be a reduction of 1% or 2%. Although aqueous-phase chemistry can lead to a strong reduction of the net-O_3 formation in the cloud the overall effect is limited by the cloud volume fraction, liquid water content and pH. The aqueous-phase reaction with iron and copper lead to an additional in-cloud reduction of O_3 by 25%.

ACKNOWLEDGMENTS

The study is sponsored by TNO and University Utrecht, and by the Dutch Ministry of Housing, Physical Planning and the Environment, the latter in the framework of the TOR-project.

REFERENCES

Builtjes, P.J.H., 1992, The LOTOS-long term ozone simulation project. Summary report, TNO-Rap. nr. IMW-R-92/240, TNO Delft, the Netherlands.

Dentener, F.J., P.J. Crutzen, 1993, Reaction of N_2O_5 on tropospheric aerosols: impact on the global distributions of NO_x, O_3, and OH, J. Geophys. Res. 98, 7149-7163.

Jacob, D.J., 1986, Chemistry of OH in remote clouds and its role in the production of formic acid and peroxymonosulfate, J. Geophys. Res. 91, 9807-9826.

Jacob, D.J., J. Liang, 1995, Effect of Aqueous-phase Radical cloud chemistry on tropospheric Ozone, J. Geophys. Res. (submitted).

Jonson, J.E., I.S.A. Isaksen, 1993, Tropospheric ozone chemistry. The impact of cloud chemistry, J. Atm. Chem. 16, 99-122.

Lelieveld, J., P.J. Crutzen, 1990, The influences of cloud photochemical processes on tropospheric ozone, Nature 343, 227-233.

Matthijsen, J., 1995, Modelling of tropospheric ozone and clouds, Ph-D thesis University Utrecht, ISBN 90-393-0559-5.

Matthijsen, J., P.J.H. Builtjes and D.L.Sedlak, 1995, Cloud model experiments of the effect of iron and copper and tropospheric ozone under marine and continental conditions, Meteor. Atmos. Phys. in press .

Matthijssen, J., P.J. Builtjes and E.W. Meyer, 1996. Modelling cloud effect on ozone over Europe, a case study, in preparation.

Schwartz, S.E., 1984, Gas- and aqueous-phase chemistry of HO_2 in liquid water clouds, J. Geophys. Res. 89, 11589-11598.

Van Weele, M., P.G. Duynkerke, 1993, Effect of clouds on photo-dissociation of NO2: observations and modelling, J. Atm. Chem. 16, 231-255.

Walcek, C.J., Hong-Hsee Yuan, W.R. Stockwell, 1993, The Influence of Heterogeneous Atmospheric Chemical Reactions on the Formation of Ozone in Polluted Air, presented at 86th Annual Meeting & Exhibition, Denver, Colorado, June 13-18.

DISCUSSION

J. CHING: With respect to the plot of the difference between model and observations for the entire year, how well correlated were the cloud field with the deviation.

P. BUILTJES: I did not look into detail yet to the correlation between differences and for example cloud cover. Previous sensitivity studies showed that especially mixing height is a sensitive parameter to ozone concentrations.

TURBULENT MIXING OF REACTIVE GASES IN THE CONVECTIVE BOUNDARY LAYER

Gé Verver

Royal Netherlands Meteorological Institute (KNMI)
P.O. Box 201
3730AE De Bilt
The Netherlands

ABSTRACT

A second order closure model is applied to the transport of chemical reactive gases in the convective boundary layer. The importance of concentration correlations on the mean transformation rate is studied. The relevance of chemistry terms in the flux equation and covariance equations is assessed. Two different cases are shown: case 1 with two reactive species released from the surface, and case 2 with a top-down and a bottom-up tracer. For case 1 it was found that the covariance of both reactive species is either positive or negative, increasing or reducing the effective transformation rate depending on the Damköhler number (the ratio of the turbulent timescale and the chemistry timescale). In case 2 the concentration covariance reduces the effective transformation rate by 30 % for slow and by 75 % for fast chemistry. A significant influence of chemistry on the flux is found in both cases, mainly due to chemistry effects on the buoyancy term.

1 INTRODUCTION

The transport of chemically reactive species in the atmosphere is usually modelled as an extension to the transport of inert, conservative species. Chemical transformation in these models is handled in a separate module that considers each grid cell as a well mixed box of reactants. Some chemical transformations are considered infinitely fast and the concentrations of the rapidly formed gases are diagnostically derived without taking into account transport. For gases that react relatively slowly the change of concentration due to both transport and

Air Pollution Modeling and Its Application XI, Edited by
Gryning and Schiermeier, Plenum Press, New York, 1996

chemistry is accounted for. In both cases the well-mixed assumption might not be correct for all gases.

To illustrate the interactions of turbulent transport and chemistry consider a species i that is depleted by a reaction with species j. The Reynolds averaged equation for the time rate of change of concentration in horizontal homogeneous conditions reads:

$$\frac{\partial C_i}{\partial t} = - \frac{\partial \overline{c_i w}}{\partial z} - k(C_i C_j + \overline{c_i c_j}) \qquad (1)$$

where C_i and c_i are the mean and fluctuating parts of the instantaneous concentration of species i and k is the reaction constant. The well-mixed assumption implies that $\overline{c_i c_j}$ can be neglected. The importance of $k\overline{c_i c_j}$ over the mean chemistry term is reflected in the intensity of segregation I_s, defined as:

$$I_s \equiv \frac{\overline{c_i c_j}}{C_i C_j} \qquad (2)$$

Schumann (1993) showed with large-eddy simulations that due to the negative correlation of two tracers that were released near the top and near the bottom respectively, I_s could reach a value as low as -0.7, thus diminishing the effective (or "well-mixed") reaction rate by 70%.

The introduction in the equation for the mean concentration (equation 1) of an extra second-order term, i.e. the covariance of concentration, needs some closure. As will be shown, all second order quantities involving the concentration of a reactive species i contain chemistry terms that may or may not be taken into account when a certain closure is chosen. The equations for the second moments including the chemistry terms are given below. The equation for the concentration covariance reads:

$$\frac{\partial \overline{c_i c_j}}{\partial t} = - \overline{c_i w}\frac{\partial C_j}{\partial z} - \overline{c_j w}\frac{\partial C_i}{\partial z} - \frac{\partial \overline{c_i c_j w}}{\partial z} - \nu_c \overline{\left(\frac{\partial c_i}{\partial x_k}\right)\left(\frac{\partial c_j}{\partial x_k}\right)}$$
$$-k(C_i\overline{c_i c_j} + C_j\overline{c_i c_j} + C_i\overline{c_j^2} + C_j\overline{c_i^2} + \overline{c_i^2 c_j} + \overline{c_j^2 c_i}). \qquad (3)$$

The flux equation:

$$\frac{\partial \overline{c_i w}}{\partial t} = - \overline{w^2}\frac{\partial C_i}{\partial z} - \beta\overline{c_i \theta_v} - \frac{\partial \overline{c_i w^2}}{\partial z} - \frac{1}{\rho}\overline{c_i \frac{\partial p}{\partial z}} - k(C_i\overline{c_j w} + C_j\overline{c_i w} + \overline{c_i c_j w}). \qquad (4)$$

For diabatic boundary layers the buoyancy term is given by:

$$\frac{\partial \overline{c_i \theta_v}}{\partial t} = - \overline{c_i w}\frac{\partial \Theta_v}{\partial z} - \overline{w\theta_v}\frac{\partial C_i}{\partial z} - \frac{\partial \overline{c_i w \theta_v}}{\partial z} - (\nu_c + \nu_{\theta_v})\overline{\left(\frac{\partial c_i}{\partial x_k}\right)\left(\frac{\partial \theta_v}{\partial x_k}\right)}$$
$$-k(C_i\overline{c_j \theta_v} + C_j\overline{c_i \theta_v} + \overline{c_i c_j \theta_v}), \qquad (5)$$

where Θ_v and θ_v are the mean and fluctuating part of the virtual potential temperature. It is seen from equations (1)-(5) that the closure problem also appears through the chemical transformation terms: the second moment equations now containing third moments, the equation for the third moment containing fourth moments and so on. Note also that the equations for the second moments are coupled due to the chemistry terms: the second moments of species i depend on second moments of species j.

Up to now previous studies focused mainly on the surface layer (Fitzgarralt and Lenschow, 1983; Vilà-Guerau de Arellano and Duynkerke,1992). Large-eddy model studies by Schumann (1993) and Sykes et al. (1992, 1993) studied the effect of chemical reactions in the convective boundary layer. We do not know of a higher order closure model of the diabatic boundary layer that incorporates the second and higher moments in the chemical transformation terms.

In this paper we focus on the convective boundary layer. A second order closure model is used to study the relevance of the second and third order chemistry terms for a simple irreversible bimolecular chemical reaction. In the next section a brief description of the model is given. Section 3 describes the experiments that are done. Two typical cases are distinguished. In the first case species i is entrained near the top of the boundary layer diffusing downward, and species j is released near the ground diffusing upward. In the second case both reactive gases are release at the ground. In section 4 results are given of these two cases. Section 5 contains the discussion and conclusions.

2 BRIEF DESCRIPTION OF THE MODEL

The model solves the equation for the mean concentration of two reactive species given by equation (1), and their second moments, $\overline{c_i c_j}$, $\overline{c_i w}$ and $c_i\theta_v$ given by equations (3)-(5). Since we are only interested in the dispersion of gases, the profiles of $\overline{w^2}$, θ_v^2, Θ_v and $w\theta_v$ are taken from literature. For $\overline{w^2}$ we use an expression from Lenschow et al. (1980), θ_v^2 is taken from Sorbjan (1989) and Θ_v and $w\theta_v$ are parameterized following Holtslag and Moeng (1991). It is asumed these profiles remain constant in time.

Closure

To close the flux and (co-)variance equations the following parameterizations are used:

- The molecular dissipation of scalar (co-)variances is parameterized conform Mellor and Yamada, 1974.
- The pressure term in the flux equations is parameterized by a Rotta return-to-isotropy term and a rapid term (André et al, 1976; Moeng and Wyngaard, 1989).
- The third moments that appear in the second moment equations are: $c_i w^2$, $c_i^2 w$, $c_i w\theta_v$, $c_i^2\theta_v$, $c_i\theta_v^2$ and $c_i^2 c_j$. The following assumptions are used to obtain the third moments:

  -The quasi-normal approximation is applied to the fourth moments (André et al., 1978)

  -The pressure fluctuation terms are parameterized conform André et al. (1982).

-Dissipation of the triple scalar products ($\overline{c_i^2\theta_v}$, $\overline{c_i\theta_v^2}$ and $\overline{c_i^2c_j}$) is modelled as a decay process (André et al.,1978).

-The time rate of change of the third moments are negligible.

-All terms of the form $M_3\,\partial M_1/\partial z$, where M_1 and M_3 are the averaged first and third moments respectively, are neglected.

-Chemistry terms are neglected.

With these assumptions the equations for the third moments reduce to diagnostic expressions in terms of second moments and their vertical derivatives.

Timescales

It is assumed that all timescales introduced through the parameterisation of variance dissipation and pressure fluctuation terms, are proportional to one reference timescale τ. This timescale is usually taken to be the dissipation timescale of turbulent kinetic energy (TKE) (τ_{TKE}= TKE/ε_{TKE}, where ε_{TKE} is the dissipation rate of TKE). The timescale is parameterized as:

$$\tau_{TKE}=\frac{1}{a_{TKE}}\frac{l}{\sqrt{\overline{w^2}}},\qquad(6)$$

where l is a length scale and a_{TKE} is the closure constant for the turbulent kinetic energy. For the length scale we take the simple expression:

$$l=\kappa z\left(1-\frac{z}{z_i}\right)\qquad(7)$$

where κ the Von Kármàn constant. To get a timescale of approximately z_i/w_* in the middle of the CBL (Isaka and Guillemet, 1983; Moeng and Wyngaard, 1989) we have put the constant a_{TKE} to a value of 0.1. The proportionality factors for all other timescales are taken from André et al. (1978, 1982), where we use for temperature and concentration the same constants.

Numerical method

The model uses a staggered grid, where the turbulent fluxes and (co-)variances are calculated on 41 flux levels, and the concentrations and third moments are calculated on the 40 concentration levels in between. The lowest level (at $0.01z_i$) and the highest level (at $0.95\,z_i$) are both flux levels where the boundary conditions are applied. The second moments that appear in the equations for the third moments (that are solved on concentration levels) are obtained by linear interpolation of the values at the surrounding flux levels. For the time integration, the model uses an explicit numerical scheme with timesteps of 0.5 seconds.

Initial and boundary conditions

The prognostic model variables, i.e. the first and second moments, are given initial values that are uniform with height. All second moments are put to zero, except at the boundaries (the highest and the lowest flux level). Both prognostic (first and second order) and diagnostic (third order) quantities are described by first order (partial) differential equations, each of which need one boundary condition. As upper boundary condition we define the entrainment fluxes and assume that the gradient of the third moment vanishes. At the lower boundary we define the surface fluxes and use similarity expressions for the variances and covariances given by André et al.(1978). In this setup having defined fluxes at both boundaries, no boundary condition is needed for the concentrations.

3 DESCRIPTION OF THE EXPERIMENTS

The experiments are designed to study the intensity of segregation I_s and other higher order chemistry terms in a number of different circumstances. Fast, moderate and slow chemistry are defined in terms of the bulk Damköhler numbers, the ratio of the turbulent timescale and the chemistry timescale:

$$Da_i \equiv k < C_i > \frac{z_i}{w_*} \tag{8}$$

where $\langle C_i \rangle$is the average concentration of species i in the boundary layer. When $Da_i>1$ the chemistry is relatively fast, when $Da_i<1$ it is slow and $Da_i \cong 1$ is called moderate chemistry. The different Damköhler numbers are obtained by prescribing different fluxes at the model boundaries. The reaction rate k is kept at the constant value of 0.0004 $ppb^{-1}s^{-1}$ (approximately the reaction rate of $NO + O_3 \rightarrow NO_2 + O_2$).

Two cases with different source configurations are studied:

- Case 1: Both tracers i and j are released in equal amounts near the surface, with zero correlation $(\overline{(c_i c_j)}_{surface}=0)$.
- Case 2: A bottom-up tracer is released near the surface with zero flux near the boundary layer top. A second (top-down) tracer is entrained at the top of the boundary layer, with zero flux at the surface. The flux ratio :

$$\frac{\text{entrainment flux of the top-down tracer}}{\text{surface flux of the bottom-up tracer}} \equiv 1 \tag{9}$$

The boundary layer characteristics for case 1 and case 2 are listed in table 1.

4 RESULTS

Results are shown after a three hour simulation, when a stationary situation is reached. In this situation the fluxes into the boundary layer of both species are balanced by their depletion through the reaction with the other component.

Table 1: Boundary layer characteristics for two cases. ($w_* \equiv ((g/\Theta)\overline{w\theta_v}z_i)^{1/3}$; $\overline{cw_0}$ is the surface flux and $\overline{cw_{z_i}}$ is the entrainment flux)

w_*	1 ms^{-1}		
z_i	1000 m		
CASE 1	**Slow**	**Moderate**	**Fast**
$\overline{c_iw_0}$	.5 ppb ms^{-1}	2.5ppb ms^{-1}	6.0 ppb ms^{-1}
$\overline{c_iw_{z_i}}$	0. ppb ms^{-1}	0.ppb ms^{-1}	0. ppb ms^{-1}
Da_i	.5 1.0	1.9	
CASE 2	**Slow**	**Moderate**	**Fast**
$\overline{c_{bu}w_0}$	.4ppb ms^{-1}	1.0ppb ms^{-1}	4.0ppb ms^{-1}
$\overline{c_{bu}w_{z_i}}$	0.ppb ms^{-1}	0.ppb ms^{-1}	0.ppb ms^{-1}
$\overline{c_{td}w_0}$	0.ppb ms^{-1}	0.ppb ms^{-1}	0.ppb ms^{-1}
$\overline{c_{td}w_{z_i}}$	.4ppb ms^{-1}	1.0ppb ms^{-1}	4.0ppb ms^{-1}
Da_1	.5	1.1	3.8
Da_2	.5	1.1	3.8

Case 1

Results of this case with two identical bottom-up tracers with equal surface sources, are plotted in figures 1 and 2. Figure 1A gives the concentration in dimensionless form ($C(z)\, w_*/\overline{cw_0}$). The concentrations become lower for faster chemistry, and the counter-gradient transport that is obtained for the slow chemistry case has disappeared for moderate and fast chemistry. Also in the dimensionless flux profiles ($\overline{cw_z}/\overline{cw_0}$; figure 1B) the effect of chemistry is seen by the stronger deviation from linearity for faster chemistry.
The intensity of segregation (equation 2) is depicted in figure 1C. It reveals different signs for slow chemistry and for moderate and fast chemistry. This means that the correlation between the two species may speed up as well as slow down the average transformation rate in the boundary layer. It may be explained as follows. As a lower boundary condition we have put the concentration covariance to zero near the surface. Since both species have the same vertical concentration and flux gradients a positive correlation will develop as we go to higher levels (see equation 3). However, chemistry will destroy this correlation: in the most extreme case of infinitely fast reaction only one species can be present, the other being completely depleted by chemistry, resulting in a perfect anti-correlation. The result of these two competing processes are seen in the intensity of segregation I_s. Note that in this case neglecting the covariance term will generate errors in the transformation rate up to 25% for slow, 20% for moderate and 70% for fast chemistry.

The flux divergence is influenced by chemistry through the chemistry terms in the flux equation (4) and the equation for $\overline{c\theta_v}$ (5) . In the stationary situation that is eventually established, the flux divergence is balanced by the chemistry terms:

$$\frac{\partial \overline{c_i w}}{\partial z} = -k(C_iC_j + \overline{c_ic_j}). \qquad (10)$$

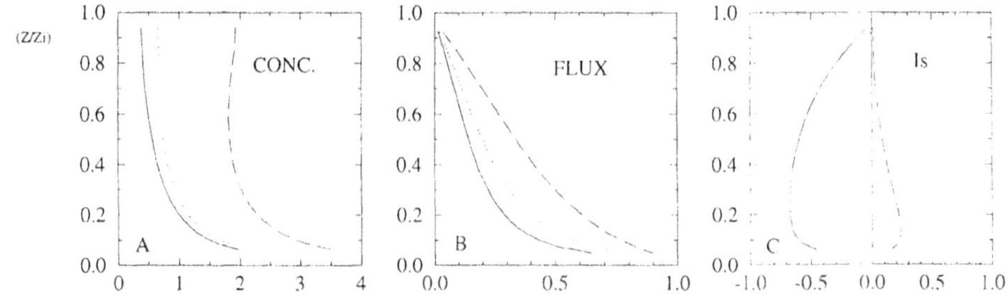

Figure 1. Dimensionless profiles of concentration, flux, and intensity of segregation (I_s), Case 1 (dashed: SLOW; dotted: MODERATE; solid: FAST chemistry).

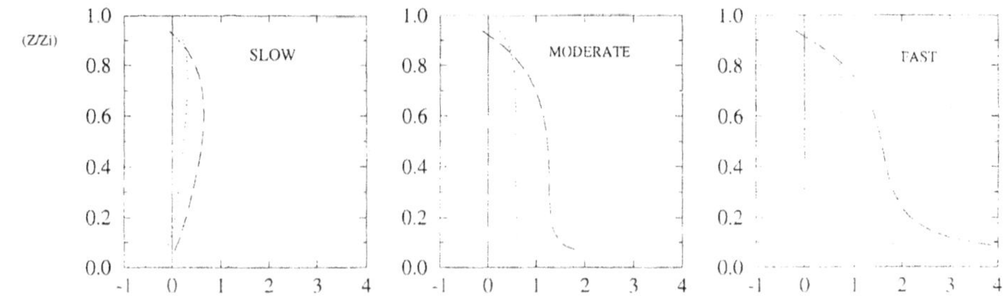

Figure 2. Chemistry contributions to the total flux divergence, Case 1 (dotted line: flux part; dashed line: buoyancy part).

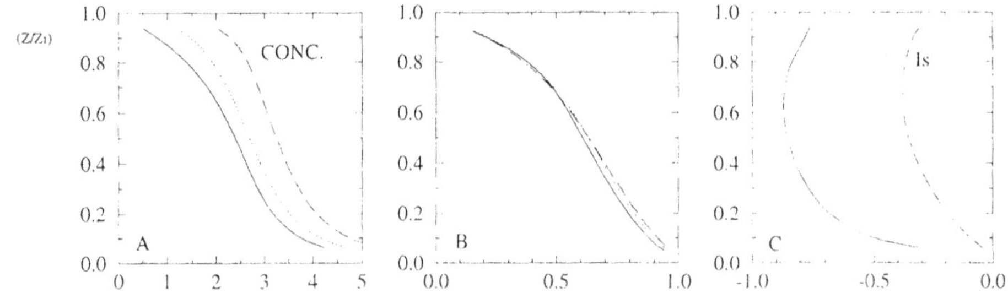

Figure 3. Dimensionless profiles of concentration, flux, and intensity of segregation (I_s), Case 2 (dashed: SLOW; dotted: MODERATE; solid: FAST chemistry).

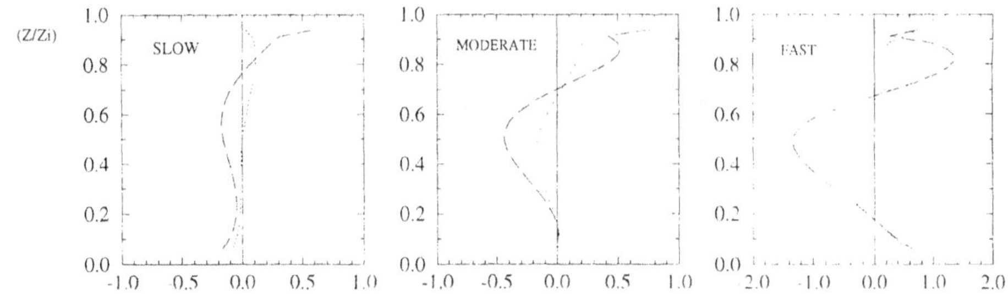

Figure 4. Chemistry contributions to the total flux divergence, Case 2 (dotted line: flux part; dashed line: buoyancy part).

In figure 2 the profile of the chemistry contribution to the flux divergence is plotted. It is made dimensionless by the total flux divergence (or equivalently by the total chemical transformation rate). It is shown that for case 1 the chemistry terms in the flux equations adds to the mean transformation rate, and is for moderate and fast chemistry even the most important term. It is also seen that for our convective boundary layer in all three cases (slow, moderate and fast chemistry) the chemical influence on the flux is mainly due to the buoyancy term.

Case 2

Results of case 2 with the top-down and bottom-up tracer are given in figures 3 and 4. Figure 3A shows the concentration profiles for both tracers. The flux profiles are not very different for the three Damköhler numbers, and deviate only little from the linear profiles. The intensity of segregation is however strongly dependent on the Damköhler numbers. In all cases the negative I_s inhibits the mean transformation rate, down to only 25% of the well mixed transformation rate in the fast chemistry run. The chemistry contributions to the total flux divergence (or total chemical transformation) is depicted in figure 4. In contrast to case 1 the contribution is much smaller, but again the chemical influence is mostly due to the chemistry terms in the $c\theta_v$ equation. In the largest fraction of the boundary layer the chemistry terms in equations (4) and (5) contribute to a decrease of the total transformation.

5 CONCLUSIONS

We focused on the transport of reactive tracers in the unstable boundary layer. It is found that the direct effects of chemical reactions on the flux is mainly due to the buoyancy term. This means that the chemical adjustment of K-theory applied in the convective boundary layer must take into account the chemistry effects on $c\theta_v$. For two initially uncorrelated reactive tracers that are released near the surface (case 1) the sign of the covariance is determined largely by the Damköhler number. In this case the intensity of segregation ($I_s \equiv \overline{c_i c_j}/C_i C_j$) is negative for fast chemistry and positive for slow chemistry. The inclusion of I_s significantly slows down or speeds up the mean reaction rate. The inclusion of the chemistry terms in the flux equation enhances the overall transformation rate. For a bottom-up tracer that reacts with a top-down tracer (case 2) the intensity of segregation is negative throughout the boundary layer, from -0.3 for slow chemistry down to -0.75 in the case of fast chemistry. The chemistry terms in the flux equation seems to be of lesser importance than in case 1.

REFERENCES

André, J. C., P. Lacarrère and K. Traoré. 1982. Pressure effects on triple correlations in turbulent convective flows. *Turbulent Shear Flows*, Ed. Bradbury et al. 3,243- 152.

André, J. C., G. De Moor, P. Lacarrère , G. Therry and R. Du Vachat. 1978. Modelling the 24-hour evolution of the mean and turbulent structures of the planetary boundary layer. *J. Atmos. Sci.* 35,1861-1883.

Fitzjarrald, D. R., and D. H. Lenschow, 1983. Mean concentration and flux profiles for chemically reactive species in the atmospheric surface layer. *Atmos. Environ.*, 17, 2505-2512.

Gao, W. and M.L. Wesely, 1994. Numerical modelling of the turbulent fluxes of chemically reactive trace gasses in the atmopheric boundary layer, *J. Appl. Meteor.*, 33, 835-847.

Garratt, J.R., 1992. *The Atmospheric Boundary Layer.* Cambridge University Press.

Holtslag, A. A. M. and C. -H Moeng. 1991. Eddy diffusivity and countergradient transport in the convective atmospheric boundary layer. *J. Atmos. Sci.* 48,(14):1690- 1698.

Lenschow, D. H., J. C. Wyngaard and W. T. Pennell. 1980. Mean-field and second-moment budgets in a baroclinic, convective boundary layer. *J. Atmos. Sci.* 37,1313-1326.

Mellor, G.L. and T. Yamada. 1974. A hierarchy of turbulent closure models for planetary boundary layers. *J. Atmos. Sci.* 31,1791-1806.

Moeng, C. -H and J. C. Wyngaard. 1989. Evaluation of turbulent transport and dissipation closures in second-order modelling. *J. Atmos. Sci.* 46,2311-2330.

Schumann, U. 1989. Large-eddy simulation of turbulent diffusion with chemical reactions in the convective boundary layer. *Atmos. Envir.*, 23, 8,pp.1713-1727.

Sorbjan, Z., *Structure of the Atmospheric Boundary Layer.* 1989, Prentice-Hall, New Jersey.

Vilà-Guerau de Arellano, J. and P.G. Duynkerke. 1992. Influence of chemistry on the flux-gradient relationships for the NO-O3-NO2 system. *Boundary-layer Met.* 61, 375-387.

DISCUSSION

I. SYKES: How did you model the third order terms in the chemistry of the second moments?

G. VERVER: The full third moment equations are simplified as follows: the quasi-normal approximation is used for the fourth moments; terms with vertical gradients of first moments (like mean temperature or concentration) are neglected; the time derivative of the third moments are neglected. These assumptions yield diagnostic expressions for the third order quantities in terms of only second order quantities and their vertical derivatives.

P. PAI: How do you propose to extend your study to three-dimensional air quality models that simulate chemistry in the atmosphere, and include many chemical species? It would be computationally impractical to numerically solve 3D equations for covariance and concentration fluctuations for so many species. Is it possible to come up with effective reaction rates based on your studies?

G. VERVER: It is not yet clear how it can be incorporated in a 3D model of transport and chemistry. A first step is to select reactions that are most sensitive to the rate of turbulent mixing. The influence of chemistry on the turbulent flux as well as the influence of turbulence on the transformation rate can then be estimated for only the relevant gases. Corrections of the exchange coefficients have already been suggested for the turbulent flux in neutral boundary layers as well as for simple chemistry in the diabatic surface layer. However, I agree that for application in large scale models further simplifications are needed.

COMPARISON BETWEEN EXPERIMENTAL DATA AND A LANGEVIN PARTICLE DISPERSION MODEL INCLUDING DRY DEPOSITION

Edvard Karlsson,[1] Erik Näslund,[1] and Sven-Erik Gryning[2]

[1]National Defence Research Establishment
S-901 82 Umeå
Sweden
[2]Risø National Laboratory
DK-4000 Roskilde
Denmark

INTRODUCTION

Random walk models are powerful tools to describe turbulent diffusion in the atmospheric boundary layer. These models describe the evolution of a passive tracers ("particles") in phase-space via a stochastic differential equation, usually the Langevin equation. They have developed from early applications to homogeneous turbulence (Lin and Reid, 1962) via applications to both inhomogeneous turbulence (Wilson et. al., 1981) and skewed turbulence (Baerentsen and Berkowitz, 1984).

Some advantages with such models are that they are grid-free, they can handle point sources and they are in total lack of artificial diffusion. Also they are easy to programme and implement on a computer.

Most particle models have so far concentrated on the transport properties and not much can be found on how to deal with secondary processes such as deposition and/or chemical interaction. A straightforward way to handle deposition is of course to use the traditional approach where the mass flux to the surface is described by deposition velocity times the concentration near the surface. This is, however, not a very attractive approach since many models are built on tracking one particle at a time and the concentration level is not always accessible during run-time.

In this paper we will investigate a model where each particle can be treated separately with no reference to concentration whatsoever. The transport and diffusion of particles are completely taken care of by the diffusion model while the decision of deposition or not only affects those particles that has reached the ground (or the level above ground where the decision process is taken place, usually the surface roughness height).

Air Pollution Modeling and Its Application XI, Edited by
Gryning and Schiermeier, Plenum Press, New York, 1996

In a number of previous papers similar approaches as taken in this paper can be found (see e.g. Boughton et al., 1987, Wilson et al., 1989). To our knowledge the proposed models have not been very well tested against experiments.

Here, we will compare the model to a number of experiments where deposition velocities were determined from concentration measurements in a box whose sides were covered with various materials. Also we show that simulation of the Prairie Grass experiments with deposition included give better agreement for the distant arcs in most cases.

THE STOCHASTIC MODEL

The evolution in phase-space of a tracer particle is, in one dimension, described by the following stochastic equations:

$$dw = a(z,w)dt + b(z,w)d\xi \tag{1a}$$

$$dz = wdt \tag{1b}$$

Here, w and z are the velocity and position of the particle, respectively, a and b are coefficients that have to be determined from physical constraints and $d\xi$ is a random velocity increment of a Wiener process with zero mean and variance dt. For two different times, t_1 and t_2 ,$d\xi(t_1)$ and $d\xi(t_2)$ are independent.

The coefficients a and b may take different forms for different kind of flows (see eg. Thomson, 1987). For inhomogeneous, Gaussian turbulence the following forms can be derived:

$$a = -\frac{w}{\tau_L} + \frac{1}{2}\left(1 + \frac{w^2}{\sigma^2}\right)\frac{\partial \sigma^2}{\partial z} \tag{2a}$$

$$b = (C_0 \varepsilon)^{\frac{1}{2}} \tag{2b}$$

where τ_L is the Lagrangian time scale, $\sigma = \sigma(z)$ is the standard deviation of w, $\varepsilon = \varepsilon(z)$ is the dissipation of turbulent kinetic energy, and C_0 is a universal constant in the range 2-8.

THE DRY DEPOSITION MODEL

We now assume that, with the additional determination of σ and ε, the transport and diffusion of tracer particles is completely determined from Equations (1-2). This means that no reference to deposition is made while the tracers are in the air. It is only those particles that reaches a surface that will be affected by the decision process whether it will deposit or not.

The starting point for the model is that the deposition velocity, v_d, somehow is known. The mass flux through a given surface during time Δt can then be written

$$F_m = Cv_d \tag{3}$$

where C is the concentration ($1/m^3$) at the surface. This flux is now compared to the actual flux of particles emerging from a known particle distribution of velocities, $f(V)$, where $V = (u^2+v^2+w^2)^{1/2}$.

The average particle speed can be written

$$\overline{V} = \int\int\int f(V)d^3V \quad (4)$$

and the mass flux through a surface in the -z-direction is

$$F_p = \int_{w<0} d^3V f(V) V \cos\theta \quad (5)$$

where θ is the angle between the particle velocity and the z-axis. These formulas are very general and may not be possible to evaluate for an arbitrary velocity distribution. A straightforward calculation (see e.g. Reif, 1965) can however be made in the important case of a Gaussian velocity distribution with zero average velocity and standard deviation σ. The result is

$$F_p = \frac{1}{4} C\overline{V} \quad (6)$$

where $\overline{V} = (8/\pi)^{1/2}\sigma$

Now, defining the probability for deposition as the quota between the two different fluxes we get for the deposition probability

$$P = \frac{F_m}{F_p} = (2\pi)^{1/2} \frac{v_d}{\sigma} \quad (7)$$

In the numerical scheme this means that, when a particle reaches the surface, a random number is drawn from a uniform [0,1] distribution which is compared to P. If P is larger than the random number, the particle gets deposited, otherwise it undergoes perfect reflection. For a surface above ground the value of v_d has to be chosen according to the influence of resistance at the surface, in the viscous sublayer and in the turbulent layer. However, at the ground, v_d should be influenced only by the surface and viscous sublayer.

COMPARISON WITH DEPOSITION EXPERIMENTS IN A TEST CHAMBER

In a previous paper (Näslund and Karlsson, 1995) results from simulation of deposition to some indoor materials were presented. Those experiments concerned the deposition of nerve agent sarin where a certain amount of the agent was injected into a test chamber of glass and concentration was measured at 10 minutes intervals. The turbulence level was chosen so that the dominant resistance for deposition came from the surface itself.

Here we will simulate more recent experiments where deposition to snow surfaces have been investigated in a similar manner (Karlsson et al., 1995). However, after the agent has been injected into the test chamber, concentration was measured every two seconds, which gave much more accuracy in the determination of the deposition velocity which is defined as

$$v_d = -\frac{1}{C}\frac{dC}{dt}\frac{Vol}{A} \quad (8)$$

where *Vol* is the volume of the box, *C* is the measured concentration and *A* is the area of the snow.

Four different measurements were chosen for the simulation, three where the snow temperature was around the melting point and one with cold, dry snow where the temperature were well below 0°C. Deposition velocities were typically 0.3 - 0.8 cm/s for melting snow, 0.05 - 0.20 cm/s for cold, dry snow and 0.1 - 0.3 cm/s for wet, old snow.

These measured values was then used in the diffusion/deposition model described above. Particles were dispersed by the Langevin equation without reference to deposition. Only when a particle hit the deposition surface it was subjected to a probabilistic judgment whether to deposit or not. In Figure 1 the results from the simulations are shown (solid lines). As can be seen they show excellent agreement with the experimental values (squares).

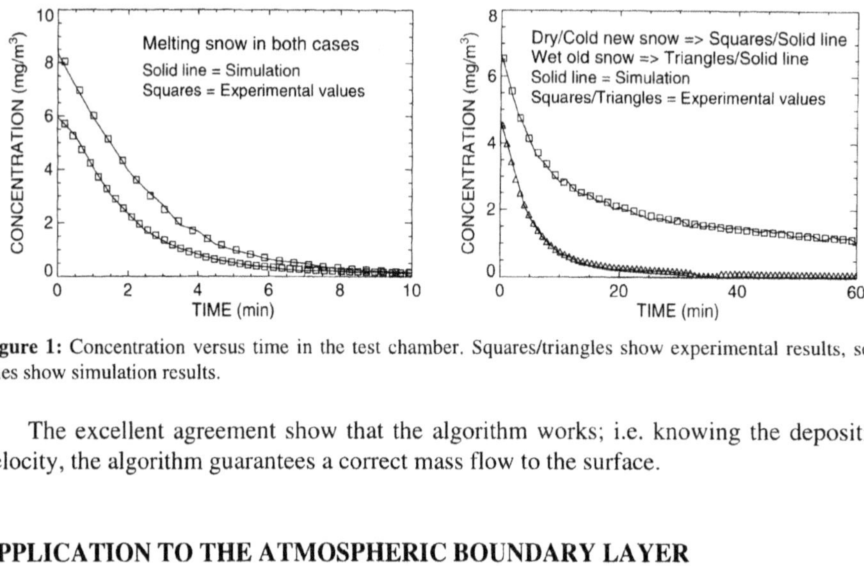

Figure 1: Concentration versus time in the test chamber. Squares/triangles show experimental results, solid lines show simulation results.

The excellent agreement show that the algorithm works; i.e. knowing the deposition velocity, the algorithm guarantees a correct mass flow to the surface.

APPLICATION TO THE ATMOSPHERIC BOUNDARY LAYER

Many attempts have been made to model the 1956 Prairie Grass experiments in Nebraska, USA (Barad, 1958). These experiments were conducted on flat grassland with a continuos release of SO_2 at a height of z = 0.46 m. Concentrations were measured on arcs 50, 100, 200, 400 and 800 m downwind.

Several modeling attempts have been quite successful, but usually one can make one major observation; the models perform better the closer to the source the comparisons are made. The explanation for this observation may be that sulfur dioxide deposits significantly, but also that the parameterization of wind and turbulent quantities are derived from profile measurements made at the 100 m arc. Although the diffusion parameters are derived from that arc, we will use those parameters for a number of runs and add deposition effects to the Langevin model. Comparisons are then made between the model calculations and the measured values at z =1.5 m.

Langevin models have previously been used for the description of some of the experiments. We will here use the parameterization used by Rodean et al., (1992) where three of the experiments (37 and 46 - stable; 45 - unstable) were successfully modeled at the 100 m arc. The mean wind profile is thus based on surface-layer similarity flux-profile relationships as follows (Dyer, 1974)

$$u(z)=\frac{u_*}{k}\left(\ln(z/z_0)+\frac{5}{L}(z-z_0)\right) \qquad \text{for } L>0 \qquad (9a)$$

$$u(z)=\frac{u_*}{k}\left(2\arctan(x)-2\arctan(x_0)-\ln\left(\frac{(1+x)(1-x_0)}{(1-x)(1+x_0)}\right)\right) \qquad \text{for } L<0 \qquad (9b)$$

where $x=(1-16z/L)^{\frac{1}{4}}$, and $x_0=(1-16z_0/L)^{\frac{1}{4}}$.

Here L is the Monin-Obukhov length, $z_0 = 0.008$ m is the roughness length, and $k = 0.4$ is the von Karman constant. The vertical variance, σ, and dissipation of turbulent kinetic energy, ε, are given by

$$\sigma^2 = u_*^2\left\{1.7\left(1-\frac{z}{h}\right)^{\frac{3}{2}} + C_1\left(\frac{z}{-L}\right)^{\frac{2}{3}}\left(1-0.8\frac{z}{h}\right)^2\right\} \qquad (10a)$$

$$\varepsilon = \frac{u_*^3}{kz}\left(1+3.7\frac{z}{L}\right)\left(1-0.85\frac{z}{h}\right)^{\frac{3}{2}} \qquad \text{if } h/L>0 \qquad (10b)$$

$$\varepsilon = \frac{u_*^3}{kz}\left\{\left(1-0.75\frac{z}{L}\right)\left(1-0.85\frac{z}{h}\right)^{\frac{3}{2}} - 0.3\frac{z}{L}\right\} \qquad \text{if } h/L<0 \qquad (10c)$$

where $C_l = 0$ for $L > 0$, $C_l = 2.4$ for $L < 0$ and h is the height of the adiabatic mixing layer that for stable/neutral cases is calculated as

$$h = d\left(L\frac{u_*}{f}\right)^{\frac{1}{2}} \qquad (11)$$

Here d is a dimensionless constant taken as 0.7 and f is the coriolis parameter. Ley and Thomson (1983) have discussed the relative arbitrariness of h for unstable cases in the range 1000-2000 m.

We have performed a straightforward application of the Langevin model with and without deposition with these input parameters for seven different runs, in each run following the trajectories of 30000 particles. The boundary-layer parameter values used in the model calculations are presented in Table 1, columns 2-4. For values of the deposition velocity we follow a discussion in Gryning et. al. (1983). Generally, v_d can be described as the inverse of atmospheric, (r_a), sublayer, (r_b), and surface,(r_s), resistance, i.e.

$$v_d = \frac{1}{r_a + r_b + r_s} \qquad (12)$$

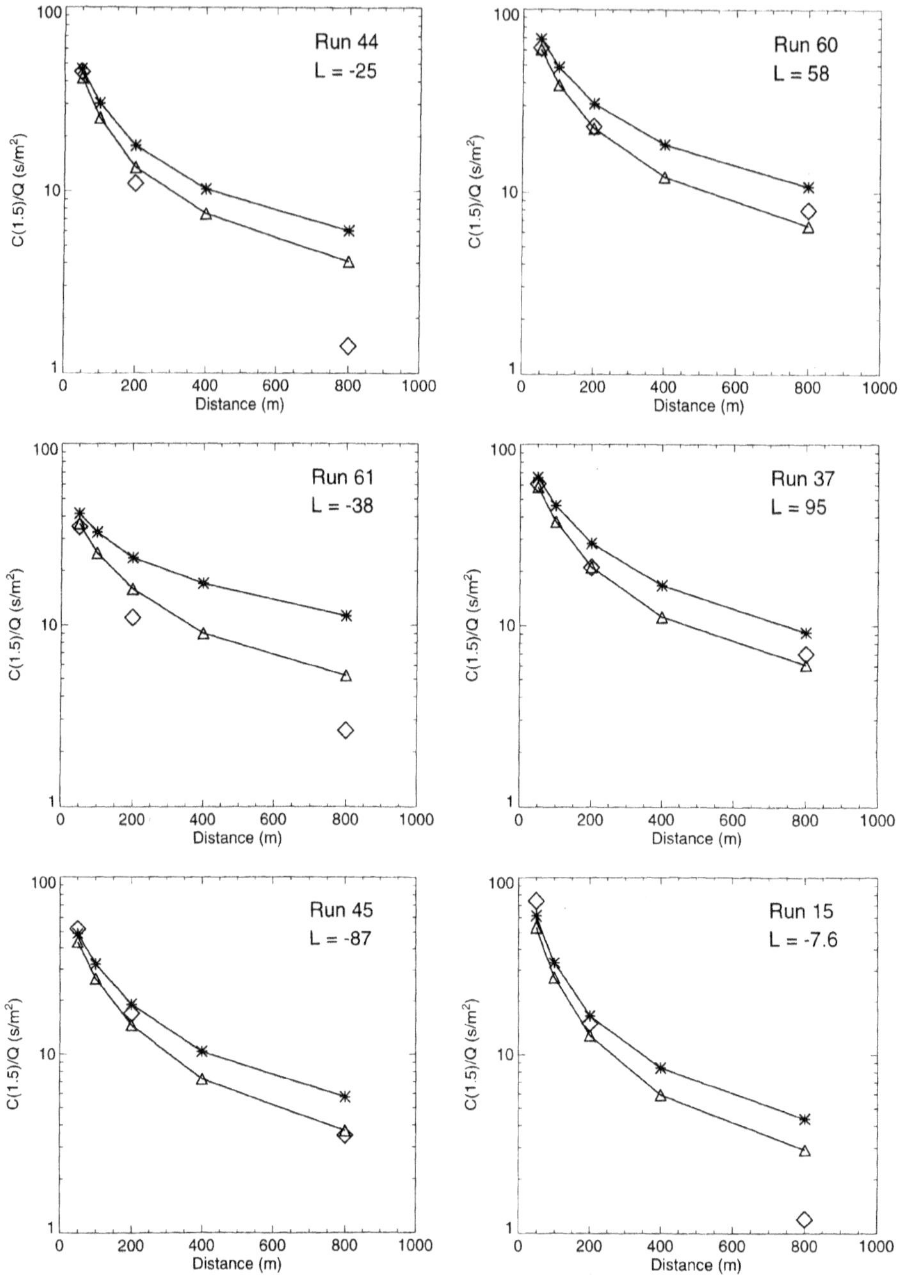

Figure 2: Simulation of some of the Prairie Grass runs. Solid line/star show simulation results without deposition. Solid line/triangles show simulation results with deposition effects included. Squares show experimental values at 50, 200 and 800 m distance from source.

where, for the Prairie Grass runs, $r_s = 45\ \mathrm{sm}^{-1}$ (Garland, 1977; Chamberlain, 1966), $r_b = 1/(Bu_*)$ and $r_a = (\ln(z_r/z_0) - \Psi_c)/(ku_*)$,, where $\Psi_c = 0$ in neutral conditions, $\Psi_c = -5z/L$ in stable conditions and $\Psi_c = 2\ln(0.5(1+(1-16z/L)^{0.5}))$ in unstable conditions. $B = 0.2$ is the sublayer Stanton number and z_r is the level where the decision of reflection or deposition is taken place. Possibly the proposed value of r_s should be different for day and night because the stomata depend on the insolation. Also the wetness of the ground (rain, dew) may influence r_s.

The resulting values for v_d, used in the simulation, are presented in Table 1, column 5.

Table 1: Surface-layer and boundary-layer parameters. Deposition velocities.

	u_* *(m/s)*	*L (m)*	*h(m)*	v_d *(cm/s)*
Run 37	0.33	95	426	1.38
Run 60	0.32	58	327	1.40
Run 15	0.26	-7.6	1000	1.14
Run 44	0.46	-25	1500	1.39
Run 61	0.58	-38	1500	1.51
Run 45	0.45	-87	1500	1.38

Concentrations at the 1.5 m level have been calculated and compared to the experimental values presented in van Ulden (1978) for the 50, 200 and 800 m arcs. The results are summarized in Figure 2, where plots of concentration vs. distance are shown.

It can be seen that the model runs where deposition is included give a better fit for the distant arcs while for the nearby arcs the effect of deposition is not as significant. For some of the runs (60 and 37) deposition even leads to a slight underprediction at the 800 m arc. One can however argue that the diffusion parameters are derived without a deposition assumption, which means that u_* (and thus σ) in reality should be somewhat smaller and all curves in Figure 2 should be shifted slightly upwards.

We also did a very stable run (Run 32) where the far arc results was really poor. This run, with a fairly low mixed layer height, gave significant underprediction for the far arcs even without inclusion of deposition. The measured values at the 800 m arc is however extremely high which may indicate an overestimation of u_* or of the Monin-Obukhov length, both leading to a too high mixed layer height. A third possibility is of course that meteorological conditions changed severely over space and time during the run.

DISCUSSION AND CONCLUSIONS

The proposed deposition model is performing very well when compared to controlled deposition experiments where diffusion parameters are well defined. This shows that the deposition algorithm gives the correct mass flux to the surface. When looking at the Prairie Grass runs deposition, for most runs, improves the quality of the simulation. However, since there are so many parameters that are poorly defined, it is no surprise that we can find some cases where the deposition extension does not lead to an improvement. A number of possible explanations for that have been given above, but the issue will be further discussed in a forthcoming paper where more Prairie Grass runs will be simulated.

REFERENCES

Baerentsen, J. H. and Berkowicz, R., 1984, Monte Carlo simulation of plume dispersion in the convective boundary layer, *Atmos. Environ.*, 18:701-712.

Barad, M. L., 1958, *Project Prairie Grass, a Field Program in Diffusion* , Report AFCRLTR235 (2 volumes).

Boughton, B. A., Delaurentis, J. M. and Dunn, W. E., 1987, A stochastic model of particle dispersion in the atmosphere, *Boundary-Layer Meteorol.* 40:147-163.

Chamberlain, A. C., 1966, Transport of gases to and from grass and grass-like surfaces, *Proc. R. Soc. Lond. A.* 290:236-265.

Dyer, A. J., 1974, A review of flux profile relationships, *Boundary-Layer Meterol.* 7:363-372.

Garland, J. A., 1977, The dry deposition of sulphur dioxide to land and water surfaces, *Proc. R. Soc. Lond. A.* 354:245-268.

Gryning, S-E., van Ulden, A. P. and Larsen, S. E., 1983, Dispersion from a continous ground-level source investigated by a *K*-model, *Quart. J. R. Met. Soc.* 109, 355-364.

Karlsson, E., Berglund, T. and Nordstrand, M., 1995, Deposition of sarin on snow, *Proceedings of the Fifth International Symposium on Protection against Chemical and Biological Warfare Agents*, Stockholm, Sweden, pp 409-415.

Ley, A. J. and Thomson, D. J., 1983, A random walk model of dispersion in the diabatic surface layer, *Quart. J. R. Met. Soc.* 109:867-880.

Lin, C. C. and Reid, W. H., 1962, Turbulent flow, *Handbuch der Physik,* VIII/2, Springer, Berlin. pp. 438-523.

Näslund, E. and Karlsson, E., 1995, Dry deposition in a model for stochastic particle dispersion, *Air Pollution 95*, Porto Carras, Greece (in press).

Reif, F., 1965, *Fundamentals of Statistical and Thermal Physics*, McGraw-Hill, pp. 269-273.

Rodean, H. C., Lange, R. L. and Nasstrom, J. S., 1992, Comparison of two stochastic models of scalar diffusion in turbulent flow, *Tenth Symposium on Turbulence and Diffusion*, Portland, Oregon, pp 106-109.

Thomson, D. J., 1987, Criteria for the selection of stochastic models of particle trajectories in turbulent flows, *J. Fluid Mech.* 180:529-556.

van Ulden, A. P., 1978, Simple estimates for vertical diffusion from sources near the ground, *Atmos. Environ.* 12:2125-2129.

Wilson, J.D., Thurtell, G.W. and Kidd, G.E., 1981, Numerical simulation of particle trajectories in inhomogeneous turbulence II: Systems with variable turbulence velocity scale, *Boundary-Layer Meteorol.* 21:423-441.

Wilson, J. D., Ferrandino, F. J. and Thurtell, G. W., 1989, A relationship between deposition velocity and trajectory reflection probability for use in stochastic lagrangian dispersion models`, *Agricultural and Forest Meteorol.* 47:139-154.

DISCUSSION

B. KARGIN: I didn't understand your consideration about probability of deposition. It may be so that the quantity which you call probability $\sqrt{2\pi}v_d/\sigma_w$ becomes greater than 1. But it's impossible by definition of probability.

E. KARLSSON: Normally the "probability" $\sqrt{2\pi}v_d/\sigma_w$ will be much less than 1.0 even at weak turbulence. As example $v_d = 0.01\ ms^{-1}$ and $\sigma_w = 0.1\ ms^{-1}$ will give $\sqrt{2\pi}v_d/\sigma_w = 0.25$. However if the algorithm is applied to particles with significant settling velocity $\sqrt{2\pi}v_d/\sigma_w$ may be larger than 1. This case is now excluded in the derivation but can be taken into account by allowing a mean vertical transfer velocity equal to the settling velocity (v_s) giving the "probability" equal to $\sqrt{2\pi}v_d/(\sqrt{2\pi}v_s + \sigma_w)$ which is ≤ 1, since $v_s = v_d$ for large v_s. Of course it also would be nicer to scale our "probability" so it always is strictly ≤ 1.

R. MERONEY: In Figure 1 of your paper you show amazing replication of your test chamber results with your model. But the model requires specification of v_d which the text indicates varies by a factor 3 to 4 for a given surface. If you vary v_d at every moment in time it would seen agreement is almost guaranteed. In the field case v_d appears to be chosen constant.

E. KARLSSON: In order to test our deposition algorithm, v_d is varied with time in figure 1 in accordance with the measured values of v_d. If the algorithm works, then an exact agreement between the model and the concentration in the test chamber should be expected. In the Prairie Grass case we have no estimate of the time variation of v_d, and therefore a constant value is used. Some of the differences between model and measurements may depend on variation of v_d with time.

PRIME—AN IMPROVED DOWNWASH MODEL

Richard H. Osa
Science & Technology Management, Inc.
2511 N. 124th Street
Brookfield, WI 53005

Charles Hakkarinen
Electric Power Research Institute
3412 Hillview Ave.
Richard H. Osa

INTRODUCTION

Electric utilities need the capability to reliably simulate the ambient impact of power plant emissions in order to meet corporate environmental objectives and support regulatory permit applications. Current regulatory models tend towards conservatism in their calculations (i.e., overpredict actual concentrations) to compensate for modeling and input data uncertainties. The scientific bases of the plume rise and building downwash routines of existing regulatory dispersion models are of particular concern since they are based on rather sparse sets of wind tunnel observations. These models also suffer from multiple discontinuities in their treatment of stack height—a vestige of their evolution by superimposition of independently-derived downwash algorithms (Corio, 1990).

The Electric Power Research Institute (EPRI) and a consortium of electric utilities have sponsored a program to develop additional observational data for developing and testing improved plume rise and downwash model formulations. A new model has been developed, the Plume Rise Model Extensions (PRIME). The EPRI program is continuing at this time with an independent evaluation comparing the performance of PRIME with that of a widely-used regulatory model (ISC3).

Research Program Design

The research project design was established consistent with published U.S. Environmental Protection Agency (USEPA) guidance on its process for selecting preferred regulatory models (USEPA, 1984). Further direction was provided in discussions with USEPA staff regarding mutual concerns over the adequacy of existing downwash models. These discussions provided insights into those attributes that would most recommend a candidate to replace the existing formulation. Foremost among these were scientific rigor and a grounding in observational measurements.

Since this is very much in keeping with EPRI's mission, the project appeared to be valuable, feasible, and adequately funded—it was
on its way.

Technically, the project was structured around data gathering (new and historical; field and laboratory), model development, and evaluation (see Figure 1). These activities were guided by an advisory committee consisting of representatives of each of the sponsors. The advisory committee participated in all activities from the initial request for proposals, through project field and laboratory activities, and acceptance of final reports.

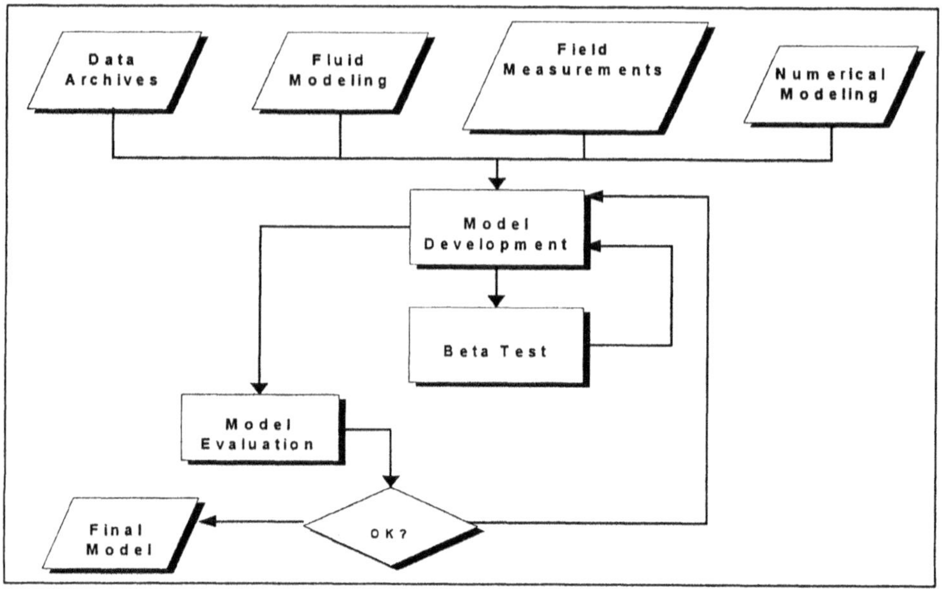

Figure 1. Plume Rise/Downwash Model Development Paradigm.

DATA GATHERING

Obtaining a rich and varied observational data base is a prerequisite to constructing a mathematical model to simulate a complex phenomenon such as atmospheric dispersion. Previous downwash models (e.g., Huber and Snyder, 1982; Schulman and Hanna, 1986) relied almost exclusively on laboratory data provided by wind tunnel experiments. Physical and scaling constraints limited such experiment to moderate and high wind speeds, and neutral stability. Regulatory needs have led to these algorithms being applied well outside the domain in which they were developed. Of particular concern is the common modeling result that critical concentrations, for regulatory purposes, often occur during simulation of light wind, stable conditions—conditions far removed from those under which the models were developed.

The EPRI data-gathering program was designed to overcome these limitations, while exploiting the strengths of wind tunnel modeling. Significant resources were devoted to carrying out a field study of plume rise and downwash from a combustion turbine and steam boiler facility.

The field study was conducted in February and March of 1994 at Jersey Central Power & Light's Sayreville Power Plant in New Jersey, USA. Power plant load during field experiments was varied over a substantial range. Unit operations and emission characteristics were monitored on a continuous basis by the host utility. Meteorological parameters were monitored using the National Center for Atmospheric Research's Atmospheric Surface Turbulent Exchange Research facility (ASTER) for micrometeorological measurements, as well as an auxiliary meteorological tower operated by ENSR Corp., and a Doppler sodar operated by Jersey Central. A limited number of tethered balloon soundings supplemented the ground-based instruments.

Plume concentrations were mapped three-dimensionally by means of mobile aerosol backscatter lidar operated by SRI, International that measured the concentration of fine smoke particles within the plume. Concentrations in the building cavity were measured by a series of NO_x monitors.

Despite uncharacteristically harsh weather, the field effort yielded a significant data base. Table 1 summarizes the observational data base. In addition to field measurements carried out as part of the EPRI study, several additional data sets obtained as part of the independent evaluation task (but not needed for the evaluation) were made available to the model development team.

Table 1. Summary of Measurements at JCPL Sayreville

Date	Start Time	End Time	Type Test	Remarks
Feb. 14	1145	1615	Downwash	Lidar on line at 1415, NW wind
Feb. 16	1000	1630	Downwash	Good test, moderate wind speeds, NW
Feb. 17	1030	1330	Plume rise	Light W wind, slight inversion
Feb. 18	1530	2015	Plume rise	Light wind, stable near ground, NE winds
Feb. 19	1030	1330	Plume rise	Temperature inversion above 50 m, NW wind
Feb. 22	0900	1045	Abort	Turbine malfunction
Feb. 26	1300	1830	Downwash	After end of storm, NW wind
Feb. 27	0900	1530	Downwash	Strong NW winds
Feb. 28	0800	0915	Downwash	Intermingled plumes from 2 steam units and 3 CTs, NW wind
Mar. 4	1100	1600	Steam Boiler	Moderate NW winds
Mar. 5	0900	1300	Steam Boiler	Modest NW wind

Supplementing the field measurements were several sets of wind tunnel studies. The original study design called for the laboratory modeling to be benchmarked against actual field observations to validate the physical model, and then to be used to extend observations to meteorological conditions not encountered in the field (e.g., an infrequent wind direction.) This strategy was carried out, with valuable information provided by collaborators at Monash University in Australia (Taylor and Melbourne, 1996) (see Figure 2).

Figure 2. Monash wind tunnel simulation of Sayreville field experiment.

The project also benefitted greatly from a set of high resolution experiments carried out by William Snyder at USEPA's Fluid Modeling Facility prior to the field mobilization (Snyder, 1993 and Snyder and Lawson, 1994). The Snyder work included concentration measurements from simulated downwash conditions as well as detailed flow and turbulence observations about the obstructing structure which was of simple, block shape.

The third data source was provided by collaborating scientists at Washington State University who adapted a finite element turbulence model to a domain useful in examining flow about a typical structure. This numerical model was exercised for a number of scenarios specified by the model developers (EARTH TECH/Sigma Research Division). While such a research-grade code solution in impractical for regulatory modeling, the fine detail available from the numerical model provided the analytical model developers insights into the physical processes at work and helped identify candidates for parameterization.

MODEL DEVELOPMENT AND FORMULATION

Model development was performed by researchers at EARTH TECH/Sigma Research Division.

Model formulation followed a process of:

- identifying the important physical processes that must be considered in order to adequately reproduce plume behavior,
- isolating those key variables that together largely determine the plume dynamics and position,
- defining and developing simplifying concepts (such as the building "cavity") that can serve to reduce the computational demands of the model while enhancing the user's ability to conceptualize the physical process being simulated, and
- implementing the new model through a computer code.

The physical processes included in PRIME include the perturbation to wind flow streamlines caused by the presence of an obstructing structure. A plume's downwash trajectory is simulated using an empirical relationship that varies with the shape of the building and the wind angle, coupled with a numerical plume rise plume rise formulation. This approach allows PRIME to determine the change in plume centerline location with downwind distance.

The method of Wilson (1979) and Fackrell (1984) is used to define the location and height of the recirculation cavity downwind of the building. This cavity structure is parameterized and figures into the computation of the slope of the downwind streamline descent.

The numerical plume rise routine uses mass, energy, and momentum conservation laws after the method of Zhang and Ghoniem (1993). As implemented in PRIME, this approach allows inclusion of vertical wind shear effects which are important for many buoyant releases from short stacks. A detailed description of the model and its development may be found in Schulman and Scire (1996).

MODEL EVALUATION

Recognizing the importance of a scientifically rigorous, independent assessment of the new model's performance, EPRI contracted with ENSR Corp. to identify and prepare existing data sets, and perform a comparative evaluation of PRIME and the current USEPA regulatory model, ISC3. A particularly challenging aspect of this assignment was the development of an evaluation protocol applicable to both short term, high resolution data sets, typical of tracer studies, and long term data sets, such as provided by compliance monitoring networks (Paine, 1996). The protocol has benefited from the review and input of the project's advisory committee, the State and Regional Modelers' Ad Hoc Committee on Downwash, and USEPA's Office of Air Quality Standards and Planning.

Paine has adapted the "Combined Model Comparison Measure" (CMCM) for the evaluation, using the comparison statistics and tests developed by Cox and Tikvart (1990) and Lee and Irwin (1995). The CMCM incorporates a weighting function designed to minimize the influence of data bases with a low confidence in the determination of the test statistics. The magnitude and sign of the CMCM indicate the relative performance as a pair of models, as evaluated against diverse sets of observational data.

The evaluation data sets have been assembled and pre-processed into a form compatible with both PRIME and ISC3 input requirements. Evaluation activities are currently underway, with preliminary results expected by the end of December.

While it is uncertain whether PRIME will outperform the ISC3 downwash algorithms, it is certain that its developers have had the advantage of the most extensive downwash data set ever assembled, and that the evaluation protocol has received scientific acceptance. Regardless of the fate of PRIME, the foundation upon which it is built is destined to lead to improvements in the modeling of the critical plume rise and downwash phenomena.

ACKNOWLEDGMENT

The Plume Rise and Downwash Project (EPRI RP3527) was carried out with the support of a consortium of funding organizations and in-kind contributors. The cooperation of Jersey Central Power & Light was crucial to the success of the field measurements program and is much appreciated. USEPA's cooperation in providing Bill Snyder's meticulous wind tunnel data sets was immensely helpful in illuminating the downwash process.

REFERENCES

Corio, L.A., 1990, Discontinuity in the Calculation of Ground-Level Concentrations Under Downwash Conditions in the ISC Model. Paper No. 90-79.3, Presented at the 83rd Annual Meeting of the Air & Waste Management Association, Pittsburgh, PA.

Cox, W.M. and Tikvart, J.A., 1990, A Statistical Procedure for Determining the Best Performing Air Quality Simulation Model. *Atm. Env.* 24A(9): 2387-2395.

Huber, A.H. and Snyder, W.H., 1982, Wind Tunnel Investigation of the Effects of a Rectangular Shaped Building on Dispersion of Effluents from Short Adjacent Stacks. *Atm. Env.* 16:2837-2848.

Lee, R.F. and Irwin, J.S., 1995, A Methodology for a Comparative Evaluation of Two Air Quality Models. In Proceedings of the Third Workshop On Atmospheric Dispersion Modeling for Regulatory Purposes.

Paine, R.J., 1996, Design of Evaluation of Building Downwash Models. Paper 6.2 in *Proceedings of the Ninth Joint Conference on the Applications of Air Pollution Meteorology*, Atlanta, GA. American Meteorological Society, Boston, MA.

Schulman, L.L. and Scire, J.S., 1996, The Development of the Plume Rise Model Enhancement (PRIME): The EPRI Plume Rise and Downwash Modeling Project. Paper 6.1 in *Proceedings of the Ninth Joint Conference on the Applications of Air Pollution Meteorology*, Atlanta, GA. American Meteorological Society, Boston, MA.

Schulman, L.L. and Hanna, S.R., 1986, Evaluation of downwash modifications to the industrial source complex model. *JAPCA*, 36: 258-264.

Snyder, W.H., 1993, Downwash of plumes in the vicinity of buildings: A wind tunnel study, in *NATO Advanced Research Workshop, Recent Advances in the Fluid Mechanics of Turbulent Jets and Plumes,* Viano so Castelo, Portugal.

Snyder, W.H. and Lawson, R.E., 1994, Wind-tunnel measurements of flow fields in the vicinity of buildings, *8th AMS Conf. on Appl. Air Poll. Meteorol.*, with AWMA, Nashville, TN.

Taylor, T.J. and Melbourne, W.H., Do building wakes increase ground level concentrations? Paper 6.3 in *Proceedings of the Ninth Joint Conference on the Applications of Air Pollution Meteorology*, Atlanta, GA. American Meteorological Society, Boston, MA.

U.S. Environmental Protection Agency, 1984, *Interim Procedures for Evaluating Air Quality Models* (Revised). EPA-450/4-84-023. Office of Air Quality Planning and Standards, Research Triangle Park, NC.

Zhang, X., and A.F. Ghoniem, 1993, A Computational Model for the Rise and Dispersion of Wind-blown, Buoyancy-driven Plumes - I. Neutrally Stratified Atmosphere, *Atmos. Environ.*, 15, 2295-2311.

DISCUSSION

P. PAI: Are there any plans to incorporate PRIME in AERMOD?

R. OSA: Yes. The PRIME development team and project management staff have been communicating with the AERMOD developers to enable PRIME to easily be accommodated within AERMOD. While there is no established schedule for inserting PRIME within AERMOD, it is considered quite likely that if the PRIME scientific evaluation shows that it is substantially superior to current regulatory model downwash techniques, AERMOD will be upgraded to take advantage of this improvement.

R. MERONEY: I suggest caution in applying the Schulman/Scire algorithm to near-building predictions and to non-flat top roof buildings in surface to surface pollution situations. In a report to EPA on wind tunnel radon plume modeling about domestic-size buildings, I found that whereas the Huber/Snyder model predicted far-wake behavior very well, the near-wake, i.e. x<building height, routinely predicted values one order of magnitude greater than those measured.

R. OSA: I appreciate the caution regarding non-flat top roof buildings. There is no doubt that building geometry can substantially influence the degree of downwash that might occur, as we saw in the experiments comparing maximum concentration with the wind at right angles to the building face and with a 45 degree angle of attack. However, I would like to point out that the Schulman/Scire algorithm was not intended to be used in describing dispersion within the cavity region, therefore it is not surprising that model predictions differ greatly from those observed in the laboratory fluid modeling.

IMPROVED ALGORITHMS FOR ESTIMATING THE EFFECTS OF POLLUTION IMPACTS FROM AREA AND OPEN PIT SOURCES

William B. Petersen and Steven G. Perry[1]

Atmospheric Sciences Modeling Division
Air Resources Laboratory
National Oceanic and Atmospheric Administration
Research Triangle Park, NC 27711

INTRODUCTION

The increasing environmental concern over hazardous emissions from landfills, landfarms, settling ponds, agricultural fields, surface mines, and surface impoundments has renewed interest in improving concentration estimates downwind from area sources. Simplistic and numerically efficient methods (for estimating area source impacts) are the virtual point source approach (Turner, 1970) as used in the models LONGZ (Bjorklund and Bowers, 1979) and VALLEY (Burt, 1977), and the finite line source approach as used in ISCST (Bowers et al., 1979). The virtual point source and finite line segment algorithms are most appropriate for estimating concentrations at downwind distances significantly larger than the side width of the area source (Hwang, 1986 and Weber, 1982). However, these constructs fail for concentration estimates within or near the area source. Thistle and Londergan (1989) showed that these algorithms behaved in a manner inconsistent with mathematical and physical principles. They demonstrated that models that use numerical techniques to approximate the Gaussian point source dispersion function over the area source are more realistic although computationally intensive. One of the models that they found to be mathematically and physically consistent was PAL (Petersen and Rumsey, 1987).

Shortly after the Thistle and Londergan study, the Office of Air Quality Planning and Standards in the U.S. Environmental Protection Agency (EPA) decided to update the area source algorithm in ISC2 (U.S. EPA, 1992) using the approach contained in PAL. At about the same time the Office of Research and Development in EPA was involved in a project to improve the computational efficiency of the area source algorithm in PAL. It is this more efficient version of the PAL area source algorithm that has been implemented in ISC2 and will be described in this paper. In addition, the improved area source algorithm is utilized in a new approach to modeling open-pit sources.

Title II, Part B, Section 234 of the Clean Air Act Amendments of 1990 provide for a reexamination of the current Environmental Protection Agency's methods for modeling fugitive particulate (PM_{10}) from open-pit surface coal mines. ISC2 is specifically named as the method that needs further study.

---

[1] Both authors are on assignment to the National Exposure Research Laboratory, U.S. Environmental Protection Agency.

To address the requirements of the Act, an open pit algorithm has been developed. Wind tunnel experimentation (Thompson, 1993, 1994a and Perry et al., 1994) suggests that the emissions within the pit are not uniformly released from the pit opening. Rather, they show a tendency to be emitted primarily from an upwind sub-area of the pit opening due to the general recirculation flow within the pit. The algorithm uses the pit depth and dimensions, height of emissions above the pit floor, and the wind direction to 1) estimate the fraction of particulate that leaves the pit (escape fraction) and 2) to calculate the sub-area dimensions. The new integrated area-source algorithm is used to model emissions from this sub-area. Initial vertical dispersion due to turbulence in the pit is also considered. Although the development work for the area source and open pit algorithms was done with ISC2, both are incorporated in ISC3. In this paper we will refer to the model incorporating these algorithms as ISC3.

NEW AREA-SOURCE ALGORITHM

Based on the results of an evaluation of area source algorithms performed for EPA (Thistle and Londergan, 1989), the finite line segment algorithm used in the ISCST model gives physically unrealistic results for receptors located near the edges and corners of the area. This is not so much a deficiency in the algorithm as it is a problem in application. The finite line segment was developed as a simple approximation to the contribution at a receptor from an area source. Conceptually, it is an improvement over the virtual point source approach without the expense of numerically approximating the point source function over an area source. It was never the intent of the developers that the area source algorithm would be applied for nearby receptors. However, recent concern for environmental impacts from nearby area sources and advancements in high speed data processing for PC's make it reasonable to revisit the area source problem. This section provides a brief description of the integration technique used in ISC2.

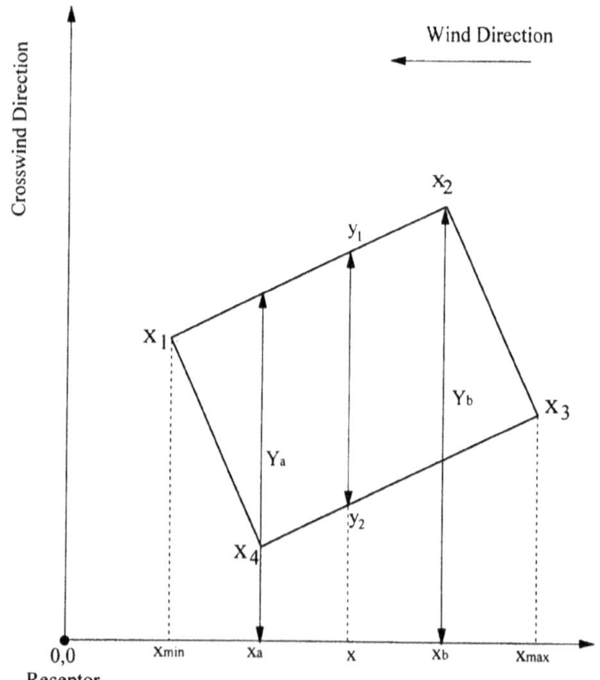

Figure 1. x_{min} and x_{max} are the distances to the closest and furthest vertices from the area source. y_1 and y_2 are the crosswind distances at an arbitrary distance x upwind.

The area source algorithms in PAL and ISC3 are functionally the same; the pollutant impact from an area source is approximated by a number of finite crosswind line sources. However, the integration techniques are different. To illustrate the integration technique we apply it for a rectangular area source. Let x be the upwind distance and y the crosswind distance. For a given source receptor pair, there is a x_{min} and a x_{max} (see Figure 1) which coincides with the x of a vertex. That is, a vertex is represented by (x_i, y_i) with $x_{n+1} = x_1$, for closure. If we let $f(x)$ be the dispersion from a point source at a distance x and let $y(x)$ be the crosswind distance to the ith side, then the integration proceeds as follows: In Figure 1, x_{min}, x_a, x_b, and x_{max} are the upwind distances from the nearest to furthest vertex from a downwind receptor and the four vertices are X_1, X_2, X_3, and X_4. The concentration can be expressed as;

$$\chi = \int_{x_1}^{x_4} f(x)\,e(p_1)\,dx - \int_{x_1}^{x_2} f(x)\,e(p_2)\,dx + \int_{x_4}^{x_3} f(x)\,e(p_1)\,dx - \int_{x_2}^{x_3} f(x)\,e(p_2)\,dx$$

where; $p_1 = \frac{y_1}{\sigma_y}$ $p_2 = \frac{y_2}{\sigma_y}$,

and $e(p_1)$ and $e(p_2)$ are defined as follows,

$$e(p_1) = \int_{-\infty}^{p_1} \frac{1}{\sqrt{2\pi}} \exp\left[-0.5\,p^2\right] dp, \quad and\ e(p_2) = \int_{-\infty}^{p_2} \frac{1}{\sqrt{2\pi}} \exp\left[-0.5\,p^2\right] dp$$

The integrals from $-\infty$ to p_1 and $-\infty$ to p_2 are the cumulative distribution of the standardized normal distribution and can be determined from tables. Each side of the area source can now be integrated independently. Figure 2 provides a simplified graphical representation of the integration by sides. In the actual integration only those portion of the sides that contribute to the concentration at the receptor are included, that is $|y(x)/\sigma_y(x)| \leq 4.0$. This criteria is a function of source-receptor geometry, wind direction, and atmospheric stability. $f(x)$ has the following forms: For stable conditions or unlimited mixing:

$$f(x) = \frac{Q}{\sqrt{2\pi}\,u\,\sigma_y\,\sigma_z}\left(\exp\left[\frac{-0.5(z-H)^2}{\sigma_z^2}\right] + \exp\left[\frac{-0.5(z+H)^2}{\sigma_z^2}\right]\right)$$

In unstable or neutral conditions and if σ_z is greater than 1.6 times the mixing height, L, the distribution below the mixing height is uniform with height.

$$f(x) = \frac{Q}{\sqrt{2\pi}\,u\,\sigma_y\,L}$$

In all other unstable or neutral conditions, that is if σ_z is less than 1.6 time L:

$$f(x) = \frac{Q}{\sqrt{2\pi}\,u\,\sigma_y\sigma_z}\sum_{N=-\infty}^{\infty}\left(\exp\left[\frac{-0.5(z-H+2NL)^2}{\sigma_z^2}\right] + \exp\left[\frac{-0.5(z+H+2NL)^2}{\sigma_z^2}\right]\right)$$

This infinite series converges rapidly, and evaluation with the integer, N, varying from -4 to +4 is usually sufficient.

AREA REPRESENTATION ALGORITHM FOR OPEN PIT MINING

The method for modeling particulate impacts from open-pit, surface coal-mine activities is based on the analysis by Perry et al. (1994) and Thompson (1994a, 1994b) of two wind-tunnel studies recently performed at the EPA Fluid Modeling Facility in which the impacts of emissions

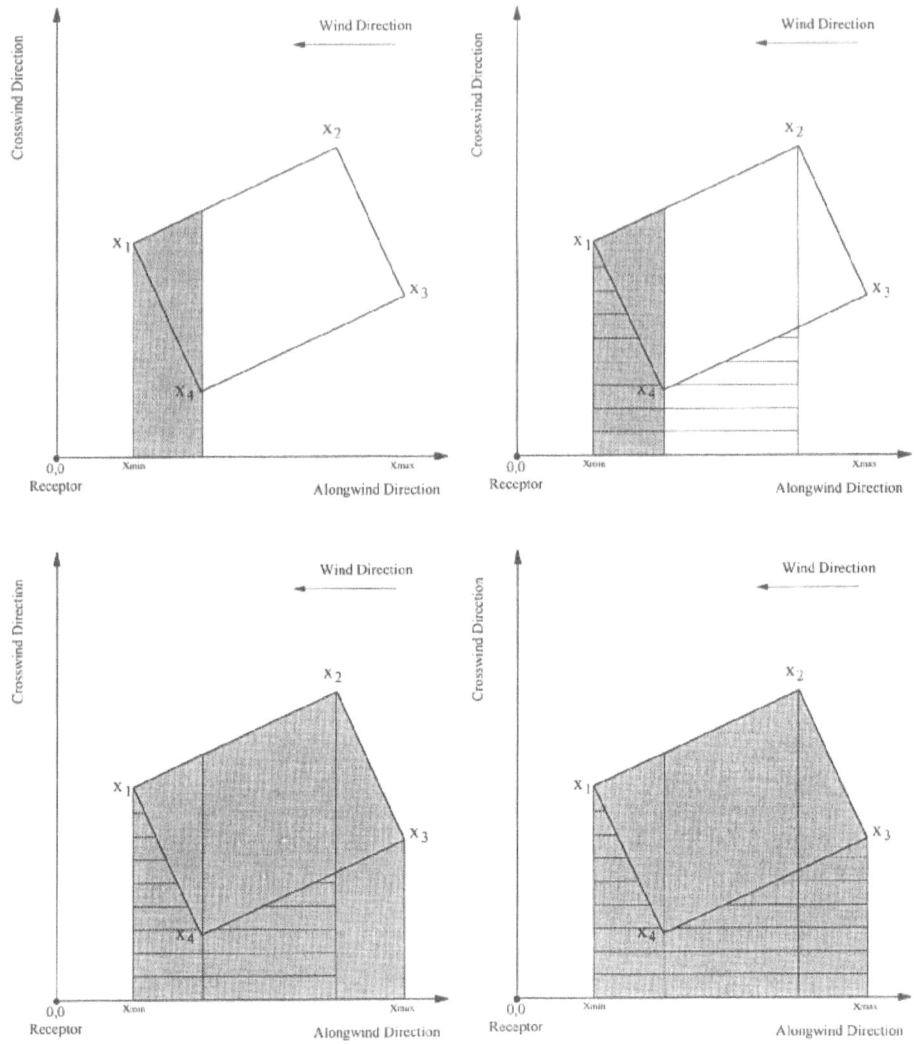

Figure 2 Graphical representation of the integration by sides. The shaded areas represent a positive contribution. The horizontal-line pattern represents a negative contribution. The positive and negative contribution cancel everywhere except the area source.

from a variety of scale model open pits were examined. Perry et al. concluded that, due to recirculations within the pit, particles that escape show a tendency to be emitted toward the upwind portion of the pit opening. The pit-mining algorithms determine the escape fraction (for each particle size group); and establish the location and size of the sub-area from which the escaped particulate is modeled with ISC3. It is important to preface that the open-pit algorithms are designed only for the modeling of impacts outside of the pit caused by emissions within the pit, that mines are assumed generally rectangular in shape, and that the emissions within the pit are assumed to be at least somewhat scattered.

ISC3 first estimates the escape fraction as a function of incident wind speed and the particle size distribution of material released within the pit (Thompson, 1994b). The computation of escape fraction is particularly important for modeling large particles susceptible to gravitational settling and less important for PM_{10} analyses where the vast majority of material escapes. As implemented in ISC3, for each of up to 20 particle size categories, an escape fraction (ε) is computed as:

$$\varepsilon_i = 1.0 \; / \; \left(1 + \frac{v_i}{\alpha \, U_r}\right)$$

where,

- v_i is the settling velocity,
- U_r is the approach wind speed at a height of 10m,
- α is the proportionality constant in the relationship between flux from the pit and the product of U_r and concentration in the pit (Thompson, 1994a).

v_i is already computed in ISC3 (as is done in ISC2) for each size category. Thompson (1994a) used laboratory measurements of pollutant residence times in a variety of pit shapes and determined that a single value of $\alpha = 0.029$ worked well for all pits studied.

The adjusted emission rate (q_i) for each particle size category is then computed as:

$$q_i = \varepsilon_i \bullet f_i \bullet q$$

where q is the total emission rate (for all particles) within the pit, f_i is the original mass fraction for the i-th size category, and ε_i is calculated from the above equation. The adjusted total emission rate, for all particles escaping the pit, is the sum of the q_i. In summary, the amount of particulate that leaves the open pit mine and is available to impact air quality downwind is a function of the total emissions within the pit, wind speed, and the particle size distribution of emitted material.

Secondly, ISC3 computes the location and size of a sub-area of the actual pit opening from which the pit emissions will be modeled. This sub-area source is assumed to be at surface level. Given an arbitrary rectangular-shaped pit with an arbitrary wind direction as shown in Figure 3, ISC3 determines the sub-area source in the following way. Based on the wind direction and pit orientation, the model determines the upwind sides of the pit; in the case of Figure 3, the upwind sides are 1 and 2. The model then computes the along wind length of the pit (l). l varies between the lengths of the two sides of the rectangular pit as follows:

$$l = L \bullet (1 - \theta / 90) + W \bullet (\theta / 90)$$

where θ is the wind direction relative to the long axis of the pit (therefore θ varies between 0° and 90°). l is the scaling factor that is used to normalize the depth of the pit.

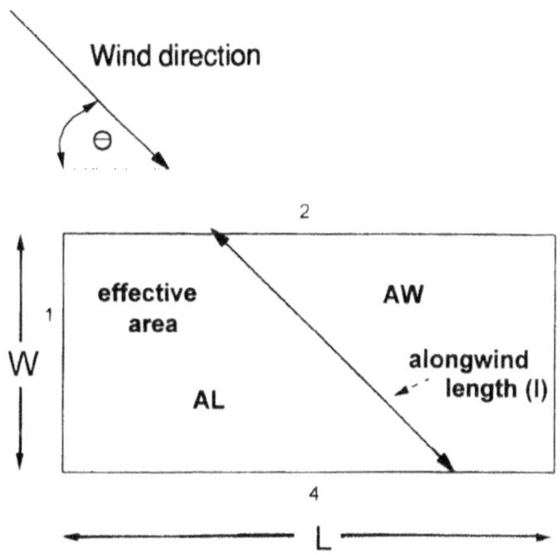

Figure 3. The effective area for a wind direction of about 45 degrees from the long axis of the pit. AL and AW (and thus the area) vary with wind direction.

The model user must specify the average height (H) of emissions from the floor of the pit and the pit volume (V). With these, ISC3 calculates the effective pit depth (d_e) and the relative pit depth (d_r) as:

$$d_e = V / (L \bullet W)$$
$$d_r = (d_e - H) / l$$

Based on observations and measurements in the wind tunnel study, it was clear that the emissions within the pit are not uniformly released from the pit opening. Rather, the emissions showed a tendency (based on smoke releases) to be emitted primarily from an upwind sub-area (or effective area, A_e) of the pit opening and the sub-area size and location depended on the relative pit depth and the wind direction. The wind tunnel data indicate that if $d_r \geq 0.2$, then the effective area is about 8% of the total opening of the mine (i.e. A_e=0.08). If $d_r < 0.2$, then the fractional area increases as:

$$A_e = (1.0 - 1.7\, d_r^{1/3})^{1/2}$$

When $d_r = 0$, the effective area is equal to the total area of the mine opening (i.e. A_e=1.0).

The specific dimensions of this assumed rectangular sub-area are calculated as a function of θ such that (see Figure 3):

$$AW = A_e^{(1 - \cos^2\theta)} \bullet W$$

and,

$$AL = A_e^{(\cos^2\theta)} \bullet L$$

W is defined as the short dimension of the pit and L is the long dimension; AW is the dimension of the effective area aligned with the short side of the pit and AL is the dimension of the effective area aligned with the long side of the pit. In the way of a few examples, if $A_e = 0.36$ at $\theta = 45°$, then AW = 0.6·W and AL = 0.6·L; at the extreme of $\theta = 0°$ (wind along the long axis of the pit), AW = W and AL = A_e·L; at $\theta = 90°$ (wind along the short axis of the pit), AW = A_e·W and AL = L.

The emission rate, q_e, for the effective area is such that;

$$q_e = q_a / A_e$$

where q_a is the emission rate per unit area from the pit if the emissions were uniformly released from the actual pit opening (with an area of L·W).

The model includes some initial vertical dispersion to simulate the high level of turbulence and initial mixing in the mine. Using the pit depth, d_e, as the representative dimension over which the pollutant is vertically mixed, ISC3 calculates an initial vertical dispersion value, σ_{zo}, equal to $d_e/4.3$ (Turner, 1970). $4.3 \cdot \sigma_{zo}$ represents about 90% of a Gaussian plume (in the vertical). For example, when d_e = 45m, σ_{zo} = 10.5m. Thus, the ISC3 model includes the pit-induced initial dispersion with the ambient dispersion as:

$$\sigma_z(x) = \left(\sigma_{zo}^2 + \sigma_z^2(x)\right)^{1/2}$$

For modeling impacts close to the pit (small x), the initial dispersion value can be particularly important.

In summary, ISC3 models the emissions from a rectangular mine as a sub-area source on the upwind edge of the rectangular pit. The size of the sub-area is a function of the relative pit depth (which is in turn a function of the actual pit depth, alongwind pit dimension, height of the emissions above the pit floor, and wind direction). The total emissions from the sub-area are equal to the total emissions (adjusted for escape fraction) from within the pit. Initial vertical dispersion as a function of the actual pit depth is modeled. The required inputs for simulating open pit sources are: the wind speed, wind direction, location and dimensions of the rectangular pit opening, pit volume, total emissions within the pit, height of emissions above the pit floor, and mass fraction and settling velocities of up to 20 particle sizes.

An example comparison of the ISC3 model with wind tunnel data is shown in Figure 4. This is a case that represents emissions that are spread throughout the floor of the pit (as may be expected in an actual mining operation with activities related to drilling, blasting, shoveling, loading, and transporting coal and topsoil throughout the pit). This example involves a pit with full scale dimensions of 225 m in the along flow direction and 450 m in the cross flow direction (the wind is perpendicular to the long dimension of the pit, i.e. θ = 90 degrees). The pit is 45 m deep and the emissions are from the pit floor (H = 0). Therefore the effective pit depth, d_e , is simply 45 m and the relative pit depth, d_r , is equal to 45/225 = 0.2. With this relative depth, the effective area for emissions, A_e , equals 0.08 (8 %), and the dimensions of the area are 18 m in the alongwind direction and 450 m in the crosswind directions. In addition, the pit-induced initial dispersion is equal to 45 m/ 4.3 = 10.5 m.

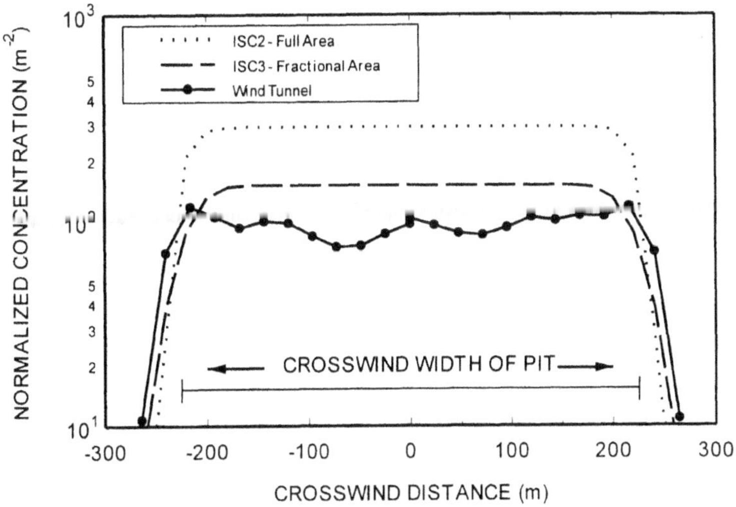

Figure 4. Horizontal concentration distributions at the downwind edge of a 225m X 450m pit with effective depth of 45m and source at the pit floor. ISC2 Full Area is the distribution from the ISC2 model and ISC2 Partial Area is from ISC3.

The concentration distribution at the downwind edge of the pit for the wind tunnel measurements and two modeling scenarios are shown in Figure 4. The results labeled FULL-AREA are those provided by the ISC2 model. The results labeled FRACTIONAL AREA are those provided by ISC3. For ISC2 where it is assumed that the entire opening of the rectangular pit acts as a surface level area source with emissions uniform over that area, the modeled distribution is nearly three times larger than that observed.

In contrast, the ISC3 model is releasing the same total emissions over a much smaller area on the upwind edge of the actual pit opening (only covering 8% of the pit opening); the plume is initially dispersion by pit generated turbulence. ISC3 is showing a slight tendency to overpredicting the distribution particularly near the center of the pit. However, it is clear that the modified area-source approach is a great improvement over that of ISC2 for impacts at the downwind edge of the pit where concentrations are expected to be highest. For large downwind distance (many times the dimensions of the pit), the ISC2 and ISC3 approaches are not expected to give very different results.

ACKNOWLEDGMENTS

The authors would like to acknowledge Richard A. Strelitz who did the original development work on the integration technique used for the new area source algorithm and Roger W. Brode who provided review and refinements to the area source and open pit algorithms and implementation in ISC3.

DISCLAIMER

This paper has been reviewed in accordance with the U.S. Environmental Protection Agency's peer and administrative review policies and approved for presentation and publication. Mention of trade names or commercial products does not constitute endorsement or recommendation for use.

REFERENCES

Bjorklund J.R. and Bowers J.F., 1979, User's instructions for the Shortz and Longz computer programs (Volume I), TR-79-131-01, H.E. Cramer Company, Inc., University of Utah Research Park, POB 8049, Salt Lake City, Utah.

Bowers, J.F., Bjorklund J.R., and Cheney C.S., 1979, User's guide for the Industrial Source Complex (ISC) dispersion models, Volume I -User Instructions, EPA-450/4-79-030, U.S. Environmental Protection Agency, Research Triangle Park, North Carolina.

Burt E.W., 1977, Valley model user's guide, EPA-450-2-77-018, U. S. Environmental Protection Agency, Research Triangle Park, North Carolina.

Hwang S.T., 1986, Methods for estimating onsite ambient air concentrations, ACS Annual Meeting Anaheim, California.

Perry S.G., Thompson R.S., and Petersen W.B., 1994, Considerations for modeling small-particulate impacts from surface coal mining operations based on wind-tunnel simulations, Eight Joint AMS/AWMA Conference on Applications of Air Pollution Meteorology, Jan. 23-28, 1994, Nashville, Tennessee, American Meteorological Society, Boston, Massachusetts.

Petersen W.B. and Rumsey E.D., 1987, User's guide for PAL 2.0 - A Gaussian-plume algorithm for point, area, and line sources, EPA-600-8-87-009, U.S. Environmental Protection Agency, Research Triangle Park, North Carolina.

Thistle H. and Londergan R.J., 1989, Review and evaluation of area source dispersion algorithms for emission sources at superfund sited, EPA-450-4-89-020, U.S. Environmental Protection Agency, Research Triangle Park, North Carolina.

Thompson, R. S., 1993, Wind tunnel simulations of dispersion from surface coal mines. Internal Data Report, Fluid Modeling Facility, U.S. Environmental Protection Agency, Research Triangle Park, North Carolina.

Thompson, R. S., 1994a, Residence time of contaminants released in surface coal mines - a wind-tunnel study. Eighth Joint AMS/AWMA Conference on Applications of Air Pollution Meteorology, Jan. 23-28, 1994, Nashville, Tennessee, American Meteorological Society, Boston, Massachusetts.

Thompson, R. S., 1994b, Personal Communication.

Turner D. B., Workbook of atmospheric dispersion estimates, U.S. Environmental Protection Agency, Research Triangle Park, North Carolina, pp 84.

U.S. Environmental Protection Agency, 1992, User's guide for the Industrial Source Complex (ISC2) dispersion models, Volume I -User Instructions, EPA-450-4-92-008a, U.S. Environmental Protection Agency, Research Triangle Park, North Carolina.

Weber E., 1982, Air Pollution Assessment Methodology and Modeling, Plenum Press, New York.

DISCUSSION

R. MERONEY:

Stable stratification (which may result in worst case pollution both in the coal pit and downwind) will change the circulation in an open pit completely. A wind tunnel simulation of flow and dispersion over a South Australian open pit mine revealed that primary release areas may be at the downwind edge of the area not upwind. See paper by Grainger and Meroney (BLM, 63, 117-140).

S. PERRY and W. PETERSEN:

We thank Dr. Meroney for his question and for his insight into the problem of modeling flow and dispersion within and around open pit coal mines. We absolutely agree with the comment that strongly stable stratification completely changes the circulation within the pit. Also we agree that the worst case concentrations of particulate within the pit likely occur when there is a near stagnant separated stable layer in the pit. These are situations that we did not simulate in the neutral wind tunnel studies that we performed as a basis for our model, and we appreciate that Dr. Meroney pointed out these important situations. In separated, stagnant conditions the residence time of the plume within the pit will certainly increase, thus increasing the in-pit concentrations and the opportunity for deposition. This increased deposition will result in less emissions from the pit than is assumed by our model.

The intent of our model was not to simulate the particulate concentration within the pit, but instead to determine impacts outside and downwind of the pit operation. Dr. Meroney goes on in his comment to say that in contrast to neutral conditions (where we found the primary emission of material from the pit to be near the upwind edge), in stable conditions where the pollutant cloud may be initially trapped in an inversion within the pit, he would expect that there would be a skimming and eventually sweeping flow, which would cause the emissions to come primarily from the downwind edge of the pit.

Based on the results of Grainger and Meroney (1993) in their inverted, heated wind-tunnel experiments, it appears from their visualizations that as the pit Froude number increases above 1.3, the pollutant cloud previously trapped by the inversion layer is washed out of the pit. These results were also supported by Bell and Thompson (1980), who reported "sweeping" flows in numerical and physical modeling of flow in an eastern U.S. valley. Therefore, for these gentle to moderately sloped valley and pit studies, this tendency to skim or sweep down into the valley and flush the pit of pollutant previously trapped by a pit inversion may, in fact, result in significant emissions from the downwind edge of the pit. This is where the contrast with our study and model can be made.

The surface coal mine pits that were examined in the U.S. prior to designing our wind tunnel study could be characterized as having rather steep (in some cases near vertical) walls. This contrasts with the relatively moderate slope of the previously mentioned studies. In fact, Grainger and Meroney cite another study on the evacuation of pollutants from stable layers in mountain valleys by Cunningham and Bedard (1992). In this study the valley walls were very steep. Cunningham and Bedard describe an "erosion" process related to the separation of the flow near the surrounding mountain tops and the effects of valley oscillations. At what they refer to as an intermediate stage in this erosion process, the leeside eddy (leeside of the upwind mountain forming the valley) continues to erode the dense valley fluid, paradoxically moving the dense fluid toward the upstream slope, where it is driven upward until it reaches the mountain top and is driven downwind. Thus, it appears that in the case of steep sloped pits or valleys, that the flow separation in stable conditions may in fact provide a significant release of pollutant near the upwind side of the pit, even in some stable conditions. Certainly, we feel that further study of stable and inversion conditions for steep-walled pits would be desirable. Once again, we thank Dr. Meroney for his comment which helped to clarify the intent of our model.

NEW DEVELOPMENTS IN THE CALPUFF NON-STEADY-STATE MODELING SYSTEM

Joseph S. Scire, David G. Strimaitis, and Mark E. Fernau

Earth Tech, Inc.
196 Baker Avenue
Concord, MA 01742

INTRODUCTION

CALPUFF was developed in the late 1980s and early 1990s as a generalized non-steady-state air quality modeling system for regulatory use (Scire et al., 1990a, 1990b). The modeling system consists of three components: (1) a meteorological modeling package with a boundary layer model and both diagnostic (Douglas and Kessler, 1988) and prognostic (Kessler, 1989) wind field generators; (2) a puff dispersion model with turbulence-based dispersion coefficients, chemical transformation, wet and dry deposition, complex terrain algorithms, building downwash, plume fumigation, and overwater and coastal interaction effects; and (3) postprocessing programs for the output fields of meteorological data, concentrations, and deposition fluxes.

Over the past two years, the CALMET and CALPUFF models have been enhanced as part of work sponsored jointly by government agencies and private industry in the United States and Australia. The improvements to CALMET include modifications to use coarse-scale gridded wind fields derived from the Penn State/NCAR Mesoscale Model (Anthes et al., 1987) with four dimensional data assimilation (MM4-FDDA) as a supplement to observational data in the CALMET fine-scale diagnostic wind model, the ability to compute Lambert conformal map factors, an option to use similarity theory (van Ulden and Holtslag, 1985) to vertically extrapolate surface wind observations, an improved technique for estimating three-dimensional temperature fields over water bodies, a modified mixing height formulation, an improved precipitation interpolation scheme, optional use of satellite cloud data, and an optional PC-based Graphical User Interface (GUI). Improvements to CALPUFF include a new non-Boussinesq numerical plume rise algorithm to predict plume rise from buoyant area sources (such as forest fires), an improved area source dispersion algorithm, new capabilities to treat buoyant line sources (e.g., aluminum reduction facilities) and volume sources, simplified treatment of plume interaction with large-scale terrain features, a subgrid-scale terrain algorithm, enhanced treatment of wind shear effects through puff splitting, a visibility module to calculate light extinction coefficients, and an optional GUI. CALPUFF has been coupled to the Emissions Production Model (EPM), developed by the U.S. Forest Service, through an interface processor. EPM provides time-dependent emission rates and heat release data for use in modeling wildfires and prescribed burns.

Air Pollution Modeling and Its Application XI, Edited by
Gryning and Schiermeier, Plenum Press, New York, 1996

Recently, the Interagency Workgroup on Air Quality Modeling (IWAQM) reviewed various modeling approaches suitable for estimating pollutant concentrations and air quality related values (AQRVs), such as visibility, within the United States' National Park system. IWAQM consists of representatives from the U.S. Environmental Protection Agency (USEPA), U.S. Forest Service, National Park Service, and the U.S. Fish and Wildlife Service. In their Phase II (refined) recommendations, IWAQM proposed the use of CALMET and CALPUFF for evaluating air quality impacts and AQRVs (USEPA, 1995a). In this paper, an overview of the CALMET and CALPUFF models is provided, and several of the key modifications to the models are described.

METEOROLOGICAL MODEL

The CALMET meteorological model consists of a diagnostic wind field module and micrometeorological modules for overwater and overland boundary layers. The diagnostic wind field module uses a two-step approach to the computation of three-dimensional wind fields (Douglas and Kessler, 1988). In the first step, an initial guess field is adjusted for kinematic effects of terrain, slope flows, and terrain blocking effects to produce a Step 1 wind field. The second step consists of an objective analysis procedure to introduce observational data into the Step 1 wind field to produce a final wind field.

CALMET also contains two boundary layer modules for application to overland and overwater grid cells. Over land surfaces, the energy balance method of Holtslag and van Ulden (1983) is used to compute hourly gridded fields of the sensible heat flux, surface friction velocity, Monin-Obukhov length, and the convective velocity scale. Mixing heights are determined using a modified Carson (1973) method based on Maul (1980). Gridded fields of stability class and, as an option, hourly precipitation rates are also determined by the model.

The aerodynamic and thermal properties of water surfaces suggest that a different method is best suited for calculating the boundary layer parameters in the marine environment (Hanna et al., 1985). A profile technique, using air-sea temperature differences, is used in CALMET to compute the boundary layer parameters in the marine boundary layer.

Diagnostic Wind Field Module

In the calculation of the Step 1 wind field, the method of Lui and Yocke (1980) is used to evaluate kinematic effects of terrain. The initial guess winds are used to compute a terrain-forced vertical velocity, subject to an exponential, stability-dependent decay function. The kinematic effects of terrain on the horizontal wind components are evaluated by applying a divergence-minimization scheme to the initial guess wind field. The divergence minimization scheme is applied iteratively until the three-dimensional divergence is less than a threshold value.

Slope flow effects are based on the empirical scheme of Allwine and Whiteman (1985). The slope flow is parameterized in terms of the terrain slope, terrain height, temperature lapse rate, and time of day. The slope flow wind components are added to the wind field adjusted for kinematic effects.

The thermodynamic blocking effects of terrain on the wind flow are parameterized in terms of the local Froude number (Allwine and Whiteman, 1985). If the Froude number at a particular grid point is less than a critical value, and the wind has an uphill component, the wind direction is adjusted to be tangent to the terrain.

The Step 1 wind field resulting from the terrain adjustments described above is then modified by the introduction of observational data through an objective analysis procedure. An inverse-distance squared interpolation scheme is used, which weights the observations heavily in data rich areas while letting the Step 1 wind field dominate the interpolation in data sparse areas. The resulting wind field is subject to smoothing, an optional adjustment of vertical velocities based on the O'Brien (1970) method, and divergence minimization to produce the final (Step 2) wind field.

CALMET has been modified to allow gridded wind fields produced by the MM4-FDDA model to be introduced in one of three ways:

- as a replacement for the initial guess wind field,
- as a replacement for the Step 1 field, or
- as pseudo-observations in the objective analysis procedure.

The weight given to the MM4-FDDA winds is weakest when they are used as the initial guess field, and strongest when used as pseudo-observations. It is expected that for large ratios of MM4-FDDA to CALMET horizontal grid size (i.e., > ~ 5), it may be more appropriate to use the MM4-FDDA winds as the initial guess field. In this way, the CALMET diagnostic terrain adjustments resolved by the fine-scale CALMET grid are incorporated into the wind fields. For moderate grid ratios, the use of MM4-FDDA winds as the Step 1 field appears to be best (USEPA, 1995b). When used in this way, no additional diagnostic terrain adjustments are applied to the MM4-FDDA winds by CALMET. The terrain effects are assumed to be represented by the interpolated MM4-FDDA winds. However, observational data are used to modify the MM4-FDDA-based Step 1 winds. When MM4-FDDA winds are used as pseudo-observations, the resulting wind field is a combination of the interpolated MM4-FDDA winds and the observed winds, with no additional (fine-scale) diagnostic terrain adjustments.

In comparisons of model performance against nested grid MM4-FDDA simulations at grid resolutions of 18-km and 54-km (USEPA, 1995b) and with tracer data collected as part of the Cross-Appalachian Tracer Experiment (CAPTEX) (Irwin et al., 1996), the use of 80-km MM4-FDDA winds resulted in improved characterization of the wind fields and trajectory patterns. As a result of these favorable results, a one-year data base of MM4-FDDA winds on an 80-km grid covering all of the continental United States, southern Canada, and northern Mexico is being made available on CD-ROM by the USEPA for a multitude of modeling applications. CALMET has been designed to be compatible with the CD-ROM MM4-FDDA data base.

Vertical Extrapolation of Surface Wind Observations

CALMET includes an option to vertically extrapolate surface observations of wind speed and wind direction based on the method of van Ulden and Holtslag (1985). The wind speed is extrapolated up to 200 meters above the ground or to the mixing height, whichever is greater. The turning of the wind with height is given by:

$$D(z)/D(h) = d_1 \, [1 - \exp(-d_2 z/h)] \qquad (1)$$

where D(z) is the turning angle at the layer height center z, D(h) is the turning angle at a reference height, h, and d_1, d_2 are empirical constants (1.58 and 1.0, respectively). A table of D(h) as a function of Monin-Obukhov length is provided in van Ulden and Holtslag (1985).

In the implementation of the scheme in CALMET, first the mixing height and Monin-Obukhov length at each surface station are determined. The amount of turning of the wind direction at the reference height of 200 m is determined by interpolating in inverse Monin-Obukhov length (1/L) between the tabulated observed values of D(h). The reference angle is then used in Eqn. (1) to yield the turning angle (D_1) at the CALMET grid point height, z. Eqn. (1) is applied a second time with the same h = 200 m and D(h) with z equal to the anemometer height of the observational station in order to obtain the turning angle (D_2) from the ground to the anemometer height. The wind direction correction at CALMET height z from the anemometer height is then computed as D_1 - D_2. In the Northern Hemisphere, the turning angle is added to the surface wind (i.e., winds veer clockwise), and subtracted in the Southern Hemisphere (i.e., wind back counterclockwise).

The wind speed profile calculations are based on Monin-Obukhov similarity theory for the surface layer:

$$U(z) = U(z_1) \frac{\left[\ln\left(\frac{z}{z_o}\right) - \psi_M\left(\frac{z}{L}\right)\right]}{\left[\ln\left(\frac{z_1}{z_o}\right) - \psi_M\left(\frac{z_1}{L}\right)\right]} \qquad (2)$$

where U(z) is the wind speed at the center of the CALMET layer, $U(z_1)$ is the wind speed at the anemometer height, z_o is the roughness length, z_1 is the anemometer height, and ψ_M is the stability function. The stability function varies with z/L.

The similarity-based extrapolation of surface winds in CALMET was found to significantly improve the ability of photochemical models to reproduce observed ozone patterns in the northeastern United States over wind fields not using the extrapolation procedure (Mayes et al., 1996).

Three-Dimensional Temperature Fields

CALMET produces three-dimensional temperature fields based on twice-daily upper air soundings, hourly surface temperature data and estimates of the local convective mixing heights. An initial 3-D temperature field is first computed from a time- and space-interpolation of the twice-daily soundings. A surface temperature field is then computed from a spatial interpolation of the hourly surface temperature data, and is used to replace the surface temperature field derived from the twice-daily sounding data. At each grid point, the surface temperature is extrapolated vertically up to the current convective mixing height assuming an adiabatic lapse rate. Above the mixing height, the interpolated temperature values from the sounding data are used. Once this initial temperature field is derived, advection effects are evaluated by applying an upwind-looking spatial smoother to the temperature field.

Over sizable water bodies, a separate interpolation scheme can be used to determine the temperature field. CALMET allows overwater data (e.g., from buoys) to be entered separately as distinct data files. The spatial interpolation of overland temperature measurements is applied only over land surfaces. The initial temperature field over water surfaces is derived from the overwater air temperature measurements. User-specified overwater temperature gradients are applied to determine the temperatures above the surface layer. Once the initial overwater temperature field is obtained, the upwind-looking spatial averaging is performed to allow, for example, coastal overland temperatures to be modified by the water body when on-shore wind conditions exist. Likewise, the overwater temperature field will be influenced in the coastal region by overland temperatures during off-shore flow conditions.

DISPERSION MODEL

CALPUFF is a multi-layer, multi-species non-steady-state puff dispersion model. It can simulate the effects of time- and space-varying meteorological conditions on pollutant transport, dispersion, chemical transformation, and removal. CALPUFF can use the three dimensional meteorological fields developed by CALMET, or simple, single station profile data (e.g., as in CTDM) or single layer data (e.g., as in ISC3). CALPUFF is intended to operate on source-receptor distances from tens of meters to hundreds of kilometers. It includes algorithms for near-source effects such as building downwash, transitional plume rise, partial plume penetration, subgrid scale terrain interactions as well as longer range effects such as pollutant removal (wet scavenging and dry deposition), chemical transformation, vertical wind shear effects, overwater transport and coastal interaction effects. It can accommodate arbitrarily-varying point, volume or area source emissions. Buoyant line sources with constant emissions can also be included. Most of the algorithms contain options to treat the physical processes at different levels of detail depending on the model application.

Area Source Plume Rise

One of the modifications made to CALPUFF is the addition of a buoyant area source algorithm. The treatment of plume rise from large buoyant area sources requires special considerations that include effects of vertical wind shear, large initial plume size, and potentially large density differences between the plume and ambient air.

The plume rise equations in the CALPUFF area source algorithm are similar to those in Weil (1988) except that the Boussinesq approximation has not been applied. The Boussinesq approximation simplifies the plume rise equations by assuming that the plume density is close enough to the ambient density that density variations can be neglected, except in the buoyancy term. The plume rise equations are solved numerically using a second order predictor-corrector scheme.

The area source plume rise model in CALPUFF is formulated to calculate the rise of buoyant plumes resulting from forest fires, the burning of leaking oil, and other types of buoyant area sources. The model is designed to be general, and has the following characteristics:

allows arbitrary ambient temperature stratifications;

· allows arbitrary wind stratifications;
· allows any arbitrarily-sized emission sources;
· includes the effects of plume radiative heat loss; and
· allows high temperature releases to be treated with its non-Boussinesq formulation.

All these factors may be important in large forest fire plumes. Due to the complex mountainous terrain in many forested areas and the strong influence it has on meteorological conditions, complex temperature and wind patterns may exist. The ambient temperature stratification is usually more complicated than the linear stratification that is normally assumed in most plume rise models. Wind shear is important because the forest fire plume starts at the ground where there is a zone of large velocity gradients in the vertical. Therefore, it is necessary for arbitrary profiles of winds and temperatures, including potential stability and wind reversals with height, to be accounted for in the plume rise. The initial fire size may be large, of the order of ten or more kilometers in radius. Since the plume temperature near the burning source is much higher than the ambient air temperature, up to 1600 °K, radiative heat loss will reduce the heat flux that is carried by the plume along its trajectory. This reduction of heat flux also reduces the buoyancy flux and thus eventually reduces the final plume rise. Also, since the initial temperature of the plume is high, and the initial density difference between the plume and ambient air is large, the application of the Boussinesq approximation becomes questionable.

The source parameters of a forest fire are usually not constant. The life cycle of a forest fire includes an initial developing stage with large increases in heat generation and pollutant emission rates, followed by a stage of decreasing values. The variation in heat generation and emission rates may be two orders of magnitude over the course of a burn. The resulting time dependency in the plume rise can be calculated by assuming that the plume motion is quasi-steady so that the source condition inputs are time-dependent but the time-derivatives in the governing equations are neglected. This assumption is reasonable because the time scale for plume rise is much shorter than that of the fire life-span. CALPUFF is designed to interface to the Emissions Production Model (EPM), a U.S. Forest Service model that estimates emission rates and heat release rates from various types of fires.

Puff Splitting

Vertical wind shear can be an important factor affecting plume transport and dispersion. The change of wind speed and wind direction with height causes a differential advection of pollutant material emitted at different heights. Even for material emitted at a given height, across-plume shear may transport the upper portion of a plume in a different direction than the lower portion when plumes become large enough. When vertical mixing brings the entire plume to the ground, the effective horizontal dispersion of the plume may be significantly enhanced as a result of the differential transport. CALPUFF explicitly models wind shear effects on different puffs by allowing each puff to be independently advected by its *local* wind speed and direction, and independently mixed vertically to the ground. For example, puffs emitted from two sources co-located in the horizontal, but with different release heights will be transported in CALPUFF in different directions and at different speeds if the wind fields indicate such a shear exists. Shear across a single puff is handled in CALPUFF by allowing the puff to split into two or more pieces when across-puff shear becomes important. Each portion of the puff is then independently transported and dispersed. A single puff may be split multiple times if it remains in the modeling domain long enough.

Alternate Approach to Terrain Adjustments

CALPUFF includes partial-height correction schemes as found in ISC3 to simulate terrain effects on ground-level concentrations, and it also allows a CTDM-like treatment of terrain effects for specific subgrid-scale terrain features (Perry et al., 1989). A third approach, more sophisticated than the partial height correction schemes but less computationally demanding than the CTDM-like approach, was recently implemented. The method has several distinctive attributes:

- Adjustments are formulated as analytic functions of readily-available physical properties of the flow, dispersion, and terrain.
- Adjustments are based on local properties of the terrain, so that heights relative to particular "stack-base" elevations or "base-plume" evaluations are not required.
- Adjustments are related to concepts used in the CTDM model.
- Puffs from different sources that happen to coincide at some point are treated alike as they interact with subsequent terrain features.

The vertical diffusion of a puff is adjusted on the basis of the local strain to the flow imparted by the underlying terrain, using a simplification of the theory on which CTDM is based. The local strain is estimated from the slope of the underlying terrain. Terrain slopes are obtained from a gridded field of terrain elevations, and apply to the segment of the trajectory over which a puff travels during one time-step.

For very stable conditions (Froude number less than 1), we assume that the wind field model will force much of the flow to be parallel to contours of terrain-height, so that puffs will directly encounter relatively minor terrain. For such minor terrain, the strain-induced adjustments described above will also apply. But in addition, we allow the lateral distribution of the puff to remain level, by placing receptors on "poles" whose heights are relative to the elevation of the terrain beneath the center of the puff. This allows the "edges" of a puff to brush against the side-walls of a valley or channel, producing "impingement" concentrations.

GRAPHICAL USER INTERFACES

CALMET, CALPUFF, and its postprocessing program have PC-based Graphical User Interfaces (GUIs) that can be used to prepare model input files, execute the model, or conduct file management functions. The GUIs contain a Help System that makes much of the information in the model user's guides available on-line. The GUIs can simplify the setup of the model input files through their click-and-point menus and internal quality assurance checking features. Although the models can be set up and run entirely within the PC-based user interface system, the GUI always creates an ASCII intermediate input file so the input files and model can be transferred to a high-speed workstation for execution, if desired.

REFERENCES

Allwine, K.J. and C.D. Whiteman, 1985: MELSAR: A mesoscale air quality model for complex terrain: Volume 1 -- Overview, technical description and user's guide. Pacific Northwest Laboratory, Richland, WA.

Anthes, R.A., E.Y. Hsie, and Y.H. Kuo, 1987: Description of the Penn State/NCAR Mesoscale Model Version 4 (MM4). NCAR Tech Note, NCAR/TN-282+STR, Available from the National Center for Atmospheric Research, P.O. Box 3000, Boulder CO 80307, 66 pp.

Carson, D.J., 1973: The development of a dry, inversion-capped, convectively unstable boundary layer. *Quart. J. Roy. Meteor. Soc.*, 99, 450-467.

Douglas, S. and R. Kessler, 1988: User's guide to the diagnostic wind field model (Version 1.0). Systems Applications, Inc., San Rafael, CA 48 pp.

Hanna, S.R., L.L. Schulman, R.J. Paine, J.E. Pleim and M. Baer, 1985: Development and evaluation of the Offshore and Coastal Dispersion model. *J. Air Pollut. Control Assoc.*, 35, 1039-1047.

Holtslag, A.A.M. and A.P. van Ulden, 1983: A simple scheme for daytime estimates of the surface fluxes from routine weather data. *J. Clim. and Appl. Meteor.*, 22, 517-529.

Irwin, J.S., J.S. Scire, and D.G. Strimaitis, 1996: A comparison of CALPUFF modeling results with CAPTEX field data results. In: *Air Pollution Modeling and Its Applications XI*, S.E. Gryning and F. Schiermeier, eds., Plenum Publishing Corp., New York, NY.

Kessler, R.C., 1989: User's guide System Applications, Inc. version of the Colorado State University Mesoscale Model (Version 2.0). Systems Applications, Inc., San Rafael, CA 75 pp.

Lui, M.K. and M.A. Yocke, 1980: Siting of wind turbine generators in complex terrain. *J. Energy*, 4, 10-16.

Maul, P.R., 1980: Atmospheric transport of sulfur compound pollutants. Central Electricity Generating Bureau MID/SSD/80/0026/R. Nottingham, England.

Mayes, P., L.L. Schulman, M.E. Fernau, and J.S. Scire, 1996: Application of the CALMET meteorological model to UAM modeling in the Northeast. Ninth Joint Conference on Applications of Air Pollution Meteorology, Atlanta, GE. American Meteorological Society, Boston, MA.

O'Brien, J.J., 1970: A note on the vertical structure of the eddy exchange coefficient in the planetary boundary layer. *J. Atmos. Sci.*, 27, 1213-1215.

Perry, S.G., D.J. Burns, L.H. Adams, R.J. Paine, M.G. Dennis, M.T. Mills, D.G. Strimaitis, R.J. Yamartino and E.M. Insley, 1989: User's Guide to the Complex Terrain Dispersion Model Plus Algorithms for Unstable Situations (CTDMPLUS) Volume 1: Model Description and User Instructions. EPA/600/8-89/041, U.S. Environmental Protection Agency, Research Triangle Park, NC.

Scire, J.S., D.G. Strimaitis, and R.J. Yamartino, 1990a: Model formulation and user's guide for the CALPUFF dispersion model. Sigma Research Corporation, Concord, MA.

Scire, J.S., E.M. Insley, and R.J. Yamartino, 1990b: Model formulation and user's guide for the CALMET meteorological model. Sigma Research Corporation, Concord, MA.

U.S. Environmental Protection Agency, 1995a: Sixth Conference on Air Quality Modeling. August 9-10, 1995, Washington, D.C.

U.S. Environmental Protection Agency, 1995b: Testing of meteorological and dispersion models for use in regional air quality modeling. U.S. Environmental Protection Agency, Research Triangle Park, NC.

van Ulden, A.P. and A.A.M. Holtslag, 1985: Estimation of atmospheric boundary layer parameters for diffusion applications. *J. Clim. and Appl. Meteor.*, 24, 1196-1207.

Weil, J.C., 1988: Plume Rise. In: *Lectures on Air Pollution Modeling*. A. Venkatram and J. Wyngaard Eds. 119-166.

DISCUSSION

P. PAI: How does CALMET reconcile between estimates of cloud parameters from a prognostic meteorological model and cloud observations from satellite?

J. SCIRE: CALMET does not use the cloud estimates from the prognostic model (MM4-FDDA). At this time, the MM4-FDDA cloud predictions are not considered as reliable as the satellite data, so CALMET gives preference to the use of the satellite data in developing the cloud field.

LOCAL-SIMILARITY DESCRIPTION OF TRAJECTORIES OF PLUMES AND JETS IN NEUTRALLY STRATIFIED TURBULENT SHEAR FLOW

Eugene Genikhovich[1]

Atmospheric Sciences Modeling Division
Air Resources Laboratory
National Oceanic and Atmospheric Administration
Research Triangle Park, NC 27711

INTRODUCTION

For neutral and unstably stratified atmospheres, the "active" part of trajectories of bent-over plumes (buoyancy-dominated rise) and jets (momentum-dominated rise) in crosswind flow is usually described by the formulae (Briggs,1984; Arya and Lape, 1990):

$$\Delta z(x) = 1.6\,(F_b/U_H^3)^{1/3}\cdot x^{2/3} \tag{1}$$

and

$$\Delta z(x) = 2\,(F_m/U_H^2)^{1/3}\cdot x^{1/3} \tag{2}$$

in which $\Delta z(x) = z(x) - H$ is the height of the plume or jet above the stack, x is the distance from the point source (along-wind direction), U_H is the wind speed at the stack height H, F_m and F_b are "fluxes" of momentum and buoyancy which depend on the stack diameter D, effluent velocity W, effluent temperature T_s, ambient air temperature T_a, and gravitational acceleration g. Fluxes F_m and F_b are determined from $F_m=0.25W^2D^2T_a/T_s$ and $F_b=0.25gWD^2(T_s-T_a)/T_s$. Equations (1) and (2) may be rewritten as $\Delta z(x)/l_b = 1.6(x/l_b)^{2/3}$ ("two-thirds law") and $\Delta z(x)/l_m = 2.0(x/l_m)^{1/3}$ ("one-third law") using length scales for buoyancy $l_b = F_b/U_H^3$ and for momentum $l_m = (F_m)^{1/2}/U_H$.

There is a general agreement that Eqs. (1) and (2) work reasonably well near the source, but some theoretical and experimental works (Murthy, 1970; Schulman and Scire, 1980; and

[1]Visiting Scientist from Voeikov Main Geophysical Observatory, St. Petersburg, Russia; Senior Research Associate of the US National Academy of Sciences

others) questioned their applicability farther from the source, because Eqs. (1) and (2) do not reflect explicitly the wind shear influence. In particular, Genikhovich and Snyder (1995) have shown that plume trajectories in the meteorological wind tunnel, observed in the EPR tests (Snyder, 1993), can be better described for $390 < x/l_b < 15300$ by the expression

$$\Delta z/l_b = 4.0\ (x/l_b)^{0.5} \tag{3}$$

which is also in agreement with published results of the field measurements. This paper is to present a local-similarity description of plume and jet trajectories in shear flows.

THEORETICAL DESCRIPTION

Let's show first that descriptions of trajectories plumes and jets in the simplest case of the neutrally stratified uniform flow, which is characterized by a constant wind speed U_0, can be based (at least, partially) on dimensional analysis. In the case of the plume, for example, Δz may depend only on x, U_0, and F_b. Both variables, Δz and x, have the same dimension of length ("m"), and U_s and F_b have dimensions ms^{-1} and m^4s^{-3}, respectively. Standard dimension analysis shows that the only combination of parameters U_0 and F_b , which has the dimension of length, is $l_b=F_b/U_0^3$. Thus, the ratio of Δz to l_b is a function of the only one argument x/l_b: $\Delta z/l_b=f_1(x/l_b)$ where f_1 is an universal function for plumes. Similarly, one can show in the case of the jet that $\Delta z/l_m=f_2(x/l_m)$ where $l_m = (F_m)^{1/2}/U_0$ and f_2 is an universal function for jets (here, the dimension of F_m is m^2s^{-2}. We can simplify more, if we note that Δz should depend independently not on the three aforementioned variables but rather on two of their combinations: x/U_0 (which is the travelling time) and F_b/U_0 (which is the buoyancy "flux from the unit of length"; the similar combination Q/U, where Q is the emission rate, has appeared in atmospheric diffusion formulae). The only combination of x/U_0 and F_b/U_0 with the same dimension as Δz is $(F_bx^2/U_0^3)^{1/3}$ which corresponds to Eq. (1). Similarly, the only combination of x/U_0 and F_m/U_0 with the same as Δz dimension is $(F_mx/U_0^2)^{1/3}$ which corresponds to Eq. (2). In other words, functions f_i (with i = 1 or 2) as functions of the non-dimensional distance to the source s_i, can be approximated by power laws :

$$f_i = A_i s_i^{\alpha_i} \tag{4}$$

where $s_1 = x/l_b$, $s_2 = x/l_m$, $A_1 = 1.6$, $A_2 = 2.0$, $\alpha_1=2/3$, and $\alpha_2= 1/3$.

Constants A_1 and A_2 should, in fact, depend on characteristics of the turbulence (*e.g.*, intensity of turbulence at the stack top, ι, or its dissipation rate, ε;). Briggs (1984) supposed that plume rise and jet rise in the low-wind turbulent flow are proportional to $\varepsilon^{-2/5}$ and $\varepsilon^{-1/7}$, respectively; ε, in turn, is proportional to ι^3, where ι is the turbulent intensity (for example, its horizontal component). Thus, A_1 and A_2 in this model are proportional to $\iota^{-6/5}$ and $\iota^{-3/7}$, respectively. Unfortunately, these results were not proved experimentally for plumes and jets in the cross-wind flow. Note that this implies that plume rise is more sensitive to the outer turbulence than jet rise, although there is no direct physical evidence that this is correct. Here, we will use a dimensional approach to describe the effects of turbulence supposing, for the sake of simplicity, that, in the presence of turbulence, parameters $F_b/(U_0+d\sigma_u)$ and $F_m/(U_0+d\sigma_u)$ should be used instead of F_b/U_0 and F_m/U_0 (here σ_u is the standard deviation of the horizontal component of the wind speed). In this case,

$$A_1 = A_1^0/(1+d\iota)^{1/3}; \quad A_2 = A_2^0/(1+d\iota)^{1/3} \tag{5}$$

where $A_1^0 = 1.6$ and $A_2^0 = 2.0$, respectively; the value of the constant d will be specified later.

In the horizontally-uniform atmosphere, however, wind speed U(z) is a function of the height. We will use in this case the concept of the local similarity, that is, that one can describe nonuniform fields with the same similarity laws as uniform ones using scaling with local values of governing parameters rather than with their global (constant) values. As applied to our problem, length scales l_{bz} and l_{mz} for plumes and jets should be introduced as functions of z as follows:

$$l_{bz} = F_b/U^3(z)\,; \qquad l_{mz} = \sqrt{F_m}/U(z) \tag{6}$$

For the sake of simplicity, a power-law wind profile with an exponent p will be used in the following transformations; the representation $U(z) = U_H\,(z/H)^p$ seems to be more convenient here than the more common formula $U(z) = U_1\,(z/z_1)^p$ where U_1 is a value of the wind speed at given height z_1 (certainly U_H can be expressed via U_1, z_1, and H). Using this wind profile, length scales can be rewritten as

$$l_{bz} = l_{bH}(H/z)^{3p}; \qquad l_{mz} = l_{mH}(H/z)^{p} \tag{7}$$

where l_{bH} and l_{mH} are calculated from Eq. (6) at z = H. Thus, a general expression for the length scale l_z (it may be either l_{bz} or l_{mz}) can be written as

$$l_{zi} = l_{Hi}(H/z)^{\gamma_i} \tag{8}$$

where $l_{H1} = l_{bH}$; $l_{H2} = l_{mH}$; $\gamma_1 = 3p$; $\gamma_2 = p$. Using variable scales, one can keep the same universal functions (4) as in the uniform flow; it yields the equation for the plume/jet centerline in a form $(z-H)/l_{zi} = f_i(x/l_{zi})$. An inverse function x(z) to the plume/jet centerline height z(x) can be derived from this equation as follows

$$x' = \left[\frac{(z')^{\gamma_i(1-\alpha_i)}\,(z'-H')}{A_i\,(H')^{\gamma_i(1-\alpha_i)}}\right]^{\frac{1}{\alpha_i}} \tag{9}$$

where $z' = z/l_{Hi}$; $x' = x/l_{Hi}$; and $H' = H/l_{Hi}$ are dimensionless height of the centerline, distance to the source and height of the source, respectively.

In general case, the sought function z(x) can be obtained from Eq. (9) numerically; this equation can be used also as a generator of analytical asymptotics of z(x). Replacing z'-H' with $\Delta z' = z'-H'$ and z' with H, in the right side, one can see, in particular, that, if x' tends to zero (simultaneously, z' tends to H'), then

$$\Delta z' \sim A_i\,(x')^{\alpha_i} \tag{10}$$

Comparison of Eqs. (10) and (4) shows that shear effects do not play any significant role near the source and that plume/jet rise near the source does not depend significantly on the stack height. Another asymptotic of z', which correspond to $x' \to \infty$ (in this case, $z' \to \infty$ too), follows from Eq. (9) after replacement z'-H' with z' in its right side:

$$z' \sim \left[A_i\,(H')^{\gamma_i(1-\alpha_i)}\,(x')^{\alpha_i}\right]^{\overline{1+\gamma_i(1-\alpha_i)}} \tag{11}$$

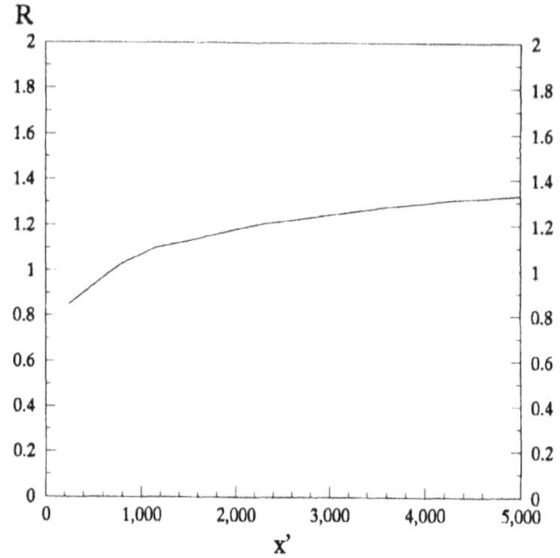

Figure 1. Ratio of the plume rises calculated with Eqs. (3) and (9).

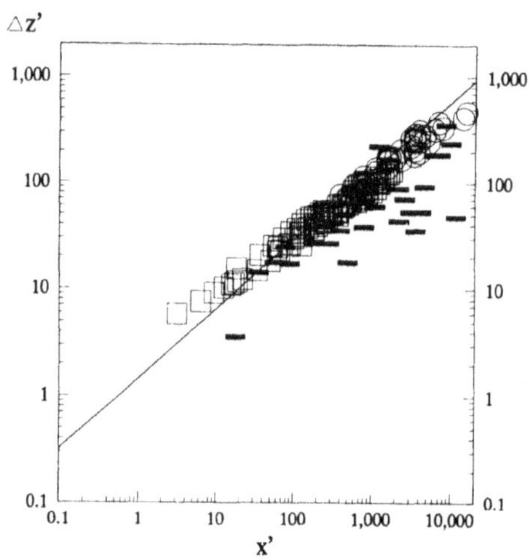

Figure 2. High-turbulence laboratory measurements in comparison with Eq. (9).

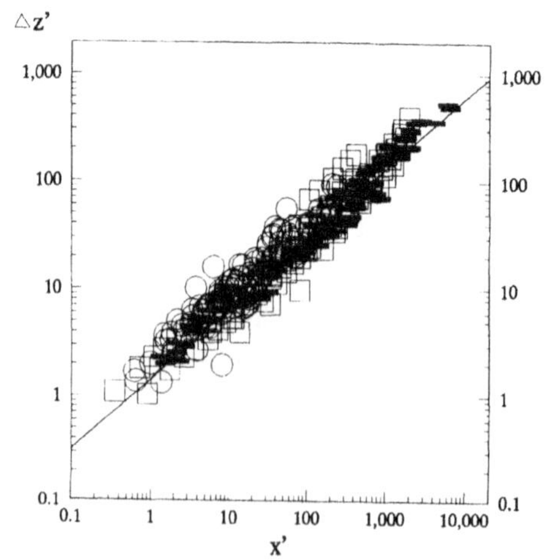

Figure 3. Field measurements in comparison with Eq. (9).

Eq. (11) shows that, asymptotically, plumes and jets rise in shear flow more slowly than in uniform flow because the exponent $\alpha_i/[1+\gamma_i(1-\alpha_i)]$ is less than α_i, if $\alpha_i < 1$ and $\gamma_i > 0$. This equation can be rewritten for plumes and jets separately:

$$\Delta z/l_{bH} \sim [A_1 (H/l_{bH})^p]^{\frac{1}{1+p}} (x/l_{bH})^{\frac{2}{3(1+p)}}$$

$$\Delta z/l_{mH} \sim [A_2^3 (H/l_{mH})^{2p}]^{\frac{1}{3+2p}} (x/l_{mH})^{\frac{1}{3+2p}} \qquad (12)$$

Eq. (12), which is applicable only at large distances from the source where z >> H, shows that, asymptotically, plume/jet rise depends on the stack height, but this dependence is comparatively weak: an increase of H in an order of magnitude corresponds to only a 35% increase in the plume rise and 23% increase in the jet rise (the value p=0.15 has been used in these estimates).

VALIDATION AND DISCUSSION

Numerical solution of Eq. (9) can be compared with empirical Eq. (3). Typical full-scale values of the parameters in the EPR tests ($H_s \propto 50$ m; p = 0.16; $l_b \propto 0.2$ - 0.3 m; $l_m \propto 5$ m; turbulent intensity at the stack top ι = 0.1) and constant d = 5 are used in these calculations. The ratio R of the plume rises, calculated with Eq. (9) and Eq. (3) at different dimensionless distances x', is shown on Fig. 1. One can see that differences between the predictions of these two formulae are within the 30% error, and this agreement seems to be reasonable. Note that such differences should not be visible, if one would plot in the logarithmic scale measured plume trajectories which have rather large scatter.

Predictions of Eq. (9) (solid line) are compared on Fig.2 with the results of the high-turbulence laboratory measurements carried by Murthy (1970), Petersen and Cermak (1979), and Snyder (1993) (squares, bars, and circles on the figure, respectively). Similar comparison with results of field measurements carried out by Balantine et al. (1983) and Spillane (1983) as well as with field measurement data published by Weil (1983) is given on Fig. 3 (circles, squares, and bars, respectively). These figures could be compared with Figs. 5 (b) and 6 from the paper by Genikhovich and Snyder (1995) which illustrated performance of Eq. (3). One can see that Eq. (9) performs reasonably well in comparison with all these data sets. It should be noted that both experimental and theoretical results do not indicate existence of the final plume rise at dimensionless distances up to 15000 - 20000 from the source.

Evidently, a similar approach can be used in order to introduce wind shear effects in other models which describe plume and jet trajectories in uniform flows, in particular, in stratified ones. In the present version, however, there are certain limitations. First of all, it uses a not very satisfactory description of the influence of turbulence on plume/jet rise (even if this influence does not seem to be strong). Some additional laboratory experiments dealing with plume and jet rise in the uniform turbulent cross-wind flow could help to improve this description. Note also that Eq. (9), as well as Eqs. (1) and (2), may not be used for ground sources. Apparently, the friction velocity u_* should be introduced in the model in this case.

CONCLUSION

Results obtained show that local-similarity technique can be applied effectively to the description of plume and jet trajectories in shear flow. Practical applications of such a description are numerous and include, in particular, modeling of atmospheric diffusion in the absence and (after some additional correction) in the presence of buildings. Preliminary results have shown that corresponding models can be used for predicting of combined effects of atmospheric diffusion and plume rise.

Acknowledgments

The author wishes to acknowledge Roger Thompson for his help in improving the English version of this paper.

REFERENCES

Arya, S.P.S., and Lape, J.F., 1990, A comparatively study of the different criteria for the physical modeling of buoyant plume rise in a neutral atmosphere. *Atmos. Environ.*, 24A: 289

Balantine, H.W., Williamson, H.J., Zubrick, S.M., and Coerr, S.P., 1983, *Plume Rise From Gas Turbines and Reciprocating Internal Combustion Engines in Gas Transmission*, Radian Corporation

Briggs, G.A. 1984 Plume rise and buoyancy effects, in: *Atmospheric Science and Power Production*, D. Randerson, ed., US Department of Energy/ TIC-27601

Genikhovich, E.L., and W.H. Snyder, 1995, Trajectories of plumes and jets in turbulent shear flow under near-neutral conditions, in *Ninth Symposium on Meteorological Observations and Instrumentation, March 27 - 31, Charlotte, NC.* AMS

Murthy C.R., 1970, On the mean path of a buoyant chimney plume in non-uniform wind, *Journ. Appl. Meteorol.*, 9: 603

Petersen, R.L., and Cermak, J.E., 1979, Plume rise and dispersion for varying ambient turbulence, thermal stratification and stack exit conditions - a numerical and laboratory evaluation", in *Wind Engineering*, J.E. Cermak, ed., Pergamon Press, Oxford.

Spillane, K.T., 1983, Observations of plume trajectories in the initial momentum influenced phase and parametrization of entrainment, *Atmos. Environ.*, 17:1207

Schulman, L.S., and Scire J.S., 1980, *Buoyant Line and Point Source (BLP) Dispersion Model User's Guide*, P-7304B, Environment Research and Technology, Inc., Concord, MA

Snyder, W.H., 1993, Downwash of plumes in the vicinity of buildings: a wind-tunnel study, in: *Recent Research Advances in the Fluid Mechanics of Turbulent Jets and Plumes*, P.A.Davies and M.J.Valente Neves,ed., NATO ASI Series E 255, Kluwer Academic Publishers, Dordrecht

Weil, J.C., 1988, Plume rise, in: *Lectures on Air Pollution Modeling*, A. Venkatram and J.C. Wyngaard, ed.,, Am. Met. Soc., Boston

DISCUSSION

G. SCHAYES: Your theory has been developed for neutrally stratified shear flow. Most applications will require non-neutral stratifications. Is your theory extendable or usable also for non-neutral flows?

E. GENIKHOVICH: Stratification effects have not been included in this model yet. It is widely believed, however, that trajectories of plumes and jets in unstably stratified flows should be close to those in neutral flows (because enhanced turbulence should compensate, at least partially, the dynamic effects).

D. ANFOSSI: You presented some comparisons between observed plume trajectories and ground-level concentrations (GLC's) computed by your new plume rise formula. What are the relative differences (percentage of improved agreement) between the results obtained by using your new formula and those obtained with the classic ("2/3" and "1/3") ones?

E. GENIKHOVICH: It is evident that wind-shear effects should reduce plume rise. For buoyant plumes, for example, the ratio R of plume rises predicted with the new formula and with the "2/3 law" is equal to 0.83 and 0.62 at the dimensionless distances 700 and 4000 from the source respectively. As a result, maximum GLC's are increased because they are, roughly speaking, inverse proportional to the square of the effective stack height.

BREAKDOWNS OF THE NIGHTTIME PLANETARY BOUNDARY LAYER OVER COMPLEX TERRAIN

Carmen J. Nappo and Richard M. Eckman

Atmospheric Turbulence and Diffusion Division
NOAA/ARL
456 South Illinois Avenue
Oak Ridge, Tennessee 37831

INTRODUCTION

Turbulence and diffusion in the stable planetary boundary layer (SBL) are often intermittent and sporadic, and this strongly affects air quality modeling. During intermittent turbulent breakdowns, pollutants in the upper levels of the SBL can be brought to the ground with relatively little dilution. Pollutants released near the surface can, in contrast, be rapidly mixed by a breakdown event, causing an unexpected dilution. These breakdowns can also have significant effects on the surface fluxes of momentum, heat, and water vapor. An understanding of these breakdowns is a necessary step in the development of improved parameterizations of turbulence and diffusion in the SBL.

Several mechanisms are possible for generating SBL breakdowns, including shear instability, the interactions of gravity waves and turbulence, and streamline deformation caused by surface obstacles and small scale terrain features. Revelle (1993) and McNider *et al.* (1995) performed numerical investigations of intermittence in the SBL, and Nappo and Andren (1995) demonstrated that breaking gravity waves can cause significant stress in the upper regions of the SBL. However, it is not yet clear how the results of these theoretical investigations can be used in the context of air quality models.

In this paper, we analyze breakdowns of the SBL over complex terrain using data from a tower network near Oak Ridge, Tennessee. A technique based on the band-pass filtering of wind speed and temperature fluctuations is used to detect breakdown events. The analysis is supplemented with a technique based on wavelet transforms. It will be demonstrated that both techniques provide complementary information on the breakdowns.

FIELD DATA

Data for this study came from the U.S. Department of Energy's Oak Ridge Site Survey (Eckman *et al.*, 1992) and from a 100 m tall tower operated by Martin Marietta

Energy Systems, Inc. Figure 1 shows the terrain and the locations of the meteorological towers used in the analysis. This part of the eastern Tennessee River Valley consists mostly of parallel ridges aligned along the valley axis. These ridges are typically 50-100 m high and are mostly forested. Towers 4, 6, 8, 15, 20, and 22 in Fig. 1 were located in valley bottoms, and towers 9, 12, 13, and 19 were located on ridge tops. The instruments on the valley-bottom towers were 10 m above the local ground, whereas the ridge-top instruments were either at 21 m (towers 9, 12, and 13) or 45 m (tower 19). All the towers collected one-minute averages of wind speed and direction, temperature, relative humidity, and rainfall. The breakdown analysis included over 150 nights of data between 12 February and 28 July 1990, with the analysis period starting at 2200 EST and ending at 0500 EST the following morning.

On the 100 m Martin Marietta tower, hourly-averaged values of wind speed and direction as well as temperature were collected at 10, 30, and 100 m, and temperature differences were measured between 10 m and 100 m. Data from this tower were used to calculate hourly-averaged gradient Richardson numbers between 10 m and 100 m.

DATA ANALYSIS

Band-Pass Filtering

A simple conceptual model of the breakdown process is used in this analysis. We consider a stably stratified boundary layer with a positive wind-speed shear. During a breakdown or overturning of the SBL, low-speed cool air is mixed upward from near the ground and high-speed warm air is mixed downward from aloft. We quantify this model

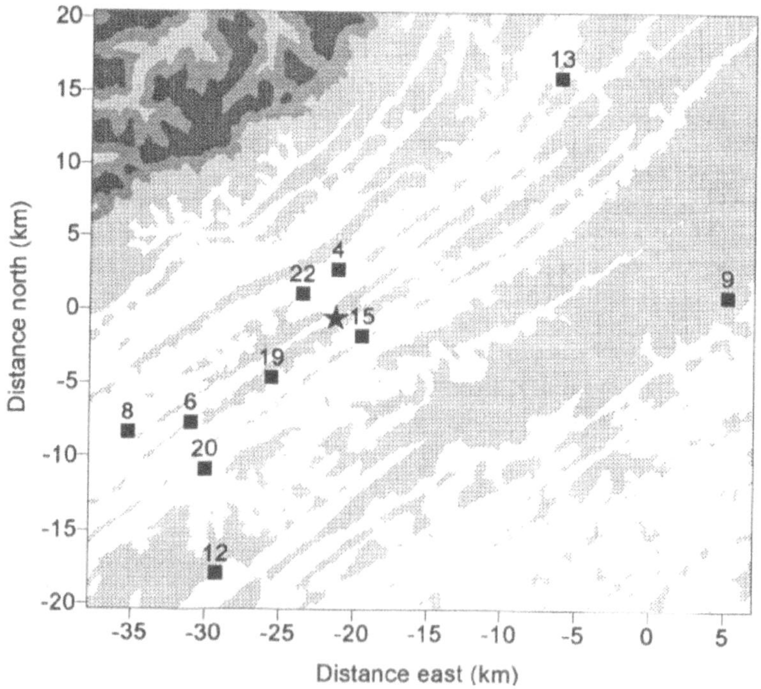

Figure 1. Map of Oak Ridge area showing the locations of the site-survey towers (squares) and the 100 m Martin Marietta tower (star). The shading represents terrain elevations above 300 m (lightest), 500 m, and 700 m MSL (darkest). The coordinate (0,0) is at 36° N latitude and 84° W longitude.

with a covariation time series defined by (Nappo, 1991)

$$C_{UT}(t) = \left(U(t) - \bar{U}(t)\right)\left(T(t) - \bar{T}(t)\right) , \tag{1}$$

where t is time, $U(t)$ and $T(t)$ are tower measurements of wind speed and temperature respectively, and the overbar represents a 19-minute running average. The covariation is in effect the product of horizontal wind-speed and temperature perturbations detected at the measurement site. The simple overturning described above tends to produce positive values of C_{UT}, since the speed and temperature perturbations are positively correlated. A breakdown event is considered to occur during periods when C_{UT} exceeds a limiting value of $0.005\ \mathrm{K\,m\,s^{-1}}$, which approximates the observed nighttime average value of C_{UT} in the Oak Ridge tower data. If we take C_{UT} to represent in some sense a vertical kinematic heat flux, then $0.005\ \mathrm{K\,m\,s^{-1}}$ corresponds to a downward heat flux of about $-6\ \mathrm{W\,m^{-2}}$, which is a typical value for the nighttime average heat flux.

Wavelet Transform

To supplement the covariation approach, we also use wavelet transforms to characterize breakdown events in the Oak Ridge data set. Wavelet transforms are more suitable than Fourier transforms because of the intermittent and nonstationary nature of turbulence in the SBL. The continuous wavelet transform of a signal $s(t)$ is defined as (Meneveau, 1991; Farge, 1992)

$$\tilde{s}(t,a) = a^{-1/2} \int_{-\infty}^{\infty} s(t')\, g^{*}\!\left(\frac{t'-t}{a}\right) dt' . \tag{2}$$

The variable t is time, a is the dilation time scale, and $g(t/a)$ is a basis function for the transform. (The star denotes a complex conjugate.) For the Oak Ridge analysis, we use a version of the complex-valued Morlet wavelet:

$$g(x) = \exp(2\pi i x)\exp(-x^2/2) . \tag{3}$$

From the basic definition of the wavelet transform it is useful to define the wavelet energy density $W_{ss}(t,a)$ as

$$W_{ss}(t,a) = \frac{1}{c_g}\,\frac{\tilde{s}(t,a)\,\tilde{s}^{*}(t,a)}{a^2} , \tag{4}$$

where c_g is a constant that is determined by the basis function g. The total energy of the signal is recovered by integrating W_{ss} over all times and dilation scales:

$$\int_{-\infty}^{\infty} dt \int_{0}^{\infty} W_{ss}(t,a)\, da = \int_{-\infty}^{\infty} s^2(t)\, dt . \tag{5}$$

We want to use the wavelet energy density to detect breakdown events. During these events, relatively rapid fluctuations in both the wind speed $U(t)$ and temperature $T(t)$ are expected. Hence, both the speed energy density $W_{UU}(t,a)$ and the temperature energy density $W_{TT}(t,a)$ should have peaks at small dilation scales during breakdowns.

For detecting breakdown events, we will integrate W_{UU} over scales between 2 and 8 min:

$$E_{UU}(t) = \int_{2\,\mathrm{min}}^{8\,\mathrm{min}} W_{UU}(t,a)\,da \ . \tag{6}$$

The lower limit of 2 min is the minimum scale that can be detected with one-minute-average data. The upper limit of 8 min was chosen because larger scale fluctuations are unlikely to represent turbulence. A breakdown is assumed to occur during periods when E_{UU} exceeds a value of 0.01 $m^2 s^{-2}$. This threshold represents about 10% of the typical values of E_{UU} that were observed during the afternoon at the Oak Ridge towers.

RESULTS

Figure 2 shows the fractions of the nighttime occupied by breakdowns, computed using the covariation technique at each ridge-top and valley-bottom tower, plotted against the median nighttime Richardson number Ri from the 100 m tower. For both ridge and valley towers, considerable scatter exists for Ri less than about 0.4. For the valley towers, the scatter reduces with increasing Ri, and a somewhat constant value of about 0.06 is reached for $Ri > 1$. Thus, during strongly stable conditions breakdowns occur about 6% of the nighttime period at valley bottoms. The ridge-top towers show considerable scatter. For $Ri > 1$, the percentage of nighttime breakdowns seems to increase, reaching a value of

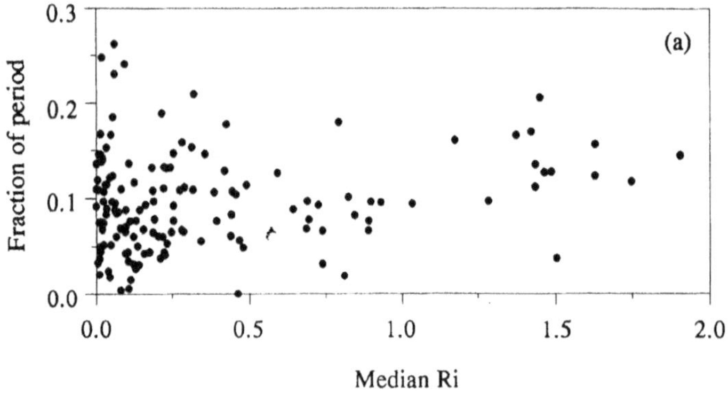

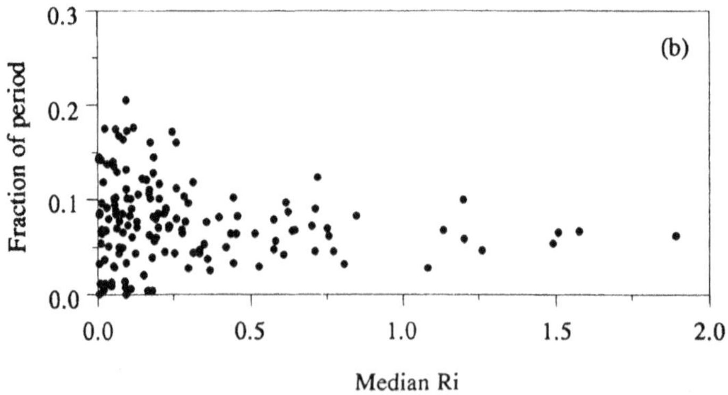

Figure 2. Fraction of nighttime period occupied by breakdowns, computed with the covariation technique, versus median Richardson number. (a) is a composite of the 4 ridge-top towers, and (b) a composite of the 6 valley-bottom towers.

about 0.14. Thus, breakdowns occur at the ridge tops about 14% of the nighttime period in strongly stable conditions.

The relative probabilities of breakdown duration and the number of breakdowns per night are plotted in Figs. 3 and 4 respectively. These histograms were compiled by ensemble averaging the breakdown data for the 6 valley and 4 ridge towers. The distributions of breakdown durations are essentially identical for ridge top and valley bottom towers, and appear lognormal. The mode is 4 min, which suggests that the majority of breakdowns are relatively short lived; however, the strongest breakdowns with large values of C_{UT} have durations between 4 and 10 min. From Fig. 4, we see a difference between the ridges and valleys in the number of breakdowns per night. Statistically, a higher frequency of breakdowns is observed at ridge tops than in valley bottoms. The average number of breakdowns per night is 12 for ridge tops and 9 for valley bottoms.

The wavelet analysis was applied only to the ridge-top tower 19 and the valley-bottom tower 15. Figure 5a shows the fraction of the analysis period during which E_{UU} exceeded its threshold value at tower 19, plotted as a function of the median nighttime Richardson number obtained from the 100 m tower. For $Ri \leq 0.1$, the fraction is nearly unity, indicating that speed fluctuations in the 2-8 min range were almost always present. Beyond $Ri \approx 1.0$, the fraction is close to zero. The figure thus suggests that at tower 19 intermittent breakdown events primarily occur when Ri is between about 0.1 and 1.0. Within this range, the fraction of the analysis period occupied by breakdowns decreases roughly logarithmically with increasing Ri.

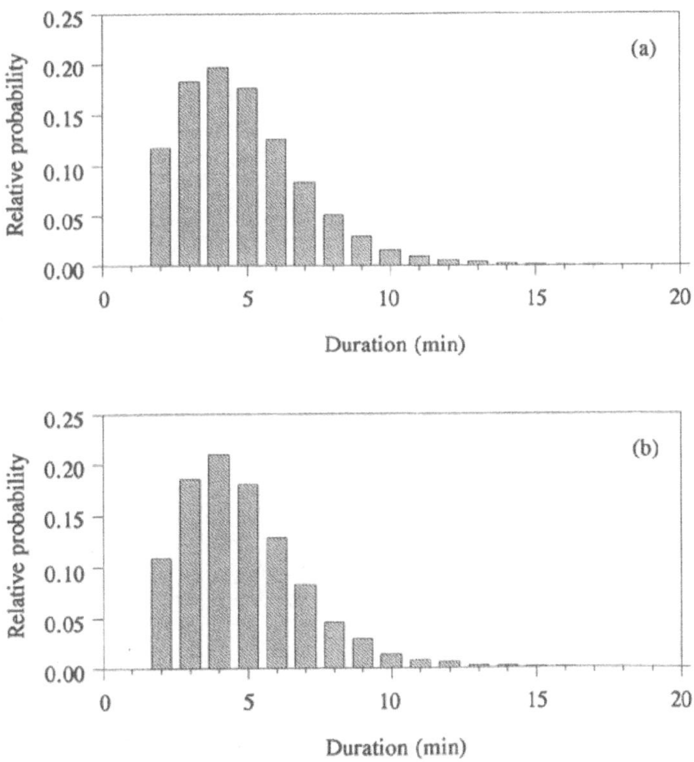

Figure 3. Histogram of breakdown duration based on covariation technique. (a) is for the ridge-top towers, and (b) is for the valley-bottom towers.

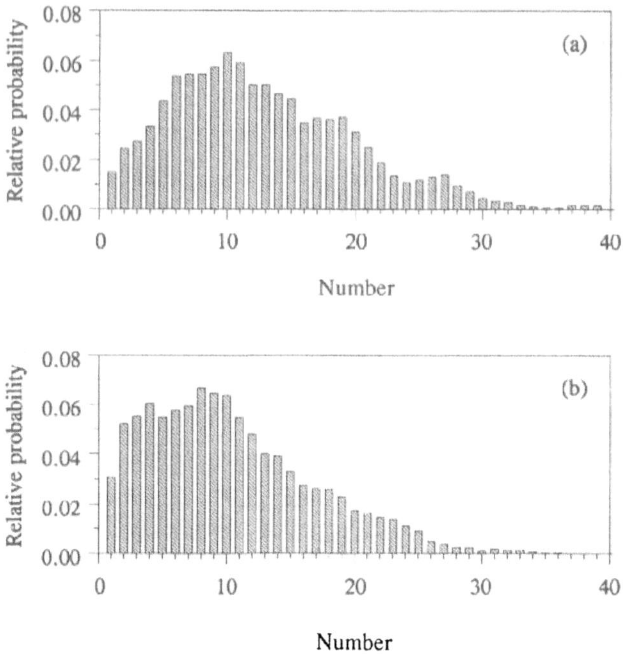

Figure 4. Histogram of number of breakdowns based on covariation technique. (a) is for the ridge-top towers, and (b) is for the valley-bottom towers.

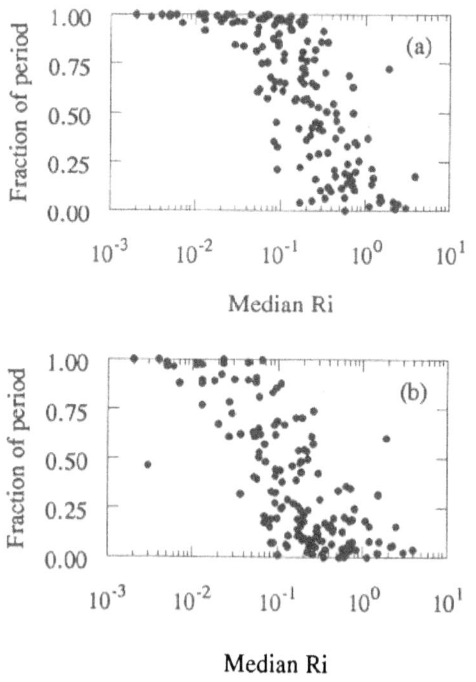

Figure 5. Fraction of nighttime period occupied by breakdowns, computed with the wavelet technique, versus median Richardson number. (a) is for tower 19, and (b) is for tower 15.

Figure 5b shows the fraction of the analysis period in which E_{UU} exceeded its threshold at tower 15. The plot looks similar to Fig. 5a, except that the data points are shifted towards smaller values of Ri. Breakdowns tend to occur mostly over the range $0.02 \leq Ri \leq 0.4$.

The duration of individual breakdown events was estimated simply by computing the length of time that E_{UU} remained above the threshold value. For each night of data, a mean breakdown duration was computed. Figure 6 shows the mean breakdown duration at the two towers plotted as a function of the median Ri. At both sites, the duration approaches its maximum value of 420 min. (*i.e.*, the total length of the measurement period) for small values of Ri. This simply indicates that more or less continuous mixing was present at these values of Ri. As is expected, this continuous mixing regime spans a larger range of Ri at the ridge-top tower than it does at the valley-bottom tower. At large values of Ri, the duration decreases to just under 10 minutes at both towers. The duration may be somewhat overestimated at these values of Ri, because the shape of the basis function $g(x)$ may affect the results when the duration is about the same size as the upper integration limit in Eq. (6).

CONCLUSIONS

The analysis of breakdown events based on the covariation C_{UT} shows that for nights with median $Ri > 1$, about 6% and 14% of the nighttime period is occupied by

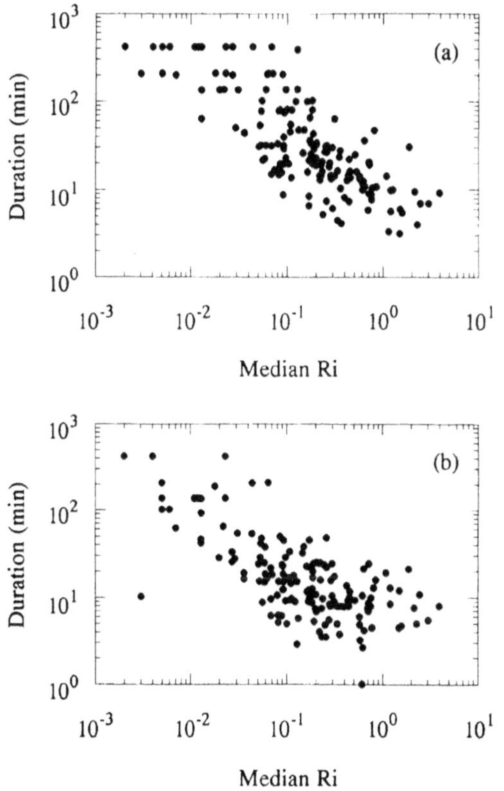

Figure 6. Breakdown duration computed with the wavelet technique plotted against the median Richardson number. (a) is for tower 19, and (b) is for tower 15.

breakdowns at the valley bottoms and ridge tops respectively; however, for $Ri < 1$, there is considerable scatter in the data, and it is not clear that a relation exists between breakdowns and median Ri. It is possible that for $Ri < 1$, breakdowns are initiated by nonlocal causes such as gravity waves or terrain-generated deformation of streamlines. In that case, finding a unique relation between breakdowns and SBL stability may be impossible.

The statistical distributions of breakdown durations and strengths are essentially the same at ridge tops and valley bottoms. The most probable breakdown duration is about 4 min; however, the durations of the strongest breakdowns range between 4 and 10 min. There is a difference between valley and ridge stations in the number of breakdowns per night. More breakdowns per night are seen at the ridge tops than in the valley bottoms. The average number of breakdowns per night is 9 for valley towers and 12 for ridge towers.

The wavelet analysis of the breakdown events at two towers indicates that the general structure of the breakdowns at the ridge-top and valley-bottom sites are similar. The main difference between the two towers is the range of Richardson numbers for which intermittent breakdowns dominate. At the valley-bottom tower, the range of Ri with intermittent breakdown events is lower ($0.02 \le Ri \le 0.4$) than the corresponding range at the ridge-top tower ($0.1 \le Ri \le 1.0$). In both cases, the fractional coverage of the breakdown events decreases roughly logarithmically with increasing Ri. The mean duration of the breakdown events is also similar at the two stations, with a duration of just under 10 min at large values of Ri.

REFERENCES

Eckman, R.M., Dobosy, R. J., and Pendergrass, W.R., 1992, Preliminary analysis of wind data from the Oak Ridge site survey. NOAA Tech. Memo. ERL ARL-193. NOAA Air Resources Laboratory, Silver Spring, MD, 45 pp.

Farge, M., 1992: Wavelet transforms and their applications to turbulence. *Ann. Rev. Fluid Mech.*, 24:395.

McNider, R.T., England, D.E., Friedman, M.K., and Shi, X., 1995, Predictability of the stable atmospheric boundary layer. *J. Atmos. Sci.*, 52:1602.

Meneveau, C., 1991, Analysis of turbulence in the orthonormal wavelet representation. *J. Fluid Mech.*, 232:469.

Nappo, C.J., 1991, Sporadic breakdowns of stability in the PBL over simple and complex terrain. *Bound.-Layer Meteor.*, 54:69.

Nappo, C.J. and Andren, A., 1995, A parameterization of subgrid scale gravity-wave-generated turbulence in a mesoscale boundary layer model. *11th Symposium on Boundary Layers and Turbulence*, 27-31 March, Charlotte, NC., American Meteorological Society, Boston, MA. 341.

Revelle, D.O., 1993, Chaos and "bursting" in the planetary boundary layer. *J. Appl. Meteor.*, 32:1169.

DISCUSSION

M. PIRINGER: Do you think your results can be generalized or classified with respect to valley structures or meteorological conditions?

C. NAPPO: I doubt that the statistics of PBL breakdowns can be related to valley structure; however, it is hoped that a relation between breakdowns and mean flow conditions can be found. Our results show some sensitivity to bulk Richardson number. I suspect that an analysis similar to ours but based on turbulence kinetic energy would show more correlation with meteorological condition.

W. KLUG: Why do you take for your correlation the median value of the Richarsdon number during the night and not the actually measured one?

C. NAPPO: We used the median value because we wanted to classify the mean nighttime stability of the PBL. We note that often the Richardson number varies greatly from hour to hour, and this we believe is due to the large vertical separation between wind speed and temperature measurements. We felt that the median value would be a more representative measure of the nighttime stability of the PBL.

S. ZILITINKEVICH: What could you say about surface flux calculations in very stable stratification in the presence of breakdowns of the stable boundary layer? Do you expect that conventional Monin-Obukhov similarity theory approach is applicable in such cases?

C. NAPPO: Our results suggest that during very stable conditions breakdowns of the PBL are the primary mechanism for exchange of heat, momentum. and trace gases between the free atmosphere and the ground surface. The often used assumption that the surface-layer fluxes are continuous functions of time may not always be applicable. It is for this reason that conventional Monin-Obukhov similarity theory might not be applicable when the vertical mixing is due to breakdowns. Indeed, a main assumption in the similarity theory is that the turbulence is stationary, but this is not the case when breakdowns are present. Thus I suspect that similarity fluxes probably underestimate the actual flux when breakdowns are present.

A NUMERICAL SIMULATION OF THE EDDY MOTION AND CLOUD DISTRIBUTIONS OVER THE SANTA BARBARA CHANNEL

Ted Yamada

President
Yamada Science & Art Corporation
Santa Fe, NM 87501

INTRODUCTION

A numerical simulation of airflows was performed over the Santa Barbara Channel, CA. The purpose of the study was to examine by using a mesoscale model how the eddy motions frequently observed in the satellite images were generated. Kessler and Douglas (1991) simulated the diurnal variations of wind distributions over the Santa Barbara Channel. Their model results produced during early morning hours a circulation which resembled the "midchannel eddy" (Smith et al., 1983; Kessler and Douglas, 1991). Satellite cloud images over the same area often show cyclonic eddies whose diameters are smaller than the circulations reproduced by Kessler and Douglas and the eddy appeared in the late morning hours.

We extended the work by Kessler and Douglas by using a nested grid version of HOTMAC* (Yamada and Bunker, 1988) where horizontal grid spacing was 2 km compared to a 10 km grid spacing used by Kessler and Douglas. HOTMAC is a three-dimensional mesoscale model based on second-moment turbulence-closure equations. The main objective of the present study was to simulate the eddy circulations presented in the satellite cloud images over the Santa Barbara channel.

MODEL EQUATIONS

Following Betts (1973), we define a liquid water potential temperature Θ_l as

$$\Theta_l = \Theta - (<\Theta> / <T>)(L_v/C_p)Q_c$$

*Higher Order Turbulence Model for Atmospheric Circulation

where Θ, T and Q_c are the potential temperature, absolute temperature, and the mixing ratio of cloud water respectively. In the above expressions, $< >$ indicates an average over a horizontal surface.

The mixing ratio of total water content, Q_w, is given as

$$Q_w = Q_v + Q_c + Q_r \tag{1}$$

where Q_v is the mixing ratio of water vapor. According to Betts (1973), "Θ_l represents the potential temperature obtained by evaporating all the liquid water in an air parcel through reversible wet adiabatic descent."

A terrain-following vertical coordinate system is used in order to increase the accuracy in the treatment of surface boundary conditions over complex terrain.

$$z* = \overline{H}\frac{z - z_g}{H - z_g}, \tag{2}$$

where $z*$ and z are the transformed and Cartesian vertical coordinates, respectively; z_g is ground elevation; $\overline{H}$ is the material surface top of the model in the z^* coordinate; and H is the corresponding height in the z-coordinate.

The equations of motion, following the coordinate transformation, are (Yamada and Bunker, 1988):

$$\begin{aligned}\frac{DU}{Dt} &= f(V - V_g) + g\frac{\overline{H} - z^*}{\overline{H}}\left(1 - \frac{<\Theta_l>}{\Theta_l}\right)\frac{\partial z_g}{\partial x} \\ &+\frac{\partial}{\partial x}\left(K_x\frac{\partial U}{\partial x}\right) + \frac{\partial}{\partial y}\left(K_{xy}\frac{\partial U}{\partial y}\right) + \frac{\overline{H}}{H - z_g}\frac{\partial}{\partial z^*}(-\overline{uw})\end{aligned} \tag{3}$$

$$\begin{aligned}\frac{DV}{Dt} &= -f(U - U_g) + g\frac{\overline{H} - z^*}{\overline{H}}\left(1 - \frac{<\Theta_l>}{\Theta_l}\right)\frac{\partial z_g}{\partial y} \\ &+\frac{\partial}{\partial x}\left(K_{xy}\frac{\partial V}{\partial x}\right) + \frac{\partial}{\partial y}\left(K_y\frac{\partial V}{\partial y}\right) \\ &+\frac{\overline{H}}{H - z_g}\frac{\partial}{\partial z^*}(-\overline{vw})\end{aligned} \tag{4}$$

$$\frac{\partial U}{\partial x} + \frac{\partial V}{\partial y} + \frac{\partial W^*}{\partial z^*} - \frac{1}{H - z_g}\left(U\frac{\partial z_g}{\partial x} + V\frac{\partial z_g}{\partial y}\right) = 0, \tag{5}$$

where

$$W^* \equiv \frac{\overline{H}}{H - z_g}W + \frac{z^* - \overline{H}}{H - z_g}\left(U\frac{\partial z_g}{\partial x} + V\frac{\partial z_g}{\partial y}\right), \tag{6}$$

and

$$\frac{D()}{Dt} \equiv \frac{\partial()}{\partial t} + U\frac{\partial()}{\partial x} + V\frac{\partial()}{\partial y} + W^*\frac{\partial()}{\partial z^*}. \tag{7}$$

The second terms on the right-hand side of Eqs. 4 and 5 indicate the effects of ground slope. For simplicity, H is specified as

$$H = \overline{H} + z_{gmax}, \tag{8}$$

where z_{gmax} is the maximum value of the ground elevation in the computational domain.

The geostrophic winds, U_g and V_g, are computed by Yamada (1981), i.e.,

$$fU_g \equiv fU_g(\overline{H})\frac{<\Theta_l>}{<\Theta_l(\overline{H})>} + g\frac{\overline{H}-z_g}{\overline{H}}\int_{z^*}^{\overline{H}}\frac{1}{<\Theta_l>}\frac{\partial}{\partial y}\Delta\Theta_l dz' - \frac{g}{\overline{H}}\frac{\partial z_g}{\partial y}\int_{z^*}^{\overline{H}}\frac{\Delta\Theta_l}{<\Theta_l>}dz'. \tag{9}$$

and

$$fV_g \equiv fV_g(\overline{H})\frac{<\Theta_l>}{<\Theta_l(\overline{H})>} - g\frac{\overline{H}-z_g}{\overline{H}}\int_{z^*}^{\overline{H}}\frac{1}{<\Theta_l>}\frac{\partial\Delta\Theta_l}{\partial x}dz' + \frac{g}{\overline{H}}\frac{\partial z_g}{\partial x}\int_{z^*}^{\overline{H}}\frac{\Delta\Theta_l}{<\Theta_l>}dz', \tag{10}$$

where $\Delta\Theta_l \equiv \Theta_l - <\Theta_l>$, and the abbreviated symbols $U_g(\overline{H}) \equiv U_g(x,y,\overline{H},t)$, $V_g(\overline{H}) \equiv V_g(x,y,\overline{H},t)$, are used. The derivations of Eqs. 4–6, 10 and 11 are given in the Appendix of Yamada (1981).

A turbulence kinetic energy equation is given by

$$\frac{D}{Dt}\left(\frac{q^2}{2}\right) = \frac{\partial}{\partial x}\left[K_x\frac{\partial}{\partial x}\left(\frac{q^2}{2}\right)\right] + \frac{\partial}{\partial y}\left[K_y\frac{\partial}{\partial y}\left(\frac{q^2}{2}\right)\right] + \left(\frac{\overline{H}}{\overline{H}-z_g}\right)^2\frac{\partial}{\partial z^*}\left[qlS_q\frac{\partial}{\partial z^*}\left(\frac{q^2}{2}\right)\right] - \frac{\overline{H}}{\overline{H}-z_g}\left(\overline{uw}\frac{\partial U}{\partial z^*} + \overline{vw}\frac{\partial V}{\partial z^*}\right) + \beta g\overline{w\theta_v} - \frac{q^3}{B_1 l}, \tag{11}$$

and a turbulence length scale l is obtained from

$$\frac{D}{Dt}(q^2 l) = \frac{\partial}{\partial x}\left[K_z\frac{\partial}{\partial x}(q^2 l)\right] + \frac{\partial}{\partial y}\left[K_y\frac{\partial}{\partial y}(q^2 l)\right] + \left(\frac{\overline{H}}{\overline{H}-z_g}\right)^2\frac{\partial}{\partial z^*}\left[qlS_l\frac{\partial}{\partial z^*}(q^2 l)\right] - lF_1\left[\frac{\overline{H}}{\overline{H}-z_g}\left(-\overline{uw}\frac{\partial U}{\partial z^*} - \overline{vw}\frac{\partial V}{\partial z}\right) + \beta g\overline{w\theta_v}\right] - \frac{q^3}{B_1}\left[1 + F_2\left(\frac{l}{kz}\right)^2\right], \tag{12}$$

where $q^2 = \overline{u^2} + \overline{v^2} + \overline{w^2}$ is twice the turbulence kinetic energy, $\overline{w\theta_v}$ turbulence heat flux, θ_v the fluctuation part of virtual potential temperature, and (F_1, F_2, S_q S_l, and B_1) = (1.8,

1.33, 0.2, 0.2, and 16.6); empirical constants were determined from laboratory experiments (Mellor and Yamada, 1982).

The conservation equation for $\Delta\Theta_l$ is written as in Yamada and Bunker (1989),

$$\frac{D\Delta\Theta_l}{Dt} = \frac{\partial}{\partial x}\left[K_x \frac{\partial \Delta\Theta_l}{\partial x}\right] + \frac{\partial}{\partial y}\left[K_y \frac{\partial \Delta\Theta_l}{\partial y}\right] + \frac{\overline{H}}{H - z_g} \times \left[\frac{\partial}{\partial z*}(-\overline{w\theta_l}) + \frac{1}{\rho C_p}\frac{\partial R_N}{\partial z*} - W\frac{\partial \Theta_l}{\partial z}\right]. \quad (13)$$

The long-wave radiation flux $R_N/\rho C_p$ for outside the cloud is computed according to Sasamori (1968). A conservation equation for the mixing ratio of total water is given by

$$\frac{DQ_w}{Dt} = \frac{\partial}{\partial x}\left[K_x \frac{\partial Q_w}{\partial x}\right] + \frac{\partial}{\partial y}\left[K_y \frac{\partial Q_w}{\partial y}\right] + \frac{\overline{H}}{H - z_g}\frac{\partial}{\partial z*}(-\overline{wq_w}). \quad (14)$$

The turbulent fluxes in Eqs. 4, 5, 12, 13, 14, and 15 are obtained from simplified second-moment turbulence-closure equations (Yamada 1983):

$$(\overline{uw}) = -lq\tilde{S}_M\left[\frac{\partial U}{\partial z}\right] \quad (15)$$

$$(\overline{vw}) = -lq\tilde{S}_M\left[\frac{\partial V}{\partial z}\right] \quad (16)$$

$$(\overline{w\theta_l}) = -\alpha lq\tilde{S}_M\left[\frac{\partial \Theta_l}{\partial z}\right] \quad (17)$$

$$(\overline{wq_w}) = -\alpha lq\tilde{S}_M\left[\frac{\partial Q_w}{\partial z}\right] \quad (18)$$

where $\tilde{S}_M$ and α are functions of the flux Richardson number, and $\alpha(\equiv K_H/K_M$ where K_H is an eddy diffusivity coefficient and K_M is an eddy viscosity coefficient) is the reciprocal of the turbulent Prandtl number.

The expressions for $\tilde{S}_M$ and α were obtained from the level 2 model of Mellor and Yamada (1974) where temporal and spatial derivatives in Eq. 12 are neglected. The readers are referred to Yamada (1975) for further discussions of the level 2 model equations. The final expressions for $\tilde{S}_M$ and α are given in Yamada (1983) and are not repeated here.

The terms for the rate of condensation are purposely eliminated by introducing the liquid water potential temperature and mixing ratio of total water [Eqs. 1 and 2]. In order to recover the potential (or absolute) temperature and the mixing ratios of water vapor and liquid water, Gaussian cloud relations, proposed by Sommeria and Deardorff (1977) and Mellor (1977), are used. The present method has been applied to simulations of the BOMEX data (Yamada and Mellor, 1979), and GATE data (Yamada and Kao, 1986).

Surface boundary conditions for mean and turbulence variables are constructed from the empirical formulas by Dyer and Hicks (1970). The surface temperatures are obtained by solving a heat conduction equation in the soil layer. The boundary conditions are heat energy balance at the surface and the soil temperature specified at a certain depth from the surface.

SIMULATIONS

Preliminary simulations of airflows over the Santa Barbara Channel were conducted with the following initial values for wind and temperature profiles. The potential temperature at sea level was 286 K and increased with height at a rate of 0.003 K/m. Wind direction was assumed to be north-westerly and geostrophic wind speed was 5 m/s. Initial wind profiles were obtained by scaling the wind speeds at the southwestern corner of the computational domain to satisfy the mass continuity.

The outer computational domain was 282 km x 282 km with an 8 km grid spacing, and the inner computation domain was 96 km x 96 km with a 2 km grid spacing (Fig. 1).

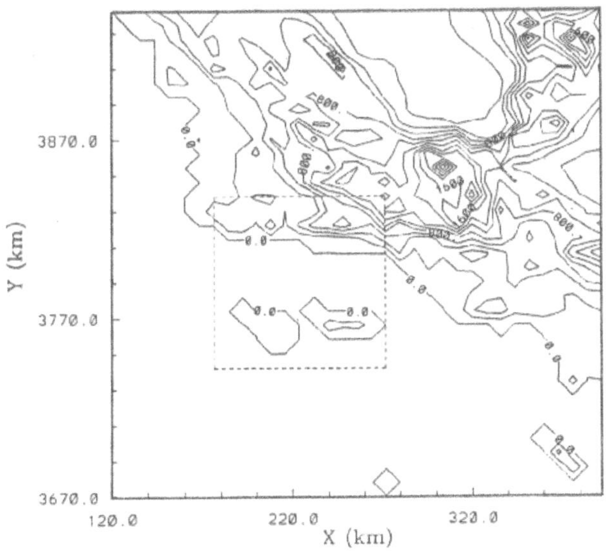

Figure 1. Outer computational domain is 282 km x 282 km. Inner domain is 96 km x 96 km and located over Santa Barbara Channel. Ground elevation is contoured with a 200 m increment.

Integration began at 1500 lst (local standard time) on Julian day 132 (May 12) and continued for 33 hours. Figure 2 shows the modeled horizontal wind vectors in the outer grid area (Fig. 1) at 180 m above the ground at 1500 lst, 24 hours after the integration started. Strong upslope flows developed which converged along the ridge lines. Figures 3 and 4 show the wind vectors in the vertical cross section through y = 3838 km (Fig. 2) at 1500 lst and 0100 lst, respectively. Verical velocities as large as 1.5 m/s were obtained (Fig. 3). The surface wind direction changed from upslope to downslope after the sunset. The upward motions subsided considerably but large disturbances remained over the ridges throughout the nights (Fig. 4). Figure 5 shows the modeled vertical profiles of mixing ratios of total water and cloud liquid water at 1500 lst. Moist air was transported from the ocean and lifted over the ridges by the upward motion.

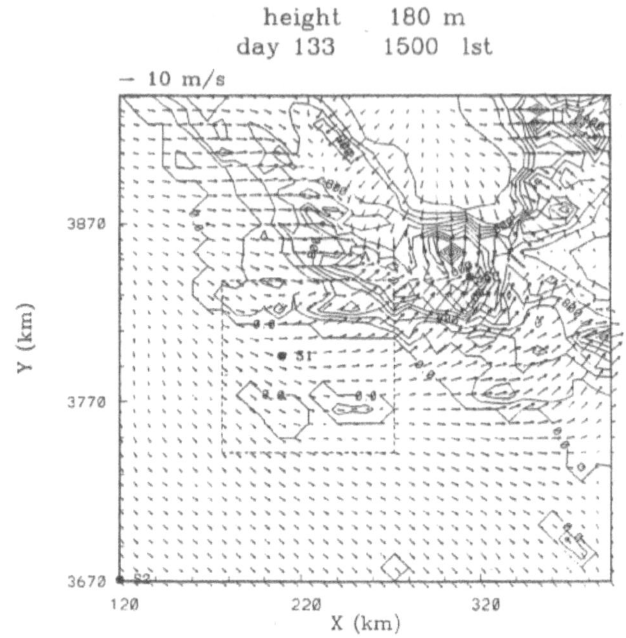

Figure 2. Modeled horizontal wind vectors in the outer grid area at 180 m above the ground at 1500 lst. Terrain is contoured by solid lines with an increment of 100 m.

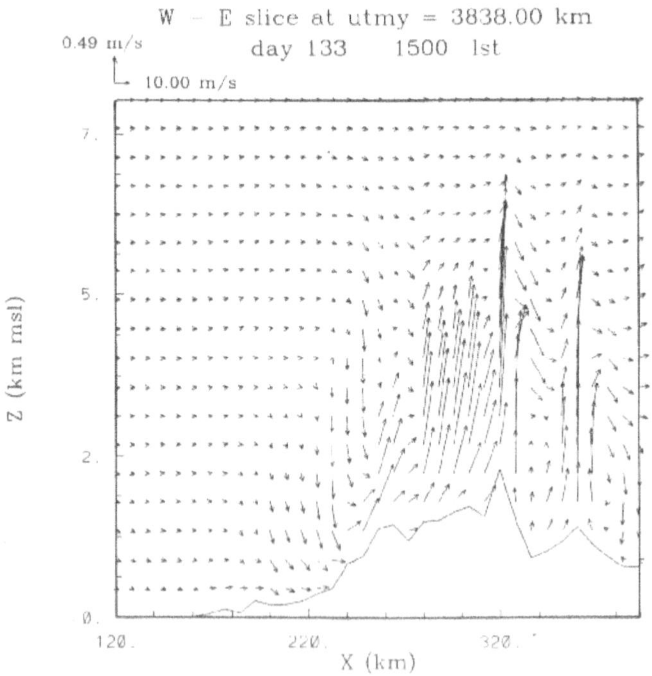

Figure 3. Modeled winds in the vertical section at y = 3838 km at 1500 lst.

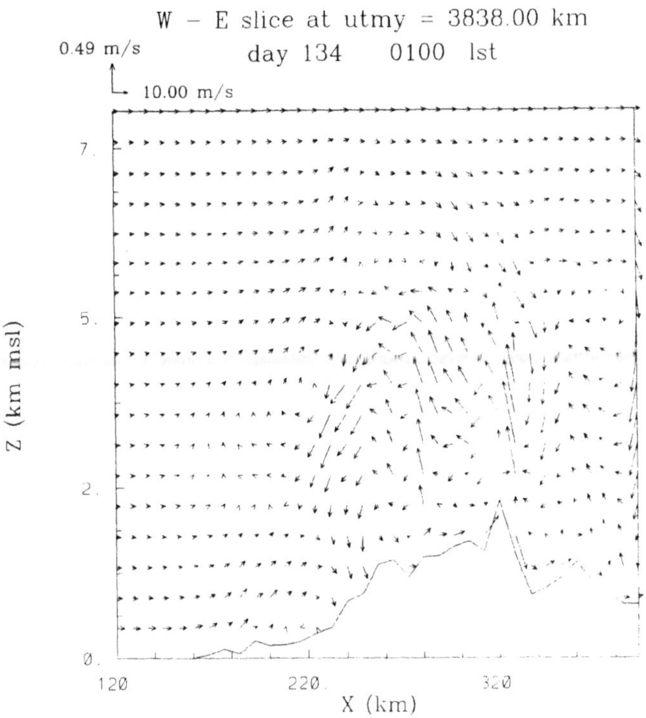

Figure 4. Same as in Figure 3 except at 0100 lst.

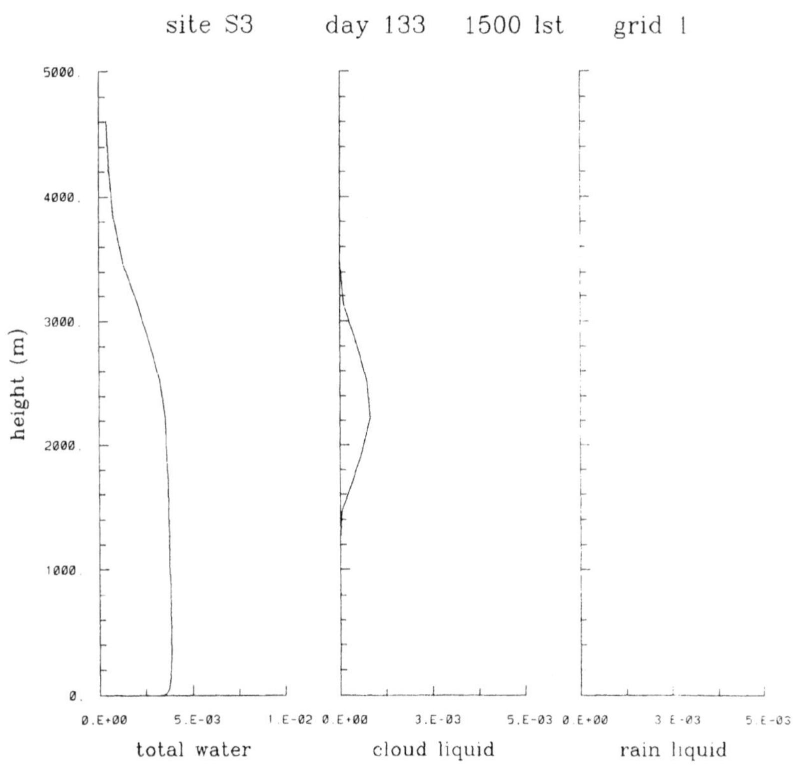

Figure 5. Vertical profiles of modeled mixing ratios of total water and cloud liquid water (g/g).

REFERENCES

Betts, A.K., 1973, Non-precipitating cumulus convection and its parameterization, *Quart. J. Roy. Meteor. Soc.* 99:178–179.

Douglas, S.G., and Kessler, R.C., 1991, Analysis of mesoscale airflow patterns in the south-central coast air basin during the SCCCAMP 1985 intensive measurement periods, *J. of Appl. Meteor.* 30:607-631.

Dyer, A.J., and Hicks, B.B., 1970, Flux-gradient relationships in the constant flux layer, *Quart. J. Roy. Meteor. Soc.* 96:715-721.

Kessler, R.C., and Douglas, S.G., 1991, A numerical study of mesoscale eddy development over the Santa Barbara channel, *J. of Appl. Meteor.* 30:633-651.

Mellor, G.L., 1977, The gaussian cloud model relations, *J. of Atmos. Sci.* 34:356–358.

Mellor, G.L., and Yamada, T., 1974, A hierarchy of turbulence-closure models for planetary boundary layers, *J. of Atmos. Sci.* 31:1791–1806, Corrigendum, *J. of Atmos. Sci.* 34:1482.

Mellor, G.L., and Yamada, T., 1982, Development of a turbulence closure model for geophysical fluid problems, *Rev. Geophys. Space Phys.*, 20:851–875.

Sasamori, T., 1968, The radiative cooling calculation for application to general circulation experiments, *J. Appl. Meteor.*, 7:721–729.

Smith, T.B., Saunders, W.D., and Shair, F.H., 1983, Analysis of Santa Barbara oxidant study, final report to California air resources board, agreement A2-086-32, [Available from Meteorology Research Inc., Altadena, CA 91001], 236 pp.

Sommeria, G., and Deardorff, J.W., 1977, Subgrid-scale condensation in models of nonprecipitating clouds, *J. of Atmos. Sci.*, 34:344–355.

Yamada, T., 1975, The critical Richardson Number and the ratio of the eddy transport coefficients obtained from a turbulence closure model, *J. Atmos. Sci.*, 32:926–933.

Yamada, T., 1981, A numerical simulation of nocturnal drainage flow, *J. Meteor. Soc. Japan*, 59:108–122.

Yamada, T., 1983, Simulations of nocturnal drainage flows by a q^2l turbulence closure model, *J. of Atmos. Sci.*, 40:91–106.

Yamada, T., and Bunker, S., 1988, Development of a nested grid, second-moment turbulence-closure model and an application to the 1982 ASCOT brush creek data simulation. *J. of Atmos. Sci.*, 27:562-578.

Yamada, T., and Bunker, S., 1989, A numerical model study of nocturnal drainage flows with strong wind and temperature gradients, *J. of Appl. Meteor.*, 28:545–554.

Yamada, T., and Kao, C.-Y.J., 1986, A modeling study on the fair weather marine boundary layer of the GATE, *J. of Atmos. Sci.*, 43:3186–3199.

Yamada, T., and Mellor, G.L., 1979, A numerical simulation of the BOMEX data using a turbulence closure model coupled with ensemble cloud relations, *Quart. J. Roy. Meteor. Soc.*, 105:915–944.

SCIPUFF - A GENERALIZED DISPERSION MODEL

R. I. Sykes, S. F. Parker, D. S. Henn, and R. S. Gabruk

ARAP Group, Titan Research and Technology
Princeton, NJ 08543-2229

INTRODUCTION

Gaussian puff models provide a flexible description of atmospheric dispersion phenomena within a practical framework. Puff models are more efficient than Eulerian grid representations for localized sources and can also describe dispersion over a wide range of scales without grid resolution restrictions or the associated numerical diffusion. Representation of the dispersing contaminant field as a collection of Gaussian puffs requires more computation than steady-state Gaussian plume models, but provides a much more general capability for describing multiple sources with arbitrary time-dependence and three-dimensional, unsteady meteorology. However, traditional Gaussian puff models have been limited by their inability to represent complex flows with shear distortions or nonlinear phenomena, such as buoyant rise dynamics or chemical reaction. Accurate calculation of shear distortions can be modeled using Lagrangian particle methods, while nonlinear interactions have required an Eulerian calculation.

As part of the development of a generalized hazardous material dispersion prediction capability, the Gaussian puff model, SCIPUFF (Second-order Closure Integrated Puff), has been significantly enhanced to remove some of the existing restrictions and widen its applicability. SCIPUFF was originally developed to describe dispersion from tall stack emissions under the sponsorship of the Electric Power Research Institute (Sykes et al., 1989). The turbulent diffusion representation is based on a Reynolds stress turbulence closure model, as described by Lewellen (1977), and includes a prediction of the concentration fluctuation variance. The variance can be used to provide a quantitative estimate of the statistical uncertainty in an observed concentration value arising from the chaotic nature of the turbulent diffusion process. Buoyant rise dynamics were also included in the original tall stack model, based on a set of conservation equations for the integrated vertical momentum and buoyancy due to each puff.

Under the sponsorship of Defense Nuclear Agency, SCIPUFF has been greatly extended to describe different material properties as they affect dispersion over a wide range of scales. Gaseous materials or solid particles with multiple size groups are described, and liquid droplets with evaporation and generation of vapor are also included. SCIPUFF represents the gravitational settling of particles and droplets and also turbulent dry deposition processes. The turbulence closure diffusion framework has been extended to describe long range transport, using the mesoscale turbulence representation of Gifford (1988), and has been compared with the continental-scale dispersion data from the ANATEX experiment (Sykes et al., 1993). Multiple sources in arbitrary combinations of instantaneous clouds and continuous plumes can be specified, and an option is provided to prescribe a fixed velocity for moving continuous sources. The SCIPUFF model has also been developed to be more user-friendly, with a graphical user interface (GUI) to allow interactive problem description and output display. In the following sections, we briefly describe the generalized Gaussian puff framework, and the extension of the concentration fluctuation variance representation necessary for a generalized puff model. Some examples of SCIPUFF calculations are presented in the final section.

GAUSSIAN PUFF REPRESENTATION

The simplified Gaussian puff representation, which usually considers only a horizontal and a vertical spread, has been generalized in SCIPUFF to contain the full tensor moment description. The concentration field due to an individual puff is described by the Gaussian form

$$c(\boldsymbol{x}) = \frac{Q}{(2\pi)^{3/2}(Det(\sigma))^{1/2}} \exp\left[-\frac{1}{2}\sigma_{ij}^{-1}(x_i - \bar{x}_i)(x_j - \bar{x}_j)\right] \tag{1}$$

where the moments are defined as follows

$$Q = \langle c \rangle \tag{2}$$

$$Q\,\bar{x}_i = \langle c x_i \rangle \tag{3}$$

$$Q\,\sigma_{ij} = \langle c(x_i - \bar{x}_i)(x_j - \bar{x}_j) \rangle \tag{4}$$

representing the puff mass, centroid location, and spatial spread respectively. The angle brackets denote a spatial volume integral over the puff distribution.

Integration of the scalar conservation equation provides the evolution equations for the Gaussian moments, which can be written

$$\frac{dQ}{dt} = 0 \tag{5}$$

$$\frac{d\bar{x}_i}{dt} = \bar{u}_i(\bar{x}) + \frac{\langle \overline{u_i'c'} \rangle}{Q} \tag{6}$$

$$\frac{d\sigma_{ij}}{dt} = \sigma_{ik}\frac{\partial \bar{u}_j}{\partial x_k} + \sigma_{jk}\frac{\partial \bar{u}_i}{\partial x_k} + \frac{\langle x_i \overline{u_j'c'} \rangle}{Q} + \frac{\langle x_j \overline{u_i'c'} \rangle}{Q} \tag{7}$$

where $u_i(\boldsymbol{x},t)$ is the turbulent velocity field. Here, the ensemble average is denoted by an overbar and the turbulent fluctuation from the average is denoted by a prime. The details of the second-order closure model for the turbulent correlations is described by Sykes et al. (1986).

The additional shear terms in (7) describe the distortion of the Gaussian in the presence of a linearly varying velocity field. The linear representation is only valid locally, however, and a puff splitting scheme is required to provide accurate solutions for an arbitrary inhomogeneous wind field. The details of the shear treatment, and puff splitting and merging algorithms are described by Sykes and Henn (1995). Figure 1 shows a comparison between the Gaussian puff calculation and the analytic solution of Staniforth et al. (1987) for the two-dimensional distortion flow of Smolarkiewicz (1982). This idealized flow is non-diffusive

Figure 1. SCIPUFF comparison with the deformation flow solution after about 5 rotations of the vortex; the puff result is shown on the left. Details of the calculation can be found in Sykes and Henn (1995).

and provides a severe test of the numerical advection scheme since the increasing distortion produces smaller and smaller length scales as time increases. The Lagrangian nature of the puff model and the flow-adaptive splitting scheme allows SCIPUFF to follow the increasing distortion of this shear flow for much longer than a conventional Eulerian calculation.

The ability to distort puffs in response to the local shearing motions and split puffs when they become too large for a purely local representation allows SCIPUFF to describe complex flow conditions, which have traditionally been considered beyond the capability of a Gaussian puff model. The distortion flow test above is unrealistic, since there is no diffusion involved, but clearly demonstrates the capability of the puff model. A more relevant example of flow over topography will be presented below.

CONCENTRATION FLUCTUATION VARIANCE DESCRIPTION

Turbulent dispersion is an essentially random process, so any concentration prediction involves some uncertainty. The conservation equations are averaged in an ensemble sense to provide a prediction of the mean, or average, value but the concentration itself is a random variable with a statistical distribution. Any assessment of the uncertainty requires an estimate of the statistical variance in addition to the mean value. A more detailed description is necessary to define the probability distribution, but the variance is the minimal requirement. The second-order closure framework provides a natural framework for predicting the concentration fluctuation variance, since the basis of the method is the development of transport equations for the second-order correlations. Previous work has applied the technique to the fluctuation prediction for plume dispersion, but we shall show the model requires generalization for application in the puff model.

The conservation equation for the mean-square concentration is obtained by multiplying the conservation of mass equation (1) by c, and averaging. Since the advection terms only transport $\overline{c^2}$ spatially, the volume integrated evolution equation for a puff can be written as

$$\frac{d}{dt}\left\langle \overline{c^2} \right\rangle = -\langle \varepsilon_c \rangle \tag{8}$$

where ε_c is the local dissipation rate for the fluctuation variance $\overline{c'^2}$. We assume that molecular dissipation is negligible, and that ε_c can be represented as

$$\varepsilon_c = \frac{\overline{c'^2}}{\tau_c} \tag{9}$$

where τ_c is a turbulent dissipation timescale.

The second-order closure model for the concentration fluctuation variance is based on the work of Sykes et al. (1984), who showed that a length scale for the scalar fluctuations, Λ_c, was required to describe the dissipation timescale. The inverse timescale was defined as

$$\tau_c^{-1} = \frac{q_c}{\Lambda_c} \tag{10}$$

where q_c is a measure of the velocity standard deviation on the length scale Λ_c. The empirical constants in the closure model were determined from the laboratory plume measurements of Fackrell and Robins (1982a).

In the general case of anisotropic mixing rates and arbitrary release conditions that can be described by a puff model, the empirical constant in the definition of the fluctuation dissipation timescale cannot be universal. This can be demonstrated by means of the idealized meandering plume model of Gifford (1957). First we note that the closure model prediction for the early time behavior of τ_c is a linear growth with time, as shown by Sykes et al. (1984). If we assume $\tau_c = \alpha t$, and further assume that $\overline{c'^2} >> \overline{c}^2$, then the variance equation becomes

$$\frac{d}{dt}\left\langle \overline{c'^2} \right\rangle = -\frac{\overline{c'^2}}{\alpha t} \tag{11}$$

and it is clear that the variance will decay like t^{-p}, where $p = \alpha^{-1}$.

The basis of Gifford's (1959) model is an instantaneous plume size, σ_i, randomly meandered over an ensemble average size, σ_e. Gifford analyzes the case where the ensemble spreads in the lateral and vertical directions are unequal, but this is not necessary for our purposes here. Fackrell and Robins (1982b) showed that Gifford's model gave good agreement with the laboratory data on the fluctuation intensity, using a spectral intensity model for the internal and ensemble spreads. This model gives the early time result that $\sigma_i \sim t^{3/2}$ and $\sigma_e \sim t$. Now the meandering plume model gives

$$I = \frac{\left(\sigma_e^2 - \sigma_i^2\right)^2}{\sigma_i^2\left(2\sigma_e^2 - \sigma_i^2\right)} \tag{12}$$

where I is the plume centerline fluctuation intensity, $\overline{c'^2}/\overline{c}^2$, and for $\sigma_i << \sigma_e$ this implies that $I \sim t^{-1}$.

The time dependence in (12) is determined by the dimensionality of the dispersion mechanism, which is 2 for the case of the plume. However, an instantaneous spherical cloud will diffuse in three dimensions, and the Gifford model is readily shown to give $I \sim t^{-3/2}$ in this case. Generally, the instantaneous concentration level, c_i, will be proportional to σ_i^{-d} and the ensemble average concentration, $\overline{c}$, will vary like σ_e^{-d}, where d is the number of dimensions in which the material is diffusing. The probability, p_i, of encountering a non-zero instantaneous concentration is the ratio of the two volumes $\left(\sigma_i/\sigma_e\right)^{-d}$, so $\overline{c'^2} \sim t^{-5d/2}$ since $\overline{c'^2} \sim p_i c_i^2$ and hence $I \sim t^{-d/2}$. We therefore see that the coefficient α in the dissipation timescale must depend on the dimensionality of the dispersion mechanisms.

The solution to this problem in SCIPUFF has been achieved through the use of three internal fluctuation scales, representing each of the three possible dimensions, two horizontal and one vertical. Thus, we define

$$\tau_c^{-1} = q_c \left(\frac{1}{\Lambda_c} + \frac{1}{\Lambda_{c2}} + \frac{1}{\Lambda_{cV}} \right) \tag{13}$$

where Λ_c and Λ_{c2} are used to describe the two horizontal directions, and Λ_{cV} represents the vertical scale. The three scales are initialized as the appropriate cloud dimensions for an instantaneous release, but a continuous source is specified with a modified horizontal scale. Λ_c and Λ_{cV} are set to the plume source dimension but Λ_{c2} must represent the streamwise scale of the plume. For a continuous source, where the source duration is long compared with turbulence time scales, Λ_{c2} is initialized as the integral length scale of the turbulence. When the source size is much smaller than the turbulence scale, the dissipation timescale in (13) effectively uses a factor of 2 for a plume, in comparison with a factor of 3 for an instantaneous cloud. The generalized dissipation timescale in (13) will therefore provide the correct dependence on the dimensionality of the diffusion process.

SAMPLE SCIPUFF CALCULATIONS

Flow Over Terrain

The puff distortion and splitting technique in SCIPUFF permits the calculation of dispersion under complex flow conditions. A simplified example is presented here, consisting of a single continuous source upstream of an isolated circular hill. The flow field is prescribed from a mass-consistent model, with a uniform wind as the initial velocity field. Two flows are specified: the first uses equal weighting on vertical and horizontal velocity components to define a potential flow over the hill, the second allows no vertical velocities in the mass adjustment procedure, and represents a strongly stable flow where the velocity is constrained to the horizontal plane. Turbulence profiles are prescribed as appropriate for a neutral boundary layer for the first flow. The stable layer turbulence profile is specified with a small vertical velocity variance ($0.01 m^2 s^{-2}$), but a larger horizontal variance ($0.25 m^2 s^{-2}$) to represent horizontal meandering motions.

The flow field is defined in a terrain-following grid coordinate system, but the Lagrangian puff calculation is carried out in a Cartesian framework. Locally normal reflections are applied to determine the concentration at the terrain surface and also to prevent the puff splitting algorithm from creating puffs below the surface.

Figure 2 shows perspective views of a particular isosurface for the plume concentration for the two flow conditions. Under neutral conditions, the plume simply follows the mean streamline over the top of the hill, but stable flow calculation is more complex. The suppression of the vertical motion in the stable case causes the plume to impact the hill on the upstream slope. As puffs approach the stagnation point, they are stretched laterally and forced to split around the sides of the hill. The lateral meandering spreads the plume in the wake and the puffs are recombined as they move downstream, as shown in the lower panel of the figure. This plume is represented by about 180 puffs and the calculation takes about 45s on a Pentium P120 to reach the illustrated steady state.

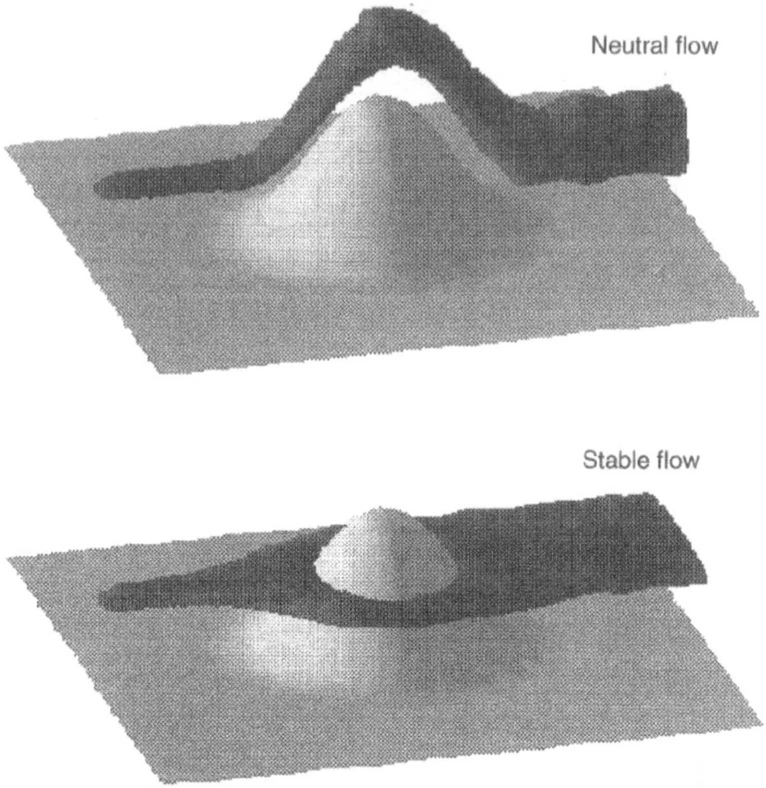

Figure 2. SCIPUFF results for plume flow over an isolated hill with flow from left to right. The upper panel represents neutral flow conditions while the lower panel represents a strongly stable flow.

Buoyant Plume Rise

The second example demonstrates the buoyant rise capability of SCIPUFF. Figure 3 shows a vertical section in the flow direction for SF_6 emission from the Kincaid power plant stack for stable early morning conditions on June 1, 1981, one of the intensive observation days during EPRI's Plume Model Validation Experiment. The puff source is specified at the stack exit, with an initial spread of 4.5m, a vertical velocity of $17ms^{-1}$, and a temperature excess of 130°C. Puffs are emitted at the stack elevation of 187m, and the ambient meteorology is specified from the model studies of Sykes et al. (1989). The time step for puffs at the stack exit is 0.2s, but an adaptive scheme allows the time step to increase as the puffs move downstream and their evolution slows down. It is somewhat inefficient to calculate the plume rise as a time-dependent problem, but this demonstrates the generality of the model and the ability to describe a wide range of sources.

Figure 3 illustrates the initial buoyant rise, which is roughly 100m in the first 2km. The horizontal axis shows UTM Easting and Northing coordinates in kilometers. The stable

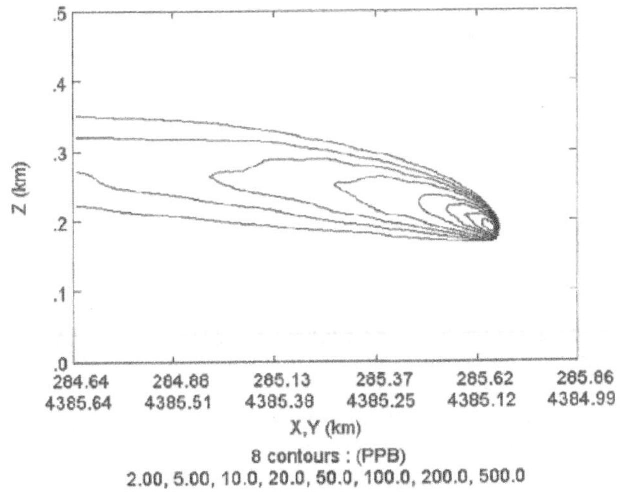

Figure 3. Vertical section along the streamwise direction through the SCIPUFF prediction for the Kincaid stack SF_6 emission.

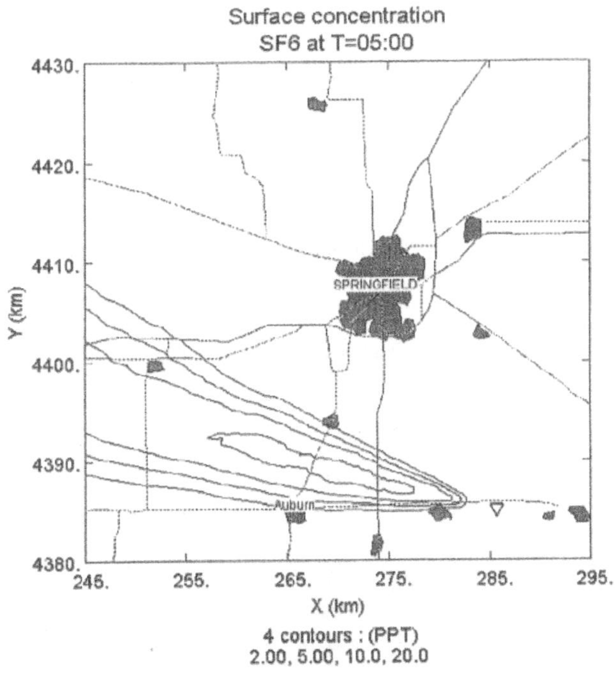

Figure 4. Surface SF_6 concentration from the SCIPUFF prediction for June 1, 1981. The source location is indicated by the inverted triangle.

temperature gradient limits the vertical rise of the plume, and equilibrium height is reached within a few kilometers. Figure 4 shows the surface concentration contour output from the interactive display on the PC. The background maps are derived from the Digital Chart of the World, a Defense Mapping Agency product, and can be selected from the Graphical User Interface.

SUMMARY

An advanced puff dispersion model, SCIPUFF, has been described with broad capability to describe complex source characteristics and material properties. The extended Lagrangian puff framework allows the treatment of arbitrary velocity fields, and an efficient numerical scheme has been implemented to provide an accurate calculation of flow distortions. SCIPUFF uses a second-order turbulence closure scheme to relate the dispersion rate to measurable turbulent velocity statistics which provides a generalized framework for representing diffusion over a wide range of scales. The closure scheme also provides a prediction of the concentration fluctuation variance due to the random velocity field, which can be used to provide quantitative estimates of the uncertainty in the mean value prediction. The concentration fluctuation model has been generalized from the earlier plume-based model so that it can properly account for instantaneous sources.

Further capability to describe buoyant rise dynamics has been demonstrated, and future work will extend the model to include dense gas effects and nonlinear chemical reactions.

ACKNOWLEDGMENT

This work is supported by the U.S. Defense Nuclear Agency, under contract number DNA001-C-95-0180.

REFERENCES

Fackrell, J. E. and Robins, A. G., 1982a, Concentration fluctuations and fluxes in plumes from a point source in a turbulent boundary layer, *J. Fluid Mech.*, 117: 1.

Fackrell, J. E. and Robins, A. G., 1982b, The effects of source size on concentration fluctuations in plumes, *Boundary-Layer Met.*, 22: 335.

Gifford, F. A., 1959, Statistical properties of a fluctuating plume dispersion model, *Adv. Geophys.*, 6:117.

Gifford, F. A., 1988, A similarity theory of the tropospheric turbulence energy spectrum, *J. Atmos. Sci.*, 45:1370.

Lewellen, W. S., 1977, Use of invariant modeling, in: *Handbook of Turbulence*, ed. W. Frost and T. H. Moulden, Plenum Press.

Smolarkiewicz, P. K., 1982, The multidimensional Crowley advection scheme, *Mon. Wea. Rev.*, 110:1968.

Staniforth, A., Côté, J., and Pudykiewicz, J., 1987, Comments on 'Smolarkiewicz's deformational flow, *Mon. Wea. Rev.*, 115:1894.

Sykes, R. I., Lewellen, W. S., and Parker, S. F., 1984, A turbulent-transport model for concentration fluctuations and fluxes, *J. Fluid Mech.*, 139:193.

Sykes, R. I., Lewellen, W. S., and Parker, S. F., 1986, A Gaussian plume model of atmospheric dispersion based on second-order closure, *J. Clim. & Appl. Met.*, 25:322.

Sykes, R. I., Lewellen, W. S., Parker, S. F., and Henn, D. S., 1989, A hierarchy of dynamic plume models incorporating uncertainty, Volume 4: Second-order Closure Integrated Puff, EPRI EA-6095 Volume 4, Project 1616-28.

Sykes, R. I., Parker, S. F., Henn, D. S., and Lewellen, W. S., 1993, Numerical simulation of ANATEX tracer data using a turbulence closure model for long-range dispersion, *J. Appl. Met.*, 32:929.

Sykes, R. I. and Henn, D. S., 1995, Representation of velocity gradient effects in a Gaussian puff model, *J. Appl. Met.*, **34**, 2715-2723.

A MODELING STUDY OF VERTICAL DIFFUSION OF PASSIVE AND REACTIVE TRACERS USING LOCAL- AND NONLOCAL-CLOSURE BOUNDARY LAYER SCHEMES

Kiran Alapaty,[1] Rohit Mathur,[1] and Daewon W. Byun[2,3]

[1]Environmental Programs
MCNC–North Carolina Supercomputing Center
3021 Cornwallis Road, P.O. Box 12889
Research Triangle Park, NC 27709-2889

[2]Air Resources Laboratory
National Oceanic and Atmospheric Administration
Research Triangle Park, NC 27711

[3]On assignment to the National Exposure Research Laboratory,
U.S. Environmental Protection Agency

INTRODUCTION

Both turbulent mixing in the atmospheric boundary layer (ABL) and chemical transformation of trace-gas species have relatively short characteristic time scales compared with those of other processes considered in existing air quality simulation models. There exists a competition between these two processes in changing the chemical state of the atmosphere. Representation of these interacting processes is further complicated by the release of various trace-gas species at different altitudes in the atmosphere. Although many air quality models (AQMs) include detailed and complex reaction pathways for species transformation, the representation of ABL turbulent mixing processes in these models is still rather crude. Many Eulerian air quality modeling systems use K-theory concepts to represent the subgrid-scale vertical mixing in the ABL. An underlying assumption in this formulation is that the turbulent mixing processes are local; consequently, the mixing is confined to adjacent layers. However, in the presence of larger eddies that can exist during convective conditions (e.g., summer daytime), the above assumption may be invalid.

In meteorological modeling, it has been well documented that either a higher-order local-closure or a nonlocal-closure ABL mixing scheme is preferable to a K-theory-based scheme in simulating various atmospheric systems. Several modeling studies of chemical reactions in the ABL (e.g., Arellano and Duynkerke, 1995) have shown that the commonly used K-theory may represent turbulent mixing processes in the ABL inadequately during convective conditions. In addition, these studies have indicated that the accuracy of simulated meteorological and chemical fields is sensitive to the turbulent mixing scheme used in a model. There is a great need to study the applicability of different boundary layer mixing schemes in air quality modeling systems. An ABL mixing scheme must meet more stringent requirements when it is used for air quality modeling than meteorological modeling. Because chemical species are released at different altitudes in the atmosphere, a mixing scheme for an AQM should be able to handle a variety of different mixing scenarios in the ABL. The objective of this study is to demonstrate the effects of using different turbulent mixing schemes on the simulation of a tracer gas and several secondary pollutants produced by nonlinear chemical reactions in the atmosphere. We compared performance of two local-closure and two nonlocal-closure mixing schemes.

MODEL DESCRIPTION

We used a three-dimensional windowed model developed by Alapaty and Pleim (1995) to study the effects of different representations of atmospheric boundary layer processes on the vertical distributions of passive and reactive trace-gas species. The windowed model has 35 vertical sigma layers and a 3×3 grid cell structure in the horizontal. The model top is located at an altitude of ~4.5 km. There are two options available in the model to perform one-dimensional numerical simulations: (1) when observational data are available at more than one site, horizontal advections of horizontal momentum, temperature, the mixing ratio of water vapor, and trace-gas species are included; or (2) when observational data are available at just one site, the advections are dropped out. Because we used the second option, all grid cells in each of the layers were assumed to be horizontally homogenous. The model uses the conservation equations for horizontal winds, temperature, water vapor mixing ratio, soil moisture and temperatures for two soil layers, and trace-gas species. It can use one of four turbulence closure schemes: (1) a K-theory scheme based on the boundary layer scaling (BKT), a local-closure scheme; (2) the turbulent kinetic energy and its dissipation (TKE) scheme (Alapaty et al., 1994), a local-closure scheme; (3) the transilient turbulence parameterization (TTP) scheme (Stull, 1993), a nonlocal-closure scheme; and (4) the Asymmetrical Convective Model (ACM) (Pleim and Chang, 1992), a nonlocal-closure scheme. Processes in the surface layer are parameterized using surface layer similarity theory, while one of the above four schemes is used for the mixed layer. Alapaty et al. (1995) contains a more detailed model description.

Our model simulations used a set of ten chemical reactions (Table 1) representing the nonlinear chemical pathways related to NO_x-VOC chemistry (Hov et al., 1989). The chemistry in these experiments was defined as simply as possible to keep the number of equations to a minimum, while representing the nonlinear chemical pathways to characterize the tropospheric chemistry (Hov et al., 1989). Although meteorological conditions are realistically represented in the model simulations, the chemistry is too simple to represent routine atmospheric measurements. However, inclusion of this rather simplistic chemical scheme is sufficient for a preliminary investigation of the impact of vertical mixing on chemical calculations. The chemistry solver is based on concepts proposed by Mathur et al. (1995).

DATA AND NUMERICAL SIMULATIONS

Observational data from the First International Satellite Land Surface Climatology Project (ISLSCP) Field Experiment (FIFE) were used to perform numerical simulations. The FIFE site is located near Manhattan, KS, covering a 15×15-km area where grass prairie is the predominant vegetation. Measurements from one of the five intensive field campaigns (Sellers et al., 1992) were selected to provide initial meteorological conditions. Initial chemical conditions used in the numerical experiments are shown in Table 2. Two sets of numerical simulations were performed with each of the ABL schemes for the 12 hours starting at 0700 LT (local time) 11 July 1987. The first set of simulations was

Table 1. Chemical scheme used in model simulations*

1.	$HC + OH$	$\rightarrow$	$4RO_2 + 2HCHO$	$k_1 = 6 \times 10^{-12}$
2.	$HCHO + h\nu$	$\rightarrow$	$2HO_2 + CO$	$k_2 = 7.8 \times 10^{-5} \exp(-0.87/\cos\theta)$
3.	$RO_2 + NO$	$\rightarrow$	$NO_2 + HCHO + HO_2$	$k_3 = 8 \times 10^{-12}$
4.	$NO + HO_2$	$\rightarrow$	$NO_2 + OH$	$k_4 = 8.3 \times 10^{-12}$
5.	$NO_2 + h\nu$	$\rightarrow$	$NO + O_3$	$k_5 = 1 \times 10^{-2} \exp(-0.39/\cos\theta)$
6.	$NO + O_3$	$\rightarrow$	$NO_2 + O_2$	$k_6 = 1.6 \times 10^{-14}$
7.	$O_3 + h\nu$	$\rightarrow$	$O_2 + O^1D$	$k_7 = 1.9 \times 10^{-4} \exp(-1.9/\cos\theta)$
8.	$O^1D + H_2O$	$\rightarrow$	$2OH$	$k_8 = 2.3 \times 10^{-10}$
9.	$NO_2 + OH$	$\rightarrow$	HNO_3	$k_9 = 1.0 \times 10^{-11}$
10.	$CO + OH$	$\rightarrow$	$CO_2 + HO_2$	$k_{10} = 2.9 \times 10^{-13}$

* θ (~71.5°) is the solar zenith angle; rate constants are in units of cm^3 $molecule^{-1}$ s^{-1} (converted to ppm^{-1} s^{-1} when performing computations); photolysis rate constants are in s^{-1}.

Table 2. Initial maximum concentrations (ppb)

$HC = 4.84$	$HCHO = 5.2$	$NO = 4.84$	$NO_2 = 4.84$
$OH = 4.2 \times 10^{-6}$	$HO_2 = 4.2 \times 10^{-5}$	$RO_2 = 4.2 \times 10^{-5}$	$O_3 = 189.0$
$O(^1D) = 4.2 \times 10^{-14}$	$CO = 42.0$	$HNO_3 = 0.0$	

designed to reveal the bottom-to-top diffusion properties of the four ABL schemes. Trace-gas species were initially present in the lowest six layers of the model, representing the conditions of a stable boundary layer. Three reactive chemical species (HC, NO, and NO_2) were continuously released in these layers. All species had maximum concentrations in the lowest model layer and then decreased smoothly to specified background values in the sixth and higher layers. The second set of simulations was designed to reveal the top-to-bottom diffusion properties of each ABL scheme. Initial chemical conditions were similar to those in the first set; the main difference was that trace-gas concentrations were initialized to have larger values at higher altitudes (~800 to 1500 m), representing conditions similar to a residual boundary layer. Concentrations of all species were highest at about 1200 m altitude and decreased linearly to specified background values in the fifth layer above and below this altitude. The three reactive species (HC, NO, and NO_2) were continuously released at these higher altitudes. In addition, in both sets of simulations, a nonreactive tracer species was released at the same altitudes as the three reactive species.

RESULTS AND DISCUSSION

Meteorological parameters that can affect chemical mixing and associated reaction processes in the bottom-to-top and top-to-bottom diffusion simulations discussed below include (1) temporal variation in the depth of the ABL, (2) representation of mixing processes, and (3) resultant profiles of water vapor mixing ratio. The effects of temporal variation in other parameters such as temperature and winds are reflected through the evolution of the depth of the ABL and vertical mixing within it. Differences in the predicted meteorological fields with each of the ABL schemes are presented in detail by Alapaty et al. (1995).

Bottom-to-Top Diffusion

Comparison of the temporal and vertical evolution of a tracer species with that of a nonlinearly evolving chemical species may provide information on how turbulent mixing processes affect the nonlinear chemical reactions. Figure 1 shows time-height sections of a passive tracer simulated by the four ABL schemes. At any point in the simulation, the vertical variation in the tracer indicates the effects of turbulent mixing and the simulated depth of the ABL. The concentration of the tracer gradually decreases with time in all simulations because the increase in the simulated ABL depth caused an overwhelming dilution of the tracer as compared to the increase in concentration due to its emission. Further, simulated virtual potential temperature and water vapor mixing ratio profiles (not shown) in all simulations showed vertical gradients close to the surface, as was found in the observations. The tracer, however, does not show this behavior near the surface. This may be attributed to the relatively low emission rates used in the simulations. In addition, the decrease in the tracer concentration with the BKT scheme is more gradual than with the TKE and TTP schemes, indicating more vigorous vertical mixing in the latter two simulations. With the ACM scheme, the tracer concentration shows a gradual decrease during the first half of the simulation, and then it reflects intense vertical mixing for the rest of the simulation period.

At the starting time of simulation (initial conditions), observations indicated the presence of a nocturnal jet in the horizontal wind fields. The vertical wind shear associated with the remnants of this jet is about 3 m s^{-1} for every 100 m in the lowest 1 km of the atmosphere. The effects of this strong vertical wind shear in the stable boundary layer are well accounted for in the TKE simulation, because prognostic equations are used for estimating the turbulent kinetic energy. The TKE panel of Figure 1 clearly shows that the tracer was mixed through a larger depth (up to ~1015 m) by 0800 LT than with the other schemes. With the TTP scheme, vertical mixing is confined to lower altitudes (~815 m) at 0800 LT because terms related to the transport of turbulence are neglected. Since neither the BKT nor ACM schemes explicitly deals with turbulence generation or depletion, the effects of wind shear on the vertical turbulent mixing are absent in the early stages of the simulation, as is reflected in the vertical extent of the tracer at 0800 LT. In the TTP simulation, the presence of vertical gradients between hours 1200 and 1300 LT indicates a weaker turbulent mixing between the surface and ~415 m altitude. The maximum depths of the ABL in the BKT, TKE, TTP, and ACM schemes are about 1300, 1400, 1700, and 1600 m, respectively. Thus, the maximum variation in the depth of the simulated ABL among the schemes is about 24%. Simulated tracer gradients by the four ABL schemes are also indicative of the differing representations of mixing intensity. In general, nonlocal-closure schemes (TTP and ACM) show more vigorous mixing as compared to the local-closure schemes (BKT and TKE).

We now consider the formation of HNO_3, a nonlinearly evolving secondary pollutant (Figure 2). HNO_3 is a particularly interesting reactive species because it is formed through complicated chemical reactions in the atmosphere. Also, the precursor species for HNO_3 formation in turn undergo complex chemistry and their relative concentrations are affected by mixing processes. Figure 2 shows that during the first half of the simulation (up to ~1300 LT), HNO_3 is produced by the chemical reactions then distributed through the vertical mixing processes. Between 1300 and 1600 LT, the production becomes smaller and is overwhelmed by the dilution due to the rapid growth of the ABL. The differences in the extent of dilution among the four schemes are striking. In the BKT simulation, the HNO_3 formed between 0830 and 1130 LT has a maximum concentration of about 2 ppb. In the TKE simulation, formation of HNO_3 is limited to a maximum of about 1.25 ppb. Further, between 0700 and 1100 LT, the turbulent mixing due the nocturnal jet has resulted in weaker vertical gradients in the HNO_3 concentrations near the top of the ABL. In the TTP simulation, the maximum HNO_3 concentration (1.25 ppb) is similar to that of the TKE simulation. The ACM simulation results are similar to those from the BKT simulation until about 1400 LT, though the peak HNO_3 predicted is relatively smaller. These preliminary

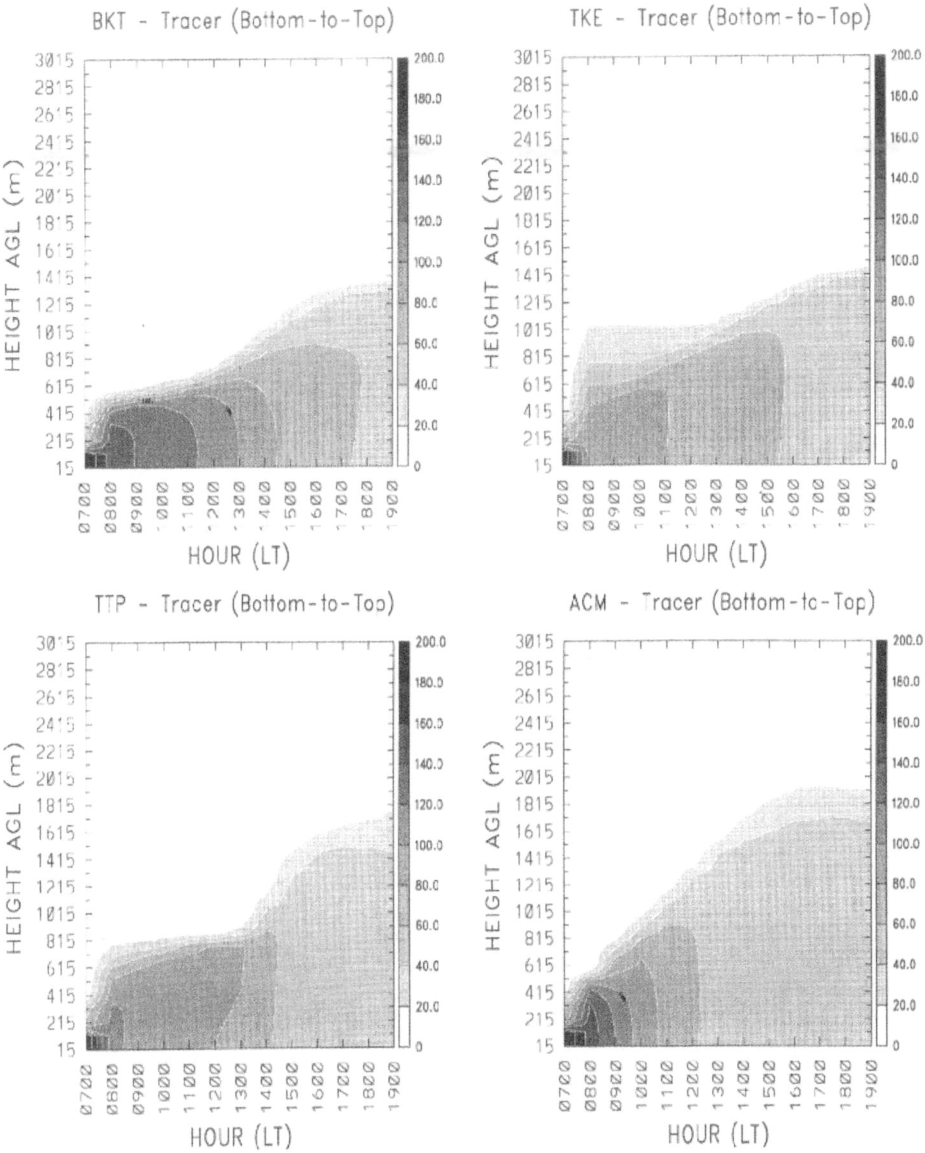

Figure 1. Temporal and vertical variation resulting from bottom-to-top diffusion of a nonreactive tracer using the four ABL schemes.

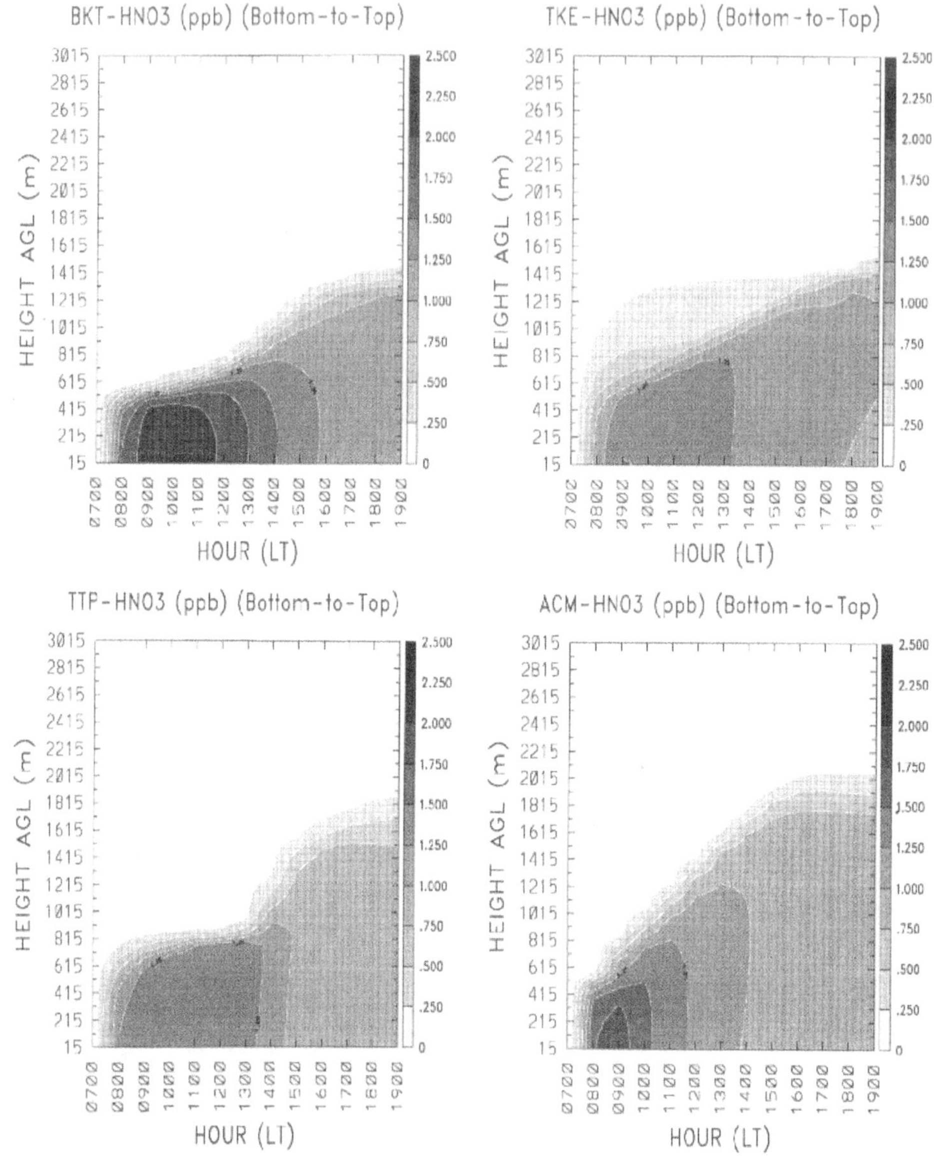

Figure 2. Temporal and vertical variation resulting from bottom-to-top diffusion of HNO_3, a nonlinearly evolving species, using the four ABL schemes.

results on HNO_3 formation indicate that the choice of turbulent mixing scheme can affect the chemical reactions.

Top-to-Bottom Diffusion

Emissions from elevated stacks are major sources of atmospheric pollutants. Also, the remnants of the previous day's mixed layer (residual boundary layer) can still contain chemical pollutants. These two sources can affect the concentration of chemical species in the ABL during the daytime. Even though the top-to-bottom component of turbulent diffusion is relatively weak compared to the bottom-to-top component, it is interesting to explore the differences that could result from using differing representations of turbulent mixing processes. The results discussed in the previous section showed that both primary and secondary pollutants are confined to a finite depth of the lower troposphere, the depth of the ABL. In the absence of moist or dry convection, a capping inversion traps pollutants in the ABL. Also, pollutants emitted above the stable boundary layer or in the residual boundary layer get concentrated during the nighttime and can then be brought down to the surface as the ABL evolves during the daytime. The representativeness of different ABL schemes under such conditions are examined and discussed here using a top-to-bottom diffusion study.

We now consider the time-height section of the nonreactive tracer released at higher altitudes (centered at ~1200 m). When the tracer is emitted well above the top of the ABL (stable boundary layer at starting time), the tracer profile is primarily determined by the mechanical mixing induced by the strong vertical wind shear. As the evolving ABL reaches this emission altitude, the tracer fumigates into the boundary layer. Figure 3 shows that the downward diffusion of the tracer is strikingly different in each of the simulations. The tracer reached the ground at a different time for each scheme. Further, the descent of the tracer to the ground is almost instant (no vertical gradient) only with the TTP scheme, while the tracer shows a gradual descent (vertical gradient) with the other three schemes. Since the downward diffusion component is local in the ACM, the vertical gradient is similar to those in the BKT and TKE simulations.

Figure 4 shows the temporal and vertical variation of HNO_3 for the top-to-bottom case. In all simulations, the descent of the HNO_3 to ground level is somewhat similar to that of the tracer. In the BKT simulation, the vertical gradient in the HNO_3 is unchanged until the top of the evolving ABL reaches the 1200 m altitude. In the TKE simulation, weaker vertical gradients indicate a stronger turbulent mixing than in the BKT simulation. In the TTP simulation, vertical turbulent mixing is weak until about 1300 LT. This leads to the maximum formation of HNO_3 (~15 ppb) at the altitudes where sources are released; after 1300 LT, stronger vertical mixing in the ABL results in well-mixed HNO_3 profiles. In the ACM simulation, turbulent mixing at the 1200 m altitude is somewhat stronger than that in the other simulations. It is interesting to note that even though the BKT and ACM schemes use the same formulation to represent turbulent mixing in the free atmosphere, simulated concentrations of HNO_3 are quite different. This is because the estimated altitudes of the top of the ABL in the BKT and ACM schemes are different, and consequently, the resulting concentrations of HNO_3 near the top of the ABL are different. Also, the near-surface (~10 m) maximum concentrations of HNO_3 with the BKT and TKE schemes during the entire simulation period are about 1.5 to 3 ppb. However, with the nonlocal-closure schemes (TTP and ACM), the near-surface maximum concentrations of HNO_3 are higher (3 to 4.5 ppb) and "touchdown" of this maximum occurs much earlier (at 1500 and 1300 LT, respectively).

These preliminary results indicate that simulated concentrations are sensitive to the type of mixed layer parameterization scheme used in the simulations. Detailed analysis of other secondary species and extension of the model to simulate realistic chemical conditions is in progress.

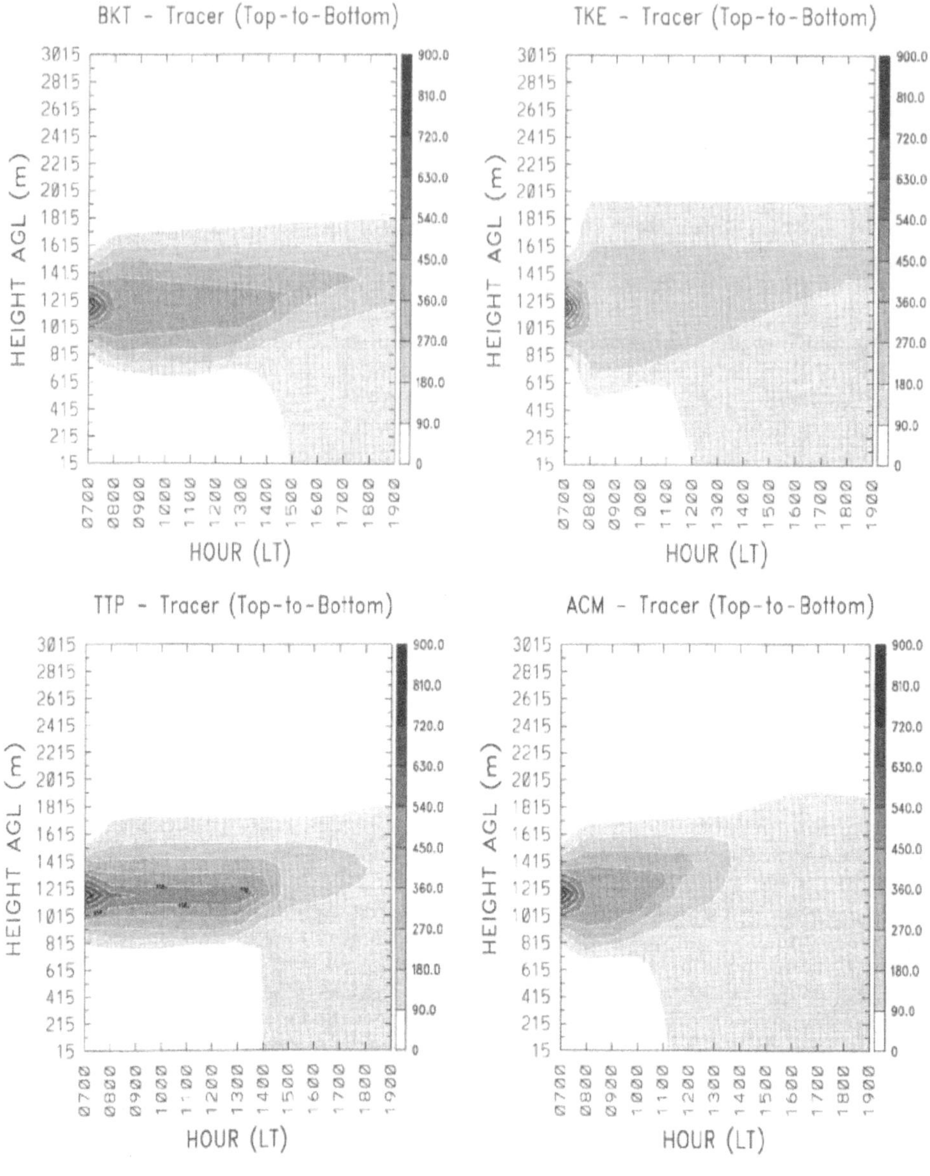

Figure 3. Temporal and vertical variation resulting from top-to-bottom diffusion of a nonreactive tracer using the four ABL schemes.

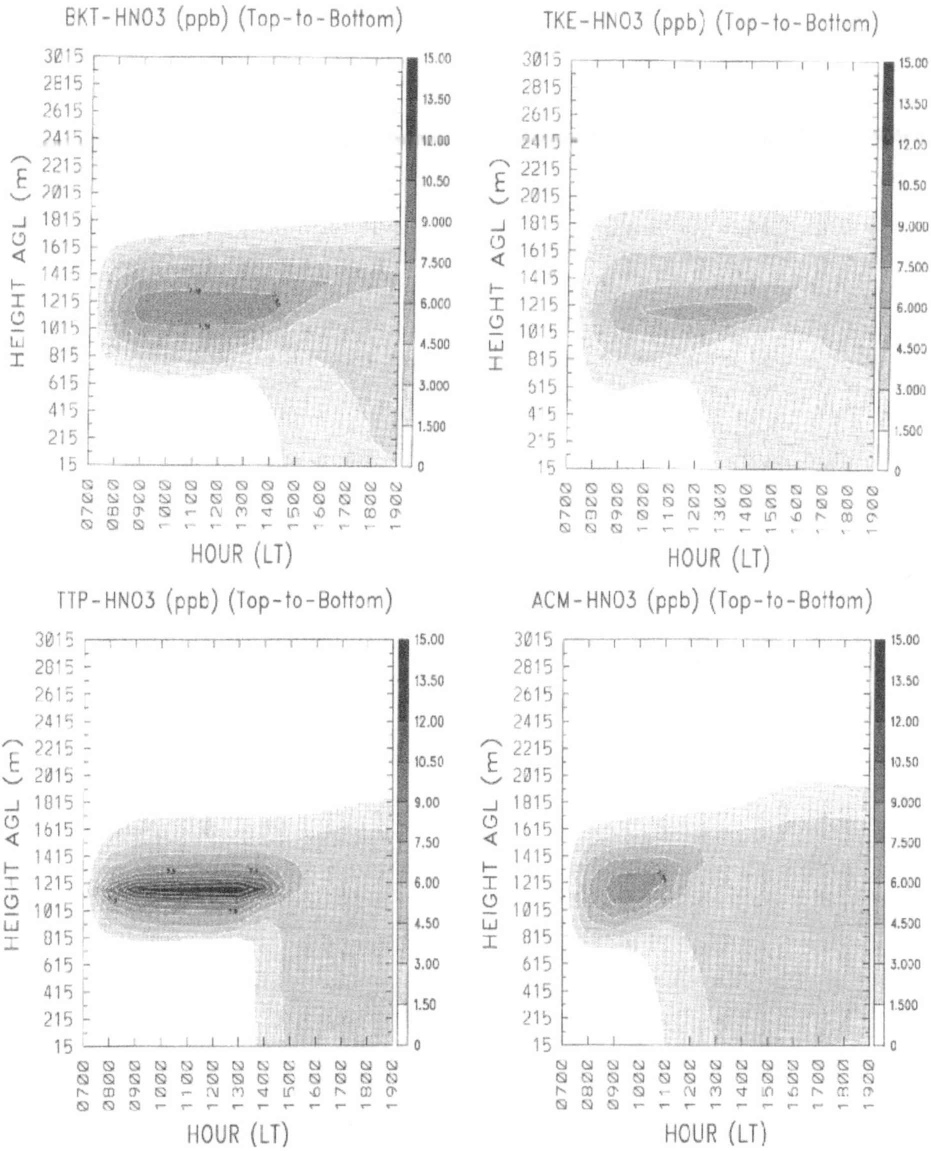

Figure 4. Temporal and vertical variation resulting from top-to-bottom diffusion of HNO_3, a nonlinearly evolving species, using the four ABL schemes.

NOTICE

This paper has been reviewed in accordance with the U.S. Environmental Protection Agency's peer and administrative review policies and approved for presentation and publication. It has also been approved for publication by MCNC. Mention of trade names or commercial products does not constitute endorsement or recommendation for use.

REFERENCES

Alapaty, K., and Pleim, J.E., 1995, Evaluation of PBL schemes using the First ISLSCP Field Experiment (FIFE) data. *11th Symposium on Boundary Layers and Turbulence,* American Meteorological Society, Charlotte, NC.

Alapaty, K., Pleim, J.E., Raman, S., Niyogi, D.S., and Xiu, A., 1995, Simulation of atmospheric boundary layer processes using local- and nonlocal-closure schemes. In preparation.

Alapaty, K., Raman, S., and Madala, R.V., 1994, Investigation of the role of the boundary layer processes in an active monsoon using a mesoscale model. *Bound.-Layer Meteor.* 67:407-426.

Arellano, V.D., and Duynkerke, P.G., 1995, Atmospheric surface layer similarity theory applied to chemically reactive species. *J. Geophys. Res.* 100(D1):1397-1408.

Hov, O., Zlatev, Z., Berkowicz, R., Eliassen, A., and Prahm, L.P., 1989, Comparison of numerical techniques for use in air pollution models with nonlinear chemistry. *Atmos. Environ.* 23:967-983.

Mathur, R., Young, J.O., and Schere, K.L., 1995, A comparison of numerical techniques for solution of mass action kinetics of the troposphere. In preparation.

Pleim, J.E., and Chang, J.S., 1992, A non-local closure model for vertical mixing in the convective boundary layer. *Atmos. Environ.* 26A:965-981.

Sellers, P.J., Hall, F.G., Asrar, G., Strebel, D.E., and Murphy, R.E., 1992, An overview of the First International Satellite Land Surface Climatology Project (ISLSCP) Field Experiment (FIFE). *J. Geophys. Res.* 97(D17):18,345-18,371.

Stull, R., 1993, Review of transilient turbulence theory and nonlocal mixing. *Bound.-Layer Meteor.* 62:21-96.

DISCUSSION

R. BORNSTEIN: Would you include the residual nighttime elevated mixed layer in your definition of the mixing depth and mixing lengths?

D. BYUN: Defining mixing depth for nighttime is somewhat tricky and there are many remaining issues needed to be addressed. We have seen some of this in Dr. Zilitinkevich's presentation on the definition of stable boundary layer height. Although we have not given enough attention on the issue whether we should include the residual elevated mixed layer as a part of the mixing depth, the ramification of this assumption is large. For an elevated source which is within the residual layer, the pollutant may disperse as if it is in neutral condition. On the other hand the dispersion of the surface emission will be limited by the stable boundary layer mixing depth. I suspect that these pollutants may not mix together until the fumigation happens after the sunrise.

ADVANCED MESOSCALE DISPERSION MODELING USING KINEMATIC SIMULATION

Robert Yamartino[1], David Strimaitis[1], and Arno Graff [2]

[1]Earth Tech Inc., 196 Baker Avenue, Concord, MA 01742
[2]im Umweltbundesamt, Bismarckplatz 1, D-14191 Berlin, Germany

INTRODUCTION

The principal urban-scale regulatory model (i.e., AUSTAL '86) in use in Germany today and the eight Guideline models (i.e., BLP, CALINE3, CDM 2.0, RAM, ISC2, MPTER, CRSTER, and OCD) currently recommended by the U.S. EPA (1987) for non-reactive pollutants, are all based on the Gaussian plume formulation. Despite its practical utility and its coverage of a wide range of relevant cases in regulatory practice, the Gaussian plume model formulation does not explicitly consider phenomena such as: non-steady state emissions or dispersion meteorology conditions, wind direction and/or speed shear with height, calm conditions, recirculating flows such as sea- and lake-breezes, the asymmetric vertical dispersion (i.e.,top-down vs bottom-up) conditions accompanying convective mixing, and spatial inhomogeneities in terrain, land use or key meteorological variables such as stability and mixing depth. These phenomena become increasingly important when modeling mesoscale (i.e., > 20 km.) regimes, where the growing use of data assimilation techniques in the meteorological models has led to greatly improved representation of the mean flow. Ideally, one would prefer a dispersion model capable of simultaneously simulating phenomena on a broad range of scales ranging from microscale through mesoscale and optimally utilizing the information produced by modern mesoscale meteorological models.

Modern Gaussian puff models, such as CALPUFF (Scire et al.; 1990, 1995--this meeting), and Langevin-equation-based, Lagrangian particle models, such as LASAT (Janicke; 1990, 1995), provide alternative approaches to modeling ensemble-average concentration, dose, and surface deposition; however, these approaches cannot predict short-term concentration fluctuations without additional assumptions about the probability distribution function (PDF) of concentrations.

Realizing that for certain situations it would be advantageous to include the above mentioned phenomena explicitly, the German Federal Environmental Agency (UBA) funded the development and preliminary testing of an advanced atmospheric dispersion model, referred to as the Kinematic-Simulation-Particle (KSP) Model and based on the kinematic simulation (KS) of atmospheric turbulence coupled to a Lagrangian particle approach. The KSP model provides the capability to assess the consequences of air pollution emissions over a broad range of spatial and temporal scales, terrain and land use environments, and non-steady meteorological and release conditions. More importantly, the KSP model has the intrinsic capacity to predict the short-term concentration fluctuations that are critical to the assessment of odor and hazardous chemical exposure problems.

The Kinematic-Simulation-Particle (KSP) model (Yamartino et al.,1993 and Strimaitis et al., 1995) is a particle-based model for transport, dispersion, and deposition, that relies on explicit simulation of the larger transportive and dispersive, sub-grid-scale (SGS) eddy structure of the flow field to advect particles. This approach is in contrast to traditional Lagrangian particle model approaches, such as the MC-LAGPAR II model of Zannetti (1986) or the LASAT model of Janicke (1990, 1995), which use independent Langevin equations for each of a particle's three degrees of freedom. Although fully responsive to the local 3-d flow and turbulence fields, the independent motions of the particles inhibits these models from predicting the higher temporal moments (e.g., C^n) and variability measures of concentrations unless additional assumptions are made concerning the expected PDF of concentrations.

Kinematic simulation avoids additional difficulties created by the use of independent Langevin equations for each particle's degree of freedom. For example, the independent velocity component approach leads directly to the problem of particle accumulation in low turbulence zones. This problem then forces one to hypothesize drift velocities to inhibit such accumulation. Other corrections are needed to recover velocity correlation terms (e.g., $u'w' \neq 0$). KS avoids these problems of particle aggregation and loss of velocity correlations by constraining the explicit SGS flow fields to satisfy the kinematic constraint of divergence-free behavior, that is

$$\underline{\nabla} \bullet \underline{V} = 0 \quad or \quad \underline{\nabla} \bullet \rho\, \underline{V} = 0 \quad ,$$

and hence the designation as kinematic simulation or synthetic turbulence models. Also, KS modeling of the SGS flow fields enables Eulerian velocity field statistics to be predicted, and thus KS serves as a complete bridge between the Lagrangian modeling frame and the Eulerian measurement frame.

A major challenge to the development of a practical KS model was the generalization of the uniform wind, isotropic turbulence formulation of Fung et al. (1990) to non-isotropic, stable through convective, turbulence in arbitrarily sheared atmospheric flows. The present paper provides 1) a brief overview of the modeling system; 2) a review of some on the model's novel features; and 3) our generalization the KS formalism to non-isotropic turbulence in arbitrarily sheared atmospheric flows.

OVERVIEW OF THE KSP MODELING SYSTEM

The "system" is comprised of a control file and input data files, meteorological processing programs, the KSP dispersion model, and a postprocessor program. Perhaps the most important input data involves the 3-d gridded mesoscale wind fields and accompanying 2-d fields specifying boundary layer quantities of interest (e.g., mixed layer depths, friction velocities, heat fluxes, convective velocity scale). These data are generally provided by an external wind field/PBL model. However, in the case where only single-point measures of wind and stability are available, an internal flow and turbulence generator creates a vertical profile which emulates the dispersion conditions specified in the German regulatory guideline (TA-Luft). The current KSP model also accesses data produced by the CALMET meteorological model (Scire et al., 1990; and U.S. EPA, 1995), which may be initialized using observations and/or outputs from other meteorological models, such as MM-4/5 (Hsie and Kuo, 1987; Grell et al., 1994), or GESIMA (Kapitza and Eppel, 1992; Eppel et al., 1994). A new interface to GESIMA developed at the Free University of Berlin (Reimer et al., 1994) also permits the modeled TKE budgets to be used in the computation of turbulence fields.

Specifications for a particular application of the KSP model are supplied via a control file prepared by the user. Its structure is self-documenting, ensuring relatively uncomplicated, straightforward use of the model. The user needs to specify the parameters associated with the sources (e.g., coordinates, stack heights, release rates) and sampling locations of interest; however, supplementary detailed information such as geophysical data on terrain and vegetation is included in additional data files that are generally prepared well in advance of applications.

KSP MODEL FEATURES

The dispersion model includes modules for particle release, SGS flow field generation, particle transport via mean and SGS flow fields, dry and wet removal processes, sampling of pollutant concentration and deposition flux fields, computation of statistical measures (e.g., means, highs, lows, standard deviations) for concentrations and deposition fluxes, and output of fields for subsequent graphical and statistical postprocessor programs. We now consider several of the KSP's modules that are unique or specific to this modeling approach.

Particle Mass

KSP includes algorithms for releasing and tracking particles from point and area sources. The pollutant mass that is chosen to be represented by each particle is based on considerations of statistical uncertainty in resolving concentrations that are near a regulatory standard. From the point of view of reducing computer resource requirements, one would like to put as much mass as possible into each particle, thereby limiting the number of particles that must be tracked. However, with fewer particles, sampling or counting uncertainties increase. In addition, any mass variability of the sampled particles will always lead to degraded, effective particle counting statistics as compared with an equal-particle-mass sample. As concentrations that are at or near a regulatory limit or standard are generally of greatest importance, the mass per particle is computed in a way that provides a user-selectable, statistical counting uncertainty for estimating instantaneous concentrations that equal or exceed the regulatory standard.

Denote the regulatory concentration for pollutant species s as C_s, the mass per particle for this species as M_s, and the receptor volume as V. Then the mass per particle that is consistent with resolving a concentration C_s to within a fractional uncertainty f is

$$M_s = C_s V f^2$$

The value of f is set sufficiently small that the scaled concentration variance associated with counting particles is small when compared to the natural or stochastic fractional concentration variance. The observed stochastic uncertainty is typically of order 1, which suggests that the variance contributed by counting uncertainty should be of order 0.1 or less to be considered small. Recognizing that the fractional variance due to counting particles is f^2, the above condition is equivalent to requiring that f be no larger than 0.3. Thus, M_s represents the maximum mass of that pollutant that may be assigned to a particle, regardless of the mass emission rate of that pollutant from any particular source.

Plume Dynamics

Plume rise in KSP is simulated by adding a vertical velocity to the advective wind field experienced by specific, stack-emitted particles. This rise velocity, w_r, decreases to zero over time after emission due to entrainment. The decrease in w_r is modeled simply as a decaying exponential function of time (or particle age), with some initial velocity. An exponential decay function for the rise velocity was derived recently by van Dop (1993) in his study of plume rise in a Lagrangian framework. Taking the leading order terms in his result, we have the approximation:

$$w_r = (B_o t + w_{exit}) e^{-t/t_r}$$

where,

B_o $= (g/T)\ (\Theta_o - \Theta_a)$
Θ_o = initial plume temperature
Θ_a = ambient temperature
w_{exit} = exit velocity of the plume

Imposing the constraint that the final rise equal the integral of $w_r(t)dt$ over all time, the decay timescale becomes

$$t_r = \frac{w_{exit}}{2 B_o}\left[\sqrt{1 + \frac{4 B_o \Delta h_r}{w_{exit}^2}} - 1\right]$$

where the plume rise, Δh_r, is computed from the Briggs (1975) plume rise equations.

Several other plume dynamics mechanisms are also included in the KSP model including stack downwash (Briggs, 1973), partial plume penetration into an elevated inversion (Manins, 1979; Hanna and Chang, 1991), and several building downwash algorithms developed for CALPUFF (Scire et al.,1995).

Particle Transport, Depletion, and Sampling

Consistent with the requirements of the German TA-Luft regulations, KSP is designed to produce estimates of area-averaged concentrations and depositions for a gridded receptor field. To this end, particles are tracked within the modeling domain and counted whenever they enter a volumetric sampling cell. The mass associated with each particle that is counted in a cell is a function of the pollutant species, and the deposition history experienced by the particle. A reduced probability of survival is computed whenever a particle enters the deposition layer and is subject to a non-zero deposition velocity. Survival probabilities less than unity are equivalent to a proportional loss of species mass from the particle. As a result, it is the survived mass of each pollutant contributed by a particle that is counted within a volumetric sampling cell to obtain a concentration, and it is the change in this survived mass (i.e., per time step per cell area) that is counted within a cell to obtain deposition flux.

Particles contribute to the average mass concentration in a receptor-box whenever they are carried into it by the flow. Sampling methods that only count particles within the receptor-box have been referred to as "box-car" methods. Because a particle is treated mathematically as a point (and tracked in KSP with a spatial resolution of 1 mm) while it carries the mass of a pollutant that is associated with a significant volume, the "box-car" method will generally produce an unrealistically "noisy" count of the mass in a receptor-box, as particles pass through the lid of the box. Therefore, the sampling method used in KSP employs a tapered sampling zone above the receptor height, in which particles are weighted via a smoothly-decaying function of their height above the top of the receptor-box.

Given the underlying KS eddy formulation and other sources of fluctuations, the concentration of pollutant s varies from one timestep to another. As these fluctuations are of interest, the maximum, minimum, and standard deviation of receptor-volume-averaged concentrations are computed at each timestep (e.g., typically about 3 seconds) in addition to the mean concentration for the period.

The loss of pollutant mass from the atmosphere to the ground as a result of deposition is modeled as a decrease in the species-dependent survival probability associated with individual particles. This change in survival probability is a function of particle height above the surface and a species-dependent deposition velocity, v_d. As individual mathematical particles represent all gaseous or particulate pollutants (i.e., those not subject to significant gravitational settling) released from a source, the mathematical particle, or marker, represents the location of pollutants released at a particular time from a certain source. Thus, while the species mass symbolized by the particle is depleted as it approaches the surface, the mathematical particle itself is not captured or removed.

The survival probability is modeled as it is in a grid model. The first-order differential equation (D.E.) for material loss within a cell of area A_{dep}, volume V_{dep}, and depth H_{dep}, is

$$V_{dep} \frac{dC}{dt} = -F\, A_{dep} \quad ,$$

where C is the concentration in the cell, and F is the deposition flux which is expressed as the product of a deposition velocity, v_d, times the concentration. For particles within the deposition layer, the solution of this first-order D.E. allows the survival probability for a particle's species mass at the end of the timestep to be written as

$$s_p(t) = s_p(t - \Delta t)\, e^{-v_d \Delta t / H_{dep}} \quad ,$$

where Δt is the length of the timestep. Mass which does not survive is assumed to be deposited to the surface within the timestep.

Because the distribution of particles within the deposition layer is not uniform and the mass associated with each particle can be different, these equations only approximate the removal process in a bulk sense. Adopting this bulk representation, we require that v_d, be evaluated at height $H_{dep}/2$ in order to be "representative" of the layer as a whole.

KSP allows a diurnal cycle of 24 hourly values of v_d to be specified in the input file, or it computes v_d internally as a function of meteorological and land use data using a resistance model. The subroutines for the internal calculation of v_d are identical to those developed for use in grid (i.e., CALGRID) and puff (i.e., CALPUFF) counterpart models.

THE KS SPECTRAL FORMULATION OF THE MODEL

Although conceived 25 years ago by Kraichnan (1970) and Drummond et al. (1984), the KS approach initially considered for the KSP model was formulated by Fung et al. (1990) for isotropic turbulence. Early in this study, it was found that this isotropic KS formulation is quite useful for simulating the type of dispersion found in simple, uniform wind and turbulence field, Gaussian plume models. The KS formalism was then extended to the more general atmosphere containing height dependent variations in wind velocity (i.e., shear), turbulence and turbulence anisotropies. This was accomplished by first transforming Fung's 3-d isotropic eddies into component 2-d eddies lying in the x-z, y-z, and x-y planes, respectively. The amplitudes of each of these components could then be independently varied to achieve the desired vertical and horizontal plume dispersion rates. The amplitudes of the x-z and y-z components were set equal and constrained so that the quadratic sum of amplitudes equaled the desired vertical velocity variance. Any excess of the horizontal velocity variance over the vertical velocity variance was then channeled into an x-y velocity meander component that advects all KS eddies.

The current KSP model uses eddies as small as 12.5m for all stabilities, an added 50m wavelength component for the most stable Klug (or Pasquill-Gifford-Turner (PGT)) Class, and a sum over five eddy sizes, with wavelengths as large as 1600m, for the most unstable Klug (or PGT) dispersion category. The relative admixture of the various wavelength components was determined from empirical spectra for the stable and convective boundary layers published by Moraes (1988) and Hojstrup (1982), respectively, and combined with our characterization of the KS system's spectral response to various input wavelengths and time scales. Though only a few discrete wavelengths are input, the fact that the amplitudes and phases of these various Fourier components of turbulence evolve according to Langevin equations causes the final spectrum of velocity fluctuations to smear out into surprisingly realistic shapes and have other desirable characteristics. For example, the traditional, independent-particle Langevin equation approach generally involves use of parameterized integral time-scale profiles, $T_L(z)$, which only implicitly recognize the fact that longer wavelength eddies have longer time scales than the shorter components. In the KS approach, each spectral wavelength component has a single, fixed time scale T_L; however, as the mix of spectral components varies with height, one obtains an effective $T_L(z)$ profile.

DISCUSSION AND CONCLUSIONS

The KSP modeling system is currently being subjected to an extensive evaluation process to determine any biases and to quantify modeling uncertainties. Such an evaluation process is also the key model acceptance criteria to emerge from the series of three European Community sponsored model 'harmonization' workshops held to date. Intercomparisons with data obtained during the Copenhagen (Grønskei, 1990) and Lillestrøm (Haugsbakk and Tønnesen, 1989) tracer experiments are now being conducted and future efforts will focus on intercomparison with the Electric Power Research Institute (EPRI) sponsored Kincaid tracer study. Although the Kincaid site offers simple terrain, the large power plant places the plume well above the surface, so that the data test the longer wavelength, convective, KS eddy formulations in the upper part of the boundary layer. During the day, these convective eddies promote unsteady plume behavior resulting in complex patterns of ground-level concentrations.

In summary, the KSP model provides the short-range (i.e., < 30 km), multi-source simulation capabilities of the TA-Luft regulatory model, AUSTAL '86, in current use, plus it provides the:

- ability to function on considerably larger mesoscale domains (e.g., to 10^3 km) encompassing wide variations in meteorology, land use and terrain;
- ability to interface to several diagnostic/prognostic meteorological models that reflect the state-of-the-science of flow and turbulence field simulation;
- capability of computing instantaneous concentration fields, higher concentration moments, $C^n(t)$, and concentration variability measures (e.g., variances, intermittency) as well as the more traditional time/ensemble average concentration measures; and
- flexibility to permit new developments in turbulence, dry and wet removal, and other scientific areas to be rapidly incorporated through a highly modular model structure.

ACKNOWLEDGEMENT

This work was supported by the German Federal Environmental Agency (Umweltbundesamt) as part of the Environmental Research Plan of the Federal Ministry of Environment, Nature Conservation, and Reactor Safety.

REFERENCES

Briggs, G.A., 1973: Diffusion estimates for small emissions (Draft). Air Resources Atmospheric Turbulence and Diffusion Laboratory. ATDL No. 79.

Briggs, G.A., 1975: Plume rise predictions. In: *Lectures on Air Pollution and Environmental Impact Analyses*. American Meteorological Society, Boston, MA, pp. 59-111.

Drummond, I.T., S. Duane and R.R. Horgan, 1984: Scalar diffusion in simulated helical turbulence with molecular diffusivity. *J. Fluid Mech.*, **138**, 75-91.

Eppel, D., H. Kapitza, M. Claussen, D. Jacob, L. Levkov, H.T. Mengelkamp, and N. Werrmann, 1994: The non-hydrostatic mesoscale model GESIMA. Part II: parameterizations and application. To be published in *Beitr. Phys. Atmosph.*

Fung, J.C.H., J.C.R. Hunt, N.A. Malik and R.J. Perkins, 1992: Kinematic simulation of homogeneous turbulent flows generated by unsteady random Fourier modes. *J. Fluid Mech.*, **236**, 281-318.

Grell, G.A., J. Dudhia and D.R. Stauffer, 1994: A Description of the Fifth-Generation Penn. State/NCAR Mesoscale Model (MM5). National Center for Atmospheric Research Technical Note NCAR/TN-398+STR., 138 pp.

Grønskei, K.E., 1990: Variation in dispersion conditions with height over urban areas - results of dual tracer experiments, 9th AMS Symposium on Turbulence and Diffusion.

Hanna, S.R. and J.C. Chang, 1991: Modifications of the Hybrid Plume Dispersion Model (HPDM) for urban conditions and its evaluation using the Indianapolis data set, Volume I: User's Guide for HPDM-URBAN. EPRI Project No. RFP-02736-1, Palo Alto, CA 94303.

Haugsbakk, I. and D.A. Tønnesen, 1989: Atmospheric Dispersion Experiments at Lillestrøm, 1986-1987 Data Report, Lillestrøm, Norwegian Institute for Air Research (NILU OR 31/89).

Hojstrup, J., 1982: Velocity spectra in the unstable planetary boundary layer. *J. Atmos. Sci.*, **39**, 2239-2248.

Hsie, E.-Y. and Y.H. Kuo, 1987: Description of the Penn. State/NCAR Mesoscale Model Version 4 (MM4). National Center for Atmospheric Research Technical Note NCAR/TN-282+STR., 66 pp.

Kapitza, H., and D. Eppel, 1992: The non-hydrostatic mesoscale model GESIMA. Part I: dynamic equations and tests. *Beitr. Phys. Atmosph*, **65**, 129-146.

Janicke, L., 1990: Ausbreitungsmodell LASAT-C: Handbuch Version 1.00. (In German)

Janicke, L., 1995: Ausbreitungsmodell LASAT: Referenzbuch zu Version 2.51, Primelweg 8, D-88662, Überlingen, FRG. (In German)

Kraichnan, R.H., 1970: Diffusion by a random velocity field. *Phys. Fluid*, **13**, 22-31.

Manins, P.C., 1979: Partial penetration of an elevated inversion layer by chimney plumes. *Atmospheric Environment*, **13**, 733-741.

Moraes, O.L.L., 1988: The velocity spectra in the stable atmospheric boundary layer. *Boundary Layer Meteorol.*, **43**, 223-230.

Reimer, E., B. Scherer, W. Klug, and R. Yamartino, 1994. A Meteorological Data Base System for Next Generation Dispersion Models and a Lagrangian Particle Model Based on Kinematic Simulation Theory. *Proceedings of the Workshop on Intercomparison of Atmospheric Dispersion Modeling Systems*, Mol, Belgium.

Scire, J.S., E.M. Insley and R.J. Yamartino, 1990: Model formulation and user's guide for the CALMET meteorological model. Prepared for the California Air Resources Board. Sigma Research Corporation, Concord, MA.

Scire, J.S., D.G. Strimaitis and R.J. Yamartino, 1990: Model formulation and user's guide for the CALPUFF dispersion model. California Air Resources Board, Sacramento, CA.

Scire, J.S., D.G. Strimaitis, R.J. Yamartino and X. Zhang, 1995: A User's Guide for the CALPUFF Dispersion Model. EARTH TECH Document 1321-2 prepared for the USDA Forest Service, Cadillac, MI.

Strimaitis, D., R. Yamartino, E. Insley, and J. Scire, 1995. A User's Guide for the Kinematic Simulation Particle Model. EARTH TECH Document 1274-2 prepared for the Free University of Berlin and the Umweltbundesamt, Berlin, FRG.

U.S. EPA, 1987: Supplement A to the *Guideline on Air Quality Models (Revised)*. EPA-450/2-78-027R.

U.S. EPA, 1995. A User's Guide for the CALMET Meteorological Model. EPA-454/B-95-002.

van Dop, H., 1993: Buoyant plume rise in a Lagrangian framework. *Boundary Layer Meteorol.*,**62**,97-105.

Yamartino, R., D. Strimaitis, M. Spitzak, and W. Klug, 1993. Development and Preliminary Evaluation of a Dispersion Model Based on Kinematic Simulation Theory. *Proceedings of the Workshop on Intercomparison of Atmospheric Dispersion Modeling Systems*, Manno, Switzerland.

Zannetti, P., 1986: Monte-Carlo simulation of auto- and cross-correlated turbulent velocity fluctuations. MC-LAGPAR II model, *Environ. Software*, **1**, 26-30.

DISPERSION IN IDEALIZED URBAN STREET CANYONS

Robert N. Meroney [1] Stillianos Rafailidis,[2] and Michel Pavageau[2]

[1]Civil Engineering Department
Colorado State University
Fort Collins, CO 80523

[2]Meteorology Institute
University of Hamburg
D=20146 Hamburg, Germany

INTRODUCTION

Urban street canyon geometry and pollution source configuration play important roles in determining the level of peak levels of air pollution due to transportation sources. Urban air pollution specialists anticipate that street canyon depth and width, intersection locations, canyon orientation to the dominant wind directions and local building geometries will determine most of the peak pollution incidents which determine air pollution control strategies. Most relevant urban air pollution models are based on limited measurements made at field and laboratory scales. Uncertainty still exists concerning the level of complexity which must be contained in numerical models to reproduce critical aspects of canyon motions, especially at lower wind speeds. Past laboratory measurements may have been biased by inaccuracies introduced by the line source designs used in the experiments. This paper reports the results of visualization, wind field, and dispersion measurements made in a meteorological wind tunnel using an improved line-source configuration. Results suggest that mean concentrations vary inversely with wind speed. Observations also confirm hat the flow within the canyon is highly intermittent, and the street level vortex periodically lifts out of the canyon and then reforms.

In complicated roughness areas like large cities pollutant dispersion is often poor and can deteriorate further when unfavorable meteorological conditions exist (weak winds, cloudy skies, low inversion capping layer). Fortunately, building and street patterns exist for which ventilation of street canyons (a road and its flanking buildings) is significantly improved (Oke, 1988). For example wind tunnel studies have found that entrapment of pollution may be alleviated by distributing higher rise structures along a street which generates corner vortices that mix pollution upward (Hoydysh and Griffiths ,1974; Wedding et al, 1977; Rafailidis, 1995).

In this work we chose to start with very simple geometric configurations which should exclude three dimensional effects. The experiments were performed in an open-circuit wind tunnel under neutral stratification conditions. The wind was directed perpendicular to the street canyons to preserve two-dimensionality.

Considerable effort was devoted to establish experimental procedures which give reliable results. Initially, emphasis was put on line source design since this is a key component of the experimental apparatus. Meroney et al. (1995) provide results from some tests which show that important errors may arise from inadequate line source design. Then we discuss the test case of an isolated street canyon in open country. The results are compared with those obtained with a street canyon amidst urban roughness.

EXPERIMENTAL SET-UP

The Atmospheric Boundary Layer Wind Tunnel

The experiments were performed in the atmospheric boundary layer wind tunnel (BLASIUS) of the Meteorological Institute of Hamburg University. The effective working section is 1 m high, 1.5 m wide and 4 m long following a 7.5 m long development section just downstream of the boundary layer stimulation system.

Surface roughness for an open country situation was provided by blocks made of two stacked Lego™ elements, placed on the floor in a regular array following a staggered pattern. The boundary layer generated is about 0.45 m thick. The vertical velocity distribution in the region where the boundary layer is fully developed may be described by a power law, displacement height is about 2 mm, and the vertical wind profile exponent α was estimated to 0.28. The turbulent intensity profiles are independent of approach velocity when it is above 0.5 m/s. The vertical turbulent momentum flux is constant in the lower 10 cm of the boundary layer. In this layer, the ratio between friction and free stream velocities $U^*/U(\delta)$ equals 0.065 and is independent of the Reynolds number, $Re^* = U(\delta)\delta/\nu$. The roughness length z_0, determined from a logarithmic profile up to 12 cm above the floor is about 1.5 mm. The roughness Reynolds number (Jensen number) U^*z_0/ν is larger than 2.5 as recommended by Snyder (1972). Comparing these tunnel characteristics with those found in the nature we conclude that this arrangement corresponds to a simulation of the atmospheric surface layer at a scale of 1:500. Different ceiling positions were used to ensure negligible longitudinal pressure gradients (~0.25 Pa/m) for all configurations studied.

Measurement techniques

Velocity. The wind velocity was continuously recorded using both a Prandtl pitot-static tube and a hot-film anemometer. The reference height was chosen to be 650 mm above the floor in the free-stream region of the flow ($z_{ref} \approx 11$ H).

Concentrations. A neutrally buoyant gas mixture of ethane and air was used to simulate the dispersion of pollution in the model street canyon. A University of Hamburg design sampling system allows simultaneous collection of 20 individual gas samples in inflatable balloons over prespecified sampling periods. Each sample is analyzed afterwards using a flame-ionization type hydrocarbon analyzer Model 400A from Rosemount Analytical Inc. The concentration sampling system was calibrated against a laboratory prepared mixture of hydrocarbon gas. The between sample variation was minimal, and the sample accuracy was determined to be ± 0.8 % of true concentration. The overall balloon sampling and concentration detection system error is about ± 5 %. The concentration measurements are presented in terms of the ratio $K = CU_{ref}HL/Q_e$, unless indicated otherwise, where

C	: actual concentration [ppm]
U_{ref}	: free-stream wind approach velocity at a height of 0.65 m above the floor level [m/s]
H	: height of physical model of building [m]
Q_e/L	: line source strength [m^2/s]
	Q_e denotes ethane flow rate and L is the source length (0.90 m)

Street canyon design

Wooden 60 mm x 60 mm bars were used to model two-dimensional multi-story flat-roofed buildings. We considered the two dimensional case which corresponds to a street canyon completely spanning the width of the tunnel and perpendicular to the wind direction. Blockage of the test section from the bars did not exceed 6 %. The approach velocity was chosen large enough so that the Reynolds number, $U_H H/\nu$, exceeds the value of 3400 suggested by Hoydysh et al (1974) to ensure that the flow pattern in the street canyon was independent of viscous effects.

The bars were equipped with flush mounted sampling ports located on a vertical plane on the tunnel axis, Figure (1). In addition to the 14 sample ports on the vertical plane, five additional holes were provided along a horizontal line at the level of port #14 to evaluate the two-dimensionality of the tests. Sample ports #15 to #19 were placed at the positions y_i equal to -75, -50, -25, +25, and +75 mm relative to hole #14.

In the following discussion "open country" refers to an isolated street canyon amidst Lego roughness. The "urban roughness" case is an approximation of the urban fabric and is obtained by replicating similar street canyons, parallel to the test canyon, upstream and downstream. Hoydysh et al (1974) determined that an upwind fetch of 8 to 10 street canyons is required before street canyon flow and dispersion are independent of upwind fetch distance. Therefore, in this study the urban fetch consisted of 8 bars downstream and up to 20 bars upstream of the test canyon.

The drag introduced by the street canyon roughness is a function of building and street dimensions. The resulting shear profile may be categorized depending on the nature of separation and attachment of streamline flow over the bars, i.e., skimming flow, wake interference, isolated roughness (Oke, 1988b). The tunnel wind velocity profiles measured for different street canyon configurations as B/H increased from 1 to 8 became fuller in shape. Power law coefficient, α, increased from 0.20 to 0.25 as the flow changed from skimming to wake interference, then remained at 0.28 for isolated roughness conditions. Displacement height varied from 57 to 40 mm as B/H increased. Surface drag coefficient, u^*/u_{ref}, varied from 0.05 to 0.10 and surface roughness, z_o, ranged from 0.25 to 3.00 mm as B/H increased.

Line source

Line sources are used to simulate exhausts from vehicles queuing along a street. Our experience suggests consistent measurements require a laterally homogeneous release; hence, Meroney et al. (1995) describe the line source and tests performed to confirm its reliability. The final design is depicted in Figure (2). The small tube diameter produced substantial pressure drop of the order of 450 Pa for a flow rate of 100 liter/hr/m; thus rendering the gas flow insensitive to local pressure fluctuations in the model street canyon above the line source. The design flow rates required during the experiments resulted in discharge velocities from such small holes of the order of 1 m/s, but any initial vertical gas momentum was deflected laterally using a capping plate.

With the line source design shown in Figure (2) the overall standard deviation of the downstream plume concentration from the mean value does not exceed ± 2.5 %. On the other hand, the maximum variability range relative to the average does not exceed 7.3 %. In the presence of a street canyon, the gas discharged from the line source becomes more evenly distributed laterally making the lateral inhomogeneity of the line source even smaller than ±2.5%.

THE TEST CASE OF STREET CANYONS

To evaluate dispersion in a model urban street canyon, we ran two series of tests. The case of an isolated street canyon in open country was examined first. The same street canyon geometry was subsequently studied in an urban environment, i.e., with additional canyons of similar geometry upstream and downstream of the test section.

In each case various street canyon aspect ratios were tested under different wind conditions. Each measurement was repeated twice to verify that an equilibrium in the canyon flow and the gas dispersion existed.

Open country case

Street canyon flow and the effect of wind speed. The effect of varying wind speeds on the concentration of pollutants within the canyon will first be discussed for the reference case B/H = 1, Figure (3). The general characteristics of the typical concentration profiles measured are discussed below.

- Visualization using smoke and a vertical light sheet revealed that clean air is drawn into the canyon by an intermittent eddy circulating down into the canyon. This eddy circulates upwind at street level. A roof top eddy which begins at the upwind upstream building roof corner sucks canyon gases onto the roof

- Significant pollution concentrations are measured on the roof of the upstream building, indicating that gas from the canyon is *sucked* inside the roof bubble before eventually being carried away by the oncoming flow. On the leeward side of the upstream building concentrations decrease from a maximum at the base of the building to lower values along the roof line (port # 7 to port #4). On the windward face of the downstream building, the concentration increases from the roof to the street (ports #11 down to #8). At positions #9 and #8 an equilibrium was apparently reached. This was consistent in all measurements, despite the fact that these positions differed vertically by 10 mm.

- The characteristics described above do not change significantly with different wind approach velocities. Any influence appears to be marginal, indicating that, at least for an aspect ratio of B/H = 1, the flow inside the canyon remains largely unchanged. It may be concluded that the canyon dispersion behavior was Reynolds number-independent.

Different street canyon configurations. The influence of different street canyon aspect ratios was studied for two different line source locations:

- The line source was placed at a fixed distance x_s = H/2 downstream of the leeward wall of the upstream bar, and its position was left unchanged while the downstream block was moved to represent street widths B = 1H, 2H, 3H, 4H, 8H and ∞ downstream, or

- The line source was placed midway between the downstream and upstream blocks, while the street widths varied, again, as B = 1H, 2H, 3H, 4H and 8H.

These tests were performed for a wind approach velocity of U_{ref} = 2 m/s. The results are presented on Figures (4) and (5). The concentrations measured reveal that:

- Substantial concentrations are always measured on the roof of the upstream building. This indicates that some pollutant gases are drawn into the roof separation bubble, before being vented away into the main flow.

- High tracer concentrations are observed at the leeward face of the upwind building. The concentrations increase from top to bottom, with the highest measurement at the intersection between the wall and the street.

- In all cases the horizontal distribution of tracer gas from the line source are homogeneous horizontally, both up- and down-stream of the source (sample ports #8 and, #15 to #19).

- In terms of the non-dimensional coefficient, $K = CU_{ref}HL/Q_s$, little variation is observed by increasing the approach velocity from 2 to 5 m/s while the canyon aspect ratio remained constant.

- Varying the canyon geometry has small effect on the concentrations on the roof of the upwind building when the source is placed at x_s = H/2. This was expected due to the relative proximity of the source and the insensitivity of the upstream roof separation bubble to the flow characteristics downstream. Concentrations at the rest of the sample ports decrease as the canyon widens.

The above observations are appropriate when the line source is close to the upstream building (x_s = H/2). Similar considerations, however, apply when the line source is halfway between the two bars.

- The roof concentrations are more heavily affected by widening the canyon than the other positions, because the source moves further away from the upstream roof than when x_s = H/2.

- Moreover, as B/H increases smaller dimensionless concentrations occur at all locations.

- Consideration of Figures (4) and (5) suggests that the flow inside the canyon can not be in the form of a permanently recirculating eddy. If a permanent eddy existed then concentrations would be indifferent to where the eddy was fed with tracer gas. Instead, the figures corroborate the hypothesis of intermittent streaming of pollutant from the source to the side walls of the canyon. Under such conditions, the geometry between the source and the walls becomes critical, influencing directly the routing of gas plume from the source to the sample ports. Thus, moving the upstream sample ports #4 to #7 farther from the source results in a drop in the concentrations measured.

Urban roughness case

Experiments were also performed on the dispersion within extended urban roughness associated with additional up- and down-stream bars. For this part of the study the line source was positioned in the middle of the street (x_s = B/2). The results are summarized in Figure (6) for U_{ref} = 2 m/s. The overall characteristics discussed for the open country case were identified also in canyons amidst large urban roughness, but some significant differences were observed.

- At both wind velocities studied (2 and 5 m/s) the pollution concentration on the roof of the upstream building is almost zero. This suggests that the roof recirculation bubble into which pollution may be drawn from inside the canyon has totally disappeared. Elsewhere in the canyon the pollution increased, in absolute terms, compare Figure (5) with (6). This suggests that intermittent ventilation of a canyon in the middle of the city by the already well-developed internal boundary layer is inhibited more than if the canyon had been surrounded by open country.

- To clarify the above description we provide sketches (7a) and (7b) for the case B/H = 1. These drawings summarize observations during a visualization study made on the circulation inside the canyon under varying

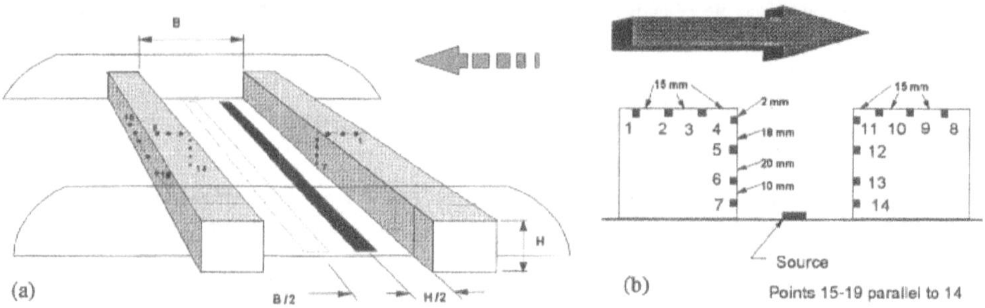

Figure 1. (a) Schematic of model street canyon, line source and sample port locations as installed in Blasius Wind Tunnel, U. of Hamburg. (b) Location of sample ports on surface of model street canyon buildings.

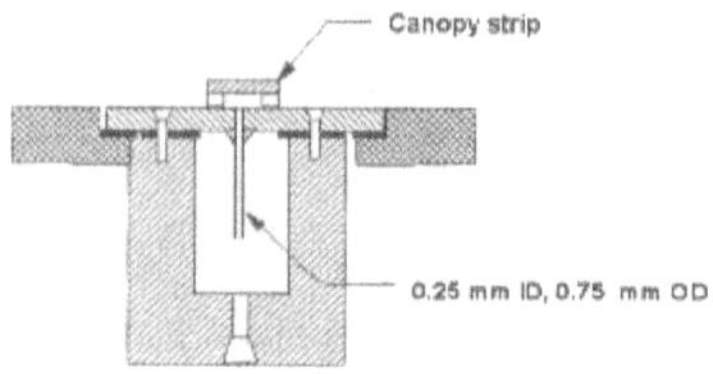

Figure 2 Schematic of line source design. Line source used 302 hypodermic tubes over a 1 meter length.

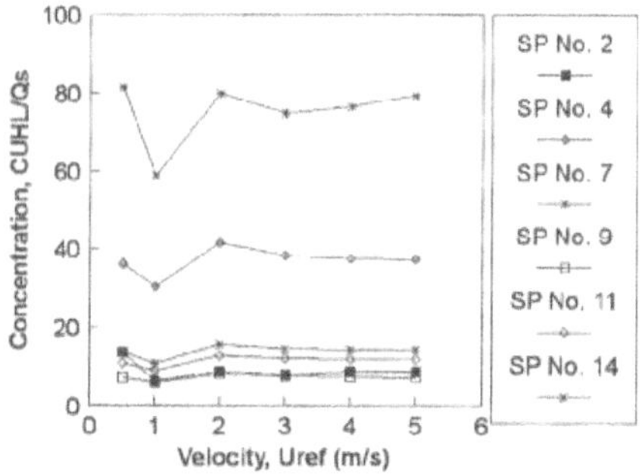

Figure 3 The effect of different wind velocities on pollutant concentrations in a model street canyon at selected sample port locations. (Open country roughness, B/H =1, Q_{air} = 100 l/h, Q_{ethane} = 4 l/h.)

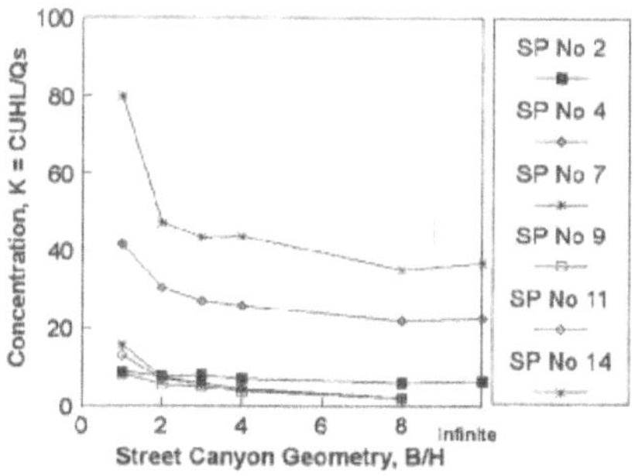

Figure 4 Concentrations in the street canyon at selected sample ports for different canyon aspect ratios. (Open country roughness, X_s = H/2, U_{ref} = 2 m/s, Q_{air} = 100 l/h, Q_{ethane} = 4 l/h).

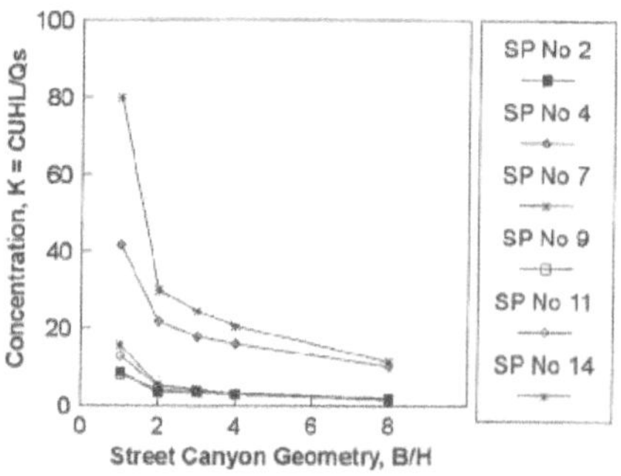

Figure 5 Concentrations in the street canyon at selected sample ports for different canyon aspect ratios. (Open country roughness, X_s = B/2, U_{ref} = 2 m/s, Q_{air} = 100 l/h, Q_{ethane} = 4 l/h).

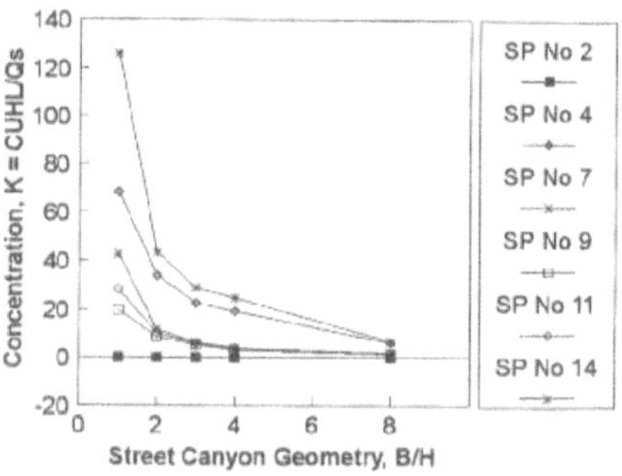

Figure 6 Concentrations in the street canyon at selected sample ports for different canyon aspect ratios. (Urban roughness, X_s = B/2, U_{ref} = 2 m/s, Q_{air} = 100 l/h, Q_{ethane} = 4 l/h).

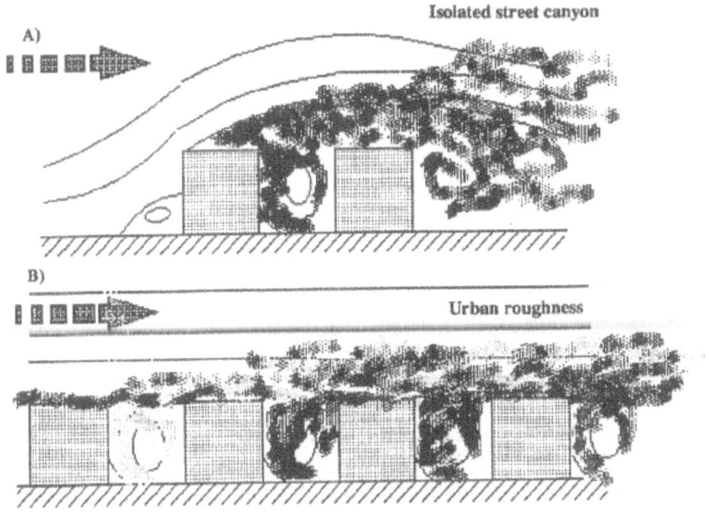

Figure 7 a) Streamlines and dispersion patterns from line source in street canyon placed in open country roughness, compared to **b)** multiple sources in street canyons placed in urban roughness.

operating conditions. Flow visualization was performed using an illuminating laser sheet and a non-buoyant mixture of air and oil smoke injected at street level along the street canyon. The beam of a 200 mW laser was spread by a semi-cylindrical lens to a light sheet of a constant thickness of the order of 2 mm. Photographs were taken using exposure times of about 5 s.

- The flow inside the canyon in open country is much more nonstationary than in the urban canyon case. A recirculating eddy forms only intermittently in the former, whereas it is clearly much more stable amidst urban roughness.

In the open country case, the first building alters the flow locally quite dramatically, promoting separation from the floor. On the other hand, in the urban roughness case the surrounding buildings raise the displacement height to the urban canopy level, almost to the roof of the buildings. The shear zone thus formed at the ceiling of the canyon induces a permanent eddy recirculating inside the canyon. As a result, vertical mixing across this shear layer is suppressed and pollution is trapped inside for longer times, until upstream turbulence in the main flow may cause sufficient disturbance to break down the recirculating eddy.

CONCLUSIONS

From the experimental work described above a number of conclusions may be drawn:

- The use of a stable and laterally homogeneous line source is one of the key requirements for simulation of vehicular pollution in an urban model study. The important design parameter is found to be the pressure difference across the discharge holes.

- An isolated street canyon in the middle of open country roughness and another in urban environment display different flow characteristics. In the canyon amidst open country the canyon vortex is unstable and is discharged regularly upwards. Inside the urban canyon a stable rotating vortex develops, street ventilation is suppressed, and pollution is trapped inside the canyon. There is a total absence of an upstream roof bubble; hence, no pollution is measured on the roof.

Acknowledgments This work was made possible through financial support by the European Union in the form of two Human Capital and Mobility Fellowships (M. Pavageau and S. Rafailidis) and by the Alexander Von Humboldt Foundation in the form of a Senior Fellowship (R.N. Meroney). The authors also extend sincere thanks to the following individuals for their technical competence, useful advice and permanent care in assisting this work: Jochen Donat , Thomas Glanert, Gopal Krishan and Dr. Michael Schatzman.

REFERENCES

Hoydysh W.G., Griffiths R.A., Ogawa Y., 1974, A scale model study of the dispersion of pollution in street canyons, APCA paper no. 74-157, *67th Annual Meeting of the Air Pollution Control Association*, Denver, Colorado, June 9-13 1974.

Hoydysh, W.G. and Dabberrdt, W.F., 1986, Kinematics and Dispersion Characteristics of Flows in Asymmetric Street Canyons, Atmospheric *Environment*, Vol. 22, pp. 2677-2689.

Meroney, R.N., Pavageau, M., Rafailidis, and Schatzmann, M. (1995), Study of line source characteristics for 2-d physical modelling of pollutant dispersion in street canyons, *J. Wind Engrg. & Ind. Aero.*, submitted August 1995, 35 pp.

Oke T.R., 1988a, Street design and urban canopy layer climate, *Energy and Buildings* 11, pp. 103-113.

Oke T.R., 1988b, *Boundary layer climates*, Methuen, New York, 2nd Edition.

Plate E., 1982, Engineering *meteorology*, Vol. 1: Fundamentals of meteorology and their application to problems in environmental and civil engineering, chap. 13, Elsevier Scientific Publishing Company.

Rafailidis S. and Schatzmann M.., 1995, Physical modelling of car exhaust dispersion in urban street canyons, *Proc. 21st Int. Meeting on Air Pollution Modelling and Its Applications*, Baltimore, Nov. 6-10, 1995.

Snyder W.H., 1972, Similarity criteria for the application of fluid models to the study of air pollution meteorology, Boundary *Layer Meteorology*, 3, 113.

Wedding, JB, Lombardi, D.J. and Cermak, J.E.,1977, A Wind Tunnel Study of Gaseous Pollutants in City Street Canyons, *Air Pollution Control Association Journal*, Vol. 27, No. 6, pp. 557-566.

DISCUSSION

R. BORNSTEIN: Does your result showing minimal mixing out of the urban canyon make it harder or easier for a mesoscale photchemical model to expect to obtain good results.

R. MERONEY: The mixing phenomena along the top of urban street canyons is a local behavior, and flows within the canyon are typically of low velocity. Since most mesoscale models currently perform at grid scales larger than the street canyon itself this suggests that the mesoscale photocemical model must treat the boundary layer source through some sort of average entrainment or mixing condition. I would expect that most mesoscale models would thus fail to predict correctly the local hot spots within the canyons themselves. On the otherhand since photo-chemical reactions involve fairly long time scales I would expect most reactions to occur over volumes large with respect to the canyon scale, so perhaps it does not matter.

ACCIDENTAL RELEASES

chairman: J. Walmsley

rapporteur: J. Pretel

VALIDATION OF A COMBINATION OF TWO MODELS FOR LONG-RANGE TRACER SIMULATIONS

Jørgen Brandt[1,2], Thomas Ellermann[1], Erik Lyck[1],
Torben Mikkelsen[2], Søren Thykier-Nielsen[2], and Zahari Zlatev[1]

[1]Department of Atmospheric Environment,
National Environmental Research Institute, Roskilde, Denmark
[2]Department of Meteorology and Wind Energy,
Risø National Laboratory, Roskilde Denmark

INTRODUCTION

The accident at the Chernobyl nuclear power plant on April 25, 1986 initiated great interest in development of warning systems that could be used as a part of the emergency planning in connection with radioactive and toxic releases due to major industrial accidents. Forecasts of the transport and dispersion of the releases by use of long-range transport models are used as an important tool in the emergency planning and it is therefore crucial to test the quality of the models, for example, by comparison of model simulations with full scale atmospheric tracer experiments.

With this background the National Environmental Research Institute (NERI) and Risø National Laboratory have developed and tested a new long-range transport model, the Danish Rimpuff and Eulerian Accidental release Model (DREAM)[1], which simulates transport and dispersion of air pollutants from strong single sources and describes the subsequent transport on a regional scale, for example Europe. The model is a combined model based on a Lagrangian meso-scale model, RIMPUFF[2,3], and a Eulerian long-range transport model, the Danish Eulerian Model[4,5,6]. RIMPUFF is used to describe the initial dispersion of the air pollutant in a sub-domain close to the source (about 300 km) while the Danish Eulerian Model is used to simulate the long-range transport up to several thousand kilometres. The combination of the two models is employed in order to gain advantage of the strengths of the two different types of models. The performance of the combined model was tested by use of a simple modification of the well-known rotation puff test and by simulation of the transport and dispersion from the Chernobyl accident. The Chernobyl simulations were compared with measured transport and dispersion of the emissions and with other long-range transport models, e.g. the EURAD model[7].

Moreover, the model has been used to study the two tracer releases of the European Tracer EXperiment, ETEX[8], and compared to tracer measurements carried out at Roskilde, Denmark located 1400 km north-east of the release position in Brittany, France. NERI has for decades carried out full scale atmospheric dispersion experiments using SF_6 as tracer[9,10,11] and recently the laboratory was supplemented by facilities for air sampling (saran bags and ETEX sample tubes[12]) and analysis of perfluoro-carbon, PFC, tracers. This equipment was used for sampling of ambient air outside and upwind of the laboratory during the days after the ETEX releases on October 23, 1994, ETEX-1, and November 14, 1994, ETEX-2[8]. The air samples were analysed for the tracer released during ETEX-1 (perfluoro-methyl-cyclohexane; PMCH), the tracer released during ETEX-2 (perfluoro-methyl-cyclopentane; PMCP), and in addition perfluoro-dimethyl-cyclohexane (PDCH).

In this presentation we give a brief introduction to the tracer model and the procedure employed to couple RIMPUFF and the Danish Eulerian Model (for further details see Brandt et al.[1]). Results from the rotation test of the tracer model will be presented together with the simulations of the two ETEX releases. Moreover, we present our tracer measurements carried out during the two ETEX campaigns and compares the tracer data with the results of the model simulations of the ETEX releases.

EXPERIMENTAL

Coupling of two models for tracer simulations

While the Lagrangian and Eulerian descriptions of the advection in principle are equivalent, the two kind of models can give different results when solved numerically. The traditional Eulerian models have difficulties in handling sharp gradients from a single source which results in unwanted oscillations[1]. The use of a finer grid does not solve this problem when transport and dispersion from a single and strong source is simulated. A finer grid will increase the gradients at the source point and will only give better results after some initial smoothing of the concentrations. Lagrangian models, on the other hand, do not have these problems with sharp gradients but have problems with uncertainties in the trajectories on large scales[13].

The performance of Eulerian models could be improved if some kind of filtering and/or dispersion is used. The effect of filtering is, however, an artificial solution and it is preferable to use a solution based on physical arguments, like a reliable parameterisation of the dispersion near a single source. The K-approach used in the Danish Eulerian Model gives unsatisfactory results close to the source[14] and another parameterisation should be used. The dispersion algorithm used in the RIMPUFF model, $\sigma = ax^b$ (where a and b are stability dependent parameters and x is the downwind distance[3]), has been verified by many experiments.

A combination of the two different types of models is therefore employed in order to gain advantage of the strengths of the two models: RIMPUFF is used to describe the initial dispersion of the air pollutant in a sub-domain of 65 x 65 grid-squares where the source is located in the centre. The size of each grid-square in the sub-domain is 5 km x 5 km. The Danish Eulerian Model is used to simulate the long-range transport in the whole domain. The size of a grid-square in the Danish Eulerian Model is 25 km x 25 km and the number of grid-squares depends on the meteorological input data. A 112 x 112 grid has been used in simulations of ETEX, and a 192 x 192 grid has been used in the rotation test and in simulations of Chernobyl. The present simulations are performed only in two dimensions, but a three-dimensional version of the model is under development.

The RIMPUFF model is a Lagrangian puff model. This means that the plume calculated by RIMPUFF consists of a series of Gaussian shaped puffs, that are released from the source-point with a certain time between two successive puffs. Each puff is advected and dispersed individually along trajectories. After a certain travel time the puffs have expanded to a size where they effectively overlap and together form a plume. Several different advection schemes can be used in the Danish Eulerian Model. A linear finite element scheme was employed in this context, because it is a fast scheme compared to other Galerkin methods.. This is important when used operationally. Other advection schemes can be used.

The main idea used during the coupling of the two models is as follows: The single puffs are incorporated individually into the Danish Eulerian Model when the puffs reach the boundary of the sub-domain. After incorporating the puff it is removed from the RIMPUFF model. This coupling procedure has been named "puff-coupling" and is illustrated in Figure 1.a which shows a single puff at the boundary of the sub-domain. The concentrations calculated by the RIMPUFF model are modified for the coarser grid by taking mean values over 25 grid-squares of the finer grid corresponding to one grid-square of the coarser grid. In every time-step the sub-domain is checked for puffs which have reached the boundary of the sub-domain.

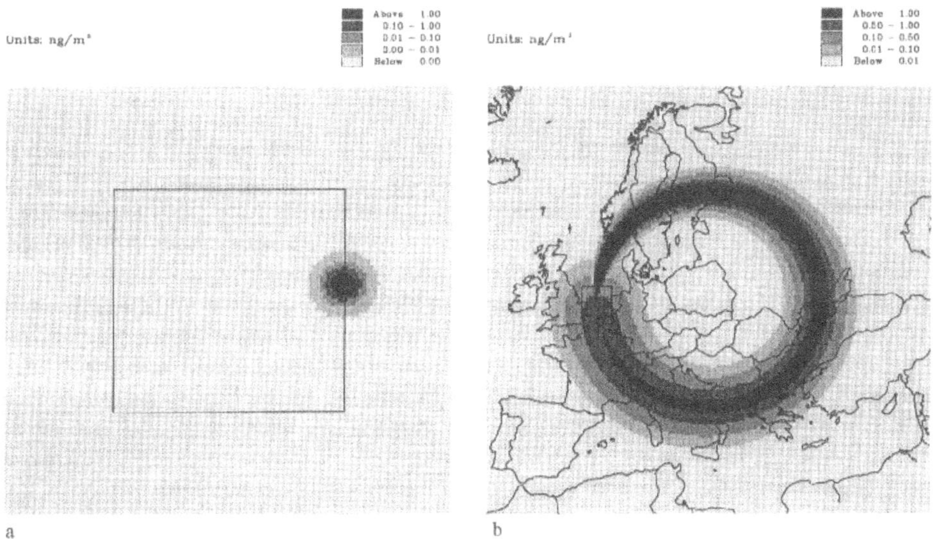

Figure 1: (a) Illustration of puff-coupling. The square represents the sub-domain where RIMPUFF is operating. (b) Rotation of a plume using the combined model.

The combined model has been tested by use of a modification of the well-known rotation test which is illustrated in Figure 1.b. Oscillations from the sharp gradient in the source point have been eliminated. The square in the figure indicates the sub-domain where RIMPUFF is operating. The combined tracer model has also been used for simulations of transport and dispersion of the emissions from the Chernobyl accident[1]. Comparison of results with measurements and other model simulations[7] indicates good performance of the model.

Tracer analysis and air sampling

The PFC tracers, PMCP, PMCH, and PDCH, are analysed by use of a gas chromatograph with electron capture detector, GC-ECD (John Booker & Co., Austin, USA). Initially the PFC content of a given air sample is up-concentrated at room temperature onto a 6.72" trap packed with 75 mg Ambersorb XE-348. The majority of oxygen is then removed by purging with a carrier gas. Finally the trap is quickly heated to 300 °C to desorb the PFC's and inject them onto the precolumn and column (0.1% SP-1000 on 80/100 Carbopak C). Prior to injection the gasses pass a catalytic converter, which removes remaining oxygen and airborne halocarbons other than the fully fluorinated substances. The reduction in the catalytic converter requires that the carrier gas consists of a mixture of 5% H_2 in N_2 (both 99.999% pure, AGA, Denmark).

The GC-ECD is calibrated towards standard gas mixtures prepared in Saran bags on the basis of permeation tubes (VICI, Santa Clara, USA) and by use of a Dynacalibrator (VICI). Permeation rates are determined regularly by weighing of the permeation tubes. Dilution gas was N_2 (99.999% pure, AGA, Denmark).

The ambient air samples were collected by use of Saran bags and ETEX sample tubes[12]. The Saran bags are used in combination with our home build sample units[10,11]. These consist basically of a time regulated pump, which pumps about 3-4 litre of air into the Saran bag. The flow rate is constant throughout the one hour sampling period. 2.2 litre air is subsequently analysed by the GC-ECD.

The ETEX sample tubes are small cylindrical tubes containing 150 mg carboxen as an adsorbant. The PFC concentration of a 30 litre air sample is up-concentrated by pulling air through the tube with a pump for 6 minutes. A Dani Sequential Tube Desorber (DANI, Monza, Italy) is coupled to the GC-ECD for handling and desorption of the sample tubes. All flow rates were controlled by employment of a DryCal DC-1 flow calibrator (Bios Int. Coop., Pompton Plains, USA).

During ETEX-1 we began the air sampling on October 24 at 10:00 GMT, i.e., 18 hours after start of the release. Saran bags were used to collect pooled one hour air samples giving hourly mean tracer concentrations after analysis. The sample tubes were used to collect 6 min mean value. Although the sample times are different we do expect that the results from bags and tubes are comparable, because of the long transport time from the release position to the sampling position at Risø. Both samplers were started every full hour until 88 hours after start of the release. During ETEX-2 we used same procedure, but began the air sampling on November 15 at 05:00 GMT, i.e., 13 hours after start of the second release. The only change was that the sample time for the sample tubes was increased to one hour by decrease of the flow. The air volume was still kept at 30 litre. Air sampling stopped 65 hours after start of the release.

The ambient concentrations of the three PFC tracers were determined prior to the ETEX releases in order to test our ability to sample and analyse for the tracers at concentrations down to a few hundredths of a pptv. The ambient air concentration was determined to 0.0053, 0.0039, and 0.022 pptv for PMCH, PMCP, and PDCH, respectively. The values given for PDCH represents the sum of the isomers. The uncertainties on the ambient concentrations are estimated to ±25% for PMCH and PMCP and ±30% for PDCH (due to the isomers). Our values are in good agreement with the values reported in 1986 by R.N. Dietz (PMCH: 0.0046 pptv; PMCP: 0.0032 pptv; PDCH: 0.022 pptv)[15].

RESULTS

Simulation of ETEX

Figure 2.a-d shows the model simulation of the transport and dispersion of the plume from the two ETEX releases. Figure 2.a shows the situation on October 24, 04:00 GMT, i.e., 12 hours after start of the first release, ETEX-1, and at the end of the release period. A low pressure system is situated over the North Sea which gives a transport of the plume to the north-east towards central Europe. Figure 2.b shows the situation on October 25, 16:00 GMT, i.e., 48 hours after start of the release. The plume has passed France and is now located over central/eastern Europe and parts of Scandinavia.The situation on November 15, 03:00 GMT, 12 hours after start of release of the second experiment, ETEX-2 is shown in Figure 2.c. The plume was transported very fast towards central Europe due to a westerly storm. At 48 hours after start of release on November 16, 15:00 GMT (see Figure 2.d) the plume is covering southern and eastern parts of Europe and is entering Russia. The meteorological wind fields used in these simulations are from the 925 hPa level and are provided by the Canadian Meteorological Centre, Canada.

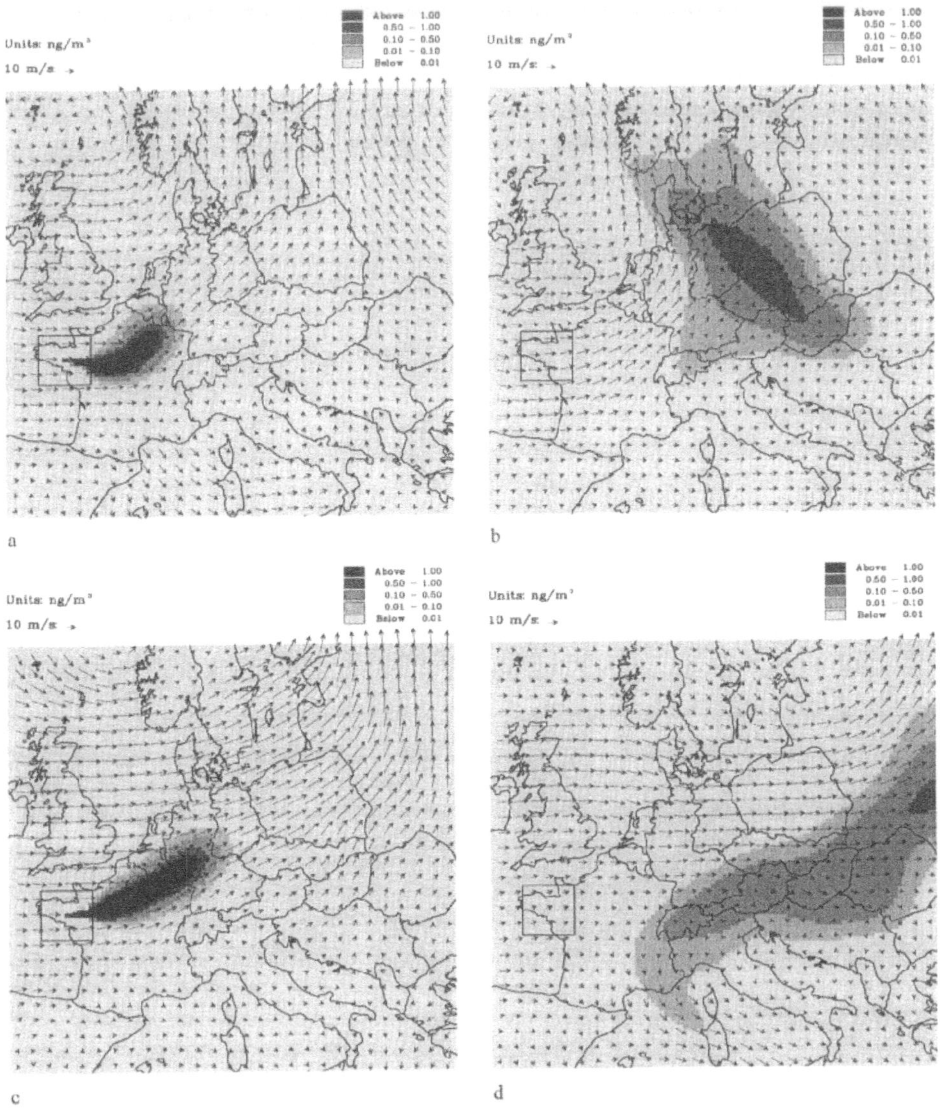

Figure 2: Simulations of the ETEX tracer plumes. (a) 12 hours after start of the release, ETEX-1. (b) 48 hours after start of the release, ETEX-1. (c) 12 hours after start of the release, ETEX-2. (d) 48 hours after start of the release, ETEX-2.

Tracer measurements during ETEX

Chromatograms obtained by analysis of air samples from ETEX-1 collected at the arrival of the front of the tracer plume at Risø are shown in Figure 3. The first chromatogram obtained on October 25, 13:30 GMT shows a PMCH concentration only slightly higher than observed prior to ETEX. During the next hours the PMCH concentration increased sharply showing clearly the arrival of the tracer plume. Note that the PDCH peak areas were constant throughout the analysis indicating that no significant change of the efficiency of the air sampling or the sensitivity of the GC-ECD could be observed during our ETEX-1 measurements. Figure 4 shows the PMCH results obtained during the passage of the cloud. The passage of the tracer plume at Risø can easily be observed with an arrival time of about 44 hours after start of the release and a passage time of 20 hours. The tracer concentration varied considerable during the passage. Two peaks were observed with maxima of about 13 and 5 times the ambient concentration. This structure was observed both in the data from the saran bags and the ETEX sample tubes (data not shown). After the passage of the tracer plume the concentrations returned to the typical ambient concentration observed before the release.

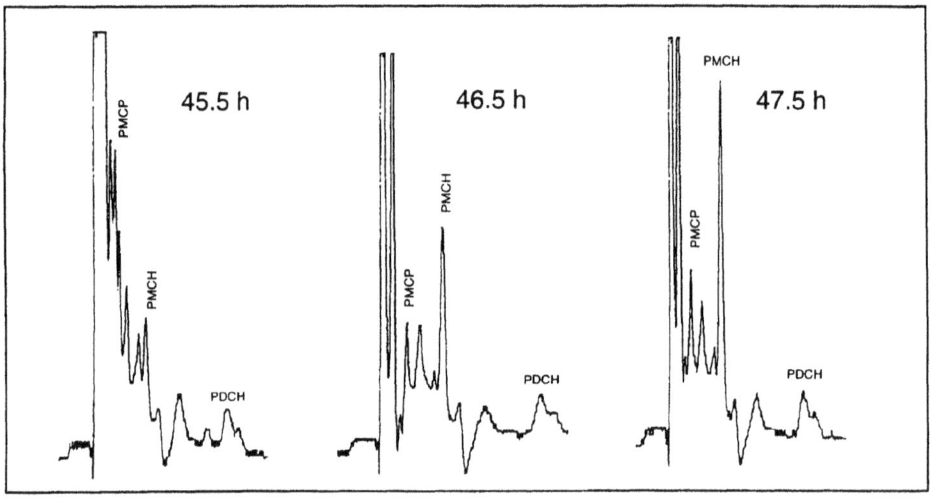

Figure 3: Gas chromatograms obtained by analysis of one hour mean air samples collected in Saran bags at the arrival of the tracer plume in ETEX-1. The inserted time gives the mean time of the sampling period after start of the release. The gas chromatograms show a four times increase of the concentration (proportional to peak area) of the ETEX-1 tracer (PMCH). The concentration of PMCP and the isomers of PDCH do not change significantly.

The calculated tracer concentrations at Risø during ETEX-1 have also been included in Figure 4 in order to compare measured and simulated data. Both the arrival time of the tracer plume and the duration of the passage is perfectly simulated by the model. The level of the tracer concentration is also in fair agreement with the measured data, when an estimated ±30% uncertainty of the measured PMCH concentration is taken into account. Moreover, at present the parameterization of the mixing height is very simple and a lower mixing height would give higher simulated concentrations. The model could not simulate the two peaks of the tracer cloud observed at Risø. Simulations of ETEX-1 using wind fields at the 850 and 1000 hPa level indicate that the reason for this discrepancy may be that the model so far is only two dimensional.

During ETEX-2 we had problems with the analysis and in many of the air samples we could not separate the peak of the tracer, PMCP. The reason is most likely the heavy rain falling throughout the sampling period. However, the data that we obtained, shows no significant change in the PMCP concentration relative to the ambient concentration prior to ETEX. It can therefore be concluded that the ETEX-2 tracer plume did not pass Risø, which is in agreement with the model simulations (see Figure 2.c-d).

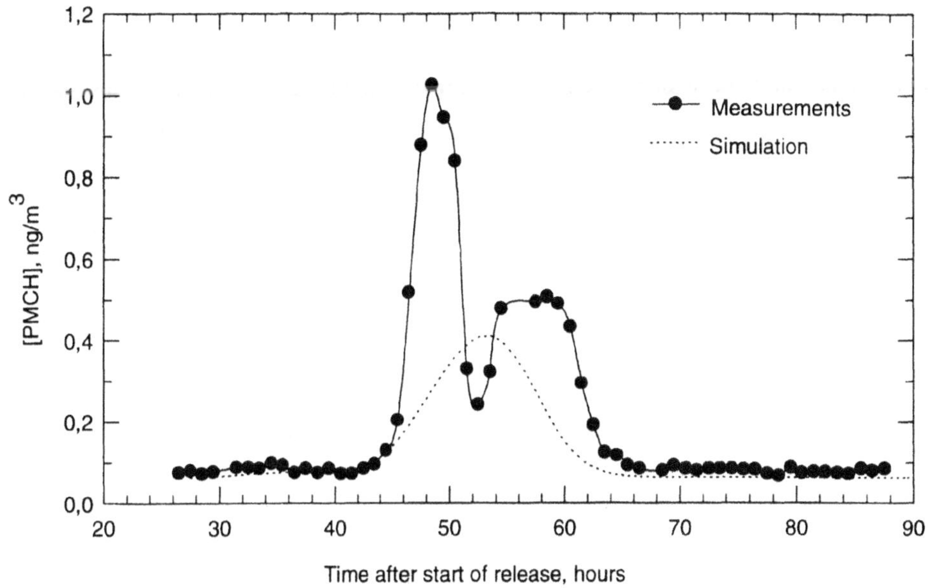

Figure 4: Measured and simulated concentration of ETEX-1 tracer (PMCH) at Risø. Measured concentrations represents the one hour mean concentration at ground level. Simulated concentrations are one hour mean values.

CONCLUSION

All of all our work has shown that the combination of a Lagrangian and a Eulerian model seems to eliminate the problems of sharp gradients in the traditional Eulerian models and the uncertainties in trajectory calculations in Lagrangian models. Furthermore, the combination of two models has the advantage that it is possible to describe the transport and dispersion both on meso-scale and on long-range using different parameterisation of the dispersion. Rotation tests and validation of the model with simulations of ETEX (and Chernobyl[1]) indicate that the combined model and the coupling procedure works satisfactorily. However, at present the model is only two dimensional, but a three dimensional version of the model is under development. Parameterisation of vertical dispersion, cloud processes, dry and wet deposition, etc, will be included in this version. The use of three dimensional models is important, especially when simulating accidental releases from nuclear power plants where there is a quick vertical distribution of concentrations from the source because of strong convection.

Moreover, the tracer measurements have shown that the passage of the tracer plume could be clearly detected even at distances of 1400 km from the release position and even though the release rate of the tracer was only about 10 g/s [8]. The measurements shows also that our sampling and analysis technique are suitable for tracer measurements down to extreme low concentrations levels, i.e., as low as a few hundreths of a pptv.

Finally, we would like to emphasize the importance of tests of models by use of full scale dispersion experiments. So far, we have only been able to compare the simulations with the tracer data obtained at Risø and we look therefore forward to evaluate the performance of the model towards the data from the ETEX measurement network.

ACKNOWLEDGEMENT

Francesco Girardi, Giovanni Graziani, Katrin Nodop, and Richard Connolly, JRC-Ispra, Italy are gratefully thanked for cooperation on ETEX and for lending us the ETEX sample tubes. Real D'Amours, Canadian Meteorological Centre, Canada is gratefully thanked for providing us with the wind fields used in the ETEX simulations. This work has partly been supported by the Danish Research Academy and the Danish Environmental Research Programme.

REFERENCES

1. J. Brandt, T. Mikkelsen, S. Thykier-Nielsen, and Z. Zlatev, Using a combination of two models in tracer simulations, Accepted, Mathematical and Computer Modelling (1995).
2. T. Mikkelsen, S.E. Larsen, S. Thykier-Nielsen. Description of the Risø puff diffusion model, Nuc. Techn., 67:56 (1984).
3. S. Thykier-Nielsen and T. Mikkelsen, RIMPUFF - Users guide, Version 20, Report No. Risø-M-2673, Department of Meteorology and Wind Energy, Risø National Laboratory, DK-4000 Roskilde, Denmark (1988).
4. Z. Zlatev, J. Brandt, A. Eliassen, and Ø. Hov, Comparison of results obtained by a large-scale Eulerian model with measurements, in: Proceedings of the EMEP Workshop on the Accuracy of Measurements, Passau, Germany, T.Berg and J. Schaug, ed., EMEP, NILU, Kjeller, Norway (1994).
5. Z. Zlatev, J. Christensen, and Ø. Hov, A Eulerian air pollution model for Europe with non-linear chemistry, J. Atmos. Chem., 15:1 (1992).
6. Z. Zlatev, Computer Treatment of Large Air Pollution Models, Environmental Science and Technology Library, Vol. 2, Kluwer Academic Publishers, Dordrecht (1995).
7. H. Hass, M. Memmesheimer, H. Geiss, H.J. Jacobs, M. Laube, and A. Ebel, Simulations of the Chernobyl Radioactive cloud over Europe using the EURAD model, Atmos. Environ., 24A:673 (1990).
8. K. Nodop, JRC-Ispra, Italy, presonal communication (1994).
9. E. Lyck and H.Rørdam Olesen, The Øresund Experiment, A Nordic Mesoscale Dispersion Experiment, SF_6 -tracer measurements performed by the Air pollution Laboratory, MST Luft-A109, National Agency of Environmental Protection, Air Pollution Laboratory, Roskilde, Denmark (1986).
10. E. Lyck, A tracer technique for full-scale atmospheric disperison experiments, in: Preprint of the Ninth Symposium on Turbulence and Diffusion, April 30 - May 3, 1990, Roskilde, Denmark, American Meteorological Society, Boston (1990).
11. R.D.A. Scholten, J.J. Erbrink, H.E. Jørgensen, T. Mikkelsen, E. Lyck, P. Løfstrøm, and T. Ellermann, Atmospheric Dispersion at a Nuclear Power Plant Building: - A Tracer Experiment and Model Evaluation, KEMA Environmental Services, Arnhem (1994).
12. K. Nodop and R. Connelly, JRC-Ispra, Italy, personal communication (1994).
13. J.D. Kahl, J.M. Harris, G.A. Herbert, and M.P. Olson, Intercomparison of long-range trajectory models applied to arctic haze, in: Air Pollution Modelling and its Application VII, H. van Dop, ed.,Plenum Press, New York (1989).
14. R.B. Stull, Review of non-local mixing in turbulent atmospheres: transilient turbulence theory, Boundary-Layer Meteorology 62, pp. 21-96, Kluwer Academic Publishers, Dordrecht (1993).
15. R.N. Dietz, Perfluorocarbon tracer technology, in: Regional and Long-Range Transport of Air Pollution, Lectures of a course held at the Joint Research Centre, Ispra, Italy, S. Sandroni, ed., Elsevier Science Publishers B.V., Amsterdam (1986).

DISCUSSION

D. BYUN: Is your advection scheme positive definite? If it is, how did you achive it?

J. BRANDT: The advection scheme used in the Eulerian model, the finite element scheme, is in principle not positive definite, but negative concentrations can be removed by using e.g. a Forrester filter. This is, however, not necessary in the model presented here, because the dispersion of initial concentrations by the Lagrangian puff model smoothes the concentration field before incorporating the concentrations into the Eulerian model. In this way the Eulerian model is not handling sharp gradients and that eliminates the unwanted oscillations and therefore the negative concentrations.

SEVERE NUCLEAR ACCIDENT PROGRAM (SNAP)
A REAL TIME DISPERSION MODEL

Jørgen Saltbones, Anstein Foss and Jerzy Bartnicki

Norwegian Meteorological Institute
P.O. Box 43 Blindern
N-313 Oslo, Norway

INTRODUCTION

The Chernobyl nuclear accident in April/May 1986 showed that Norway was not well prepared for this kind of emergency situation[1]. Scientists who had relevant information felt frustrated because there was no clear answer to the question: where and how their information should be delivered, assessed and synthesized[1]. In addition, there was no dispersion model implemented in an operational environment, ready to simulate and forecast atmospheric transport and deposition of the radioactive pollutants over Europe based on real-time meteorological data. This was also the situation in most countries in Europe[2].

Norway is better prepared to handle such an emergency situation today[3]. New tools have been developed and a new administrative structure has been set up to handle the large flow of information and bring it to the attention of decision makers. Severe Nuclear Accident Program (SNAP) is one of these new tools. A first version of the SNAP model has been developed and implemented at the Norwegian Meteorological Institute[4,5]. It is based on the same theoretical assumptions and has the same basic architecture as the UK model NAME[6]. Although the first version of SNAP is in operational use at DNMI today, it is still under development as part of DNMI's contribution to the Norwegian Application of MEMbrain, the Major Emergency Management - brain, as part of EUREKA (EU-904). This is a computer based decision support system for emergency management in case of nuclear accidents. A/S Quasar Consultants is responsible for the project management in Norway. The user of the system will be Norwegian Radiation Protection Authority (NRPA) in its function as secretariat for the Nuclear Emergency Preparedness Organization in Norway. The development of the system have been given a financial support from EKSPOMIL and Norwegian State Authorities.

The main objective of SNAP is to provide decision makers and Government officials with a tool for real-time simulation of large accidental releases of radioactivity to the atmosphere in case of nuclear accidents. SNAP is now fully operational at DNMI, which means the meteorologist on duty, on request from the NRPA, can start a simulation at any time.

SNAP MODEL

SNAP is developed in the Lagrangian framework, in which atmospheric transport is simulated by emitting a large number of 'particles' (a mathematical concept, - see p.3) from the source point. The main advantage of the Lagrangian approach is a possibility of precise parameterization of advection processes. After release, each particle carries a given mass of radioactive nuclides, which can be in the form of gas or aerosol. During the atmospheric transport, the radioactive mass of each particle is reduced by dry deposition, wet deposition and radioactive decay. SNAP can be used to predict the transport and deposition of a radioactive cloud in the future (up to 48 hours, in the present version) or to analyse the transport/diffusion behaviour of the cloud in the past. It is also possible to run the model in the mixed mode (partly analyses and partly forecasts).

Model Domain

The grid system most frequently in use for SNAP simulations is shown in Figure 1, the grid system for LAM50S (Limited Area Model), currently DNMI's main operational weather prediction model[7,10]. The model domain is covered by the numerical grid with a mesh size of 50 km at 60^{0}N in the Polar Stereographic Projection.

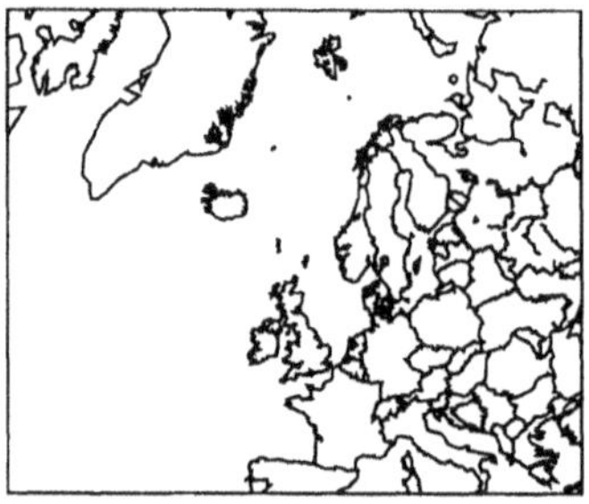

Figure 1. Model domain most frequently used for SNAP computations.

The vertical structure of SNAP is made compatible with LAM50S, using data from a subset (14 layers) of the 31 layers available. This choice is determined by the meteorological data available from LAM50S as input to SNAP simulations. Concentration and deposition fields computed by SNAP are stored/presented in the same grid as shown in Figure 1.

There are two classes of meteorological input data for SNAP. The first is produced directly by LAM50S and includes: wind, temperature and precipitation. The second includes the mixing height, which defines the height of the Atmospheric Boundary Layer (ABL). This height is computed from NWP model output, but it stands out as a separate class; both due to its importance for SNAP as input, but also because the procedure to determine the mixing height is so sensitive for the outcome of the simulation. This procedure is described in[6].

The most important meteorological elements provided by LAM50S model are:

- horizontal wind (u,v) - in 14 layers
- vertical velocity $\dot{\sigma}$ - in 14 layers
- temperature - in 14 layers
- precipitation and pressure - at the ground

Source Term Parameterization

Considering the importance for potential off-site consequences of a nuclear accident, according to Stokke[8], approximately 60 radionuclides can give harmful radiation if released to the environment/atmosphere. Fission products can be grouped in different release categories according to their volatility, because this determines the sequence of release. Noble gasses form a special group because there are few retention mechanisms restricting their release once core damage has occurred. Concerning off-site consequences, the most important elements can be grouped in four release categories presented in Table 1.

Table 1: Release categories according to volatility

Group	Elements	Compounds
Noble gases	Xe, Kr	None
Volatile species	I, Cs	CsI, CsOH, Cs uranates
Semivolatile species	Te, Ru	Oxides
Low-volatility species	Sr, Ce, Ba, La	

The amount of radioactivity released to the atmosphere in case of an accident depends on the seriousness of the accident (accident class), reactor type and power involved in reactor operation at the time of accident.

To simplify the source term for SNAP, we have grouped radionuclides according to physical properties.

a: Noble gases. They are neither wet nor dry deposited.

b: Different isotopes released as gas (example I-131).

c: Different isotopes released as aerosols (example: Cs-137).

The mass initially allocated to each of these classes, is treated somewhat differently with regard to the depletion process. At the early phase of an accident, we do not expect to have this detailed information available and we must run the model under default (or preset) release conditions. Two standard vertical release profiles can be specified by the user. In the first profile all particles are emitted into the ABL. In the second, 70% is emitted into the ABL and 30% above. User can also specify two standard time profiles for the release; emission intensity constant during the release time or decreasing linearly at the end of the release.

Advection of particles

Emission of radionuclides into the atmosphere is simulated by releasing a large number of 'particles'. 'Particle' is here a purely mathematical concept, being the 'vehicle' that carries the information (mass) from the source point. Part of this mass can be dumped off the vehicle as a result of depletion processes (wet or dry). We release up to 900 particles for each timestep (15 min.) in the model computations. These particles are advected by wind at different levels of the atmosphere.

The displacement of each particle is calculated for every model time step. In this calculation, three-dimensional velocity is interpolated to particle position from the eight nearest nodes in the model grid. Linear interpolation in space is applied to the components of the velocity field. In addition, a linear interpolation in time is applied between six-hourly meteorological input fields provided by LAM50S. The displacement of each particle in one time step is calculated according to the following equation:

$$\vec{x}'_{t+\Delta t} = \vec{x}_t + \vec{v}(\vec{x}_t)\,\Delta t \tag{1}$$

where $\vec{x}_t = (x, y, \sigma, t)$ is the position of the particle and $\vec{v}_t = (u, v, \dot{\sigma}, t)$ is the velocity field at time t, and the position of the particle after advection procedure is denoted by the vector $\vec{x}'_{t+\Delta t}$.

Horizontal and vertical diffusion

In the model computations, the processes - advection and diffusion - are separated so that advection is first performed, then followed by the diffusion process in a stepwise manner. A 'random walk' approach is used to parameterize horizontal and vertical diffusion, with slightly different parameterization for particles located within the ABL and for those above, but it can be described by the same equations. The new particle position is:

$$\begin{aligned} x'' &= x' + r_x l \\ y'' &= y' + r_y l \\ \sigma'' &= \sigma' + r_\sigma l_\sigma \end{aligned} \tag{2}$$

where $\vec{x}''_{t+\Delta t} = (x'', y'', \sigma'')$ is the particle position after the horizontal diffusion operator has been applied; r_x, r_y and r_σ are the randomly sampled values from the range <-0.5, +0.5>, generated from the uniform distribution; l and l_σ are the length scale parameters for horizontal and vertical turbulent motion. For constant $\vec{v}(\vec{x}, t)$, this formulation gives a parabolic spread with standard deviation s_y:

$$s_y = \sqrt{2nK_h\Delta t} \tag{3}$$

where $n\Delta t$ is the elapsed time since the release of the particle.

Horizontal diffusion in SNAP is parameterized in the same way for the particles within the ABL and for those above the ABL, but the numerical value of the coefficient of proportionality is different for the two regions. We assume a horizontal length-scale for the turbulent motion, and defines horizontal diffusion as:

$$l = ax^b \tag{4}$$

where $x = |\vec{v}|\Delta t$; $|\vec{v}| = \sqrt{u^2 + v^2}$ is the wind speed (with $|\vec{v}|$ units being (m/s) and Δt units being (s); b = 0.875 and a = 0.5 within the ABL and a = 0.25 above). In case of vertical diffusion, parameterization is different within the ABL and outside. The timestep of 15 min. in SNAP corresponds to the turn-over time of large eddies within the ABL: $\Delta t \approx z_i/w_c \approx 15min.$, where z_i is the ABL height and w_c is a characteristic convective velocity. This means that after approximately 15 min., a particle does not 'remember' its previous position within the ABL. Therefore, vertical diffusion in this region is parameterized as a random relocation of the particles between surface and z_i, applied after each timestep. To allow a vertical exchange of particles between the ABL and the free atmosphere above, the mixing zone is extended above the ABL from z_i to $z_i + \Delta z_i$. In σ-coordinates, vertical diffusion including the exchange mechanism is described by the following equation:

$$\sigma'' = 1 - (1 - \sigma_H - \Delta\sigma)\,(r_\sigma + 0.5) \qquad (5)$$

where σ_H is the vertical position of the ABL top (in σ-coordinates) and $\Delta\sigma = 0.005$. Above the ABL, vertical diffusion is parameterized by Equation (2) with $l_\sigma = 0.04$.

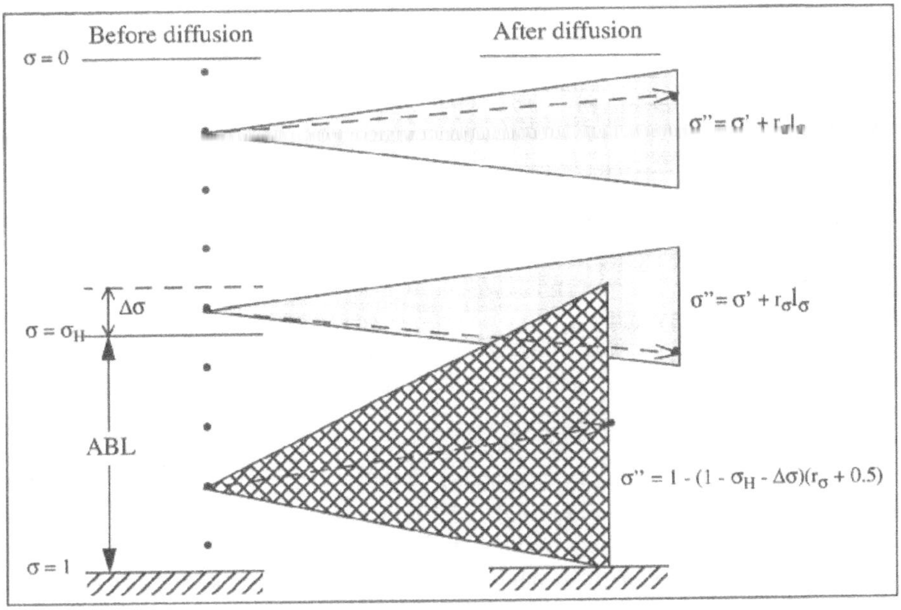

Figure 2. Diffusion in the well mixed ABL and above.

Figure 2 shows how the vertical diffusion process works at different levels.

a: For particles well above the ABL: After a time Δt they will find themselves at the new vertical position: $\sigma_p(t+\Delta t) = \sigma_p(t) + r_\sigma \cdot l_\sigma$ where $r_\sigma \in [-0.5, 0.5]$ is randomly sampled in the given interval, and l_σ gives the scale of the vertical turbulent transport.

b: For particles slightly above the ABL: They are treated in the same way as under a: - above. As we can see, this formulation will create an entrainment flux of particles into the ABL from above, - depending on l_σ.

c: For particles in the ABL: We assume that after a time Δt, a full turn-over in the ABL has taken place so that the particle can be found with equal probability at any height in the ABL or slightly above. If the height of the ABL is $(1 - \sigma_H)$ in σ-units, we add an increment (-Δσ) to it and let the random walk process (relocation) work inside the range $(1 - \sigma_H - \Delta\sigma)$ for all particles found inside the ABL. This will create a compensating upward flux of particles from the ABL to the layers above. Δσ must be adjusted to mach the effect of the entrainment of particles from above, - see b: above.

Dry Deposition

A particle carry an initial mass - $M_p(t_0)$ - from the source. The mass $M_p(t_0)$ is subdivided into smaller units. Particles located above the atmospheric boundary layer, are not subject to the dry deposition process. For particles in the boundary layer, we assume: The boundary layer consists of - n - 'fictitious layers' or $n=z_b$, where z_b is the number of 'meters' of the boundary layer thickness. The particle is randomly 'assigned' a number - n_p - from the number set $<z_b>$. The number n_{crit} is prescribed, being an element in the lower part of the number set $<z_p>$. If $n_p > n_{crit}$, no mass is lost by this particle. If $n_p \leq n_{crit}$, a certain fraction of the particle's mass - ΔM_{pd} - is deposited onto the ground.

The probabilistic parameters in this process, are given numerical values corresponding to a rate of efficiency that can be compared with the more commonly used parameterization of the process, for example, the concept 'dry deposition velocity'. The values of the 'key' parameters n_{crit} and ΔM_{pd} determine the efficiency of the dry deposition process.

Parameterization of Wet Deposition

In the parameterization of the wet deposition process, we use 'probability curves'. A 'probability curve' can be explained as follows: If we choose an arbitrary point (x, y, t) in a 'grid-volume' (Δx, Δy, Δt), a 'probability curve' will tell us: "What is the probability for 'real' precipitation to occur in this specific point in space and time, - given we have access to the amount of precipitation (Prec) predicted by DNMI's operational NWP model". These ideas and concepts have been developed and used in the models at DNMI[9], describing transport/deposition of acidifying substances.

For Prec = 0, no wet process acts on the particles. For Prec > 0, we define a two-step process. First step: We read from the 'probability curve' for the precipitation amount in question, what the probability for 'real' precipitation to occur is at the 'space-time' point in question: $P_w = Prob(prec)$. Next step: We randomly assign '0' or '1' to each particle. The '1' value is chosen with a probability P_w, and '0' value with a probability $(1 - P_w)$. This means, if the total number of particles is N in the grid-volume in question, then P_w N particles are affected by the wet deposition in a timestep Δt, and a certain fraction of the 'wet' particles' mass is deposited to the ground.

The probabilistic parameters for the process, are given numerical values corresponding to a rate of efficiency that can be compared with more common parameterization of wet deposition, for example, the 'scavenging ratio' concept. The 'probability curves' will be the 'key' in determining the efficiency of the process. The 'probability curve' worked out by P.E. Haga[9] is shown in Figure 3.

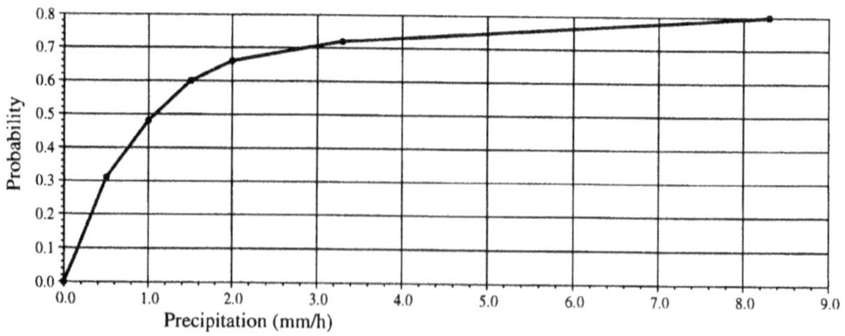

Figure 3. Probability for the point in a grid square to be affected by 'real' precipitation, as a function of 'model precipitation intensity'.

APPLICATION OF SNAP

The main application of SNAP is real-time simulation of the transport of radionuclides released during a nuclear accident (operational application). In addition SNAP can be used to simulate events from the past, for which meteorological data are available in archives, and special events related to model intercomparisons and tests.

SNAP is now fully operational. This means that it can be run by the meteorologist on duty at any time. During the simulation of a nuclear accident, SNAP produces different types of output. Part of this output is graphically displayed directly on the screen during the simulation. Another part is stored in public accessible disc files for further processing.

As an example of on-line screen graphics during simulation, we show a screen-dump in Figure 4. This shows individual particles after 60 hours of continues release from a source point in UK. During the model run, user can display on-line the following parameters:

- particles effected by precipitation - individual locations
- particles not effected by precipitation - individual locations
- precipitation amount for the last three hours - isolines
- concentration of mass/activity in the ABL - isolines
- accumulated wet deposition - isolines
- accumulated dry deposition - isolines
- accumulated total (wet + dry) deposition - isolines

Hard copies of the screen can be printed during the model run. When the simulation is finished, the fields listed in the previous paragraph (except particles) can repeatedly be seen on the screen with better resolution and hard copies of these fields can be printed.

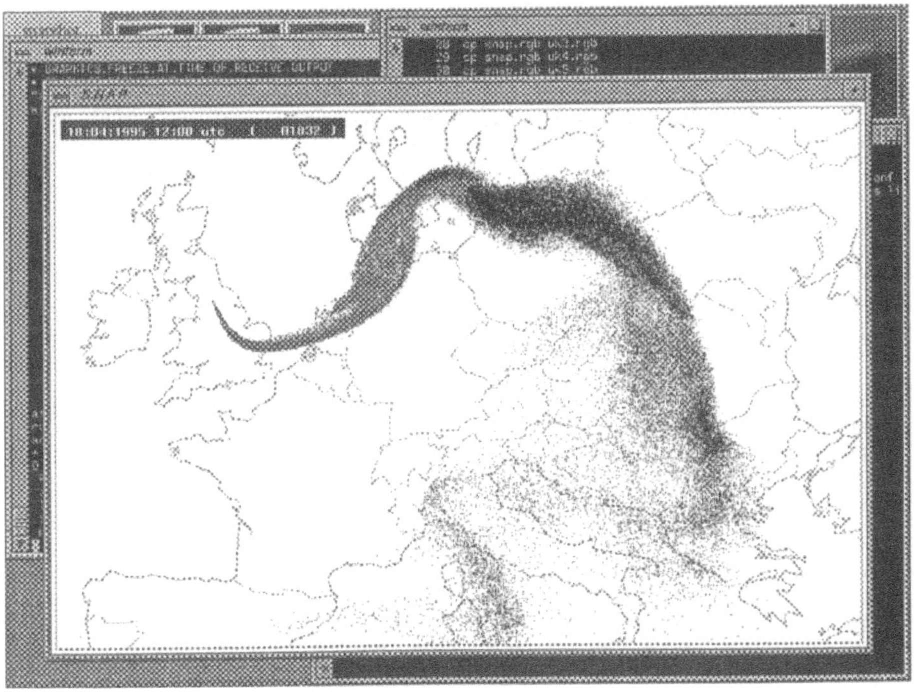

Figure 4. Particles after 60 hours of continuous release from a source in UK (Screen dump).

Direct access to the output files from the SNAP simulations is restricted to DNMI staff. However, the user of the Norwegian Application of MEMbrain, (NRPA), will also have remote access to the results, written in agreed format on a special disc at DNMI which is available for the outside users.

Besides the operational applications, SNAP is often used for special tasks. For example, SNAP was used in the ETEX experiment. Also, the Chernobyl accident in April 1986 has been simulated using SNAP. In this simulation, special attention was paid to atmospheric transport of radionuclides to Norway. The results from SNAP have also been compared with outcome from other real-time dispersion models in use in the Weather Services in Denmark, Finland and Sweden in the frame of Nordic collaboration in functional exercises. However, despite all these applications, the operational application of the model is the most important, and the future development of SNAP will mainly be directed towards operational aspects.

CONCLUDING REMARKS

SNAP has reached a state of development where useful results can be produced. A state of preparedness has been reached (implemented in operational service) at DNMI, when SNAP can be used at any time (day and night) by the meteorologist on duty.

REFERENCES

1. NOU 24A, 1986. Experience from the nuclear accident in Chernobyl. (NOU: Norges Offentlige Utredninger - Norwegian Official Reports)
2. Smith F. B. and M. J. Clark (1989). The transport and deposition of radioactive debris from the Chernobyl nuclear power plant accident with special emphasis on consequences to the United Kingdom. *Meteorological Office Scientific Paper No. 42*, HMSO London.
3. NOU 5, 1992. Tiltak mot atomulykker. Anbefalinger om videre styrking av norsk beredskap mot atomulykker - Efforts Against Nuclear Accidents. Recommendations for Further Strenghtning of Norwegian Preparedness Against Nuclear Accidents, (NOU: Norges Offentlige Utredninger - Norwegian Official Reports)
4. Saltbones, J. (1994). Real-Time Dispersion Model Calculation as a Part of NORMEM - WP19. To be published in the journal: Safety Science, Elsevier Science B. V.
5. Saltbones J. and A. Foss. (1994). Real-Time Dispersion Model Calculations at DNMI of Radioactive Pollutants. Part of Norwegian Preparedness Against Nuclear Accidents. Preprints for the XIX Nordic Meteorologists Meeting, (6-10)/6-1994, Kristiansand, Norway.
6. Maryon R.H., J.B. Smith. B.J. Conway and D.M. Goddard (1991). The United Kingdom Nuclear Accident Model. *Prog. Nucl. Energy*, **26**: 85-104.
7. Grønås S. and K.H. Midtbø (1994). Operational Multivariate Analysis by Successive Corrections. *J. Meteo. Soc. Japan*, **65**: 61-74.
8. Stokke, E. (1994). Technical note about the source-term parameterization from IFE to DNMI. (available within the NORMEM Project)
9. Haga P.E. (1991) Hvordan influerer nedbørprosessers tids- og romskala på langtransport av svoveldioksyd og partikulært sulfat? (in norwegian). Thesis, Oslo University.
10. Nordeng T.E. (1986) Parameterization of Physical Processes in a Three-Dimensional Numerical Weather Prediction Model. DNMI Technical Report No. 65.

DISCUSSION

H.-J. PANITZ:

Your parameterization of the particle behaviour close below and above the top of the mixing height layer does it avoid the accumulation of particles in this area or do you also need a drift velocity term?

J. BARTNICKI:

In the SNAP model, we compensate downward particle flux at the top of the ABL by introducing the entrainment layer just above the mixing height. The depth of the entrainment zone is related to the length scale of the vertical turbulent motion above the ABL. This parameterization completely avoids the accumulation of the particles in the mixing layer. Therefore, we do not use a drift velocity term in the model.

STUDIES OF DENSE GAS DISPERSION FROM SHORT-DURATION TRANSIENT RELEASES OVER ROUGH SURFACES DURING STABLE CONDITIONS

Steven R. Hanna[1] and Kenneth W. Steinberg[2]

[1]Earth Tech, 196 Baker Ave., Concord, MA 01742
[2]Exxon Research and Engineering, 180 Park Ave., Florham Park, NJ 07932

INTRODUCTION

Emerging legislation in the United States and Europe has brought renewed emphasis to the analysis of hypothetical, accidental releases of hazardous substances at industrial facilities. Most dense gas models have been developed and evaluated using field and wind tunnel data from idealized experiments in which the source emissions are either continuous or instantaneous, the underlying surface is relatively smooth, and the ambient stability is nearly neutral. In contrast, most accidental releases of hazardous chemicals are likely to involve source emissions that are transient (i.e., the release rate is controlled or mitigated so that it drops to zero within a minute or two), underlying surfaces that are characterized by a rough surface including some 10-m to 20-m high obstacles (typical of an industrial facility), and worst-case meteorological scenarios that would be defined by stable light-wind ambient conditions. Evaluations of the HEGADAS, DEGADIS, and SLAB models with limited existing field and wind tunnel data suggest that they are biased (by as much as a factor of two to four) if one or more of these conditions is satisfied (Hanna et al., 1993).

The research deficiencies listed above are being addressed as part of a cooperative research project cosponsored by the Petroleum Environmental Research Forum (PERF) Dispersion Modeling Project 93-16, the U.S. Department of Energy (DOE), and the U.S. Environmental Protection Agency (EPA). PERF was established in 1986 to provide a stimulus for cooperative research and development of technology for environmental pollution control for the petroleum industry and to provide a forum to discuss and develop proposals for cooperatively-funded projects in pollution control. The projects are funded and carried out by interested parties including non-members of PERF.

Air Pollution Modeling and Its Application XI, Edited by
Gryning and Schiermeier, Plenum Press, New York, 1996

The study has three main components:

(1) A field study was carried out in summer 1995 at the Nevada Test Site. Dense gas (CO_2) was released for 5-minute (continuous) and 20-second (short-duration transient) periods, including both neutral and stable conditions. The desert surface was artificially roughened using combinations of obstacles in order to simulate the roughness of an industrial site.

(2) Wind tunnel studies have been carried out in which continuous and transient dense gas releases were made over rough surfaces. A separate wind tunnel study involves careful attention to properly simulating the stable boundary layer.

(3) Previous and current field and laboratory data are being analyzed and further research carried out in order to modify algorithms in dense gas models so that the predictions have minimum bias and scatter.

An overview of the status of the three research components is given below, with emphasis on new findings concerning technical issues such as entrainment into dense gas plumes and variations of dense gas transport and dispersion as the surface roughness increases. At the completion of this study, it is expected that revised versions of state-of-the-art heavier-than-air dispersion models will be available that are able to more accurately simulate dense gas releases with short durations at typical industrial sites.

TECHNICAL ISSUES

The wind tunnel and field experiments and the data analysis and model development activities are intended to improve our knowledge of six primary technical issues, as listed below.

Vertical Entrainment into Dense Gas Plumes

Current dense gas dispersion models account for entrainment of ambient air into ground-based dense gas plumes by assuming that the vertical entrainment velocity, u_e, is a function of the ambient friction velocity, u_*, and the plume Richardson number, Ri. The plume Ri can be defined as $Ri = g(\rho_p - \rho_a)h/\rho_a u_*^2$, where g is the acceleration of gravity, h is the local plume depth, ρ_p is local plume density, and ρ_a is ambient density. Unpublished wind tunnel data collected by McQuaid were plotted in the form of u_e/u_* versus Ri to derive the relation in current use in most models:

$$u_e/u_* = a/(1 + b(Ri)^c) \qquad (1)$$

where a is a "constant" ranging from about 0.4 to 0.8, b is a constant on the order of 1.0, and various models use the power, c, in the range from about 0.5 to 1.0 (e.g., Havens and Spicer, 1985; Ermak, 1990; Witlox, 1994). It is also assumed in all models that the dense gas cloud does not alter the ambient boundary layer, so that the ambient u_* is constant in space. The PERF project is coordinating/supporting wind tunnel tests at three institutions (EPA Fluid Modeling Facility, University of Arkansas wind tunnel, and University of Surrey wind tunnel) and a field test at the Nevada Test Site in order to better define the entrainment relation in equation (1). A wide range of roughness types, including those typical of an industrial site, are included in the tests. The experiments during stable conditions at the University of Surrey and at the Nevada Test Site will allow the effects of ambient stability to be parameterized as functions of the ambient Richardson number or the Monin-Obukhov length L. Thus the modified version of equation (1) should be valid for all expected surface roughness and stabilities.

Along-Wind Dispersion Coefficient σ_x

When the time duration of the release, T_d, is less than the time of travel to the receptor, t = x/u, the cloud looks more like an instantaneous puff, for which the along-wind dispersion coefficient, σ_x, must be known. Because of the tilting or stretching effect of windshear, σ_x is typically two to four times the lateral dispersion coefficient, σ_y (Van Ulden, 1992). At downwind distances where $T_d u_c/\sigma_x$ is about unity (where u_c is the cloud advection speed), the predicted concentrations can be in error by as much as a factor of four if σ_x is in error by a factor of two (Petersen, 1992). For the typical range of values of T_d (about 1 min) and x (about 1000 m) for accidental releases, the ratio $T_d u_c/\sigma_x$ is often close to unity, implying that σ_x must be accurately specified. Current models use σ_x formulas based on 15 to 25 year-old theoretical analyses (e.g., Chatwin, 1968; or Wilson, 1981). More recent theories (e.g., Van Ulden, 1992) suggest that σ_x should be larger. The new theories account for the effects of large horizontal eddies as well as the effects of small eddies and wind shear. Fortunately the effects of stability are reduced by the tendency for wind shear and vertical diffusion to be negatively correlated. The PERF wind tunnel and field tests emphasize release durations and distances such that the ratio $T_d u_c/\sigma_x$ is in the critical region near unity where model improvements are needed.

Cloud Advection Speed u_c

The cloud advection speed, u_c, for ground-based plumes is calculated in most models by means of a concentration-weighted vertical integration of the wind profile. Sometimes u_c is set equal to the wind speed at a height equal to a set fraction (say, 0.6) of σ_z. This speed can be calculated from observations of concentration time series in wind tunnel or field experiments. For example, analysis of Petersen's (1992) wind tunnel data suggest that u_c is being overestimated by the HEGADAS model (Witlox, 1994). As a result, the ratio, $T_d u_c/\sigma_x$, is overestimated and the modeled transition from a steady-state plume to puff dispersion is delayed too long. The speed u_c can be calculated from the observations in the PERF experiments since high-resolution concentration time series are available from all monitoring locations.

Effects of Surface Roughness

Most dense gas effects are strong functions of friction velocity, u_*, which is itself a function of surface roughness. For example, the source Richardson number Ri_o, is defined as follows:

$$Ri_o = \frac{g(\rho_{po}-\rho_a)q_o}{u_*^3 D_o} \qquad (2)$$

where ρ_{po} is the initial cloud density, q_o is the source volume flux, and D_o is an initial dimension of the source (say, the diameter of a ground-level area source). If $Ri_o >> 1$ then dense gas effects must be accounted for. But because u_* is cubed in equation (2), a factor of two change in u_* causes a factor of eight change in Ri_o. Similarly, the rate of lateral gravity spread of the dense gas cloud is inversely proportional to u_*, and the distance a dense cloud spreads upwind is inversely proportional to u_*^3. Because u_* is a function of surface roughness ($u_* \propto (\ln z_o)^{-1}$), as the underlying surface changes from grass ($z_o \approx 0.01$ m) to an industrial site ($z_o \approx 1$ m), u_* can increase by a factor of two or three, thus greatly reducing the possibility of gravity slumping of an dense gas cloud. Ground level concentrations can be a factor of five to ten less over an industrial surface than over a flat rural surface. For this reason, the PERF experiments cover a wide range of surface roughness, including those typical of an industrial site.

Effects of Stability

All previous wind tunnel tests and most field tests of dense gas plume dispersion have involved neutral boundary layers with moderate to strong winds (Hanna et al., 1993). However, the "worst-case" meteorological conditions assumed for hazards assessments are usually assumed to be stability class F with a wind speed of 1 to 2 m s, for which concentrations are expected to be highest. In order to address this deficiency, the University of Surrey wind tunnel tests and the Nevada Test Site field tests are focussing on stable, light wind conditions. As the ambient boundary layer stabilizes, the effects of ambient turbulence may be reduced in comparison to turbulence generated inside the cloud, and the parameterizations in equation (1) may need refinement.

Effects of Obstacles

The height of the dense gas cloud in an industrial area may be less than the height of the obstacles (buildings, tanks, pipe racks, etc.). In this case, the cloud will be influenced by local flows around the obstacles and the general entrainment equation (1) may no longer be valid. It is especially important to be able to parameterize the effective wind speed within the obstacles and the averaged entrainment rate. The wind tunnel studies by the EPA and the University of Arkansas include some special cases when the cloud height is less than the obstacle height. Also, because the field study at the Nevada Test Site is concerned with ground-level dense gas releases within an array of obstacles 2.4 m high, it is expected that the cloud height will be less than the obstacle height at the first two downwind monitoring arcs (at x = 25 m and 50 m). These experiments will also provide observations in the critical regime as the cloud grows through a range such that its height is about equal to the obstacle height.

WIND TUNNEL EXPERIMENTS

A coordinated set of wind tunnel experiments is included in the PERF program in order to provide answers to some of the technical issues mentioned above. Petersen's (1992) initial investigations of the dispersion of finite duration dense gas releases over rough surfaces suggested that improved formulations for entrainment and along-wind dispersion were needed. Several studies have been undertaken or are underway as part of the PERF cooperative study with the EPA and the DOE.

EPA Wind Tunnel Study of How Roughness Elements should be Efficiently Designed.

Snyder (1995) carried out an investigation in the EPA wind tunnel of the effective roughness length generated by various shapes and spacings of rectangular roughness elements. The results of this study were to be used to design the roughness elements to be placed on the desert floor at the Nevada Test Site for the field experiment. Snyder found that rectangular plates 4 times as wide as they are high, with spacings of 12 plate heights (staggered from one row to the next), were the most efficient roughness generators. Their effective roughness length, z_o, was about 1/14 of their height. Snyder also determined that if a row of tall triangular-shaped Irwin spires (turbulence generators) were installed on the upwind edge of the roughness array, the equilibrium boundary layer over the array would develop more rapidly.

CPP Wind Tunnel Study of How Roughness Elements Should be Designed to Represent an Industrial Site

The CPP wind tunnel facility had on hand some scale models of industrial sites from previous studies. Petersen and Cochran (1995) carried out an investigation with various shapes and spacings of flat square elements in order to determine which combination best represents the surface roughness and turbulence over the industrial sites. They concluded that the square elements with height H should be set in staggered rows 2.4 H apart with spacing of 3.4 H between centers of elements along a row. These results were also used to design the field experiments.

EPA and University of Arkansas Wind Tunnel Studies of Entrainment into Two-Dimensional Dense Gas Plumes during Neutral Conditions

As mentioned earlier, a fundamental technical issue concerns the parameterization of entrainment rate in dense gas models (e.g., see equation (1)). This issue is being addressed in the fall of 1995 in coordinated wind tunnel studies at the EPA and the University of Arkansas wind tunnels. The University of Arkansas wind tunnel, operated by J. Havens, has been recently upgraded under support of the DOE and the Gas Research Institute.

To simplify the analysis, the experiments in the two wind tunnels are being set up so they are as similar as possible. The dense gas is emitted from a cross-wind line source into a neutral ambient boundary layer. The initial density is adjusted so that six different initial cloud Richardson number Ri_o, are tested, including one case with Ri_o = 0 (i.e., a passive gas plume). Three alternative arrays of roughness obstacles are being tested: (1) width (W) to height (H) ratio of 4, with spacing of 12 H, and H = 5 cm; (2) W/H = 4, spacing of 12 H, and H = 0.85 cm, and (3) W/H = 1, spacing of 8 H, and H = 3.8 cm. These array types were tested in Snyder's (1995) earlier studies, reported above, and the W/H = 4 array with spacing of 12 H is being used in the field study. Array types (1) and (2) employ the same basic geometrics, but type (1) is larger (H = 5 cm for type (1) and H = 0.85 cm for type (2)), so that experiments with the cloud height less than the obstacle height can be included. The wind tunnel studies of the three array types are being split so that the tests at the EPA employ array types (1) and (2) and the tests at the University of Arkansas will employ array types (1) and (3). Array type (1) is being tested at both places so that the results can be compared from the two tunnels. Otherwise, the two wind tunnels are being set up with similar test section sizes and other important scaling parameters. The entrainment velocity, u_e, is assumed to be given by the product of the cloud advection speed, u_c, and the rate of growth of the cloud depth, dh/dx:

$$u_e = u_c \frac{dh}{dx} \tag{3}$$

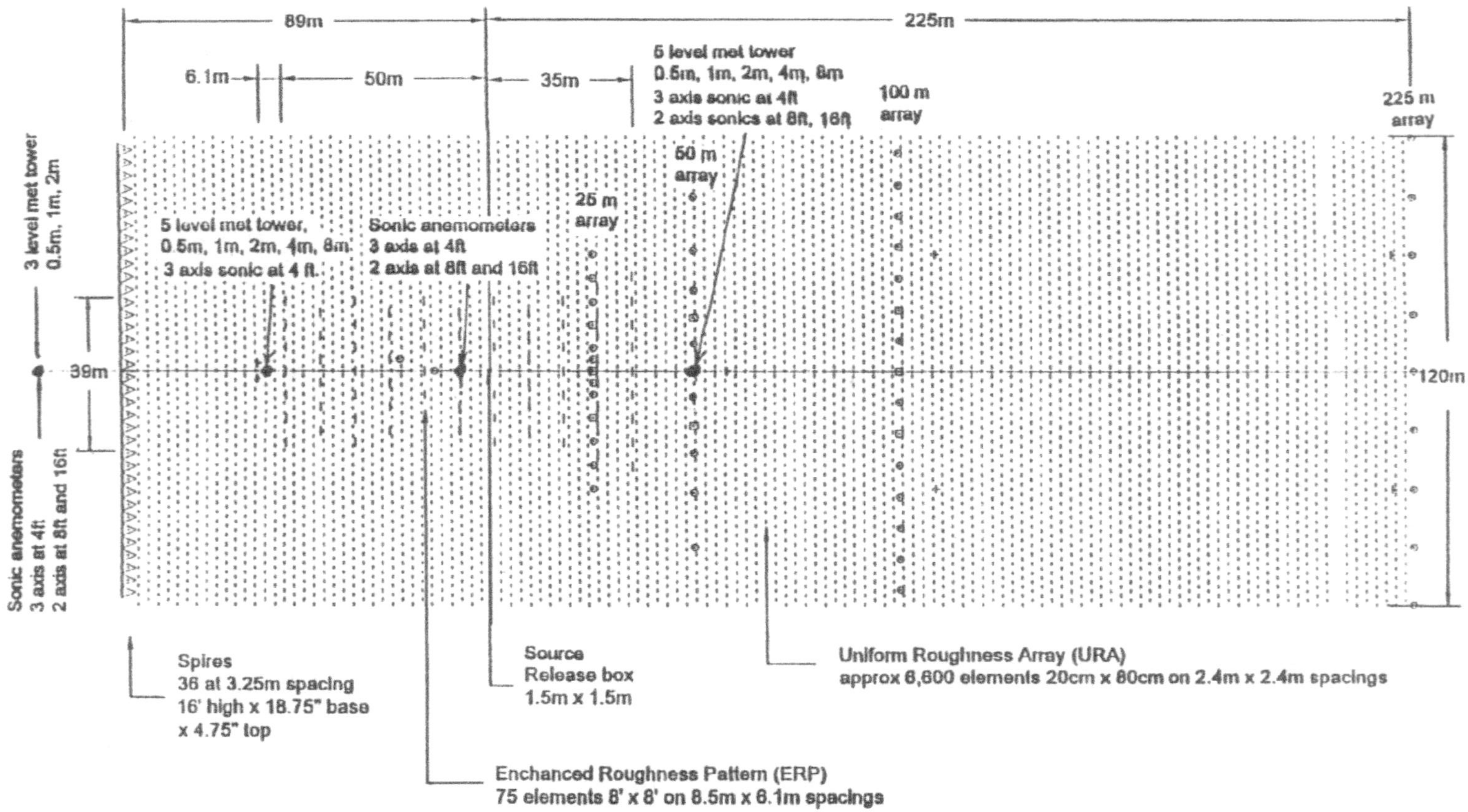

Figure 1. Plot plan of experiment at Nevada Test Site. Circles are near-ground concentration monitors; squares are tower-mounted concentration monitors; meteorological towers are marked by solid circle; wind vanes are marked by sideways "T" (figure from WRI).

University of Surrey Wind Tunnel Studies of the Stable Boundary Layer and Entrainment into Dense Gas Plumes.

Because of interest in the worst-case meteorological condition associated with light-wind stable conditions, and the lack of data concerning entrainment into dense gas clouds during these conditions, PERF is supporting series of stable wind tunnel tests in 1995 and 1996 at the University of Surrey. These stable tests, being carried out by I. Castro and A. Robins, are being coordinated with the EPA and University of Arkansas neutral wind tunnel tests described above. The first task involves a demonstration that the tunnel can simulate the turbulence structure in a stable boundary layer, since several reviewers have expressed concern about that issue. If this demonstration is successful, the second task will begin, related to entrainment into a dense cloud from a two-dimensional line source, analogous to the neutral studies at the EPA and University of Arkansas. It is expected that these data will allow equation (1) to be modified to account for changes in ambient stability.

FIELD EXPERIMENT

A joint field experiment has been conducted in August and September, 1995, at the Nevada Test Site (NTS), under support from PERF, EPA, and DOE. The operations were carried out by Desert Research Institute (DRI) and Western Research Institute (WRI). The entire experimental set-up is intended to represent about 1/10 of full-scale. The flat desert lake bed has had thousands of roughness elements installed over a 120 m x 314 m area in order to increase the roughness to a value typical of an industrial site and its surroundings (see Figure 1). The roughness arrays were designed by wind tunnel experiments by CPP and EPA (see above). The field experiment has the following characteristics (refer to Figure 1):

- Large Equivalent Roughness Pattern (ERP) (2.4 m squares) on inner 40 m x 85 m rectangle; smaller Uniform Roughness Array (URA) (0.2 m high by 0.8 m wide rectangles) on outer 120 m x 314 m rectangle; Irwin spires on upwind edge.
- CO_2 is released from a 1.5 m x 1.5 m area source in middle of ERP, with emission rate of 4 kg/s in 5 minute (continuous) and 20 second (finite duration) bursts.
- 84 fast-response concentration monitors on 4 downwind arcs (25 m, 50 m, 100 m, and 225 m), with some mounted on towers.
- Meteorological instruments (including nine sonic anemometers) on five towers, plus wind vanes at six other locations.
- Schedule: 1 week with ERP and URA in place; 1 week with URA in place; 1 week with URA and ERP removed (i.e., flat desert).

The CO_2 concentration data and the meteorological data from these experiments will be used to test the revised entrainment parameterizations and other algorithms developed as part of the PERF project.

MODEL MODIFICATIONS AND TESTING

Two operational dense gas models, HEGADAS (Witlox, 1994) and SLAB (Ermak, 1990) have been used to help design the field experiment. Model predictions of concentrations, cloud height, and cloud width were generated for a variety of source conditions, meteorological inputs, and surface roughness, in order to plan the calibration ranges for the instruments and their locations in x, y, and z. An attempt was made to account for changes in the boundary layer as it passes from one roughness type to another, and to account for reductions in wind speeds below the tops of the obstacles. It was found that the predicted concentrations and cloud dimensions were sensitive to accurate knowledge of ambient friction velocity.

The data from the wind tunnel and field experiments, as well as data from previous related experiments, are being analyzed from the point of view of the several technical issues briefly discussed above. Primary emphasis is on developing improved parameterizations for entrainment into dense gas clouds in neutral and stable ambient conditions over a wide range of surface roughness, for the advection speed of dense gas clouds, for the along-wind dispersion coefficient, and for the effects of obstacles on the transport and dispersion of dense gas clouds. These revised parameterizations will be sufficiently general that they can be used in most current dense gas models (e.g., publicly-available models such as SLAB, DEGADIS, ALOHA and HEGADAS; or proprietary models such as PHAST and TRACE). The revised models will be evaluated with the new field data and with the sets of field data used in the comprehensive evaluations by Hanna et al. (1993). It is expected that the revised models can be more confidently used to analyze the consequences of accidental releases of hazardous gases.

ACKNOWLEDGEMENTS

Exxon Research and Engineering Company (ER&E) serves as contract coordinator for this PERF 93-16 Project. The other companies financially supporting in this PERF Project include Allied-Signal Incorporated, AMOCO Corporation, Chevron Research and Technology Company, Mobil Research and Development Company, Shell Development Company, CITGO Petroleum Corporation, Clark Oil and Refining Corporation, Marathon Oil Company, and Phillips Petroleum Company. EPA is providing in-kind support through a Coordinated Research And Development Agreement (CRADA) with ER&E. DOE is providing financial support through their jointly sponsored research program with the Western Research Institute. In particular, Manny Vazquez of Allied-Signal, Doug Blewitt of AMOCO, Dave Fontaine of Chevron, John King of Shell, Gary Briggs, Bill Petersen, and Bill Snyder of EPA have made numerous technical contributions. The field design plan and field experiments are largely the responsibility of WRI, with David Sheesley as project manager. Joe Chang of Earth Tech has been responsible for the HEGADAS simulations in support of the PERF activities.

REFERENCES

Chatwin, P.C., 1968, The dispersion of a puff of passive contaminant in the constant stress region, *Quart. J. Roy. Met. Soc.*, **94:** 350.

Ermak, D.L., 1990, *User's Manual for SLAB: An Atmospheric Dispersion Model for Denser-Than-Air Releases*, UCRL-MA-105607, Lawrence Livermore National Laboratory, Livermore, CA.

Hanna, S.R., Chang, J.C., and Strimaitis, D.G., 1993, Hazardous gas model evaluation with field observations. *Atmos. Environ.,* **27A**: 2265.

Havens, J.A., and Spicer, T.O., 1985, Development of an Atmospheric Dispersion Model for Heavier-Than-Air Gas Mixtures. Final Report to the U.S. Coast Guard, CG-D-23-85, USCG HQ, Washington, DC.

Petersen, R.L., 1992, Wind Tunnel and HEGADAS Simulations of Finite Duration Releases for a Specific Refinery with Uniform Roughness. Report 91-9782 by CPP, 1415 Blue Spruce Dr., Ft. Collins, CO 80524.

Petersen, R.L. and Cochran, B.C., 1995, Wind Tunnel Determination of Equivalent Refinery Roughness Patterns, CPP Report 94-1152 (Part I dated February 1995 and Part II dated July 1995), CPP, 1415 Blue Spruce Drive, Fort Collins, CO 80524.

Snyder, W.H., 1995, Wind Tunnel Roughness Array Tests. ASMD/NOAA, Research Triangle Park, NC 27711.

Witlox, H.W.M., 1994, The HEGADAS model for ground-level heavy-gas dispersion - II. Time dependent model. Atmos. Environ., **28**: 2933.

Van Ulden, A.P., 1992, A surface-layer similarity model for the dispersion of a skewed passive puff near the ground. Atmos. Environ., **26A**: 681.

Wilson, D.J., 1981, Along-wind diffusion of source transients. Atmos. Environ., **15**: 489.

DISCUSSION

I. SYKES: Were high frequency concentration observations taken during the field experiments at the Nevada Test Site.

G. BRIGGS: We had over 80 continuous carbon dioxide monitors on four arcs (25, 50, 100, and 225 m). Their response time was about one second and the data are being reported as time series of one second averages.

EMERGING AIR QUALITY MODELING TECHNOLOGIES FOR HIGH PERFORMANCE COMPUTING AND COMMUNICATION ENVIRONMENTS

Daewon W. Byun[1§] Donald Dabdub[2], Steve Fine[3], Adel F. Hanna[3], Rohit Mathur[3], M. Talat Odman[3], Armistead Russell[4], Edward J. Segall[5#], John H. Seinfeld[6], Peter Steenkiste[4], and Jeffrey Young[1§]

[1]Atmospheric Sciences Modeling Division, Air Resources Laboratory, National Oceanic and Atmospheric Administration, RTP, NC 27711

[2]Department of Mechanical & Aerospace Engineering, University of California Irvine, Irvine, CA 92717-3975

[3]MCNC, North Carolina Supercomputing Center, Environmental Programs, 3021 Cornwallis Road, Research Triangle Park, NC 27709

[4]Department of Mechanical & Engineering, Carnegie Mellon University 5000 Forbes Avenue, Pittsburgh, PA 15213-3891

[5]School of Computer Science, Carnegie Mellon University 5000 Forbes Avenue, Pittsburgh, PA 15213-3891

[6]Department of Chemical Engineering, California Institute of Technology, Pasadena, CA 91125

INTRODUCTION

An Eulerian photochemical air quality model (AQM) simulates atmospheric processes such as advection and diffusion, pollutant emissions, and chemical reactions using a grid system defined in a fixed frame of reference, usually the surface of the earth. Because the modeling framework is convenient to incorporate many atmospheric processes affecting air quality, Eulerian models have been widely used to study the effects of emissions control strategies on air quality.

Most AQMs depend on meteorological models to provide detailed atmospheric dynamic features and simulate highly nonlinear reactions of about fifty chemical species in the atmosphere. The temporal scales of important meteorological processes range from minutes to weeks while those of chemical reactions range from seconds to months. Thus, air quality models need to operate at temporal resolutions of minutes and horizontal spatial resolutions of a few kilometers. Computational domains range from a few hundred km (urban scale) to two-thousand km (regional scale). For realistic descriptions of atmospheric flux exchange processes, the models require five to thirty vertical layers. Therefore, current problem size is typified by 20,000 coarse cells covering a regional domain to set boundary conditions for 70,000 high-resolution cells for embedded urban domains. In addition, the description of atmospheric chemistry requires approximately 50 state variables (transported species), resulting in a "problem size" of 4.5 million variables. The computational demand is dominated by the desire to obtain accurate solutions to a set of stiff ordinary differential

§ On assignment to the National Exposure Research Laboratory, U.S. Environmental Protection Agency

Currently with Department of Computer and Information Sciences, University of Delaware, Newark, DE 19716

equations (ODEs) describing the photochemical kinetics. On the other hand, environmental impact assessments need hourly modeling results for periods ranging from weeks to years. To represent the physical-chemical processes with the degree of accuracy needed to reliably support decision making, we believe the number of cells needs to be increased by a factor of at least 5 and the number of state variables increased by a factor of 3. Also, the models must be capable of providing the simulation results for many "what if" questions with reasonable turn-around times (for example, 10% of simulation time).

For the last decade there has been an explosion of developments in both computer hardware and software technologies. Faster workstations have become common place and not only vector supercomputers but also many parallel computers of different architectures, such as SIMD (single instruction multiple data), MIMD (multiple instruction multiple data), or hybrid designs are available. Also, software engineering methodologies have been improved and applied for many complex scientific software development projects. Recently, the U.S. Government initiated the High Performance Computing and Communications (HPCC) Program to encourage the development of high performance computers and networks and to assist in the dissemination of such new systems.

To demonstrate applications of the HPCC technologies in air quality models, we organized the Specialty Evening Session 1, "Emerging Air Quality Modeling Technologies for High Performance Computing and Communication Environments" as a part of the Twenty-First NATO/CCMS International Technical Meeting on Air Pollution Modelling and Its Application, 6-10 November, 1995. This paper summarizes presentations in the Specialty Session. First, we describe a new design concept for the Eulerian AQMs incorporating recent developments in computer and atmospheric sciences. Second, we present several different experiences on parallel computation approaches for achieving fast turnaround time (speed) and, at the same time, implementing sufficient scientific details (accuracy) in AQMs. Finally, we show an application demonstration of a prototype multiscale air quality model collaboratively developed by scientists and system engineers under U.S. Environmental Protection Agency's (EPA's) Models-3 and MCNC's Environmental Decision Support System (EDSS) projects.

REDESIGNING EULERIAN AQMS FOR HPCC ENVIRONMENTS

Operational experiences with many existing Eulerian air quality models have revealed many shortcomings both in scientific.algorithms and in computational efficiency (Byun et al., 1995). In addition to the improvement of scalability of scientific algorithms, the computational issues such as lengthy execution time, unnecessarily complex data movement, the outdated software design, inadequate modularity of model components, and limited flexibility and interchangeability of science modules need to be addressed for the next generation AQMs. To take advantage of the HPCC technology, we need to redesign the system framework and science algorithms on which current AQMs are based. In the context of the Models-3 and EDSS systems, a system framework is the mechanisms necessary to manage the scheduling and execution of computational models, the data produced by the computational models, the analysis and visualization tools necessary to provide understanding of their results for decision making, and the interfaces to all these capabilities. In the following we describe key design characteristics of the system framework and science models in the Models-3 and EDSS systems.

Requirements: As our knowledge of the environment improves, the linkage of natural processes will be diversified. Because no one group of scientists can be experts in many diverse air quality science areas, we promote a community modeling paradigm for the Eulerian AQMs. A community modeling concept will change the modeling culture as modelers become more open and collaborative and allow verification of modeling results by other scientists. AQM users demand that multipollutant and multiscale modeling should be done within one system. Models must be designed with highly modular characteristics because the science of many key processes is constantly improving and many alternative physical/chemical parameterizations need to be tested. AQMs should also be designed to easily incorporate advances in computational techniques including numerical solvers, systematic sensitivity/uncertainty analysis methods, optimization procedures (for decision support), processor analysis tools, and scientific visualization.

Design and Implementation: To handle the ever increasing complexity of AQMs and to meet the demand for linking AQMs with other environmental models, rigorous software development methods are adopted. Because of the inherently different tasks involved, we use different development strategies for the system framework and science models.

For the framework development, both *rapid prototyping* and *object oriented analysis* (OOA) methods are used. Rapid prototyping refers to a software development method that clarifies requirements and desirable features of the target system by developing simplified prototypes early in the development process. OOA is a method that describes the structure of a system in terms of objects and relationships (data structure and behavior). Key objects of the framework have been identified through the rapid prototyping process, and the relations among objects, their attributes, and their operations identified through OOA. As the framework must be able to adapt to advances not only in science but also in computer software and hardware technologies, we have structured the framework with isolated architectural layers. For example, the Models-3 framework consists of six processing layers: user interface, system manager, UNIX environment, computational program, data access, and data storage layers. Science models reside in the computational program layer. Each layer can be replaced independently as component computer technologies advance and standards change. In terms of functionality, the framework is divided into five major partitions–main control, strategy building (or decision support), model building, analysis, and administration. Some of the key framework objects are: (1) Input/Output Application Programming Interface (I/O API)–provides a uniform input/output system to facilitate efficient and disciplined subsystem communication; (2) Model Manager–manages source version control and compilation of the air quality model code following the modular concept described later; (3) Study Planner–plans and executes simulations by linking a set of programs and associated data files; and (4) Dataset Manager–controls and manages datasets used and produced by the system.

We use the evolutionary prototyping method with detailed data flow and model structure analyses for the development and integration of science models. Thus, we can build more complex models based on the design concepts tested on simple prototypes. In addition we can identify required data flow of the system and modify the science model design accordingly. We have found that it is important to define the levels of modularity in the AQM system. The modularity levels are classified based on the granularity of the modeling components: (1) distinction of system framework and science models; (2) division of science components into submodels such as meteorology, emissions, chemistry-transport model (CTM), and analysis and visualization subsystem; (3) distinction of driver, processor modules, assistant subroutines in a CTM (modularity in the CTM is based on the operator splitting concept of science processes); (4) division based on the computational functionality in a processor module, i.e., science parameterization, numerical solver, processor analysis, and input/output routines; and (5) isolation of sections of codes that can benefit from machine dependent optimizations. Currently, the CTM is structured based on the third level of modularity taking advantage of the state-of-the-art I/O API utilities that provide selective random access to machine portable data files. Other design considerations for science process algorithms are described in a later section describing the Models-3/EDSS AQM prototypes.

Implementation for HPCC Environment: Specific implementation considerations adopted in the Models-3/EDSS AQMs for high performance computations are:

- To enhance portability of the prototype codes across multiple UNIX workstations, vector supercomputers, and some parallel computers: We not only modularize the code at the science processor level but also distinguish science parameterizations and numerical solvers to localize coding differences in the optimized solvers for different computer architectures.
- To benefit from the architecture of high performance computers: A science module of a CTM prototype operates on the entire three-dimensional gridded concentration fields at a time within a DO-loop, exposing more data parallelism than existing models which share grid traversal responsibilities between the driver and the science modules.
- To utilize available computing resources efficiently: We are developing a framework of process and data management tools for networks of heterogeneous and distributed computers. This framework takes advantage of message passing libraries to distribute work across vector supercomputers, distributed workstations, and massively parallel architectures.

PARALLEL COMPUTATIONAL SCALABILITY ISSUES FOR THE URBAN AND REGIONAL MULTISCALE (URM) AIRSHED MODEL

We studied parallel scalability issues using URM (Kumar et al., 1995). URM is a multiscale version of the CIT Airshed model (Harley et al., 1993). It solves the atmospheric diffusion equation using operator splitting, advance in time as $c^{n+1} = L_{xy}(\Delta t/2)\, L_{cz}(\Delta t)\, L_{xy}(\Delta t/2)\, c^n$ where L_{xy} is the horizontal transport operator and L_{cz} is the chemistry/vertical transport operator. The Streamline Upwind Petrov-Galerkin (SUPG) finite element method is used for L_{xy} (Odman and Russell, 1991). The two-dimensional horizontal transport operator is a direct result of the use of multi-resolution rectangular grid, which is computationally more efficient than use of a uniform grid because it requires evaluation of L_{cz} at fewer points. For L_{cz}, the hybrid scheme of Young and Boris (1977) for stiff systems of ODEs is used. The URM was converted from a sequential to a message-based parallel program.

Distributed URM–Data Parallelism: Figure 1 shows the structure of the sequential URM hourly loop. **chem** corresponds to L_{cz} and **trans** to L_{xy}. The loop first reads hourly meteorological and emissions input data. Next, for each time step, **chem** and **trans** update the concentration array, **conc**. Finally, hourly output data are saved. Within **chem**, node loop iterations are independent, allowing data-parallel execution. Within **trans**, the layer/species loops contain two-level nested data-parallelism. Grid point iterations in **trans** are not independent. Consequently, parallelism in **chem** and **trans** requires orthogonal data distributions. **em** represents pollutant emissions input.

```
do for every hour
   call input(met,em)
   do for every timestep
      call trans(conc,met)
      call chem(conc,em)
      call trans(conc,met)
   enddo
   call output(conc)
enddo
```

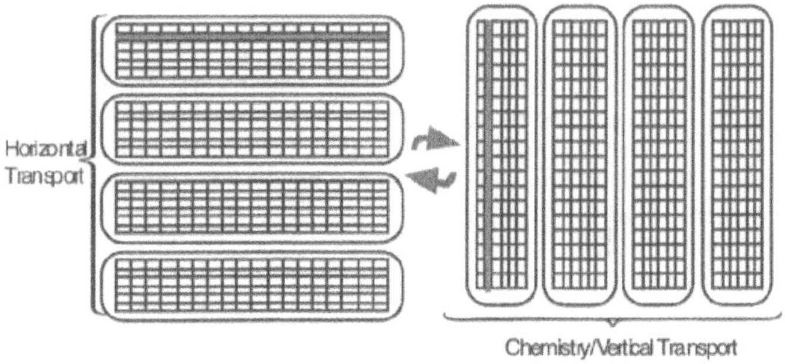

Figure 1. URM code structure (left) and main data structure (right)

We have developed a distributed version of URM that exploits outer-loop data parallelism in **chem** and **trans**. It uses PVM (Parallel Virtual Machine; Geist and Sunderam, 1991) for communication. The main source of interprocess communication is the repeated redistribution of **conc** between **chem** and **trans**. It is executable on heterogeneous combinations of parallel computing platforms, including high-speed workstation clusters, and the Intel iWarp, Intel Paragon, Cray C90, and Cray T3D heterogeneous supercomputer (Segall, et al., 1995a). It has achieved speedups ≥ 6.0 on ten workstations for both the Los Angeles and northeastern U.S. regions. Analysis suggests that 25 is about the maximum possible speedup regardless of the number of processors used (assuming no change to the size of the domain or to the solvers used). In contrast, Dabdub and Seinfeld (1994) have recently reported speedups of about 30 and 87, respectively (on a 256-processor Intel Touchstone Delta) for two different transport schemes, for a parallel version of the CIT Airshed model (using the same inputs), and have recently increased speedups to 45 and 95 (see below).

Part of the reason for this difference is the relatively low degree of outer-loop parallelism in L_{xy}, compared to that available from separate L_x and L_y operators. In compensation, multi-resolution grid methods require significantly less computation to model a given region, so time-to-solution is shorter even though speedup is lower. This illustrates the principle that when evaluating the performance of a distributed computation, not only speedup should be considered but also the algorithms used. Further, detailed comparison of different models requires a method of normalizing complexity of algorithms that exhibit different numerical behavior. Significantly, we observed that a cyclic distribution of grid points to processors results in improved load balance, by breaking up spatial locality in the load distribution. An interesting opportunity results from the observation that L_{xy} contains moderate outer-loop and vector parallelism, whereas L_{cz} contains massive outer-loop parallelism and little vector parallelism. In Segall et al.(1995b), we discuss tradeoffs involved in a heterogeneous implementation that places each phase on an architecture best suited to its solution.

Distributed URM -- Mixed Task and Data Parallelism: Other speedup-reducing factors include sequentiality in data I/O, hourly pre- and post-processing, and communication latency between **chem** and **trans**. For example, L_{xy} uses only one finite element stiffness matrix per layer (per hour): during preprocessing, outer-loop parallelism is limited to only the number of layers. Although preprocessing time is small compared to the overall sequential time, parallelization of **chem** and **trans** increases its relative significance. These factors are currently more significant than L_{xy} parallelism. They can be addressed by combining task with data parallelism (Figure 2).

I/O can be parallelized using data parallelism (input each array using a separate processor) plus pipelining (input data for hour $n+1$ while computing for hour n). For preprocessing, our current implementation factors the stiffness matrices for different layers in parallel. Pipelining can also be used, as for inputs. Critical-path communication time is ultimately a function of platform selection. Its impact is strongly dependent on problem size and the computation/communication ratio (grain size) of the solvers used.

In parallelizing URM, we have learned that multi-resolution grid models are inherently harder to parallelize than uniform grid models. Straightforward data parallelism (e.g., Fortran-90) does not lead to high parallel speedups with current generation solvers. However, the efficiency of the approach provides good time-to-solution with moderate speedups. Further, proper data distribution and mixed task and data parallelism can be used to achieve higher speedups where required. Finally, the different characteristics of the chemistry and transport solvers can be exploited to take advantage of heterogeneous distributed computing platforms.

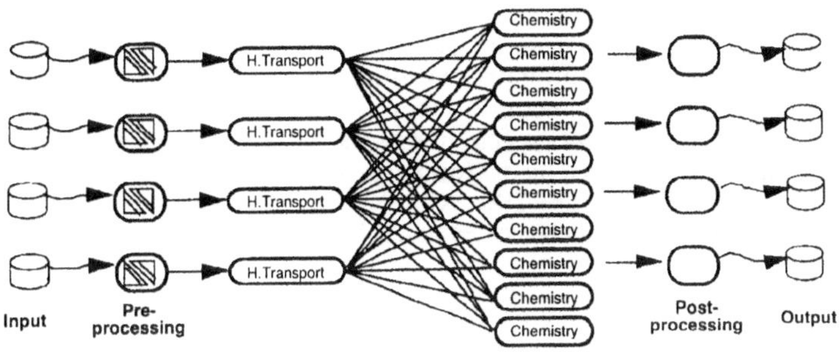

Figure 2. Mixed Task and Data Parallelism in the URM Model

OPTIMIZING GAS-PHASE CHEMISTRY SOLVERS FOR MASSIVELY PARALLEL PROCESSING ON THE CRAY T3D

Among the various physical processes simulated by air quality models, most of the computational effort is spent in solving the chemical transformations of atmospheric gaseous species. Thus efforts to significantly improve the computational performance of AQMs imply focusing on the chemistry solvers.

Testbed Core Model and Gas Phase Chemistry Solver: Successful implementation of AQMs on massively parallel processor (MPP) computers requires major shifts in the programming paradigms. In an effort to understand the interplay between computational algorithms and MPP architectures, we have developed a testbed core model (TCM) to study the effects of algorithms and data distribution on performance. In particular, we have focused primarily on the Cray-T3D MPP. The TCM, based on the Regional Oxidant Model (ROM: Lamb, 1983, 1984), simulates horizontal transport, vertical exchange between adjacent model layers with emissions and deposition, and gas-phase chemical interactions. We used a test dataset, provided by the ROM system input processors, for a 64 column (longitude) by 52 row (latitude) domain in the northeastern U.S. for one high-ozone, 24-hour period. The grid cell resolution is 1/4° longitude by 1/6° latitude. The TCM was initially optimized for a vector architecture (Cray C90) so that results from the T3D implementation could be verified and compared. The TCM uses an explicit, quasi-steady state approximation (QSSA) solver, which is summarized in Young et al. (1993), and the RADM2 chemical mechanism (Stockwell et al., 1990), as representing a current state-of-the-science mechanism used extensively in regional air quality modeling.

Implementation of TCM on Cray T3D: For the TCM implementation, the array containing the concentration field is distributed using the "CDIR$ SHARED" CRAFT directive (Cray Research, 1995), with a block distribution that allocates the number of domain columns (64) evenly among all the processor elements (PEs) and with all the rows and layers (and species) given to each PE. Note that the T3D requires users to allocate PEs by powers of 2 only. The main calling routine invokes first the horizontal transport, followed by the chemistry/vertical-exchange routines, using a technique called "shared-to-private coercion". Each of the routines called (in parallel), through their argument lists, references only their portion of the data to work on. At this level of granularity, the chemistry (and other processes) is assigned a sub-domain in which to work. Thus with this decomposition, we are essentially trying to optimize the chemistry on a single PE, which is a DEC Alpha 64-bit RISC microprocessor.

The main challenge with the Alpha microprocessor is with its cache memory system. The cache, with 256 4-word cache lines, is relatively small and direct-mapped. Therefore, the way data are laid out in memory is critical for performance. For example, cache misses result in up to a factor of 9 in performance degradation. Similarly if cache conflicts occur (memory references that hit the same cache line), serious degradation can result. There are other challenges as well, which involve the write buffers, the non-pipelined divide instructions, memory page misses, and instruction cache sensitivity.

Results: We experimented with various techniques to improve performance, such as loop unrolling, cache alignment, and read-ahead directives, and discovered widely varying results. For example, using 16 PEs, unrolling loops in the QSSA subroutines by 8 and limiting the store streams to one (for write buffer optimization), we obtained a 27% speedup (see Figure 3). The best results were obtained with more drastic code revision. By swapping the innermost loop indices from cell to species iterations, we obtained a 35% improvement. With 16 PEs we still have not achieved the performance obtained with the vector code on the C90. When 64 PEs were used, the C90 performance was exceeded. However, with approximately half the CPU time, we observe a sub-linear speedup from the 16 PE case.

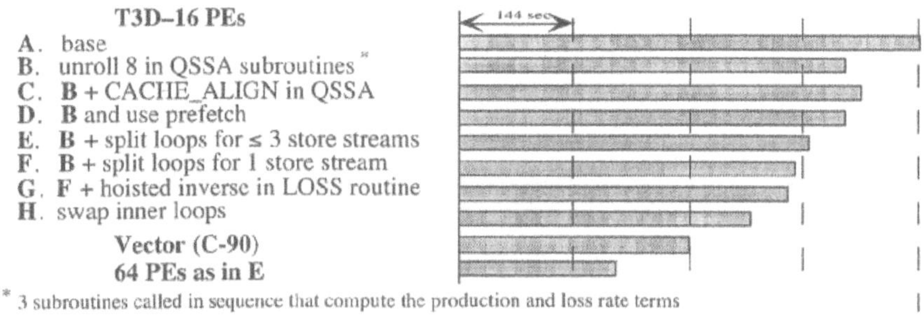

Figure 3. CPU Totals for 6 steps (3 hours starting at midnight)

PARALLEL COMPUTATION STRATEGIES FOR UNIFORM-GRID CIT URBAN AIRSHED MODEL

Here, we present some techniques used in the parallel computation of the CIT urban-scale air quality model on a massively parallel architecture. In particular, the work focuses on the design of the computational infrastructure needed to incorporate new physical and chemical phenomena in the next generation of urban- and regional-scale AQMs. Implementation issues on parallel architectures include domain decomposition strategies, algorithm design and evaluation, portability, modularity, and buffering techniques used in I/O operations. The phenomena incorporated in the studied models include gas-phase chemistry, advection, and diffusion. Here we summarize implementation experiences of the CIT urban air pollution model on distributed memory MIMD machines ranging from a cluster of workstations to a 512 node (PE) Intel Paragon. For details, refer to Dabdub and Seinfeld (1994).

To improve the portability of the code, we designed a new message passing interface and incorporated it into the parallel model. The routines are designed to run on top of message passing libraries P4 (Butler and Lusk, 1994) and NX. Other libraries such as PVM or EXPRESS (Parasoft Corporation, 1988) can be easily incorporated. Use of such routines produces a "universal" code that runs on different architectures under different basis communication libraries. A predetermined domain decomposition is used to parallelize the chemistry operator. Because of its simple data flow structure, a predetermined decomposition avoids the most common pitfalls of MIMD programming: algorithmic and cyclic deadlocks as well as process starvation. Different strategies for the parallelization of the advection operator are discussed, implemented, tested, and compared. It has been found that a dynamic data distribution presents the best performance for the high-number-of-processors regime. Implementing the I/O operations in parallel increases the speedup of the code significantly. The parallelization of the I/O operations involves distribution of the file access operations between the two hosts. In addition, the data is striped across the hard disks.

When the chemistry, transport, and I/O operators of the model are implemented in parallel, we obtained the following best speedup results: Using a Galerkin horizontal transport solver, a typical 24-hour simulation of the South Coast Air Basin requires less than 7 minutes on the Intel Paragon with 128 nodes. This corresponds to a speed-up of 45. Using the more computationally intensive Accurate Space Derivative (ASD) transport solver (Gazdag, 1973), the same run requires less than 11.5 minutes, which corresponds to a speed-up of 95. The speedup numbers reported here represent the parallel implementation of a model that was designed to run on a sequential machine. Parallel computing motivates research to find new solution strategies to the atmospheric diffusion equation (with nonlinear chemistry) not envisioned when only uniprocessors are used.

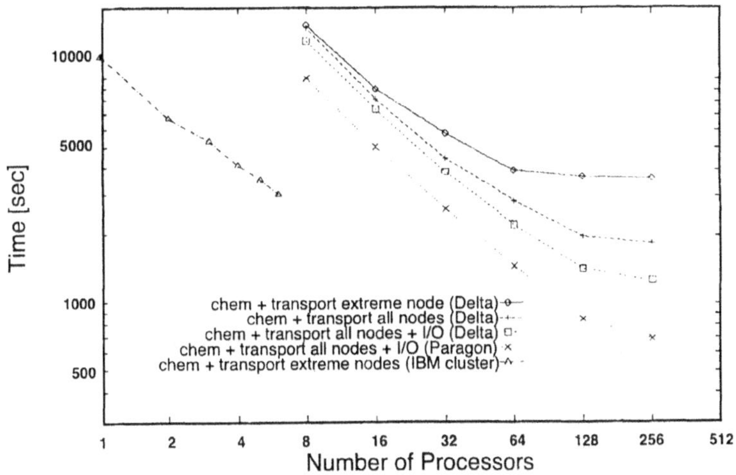

Figure 4. CPU time for a 24-hour simulation versus number of nodes for the ASD implementation of the CIT model using different parallelization strategies.

When the model is highly parallelized there is a performance decrease when adding extra processors in the high-number-of-processors regime. This effect occurs because the computation to communication ratio inherent in the model is not sufficiently high. Stated in a different manner, the task of simulating this Los Angeles Basin gas-phase air pollution episode is sufficiently large to run optimally on $O(10^2)$ nodes. If the complexity is increased with more detailed physics and chemistry, then the model will benefit from a higher number of nodes available (Figure 4).

MODELS-3 AND EDSS AIR QUALITY MODEL PROTOTYPES

Both the EPA's Models-3 and MCNC's EDSS projects have the common goals for redesigning the framework and air quality models for HPCC environments and have collaboratively developed science model prototypes. The current prototype design divides an air quality model into flexible and explicit process modules that are amenable to modification, revision, and different applications. The system also features automated process and data management and enhanced analysis tools (Byun et al., 1995).

The AQM prototypes are used as testbed systems for evaluating the performances of air quality related processes such as transport, chemistry, clouds, and aerosol/visibility. Currently, twelve different advection schemes (Odman et al., 1996), eleven test cases, and two different analysis and visualization tools are available. We have introduced several photochemical mechanisms for describing detailed chemical pathways related to acid deposition and oxidant formation. We have incorporated generalized and flexible chemistry solvers together with a chemical mechanism reader, which enables the models to easily incorporate or modify the chemical mechanisms for different applications. For cloud mixing and aqueous processing, two closely allied deep convective numerical cloud chemistry schemes based on the Kain-Fritsch convective parameterization (McHenry and Kain, 1996) have been designed and tested. We have improved algorithms for aerosol removal by dry deposition, aerosol-cloud droplet interaction and removal by precipitation, and for the binary homogeneous nucleation of the sulfuric acid/water vapor system (Binkowski and Shankar, 1995). Also, we have introduced a method of visibility estimations using simulated aerosol size parameters (Hanna et al., 1993). As described below, we are investigating grid nesting and grid refinement techniques to address the issue of scale interactions and representation of phenomena that are essentially at a subgrid level with respect to traditional model resolutions. We have developed a one-way nested grid model to accurately represent the transport and chemistry of atmospheric pollutants. Various methods for passing the coarse-grid information to the nested fine grid have been developed and tested for simple cases involving coupled transport and chemistry processes. We have also investigated the use of variable-source adaptive grids for accurate representation of emissions and the subsequent transport and reactions of atmospheric chemicals.

Advances in meteorological modeling are being pursued to improve the meteorological capability of predicting surface and atmospheric boundary layer characteristics that are critical to air quality applications. We have developed a sophisticated land-surface/boundary layer model with an advanced evapotranspiration parameterization scheme. We have also developed an initialization scheme and a data assimilation scheme for soil moisture (Pleim and Xiu, 1995). We have tested several atmospheric turbulence mixing schemes for their applicability in air quality modeling studies. In addition, the prototype employs a sophisticated process analysis method (Integrated Reaction Rate/Mass Balance), and sensitivity and uncertainty analysis methods using ADIFOR, an automatic differentiation technique (Hwang and Byun, 1995).

MULTISCALE APPLICATION OF NESTED EDSS/MODELS-3 PROTOTYPE

There are complex characteristic scale interactions between the dynamic and chemical processes that affect air quality. The lack of adequate resolution can limit our ability to accurately simulate individual processes and their interactions. On the other hand, modeling large geographic regions with uniform resolution at the smallest relevant scale is beyond the realm of current generation computers. A coarse resolution AQM will have a significant error in representing small scale processes such as cloud formation and development, precipitation, emissions from individual sources, and chemical interactions. For example, in uniform grid regional models, when plumes are injected into coarse grid cells, they are mixed immediately with the cell contents, in other words, diluted, and the details and effects of the near field chemistry are lost. Multiscale models can provide finer resolution in key regions (e.g., urban areas or major emission sources) thereby capturing the interactions between small and large scale processes in a computationally efficient way.

As part of the development of the Models-3/EDSS system, we are investigating various multiscale modeling techniques. Our goal is to develop a unified modeling system for different scales (e.g., local, urban, regional, global) with accurate description of the interactions among the relevant processes at each scale. In contrast with the multi-resolution grid method presented earlier, we describe a multiscale AQM based on a multiple nesting of uniform grids of different resolutions to resolve smaller scale processes. We will also examine the impact of grid resolution on the model predictions of pollutant distributions.

Model Formulation: The CTM is formulated using a generalized coordinate system allowing the user to select from a variety of possibilities. Depending on the application, one coordinate system may be more suitable than others (e.g., Lambert conformal map coordinates in the horizontal and normalized pressure coordinates in the vertical for regional scales; UTM coordinates in the horizontal and normalized altitude coordinates in the vertical for urban scales). The selection of the coordinate system may also be influenced by the source of input data. If meteorological data are supplied by a prognostic model that uses a certain coordinate system, the CTM can easily adapt to that coordinate system. The species continuity equation in generalized coordinates is expressed as follows:

$$\frac{\partial}{\partial t}\left(\sqrt{\gamma}\,\overline{c_i}\right)+\frac{\partial}{\partial x^j}\left(\sqrt{\gamma}\,\overline{c_i}\,\overline{v^j}\right)+\frac{\partial}{\partial x^j}\left(\sqrt{\gamma}\,\overline{c_i'\,v^{j\prime}}\right)=\sqrt{\gamma}R_i+\sqrt{\gamma}S_i\,;\quad i=1,\dots,N\,;\ j=1,2,3 \quad (1)$$

where, $\overline{c_i}$ is the concentration of species i, $\overline{v^j}$ is the velocity component in the x^j-direction (contravariant velocity), γ is the determinant of the coordinate system's metric tensor, R_i denotes the chemical production or loss, and S_i is a source/sink term for emissions and deposition processes. The expressions of γ for a variety of coordinate systems can be found in Odman et al. (1995a). The turbulent diffusion terms, $\overline{c_i'\,v^{j\prime}}$, must be parameterized; for example, in the x^3-direction (generalized vertical), the use of mixing-length theory yields

$$\overline{c_i'\,v^{3\prime}} = -\bar{\rho}\,K^{33}\frac{\partial}{\partial x^3}\left(\frac{\overline{c_i}}{\bar{\rho}}\right) \quad (2)$$

where K^{33} is the diffusivity in the x^3-direction and $\bar{\rho}$ is the air density. Equation (1) is discretized by defining a grid structure over the domain of interest. The multiscale capability can be achieved by using static grid nesting.

Static Grid Nesting: In static grid nesting, finer grids (FGs) are placed (i.e., nested) inside coarser ones (CGs). The resolution and the extent of each grid are determined *a priori* and remain fixed throughout the CTM simulation. The approach is illustrated in Figure 5. The spatial resolution of the CG is usually an integer multiple of the FG's. First, the CG solution is marched forward one time step. This solution provides boundary and initial conditions (both concentration and flux) to the FG solution that is advanced at a smaller time step (usually an integer fraction of the CG's time step). Though not necessary, it is customary to set the time step ratio equal to the grid size ratio. When the FG solution catches up with the CG solution, the former may be used to update the latter. In one-way nesting, no feedback from the FG to the CG is allowed.

In one-way nesting, the primary concern is the mass conservation at the CG-to-FG interface where boundary conditions are input to the FG using the CG solution. The advective flux at the inflow boundaries of the FG is the flux passing through this interface as determined by the CG solution. We also allow for time variation of the flux during the CG time step. This is performed by computing the departure point of the last particle passing through the interface for each FG time step. In the Lagrangian description, the mass crossing the interface is equal to the spatial integral of the concentration distribution between the departure point and the interface. Most flux-conserving advection schemes use the same Lagrangian concept to calculate the mass transfer between grid cells (e.g., Bott, 1989). During each FG time step, we meter into the FG the exact amount of mass that would have crossed the interface on the CG.

In two-way nesting, we update the concentration of each CG cell by the average of the concentrations of all the FG cells overlapping with the CG cell. Special care is required to assure strict mass conservation at the FG-to-CG interface. The mass of some species (e.g., radicals) may no longer be conserved because, when advancing the FG solution, we perform non-linear chemical transformations in addition to transport. However, the mass of the basic chemical elements such as sulfur, nitrogen, and carbon must be conserved. The FG solution is used to compute the flux of each element at the FG-to-CG interface. When conservation principles are applied to the FG-to-CG interface, as described above, the CG concentrations near the interface must be corrected (when the Courant stability limit is applied, only the first row of CG cells immediately outside the interface need correction). This is done by renormalizing the concentration of each species based on the assumption that the ratio of the species mass to element mass will remain the same before and after the correction. A similar renormalization procedure was used to make slightly non-conservative chemical solvers strictly conservative (Odman et al., 1992).

The nested grid CTM is designed to provide a high level of modularity: different processes are coded as separate modules; each module reads its required input data independently from others. This minimizes the data dependency between processes. The model is also designed to provide a high level of flexibility for multiple and multilevel grid nesting. The information about the grid is communicated to each process module through a structure that is expanded during compilation through the FORTRAN include mechanism. This allows for the use of the same process module for different grids. Customizing a nested model is as simple as preparing include files with grid dimensions for each grid and a driver with the appropriate process calling sequence. Static grid nesting conserves mass and preserves transport characteristics at the interfaces of grids with different resolutions (Odman, 1995b) It allows for effective interaction between different scales with efficient use of computing resources. The nested CTM simulates each grid with its own time step and periodically exchanges information between them.

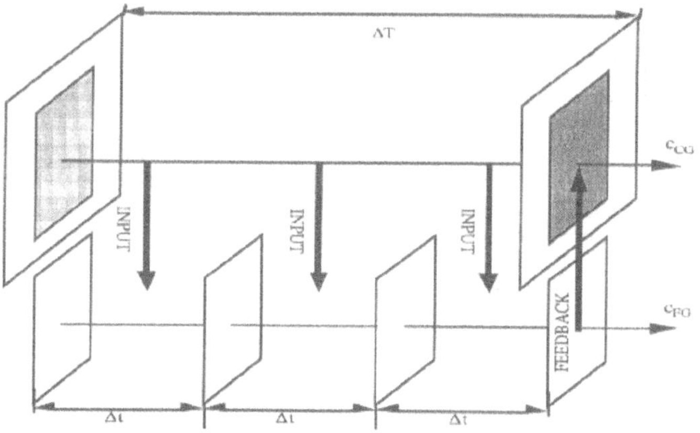

Figure 5. Static grid nesting.

Application to the Northeast: The nested grid CTM was applied to simulate the air quality in the northeastern U.S. with one- and two-way nesting. This 36-hr simulation started at 00 GMT on August 2, 1988. A CG with 54-km resolution over the eastern U.S. and southeastern Canada and an 18-km resolution FG over the Northeast were used. Both grids were simulated simultaneously (this is not required in one-way nesting); they each had approximately 100,000 cells and 58 chemical species were followed. The simulation required approximately 35 MWords of memory and took 7.5 hr on one processor of a Cray C90.

There were no significant differences between the one-way and two-way nested model predictions over the FG and the portion of the CG outside the nest. This may be related to the large size of the region covered by the FG and relatively short duration of the simulation. In Figure 6, afternoon average (12:00–18:00 LST) surface ozone predictions of the CG are compared to those of the FG. While the two solutions are in general agreement, there is more spatial variation in ozone concentrations predicted by the FG. Also, the peak ozone levels predicted by the FG are generally lower than those of the CG. This result is consistent with the general non-linear character of ozone production as related to ambient NO_x levels; the relatively more diluted NOx emissions in the CG lead to higher ozone concentration (photochemical ozone production is more efficient at lower NO_x levels).

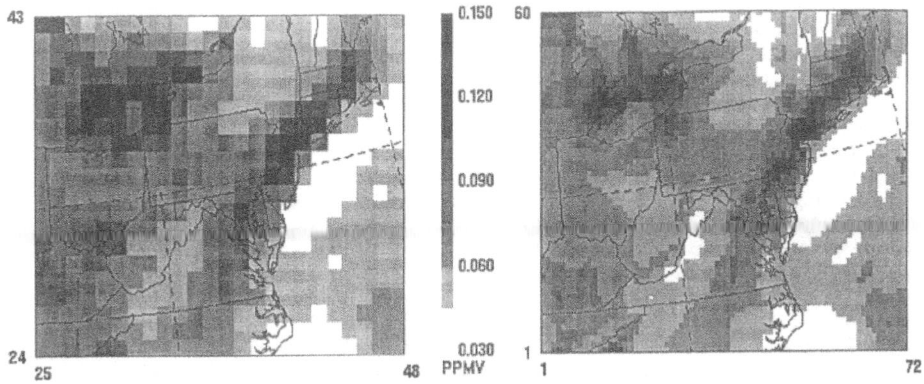

Figure 6. Afternoon averaged (August 2, 1988) surface layer ozone concentrations (in ppm) as predicted by the 54-km coarse grid (left) and the 18-km fine grid (right).

CONCLUDING REMARKS

We have presented several examples of how HPCC technology can improve performance of current air quality models. It was shown that the speedup of the parallelization depends significantly on the grid structure of model, algorithms used, and methods of parallelization. Use of advanced computational techniques will benefit air quality science tremendously. For example, larger problem sizes and more compute-intensive aqueous and aerosol-phase chemistry solvers can be handled with improved efficiency by suitable problem decomposition. Current parallelization efforts focus mostly on the data parallelism because it provides good efficiency for grid models on tens of processors. In the near future, we will test inner-loop and outer-loop parallelism of the transport layer/chemistry loop for very high scalability. Adding task parallelism will further reduce model execution time.

To incorporate recent advances in computing technology and continual developments in atmospheric sciences, we have redesigned both the framework and science submodels of AQMs using software engineering methods. A key design concept for the system framework is the layering structure of the system components to allow timely replacement of a processing layer as technology advances and standards changes. For science models, we have identified modularity levels and redesigned data flow to provide extensibility and interchangeability of codes at the science processor level. We expect this feature will allow easy incorporation of new developments in different disciplinary areas or addition of new atmospheric processes. We have introduced an EDSS/Models-3 CTM prototype with static grid nesting to demonstrate the framework and science concepts. Finally, we would like to promote the community modeling idea for the development of comprehensive air quality models.

This paper has been reviewed in accordance with the U.S. Environmental Protection Agency's peer and administrative review policies and approved for presentation and publication. Mention of trade names of commercial products does not constitute endorsements or recommendation for use.

REFERENCES

Binkowski, F.S., and Shankar, U., 1995, The regional particulate model, Part I: Model description and preliminary results. *J. Geophys. Res.* (in press)

Bott, A., 1989, A positive definite advection scheme obtained by nonlinear renormalization of the advective fluxes, *Mon. Wea. Rev.* 117:1006-1015.

Byun, D.W., Hanna, A.F., Coats, C.J., and Hwang, D., 1995, Models-3 air quality model prototype science concept development. In A&WMA Transactions, Regional Phtochemical Measurement and Modeling Studies, TR-24: 197-212.

Butler, R., and Lusk, E., 1994, Monitors, message, and clusters: The p4 parallel programming system, *Parallel Computing* 20:547-564.

Cray Research Inc., 1995, Cray MPP FORTRAN Reference Manual, SR–2504 6.2.1.

Dabdub, D., and Seinfeld, J.H., 1994, Air quality modeling on massively parallel computers. *Atmos. Enviro.*, 28:1679-1687

Gazdag, J., 1973, Numerical convective schemes based on accurate computation of space derivatives. *J. Comp. Phys.* 13: 100-113.

Geist, G.A., and Sunderam, V.S., 1991, The PVM system: Supercomputer level concurrent computation on a heterogeneous network of workstations. in Proceedings of the Sixth Distributed Memory Computing Conference, IEEE, April 1991, 258-261.

Gross, T., O'Hallaron, D., and Subhlok. J., 1994, Task parallelism in a high performance FORTRAN framework. *IEEE Parallel & Dist. Technol.* 2:16-26

Hanna, A. F., Binkowski, F. S. and Shankar, U., 1995, Analysis of regional visibility in the eastern United States using aerosol models. Proceedings of the A&WMA conference on Regional Photochemical Measurement and Modeling Studies. San Diego, California, November 7-12. 1993. (in press)

Harley, R.A., Russell, A.G., McRae, G.J., Cass, G.R., and Seinfeld, J.H., 1993, Photochemical modeling of the Southern California air quality study. *Envir. Sci. Technol.* 27:378-388

Hwang D., and Byun, D.W., 1995, Application of automatic differentiation for studying the sensitivity of numerical advection schemes in air quality models, Proceedings of the High Performance Computing '95, Phoenix, AZ, April 9-13, 1995. 52-57.

Kumar, N., Russell, A. G., Segall, E., and Steenkiste, P., 1995, Parallel and distributed application of an urban-to-regional multiscale model. *Comp. and Chem. Eng.* (in press).

Lamb, R.G., 1983, A Regional Scale (1000 km) Model of Photochemical Air Pollution, Part1: Theoretical Formulation. EPA-600/3-83-035, U.S. Environmental Protection Agency, Research Triangle Park, NC.

Lamb, R.G., 1984, A Regional Scale (1000 km) Model of Photochemical Air Pollution, Part2: Input Processor Network Design. EPA-600/3-84-085, U.S. Environmental Protection Agency, Research Triangle Park, NC.

McHenry, J. , and Kain, J.S., 1996, Development and testing of a Kain-Fritsch deep convective cloud chemistry scheme for incorporation into multiscale air quality models. Accepted for presentation in the Ninth Joint Conference on the Applications of Air Pollution Meteorology with the A&WMA , Atlanta, GA, Jan. 28- Feb. 2, 1996. (in press)

Odman, M.T., Byun, D.W., Srivastava R.K., McRae, D.S., 1995a, On air quality modeling using generalized coordinate systems. (in preparation)

Odman, M.T., Ingram C.L., and Yamartino, R.J., 1996, A comparison of advection schemes for use in air quality models. (in preparation)

Odman, M.T., Mathur, M., Alapaty, K., Srivastava R.K., McRae, D.S., and Yamartino, R.J., 1995b, Nested and adaptive grids for multiscale air quality modeling. U.S. EPA workshop on Next Generation Models Computational Algorithms (NGEMCOM), Bay City, MI, June, 1995

Odman, M.T., Kumar, N., and Russell, A.G., 1992, A Comparison of fast chemical kinetic solvers for air quality modeling, *Atmos. Environ.* 26A:1783-1789.

Odman, M.T., and Russell, A. G., 1991, Multiscale modeling of pollutant transport and chemistry. *J. Geophys. Res.* 96:7363-7370.

Parasoft Corporation, 1988, Express Version 1.0: A Communication Environment for Parallel Computers.

Pleim, J.E., and Xiu, A., 1995, Development and testing of a surface flux and planetary boundary layer model for application in mesoscale models, *J. Appl. Meteor.* 34:16-32.

Segall, E., K. Walker, Steenkiste, P., and Russell, A.G., 1995. Environmental modeling on the C90-T3D heterogeneous system. In Proceedings of the U.S. EPA Next Generation Environmental Models Computation Methods (NGEMCOM) Workshop. August 1995. SIAM Press (in press).

Segall, E., Steenkiste, E., Kumar, N., and Russell, A.G., 1995, Scalability of a portable distributed multiscale air quality model. In Proceedings of the U.S. EPA Next Generation Environmental Models Computation Methods (NGEMCOM) Workshop. August 1995. SIAM Press (in press).

Srivastava, R.K., McRae, D.S., Odman, M.T., 1995, Governing Equations of Atmospheric Diffusion. Technical Report prepared for U.S. EPA (unpublished).

Stockwell, W.R., Middleton, P., and Chang, J.S., 1990, The second generation regional acid deposition model chemical mechanism for regional air quality modeling, *J. Geophys. Res.* 95:16343-16367.

Young, J.O., Sills, E., and Jorge, D., 1993, Optimization of the Regional Oxidant Model for the Cray Y--MP, EPA/600/R-94/065, U.S. Environmental Protection Agency, Research Triangle Park, NC

Young, T.R., and Boris, J.P., 1977, A numerical technique for solving ordinary differential equations associated with the chemical kinetics of reactive flow problems. *J. Phys. and Chem.* 81:2424-2427.

MODEL ASSESSMENT AND VERIFICATION

chairman: I. Ziomas

rapporteur: R. Salvador

AIR POLLUTION IN COASTAL CITIES

D. G. Steyn

Atmospheric Science Programme, Department of Geography
The University of British Columbia
Vancouver, B.C., Canada, V6T 1Z2

INTRODUCTION

Air pollution of urban atmospheres arises from a myriad of pollutant arising from the high density of human activity found in cities. These sources range from the traditional "smokestack" industries such as metal smelting, fuel refining, and power generation, through lighter industries such as manufacturing and fabrication to the many automobiles and other mobile sources associated with transportation. The sources are located in complex spatial patterns determined by the history of the city, as well as current air quality management and zoning legislation. As the plume of pollutants moves downwind, it is quickly dispersed in the vertical, and less quickly in the horizontal. A few tens of kilometres downwind of the city, the distribution of pollutants becomes independent of source details, and depends only on gross city shape, and local meteorology. If the city is located in the coastal zone, the medium range (tens to hundreds of kilometres) distribution of pollutants will be strongly influenced by meteorological phenomena found at coastlines. If the coastal lands have complex topography, meteorological phenomena driven by terrain forcing will also be important.

For the purposes of this paper, I will take the coastal zone to be all regions within one sea-breeze horizontal length scale of the coastline as defined by Niino (1987). In most cases, this corresponds to a length between 50 and 150 km. Only cities within this distance of a coastline (of both oceans and large lakes) will be considered in this paper.

De Koning (1986), Akland et al. (1992) and WHO/UNEP (1992) catalogue polluted cities based on measured levels of sulphur dioxide and suspended particulate matter - primary pollutants generally associated with traditional heavy industries and thermal power generation. that fall into the coastal zone. Figure 1 shows the location of the subset of coastal cities catalogued in these three works as well as Vancouver. These 34 cities have approximately 250 million inhabitants (1990 figures) in both developing and developed nations. As is common with expanding metropolitan regions, some coastal cities have expanded merged (in land-use if not political terms) with adjacent (and also expanding) cities to create continuously urbanized coastal zones. Probably the best example of this is the Rotterdam/The Hague /Amsterdam complex known as "Randstad". This list of polluted coastal cities implied in Figure 1 is undoubtedly incomplete, since some polluted cities may be excluded simply because pollution monitoring in those cities has not been established for long enough or extensively enough to detect the pollution. Omitted from Figure 1 will be cities subject to significant pollution by secondary pollutants arising from transportation sources. This means that cities such as Brisbane, and Vancouver are not on the list in spite of being significantly polluted. What this brief exercise shows is that a significant fraction of the worlds population lives under the influence of air pollution whose distribution will be influenced by coastal effects.

Figure 1. World map showing coastal cities subject to significant air pollution. The cities mapped are the coastal subset of those referred to in De Koning (1986), Akland et al. (1992) and WHO/UNEP (1992). The location of Vancouver is also shown.

Drawing most of his examples from cities in the Great Lakes Basin, Lyons (1975) examined turbulent diffusion in shoreline environments, concentrating on primary pollutants emitted as detectable plumes from large point sources,. I will attempt a more global view, and intend to concentrate on regional photochemical pollution. The objectives of this paper are twofold. Firstly, I will examine meteorological and other effects that affect the distribution and severity of air pollution in coastal zones. I will give examples of these effects by referring to studies from three of the cities included in Figure 1. Secondly, I will attempt to illustrate these effects in detail by reporting results from studies of air pollution (photochemical smog in particular) in Vancouver, Canada.

PHYSICAL AND METEOROLOGICAL CHARACTERISTICS OF COASTAL ZONES

Open coastal locations are often subjected to strong, even severe wind strengths. Under these conditions, large ventilation potentials result in air pollution generally not being a significant problem. If winds are light and offshore, pollution levels over the lake or ocean surface may well be significant, but impacts will be limited since the ocean generally has a large assimilation capacity compared with pollutant loadings. The most significant cases to be considered are thus those occurring under light, onshore wind conditions.

Sea-Breeze Conditions

Under slack synoptic gradients and moderate or stronger insolation, diurnally reversing, thermally driven circulations will occur at land water interfaces. These circulations called land/sea(lake) breezes are described in Atkinson (1983), Hsu (1988), Lyons (1975) and Simpson, (1994). For simplicity, I will refer to them as land/sea breezes. Among their characteristics, the most important in the context of pollutant distribution at regional scales are light winds, shallow circulations and the possibility of both vertical and horizontal recirculation of pollutants.

Recirculation of pollutants in land/sea breeze circulations can occur in two ways. Polluted air may follow a recirculating trajectory by being advected landward in the sea breeze, be lofted by rising air at the sea breeze front, and return seaward in the upper return flow. I refer to this as vertical recirculation. Horizontal recirculation occurs when polluted air is carried landward by the sea breeze during daylight hours, and seaward by the land breeze at night.

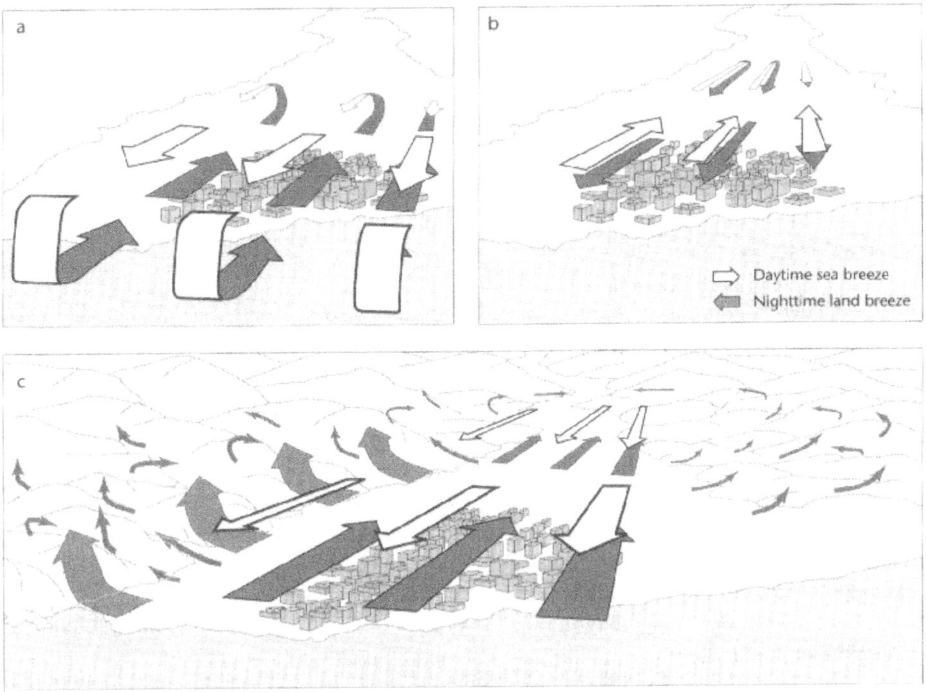

Figure 2. Possible recirculation trajectories in a sea breeze circulation. a) Vertical recirculation in sea breeze and upper return flow, b) Horizontal recirculation in daytime sea breeze and nighttime land breeze, c) Recirculation in interacting sea breeze and slope flows.

These two modes of recirculation are depicted in Figure 2, a and b. In both cases, pollutants will not in general return to their point of emission because of Coriolis effects. These effects may be overcome by topographic forcing (Steyn and Kallos, 1991). If the coastal zone is also a region of significant relative relief, topographically forced circulations will interact with land/sea breeze circulations to further complicate pollutant advection. The case of a coastal valley with associated daytime upslope flows interacting with the sea breeze is shown in Figure 2 c. In this case, pollutants advected shoreward by the sea breeze are eventually caught up in upslope flows which enhance the sea breeze strength, but also loft pollutants to greater heights than would be the case of sea breezes alone. This may result in pollutants being vented away from the city, as reported by Ulricksen and Mass (1990) for the Los Angeles basin. In other cases, this mechanism can deposit pollutants from the surface into elevated residual layers, leading to enhanced pollutant concentrations on subsequent days as reported by McKendry et al. (1996) for Vancouver.

Scaling calculations in the Appendix show that the vertical recirculation time will be in the range of 4 to 10 hours. Horizontal recirculation times will be on the order of 24 hours, the life cycleof the land/sea breeze circulation. Both types of recirculation can occur simultaneously, but vertical recirculation will be rarer, as even relatively weak onshore synoptic winds will obliterate the return flow aloft. If present, vertical recirculation can have a significant effect on regional photochemical pollution, since it will result in the advection and chemical transformation of early morning emissions, and a subsequent return to the shoreline by midday or mid afternoon. The sea breeze air will thus not be reflective of a clean marine environment, but may well contain substantial amounts of photochemical end products. If the recirculation is not completed within a 24 hour period, it will result in an elevated residual layer of polluted air at mid sea breeze level (z_s = 371 m). Fumigation of this layer to the surface will result in an accelerated production of photo-chemical end products the next morning. Horizontal recirculation of pollutants will be a much more frequent mode of recirculation, since it depends on the more robust near-surface sea- and land-breezes. This recirculation is the major cause of day-to-day intensification of photochemical episodes since it provides a mechanism for returning polluted air at low levels to the site of early morning emissions.

Thermal Internal Boundary Layers

Under conditions of light onshore winds and strong insolation, a Thermal Internal Boundary Layer (TIBL) develops in the coastal zone (Gryning and Batchvarova, 1990; Steyn and Oke, 1982). Over distances comparable to a sea-breeze horizontal length scale x_s, this layer rises from its over-water depth (typically 50 to 100 metres) to a continental depth (typically 1500 to 2000 metres) This growth occurs in an approximately quadratic way, with depths rising to roughly one third of their final value within one quarter to one third of x_s. In terms of pollutant distribution, the salient features of the TIBL are: extremely limited vertical mixing in the near shore zone, and downward fumigation of pollutants trapped in the stable marine air as it encounters the growing TIBL (Lyons, 1975). Both processes lead to enhanced potential for pollution in the coastal zone.

Land/water contrasts

In addition to meteorological effects related to the proximity of a coastline, the significant contrasts in surface properties that occur at a land/sea interface have a powerful effect on pollutant patterns in the coastal zone. The major effect in operation is that emissions of pollutants (or pollutant precursors) are zero or very low over water surfaces, and rise to relatively high values over urban surfaces. This sharp spatial change in emissions results in correspondingly steep spatial gradients in ambient pollutant concentrations. While this is not a particular difficulty when considering primary pollutants, it has a significant influence on the difficulty of adequately modelling the generation rate of secondary pollutants such as photochemical smog.

Currently operated models for photochemical smog, such as the Urban Airshed Model and Calgrid, are based on highly simplified reaction schemes. These simplifications inevitably result in a limited range of operating conditions under which the models can mimic the real world. One such limitation is the inability of the model to respond to sharp spatial changes in chemical fields. In reality, this is a limitation in a model's ability to respond to quick temporal changes in concentration of precursor substances. Of particular concern would be modelling a change from a low NOx to a high NOx regime.

SELECTED POLLUTED COASTAL CITIES

In order to reveal actual examples of the effects described in the preceding section, it is instructive to examine a few notably polluted cities. The cities selected for examination are chosen because of extensive published research on the interaction between meteorology and air pollution, and because that research shows clearly the effects under discussion.

Tokyo

Much has been written on the interaction between mesoscale meteorological phenomena and air pollutants in the Tokyo area. Of direct relevance to the present discussion is the work of Ueda, *et al.* (1988). The authors describe the profound influence of land/sea breezes (which they treat as gravity currents) and downslope winds off the mountains following sunset on the mesoscale distribution of photochemical pollutants over the greater Tokyo area. An unusual feature of their work is the generation of "nighttime smog" by down-drafts resulting from internal circulation at the head of the gravity current being responsible for the entrainment of oxidants from the upper layers. Entrained oxidants are drawn from upper boundary layer levels where they are deposited by day from vertical recirculation. Yoshikado (1992) shows how pollutants passing over Tokyo under the influence of sea breezes become concentrated in the urban-suburban area, rise in the vertical (up to 1 km) and stagnate somewhat. Once here, pollutants are either returned in the seaward flow or move inland to become part of background pollution on the following day. Both of these studies emphasize the influence of suppressed boundary layer depth on the severity of conditions.

Athens

Thanks to strong seasonal winds (the *Meltemi*), the city of Athens and surrounding communities generally do not experience particularly severe air pollution in summer. They do however suffer from notoriously poor air quality in the between seasons. Moussiopoulos, (1993) shows how most air pollution episodes are associated with the sea breeze, and that horizontal recirculation of pollutants during land/sea breezes lead to increases in pollution from day to day. In addition, strong nighttime temperature inversions isolate chemically mature air aloft, thus protecting oxidants from NOx scavenging. The re-entrainment of these oxidants into the growing TIBL on the following morning ensures high ozone production rates.

Lalas *et al.*, (1983) also show the strong links between sea breeze circulations, TIBL growth and photochemical pollution in this region. During the evening and early morning, the vertical structure of the atmosphere produces multiple inversion layers, each characterized by different pollution concentrations. Those critical to this discussion are the two found in the lowest levels (i.e. the ground based inversion and that above it.). Oscillation in height is apparent in both. By sunrise, the lower inversion traps vehicle emissions. Ozone levels are negligible in this layer as compared to concentrations in the layer directly above. Stratification of the atmosphere is produced by the complex vertical wind pattern associated with the sea breeze circulation. This results in distinct layering of pollution. Downward mixing of recirculated pollution by the land-sea breeze during the early a.m. (0500 to 0600) is restricted by the internal boundary layer at the coastline. Rather, downmixing occurs at a distance of 3-5 km from the shore at which location they observe an increase in ozone concentration.

Coastal region of Valencia and Catalonia (including Barcelona and Castellon)

The conurbation stretching up the Spanish East Coast is afflicted with seasonally severe air pollution. Millan and Artinano (1992) show how the onset of the sea breeze is accompanied by a sharp rise in ozone concentrations, which appear to be associated with the arrival of a maritime airmass. At the coastline, the observed ozone concentrations persist during the duration of the breeze period, with ozone concentrations patterns from the coast to rural inland sites following well defined and repetitive cycles every day. This study uncovers an extremely complex layering of pollutants due to interaction between land/sea breezes and TIBL structures, and shows strong influences of horizontal recirculation. Martin *et al.* (1991) show how land/sea breeze circulations can occur throughout the year in this region. Topographic interactions tend to channel sea breeze flows, especially during summertime occurrences of the Iberian Low. In all seasons, flow patterns in land/sea breezes are dominant in determining the distribution and recirculation of pollutants in the coastal zone.

ILLUSTRATIONS FROM THE LOWER FRASER VALLEY, B.C.

The Lower Fraser Valley (LFV) of British Columbia, spanning the Canada/USA border at 49 ° N, is a roughly triangular valley with its westward end being the shoreline of the Strait of Georgia (see Figure 3).

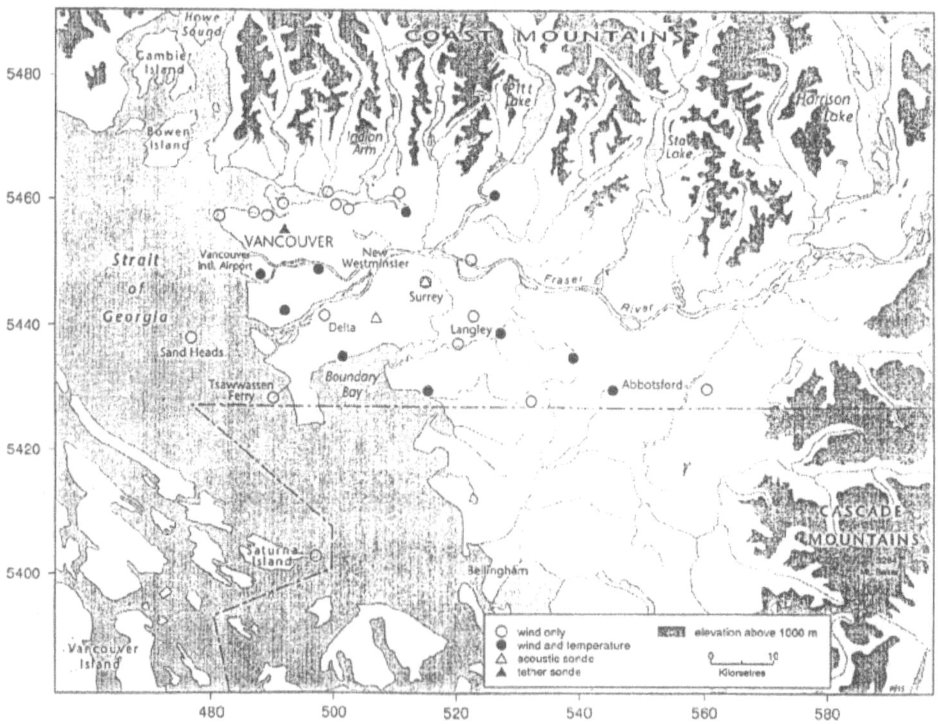

Figure 3: Map of the Lower Fraser Valley, showing, major population centres, significant geographic features and all places referred to in this paper.

The walls of the LFV consist of the Coast Mountain Ranges in the north, and the Cascade Ranges in the south. The valley floor is nearly flat, at an elevation of no more than a few hundred metres above sea level. However, within 10 km of the floor, valley walls to the north rise to 2000 metres above sea level, while walls to the south rise to 1000 metres above sea level. The LFV contains the City of Vancouver and its satellite communities with a total population of roughly 2.0 million persons. Most of the population, and current growth lies in the Canadian (northwestern) part of the valley. The LFV experiences land/sea breeze circulations on up to 17 days per month in summer (Steyn and Faulkner, 1985). As elsewhere, conditions which favour the formation of land/sea breeze circulations also lead to the formation of photochemical smog, as summarized by Oke and Hay (1994). In spite of its relatively modest population and small size, the region experiences surprisingly severe episodes. This severity is in part attributable to both land/sea breeze and TIBL effects.

Sea breeze and TIBL effects on ozone pollution in the LFV

Observations of the vertical distribution of pollutants in the LFV were taken during an extensive field campaign in 1993. McKendry *et al.* (1996) report on data derived from aircraft-borne downlooking LIDAR, tethersondes, and aircraft mounted sensors. Figure 4 indicates the complicated layering in the LFV revealed by a South-North transect of the downlooking LIDAR. The layers of strongly scattering (and presumably also polluted) air have been created by vertical advection of polluted air from low levels. This advection is driven by uplift in the land/sea breeze circulation as well as upslope flows on the slopes of the

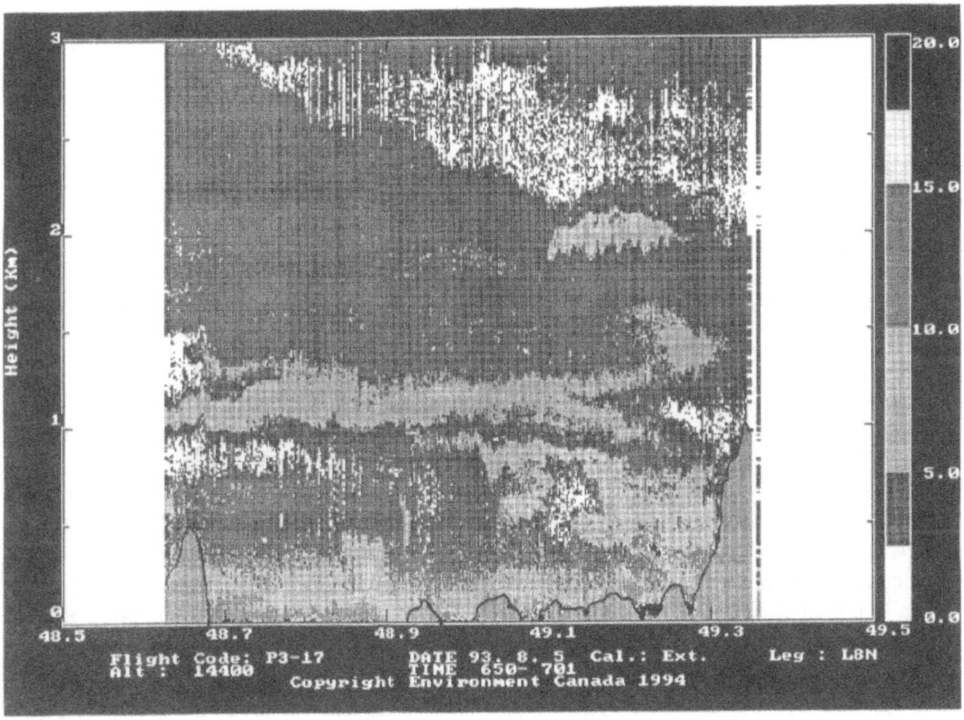

Figure 4: South-North transect of aircraft borne downlooking LIDAR taken between 0702 and 0714 PST on 1993.08.05. Aerosol backscatter is shaded in ranges of multiples of clear air backscatter. Terrain appears as a gray shaded region outlined in black. Vertical exaggeration is roughly 50:1.

Coast Mountain Ranges. McKendry *et al.* (1996) show how the residual polluted air is entrained into the boundary layer the next morning, at which time the oxidant rich air accelerates ozone production.

The powerful interaction between land/sea breezes and slope flows with both vertical and horizontal recirculation occurring during the most severe ozone episodes was investigated by Miao and Steyn (1995). They calculated forward trajectories from points within the LFV that coincided with strong emissions of ozone precursor substances. The calculations were based on output from a 3-Dimensional mesoscale meteorological model. Figure 5 shows that, due to a combination of horizontal and vertical recirculation, trajectories from a wide range of positions within the major source region all remain within the valley. Vertical representations of these trajectories show the surface released air parcels to rise to a height of roughly 650 metres, and fall under slope winds/land breezes to roughly 200 metres above ground level. These heights are consistent with polluted layers shown in Figure 4. These trajectories show the great importance of sea/land breeze circulations in recirculating pollutants in the coastal zone.

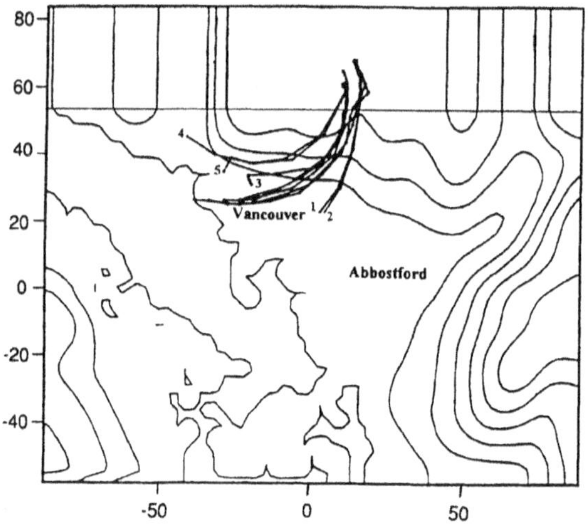

Figure 5: Plan view of three-dimensional grid-scale trajectories from five near-surface sources in the City of Vancouver. Trajectories are released at 0800, and are followed for 19 hours. Contours are strongly smoothed topography in 200 metre intervals.

A major factor controlling the severity of these ozone episodes is the very low TIBL heights in the LFV (Steyn *et al.* 1993). They provide acoustic sounder data as evidence of the very thin (often as shallow as 500 metres) layer in which the reacting urban plume from Vancouver is contained. More comprehensive pictures of the spatial variability of this thin layer are provided by the analytical model of Gryning and Batchvarova (1996) and the numerical modelling of Cai and Steyn (1995). This latter work diagnoses the TIBL depth from turbulent kinetic energy profiles calculated by a higher order closure scheme for boundary layer turbulence imbedded within a three dimensional mesoscale meteorological model. The calculated heights are shown to compare well with measured values derived from two acoustic sounders and a tethered balloon flown within the base of the LFV. Figure 6 shows the mid afternoon TIBL depth to rise from its over-water value of roughly 60 metres to a maximum of just over 900 metres at roughly 80 km from the coastline. The very complex convective boundary layer depth structure over the mountain ranges is expected, but no data were available to confirm its correctness.

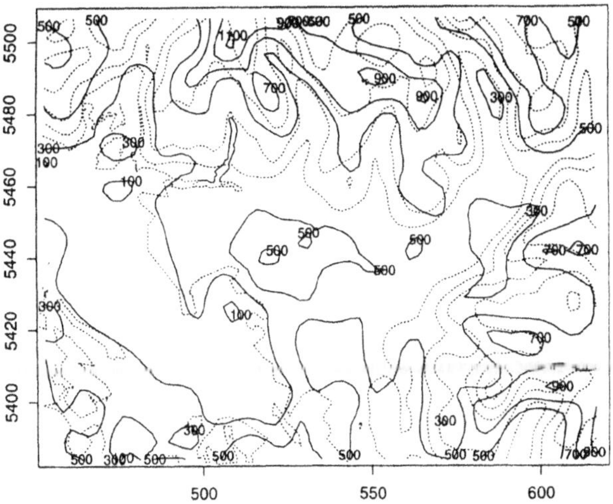

Figure 6. Mixed layer depths in the Lower Fraser Valley at 1600 PST on 1985.07.17 derived from turbulent kinetic energy profiles calculated on a 2.5 km grid.

Modelling ozone pollution in the LFV

In attempts to improve on degraded air quality in the LFV, a number of research and environmental management initiatives are underway. The research initiatives involve a combination of continuing summer pollution episode studies, occasional intensive field measurement campaigns, and ultimately modelling studies. The numerical modelling efforts are intended to provide scientifically defensible tools for guiding emissions management initiatives.

In order to model the combination of land/sea breezes and slope wind systems in steep topography, a fully non-hydrostatic mesoscale meteorological model is needed. Furthermore, since the LFV is surrounded by a region of strong relative relief, the model must be able to accommodate strong lateral domain forcing. This is best achieved through a telescoping, interactive, nested grid structure. Furthermore, in order to adequately model TIBL growth, a higher order closure scheme for boundary layer turbulence is needed. The model chosen for this study is the Colorado State University - Regional Atmospheric Modelling System (CSU-RAMS) described by Pielke *et al.* (1992). There exist many photochemical models for grid based studies of regional ozone pollution. Urban Airshed Model (UAM-V) (Systems Application International, 1995) was chosen since its grid structure allows for the high vertical resolution needed to model vertical structures evident in observations presented herein. UAM-V is based on the Carbon bond-IV photochemical mechanism, and incorporates a nested grid structure that is easily compatible with output from models such as CSU-RAMS. Having selected the models for the task, it is necessary first to define the domain over which the modelling is to be conducted, and second to provide meteorological and emissions input data for the initial, boundary and forcing conditions.

Domain selection is an important step since the domain must be selected to cover the entire polluted region, as well as being large enough to allow full expression of all important meteorological phenomena. In the LFV case, this requires extending the domain far enough to allow land/sea breeze and topographic circulations to be represented without lateral boundary effects. This is made possible by the nested grid facility. Figure 7 shows the three nested domains employed for CSU-RAMS. In order to follow synoptic scale forcing, the model is nudged towards upper level and surface data in the two outer grids, but allowed to run in purely prognostic mode on the inner grid.

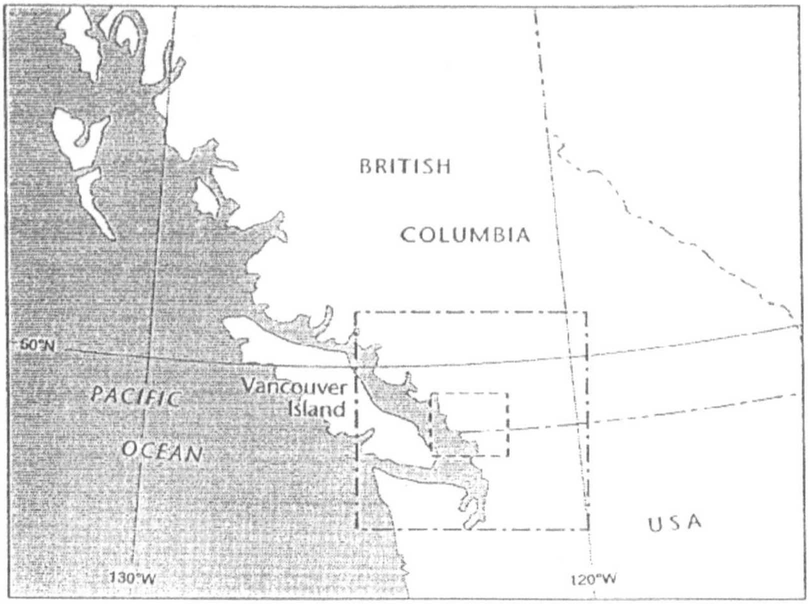

Figure 7. Domain definition and nested grids used for CSU-RAMS simulation of meteorological fields during ozone episodes in the LFV. Grid spacing in the outer/middle/inner grids are 40/10/2.5 km respectively.

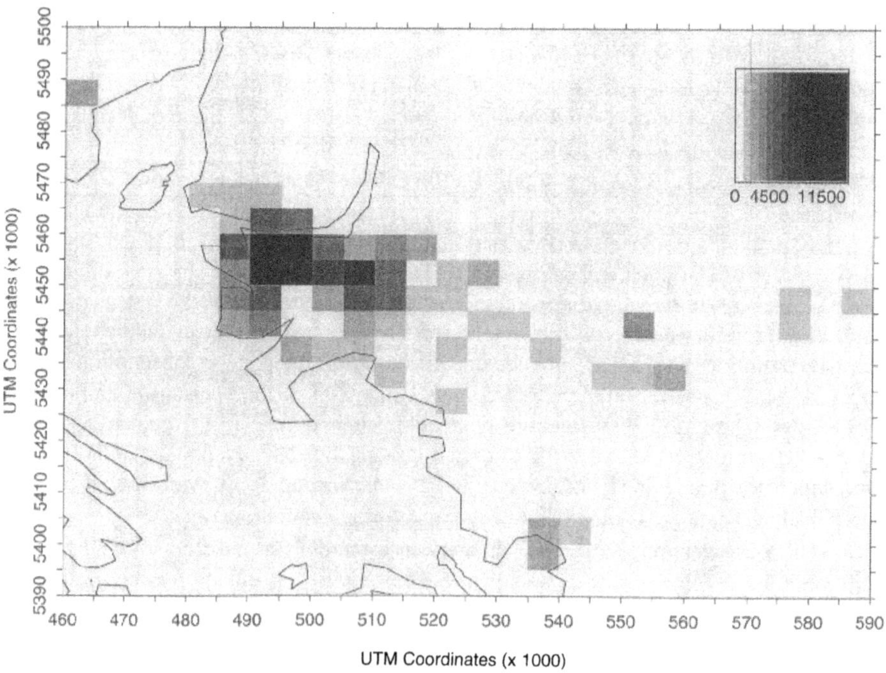

Figure 8. Emissions patterns of NO at 0800 PST on 1985.07.19 on a 5 km grid in the LFV.

Of relevance in the present context is the emissions data. As pointed out earlier, emissions fields will contain a strong discontinuity in the horizontal due to land/water contrasts. In the LFV region, anthropogenic source over the water are due to ferry and pleasure craft emissions. These are far weaker than vehicle and industrial emissions over land. Similarly, biogenic emissions of ozone precursors are much smaller over water than over land. As an example of this, Figure 8 shows the spatial pattern of NO emissions on a 5 km grid in the LFV for an hour in the morning traffic maximum. This emissions map demonstrates the very strong spatial gradients of NO, a major ozone precursor substance. To successfully model the observed ozone fields, the photochemical mechanisms must be able to accomodate the implied rates of entrainment of NO.

The modelling work is currently underway, so no detailed output fields (meteorological and chemical) are presented here.

CONCLUSIONS

It is shown that a significant portion of the world's population lives in polluted coastal cities. This review has demonstrated that meteorological conditions found in coastal zones (especially land/sea breezes and thermal internal boundary layers) have a profound effect on the distribution of pollutants emitted by cities near the coast. I have argued that the net effect in general is to exacerbate pollution levels as compared to conditions resulting from similar emissions in non coastal environments. A major consequence of environmental conditions in coastal cities is that numerical modelling of regional scale pollution problems can only be undertaken if attention is paid to the particular meteorological conditions found in coastal zones. This consideration will influence selection of meteorological and photochemical models, as well as domain size and location.

Acknowledgments

Kate Stephens helped immensely with the preparation of text and figures. Paul Jance and Eric Leinberger drafted some of the diagrams. I have drawn on the work of Xioaming Cai, Natalie Suzuki, Roy Hourston and Yuelong Miao for much of the Vancouver material. Ray Hoff kindly provided Figure 4. The National Science and Engineering Research Council of Canada and Environment Canada provided my research funding. Don Hastie and Jan Bottenheim helped cover my deficiencies in Atmospheric Chemistry. The numerical code CSU-RAMS was written by the Atmospheric Science Department at Colorado State University under funding support of the National Science Foundation and the Army Research Offices.

REFERENCES

Atkinson, B.W., 1981: *Meso-scale Atmospheric Circulations.* Academic Press 495p.

Akland, G.G., H. de Koning, D.T. Mage and G. Ozolins, 1992: Global assessment of ambient urban air quality. in D.A. Dunnette and R.J. O'Brien (Eds), 1992: *The Science of Global Change*. American Chemical Society, Washington, 498p.

Cai, X., and D.G. Steyn, 1995: *Mesoscale meteorological modelling study of the Lower Fraser Valley, B.C., Canada from July 17 to 20, 1985.* Contract report to Environment Canada. August, 1995. (also as The University of British Columbia, Department of Geography, Occasional Paper No. 40). 60pp.

De Koning, H.W., J.G. Kretzchmar, G.G. Akland and B.G. Bennett, 1986: Air pollution in different cities around the world. *Atmos. Env.*, Vol. 20, No. 1, 101-113.

Gryning, S.-E. and E. Batchvarova, 1990: Analytical model for the growth of the coastal internal boundary layer during onshore flow. *Quart. J. Roy. Met. Soc.* 116, 187-203.

Gryning, S.-E. and E. Batchvarova, 1996: Modeling the internal boundary layer height over the Vancouver area. Paper submitted to *Ninth Joint Conference on the Applications of Air Pollution Meteorology*. American meteorological Society and Air and Waste management Association. Atlanta, 28 Jan. to 2 Feb, 1996.

Hsu, S.A., 1988: *Coastal Meteorology*. Academic Press 260p

Lalas, D.P., D.N. Asimakopoulos, D.G. Deligiorgi, 1983: Sea-breeze circulation and photochemical pollution in Athens, Greece. *Atmos. Env.* 17, 1621-1632.

Lyons, W.A., 1975: Turbulent diffusion and pollutant transport in shoreline environments. in *Lectures on Air Pollution and Environmental Impact Analysis* D.A. Haugen (Ed.) American Meteorological Society, Boston, 296p.

McKendry, I.G., D.G, Steyn, R.M. Hoff, W. Strapp, K. Anlauf, F. Froude, B.A. Martin, R.M. Banta, and L.D. Olivier, 1996: Elevated pollution layers and vertical downmixing over the Lower Fraser Valley, B.C. *Atmos. Env*. Submitted.

Martin, M., J. Plaza, M.D. Andres, J.C. Bezares and M.M. Millan, 1991: Comparative study of seasonal air pollutant behavior in a Mediterranean coastal site: Castellon (Spain). *Atmos. Env.* 25A, 8, 1523-1535

Miao, Y. and D.G. Steyn, 1994: Mesometeorological modelling and trajectory analysis during an air pollution episode in the Lower Fraser Valley, B.C. *Transactions of Regional Photochemical Measurement and Modelling Studies Conference*. (September, 1994), Air and Waste Management Association, Pittsburgh.

Millan, M.M. and B. Artinano, 1992: *Mesometeorological Cycles of Air Pollution in the Iberian Peninsula*. Commission of the European Communities, Brussels, 219p.

Moussiopoulos, N., 1993: *The Athens Experience*. Moussiopoulos, Thessaloniki

Niino, H., 1987: The linear theory of land and sea breeze circulation. *J. Meteorol. Soc. Japan*, V. 65, No. 6, 901-920.

Oke, T.R. and J.E. Hay, 1994: *The Climate of Vancouver*. B.C. Geographical Series, Number 50, Vancouver, 84p.

Pielke, R.A., W.R. Cotton, R.L. Walko, C.J. Tremback, W.A. Lyons, L.D. Grasso, M.E. Nicholls, M.D. Moran, D.A. Wesley, T.J. Lee and J.H. Copeland, 1992 A comprehensive meteorological modelling system - RAMS. *Meteorol. Atmos. Phys.* 49, 69-91.

Simpson, J.E., 1994: *Sea Breeze and Local Winds*. Cambridge University Press, Cambridge, 234p.

Steyn, D. G. and D. A. Faulkner, 1986. The climatology of sea-breezes in the Lower Fraser Valley, BC, *Climatological Bulletin*, 20 (3), p. 21-39.

Steyn, D. G. and G. Kallos, 1991. A dynamical study of hodograph rotation in the sea breeze of Attica, Greece, *Boundary-Layer Meteorology*, 58, p. 215-228.

Steyn, D. G. and T. R. Oke, 1982. The depth of the daytime mixed layer at two coastal sites: a model and its validation. *Boundary-Layer Meteorology*, 24, p. 161-180.

Steyn, D. G., A. C. Roberge and C. Jackson, 1990. *Anatomy of an Extended Air Pollution Episode in British Columbia's Lower Fraser Valley*. Prepared for BC Ministry of Environment. 49 pp.

Systems Application International, 1995: *User's Guide to the Variable-Grid Urban Airshed Model (UAM-V)*. Systems Application International. San Rafael. SYSAPP-95/027.

Ueda, H., S. Mitsumoto and H. Kurita, 1988: Flow mechanism for the long-range transport of air pollutant by the sea breeze causing inland nighttime high oxidants. *Journal of Applied Meteorology*. 27, 182-187.

Ulricksen, B.L. and C.F. Mass, 1990: Numerical investigation of mesoscale circulations and pollutant transport over the Los Angeles Basin. Part II: Synoptic influences and pollutant transport. *Mon. Wea. Rev*. 2162-2184.

WHO/UNEP, 1992: *Urban Air Pollution in Megacities of the World*., Blackwell, Oxford, 230p.

Yoshikado, H., 1992: Numerical study of the daytime urban effect and its interaction with the sea breeze. *J. Applied Meteorol.* 31, 1146-1164

APPENDIX: SEA BREEZE SCALES FROM NIINO, 1987

Niino (1987) presents a set of governing scales for the sea breeze circulation. He defines a horizontal length scale (x_s), vertical length scale (z_s), and a horizontal velocity scale (u_s), where:

$$x_s = N\kappa^{1/2}\omega^{-3/2}F\left(f/\omega\right), \qquad \text{(A.1)}$$

$$z_s = \kappa^{1/2}\omega^{-1/2}, \qquad \text{(A.2)}$$

and

$$u_s = \frac{g\Delta T}{T_0 N}. \qquad \text{(A.3)}$$

where g is the acceleration due to gravity, ω is the sidereal day period, f the Coriolis parameter, N the Brunt-Vaisala frequency for overlying stably stratified air, κ the eddy thermal diffusivity, T_0 is the base state temperature and ΔT is the land-sea temperature difference. F is a universal function defined by Niino (1987.

An estimate of the time for a complete circulation around the sea breeze cell is:

$$t_c = 2(x_s + z_s)/u_s, \qquad \text{(A.4)}$$

Typical values for governing quantities, the three scales and t_c are given in Table A.1.

Table 1. Governing quantities and sea-breeze scales.

Quantity	Typical Magnitude
g	9.8 m s^{-2}
N	10^{-2} s^{-1}
ω	7.27x10^{-5} s^{-1}
κ	10 $m^2 s^{-1}$
f	0 to 1.45x10^{-4} s^{-1}
F(f/ω)	0.9 to 2.3
T	300 K
ΔT	2 K
u_s	6.5 m s^{-1}
x_s	47 to 117 km
z_s	371 m
t_c	4 to 10 h

DISCUSSION

R. BORNSTEIN: What interesting complications arise from a complex shaped terrain?

D. STEYN: Complex terrain in the coastal zone complicates things enormously. One must consider the direct channelling or diversion of sea breeze flows, as well as the interaction between sea breezes and slope flows. Obviously details of these effects depend very strongly on topographic details. One interesting and important phenomenon is the upslope venting of pollutants along sea facing slopes. This phenomenon is very important in both Vancouver and Los Angeles.

H. KAMBEZIDIS: The speaker has tackled a hot problem related to environmental problems in coastal urban areas. This is being dealt with in a NATO/CCMS pilot study. In Athens there are three sea breezes: (I) Full sea breeze development with stagnant synoptic conditions.; (ii) Semi-development of the sea breeze with opposing NE winds; (iii) Full development of the sea breeze up to a height of 400 m (height of mountains west of Athens Basin) and synoptic W winds aloft.

D. STEYN: It is gratifying to hear your expression of interest in my topic. Of course I have here restricted my consideration to sea breezes under light or no synoptic winds. The matter becomes quite different (and simpler) when synoptic flows override the return flow. Most of the air quality exacerbating effects are reduced in that case.

H. van JAARSVELD: Do you have any indication that upwind concentrations (eg. NO_2) are enhanced due to recirculation effects?

D. STEYN: I suspect it would be very difficult to find directly substantiating evidence of this. It would require profiles of NO_2 through the sea breeze cell at the coastline. I know of no such data.

TOWARD THE ESTABLISHMENT OF A COMMON FRAMEWORK FOR MODEL EVALUATION

H.R. Olesen

National Environmental Research Institute
P. O. Box 358
DK-4000 Roskilde
Denmark

1. INTRODUCTION

In this paper, a presentation will be given on current work toward the establishment of a common frame of reference for the evaluation of atmospheric dispersion models. The focus will be on issues within local-scale regulatory modelling.

First, the motivation for trying to establish a common framework will be discussed: What is unsatisfactory about the way that model evaluation is currently being practised?

Next, in Section 3, the actors on the scene will be presented. Work is being conducted internationally within the framework of several organisations in order to organise a common basis for model evaluation; some of these activities will be described.

An important activity in this respect is that of the *Initiative on Harmonisation within Atmospheric Dispersion Modelling for Regulatory Purposes*. Under this initiative, a series of workshops have been held on harmonisation of models. Model evaluation has been a key issue. In Section 4 some important experiences gained through this work will be discussed.

The concluding Section 5 proposes some responses to the question: What can be done better?

2. WHAT IS UNSATISFACTORY?

Model evaluation studies abound in the technical literature. Still, it is not easy for a decision-maker to determine whether a particular model is a good model, and whether it fits his particular purpose.

A fundamental problem in model evaluation practice can be illustrated by the following constructed example:

Let us assume that in the year 1990, a particular research group has received funding to evaluate models A, B and C, on the basis of data sets X and Y. The evaluation is performed. Some years later, model A is revised. However, the original model evaluation is not repeated. And when one day model D appears, the developers of model D do not have access to the data sets and software used in the 1990 evaluation. So nobody is able to make a qualified comparison of the revised model A and model D with the earlier models.

The basic problem is that, until now, a common frame of reference for model evaluation has to a large extent been lacking. Tools that would make it easier for modellers to adhere to standards have been few. Those that do exist have been unknown to most of the modelling community.

The problem is severe, because model evaluation is intrinsically difficult. In the model evaluation business, one often meets statements like the following:

"For model A, the fractional bias is 0.27 when using data from experiment X."

Such a statement by itself is totally inadequate as basis for a judgement about a model. For a serious judgement, one needs to know a lot of details. What data sets were used? What information is contained in the data sets, and what information is missing? Were some of the original data discarded? On what criteria? Which concentration measures were considered (e.g. area-wise maxima, arc-wise maxima etc.)? What kind of averaging was performed? What processing of data took place?

Such information is usually included in the documentation from a model evaluation, but only to a limited extent. There are so many details involved that it is very difficult to get a precise picture of the abilities of a model. Only with performance evaluation results from *several models* that are formed on *exactly the same basis,* can one draw any conclusions about model performance. This is the background that there is clearly a need for "something to be done".

3. ACTORS ON THE SCENE

Over the years, several well-organised model evaluation exercises have been conducted, e.g. the work during the EPRI Plume Model Development and Validation project (e.g. Bowne et al., 1983); the DOE/AMS Air Pollution Model Evaluation Workshop in 1984 (e.g. Kurzeja and Weber, 1985); a series of activities of the US EPA (e.g. Lee, 1993); the ATMES study following the Chernobyl accident (Klug et al., 1992); the APSIS exercise concerning the Athenian photochemical smog (e.g. a series of papers in the proceedings of the previous ITM; Gryning and Millan, 1994), and the comprehensive hazardous gas model evaluation reported by Hanna et al. (1993). The list does not pretend to be complete, especially not regarding long range modelling, but the studies mentioned are among the "classics".

Limits in past computer technology have restricted the portability of software from these studies. In most cases these studies have been stand-alone exercises with no follow-ups. However, the studies do represent a valuable resource of information that should still be exploited.

Presently, there are several initiatives involved in formulating procedures for model evaluation, the activities of which shall be described in the following subsections:

- the *Initiative on Harmonisation within Atmospheric Dispersion Modelling for Regulatory Purposes*;
- the *European Topic Centre on Air Quality* within the framework of the European Environment Agency;
- the *Model Evaluation Group* under the EC DG XII research programme on major industrial hazards.

Conclusively, in Subsection 3.4, a few other initiatives will be briefly mentioned.

3.1 The Initiative on Harmonisation within Atmospheric Dispersion Modelling for Regulatory Purposes

During the past few years, a "new generation" of models has emerged with physically more justifiable parametrisations of dispersion processes than previously. A need has been felt for these new models to be developed in a well-organised manner and turned into practical, generally accepted tools fit for the various needs of decision-makers. Therefore, in 1991, a European initiative was launched for increased cooperation and standardisation of atmospheric dispersion models for regulatory purposes: the *Initiative on Harmonisation within Atmospheric Dispersion Modelling for Regulatory Purposes.* This initiative has organised a series of workshops to promote the use of new-generation models within atmospheric dispersion modelling, and in general improve "modelling culture". Model evaluation is a key issue in this respect.

Until now, three workshops have been held (see Olesen and Mikkelsen, 1992; Cuvelier, 1994), while a fourth will take place in May 1996. An activity closely related to the workshops is the distribution of the so-called *Model Validation Kit.* This kit was originally prepared for the second of the workshops (in Manno, Switzerland, 1993). It was revised and used again for the third workshop (in Mol, Belgium, 1994). In total it has been distributed to about 80 research groups.

The *Model Validation Kit* is a collection of three experimental data sets accompanied by software for model evaluation. It is a practical tool, meant to serve as a common frame of reference for modellers. The exercises based on the kit are limited in scope (only short-range models considered, homogeneous terrain, only three data sets, etc.). It has not been attempted to perform an in-depth investigation of models; rather, the work with the kit has served as a demonstration exercise. Many useful experiences have been gained through the use of the kit. These will be discussed in Section 4.

It should be noted that the work is continuously evolving. Up-to-date information on the workshops as well as practical tools are available through the World Wide Web (see Subsection 3.5).

3.2 The European Topic Centre on Air Quality

The European Environment Agency (EEA) is an institution under the European Union. The mandate of the EEA is to provide the member states with information on the environment. The technical tasks of the EEA are often delegated to the so-called *European topic centres*. The *Topic Centre on Air Quality* is a consortium led by the Dutch RIVM.

The 1995 work plan of the *Topic Centre on Air Quality* includes the project *MA3: Harmonisation in the use of models for ambient air quality and pollution dispersion/transport*. This project has as its objective: "To increase consistency of models already in use; To develop guidance on criteria for selecting appropriate models and their application for the assessment and management of air quality". The project includes several subtasks, one of which is to establish a documentation centre and toolkits for testing models. The work is conducted in cooperation with the above mentioned *Initiative on Harmonisation*.

3.3 Model Evaluation Group

The EC DG XII (the European Commission Directorate-General for Science, Research and Development) coordinates research programmes within the EC. Within the *Major Industrial Hazards* programme, efforts are in progress to set up a more systematic framework for model evaluation. A *Model Evaluation Group* (MEG) has devised a protocol for model evaluation (Model Evaluation Group, 1994). This protocol is generic, i.e. it is a framework to be filled in for several different kinds of technical models (not just dispersion models). An ongoing activity in this framework is an evaluation of models for dense gas dispersion (Duijm, 1995).

3.4 Others

In May 1995, the council of the British Royal Meteorological Society issued a policy statement titled *Atmospheric dispersion modelling: Guidelines on the justification of choice and use of models, and the communication and reporting of results* (Royal Meteorological Society, 1995). The aim is to promote the best practice in the use of mathematical modelling for atmospheric dispersion.

The ASTM (American Society for Testing and Materials) is considering to establish standards relating to performance evaluation of atmospheric dispersion models. The ASTM is an influential, voluntary standards development organisation which is open to members from all over the world. At an ASTM workshop held in July 1995 titled *Performance evaluation of atmospheric dispersion models* it was found that it would be both possible and useful to develop some standards for performance evaluation in cooperation with similar European initiatives.

3.5 Future developments

The activities mentioned above are *not* intended to be competing with another. They are all inspired by the same problems, and they can be seen as a an attempt to sow the same kind of seed in various contexts.

Funding is a general problem for all the activities mentioned. Much work is done on a voluntary basis, so progress must be expected to be slow.

A useful point of entry to up-to-date information on these issues is the WWW pages of the *Initiative on Harmonisation within Atmospheric Dispersion Modelling for Regulatory Purposes* (http://www.dmu.dk/AtmosphericEnvironment/harmoni.htm).

4. THE MODEL VALIDATION KIT: EXPERIENCES GAINED

This section sums up some experiences gained through work with the *Model Validation Kit*. Through examples, a few essential problems will be highlighted, and it will be discussed how they can be overcome.

The scenario considered by the *Model Validation Kit* is relatively simple, namely one where a single source emits a non-reactive gas in homogeneous terrain. The kit includes data sets from three atmospheric field experiments (Kincaid, USA; Copenhagen, Denmark; Lillestrøm, Norway). The data sets are accompanied by a package of model evaluation software. The software was essentially developed by Steve Hanna and his colleagues for use with hazardous gas dispersion models (Hanna et al., 1991).

In the subsequent sections, some results based on Kincaid data will be presented. During the Kincaid experiment, the network of tracer monitors was dense; it has therefore been possible to determine maximum concentrations for crosswind arcs of monitors. Such maximum arc-wise concentrations have been compared to the computed plume centre line concentration at ground level at the same distance.

More information on the *Model Validation Kit* can be found in Olesen (1995a). The results of applying the kit on five models is discussed in depth in Olesen, 1995b.

In a paper for the previous ITM, the author analyzed a number of difficulties connected to model evaluation, and listed pertinent reactions to them (Olesen, 1994). A slightly revised version of this list of difficulties looks as follows:

Why is model evaluation difficult?

- The appropriate evaluation method cannot be uniquely defined.
- Input data sets are limited - they reflect only few of the possible scenarios.
- Processing of input data for validation is far from trivial.
- The luxury of independent data sets can rarely be afforded.
- There are inherent uncertainties.

Selected items from this list will be discussed below.

4.1 The appropriate evaluation method cannot be uniquely defined

It is important to recognise that there will never be just *one* recommended method for validating models. Just as models should be fit for purpose, so should evaluation methods. When one wants to make a judgement concerning a model, one should be convinced that the tests undertaken really correspond to the questions which should be answered.

Not only do the relevant evaluation methods depend on the context of the application; they also depend on the data sets available. There is a difference between the tests one can undertake, depending on whether one has a dense set of monitors or a sparse net.

A useful reaction to the difficulty of choosing a relevant evaluation method is to *develop an array of methods to be used in various contexts.*

4.2 Input data sets are limited - they reflect only few of the possible scenarios

One fundamental difficulty with model evaluation is that experimental data sets are limited in several respects. Usually, a data set is from a campaign with only *one* source configuration, *one* set of terrain conditions, and a limited number of meteorological scenarios. Further, the number of points where the concentration has been measured is small.

In contrast, users want models that can be used for a broad range of source configurations, for a large number of meteorological scenarios, and that can compute concentrations at all points in space. A natural, and almost inevitable reaction to this dilemma is to *extrapolate model behaviour* to conditions where the model has not been validated. This implies a problem: the extrapolation will only be credible if the model has a sound description of physical processes. Thus, the validation process should be used to *develop an understanding* of model behaviour. This implies in turn that a *diagnostic approach* to model evaluation is indispensable.

In practice, a very useful approach is to stratify data according to physical parameters, and present residuals graphically. Such plots are very easy to produce with the *Model Validation Kit.* An example of a graph showing residuals is reproduced as Figure 1 (see Olesen, 1995 for full details). Some explanation is necessary.

Arc-wise maximum concentration values have been determined from tracer data, and have been predicted by models A and B. These pairs of data, (c_{obs}, c_{mod}), have been stratified according to distance from the source. Seven subsets have been formed. E.g., there is one subset corresponding to distances between 0 and 2 km. It contains 36 pairs (N=36). For each pair, the ratio (c_{mod}/c_{obs}) has been determined.

The box plots display the distribution of this ratio. The boxes indicate percentiles 5, 25, 50, 75 and 95. Taking the left plot (A) as an example, we can consider the 36 pairs in the first group. The 95 percentile is close to unity, implying that for almost 95% of the pairs model A underpredicts. The 75 percentile is approximately 0.2, implying that for 75% of the pairs model A underpredicts by more than a factor of five.

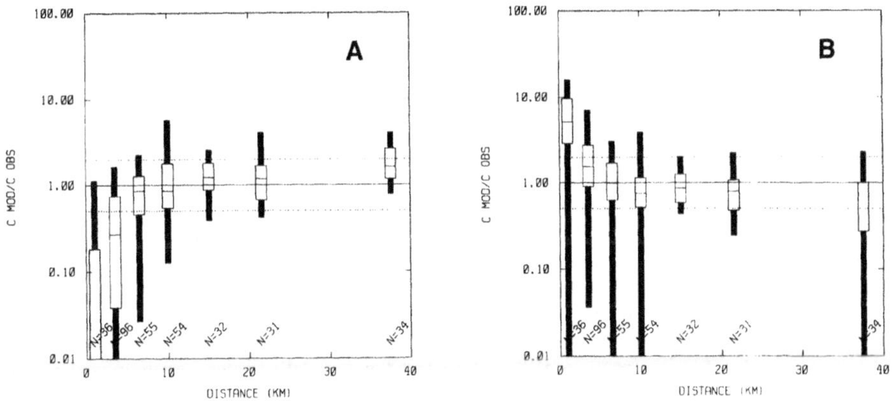

Figure 1. Model results for Kincaid data according to two models. For observations of "quality 3", the ratio of c_{mod}/c_{obs} is analyzed in terms of distance. The boxes indicate percentiles 5, 25, 50, 75 and 95. See text for further explanation.

For a physically correct model, there should be no more tendency for overprediction close to the source than far from it. Figure 1 shows that both model A and model B have problems in this respect, but behave quite differently.

Concerning data selection and processing, the graphs are based on a subset of data with well-defined arc-wise maxima (the so-called "quality 3" data). In order to reduce "noise", a filter has been imposed on the data presented in the figure: when *both* the observed and the predicted values are small (normalized concentration less than 15), the ratio is assumed to be unity.

In conclusion, one useful reaction to the problem of limited data sets is to *develop an understanding of model behaviour*, e.g. by using stratification of data as in the example.

4.3 Processing of input data for validation is far from trivial

In previous papers based on the results from the "harmonisation workshops" (Olesen, 1994 and Olesen, 1995b), it is demonstrated by many examples that processing of data is far from trivial. One important lesson to be learned is that it is extremely important to have well-defined criteria for selection of data.

The Kincaid data set lends itself as an excellent example to illustrate the problems related to quality control and rejection of data.

For the Kincaid experiment, the concentration pattern is often irregular - high and low concentrations occur intermittently along an arc. Hence, it is in many cases difficult to determine a representative maximum concentration along a cross-wind arc of monitors. Further, there are frequently gaps in the monitoring arcs. The question emerges: for cases where it is questionable whether the observed maximum is a reliable estimate of the actual maximum, should the data then be discarded from the set of arc-wise maxima?

It was found that a satisfactory solution has been to assign a *quality indicator* to each monitoring arc, indicating *how reliable* the arc-wise maximum should be considered. This quality indicator has been assigned on the basis of manual inspection of the geographical patterns of concentration distribution.

This procedure has provided a well-defined, common basis for model evaluation. If such a basis does not exist, some modellers will reject one part of the data, while others reject a different part, and any comparison will be futile.

The most important conclusion of the discussion on processing of input data is that careful work should be done. Data sets should be *prepared for ease-of-use, with their peculiarities and pitfalls well documented.* When comparing models, the models should all be run on the same data according to a common protocol.

4.4 The luxury of independent data sets can rarely be afforded

In model evaluation protocols, it is often stated that models should be independent of the data sets used for validation. At first sight this requirement may sound like common sense. However, in the author's opinion this is *not* a realistic claim in practice. A model may start its existence unbiased by existing data, but, as time passes by, no model can be expected to preserve its virginity in respect to the relatively few experimental data sets around.

A case story can serve to illustrate this, and at the same time demonstrate the capabilities of the *Model Validation Kit.*

Fig. 2 shows again results based on Kincaid data. The graphs require some explanation.

As before, arc-wise maxima are considered, and "quality 3 data" selected. The five models A-E were used to compute centerline maxima at ground level. For each model, this yields a set of modelled concentrations, which can be compared to the set of observed concentrations. The two sets have been sorted by the magnitude of concentration, so that observed and modelled values can be paired according to the rank. Finally, quantile-quantile plots as shown have been produced. Taking as an example the upper lefthand plot, it shows that for model A, the highest observed value is 319, while the highest computed value (which did not occur at the same time and arc) is 132.

Roughly speaking, a "perfect model" should be expected to produce points close to the one-to-one line*

The plots A-E were presented at the Mol workshop. There are striking differences between the models. Models A and D are not able to predict as high values as those actually observed; inversely, model B strongly overpredicts. Figure 2 is an example of a very fruitful approach to model evaluation: to run models on exactly the same basis, and then line up the results.

Actually, this graph triggered an immediate reaction when it was presented in Mol. In informal discussions it was attempted to identify the causes for the differences in model behaviour. Some potential causes were identified, and model B was revised shortly after the workshop. Therefore, now, model B should be judged not on the basis of the plot labelled B, but on the one labelled "Revised B".

This sequence of events is a perfect illustration of a mechanism, which complicates model evaluation, but must be regarded as natural and unavoidable: modellers will take advantage of the information in validation data sets, and use validation results for deducing necessary model improvements.

As a consequence, results from model validation exercises invariably become outdated. Furthermore, models will *not remain independent* of the data sets around. A suitable reaction to this problem is to validate models against many data sets.

5. WHAT CAN BE DONE BETTER?

Throughout this paper it has been emphasized that a common frame of reference is essential for model evaluation.

In the author's opinion, it would greatly enhance the productivity of the modelling community if tools for model evaluation were made generally available, so that the modelling community in practice would be able to use a common frame of reference. Such tools include *carefully prepared data sets, model evaluation software, and model evaluation protocols.* The *Model Validation Kit* is one initial step along this road.

Concerning data sets, they should be well organised, carefully checked, and with their pitfalls and peculiarities well documented. We still suffer from a lack of such data sets.

Concerning model evaluation software, we have some tools in the package of the *Model Validation Kit.* This kit is useful, but there is a need for supplementary software. An array of various model evaluation methods and corresponding software must be developed and be freely available.

Concerning model evaluation protocols, protocols for specific applications should be developed and their usability thoroughly tested.

A very real impediment to progress is that, until now, those few tools which actually did exist have been unknown to the majority of the modelling community. Now, however, recent developments in

* There are some reservations as to what one should expect: the prediction of a model is assumed to be an estimate of the *ensemble average* over many realisations in the centre of the plume. Even for a perfect model this value need not be identical to the one actually observed at the monitors, which is taken from a *single realisation.* Another complication is due to the fact that the observation is not necessarily performed in the plume centre. See Olesen, 1995b for a further discussion.

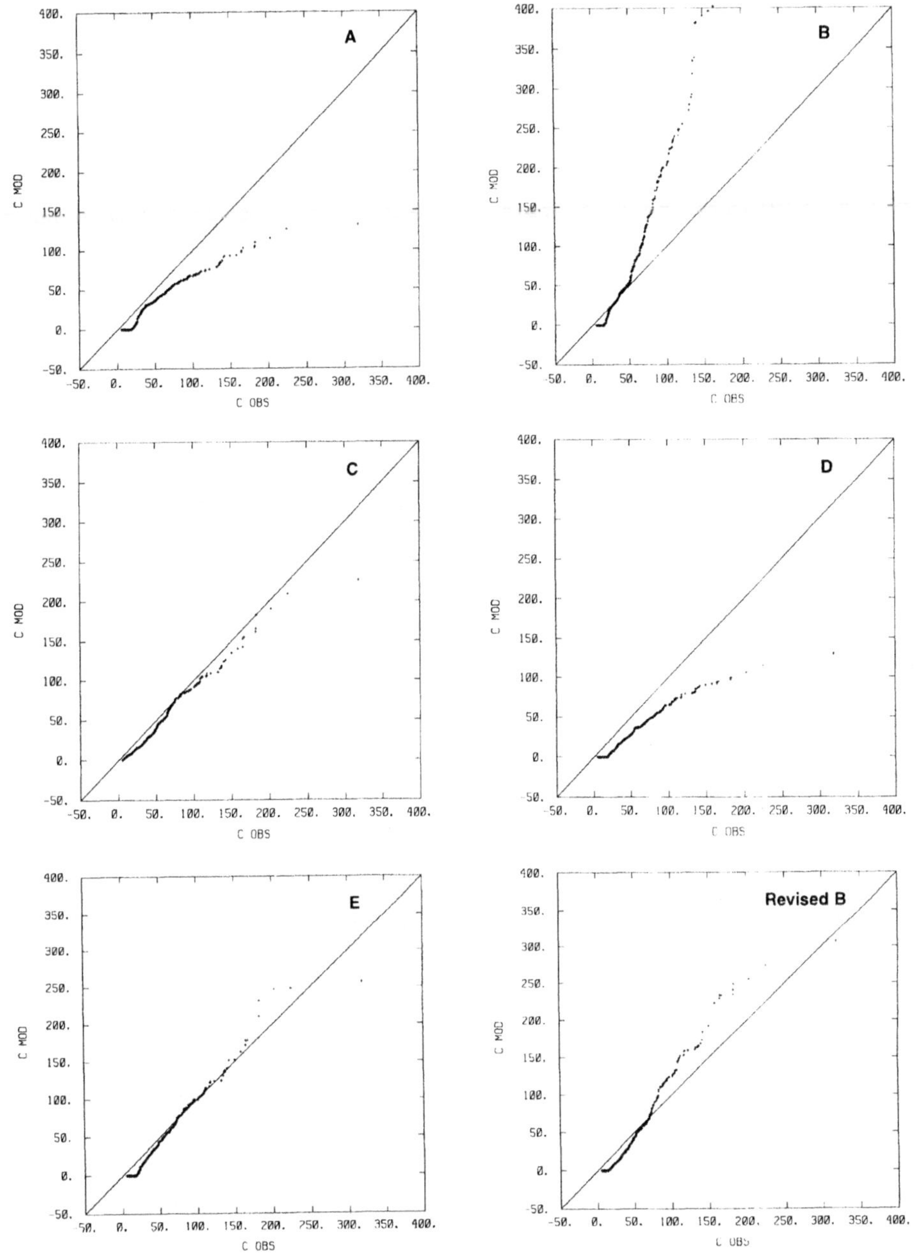

Figure 2. Kincaid data. Quantile-quantile plot of normalized concentrations according to five models. Observations of quality 3. Concentration unit is s/m^3 10^{-9}. Further explanation in text.

electronic communication have placed us in a much better position than a few years ago to share common tools - such as software and guidelines. The so-called World Wide Web (WWW) is an advanced system for accessing documents over the Internet. The most distinct feature of the WWW is its use of *hypertext* and *hyperlinks*. Some words in a document are marked, and each marked word has a link to another document or resource - which may physically be located thousands of kilometres away at another computer. Yet, the user can smoothly move between these documents without concern about their physical location.

An example of the use of the WWW in order to improve model evaluation practice has been mentioned in Subsection 3.5.

The advent of the WWW makes it possible to provide easy access to a common pool of information concerning model evaluation, such as guidelines, software and evaluation databases. A great advantage of the WWW is that it is not necessary for some central body to collect all information and continuously keep it up to date. This task would be so overwhelming for a single institution, that in practice it could never really be accomplished. On the other hand, with the use of the WWW, pooling of information can take place in a decentralized manner, so that several institutions can make contributions, each within their specific field of expertise.

REFERENCES

Bowne, N.E., Londergan, R.J., Murray, D.R. and H.S. Borenstein, 1983, *Overview, Results, and Conclusions for the EPRI Plume Model Validation and Development Project: Plains Site.* EPRI EA-3074, Palo Alto, CA.

Cuvelier, C. (editor), 1994, Proceedings of the workshop *"Intercomparison of Advanced Practical Short-Range Atmospheric Dispersion Models".* August 30- September 3, 1993, (Manno - Switzerland), CSCS (Centro Svizzero di Calcolo Scientifico). Joint Research Centre, European Commission, EUR 15603 EN. Available from C. Cuvelier, JRC Ispra, TP 690, 21020 Ispra, Italy.

Duijm, N.J., Stork, B., Nielsen, M., Ott, S., 1995, An overview of the REDIPHEM project. Workshop on Operational Short-range Atmospheric Atmospheric Dispersion Models for Environmental Impact Assessment in Europe, Mol, Belgium, Nov. 1994, *Int. J. Environment and Pollution,* Vol. 5, Nos. 4-7, pp.-

Gryning, S-E. and Millan, M.M. (Eds.), 1994, *Air Pollution Modeling and Its Application X.* Plenum Press, New York.

Hanna, S.R., Strimaitis, D.G. and J.C. Chang, 1991, *Hazard Response Modeling Uncertainty (a Quantitative Method). Vol. I: User's Guide for Software for Evaluating Hazardous Gas Dispersion Models.* Sigma Research Corporation, Westford, Ma.

Hanna, S.R., Chang, J.C. and Strimaitis, D.G., 1993, Hazardous gas model evaluation with field observations, *Atm. Env.* 27A:2265.

Klug, W., Graziani, G., Grippa, G., Pierce, D., Tassone, C., 1992, *Evaluation of Long Range Atmospheric Transport Models Using Environmental Radioactivity Data From the Chernobyl Accident. The ATMES Report.* Elsevier. Pub. EUR 14147 of the Commission of the European Communities.

Kurzeja, R.J. and A.H. Weber, 1985, *Proceedings of the DOE/AMS Air Pollution Model Evaluation Workshop. Vol. 3. Summary, Conclusions, and Recommendations.* DP-1701-3. E.I. DuPont de Nemours & Co., Savannah River Laboratory, Aiken, SC 29808.

Lee, R.F., Overview of the U.S. Environmental Protection Agency's model evaluation activities, in: *Cuvelier, 1994.*

Model Evaluation Group, 1994, *Model Evaluation Protocol.* Can be requested from DG XII/D1, Rue de la Loi 200, B-1049 Brussels, Belgium. Fax +32 2 296 3024.

Olesen, H.R. and Mikkelsen, T. (Eds.), 1992, *Proceedings of the workshop on Objectives for Next Generation of Practical Short-Range Atmospheric Dispersion Models*, Risø, Denmark. National Environmental Research Institute, P.O. Box 358, DK-4000 Roskilde, Denmark. The volume is now out of print, but it is possible on request to obtain copies of specific papers.

Olesen, H.R., 1994, European coordinating activities concerning local-scale regulatory models, in: *Air Pollution Modeling and Its Application X.* Edited by S-E. Gryning and M.M. Millan, Plenum Press, New York.

Olesen, H.R., 1995a, Data sets and protocol for model validation. Workshop on Operational Short-range Atmospheric Atmospheric Dispersion Models for Environmental Impact Assessment in Europe, Mol, Belgium, Nov. 1994, *Int. J. Environment and Pollution,* Vol. 5, Nos. 4-7, pp.-----.

Olesen, H.R., 1995b, The model validation exercise at Mol. Overview of results. Workshop on Operational Short-range Atmospheric Atmospheric Dispersion Models for Environmental Impact Assessment in Europe, Mol, Belgium, Nov. 1994, *Int. J. Environment and Pollution,* Vol. 5, Nos. 4-7, pp.-----.

Royal Meteorological Society, 1995, *Atmospheric Dispersion Modelling: Guidelines On the Justification of Choice and Use of Models, and the Commmunication and Reporting of Results.* 104 Oxford Road, Reading RG1 7LJ, England.

DISCUSSION

G. SCHAYES:

As far as I know, the present harmonization and evaluation procedure is aimed towards current operational models. As most advanced models are now based on wind fields generated by meso-scale models, I think a similar initiative would be needed to evaluate meso-scale models.

H.R. OLESEN:

I agree that it certainly is appropriate to do some work for other classes of models, including mesoscale models. The initiative on Harmonisation within Atmospheric Dispersion Modelling for Regulatory Purposes has until now been considering local-scale regulatory models and the seemingly rather uncomplicated problem of predicting dispersion from a single source in homogeneous terrain. I expect that this work will continue and be gradually extended somewhat. But more actors should - and do - enter the scene in order to cover more model classes. Thus, the European Topic Centre on Air Quality of the European Environment Agency has not confined its evaluation work to regulatory models: also, related activities for mesoscale models are going on in the framework of e.g. ETEX.

J. CHING:

Please clarify the distinction between "model validation" and "model evaluation". Please discuss in context of current evaluation projects and initiatives.

H.R. OLESEN:

There is some confusion as to what the terms "evaluation" and "validation" mean, as they have been defined and used differently within the air pollution society.

In a classic review paper from 1979, D.B. Turner suggests that "validation" implies confirmation and acceptance, and should be reserved for use in situations where a second verification substantiates a previous. Thus, in Turner's terminology the concept validation comprises much more than just comparing modelled and observed data.

Along the same lines, the American Society for Testing and Materials gives the following definitions in its Standard Guide for Statistical Evaluation of Indoor Air Quality Models from 1991 (D 5157-91): "Model evaluation" is a series of steps through which a model developer or user assesses a model's performance for selected situations. "Model validation" is a series of evaluations undertaken by an agency or organization to provide a basis for endorsing a specific model (or models) for a specific application (or applications). Thus, model validation in the sense of the ASTM is a very demanding task.

Quite a different use of the same terms is prescribed in the official "Model Evaluation Protocol" (MEP) from 1994, endorsed by the European Commission, and mentioned in Section 3.3. According to the MEP, model evaluation is a set of procedures, namely (I) Model assessment including (a) a scientific assessment (b) a user-oriented assessment (II) Verification of correct coding (III)Model validation using experimental data sets. Note that the term validation is used in a very different manner than in the ASTM guide. Validation in the sense of the MEP is a much more limited undertaking than according to Turner's and ASTM's definitions. According to the glossary underlying the MEP, "validation is the process of assessing a model so that its accuracy and usefulness can be determined. This process often involves comparison with other models or experimental data." As I see it, a main obstacle preventing agreement on a common terminology is that we lack a good term for the task of comparing model results with observations. This activity is variously called "evaluation" (ASTM) and "validation" (MEP). The use of any of these two terms can lead to confusion, because the same terms are used for the overall set of procedures involved in the evaluation process. "Model performance evaluation" has also sometimes been used, but again with some danger of misunderstandings.

As to my usage of these terms, until 1994, I used the terminology defined by Turner (I was not aware that the similar ASTM standard for indoor models existed). When the MEP was published in 1994, I switched to the MEP conventions; hence, I gave the Model Validation Kit its name, and I have used the MEP conventions in the present paper. However, I have experienced that speaking about "validation" often leads to confusion, and in future I intend to avoid using this term.

VISIBILITY EPISODES IN SOUTHWESTERN BRITISH COLUMBIA: EVALUATION OF A GAS-TO-PATICLE SCHEME USING THE REVEAL DATA SET

S.C. Pryor[1], R.J. Barthelmie[1,2], and O. Hertel[1]

[1]Climate and Meteorology Program, Department of Geography, Indiana University, Bloomington, IN 47405
[2]Department of Meteorology and Wind Energy Risø National Laboratory, Roskilde, Denmark
[3]National Environment Research Institute Frederiksborgvej 399, Roskilde, Denmark

INTRODUCTION

Elevated particle concentrations are responsible for health impacts and impaired visibility. For these and other reasons there is regulatory interest in implementing measures to reduce precursor and particulate emissions, and hence ambient particulate concentrations, in a number of environments. Atmospheric particles range in size up to tens of microns in diameter and exhibit a range of chemical composition. PM_{10} (particles of diameter $\leq 10\mu m$) and $PM_{2.5}$ (fine particles) (particles of diameters$\leq 2.5\mu m$) are sub-sets of TSP (total suspended particles). The relative abundance of particles in these size ranges; TSP, PM_{10} and $PM_{2.5}$ is highly variable and has a profound influence upon the health and visibility impact of a given concentration of aerosols. Due to the diversity of particle sources and sinks, and complexity of physical and chemical transformation processes, source-receptor relationships and the impact of emissions control measures can best be described using aerosol models.

The Atmospheric Chemistry and Deposition model (ACDEP) is based on the Carbon-Bond Mechanism IV and includes extended ammonia/ammonium chemistry (Hertel et al. 1995). ACDEP calculates vertical and horizontal transport, chemical transformations and dry and wet deposition along a trajectory. The model has previously been successfully applied to calculate the deposition of nitrogen to the sea around Denmark using meteorological fields at a grid scale of 150 km and emissions either at the same scale or, within Danish territory, at a resolution of 1 km (Asman et al. 1994).

The research described herein has two principal objectives:

I. Continued development and validation of this scheme - particularly adaptation of the scheme to use finer spatial and temporal resolution meteorological/ emissions fields.
II. Implementation of the model in a region of copious and diverse emissions and complex meteorology.

Air Pollution Modeling and Its Application XI, Edited by Gryning and Schiermeier, Plenum Press, New York, 1996

The region selected for application of ACDEP is the Lower Fraser Valley (LFV), British Columbia (BC). The LFV is a densely populated coastal river valley traversed by the border between Canada and the USA. This area is characterized by significant anthropogenic and biogenic gaseous and particulate emissions and complex meteorology (due to both topographic flows and mesoscale circulations). An intensive field campaign (REVEAL) was conducted during 1993 to characterize summertime visibility and aerosol loadings in this region (Pryor et al. 1995a,b). Analysis performed on the data from REVEAL indicates that the dominant sources of fine aerosol mass in the LFV are direct vehicle emissions, secondary aerosols and soil and road dust (Pryor et al. 1995b). Organic aerosols dominate fine aerosol mass (accounting for 35-46% of the mean fine aerosol mass) in the LFV, but ammonium nitrate and ammonium sulphate dominate particle light scattering (b_{sp}) (sulphate and nitrate accounted for 55-67% of mean b_{sp} during REVEAL) in this region (Pryor et al. 1995a). Thus, it can be seen that information regarding the formation and transport of secondary inorganic aerosols is central to understanding summertime episodes of visibility degradation in the LFV, and hence is required for input into regulatory processes designed to improve visibility in the region.

The model simulations presented herein portray conditions during a period of elevated fine aerosol concentrations and reduced visibility which has previously been described by Pryor and Barthelmie (1995). The simulation results are evaluated using aerosol and gaseous data from two sites which lie in the center of the modeling domain; one in the western reaches of the valley, just east of metropolitan Vancouver (PIME), the other in the east-central valley, in the region of maximum fine aerosol concentrations (CHIL). These sites are shown in Figure 1.

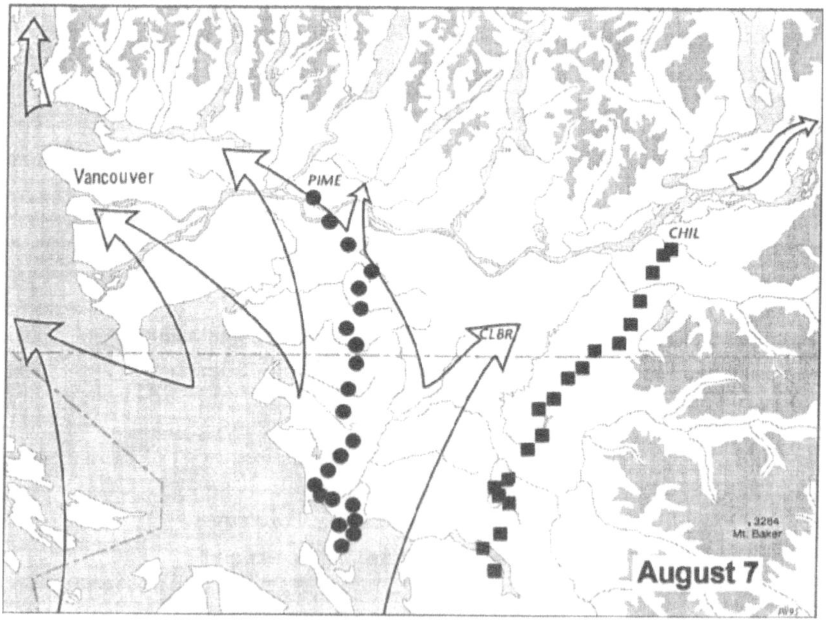

Figure 1. The LFV modeling domain. The position of 24 hour back trajectories (at each hour) to both receptor sites are denoted by solid symbols. The open arrow indicates a schematic of flow generated by Pottier (1995) based on surface wind observations.

THE ACDEP MODEL

Model description

The model is fully described in Hertel et al. (1995) and Asman et al. (1994). The main features of the model may be summarized as follows:

I. Back trajectories are calculated using a modified version of the Norwegian Meteorological Institute method (Pettersen 1940).
II. The vertical grid structure is telescopic - there are ten model layers and the depth of the layers increases logarithmically with height.
III. Vertical dispersion is calculated using the eddy diffusion coefficient which is formulated using the friction velocity and Monin-Obukhov length.
IV. Chemical transformations (including gas-to-particle conversion) are calculated using an extended version of the Carbon-Bond Mechanism IV.
V. Dry and wet deposition (in-cloud and below cloud scavenging) are computed.

Because;
(a) the modeling domain used herein is significantly reduced in comparison with the domain used in Hertel et al. (1995) (the domain is shown in Figure 1) and the back-trajectories have a duration of only one or two days,
and (b) the emissions and meteorological data are available on a one hour resolution,
it was possible to reduce the time step for the chemical mechanism to 300 secs without incurring the penalty of unfeasibly long computation times.

Atmospheric Conditions During the Model Simulation Period

Preliminary results from simulations performed for August 7 1993 (starting at midnight) are presented herein. On this day, the visual range in the central LFV was reduced below 7 km, 24 hour average measured fine mass exceeded 20 μgm-3, and up to two-thirds of hydrated fine aerosol mass was attributable to ammonium sulphate and ammonium nitrate. During the night of August 6-7, a marine stratus bank pushed into southwestern BC, ending a period of stagnation and elevated oxidant levels, and causing increased relative humidity and a marked change in the prevailing flow. Surface winds during August 7 were predominantly from the southwest, and upper level flow changed from northwesterly to southwesterly (Pryor and Barthelmie 1995).

Model Inputs - Back Trajectories

Back-trajectories from the receptor points (PIME and CHIL) are calculated using a modified version of the Norwegian Meteorological Office trajectory code (Pettersen 1940). The back-trajectories calculated for each hour of August 7 (starting at midnight) from PIME and CHIL are shown in Figure 1. The trajectory which terminates at PIME is initially relatively stagnant over the coast of Washington state (west of Bellingham) then begins to travel northward terminating at PIME at 9pm (PST). The trajectory which terminates at CHIL begins further inland and moves northeast over predominantly agricultural areas and transport routes before arriving at CHIL. The speed and direction of the modeled trajectories are in agreement with the observed wind fields and the generalized flow patterns (from the interpretation of Pottier 1995) shown in Figure 1.

Model Inputs - Meteorological Data

During REVEAL the routine meteorological monitoring network was supplemented by additional stations. Data from this expanded network are interpolated onto the same 5km by 5km grid as the emissions data for use in the trajectory module and ACDEP. The gridded meteorological fields used include upper level and near-surface wind components, upper level and near-surface temperature, relative humidity and surface pressure. Estimates of mixed layer depth, cloud cover and heat and momentum fluxes are also used. No precipitation was observed during the modeling period.

Model Inputs - Emissions Fields

For inventory purposes, the domain shown in Figure 1 is divided into 28 (West-East) by 22 (North-South) grid cells of 5km x 5km. Emissions inventories of sulphur dioxide (SO_2), oxides of nitrogen (NOx), and volatile organic compounds (VOC) (GVRD 1994, Levelton 1993a, b) were processed for a period of elevated oxidant concentrations in early August 1985 by Suzuki (1995) and are used here to represent typical summer conditions. A full inventory is not yet available for ammonia (NH_3). However, it is thought that NH_3 emissions in the LFV are dominated by agriculture, particularly livestock husbandry, and preliminary estimates of ammonia emissions generated by Barthelmie and Pryor (1995) are employed here. The spatial distribution of emissions shows:

I. NOx emissions are dominated by the mobile sector and are therefore concentrated in downtown Vancouver and the suburban core.
II. SO_2 emissions are dominated by industrial point sources both within Vancouver and to the south of the border.
II. VOC emissions are a mix of biogenic and anthropogenic emissions and hence are dominated by area sources.
IV. NH_3 emissions are principally from agricultural areas in the center/east of the LFV.

The implications of the spatial distribution of these emissions for the formation of secondary aerosols are discussed in Pryor et al. (1995b).

Within ACDEP, emissions are distributed into either the lowest two model layers for area sources (including all NH_3 emissions) or into the third and fourth layers for elevated point sources.

Figures 2a and b show emissions of NOx, SO_2, VOC and NH_3 for the grid cells over which the trajectories shown in Figure 1 pass. The emissions into the one-dimensional column vary both spatially and temporally. As shown by Figure 2a, the trajectory which terminates at PIME initially passes over dominantly mixed agricultural regions of low NOx, SO_2, and VOC emissions until hour 12 when the column passes over a localized (point) source of SO_2 emissions. The total SO_2 emissions, converted to concentrations in ppb, in the lowest two model layers are 3.8 ppb (and in layers three and four are approximately 0.4 ppb)[1]. Subsequent to hour 12 the trajectory passes over mixed agricultural/transport links/urban developments associated with higher VOC and NOx emissions. As shown by Figure 2b, in the early morning hours the trajectory which terminates at CHIL is located over mixed agricultural/rural land and transport links associated with area sources of VOC species and NH_3. After hour 9, the trajectory moves over mainly low emission areas and it is not until hour 18 that significant NOx emissions are encountered.

---

[1] The converted emissions concentrations are not shown in Figure 2a, because those values are off scale.

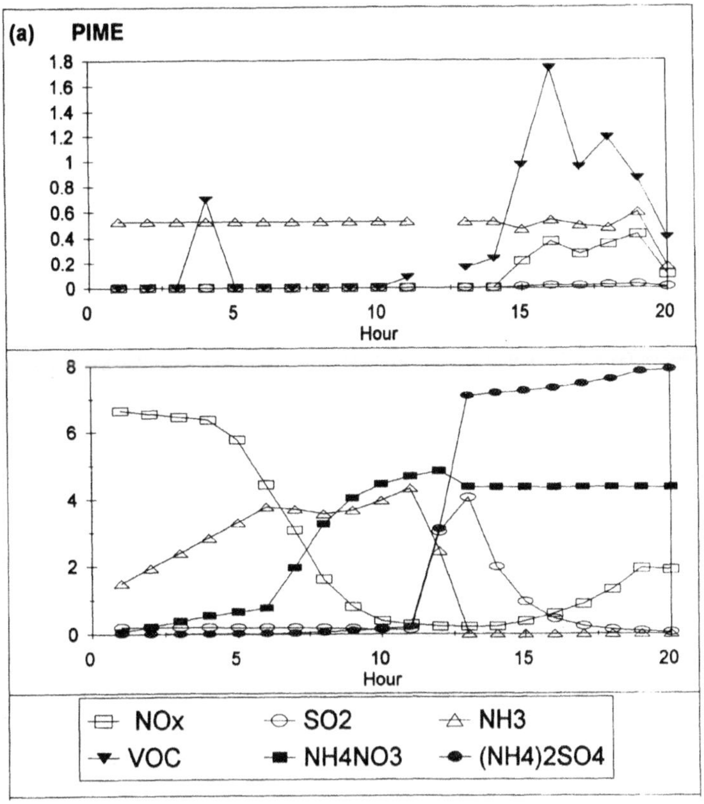

Figure 2a. Above: Emissions for each hour of the back-trajectory from PIME shown in Figure 1. Emissions of SO_2, NOx, VOC and NH_3 (converted to ppb) distributed through the lowest two layers of the model (to a height of 24m). Below: Concentrations of key chemical species in the lowest two model layers. Concentrations of aerosol species are given in μgm^{-3}, concentrations of gaseous species are given in ppb.

Model Inputs - Initial Chemical Concentrations

The relatively small domain used herein means that the lifetime of some of the reactants can exceed the duration of the trajectory. Hence it is necessary to assume initial concentrations for species with lifetimes of less than one hour. Measurements taken in the LFV are used to determine initialization concentrations of O_3, NOx, CO and SO_2 (Rogak et al. 1993) and VOCs (Dann 1994). Where measurements were not available, values taken from the literature are employed (Hertel et al. 1995).

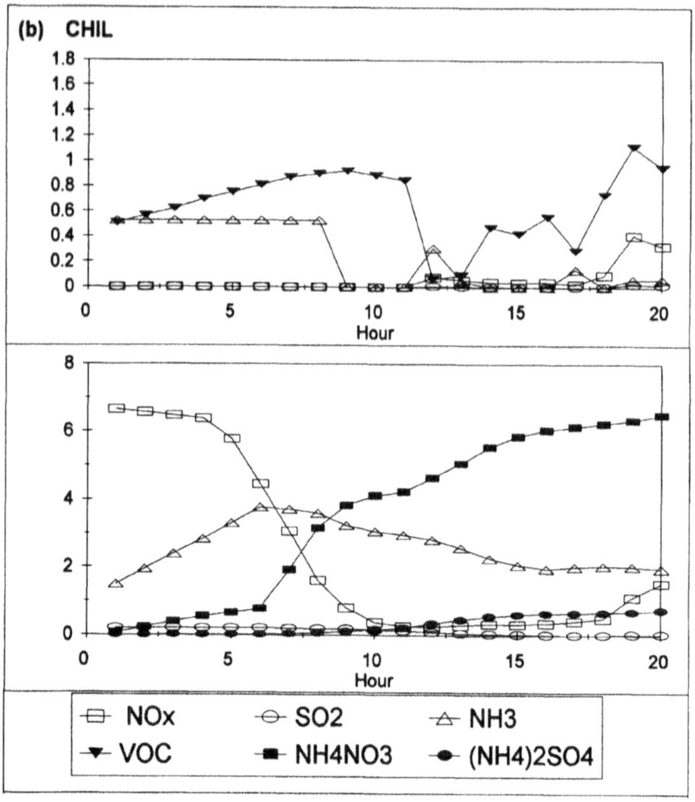

Figure 2b. Above: Emissions for each hour of the back-trajectory from CHIL shown in Figure 1. Emissions of SO_2, NOx, VOC and NH_3 (converted to ppb) distributed through the lowest two layers of the model (to a height of 24m). Below: Concentrations of key chemical species in the lowest two model layers. Concentrations of aerosol species are given in μgm^{-3}, concentrations of gaseous species are given in ppb.

PRELIMINARY RESULTS

Comparison of measured and predicted daily average concentrations of NOx, CO and O_3, at PIME and CHIL are shown in Table 1. Modeled concentrations of SO_2 are also shown. Unfortunately, there are no observations of SO_2 from these sites over this period. As shown, the observed and simulated CO concentrations show good agreement, however, both O_3 and NOx are under predicted.

Table 1. Hourly average observed and predicted concentrations of key gaseous species (ppb) at 9pm (PST) August 7.

	PIME		CHIL	
	Observed	Predicted	Observed	Predicted
NOx	N/A	1.8	16.0	1.5
O_3	34.0	24.0	N/A	20.0
CO	390	406	330	365
SO_2	N/A	0.04	N/A	0.02

N/A indicates not available.

Concentrations of key aerosol and precursor chemical species in the bottom two layers of the one-dimensional column are also shown in Figure 2a and b. As shown in Figure 2a, in the simulation for PIME, initial concentrations of NOx and NH_3 are high, but after sunrise, ammonium nitrate (NH_4NO_3) formation commences and, in the absence of fresh NOx emissions, this aerosol formation depletes ambient NOx concentrations. Ammonium sulphate ($(NH_4)_2SO_4$) concentrations are also initially low but increase rapidly subsequent to the injection of significant SO_2 emissions. Chemical concentrations in the simulation terminating at CHIL (Figure 2b) show a similar pattern to those in Figure 2a with the exception that SO_2 emissions into the column are lower than those in the simulation for PIME and hence $(NH_4)_2SO_4$ formation is greatly retarded.

Twenty-four hour averaged measurements of fine aerosol composition from PIME for the period 7am (PST) August 7 until 7am (PST) August 8 show fine aerosol SO_4^{2-} concentrations of approximately 5 μgm^{-3}, NO_3^- concentrations of 1.2 μgm^{-3}, and NH_4^+ of 1.7 μgm^{-3}. Although these data are not directly comparable to the modeled aerosol concentrations because;

I. they are 24 hour averaged concentrations and the model predicts instantaneous concentrations,

II. the measurements record fine aerosols only, while the model predicts TSP,

it is interesting to note that the relative abundance of these species is well represented by the model simulation for PIME. However, the concentration of SO_4^{2-} is considerably over-predicted. Total aerosol NO_3^- and NH_4^+ are predicted at hour 20 to be 4.4 and 1.3 μgm^{-3}, respectively. Observations of 24 hour averaged fine aerosol SO_4^{2-}, NO_3^- and NH_4^+ from CHIL for the 7-8 August indicate concentrations of 5.9, 5.4 and 3.2 μgm^{-3} respectively. Again, given the caveats listed above, the total aerosol NO_3^- concentration at hour 20 of the simulation (6.4 μgm^{-3}) shows good accord with this measurement, but SO_4^{2-}, and NH_4^+ are significantly under predicted compared to the observations. It is speculated that the overprediction of SO_4^{2-} concentrations at PIME and the underprediction at CHIL is due to:

I. The dominance of point sources of SOx in the domain. If the trajectory does not pass directly over these point sources, emissions of SOx to the column, and hence sulphate production, will be low. This will be addressed in future model runs by attention to the parameterisation of horizontal diffusion.

II. The omission of biogenic SO_2 and NH_3 sources from the inventory (see Pryor et al. 1995a). There is also considerable uncertainty in the accuracy of NH_3 emissions values, particularly in the south of the domain.

FUTURE WORK

It is clear that further evaluation of the model results is required in different seasons and over longer time scales than are shown here. This will be possible following the release of longer-term data from a one year extension of REVEAL(REVEAL II).

Further amendments to both model inputs and the model parameterisations are required to improve results in the LFV. The most readily identified improvements in terms of the emissions inventories are;

I. inclusion of land use data bases in the determination of ammonia emissions,
II. improvements in identifying and quantifying biogenic emissions,
III. further work on the temporal variability of emissions and better parameterisation of emissions from SO_2 point sources.

Development of a nested grid system to allow use of longer trajectories and improvements to the stability scheme within the model are also desirable. Additions to the chemical mechanism to take account of marine sulphur species and to parameterize organic aerosol formation would be advantageous.

ACKNOWLEDGMENTS

The authors would like to thank; Natalie Suzuki of the BCMoELP for granting permission to use emissions fields from her thesis, Sue Grimmond of IU for donating computing resources, and researchers in both the Danish NERI and the Norwegian Geophysical Institute for allowing use of the ACDEP model. We are also grateful to the people who initiated and participated in REVEAL and to Paul Jance who produced Figure 1. Finally, RJB would like to thank Sven-Erik Gryning for his encouragement and Søren Thykier-Nielsen of Risø for his assistance with computing.

REFERENCES

Asman W.A.H. et al. (1994), Marine Research from the Danish Environmental Protection Agency, No 37, Copenhagen, Denmark.
Barthelmie R.J. and Pryor S.C. (1995): The role of ammonia in fine aerosol formation and visibility impairment. *Accepted for inclusion in the Second International Conference on Atmospheric Ammonia,* Oct. 1995.
Dann T. (1994): Environmental Technology Centre of the Environmental Protection Service of Environment Canada. Report # PMD 94-1. January 1994.
Greater Vancouver Regional District (GVRD) (1994): Prepared by the GVRD, Burnaby, February 1994.
Hertel O. et al. (1995): Development and testing of a new variable scale air pollution model - ACDEP. *Atm. Env. 29* 1267-1290.
Levelton and Associates Ltd. and Western Research (1993a): Prepared for GVRD, Burnaby, B.C.. Jun 1993.
Levelton and Associates Ltd. (1993b): Prepared for Atmospheric Environment Service, Vancouver. March 1993.
Pettersen S. (1940): Weather Analysis and Forecasting. Mcgraw-Hill, New York
Pottier J. (1995): Personal communication.
Pryor S.C. et al. (1995a): Visibility and aerosol composition in the Fraser Valley during REVEAL. *Conditionally accepted for publication in JAWMA.*
Pryor S.C. et al. (1995b), REVEAL: Characterizing fine aerosols in the Fraser Valley. *Atm. Env. (in review)*
Pryor S.C. and Barthelmie R.J. (1995), Investigation of the causes of visibility degradation in the Fraser Valley, British Columbia. *1995 AWMA Conference proceedings (in press).*
Rogak S.N. et al. (1993): Report prepared for the GVRD, March 1993.
Suzuki N. (1995): Application of the Urban Airshed Model to the Fraser Valley, BC. *Thesis to be submitted to the Dept. of Geography, UBC.*

DISCUSSION

J. CHING: Please comment on under prediction of sulphate and over prediction of nitrate. In particular please discuss the role of low dispersion of the SO_2 sources in lagrangian model as a possible explanation.

R.J. BARTHELMIE: In the current formulation of the model, horizontal dispersion is assumed to be contained within the grid cells. Hence, unless the trajectory passes over a point sources (i.e. within the same grid cell), no contribution from that source is incorporated into the column. Because, sulphur dioxide emissions in the LFV are dominated by point sources, while oxides of nitrogen are emitted principally from mobile sources, the sulphur species are particularly sensitive to the parameterization of this process. Underestimation of sulphate may also reflect inaccuracies/omissions in the emission inventory. For example, biogenic sulphur emissions in the coastal waters to the west of the LFV are thought to comprise a significant sulphur source, however, these emissions are not included in the inventory (see Pryor et al. 1995a). Note also that underestimation of sulphate may lead to overestimation of nitrate since ammonium sulphate is formed preferentially to ammonium nitrate.

D. ANFOSSI: At which height over ground do your back trajectories start?

R.J. BARTHELMIE: The trajectory scheme used for the preliminary runs of ACDEP is a single layer scheme. Observed wind speeds in the boundary layer were used to calculate a bulk boundary layer average. Work is currently underway at NERI to develop a multi-layer trajectory scheme for use with the ACDEP model.

J.A. van JAARSVELD: Did you take into account any background concentrations for sulphate and nitrate? The atmospheric lifetime of sulphate is supposed to be at least 2 days, while your trajectories are only 24 hours.

R.J. BARTHELMIE: The relatively small domain used in this analysis means that the lifetime of some chemical species can exceed the duration of the trajectory. For this reason it was necessary to assume background concentrations of some species based on ambient measurements in the region. Sulphate and nitrate aerosol are among those species for which this is necessary. The following were used as background concentrations: NH_4NO_3: $0.32 \mu gm^3$, H_2SO_4: $0.40 \mu gm^3$, NH_4HSO_4: $0.47 \mu gm^3$, NH_2SO_4: $0.54 \mu gm^3$.

J. WALMSLEY: When did/will REVEAL II take place and was there/will there be a full suite of meteorological measurements?

R.J. BARTHELMIE: REVEAL II was conducted between April 1994 and June 1995. IMPROVE aerosol samplers were operated at two sites (CLBR and CHIL) on a sampling regime of 24 hour averages, two days a week. Data from co-located open chamber nephelometers, humidity, temperature and wind speed/direction measurements were also collected on an hourly time frame. Scenic information was collected by two still cameras. Standard near-surface meteorological measurements are routinely available from a network of approximately 20 stations in the domain (including airports in the region). The closest radiosonde release locations are on the Olympic Peninsula to the southwest and the coast of B.C. to the north. An acoustic sounder is also operated at Abbotsford Airport.

VALIDATION OF A MICROSCALE POLLUTION DISPERSAL MODEL

Joachim Eichhorn

Institute of Atmospheric Physics
Johannes Gutenberg - University
Mainz, F.R.G.

INTRODUCTION

The three-dimensional numerical model MISCAM (**Mi**cro **S**cale **A**ir Pollution **M**odel) has been developed to study wind flow and pollutant dispersal in densely built-up urban areas (Eichhorn, 1989). The model has been successfully applied for planning purposes by a variety of institutions in Germany. MISCAM consists of the non-hydrostatic Eulerian equations of motion and a transport equation for pollutants. Turbulence closure is carried out by means of a k-ε-model. To reduce numerical diffusion errors, Smolarkiewicz and Grabowski's (1989) scheme may be used for the calculation of advective transport. Additionally, sedimentation and dry deposition of pollutants may be taken into account.

Model simulations will be an important factor in urban planning, once the quality of model results in comparison to measurements has been proven. Therefore, a series of validation experiments has been carried out. Comparisons of MISCAM-simulations with wind tunnel data as well as results of a field experiment will be presented. Wind tunnel data are used to examine the model's capabilities to reproduce flow conditions near isolated buildings. MISCAM-simulations of height and horizontal extent of recirculation zones compare favourably to wind tunnel measurements.

During a one-week field experiment continuous measurements of air pollution due to traffic have been carried out near a four lane road at Frankfurt. The measurements cover the analysis of NO, NO_2, O_3 and SO_2. Vertical profiles have been recorded as well as time series of pollutant concentrations near the ground at a couple of sites along the road. Again, MISCAM results are in reasonable agreement with the experimental data.

Air Pollution Modeling and Its Application XI, Edited by
Gryning and Schiermeier, Plenum Press, New York, 1996

Model results imply that MISCAM may be presently applied as a useful instrument in urban planning. Nevertheless, additional work has to be done to improve and ensure the model's quality.

THE MODEL

The atmospheric flow field is calculated from the three-dimensional set of Eulerian equations of motion using the anelastic Boussinesq approximation. The basic set of equations reads

$$\frac{\partial u_i}{\partial t}+\frac{\partial u_j u_i}{\partial x_j}=-\frac{1}{\rho}\frac{\partial p}{\partial x_i}+\frac{\partial}{\partial x_j}\left[K_m\left\{\frac{\partial u_i}{\partial x_j}+\frac{\partial u_j}{\partial x_i}\right\}\right] \tag{1}$$

$$\frac{\partial u_j}{\partial x_j}=0 \tag{2}$$

Pollutant transport is modelled by solving a transport equation for arbitrary mass concentrations

$$\frac{\partial c}{\partial t}+\frac{\partial u_j c}{\partial x_j}=\frac{\partial}{\partial x_j}\left[K_h\frac{\partial c}{\partial x_j}\right]+Q_c \tag{3}$$

where Q_c denotes the sum of sinks, i.e. deposition at the ground, and sources of the mass concentration c.

The turbulent exchange coefficients are computed from a k-ε-model. Once the prognostic equations for the turbulent kinetic energy k and for the dissipation rate ε

$$\frac{\partial k}{\partial t}+\frac{\partial u_j k}{\partial x_j}=\frac{\partial}{\partial x_j}\left[K_m\frac{\partial k}{\partial x_j}\right]+P_m+P_h-\varepsilon \tag{4}$$

$$\frac{\partial \varepsilon}{\partial t}+\frac{\partial u_j \varepsilon}{\partial x_j}=\frac{\partial}{\partial x_j}\left[\frac{K_m}{\sigma}\frac{\partial \varepsilon}{\partial x_j}\right]+c_1\frac{\varepsilon}{k}(P_m+P_h)-c_2\frac{\varepsilon^2}{k} \tag{5}$$

are solved, K_m is computed from

$$K_m=c_\mu\frac{k^2}{\varepsilon} \tag{6}$$

For simplicity, K_h=1.35 K_m is assumed. Values for the empirical constants c_μ, c_1, c_2 and σ are taken from Paterson and Apelt (1989). At the surface, exchange coefficients are calculated from the external Monin-Obukhov laws as specified by Clarke (1970).

The solution of the velocity field is carried out by using the well-known procedure described by Patrinos and Kistler (1977) who introduce an auxiliary velocity field which is free from the dynamic pressure disturbance. The final velocity field is obtained by iteratively solving a Poisson equation for the pressure disturbance and substituting the results into the preliminary wind field.

Special care has been taken to simulate flow separation at building edges which has been a minor defect of earlier model versions. The method adopted in the present study is due to Paterson and Apelt (1989) and uses a zero longitudinal velocity at the upper front edges of buildings. All other grid points near solid surfaces are treated as usual.

The prognostic equations are solved on an arbitrarily scalable Cartesian grid. Upstream discretisation is used for the advection terms. In order to reduce numerical diffusion, however, the advection scheme proposed by Smolarkiewicz and Grabowski (1989) may be applied. The diffusion equations are solved by use of a standard ADI procedure. Finally, a SOR method is used to solve the Poisson equation. This method is known to be reliable but slow in comparison to newer techniques. Nevertheless, SOR has been selected because of its remarkably lower memory requirements thus enabling MISCAM to run on standard personal computers.

WIND TUNNEL SIMULATIONS

A variety of wind tunnel experiments have been carried out by Klein *et al.* (1994) to obtain a data base for the validation of microscale models. A comparison of five different models, including MISCAM, as applied to the wind tunnel data was subject to a study by Röckle and Richter (1995).

Since MISCAM has been improved in various parts, several of these simulations have been repeated by the author. The following figures show results for the flow and pollutant transport around a U-shaped building. In Figure 1, the horizontal wind field in the lowest grid level is shown. A well-pronounced pair of lee-eddies is noticed as well as the reattachment of the flow about 50 *m* downstream of the building. This gives a recirculation zone extending raughly two times the building height. The wind tunnel measurements by Klein *et al.* imply a slightly larger recirculation zone.

Pollutant dispersal has been simulated for an arbitrary gas released from a point source located at ground level outside the backyard, at the centre of the backyard and above the building's roof. Figure 2 shows a vertical cross-section of the normalized mass concentration cU/Q where U is the inflow velocity at height 10 *m*, Q is the emission rate and c the mass concentration with the source located at the centre of the yard. The simulation corresponds to the wind field shown in Figure 1. The dotted line represents the 10^{-3} m^{-2} contour line as obtained from the wind tunnel experiments showing an almost exact agreement of model result and measurement.

MEASUREMENTS OF TRAFFIC INDUCED IMMISSIONS

Wind tunnel and model studies of pollutant dispersal have also been carried out for more complex building configurations representing parts of a large chemical plant at Ludwigshafen (F.R.G.). Details are given in the report by Röckle and Richter (1995). The present study will give an impression of MISCAM's performance for a highly structured build-up by applying the model to an inner city roadway at Frankfurt. Model results will be compared to immission data that have been collected during a one-week measuring program carried out by the Centre of Environmental Research at the University of Frankfurt in conjunction with Lahmeyer International, Frankfurt.

A vertical profile of four equally spaced measuring points was taken as well as two additional ground level measurements, one near the entrance of a building facing the

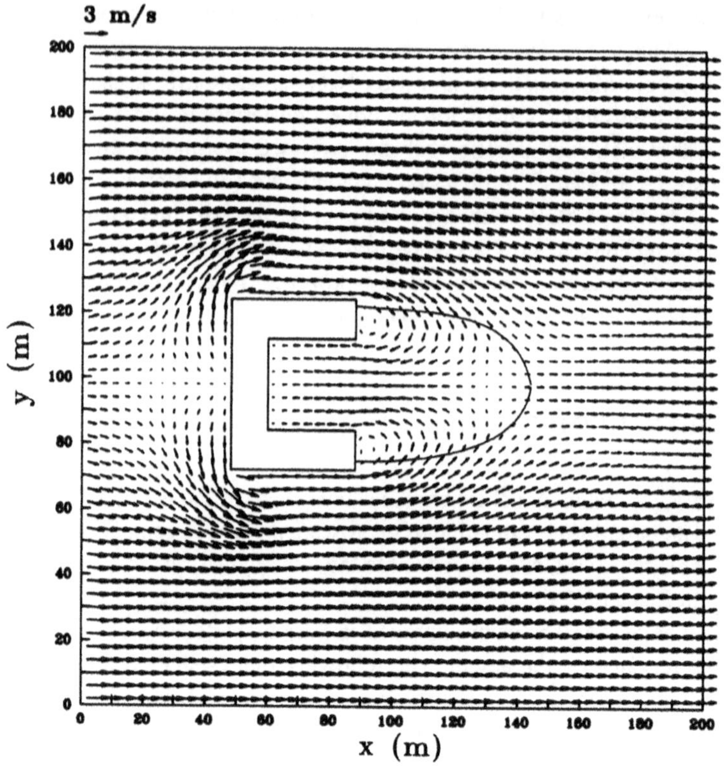

Figure 1. Flow around a u-shaped building. Horizontal wind field at a height of 2 *m*. Inflow velocity is 5 *m/s* at a height of 10 *m*, the building height is 28 *m*. The thick line corresponds to circulation zone from wind tunnel experiment.

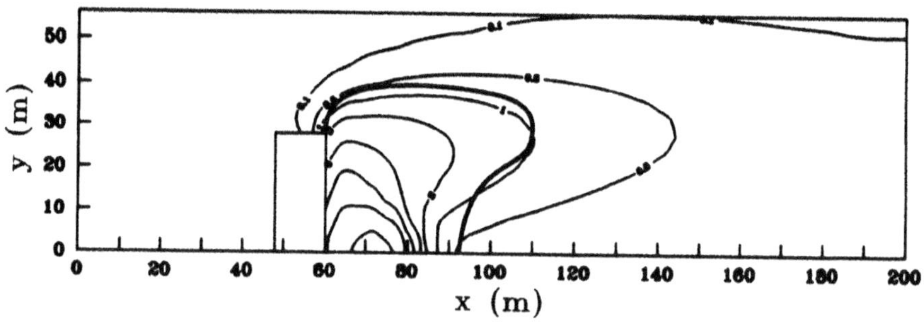

Figure 2. Distribution of a polluting gas from point source at the centre of the backyard. Normalized mass concentration ($10^{-3}m^{-2}$), the thick line indicates the 10^{-3} m^{-2} contourline as obtained from the wind tunnel experiment.

roadway, the other at the rear side of the building row, representing the undisturbed background conditions. The measuring site was located at the west side of the road. Since westerly winds were prevailing throughout the measuring period, highest immission concentrations had to be expected at the ground level data points. Therefore, the lowest point of the vertical profile was used as reference point. During peak hours (6.30 to 9.30 a.m., 3.30 to 7.30 p.m.), each of the measuring points was attached every 5 minutes. For the remaining time, an alternating scheme of 30 minutes measurements of the vertical profile and measurements at the reference point was used.

To obtain realistic emission rates as input for the numerical simulations, emission factors for different types of vehicles have been taken from a recent study (PROGNOS, 1993). Traffic data (number of vehicles per hour, fraction of different car types etc.) have been estimated from a traffic census carried out for one day of the measuring period.

MODEL RESULTS

Preliminary results of MISCAM-computations of the immission distributions as compared to the measured data will be presented below. The simulations were carried out for Dec 8, 1994, covering the peak hours from 6 a.m. to 10 a.m. Figure 3 shows the time averaged mass concentrations for NO_x at the lowest grid level (1.5 *m*).

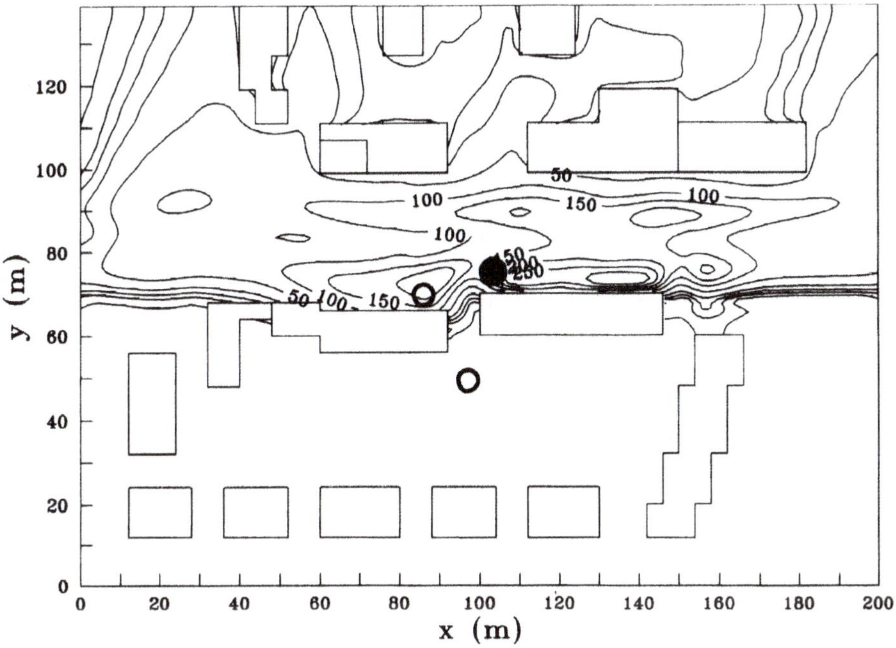

Figure 3. Computed NO_x-concentrations ($\mu g/m^3$) at a height of 1.5 *m* at Eschersheimer Landstraße, Frankfurt. Averaged values for Dec 8, 1994, 6 a.m. to 10 a.m. Open dots denote locations of ground level measuring points, the solid dot denotes the location of the vertical profile.

As expected, maximum values are found at the lee-side of the roadway. A more homogeneous structure of the immission distribution results in those parts of the model domain where the flow field is less influenced by the building structure.

Since the model runs for the complete measuring period are not yet finished, only preliminary findings can be drawn concerning the quality of the model simulations. A first examination of simulated and observed concentrations for the period evaluated in Figure 3 shows a moderate overestimation of the observed mean values at ground level by roughly 15 %, while model results for the upper points of the vertical profile are lower than the corresponding observations. This discrepancy may be removed by properly specifying the thermal stratification as input for the numerical simulations.

To obtain a comprehensive impression of MISCAM's quality, an evaluation of the complete experimental data set is required. Furthermore, model sensitivity to external parameters (thermal stratification, roughness of underlying surfaces etc.) must be examined comprehensively. Present results, however, indicate that MISCAM is a well-designed tool for practical planning purposes. This may also been concluded from validation experiments which have been carried out by other authors. For example, Bächlin *et al.* (1995), have tested various microscale models, including MISCAM, by simulating annual average concentrations of exhaust gases near an inner city road in Hannover / Germany. There has been a one-year series of measurements for the corresponding site which has been carried out by the German control organization TÜV. A comparison of observed and simulated mean values gives deviations of approximately 10 % for the MISCAM simulations which agrees well with the results of the present study.

CONCLUSIONS

To validate the three-dimensional flow and transport model MISCAM a series of numerical simulations has been carried out. Model results have been compared to wind tunnel data as well as immission data gathered at different measuring sites in German cities. An examination of the model results shows that MISCAM is capable to satisfactorily reproduce dynamic effects of single buildings and groups of buildings. Additionally, observed immission concentrations are simulated within an acceptable error range. This suggests, that MISCAM may be applied to problems in urban planning without major modifications of the code.

With some extensions, MISCAM may be used to examine highly complex configurations, e.g. tunnel exits located in densely built-up terrain. Examples of such applications are documented in another part of this volume.

ACKNOWLEDGEMENTS

The author wishes to express his gratitude to Prof. W. Jaeschke, Dr. N. Beltz and J. v. Trümbach (Centre of Environmental Research, University of Frankfurt) for carrying out the measurements and to R. Ries (Lahmeyer International, Frankfurt) for supplying emission data and for helpful discussions.

REFERENCES

Bächlin, W., Lohmeyer, A., Schädler, G., 1995, *Vergleich und Bewertung derzeit verfügbarer mikroskaliger Strömungs- und Ausbreitungsmodelle,* annual report of project PEF 349.

Clarke, R. H., 1970, Recommended methods for the treatment of the boundary layer in numerical models, *Austr. Met. Magazine,* **18**, 51-73.

Eichhorn, J., 1989, *Entwicklung und Anwendung eines dreidimensionalen mikroskaligen Klimamodells,* PhD-thesis, University of Mainz, Germany.

Klein, P., Rau, M., Wang, Z. and Plate, E. J., 1994: *Ermittlung des Strömungs- und Konzentrationsfeldes im Nahfeld typischer Gebäudekonfigurationen (Experimente),* final report of project PEF 2 92 003, Karlsruhe, Germany.

Paterson, D. A. and Apelt, C. J., 1989, Simulation of wond flow around three-dimensional buildings, *Building and Environment,* **24**, 39-50.

Patrinos, A. N. A. and Kistler, A. L., 1977, A numerical study of the Chicago lake breeze, *Boundary Layer Meteor.*, **12**, 93-123.

PROGNOS, 1993, *Entwicklung der Emissionen von Luftschadstoffen und CO_2 durch den Verkehr in Ost- und Westdeutschland bis 2010,* final report of project 90344/91, German Department of Traffic.

Röckle, R. and Richter, C. J., 1994, *Ermittlung des Strömungs- und Konzentrationsfeldes im Nahfeld typischer Gebäudekonfiguratioen (Modellrechnungen),* final report of project PEF 92/007/02, Freiburg, Germany.

Smolarkiewicz, P. K. and Grabowski, W. W., 1989, The multidimensional positive definite advection transport algorithm: nonoscillatory option, *J. Comput. Physics*, **86**, 355-375.

DISCUSSION

E. GENIKHOVICH: The vertical step in your model is about 2 m near the ground. How could you resolve the roughness parameter effects? Could you simulate the logarithmic wind profile outside the building influence area?

J. EICHHORN: MISCAM uses the roughness length to obtain turbulence variables at the surface, influencing diffusion coefficients and velocity components in the lowest grid levels. Without the presence of buildings MISCAM produces a logarithmic wind profile as long as neutral stratification is assumed.

M. PIRINGER: Can you comment on the comparison of simulations and measurements comprising the whole day? What were the meteorological conditions like on that particular day (main wind direction along or perpendicular to the street,...)? How sensitive is the model to varying meteorological conditions?

J. EICHHORN: A comprehensive examination of the complete measuring period is still under construction. Most of the time, the measuring site represented the lee-side of the road since SW was the prevailing wind direction. Of course there is a significant sensitivity to meteorological parameters such as wind speed and thermal stratification. For the simulations shown here a slightly stable atmosphere has been assumed, following data of the nearest synoptic stations.

G. COSEMANS: How accurate are the emission estimates for the hours of the measurement campaign? In other words, within what range (% less or more) can real emissions have been compared to the emission values used in the modelling?

J. EICHHORN: The estimation of emission rates is still one of the major problems in pollution modelling. Since there is a couple of unknown or uncertain parameters, an error range of 20 to 30 % seems adequate, implying that a comparison of absolute values is a questionable approach to validate the model. Another feature of the pollutant concentrations is their relative order of magnitude depending on the position of the measuring inlet, which is reproduced quite satisfactorily by the MISCAM simulations.

R. YAMARTINO: You mentioned replacing the upstream-differencing advection scheme with a higher order (less diffusive) transport scheme. Since you solve to steady state, why not use the low-computation cost scheme until reasonable convergence is achieved and the switch to then more expensive transport algorithm?

J. EICHHORN: The less-diffusive algorithm (McCormack) will prevent existing gradients from being smoothed by numerical diffusion, but it will not recreate those gradients which have already been destroyed by a couple of upstream-steps. Since McCormack's scheme may also be applied to the diffusion equation where it will replace the stable but time-consuming ADI-procedure, there will be no significant change of computation costs.

T. YAMADA: Did you include an equation for temperature to consider density stratification? Local convectional circulation is expected because building walls are heated by the sun. Without considering temperature influences, how did you get such good agreements between modeled and observed concentrations?

J. EICHHORN: A heat transport equation is not included in the model version used for the current study. Since weather conditions were mostly cloudy during the measuring period, convective effects were of minor importance. Nevertheless the influence of convection due to heated building walls is out of question. Therefore, future versions of MISCAM will be extended to model heat transport.

A NUMERICAL STUDY OF DISPERSION OVER THE GREATER ATHENS AREA DURING TWO DAYS OF HEAVY POLLUTION USING THE URBAN AIRSHED MODEL

Ioannis C. Ziomas, Paraskevi Tzoumaka, Dimitrios Balis, Dimitrios Melas, and Christos S. Zerefos

Laboratory of Atmospheric Physics, Aristotle University of Thessaloniki, 54006 - Thessaloniki, Greece

ABSTRACT

The Urban Airshed Model (UAM) is a three dimensional grid (Eulerian) model which calculates the concentrations of both inert and chemically reactive pollutants by simulating the physical and chemical processes in the atmosphere that affect pollutant concentrations. The UAM simulates the emission, advection and dispersion of pollutants within every grid cell of the modelling domain.

The aim of the present study is to use UAM in order to simulate the spatial and temporal variations of pollutants over the Greater Athens Area (GAA.) during the 14th and 15th of September 1994.

The period of the simulation are days of pollution episodes during the MEDCAPHOT - TRACE (Mediterranean Campaign of Photochemical Tracers - Transport and Chemical Evolution) project, that was carried out in GAA during the period August 20 - September 20 1994. The two days of simulation are similar in terms of the meteorological conditions, while they differ in terms of pollutant emissions, due to the traffic restrictions imposed by the responsible authorities on the 15th of September.

The meteorological data used resulted from the application of a prognostic mesoscale model developed at the Department of Meteorology in University of Uppsala (MIUU).

The simulation results of the UAM are compared with pollution measurements, performed during the same period in the frame of MEDCAPHOT-TRACE. The comparison results indicate that the UAM is capable of simulating very well the pollution concentrations and shows good agreement with the spatial and temporal variation of pollutants derived by the measurements.

INTRODUCTION

The study area is the Greater Athens Area (GAA) which is one of the largest urban agglomerations in the Mediterranean region (4 million inhabitants) and is characterised by: i) high contents of primary and secondary air pollutants frequently exceeding the European Union and the World Health Organisation air quality standards, ii) high levels of solar irradiation and temperature throughout the year stimulating the photochemical activity, iii) a geographical topography facilitating the accumulation of primary/secondary pollutants (see Figure 1). A detailed description of the Athens area is found in Ziomas et al. (1995).

Athens is considered, together with Los Angeles, as an international example of intense photochemical air pollution (Mantis et al. 1993). During the summer months the prevailing synoptic north-eastern winds, known as Etesians, ventilate the Athens basin and thus no pollution episodes occur. On the contrary, in late spring and late summer (pre-Etesian and post-Etesian periods respectively) the development of land/sea breeze circulation, in conjunction with local air pollution sources (mainly transportation, and industry) favour the appearance of extreme photochemical air pollution episodes.

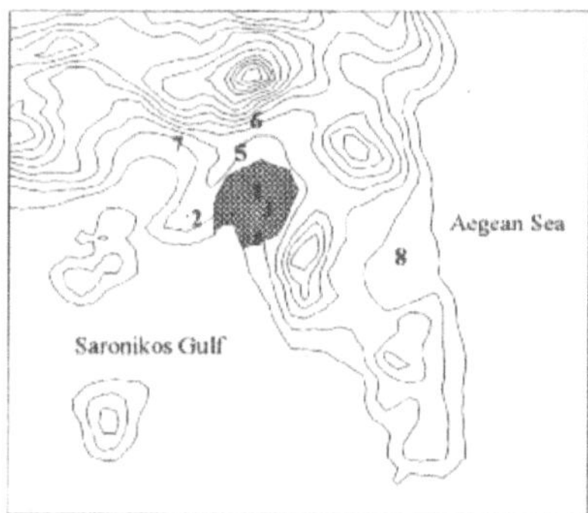

Figure 1. The Greater Athens Area. With numbers it is shown the location of the monitoring stations; 1 Nea Smirni, 2 Rentis, 3 Athinas, 4 Patision, 5 Liosia, 6 Thrakomakedones, 7 Elefsis and 8 Spata. The shaded area is the outer traffic ring of Athens.

The Mediterranean Campaign of Photochemical Tracers - Transport and Chemical Evolution (MEDCAPHOT-TRACE)

The MEDCAPHOT-TRACE experiment took place in the Greater Athens Area (GAA) during the summer of 1994 (20 August - 20 September) with the participation of many European scientific groups.

The purpose of MEDCAPHOT-TRACE was to study the chemical and meteorological evolution of ozone related trace gases (over land and sea), modulated by the various synoptic and local meteorological conditions (sea/land breeze circulations, local winds, inversion layers etc.) in one of the largest urban agglomerations in the Mediterranean region

Apart from the stations of the existing monitoring network, 18 new stations operated in the city of Athens and in the Athens basin, involving for the first time, advanced remote sensing and in-situ analytical chemistry instrumentation, installed on ground based and airborne platforms. The scientific instrumentation involved in the experiment consists of ground based air pollution measuring systems (Gas analysers and Gas chromatographers), integrated-path air pollution measuring systems (DOAS systems), ground based laser remote sensing systems (LIDAR technique), meteorological instrumentation (slow- and fast-response) and aircraft measurements (air pollution and meteorological). The study area covers the whole GAA and extends further down the Saronikos Bay.

Measurements of ozone, hydrocarbons, nitrogen oxides and PAN took place during the experiment, as well as a great variety of meteorological measurements. A detailed description of the MEDCAPHOT-TRACE as well as the scientific results of the project will be published in a special issue of the Urban Atmosphere (1996).

The photochemical model

The UAM is a three dimensional photochemical grid model which calculates the concentrations of inert and chemical reactive pollutants by simulating the physical and chemical processes in the atmosphere that affect pollutant concentrations (Scheffe and Morris 1993). The basis for the UAM is the atmospheric diffusion equation that represents a mass balance in which emissions, transport, diffusion, chemical reactions and removal processes are expressed in mathematical terms. The UAM accounts for spatial and temporal variations as well as differences in the reactivity of emissions.

The first complete application of the early version of UAM was performed in the early 70's (Reynolds et al. 1974, Roth et al. 1974). Since then UAM was continuously improved. In the present work the version of UAM used is the most recent one, which uses the Carbon Bond IV chemical mechanism (Gery et al. 1989), and the Smolarkiewitcz numerical integration scheme to solve the advection equation Smolarkiewitcz 1983).

The UAM simulates the emission, advection and dispersion of precursors and the formation of ozone within every grid cell of the modelling domain. For this purpose the following steps have taken place for the preparation of the input data :

a) The modelling domain; GAA, has been divided to a 40 x 36 grid (grid cell: 2 x 2 km^2) and the topographical and land use data have been prepared for this area.

b) An emission inventory has been also prepared according to the methodology described in Ziomas et al. (1995), for the main primary pollutants (NOx, CO and VOC). The main categories of sources considered for the preparation of the inventory are traffic and industrial installations, since there is no central heating operating during this time of the year.

c) The meteorological input data have been prepared by interpolating the results of MIUU, which is a three-dimensional mesoscale model with a terrain influenced coordinate system. A detailed description for MIUU can be found in other literature references (Enger 1986, Tjenstrom 1987).

d) The concentrations of O_3, NO_x, CO and HC measured at the various ground stations, as well as by the aircraft flights, were used, after interpolation, for the initial concentration field and for the determination of the boundary concentrations.

RESULTS

The days of the experimental campaign selected for applying the numerical models was the 14th and the 15th of September, 1994, when the highest ozone concentrations during the campaign were observed. The two days of simulation are similar in terms of the meteorological conditions, while they differ in terms of pollutant emissions, due to the traffic restrictions imposed by the responsible authorities on the 15th of September. For this reason the traffic emissions used as input to UAM for the second day of simulation were reduced by a factor of 30% within the area of the outer traffic ring of Athens (see Figure 1).

During both days, Greece was under the influence of anticyclonic conditions associated with weak pressure gradients and clear skies. The geostrophic wind was from the NNE-NNW with speeds of ~4.0 ms^{-1}. In the morning hours a westerly flow was established over Saronikos gulf while the winds over most of the Attica peninsula are low with variable directions. The surface wind pattern over Athens basin was dominated by very weak (>1 ms^{-1}) northerly flow and the air mass practically stagnated over Athens basin. During daytime, the situation changed drastically due to the development of thermally induced local circulations. Figure 2 shows the simulated horizontal wind fields at 14.00 LST for the two days of simulation. The corresponding height is approximately 8 m above the surface and for clarity of presentation, a fraction of the computational domain near the lateral boundaries is omitted. The flow over the sea was generally from the W with wind speeds ~5 ms^{-1} but there is a splitting in the vicinity of the coast of the Athens basin and part of the air moves in over the Athens basin. This is due to the sea breeze circulation which diverts a part of the flow towards the Athens basin. The winds over Athens area are from SW with wind speeds of the order of 3-4 m s^{-1}. The notable wind deceleration downwind of the shoreline is resulting from the high roughness of the city. Another sea breeze system was generated in the west side of mountain Aigaleo, at Thriassion plain, where the flow was from the SW with speeds of the order of ~3-4 ms^{-1}. In the east part of Attica peninsula, at Mesogia plain, model results show that the flow was generally from E, corresponding to a third sea breeze system. Finally, it should be mentioned that the simulated thermal structure of the atmosphere (not shown here) reveals that the convective boundary layer over the land surface was very shallow in the vicinity of the shoreline and it was increasing gradually to a height of 400 m AGL at about 10 km downwind the shoreline.

In Figure 3, predicted ozone concentrations from the application of UAM and the respective observations at the stations Nea Smirni and Rentis (southern basin), Athinas and Patision (city centre), Liosia and Thrakomakedones (northern basin), Spata (eastern Attica) and Elefsis (western Attica), for the 14th and the 15th of September, 1994 are presented. The location of the monitoring stations is shown in Figure 1. As it appears from Figure 3, the results of UAM are in rather good agreement with observations for all stations, especially concerning both the time and the amplitude of the daily maximum. The observed increasing gradient of ozone from south to north is well predicted by the UAM, as well as the differences between the areas within the Athens basin and outside it.

In some stations, and for the first hours of each day, the UAM underestimates the ozone concentrations, probably due to overestimation of the NOx emissions used in the inventory of emissions. During the midday hours a sharp increase of ozone is predicted and observed at the northern stations and the daily maximum occurs.

In Figure 4 the spatial distribution of the maximum hourly ozone concentrations, as calculated by UAM, over the whole modelling domain is presented, for both simulation days. As it appears from Figure 4, the highest ozone concentrations are observed at the northern part of the Athens basin, while ozone is higher than the 100 ppb limit even at the city centre. The areas outside Athens basin as well as the southern part of the basin do not experience ozone concentrations above this limit. It is significant to note that the 30% reduction of traffic emissions over the southern and central area of the Athens basin, during the 15th of September, 1994, did not affect the production of high ozone concentrations, which appeared again at the northern part of the basin.

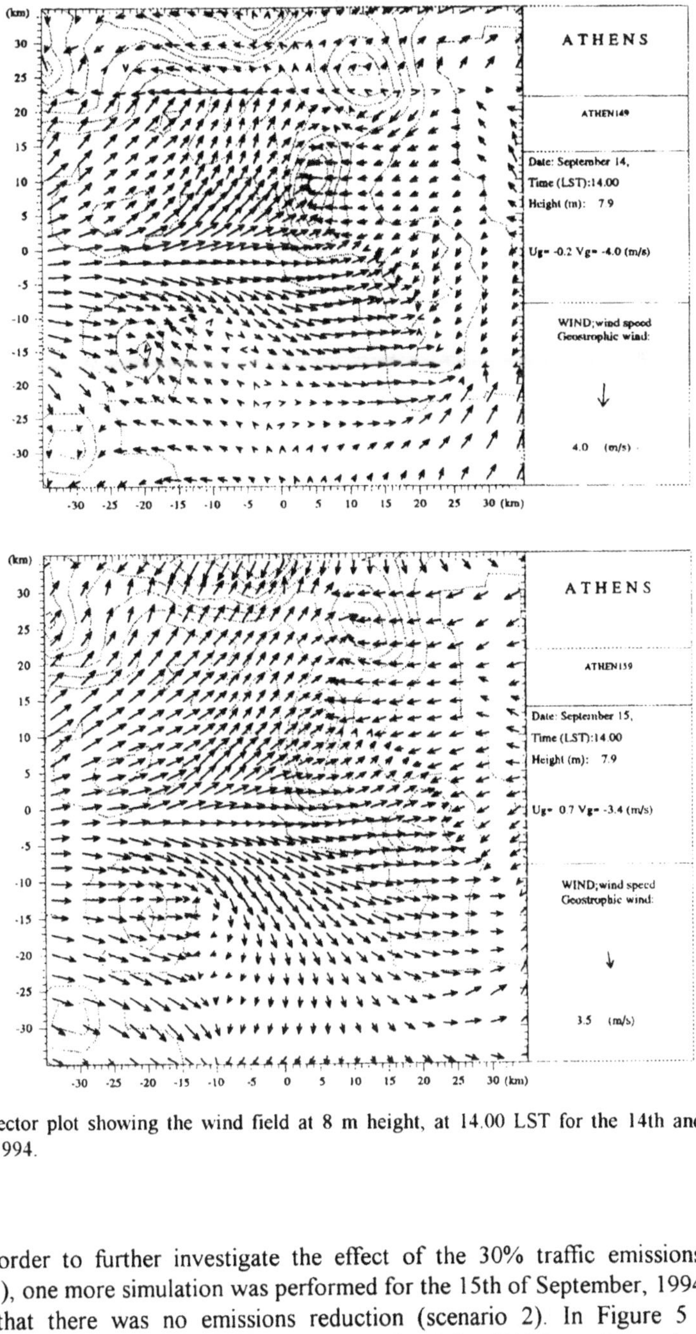

Figure 2. Vector plot showing the wind field at 8 m height, at 14.00 LST for the 14th and the 15th of September, 1994.

In order to further investigate the effect of the 30% traffic emissions reduction (scenario 1), one more simulation was performed for the 15th of September, 1994, assuming this time that there was no emissions reduction (scenario 2). In Figure 5 the spatial distribution of the differences (%) of the calculated by UAM maximum hourly ozone concentrations for both scenaria is presented. It appears that within the area where the emissions reduction was applied, the maximum ozone concentrations were lower (up to -6%), while, at the northern part of the basin, ozone increased (up to 6%). Many factors have to be considered for the causes of this phenomenon and a direct explanation cannot be given. However, since similar phenomena were observed by analysing past ozone measurements during days that similar traffic reduction measures were applied in Athens, the present theoretical finding certainly worthies further research.

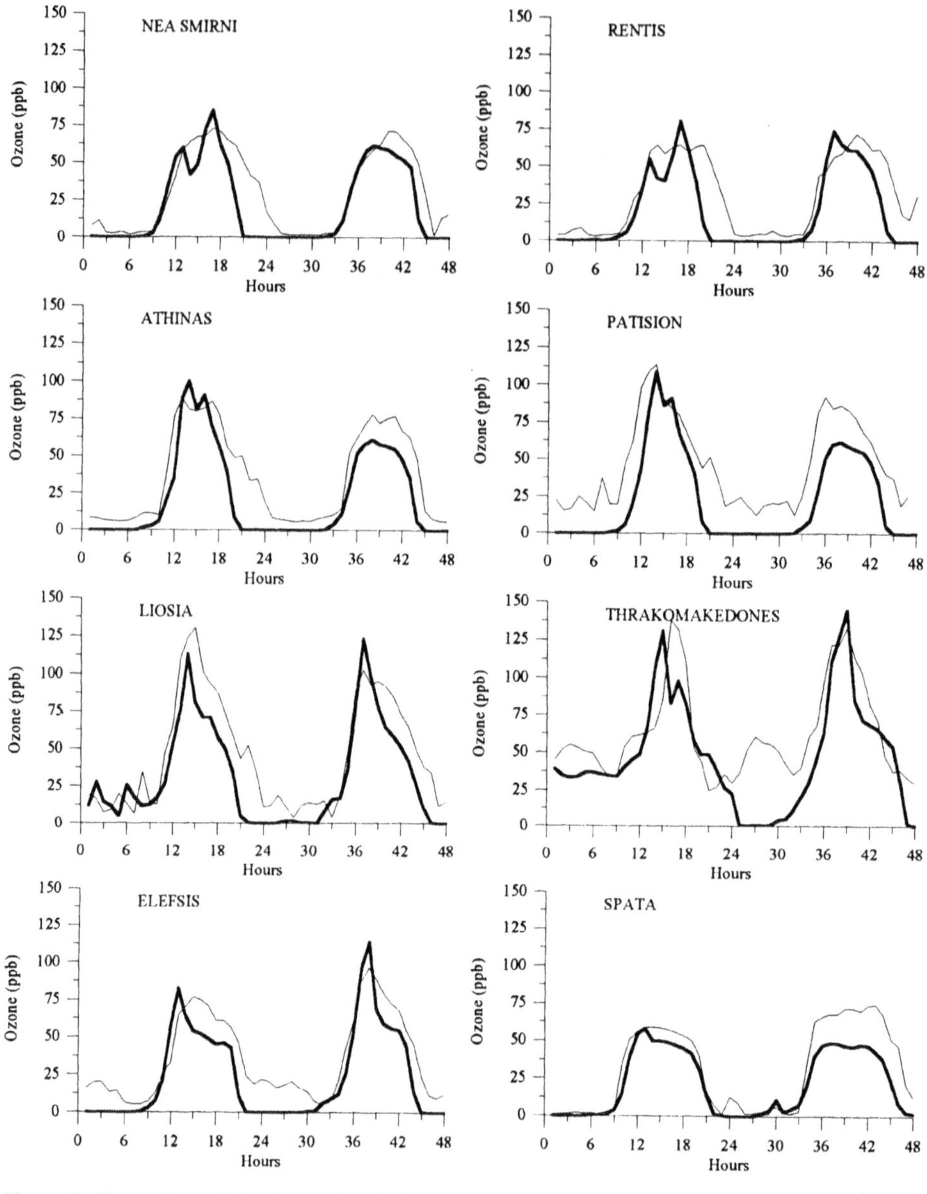

Figure 3. Comparison of the observed variation of ozone (thin line) at the various monitoring stations during the 14th and the 15th of September, 1994, with the corresponding ozone concentrations predicted by UAM (thick line).

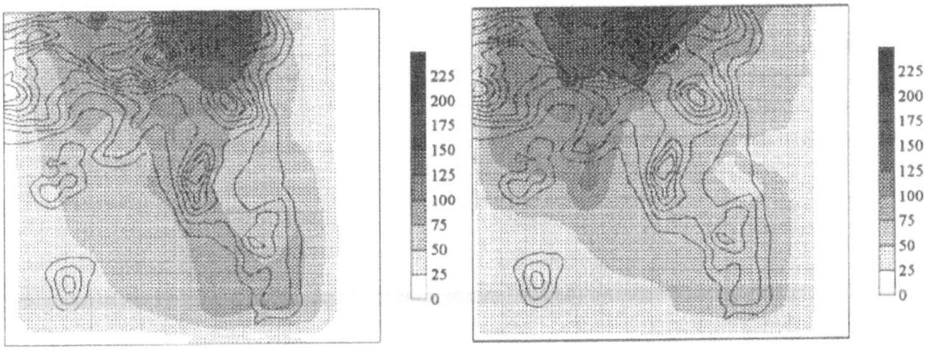

Figure 4. Spatial distribution of the maximum hourly ozone concentrations (ppb) predicted by UAM for the 14th and 15th of September, 1994.

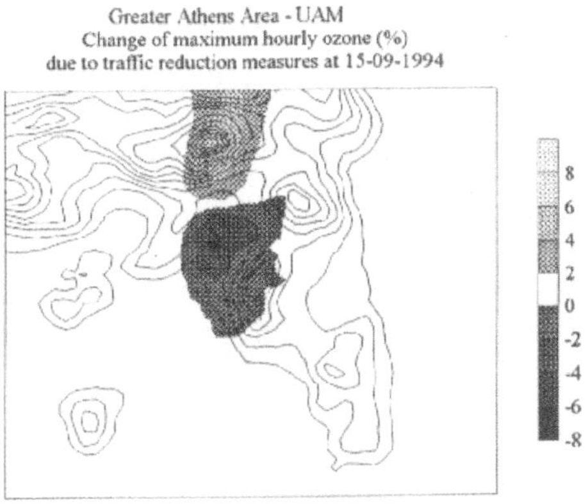

Figure 5. Spatial distribution of the differences (%) between the predicted by UAM hourly maximum ozone concentrations, for the 15th of September 1994, by using two emissions scenaria; a) traffic was as usual within the modelling domain and b) traffic emissions were reduced by a factor of 30% within the outer traffic ring of Athens.

CONCLUSIONS

The presented results of UAM, concerning the spatial and temporal distribution of ozone over the GAA during pollution episode days, are in rather good agreement with observations from the MEDCAPHOT-TRACE research project.

It appears that the UAM is capable to describe, in good agreement with observations:

- the increasing ozone gradient from the southern to the northern part of the Athens basin;
- the abscence of high ozone levels at the western and eastern areas of the Attica peninsula;

- the presence of high ozone levels (greater than the 100 ppb limit) even at the city centre.

Traffic reduction measures seem to have no effect on the maximum ozone levels observed. A theoretical investigation, by means of UAM, of the effect of traffic reduction within the outer traffic ring of Athens, show that a small decrease of ozone should appear within the traffic ring, but, at the same time, a small ozone increase should be observed at the northern part of the basin.

Acknowledgements

The authors wish to thank the staff of the Directory of Air and Noise Pollution Control, Ministry of Environment, Athens, Greece and all the participants of the MEDCAPHOT-TRACE project for providing their data. The study was supported by the European Union, DGXII, within the MEDCAPHOT-TRACE project (Contract EV5V-CT93-0312)

REFERENCES

Enger L., 1986, A higher order closure model applied to dispersion in a convective PBL. Atmos. Environ. 20, 879-894.

Gery, M.W., G.Z. Whitten, J.P. Killus and M.C. Dodge, 1989, A photochemical kinetics mechanism for urban and regional scale computer modelling. J. Geoph. Res. 94.

Scheffe, R.D. and R.E. Morris, 1993, A review of the development and application of the Urban Airshed Model. Atmos. Environ. 27B, 23-39.

Smolarkievicz, P.K., 1983, A simple positive definite advection scheme with small implicit diffusion. Mon. Weath. Rev. 111, 479-486.

Tjenstrom M., 1987, A study of flow over complex terrain using a three dimensional model. A preliminary model evaluation focusing on stratus and fog. Annales Geophysicae 5B, 469-486.

Ziomas I., P. Suppan, B. Rappengluck, D. Balis, P. Tzoumaka, D. Melas, G. Jacobi, A. Papagiannis, P. Fabian and C. Zerefos, 1995, A Contribution to the Study of Photochemical Smog in the Greater Athens Area, *Beitr. Phys. Atmosph.*, 68, 3, 191-203.

DISCUSSION

R. SAN JOSE: What percentage represents the biogenetic emissions relative to the antropogenic emissions in your emission model?

I. ZIOMAS: Biogenic emissions are not included in the inventory used.

G. KALLOS: Have you done any sensitivity test with your models, and especially with the meteorological one?

I. ZIOMAS: Several sensitivity tests, for both UAM and MIUU, have been published in the intenational scientific literature.

G. KALLOS: I agree with you on using the roof level rather than the street level as measuring point, but what is really your first model level using the UAM? Is it related to the roof level or the ground level?

I. ZIOMAS: The height of the first layer of the UAM, during the noon hours at the city center, is about 50 m. Consequently it is not possible to calculate different concentrations for the ground level and for the roof level (30 m). Also, the spatial resolution of the UAM is 4 km^2, which means that the results of the model should be compared with observations representative for the whole area of the 4 km^2.

At the station PATISION ozone measurements were available from an analyzer at the roof level and from another analyzer sampling air a few meters above ground, within a street canyon with heavy traffic load. The ozone concentration measured near the ground is much less (2-3 times) than the ozone measured at the roof level, due of course to the high NO_x emissions from the traffic within the street canyon. So the UAM results were compared with the ozone measurements at the roof level because we strongly believe, and I am very happy that you agree, that these measurements are more representative for the area.

G. KALLOS: Are you sure the aircraft was flying 100 m above ground over Athens?

I. ZIOMAS: During all 9 flights, at least one flight path of the aircraft, from SW to NE, was at an altitude ranging from 150 to 300 m above sea level.

R. BORNSTEIN: Your results in conjunction with the urban canyon results shown yesterday by Prof. Meroney show that roof top measurements are the best ones to use as model validation data. Also I believe that the N-S convergence zone through the center of downtown Athens is the cause of the high O_3 values in the area.

I. ZIOMAS: I agree with your comments.

R. YAMARTINO: Given that your model's vertical resolution is rather coarse (e.g. 50 m) how were you able to resolve the difference between street level and roof top level O_3, so that you could conclude the model is appropiate for rooftop level O_3 measurements?

I. ZIOMAS: The model results were compared only with ozone measurements performed at the roof level, for the reasons already explained to Prof. Kallos. Also, it must be noted that the average depth of the first layer of the model is 25 m, which is more close to the roof level height than to the ground.

INTERCOMPARISON OF 3-D FLOW AND PARTICLE MODELS WITH TRANSALP 1989 METEOROLOGICAL AND TRACER DATA

E. Ferrero[1], F. Desiato[2], G. Brusasca[3], D.Anfossi[4], G. Tinarelli[3], M.G. Morselli[3], S. Finardi[5], and D. Sacchetti[6]

[1]Universita', Dip. Scienze Tecn. Avanz., Alessandria, Italy
[2]A.N.P.A., Roma, Italy
[3]ENEL/CRAM - Servizio Ambiente, Milano, Italy
[4]C.N.R., Istituto di Cosmogeofisica, Torino, Italy
[5]CISE, Segrate (Milano), Italy
[6]Universita', Istituto di Fisica Generale, Torino, Italy

INTRODUCTION

In this paper two mass-consistent models coupled with dispersion models have been applied to simulate the 3-D flow and dispersion during the first TRANSALP experiment (Ambrosetti et al., 1991; T1 hereafter). The work is focused on the intercomparison between model results. TRANSALP experiments (Ambrosetti et al., submitted) were designed to study the transport of atmospheric trace constituents over the Alpine barrier from the western Po valley (Italy) to the Swiss Plateau and vice versa. They were part of TRACT, a sub-project of EUROTRAC within the European initiative EUREKA (Fiedler, 1989).

SITE AND EXPERIMENT

T1 was performed on October 19, 1989 in the Ticino river valley (Canton Ticino, Switzerland) by an international collaboration among the Joint Research Centre of Ispra, Italy, Paul Scherrer Institut of Villigen, Switzerland, Osservatorio Meteorologico of Locarno Monti, Switzerland, ENEL/CRAM of Milan, Italy and C.N.R./Istituto di Cosmogeofisica of Turin, Italy. The site (Fig. 1) is characterised by the bifurcation of the Ticino valley in the Leventina and Blenio valleys, oriented in the NW and N directions, respectively. The topography is extremely complex: the two valleys are deep and narrow and are surrounded by mountains reaching the height of 3000 m, characterised by steep slopes. A tracer release was performed at Iragna (287 m a.s.l), about 4 km upwind from the valley bifurcation, during diurnal valley breeze conditions. C_7F_{14} was released from 11 to 12 a.m. at a rate of 4 g/s and at 8 m height, while tracer concentrations within the Leventina and Blenio valleys were monitored by 40 automatic samplers until 5 p.m.. A ground-based anemometer network covered the experimental area, a sonic anemometer and one Doppler SODAR were located near the release point; two more SODARs were positioned in the Leventina valley. In addition, pilot balloon launches were performed every half an hour at the entrance of the Blenio valley (see Fig. 1).

During the experiment the synoptic situation was characterised by a high pressure system centred over Eastern Europe with moderate SW upper level winds which favoured the development of local breeze systems. The breeze drove the tracer towards the Leventina and Blenio valleys. Measured tracer concentrations indicated the occurrence of a plume bifurcation, with the larger amount of tracer (approximately in the proportion 4:1) flowing inside the Leventina valley.

The evaluation of the boundary layer thermal structure is difficult due to the lack of thermal soundings inside the valleys. From sonic anemometer measurements at Iragna, Monin-Obukhov lengths ranging from -4 to -10 m during insolation hours are derived. These values indicate unstable conditions in the surface layer. The stability can be expected to vary in space due to the effect of topography.

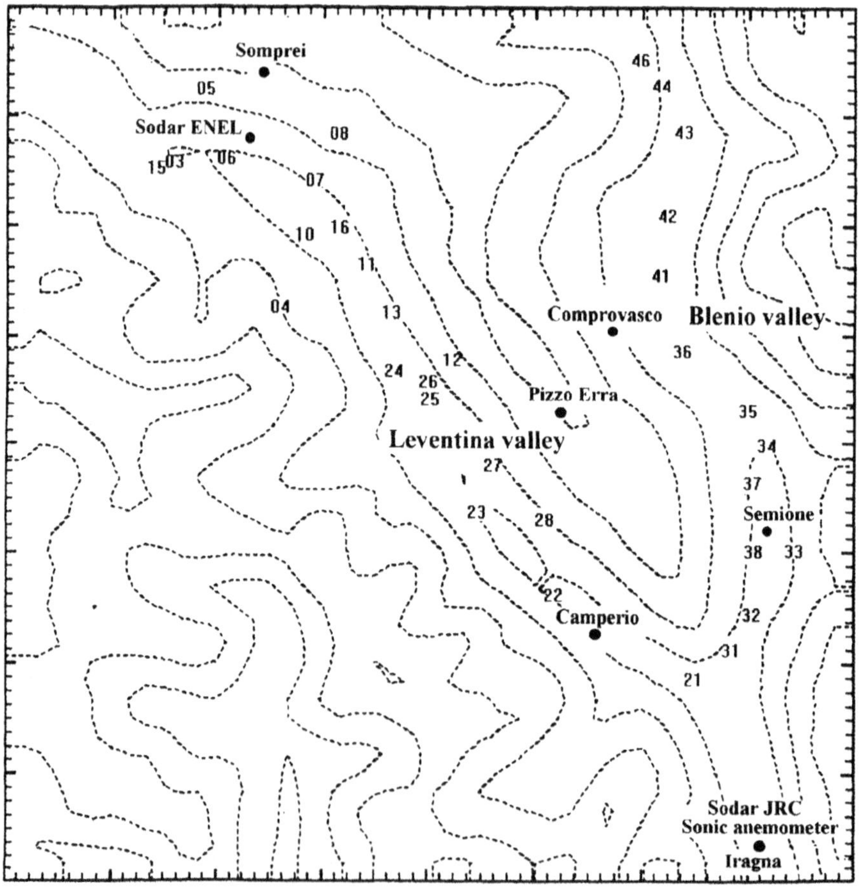

Fig 1. Map of the site of the exercise. Sites of meteorological stations referred in the text are indicated. Numbers indicate sampler locations. Contour interval is 500m.

FLOW MODELLING

The accuracy of three-dimensional wind field reconstruction is a fundamental requirement to predict pollutant dispersion over complex terrain. Diagnostic models can quickly generate wind fields satisfying the continuity equation, based on the available wind measurements. Despite of their physical limitations, mass-consistent models are useful tools due to their relative simplicity and reduced computational needs. Their accuracy is of course tied to the quality and representativeness of the measurements. Here two mass-consistent models, CONDOR (Moussiopoulos et al., 1988) and MINERVE (Geai, 1987), have been applied to simulate the three-dimensional flow during TI. The two models differ in the numerical scheme to solve the continuity equation and in the stability parameterization governing the relative amount of adjustment of the vertical and horizontal wind components.

The main inputs to CONDOR and MINERVE are the following: grid = 49 x 49 x 21, horizontal cell size = 500 m, domain top = 7500 m, wind speed and direction at 5000 m = 7 m/s and 220^0, first vertical level = 10 m. Wind measurements were grouped into half-hour intervals starting from 11.00 am, i.e. 11 wind fields were computed to cover a tracer sampling period of 5 hours.

Using mass-consistent models in complex flows requires the evaluation of wind data representativeness. Some stations, representative of the main flow, are more suitable for modelling use, while others are more influenced by local circulations, not resolvable by the model. For this reason several CONDOR and MINERVE runs were executed in order to investigate their sensitivity with respect to the number and location of wind measurements. In particular both models were run with just the sonic anemometer observations and SODAR profiles measured near the release point. In this case the horizontal variation of the wind field is the result of the mass-conservative adjustment only. The effect of topography on wind direction is evident, even if a large part of the adjustment takes place on the vertical, due to the slightly unstable conditions assumed in the boundary layer. In particular, different wind directions in the lower layer of the Leventina and Blenio valleys are obtained, which represent a fundamental characteristic of the advection field for driving part of the tracer in both valleys, as observed. Several other runs were carried out varying the number of input wind measurements. The results were evaluated by comparing the observed and predicted wind values at the stations not utilised in the model initialisation and from the comparison of the observed and computed tracer concentration patterns resulting from the dispersion models.

The introduction of two surface anemometer measurements, one at Somprei (1850 m a.s.l.) in the upper part of Leventina valley, and the other at Comprovasco (552 m a.s.l.) in the middle of Blenio valley (C and M run), substantially improved the results of the first run, and did not differ significantly from other runs with a larger number of input data. For this reason, they have been finally used to provide the advection field to the particle models.

The horizontal sections at 10 m height of wind field, simulated in C and M runs one hour after the release start, are shown in Figs. 2a and 2b.

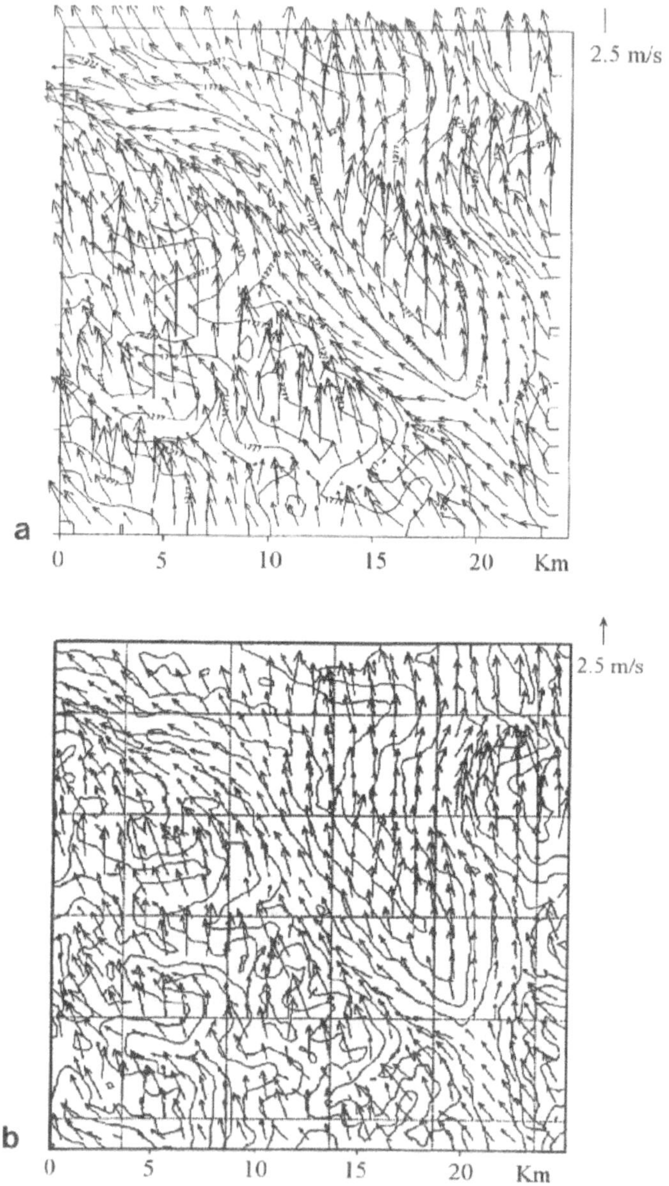

Fig. 2. Horizontal cross section of the wind field at the level of 10m above the terrain, produced by run C of CONDOR (a) and by run M of MINERVE (b).

In Tables 1 and 2 a summary of the model evaluation for all the eleven half an hour periods at three surface stations (Semione, 370 m a.s.l.; Camperio, 1420 m a.s.l. and Pizzo Erra, 2200 m a.s.l.; see Fig. 1) not used to initialize the model is presented. The poor agreement of wind direction at Semione is due to an observed rotation of about 150^{0} during the last hour, which is not observed at Iragna and Comprovasco, and therefore is not reproduced by the model. Excluding the observations of the last hour, Bias and RMSE at Semione are -18 and 21, respectively for Run C and M.

Table 1.

	d_0	d_p		σ_0	σ_p		bias		RMSE	
		C	M		C	M	C	M	C	M
Semione	230	180	164	70	24	23	-50	-66	73	84
PizzoErra	193	184	187	9	16	13	-9	-6	25	22
Camperio	194	186	160	25	31	19	8	34	25	39

Table 2.

	s_0	s_p		σ_0	σ_p		bias		RMSE	
		C	M		C	M	C	M	C	M
Semione	1.44	1.84	1.78	0.40	0.43	0.40	0.40	0.27	0.65	0.63
PizzoErra	3.48	2.78	3.05	0.51	0.42	0.69	-.70	-.43	0.91	0.68
Camperio	1.62	1.34	1.71	0.17	0.30	0.52	-.28	0.09	0.49	0.65

DISPERSION MODELLING

In order to simulate the tracer dispersion, two Lagrangian particle models, ARCO (Desiato and Lange, 1989) and SPRAY (Tinarelli et al., 1994) were applied using the 3-D flow fields produced by CONDOR and MINERVE respectively. 3-D turbulence fields were prescribed by two parameterization schemes based, as far as possible, on the data gathered during the field exercise.

ARCO (Atmospheric Release in COmplex terrain) retains the turbulence scheme used in the Monte Carlo version of ADPIC (Lange, 1990) and adopts the solution of the Langevin equation suggested by Legg and Raupach (1982). SPRAY solves three Langevin equations (for u,v and w); the values of 3-D wind fluctuations are calculated according to a numerical scheme developed by our team (Tampieri et al. 1992, Tinarelli et al. 1994), starting from Thomson (1984). In both models the mixing height h was kept constant in time over all the domain. Its value was estimated from the σ_w profiles measured by the 3-D Doppler Sodar in the Leventina Valley keeping h equal to three times the height where σ_w is maximum as observed in unstable conditions. An inspection of the time sequence of the σ_w profiles during the considered day yielded an average value of 800 m for h. ARCO and SPRAY produced a sequence of 3D concentration fields having the same chronology of the tracer sampling.

An inspection of the first sequences of particle position plots produced by ARCO and SPRAY (not shown here) shows that the simulation of the plume behaviour looks quite similar for the two models. The plume moves at an average speed of about 2 m/s towards the valley bifurcation. Most of the plume enters initially the Leventina valley, although a consistent part of the tracer starts rising over the relief dividing the two valleys. The ratio of the tracer mass entering the two valleys is correctly simulated by the models. Afterwards Leventina valley empties progressively due to the Southern wind at levels above a few hundreds metres. In fact the furthest sampling points, as the simulation time go on, show more and more average underestimation. The main reason is probably the reduced representativeness of the Sodar profile at the release point with respect to the actual profiles within the two valleys.

Fig. 3 depicts two examples of the observed concentration course recorded at two sampling stations and the corresponding values predicted by the two models. Station 22 is in the first part of Leventina valley whereas station 31 is at the entrance of Blenio valley. Fig. 3 gives an idea of the quality of the results obtained at the samplers nearest to the source. As anticipated above, the accuracy decreases with the distance from the release point. To get some quantitative insight on the simulations accuracy, a model evaluation was performed by using the statistical package MODIA (Morselli and Brusasca, 1991). The data consists of all the available (193) half an hour concentration measurements at the samplers included in the computation domain (33) and the corresponding concentrations produced by ARCO and SPRAY simulations. Mean value, standard deviation, maximum value, fractional bias (FB) and correlation coefficient have been computed. The results shown in Table 3 indicate that SPRAY captures correctly mean and maxima values while ARCO tends to overestimate. On the contrary ARCO shows a better correlation coefficient. Finally Tables 4 and 5 show the model evaluation indexes relative to the distribution of time integrated concentrations for all the samplers and of maxima at each sampler, respectively. The general results indicated in Table 3 are confirmed.

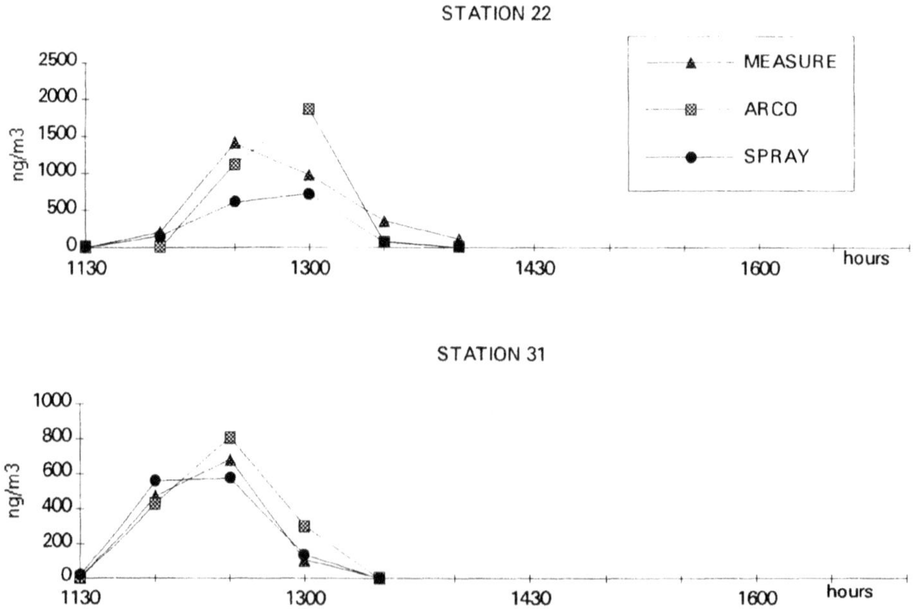

Fig. 3. Comparison between the temporal trends of ground level concentration at stations 22 and 31

Table 3.

	Mean (ng/m^3)	St.dev. (ng/m^3)	Max (ng/m^3)	Fractional bias	Correlation coefficient
Observed	113.6	216.8	1502.3		
ARCO	138.4	313.9	2221.1	-.1970	0.84
SPRAY	108.4	214.4	1553.3	0.0480	0.74

Table 4.

	Mean (ng/m^3)	St.dev. (ng/m^3)	Max (ng/m^3)	Fractional bias	Correlation coefficient
Observed	664.6	739.0	3066.5		
ARCO	809.3	918.2	4672.0	-.1970	0.87
SPRAY	633.7	679.3	3452.0	0.0480	0.67

Table 5.

	Mean (ng/m^3)	St.dev. (ng/m^3)	Max (ng/m^3)	Fractional bias	Correlation coefficient
Observed	303.7	347.5	1502.3		
ARCO	458.0	480.3	2221.1	-.4050	0.88
SPRAY	303.8	298.8	1553.3	0.0002	0.70

CONCLUSIONS

A model intercomparison exercise has been performed and presented. Two Lagrangian particle models (ARCO and SPRAY) were tested by comparing their simulation results with the data collected during TRANSALP 1989 tracer experiment held in the Alpine region. The 3-D wind field was given by two diagnostic mass-consistent models (CONDOR and MINERVE). Many runs have been performed looking for the sensitivity of the models to the number and position of input wind information. It was found that: i) - the performance of the two models is quite similar. Starting from the same input data and parameters, they converge approximately to the same 3D solution; ii) - the best results were obtained by initialising the models with the Sodar profile at the release point and two surface winds observed by anemometers located in the two valleys.

As regards the dispersion simulation, it was found that the evolution of tracer plumes inside the valleys is captured at least in its ensemble characteristics. Except during the last hour of simulation, the overall concentration field is correctly simulated by both models even if some disagreement appears for particular samplers. Taking into account both the limited available meteorological information and the extreme complexity of the site in which the tracer experiment took place the present results seem quite encouraging.

This allows to conclude that these models, considering also the computational power nowadays available on workstations, can be applicable for practical purposes.

Acknowledgements

The authors wish to acknowledge the Joint Research Centre of Ispra, Italy, Paul Scherrer Institut of Villigen, Switzerland and Osservatorio Meteorologico of Locarno Monti, Switzerland for the fruitful co-operation in organising and performing the TRANSALP 1989 campaign.

REFERENCES

Ambrosetti P., Anfossi D., Cieslik S., Gaglione P., Lamprecht R., Marzorati A., Sandroni S., Spinedi F., Stingele A. and Vogt A., The TRANSALP-89 exercise - a tracer release experiment in a subalpine valley. EUR Report 13474 EN (1991)

Ambrosetti P., Anfossi D., Cieslik S., Graziani G., Lamprecht R., Marzorati A., Nodop K., Sandroni S., Stingele A. and Zimmerman H., Mesoscale Transport of atmospheric trace gases across the central alps: the TRANSALP tracer release experiments. Submitted for publication on *Atmos. Environ.*

Anfossi D., Ferrero E., Brusasca G., Tinarelli G., Giostra U., Tampieri F., and Trombetti F., Dispersion simulation of a wind tunnel experiment with Lagrangian particle models. *Nuovo Cimento*, 15 C, 139-158 (1992).

Anfossi D., Desiato F., Tinarelli G., Brusasca G., Ferrero E., Sacchetti D., TRANSALP 1989 experimental campaign - Part II: Simulation of a tracer experiment with Lagrangian particle models. Submitted for publication on *Atmos. Environ.*

Desiato F., Finardi S., Brusasca G., Morselli M.G. TRANSALP 1989 experimental campaign - Part I: Simulation of 3-D flow with diagnostic wind fiel models. Submitted for publication on *Atmos. Environ.*

Desiato F.and Lange R., An intercomparison of atmospheric turbulence parameters and their applications to a tracer experiment using a Monte Carlo particle model. Air Pollution Model and its Applications IX, H. Van Dop and G. Kallos, ed., Plenum press (1992).

Fiedler F., Proposal of a Subproject "Transport of air pollutants over complex terrain - TRACT", EUREKA Environmental Project, Karlsruhe (1989).

Geai P., Methode d'interpolation and reconstitution tridimensionelle d'un champ de vent: le code d'analyse objective MINERVE. E.d.F. Chatou (France), Report DER/HE/34-87.03 (1987).

Lange R., A comparison of the Monte Carlo and the flux gradient method for atmospheric diffusion. Proceedings of the 18th NATO/CCMS Int. Technical Meeting on Air Pollution Modelling and its Applications. Vancouver, Canada (1990).

Legg B.J. and Raupach M.R., Markov chain simulation of particle dispersion in inhomogeneous flow: the mean drift velocity induced by a gradient in Eulerian velocity variance.*Bound. Lay. Met.*,24,3-13 (1982).

Morselli M.G. and Brusasca G., MODIA: Pollution dispersion model in the atmosphere. In: *Environmental Software Guide*, 211-216, Patron, Bologna (1991).

Moussiopoulos N., Flassak T. and Knittel G., A refined diagnostic wind field model. *Environmental Sortware*, 3, 85-94 (1988).

Tampieri F., Scarani C., Giostra U., Brusasca G., Tinarelli G., Anfossi D., and Ferrero E., On the application of random flight dispersion models in inhomogeneous turbulent flows.*Annales Geophysicae*, 10, 749-758 (1992).

Thomson D.J., Random walk modelling of diffusion in inhomogeneous turbulence. *Quart. J. Roy. Meteor. Soc.*, 110, 1107-1120 (1984).

Tinarelli G., Anfossi D., Brusasca G., Ferrero E., Giostra U., Morselli M.G., Moussafir J. Tampieri F., Trombetti F., Lagrangian particle simulation of tracer dispersion in the lee of a schematic two-dimensional hill, *Journal of Applied Meteorology*, 33, 744-756 (1994).

J. WALMSLEY: Were you able to determine whether differences in results from the two model systems were due to the meteorological or the dispersion components?

G. BRUSASCA: It is difficult to determine the contributions due to the two components but the weak points of the simulations can be evidentiated:

- for the meteorological component the problem is connected to the interpolation phase, in particular for the grid points far from measurements;
- for the dispersion component the most critical parameter seems to be the PBL height, in particular because we fix it constant in the domain.

R. BORNSTEIN: Do you think that your upper level winds show evidence of gravity wave activity?

G. BRUSASCA: I think that it is not possible to detect the presence of gravity waves using a mass-consistent model with so few wind measurements and a so complex topography.

QUANTITATIVE EVALUATION OF A MESOSCALE NUMERICAL MODEL SIMULATION USING FOUR-DIMENSIONAL DATA ASSIMILATION OF COMPLEX AIRFLOW OVER THE KANTO REGION IN JAPAN

Itsushi Uno

National Institute for Environmental Studies
Onogawa 16-2, Tsukuba, Ibaraki 305, Japan

INTRODUCTION

Summer time typical mesoscale circulation over the Tokyo metropolitan area (TMA) plays a significant role in the transport of photochemical air pollutants (e.g., Wakamatsu *et al.* 1990). The TMA is located in the Kanto Plain which extends about 100 km in the E-W and 200 km in the N-S directions. Topographically the Kanto region contains mountains (1000 - 2000 m) on the W and N sides, and Tokyo and Sagami bays and the Pacific Ocean at the center, S and E sides of the plain, respectively (see Figures 1 and 2).

Mesoscale numerical models have been applied to the Kanto region to better understanding land/sea and mountain/valley wind circulation mechanisms (e.g., Kimura and Arakawa, 1983; Kondo 1990). Such mesoscale models can be significant tools for providing accurate wind field inputs to photochemical gird models since these photochemcal grid models are very sensitive to the accuracy of simulated wind fields (e.g., Lyons *et al.*, 1995). Therefore, it is necessary to quantitatively and qualitatively evaluate mesoscale numerical model performance based on intensive 3-D meteorological observation data in the Kanto region before their application to photochemical transport models in the region.

There are two objectives of this paper. First, to qualitatively and quantitatively validate mesoscale model results with and without using four-dimensional data assimilation (FDDA) over the complex terrain of the Kanto region. And second, to examine, using a statistical method, the sensitivity of observation data density used in the FDDA. Such sensitivity analysis will help determine the reliablity of simulation results obtanined from a limited number of observation points.

MESOSCALE MODEL AND FOUR-DIMENSIONAL DATA ASSIMILATION

The Colorado State University Mesoscale Model (CSU-MM; Pielke, 1974; Mahrer and Pielke, 1978) was used for the mesoscale numerical model in this study. The model is hydrostatic and consists of the equations of motion, moisture and continuity within a three-dimensional terrain-following coordinate system. It includes a thermodynamic equation, a diagnostic equation for pressure and a surface heat budget. The modified CSU-MM version of Urlickson and Mass(1990) and Kessler and Douglas (1992) was applied in this study.

Recently the four dimensional data assimilation (FDDA) technique has been applied to include synoptic pressure system changes and to enhance model performance within the numerical domain. The FDDA option developed by Kessler and Douglas (1992) was used in this study. The general form of the prognostic equation for a variable ϕ is,

$$\frac{\partial \phi}{\partial t} = F(\phi, x, y, z, t) + G_\phi \times w_t \times w_{xyz} \times (\phi_{obs} - \phi). \tag{1}$$

The first term on the right-hand side of equation (1), F, represents all of the model's physical processes. The second term is the nudging term. G_ϕ determines the relative weight of the nudging term with respect to the model's physical processes. The variables w_t and w_{xyz} are temporal and spatial weighting factors, respectively, and the quantity ϕ_{obs} represents the "observed" value.

The "observed" value ϕ_{obs} for horizontal wind components (i.e., $\phi_{u,\,obs}$ and $\phi_{v,\,obs}$) were computed using a target wind scheme by Yamada and Bunker (1989) and Yamada and Henmi (1994). These target wind components were determined from observed wind components (u_{obs} and v_{obs}) and geostrophic winds (U_g and V_g):

$$\phi_{u,obs} = u_{obs} - \frac{f}{G_u}(v_{obs} - V_g) \tag{2}$$

$$\phi_{v,obs} = v_{obs} + \frac{f}{G_u}(u_{obs} - U_g). \tag{3}$$

The nudging coefficient G_u was set at $G_u/f = 5$ following Yamada and Bunker (1989), where f is the Coliolis factor. No nudging was applied to temperature and water vapor. Observation data for wind (surface and upper air soundings) were interpolated into the mesocale model grid system, and these interpolated values were used in the FDDA (analysis nudging).

OBSERVATION DATA AND EXPERIMENTAL DESIGN

Intensive field observation data covering the Tokyo metropolitan area were used for the quantitative evaluation of the numerical model. Three-dimensional observations were conducted from July 15 to July 17, 1981 mainly covering the south Kanto Plain. Because a summary of these observations has already been published by Wakamatsu *et al.* (1990), only the meteorological observation network will be briefly described here.

Figures 1 and 2 show the meteorological observation stations (Figure 1 indicates the model domain and elevation contours). Vertical wind profiles were observed every hour at 23 points using a pilot-balloon (P in Figure 1 and P_1 - P_{23} in Figure 2). Data were collected at each 100 m interval above ground level up to 3000 m. Also at four points (P_{06}, P_{10}, P_{11} and P_{21} in Figure 2) temperature and humidity profiles were measured by radiosonde at about 3-h intervals. The objective analysis scheme proposed by Sherman (1978) was applied to the pilot-balloon data to create a three-dimensional interpolated wind dataset to be used in the FDDA.

Besides the special observation data described above, routine observation data of Japan Meteorological Agency (JMA) were also used. This consisted of five upper air soundings (from aerological stations denoted by "A" in Figure 1) and data from Automated Meteorological Data Acquisition System (AMeDAS) stations (denoted by a solid circle in

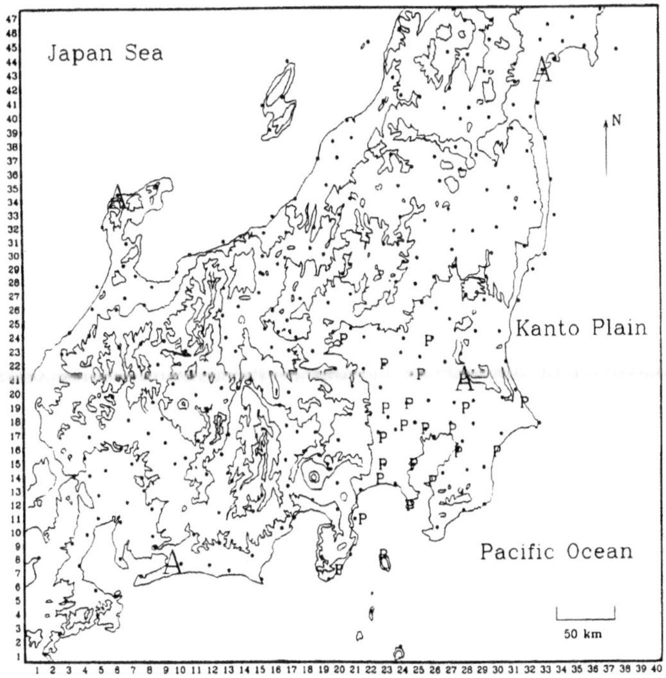

Figure 1. Map of model domain. Elevation contours are 200, 500, 1000, 2000m. Symbols shown in figure are upper air sounding stations (A), pilot-balloon observation sites (P) and AMeDAS stations (•). Axes represent numerical girds location.

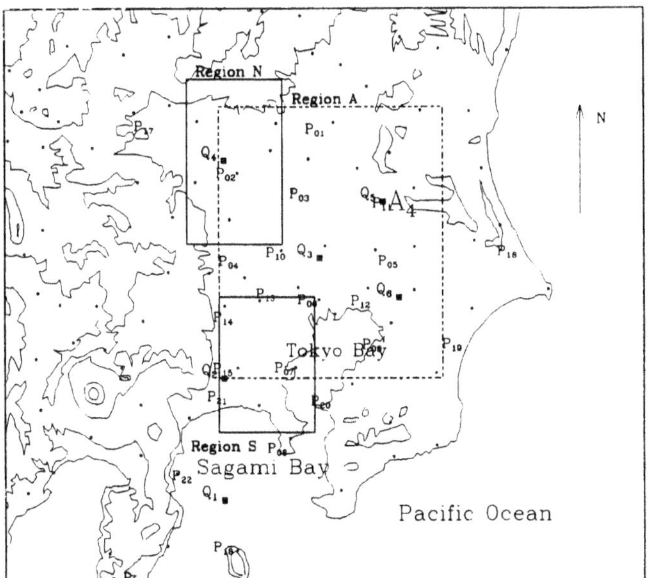

Figure 2. Detailed map of Kanto region. Symbol (•) denotes AMeDAS stations, P_1-P_{23} pilot-balloon observation sites, and Q_1-Q_6 (■) the location of vertical winds (interpolated from pilot-balloon observations) used in the FDDA. Regions S, N and A are the areas used for model evaluation.

Figures 1 and 2). Aerological data provided the wind speed, direction, temperature and humidity at specified pressure heights (1000, 900, 850, 700, 600, 500, .., 10 hPa) at 12-h intervals (6-h intervals for winds) and surface level wind and temperature were obtained from AMeDAS every hour.

Numerical simulation was started on July 16, 0300 JST and terminated on July 17, 2200 JST. The model domain was intialized by a single point upper air sounding at July 16, 0300 JST (WD was set to SW based on synoptic scale pressure analysis). Dynamic initialization (DI) was applied for three hours. Figure 1 shows the model domain. The simulation was performed on a 40 × 47 × 23 grid structure with a horizontal grid spacing of Δx=13.6 km and Δy=11.1 km. In the vertical direction, the atmosphere was divided into 23 layers (up to 8000 m) with different thicknesses from a minimum of 5 m near the surface to a maximum of 1000 m at the top of the domain. An absorbing layer was included above 4000 m. The time step for numerical integration was chosen as Δt=60 sec.

Sensitivity analysis of observation data density used in the FDDA is an important issue in order to obtain reliable results. A series of numerical experiments were designed. Table 1 shows the numerical experiment details. Experiment "000" is a pure prognostic simulation (without FDDA), "500" is FDDA simulation using only aerological data points, "5A0" is FDDA using both aerological and AMeDAS data, "5AN" is FDDA using aerological, AMeDAS and upper air soundings from Q_1 to Q_N (for location see Figure 2) interpolated from pilot-balloon observation. The FDDA time interval was 1-h except in experiments IT3 and IT6. The model performance evaluation measures proposed by Willmott (1982) and Willmotte *et al.* (1985) were used.

RESULTS AND DISCUSSION

Figure 3 shows the surface level (z*=5 m) wind fields for Experiment 5A4 from July 16, 1400 JST to July 17, 0200 JST for every 6 hour interval. The figure only shows the wind field around the Kanto district. Bold arrows in the figure show AMeDAS observation data. Observed winds indicated that the northern part of the Kanto Plain was covered by an E wind from the Pacific Ocean until late afternoon. Sea breeze penetration from Sagami Bay (south of the Kanto Plain) was blocked by this E wind zone. The pure prognostic experiment "000" showed deep penetration of the sea breeze into the northern part of the Kanto Plain. The east wind component (from surface to 2000 m) was observed at aerological station A_4 (Tateno). Inclusion of this local east wind by FDDA (Experiments 500 - 5A6) successfully simulated the blocking of the sea breeze until late afternoon. Simulated winds at night time (Figure 3c) predicted the very calm wind over the Kanto Plain. The general flow fields agreed with observations, however, the wind speed and direction are slightly different from observation.

Wind and Temperature Field Validation

Wind field analysis showed different patterns between the southern and northern areas of the Kanto Plain (see Figure 3). Two regions were selected to compare model results with observations. Region S includes the south Kanto area, while Region N includes the north Kanto area (see Figure 2)

Figure 4 (1) and (2)shows a statistical comparison of WS, WD, σ_{WS}, RMSD (root-mean squared difference) and d (agreement index by Willmotte(1982)) for Region S at z*=5 m and 200 m, respectively. Numerical results (Experiments 000, 500, 5A2 and 5A4) were compared with AMeDAS station data (five stations within Region S).

Sea breeze penetration started from 1000 JST on July 16 and continued until 2200 JST. The pure prognostic experiment "000" was clearly unsuccessful in predicting the sea breeze penetration time and overestimated the wind speed (wind speed at z*=200 m showed

Table 1 Simulation Design

Experiment[a)]	Contents	Aerological	AMeDAS	Pilot-Balloon[b)]
000	No FDDA	0	0	0
500	aerological only	5	0	0
5A0	aerological + AMeDAS	5	All	0
5A2	5A0 + 2 pibal stations	5	All	2
5A3	5A0 + 3 pibal stations	5	All	3
5A4	5A0 + 4 pibal stations	5	All	4
5A6	5A0 + 6 pibal stations	5	All	6
IT3	=5A4 (but 3 hour interval pibal observation)	5	All	4
IT6	=5A4 (but 6 hour interval pibal observation)	5	All	4

a) XYZ : X = number of aerological stations, Y=AMeDAS surface station flag (Y=0 means no AMeDAS station, =A means use of all AMeDAS stations), Z= number of pilot balloon sounding stations.

b) Pilot-balloon observation stations used in each experiment are from Q_1 to Q_N

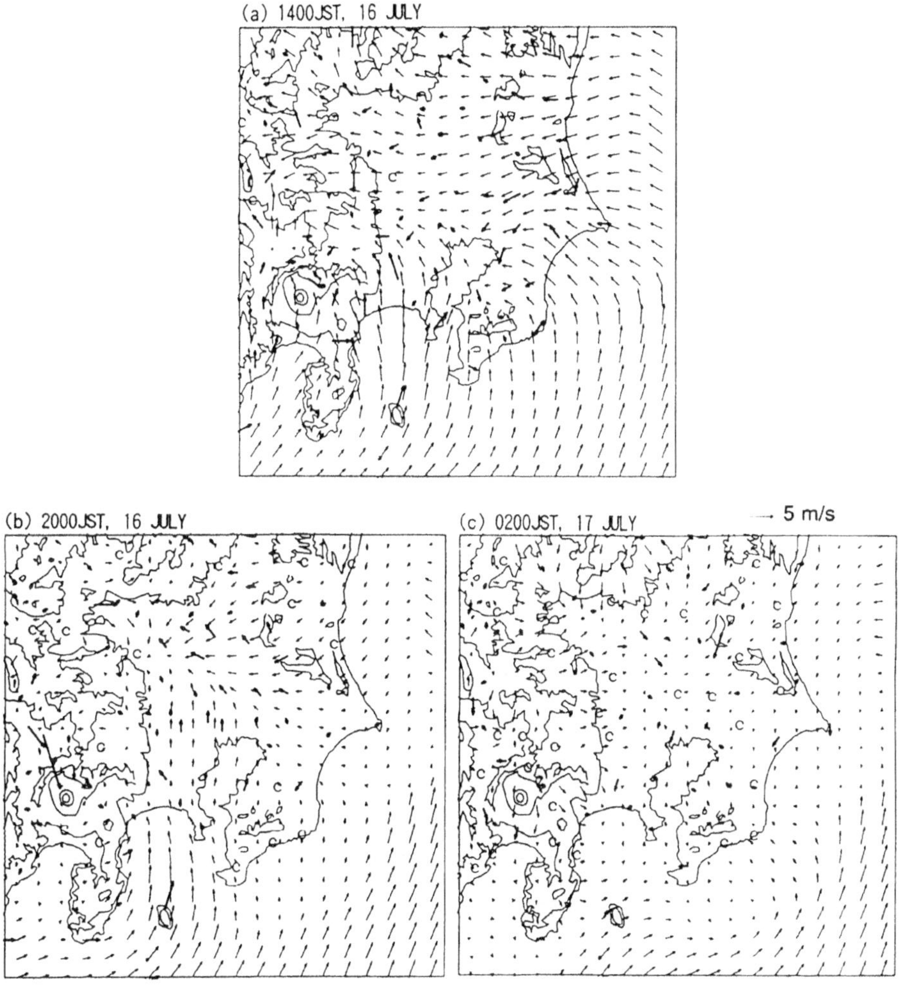

Figure 3 Results of Experiment 5A4 at z*=5 m level for (a) 1400 JST, (b) 2000 JST 16 July 1981 and (c) 0200 JST, 17 July 1981. Bold wind vectors indicates AMeDAS observed wind.

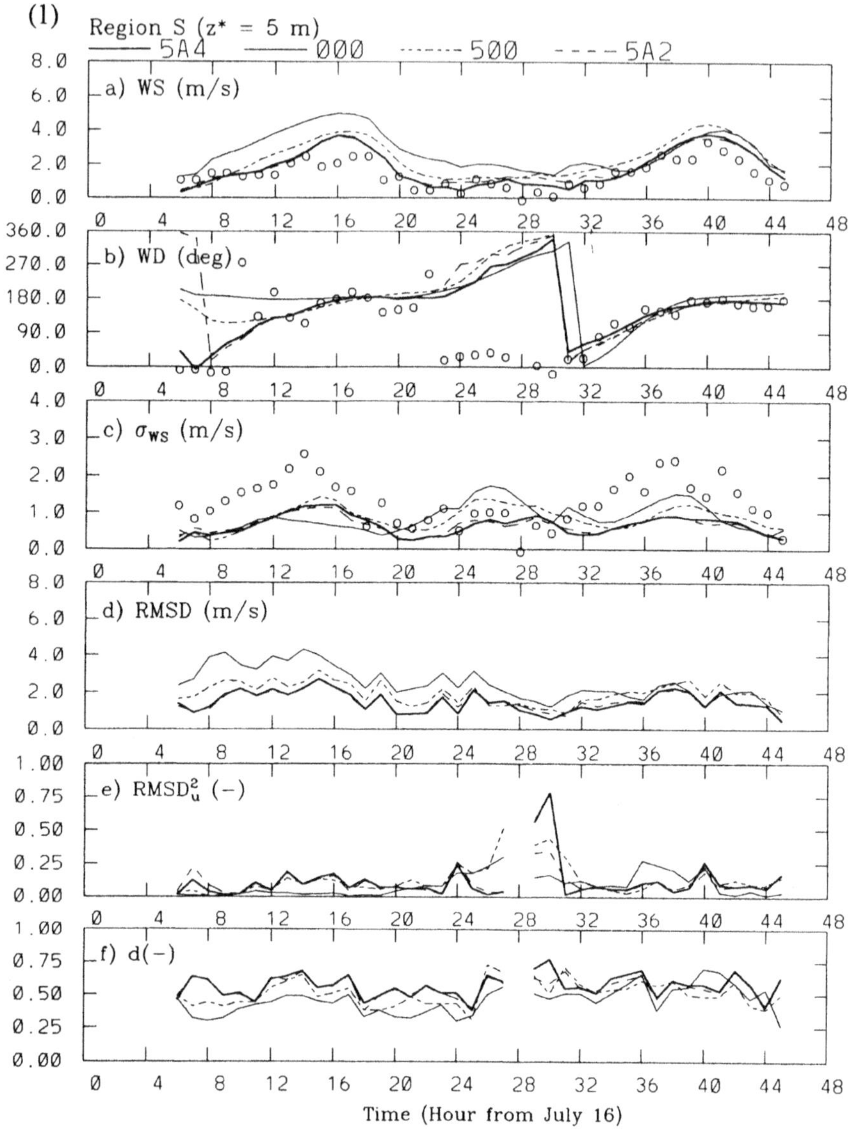

Figure 4 The evolution of modeled and observed (a) averaged WS, (b) averaged WD, (c) standard deviation of WS, (d) root mean square deviation (RMSD), (e) $RMSD_u^2$ $/RMSD^2$, the normalized unsystematic root mean square deviation ($RMSD_u$), and (f) index of agreement for Region S from 16 July to 17 July 1981. Open circles denote the averaged data from AMeDAS stations. The model results from RUN 000, 500, 5A2 and 5A4 are compared at z *=5 m (1), and z*=200 m (2).

(2)

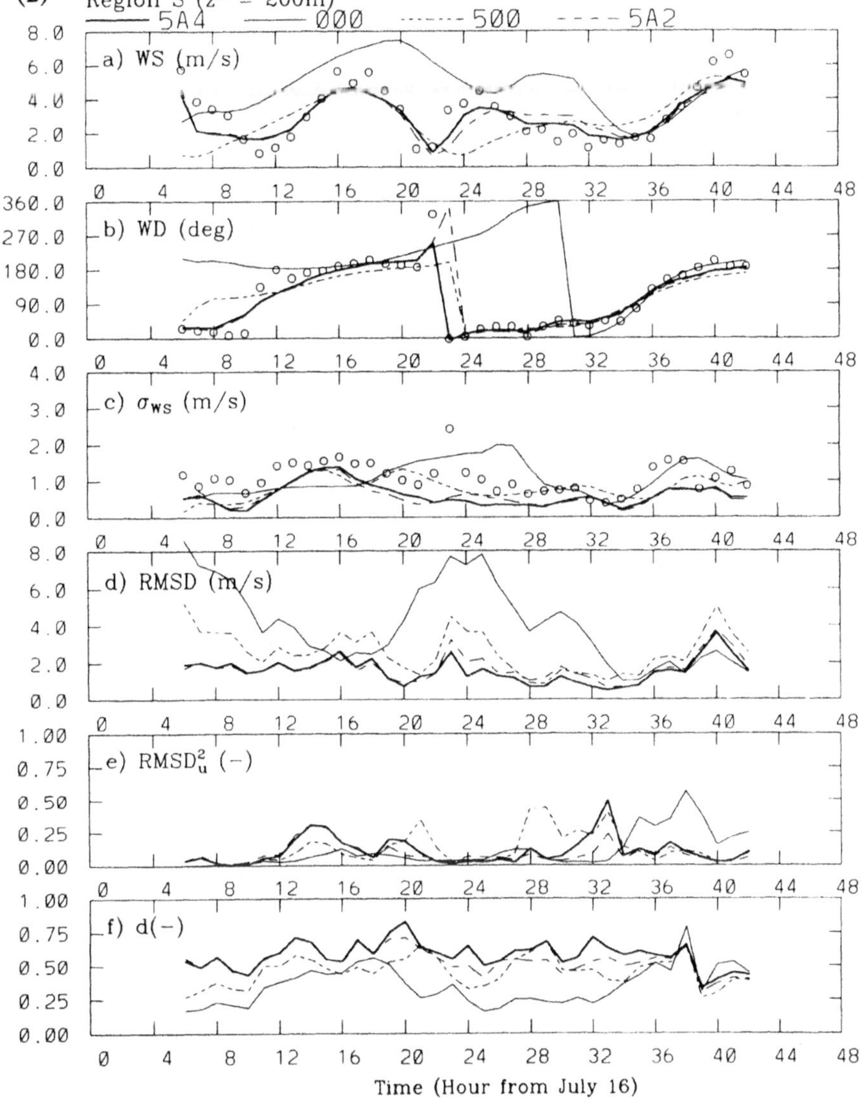

Figure 4. (cont.)

strong overprediction). FDDA experiments (500 - 5A4) showed similar results. WS and WD prediction by FDDA experiments during the daytime showed good agreement and no systematic differences were found between the experiments. Night time simulation results showed poor agreement for WD at z*=5 m.

Figure 5 shows the results of a statistical comparison of temperature at z*=5 m for Region S. The figure only shows T, σ_θ and d. The diurnal change of simulated temperature showed good agreement with observations. However, σ_θ always gave smaller values than those observed. This is due to the fact that the model does not take into account the spatial variability of surface properties. The agreement index d had a significantly smaller value in the morning hours. This indicates that the model does not correctly reproduce the observed spatial variability .

Time-Height Cross-Section Comparison

The time-height cross-section of model simulated and radiosonde observed potential temperature and wind fields were compared. Figure 6 shows a comparison near the center of Tokyo (Otemachi, Tokyo denoted as P_{06} in Figure 2). The figure shows the observations (a) and Experiments 5A4 (b), 500 (c) and 000 (d). Because the model was initialized by a single point sounding, the absolute value of the temperature comparison is not of interest (the model has about a 3K offset in temperature).

Figure 6 clearly shows the alternation of land/sea breeze. Observation indicates that the prevailing wind direction above the 1000 m level is N - NE (which is consistent with aerological station A_4). The thickness of the sea breeze layer was about 900 m on July 16 and about 1200 m on July 17. The daytime mixed layer height was approximately 1400 - 1500 m on both days. No surface inversion layer was observed and the potential temperature profile was isothermal below 350 m level in night time.

Experiment "000" simulated a strong surface inversion. The transition time from sea to land breeze was different from observation. The potential temperature variation in Experiments 500 - 5A4 was almost the same. A small difference was found in WS and WD. The results of Experiment 5A4 indicated that the simulated night time stratification was significantly different from observation even with the FDDA option. This may be due to the lack of different surface layer properties and the anthropogenic heat release in the Tokyo area both of which were ignored in the present model simulation.

The simulated mixed layer depth (in the FDDA experiments) was 1200 m for both days. This was in good agreement with observation and can be explained that the fact that the FDDA experiment predicted the sea breeze penetration from Sagami Bay correctly, which has a strong effect on the development of the mixed layer height. This good agreement between simulted and observed mixed layer height by CSU-MM was also reported by Steyn and MeKendry (1988) and Ulrickson and Mass (1990).

Horizontal Wind Fields

Sensitivity study of the number of observation points within FDDA is important. Simulated results were statistically evaluated over Region A which includes the Kanto Plain (see Figure 2). The horizontal wind at z*=200 m was examined. Figure 7 shows the results for WS, RMSD and d. Experiment 000 was unsuccessful. Experiment 500 predicted the sea breeze penentraion but the transition from sea breeze to land breeze was not correct until Experiment 5A3. Systematic improvement is obviously proportional to the number of observation data points used in FDDA. Experiment 5A3 included three vertical observation points in FDDA (Q_2 located on the south-west edge of Region A and Q_3 located at the center of Region A). These results indicate that night time simulation (stable condition) can not be completely accurate without at least one vertical observation point within the region of interest if the topography is complex.

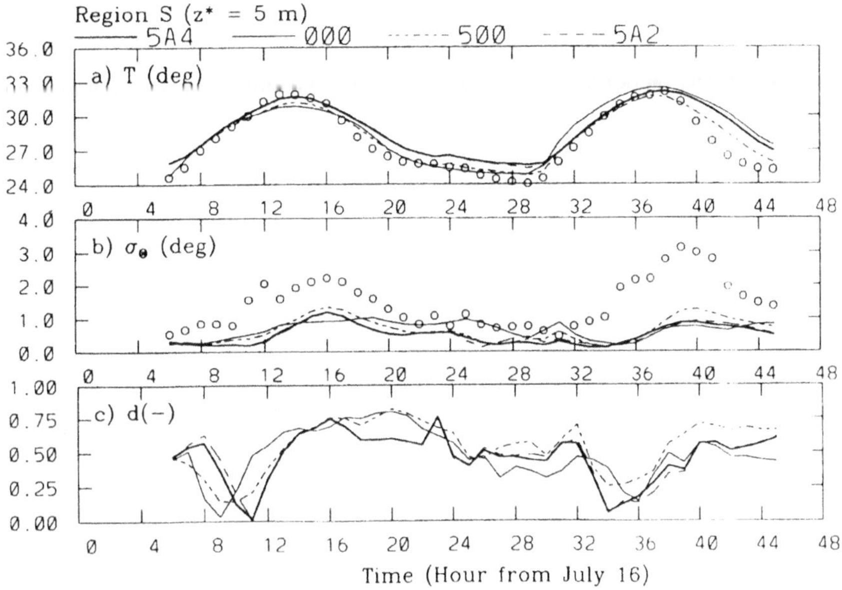

Figure 5 As in Figure 4 but for temperature.

Sensitivity of the Time Interval of Observation

Another important problem is the observation interval used in the FDDA. The statistical analyses shown in Figures 4 - 7 clearly indicate that the mesoscale model is quite accurate in simulating sea breeze penetration and the development of the mixed layer if the model was nudged toward aerological upper air sounding data correctly.

Sensitivity experiments of time interval IT3 and IT6 for Region A was also conducted. As shown in Table 1, IT3 and IT6 retrieve the same observation points as 5A4, however, IT3 (6) used three (six) -hour interval observation data points. It was shown that there are only small differences in the daytime results, however, the agreement index d detects a difference in the night time simulations among 5A4, IT3 and IT6. However, the difference between 5A3 and IT3 is very small and thus IT3 is quite successful. This result indicates that the time scale of a mesoscale phenomena of interest plays an important role in determining the observation interval.

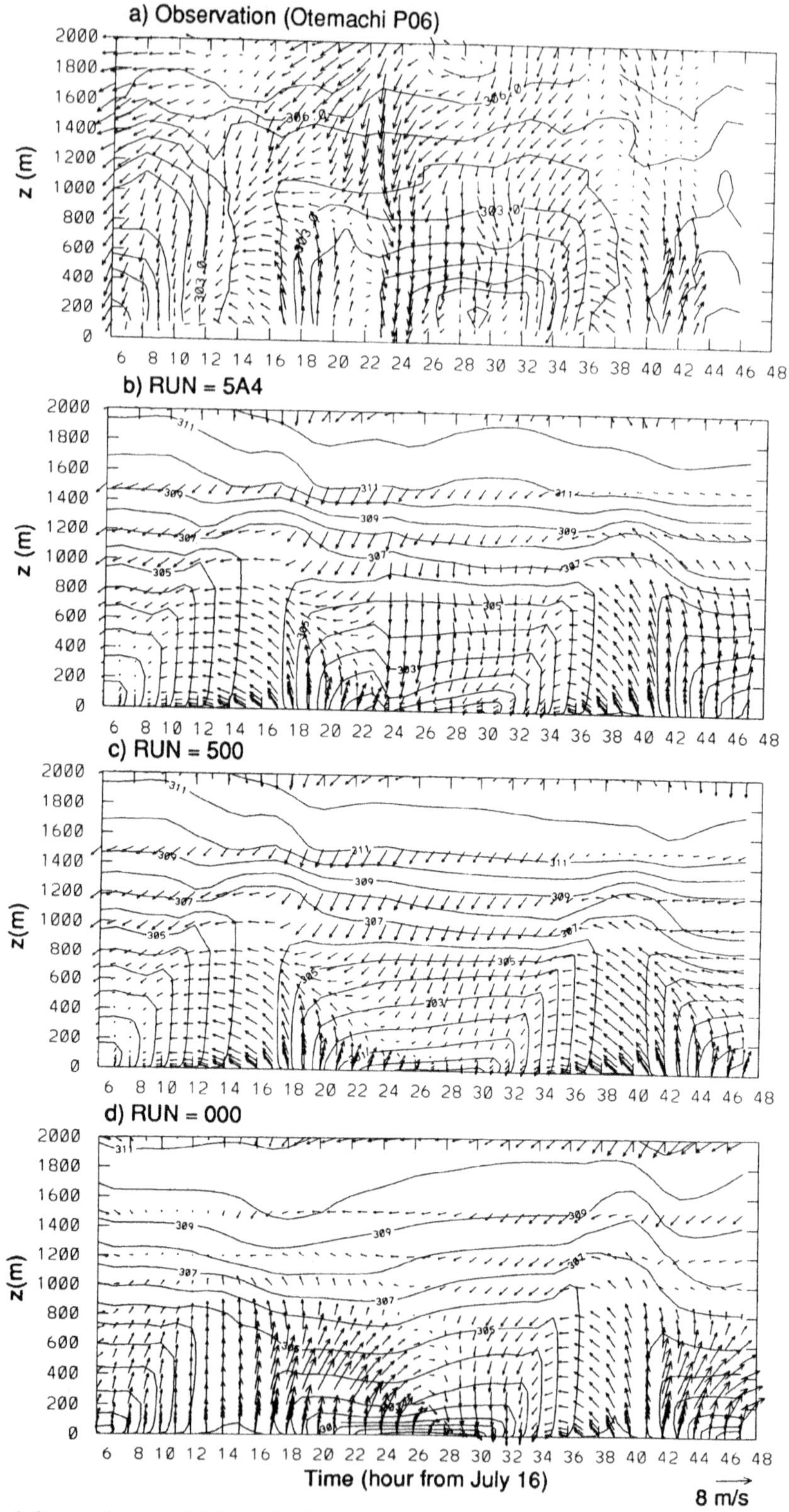

Figure 6. Observed and modeled time-height cross-sections of wind vector and potential temperature at Otemachi site (P06) from 0600 JST, 16 July to 2400 JST, 17 July 1981. Arrow indicates the horizontal plane wind vector.

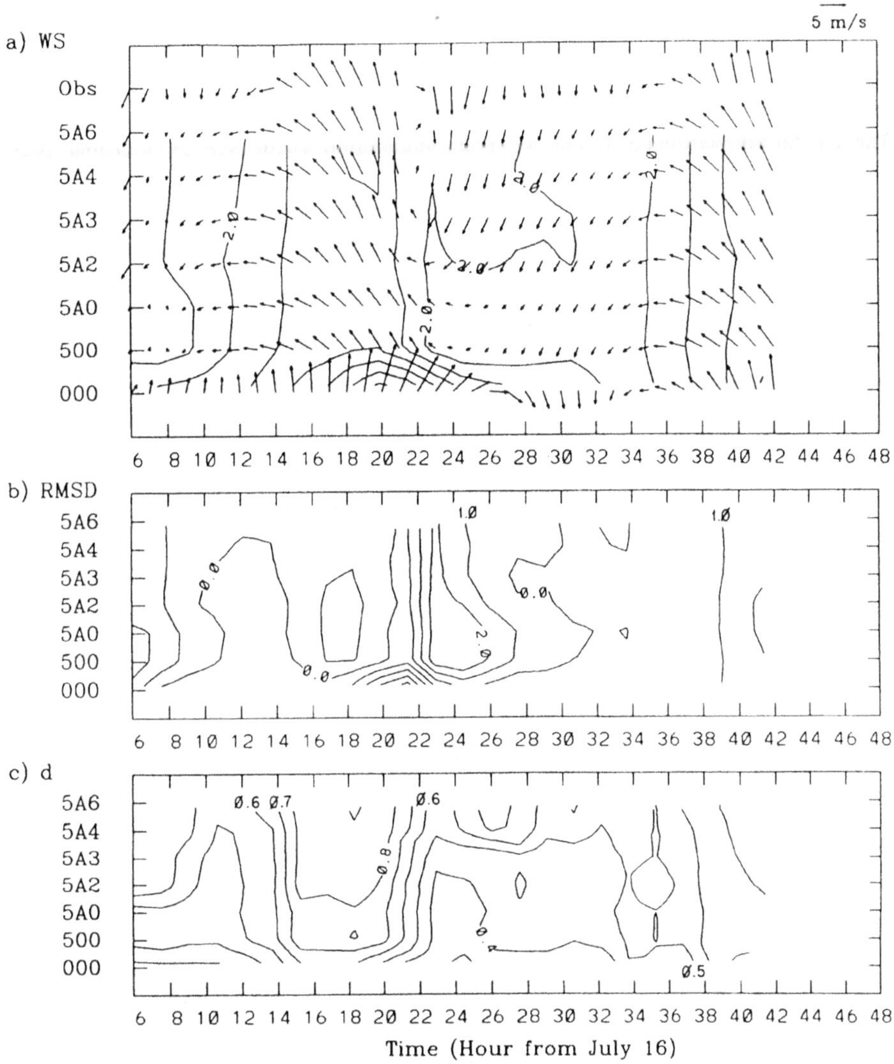

Figure 7. The evolution of a modeled and observed (a) averaged WS, (b) RMSD and (c) index of agreement for Region A from 16 July to 17 July 1981. The model results from RUN 000, 500, 5A0, 5A2, 5A3, 5A4, and 5A6 at z*=200 m are plotted. The observed(Obs) WS in (a) is averaged wind vector from balloon-pilot observation.

CONCLUDING REMARKS

A mesoscale numerical model simulation using four-dimensional data assimilation (FDDA) was conducted to quantitatively evaluate complex airflow over the Kanto region in Japan. The Colorado State University Mesoscale Model (CSU-MM) was used in this study. Model simulation using only synoptic scale geostrophic winds (without FDDA) was unsuccessful in predicting the observed wind fields. However, model simulation with FDDA using aerological data dramatically improved wind field prediction. Simulated wind speed and wind direction agreed well with observation, especially in daytime. In night time, the FDDA run using aerological and AMeDAS data simulated well typical sea breeze and land breeze alternation, however, it could not simulate observed dramatic wind changes. The standard deviations of model wind and temperature fields were smaller than those of observed. Examination of the spatial density and time interval of observation showed that model and observation agreement increased in proportional to the number of observation points, and that the effect of using FDDA was most significant in improving night time simulation.

Acknowledgements The author wishes to express his sincere thanks to Mr. J. M. Godowitch of Atmospheric Research and Exposure Assessment Lab. (U.S. EPA) and Prof. D.G. Steyn of the University of British Columbia for their help during this study.

REFERENCES

Kessler,R.C. and Douglas,S.G., 1992, *User's guide to the System Applications International Mesoscale Model (Ver. 2.0)*, System Application Int., SYSAPP-92-085.

Kimura,F. and Arakawa, S., 1983, A numerical expreiment on the nocturnal low level jet over the Kanto Plain, *J. Meteorol. Soc. Japan*, **61**, 848-861.

Kondo,H., 1990, A numerical experiment of the "Extended Sea Breeze" over the Kanto Plain, *J. Meteorol. Soc. Japan*, **68**, 419-434.

Lyons, W.A., Tremback, C.J. and Pielke, R.A., 1995, Applications of the regional atmospheric modeling system (RAMS) to provide input to photochemical grid models for the Lake Michigan ozone study (LMOS), *J. Appl. Meteorol.*, **34**, 1762-1786.

Mahrer,Y. and Pielke, R.A., 1978, A test of an upstream spline interpretation technique for the advective terms in a numerical mesoscale model, *Mon. Wea. Rev.*, **106**, 818-830.

Pielke, R.A. , 1974, A three dimensional numerical model of the sea breezes over South Florida, *Mon. Wea. Rev.*, **102**, 115-134.

Ulrickson, B.L. and Mass, C.F., 1990, Numerical investigation of mesoscale circulations over the Los Angles basin Part I A verfication study. *Mon. Wea. Rev.*, **118**, 2138-2161.

Sherman, C.A., 1978, A mass consistent model for wind fields over complex terrain. *J. Appl. Meteorol.*, **17**, 312-319.

Steyn,D.G. and McKendry, I.G., 1988, Quantitave and qualitaative evaluation of a three-dimensional mesoscale numerical model simulation of a sea breeze in complex terrain. *Mon. Wea. Rev.*, **116**, 1914-1926.

Yamada,T. and Bunker,S., 1989, A numerical model study of nocturnal drainage flows with strong wind and temerature gradients. *J. Applied Meteorol.*, **28**, 545-554.

Yamada, T. and T. Henmi, 1994: HOTMAC: Model performance evaluation by using WIND phase I and II data, *Mesoscale Modeling of the Atmosphere* (eds. R.A. Pielke and R.P.Pearce), *Meteorological Monographs*, **25**, 123-135.

Wakamatsu, S., Uno, I and Suzuki, M., 1990, A field study of photochemical smog formation under stagnant meteorological condition, Atmos. Environ., **24A**, 1037-1050.

Willmott,C.J., 1982, Some comments on the evaluation of model performance. *Bull. Amer. Meteorol. Soc.*, **63**, 1309-1313.

Willmott, C.J., S.G. Ackleston, R.E. Davis, J.J. Feddema, K.M. Klink, D.R. Legates, J.O'Donnell and C.M. Rowe, 1985, Statistics for thr evaluation and comparison of models. *J. Geophys. Res.*, **90**, 8995-9005.

DISCUSSION

R. BORNSTEIN: How do you treat both surface and PBL temperature during you dynamic initialization procedure ?

I. UNO: Dynamic initialization and the following FDDA simulation were conducted by only nudging the horizontal wind components, U and V. Trial simulation with surface temperature nudging showed the difficulty under the complex terrain conditions. Therefore no nudging was applied for surface and PBL temperature field.

WIND FLOW OVER ROLLING TERRAIN: A COMPARISON OF THREE RECENT MODELS AND FULL-SCALE OBSERVATIONS

J.M. Santabàrbara[1], J.F. Hernández[1], J. Calbó[1], L. Cremades[1], T. Mikkelsen[2], and J.M. Baldasano[1]

[1]ITEMA, Universitat Politècnica de Catalunya (UPC), Barcelona, Spain
[2]Risø National Laboratory, Roskilde, Denmark

ABSTRACT

Results from three recently developed wind flow models: *MCM* (Hernández *et al.*, 1995), *L* (Santabàrbara *et al.*, 1994) and *PROMETEO* (Calbó and Baldasano, 1995) are compared with data from the Sep.1992 *MADONA* experiment (Cionco *et al.*, 1995). The three models are representatives of three different approaches used in flow modelling. With increasing complexity, these approaches are: 1) a *mass-consistent* algorithm, 2) a diabatic extension of a *linearized* diagnostic model and 3) a primitive-equations based *prognostic* model. They are applied on a non-uniform gentle topography terrain at Porton Down (UK) and their results compared to wind measurements monitorized by 15 meteorological masts. The comparison evaluates the ability of the three models to reproduce the data set, keeping in mind the differences among the models. While the two diagnostic models compute the mean wind very fast (specially *L*, which runs on a 486 based PC in the range of seconds for this case), the prognostic model *PROMETEO*, originally designed for mesoscale scenarios is able to predict more characteristics of the flow. As far as the comparison against measurements is concerned, the behaviour of the three models is quite similar.

INTRODUCTION

The interest of modelling wind flow over non-flat terrain is mainly coming from disciplines such as atmospheric dispersion, civil engineering or wind energy resources. The specific needs in these and other fields of application have lead to different modelling concepts in search of an optimized cost benefit approach with CPU time and model accuracy as parameters. A classification of wind flow models exist according to the to the complexity of their theoretical basis. One fundamental division is between *prognostic* and *diagnostic* models.

Prognostic models allow the forecast of the wind field on the basis of some initial input information. Their theoretical fundament is the full *Navier-Stokes set of equations* which drive the physical behaviour of the air flow. Since there is no general analytical solution for this set of equations, the models solve this system numerically.

Diagnostic models compute a wind field which is simultaneous to a set of measurements. To do this, several schemes are currently used. A second classification level can be stablished (Mikkelsen and Desiato, 1993) from the simple interpolation procedures to a range of increasing complex approximations of the Navier-Stokes set of equations.

The so-called *mass-consistent* models represent one class of the diagnostic models. The principle within mass-consistent models consist of an iterative modification of a initially guessed wind field (computed by interpolation of the measurements) in order to fulfil the divergence-free condition (which results from a simplification of one of the Navier-Stokes equations).

Another type of diagnostic models have been developed under the so-called *linear theory*. The basis of this theory is an approximation the complete Navier-Stokes system where some equations have been truncated to eliminate the non-linear terms.

Evaluation of wind flow models is usually made by comparison of their results with experimental data obtained from field experiments. The aim of this paper is to compare the results of three models (*PROMETEO, MCM* and *L*) with detailed experimental data. The models can be taken as representatives of the three described types of models: prognostic, mass-consistent and linear theory-based models. Their results are compared with data from the extensive field study *MADONA*.

THE *MADONA* EXPERIMENT

Trial *MADONA* (*M*eteorology *A*nd *D*iffusion *O*ver *N*on-uniform *A*reas) was a field experiment conducted by an international consortium during September and October 1992 in the South-West of England. *MADONA* was composed of meteorological data collection and diffusion experiments using smoke, SF_6 and propylene gas during unstable, neutral and stable atmospheric conditions in an effort to obtain terrain-influenced meteorological fields, dispersion data, and concentration fluctuation measurements.

The *MADONA* field site is located on the Range area at the Chemical and Biological Defence Establishment (CBDE) at Porton Down, Salisbury, Wiltshire, UK. The topography of the area is gently rolling terrain with a ridge running southwest to northeast with higher terrain at each end. Total relief for the test area is 100 m ranging from 70 to 170 m. Figure 1 gives a perspective of the topography of the area viewed from South-West corner. The ground is comprised of gently rolling features, mainly of short sward but with patches of scrub and some stretches of mature woodland. Although the test area cannot be described as very complex terrain it is clearly not a flat, uniform terrain. As a result of this non-uniformity, the wind flow over the area is perturbed by terrain-induced effects, not only by dynamical (speed-up, lee wakes) but also thermally driven slope and drainage winds.

Fourteen cup-and-vane anemometers were positioned across the area (Figure 1) at a height of ten meters, their exposures chosen in order to reflect a representative sample of wind measurements. A fifteenth anemometer at a height of 30 m (mast 9 in Figure 1) was also recording the experimental data: the average wind speed and direction logged every six seconds for the thirty one days of the trial. Short gaps (of ½ hours duration) exist in every 24 hour's record when data downloading occurred once a day.

Upper air wind measurements were made by means of two periodical balloon soundings (up to 500 mb) and by around-the-clock SODAR data recording. Many other

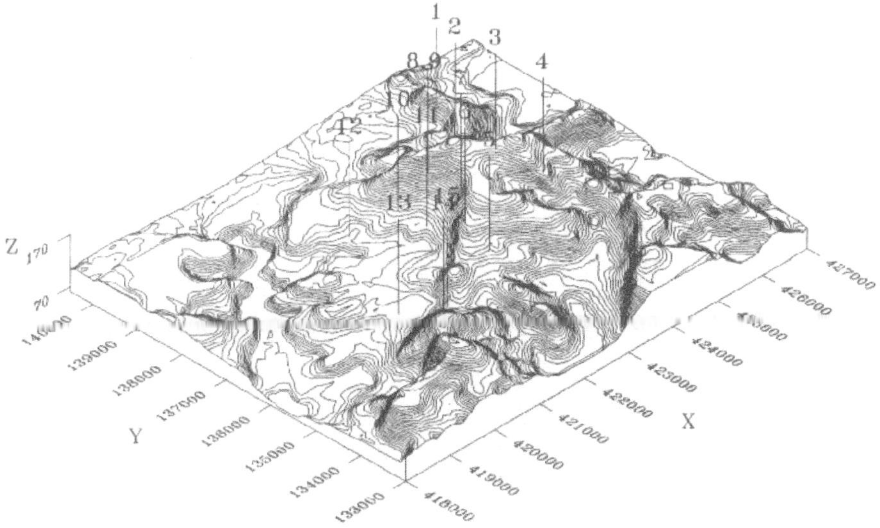

Figure 1 The MADONA terrain perspective viewed from S-W corner. Heights have been amplified by a factor 10. The positions of the 15 meteorological masts used in this study are also shown.

experimental data were recorded during the trial. Data from nine days of diffusion experiments: smoke and trace gas releases was acquired using not only traditional gas detectors but remote sensing techniques such as LIDAR or Video Imaging systems.

Daily weather conditions during the study period were generally partially cloudy to overcast skies. Winds were mostly moderate coming from southwest. Temperatures moved mainly in the range 14 to 18 ºC.

Additional information regarding the MADONA trials including a reference to an available CD-ROM media (containing all wind data among other information) can be found in Cionco *et al.* (1995), Joergensen *et al.* (1995) and Mikkelsen and Joergensen (1995).

DESCRIPTION OF THE MODELS

The mass-consistent model *MCM* (Hernández *et al.*, 1995)

This diagnostic model is based on the variational approach proposed by Sasaki (1958), Sherman (1978) and Mathur and Peters (1990). The cornerstone of the model is the variational analysis with continuity equation used as strong constrain.

The integral function defined in the variational approach is,

$$E(u,v,w,\lambda) = \iiint \left[\alpha_1^2(u-u_0)^2+\alpha_2^2(v-v_0)^2+\alpha_3^2(w-w_0)^2+ \lambda\left(\frac{\partial u}{\partial x}+\frac{\partial v}{\partial y}+\frac{\partial w}{\partial z}\right)\right] dx\, dy\, dz \tag{1}$$

where, $\boldsymbol{u}=(u,v,w)$ is the fitted wind vector, $\boldsymbol{u}_0=(u_0,v_0,w_0)$ is the corresponding interpolated wind field from observed data, $\boldsymbol{\lambda}(x,y,z)$ is the Lagrange multiplier and $\boldsymbol{\alpha}_i$ are weighting functions relating horizontal wind components with the vertical ones. It is assumed that the horizontal weighting functions are identical ($\boldsymbol{\alpha}_1=\boldsymbol{\alpha}_2$).

The Euler-Lagrange equations resulting from the minimization problem are,

$$u_i = u_{0_i} + \frac{1}{2\alpha_i^2}\frac{\partial\lambda}{\partial x_i} \quad , \; i = x,y,z \tag{2}$$

$$\frac{\partial u}{\partial x} + \frac{\partial v}{\partial y} + \frac{\partial w}{\partial z} = 0 \tag{3}$$

The fundamental equation results by combining the above expressions and imposing the natural boundary conditions is,

$$\left(\frac{\partial^2}{\partial x^2} + \frac{\partial^2}{\partial y^2} + \left(\frac{\alpha_1}{\alpha_3}\right)^2 \frac{\partial^2}{\partial z^2}\right)\lambda = -2\alpha_1^2\left(\frac{\partial u_0}{\partial x} + \frac{\partial v_0}{\partial y} + \frac{\partial w_0}{\partial z}\right) \tag{4}$$

Due to a lack of observed data for the vertical velocity component, equation (4) is solved for the Lagrangian multiplier λ by setting $w_0=0$. After substitution in (2), adjusted components are obtained through the natural boundary conditions. The final result is a mass-consistent (with minimum divergence) wind field, representative of mean flux in the domain.

Among other characteristics, some model features are: terrain-following coordinates system (x,y,σ) allowing for variable vertical grid spacing and ability to incorporate many commonly available meteorological data (surface and profiles).

The linear model *L* (Santabàrbara *et al.*, 1994)

This model is based on an extension to thermally stratified flows of the *LINCOM* code first described by Troen & De Baas (1986) which, in its turn, relates back to earlier work by Jackson & Hunt (1975). This last paper established the basis for the so-called *linear theory* also found in models such as *MS3DJH* (Taylor *et al.*, 1983) which is still undergoing development (Wamsley *et al.*, 1993) and the *FLOWSTAR* (Carruthers, 1993). The cornerstone of the linear theory is the truncation of the Navier-Stokes set of equations by eliminating the non-linear terms to lead to a set of simplified equations (Santabàrbara *et al.*, 1994) similar to the following:

$$\begin{cases} \left(U\frac{\partial}{\partial x} + V\frac{\partial}{\partial y} - K\frac{\partial^2}{\partial z^2}\right)\breve{u} + \frac{\partial \breve{p}}{\partial x} = 0 \\ \left(U\frac{\partial}{\partial x} + V\frac{\partial}{\partial y} - K\frac{\partial^2}{\partial z^2}\right)\breve{v} + \frac{\partial \breve{p}}{\partial y} = 0 \\ \left(U\frac{\partial}{\partial x} + V\frac{\partial}{\partial y} - K\frac{\partial^2}{\partial z^2}\right)\breve{w} + \frac{\partial \breve{p}}{\partial z} = g\frac{\breve{\theta}}{\langle\theta\rangle} \\ \frac{\partial \breve{u}}{\partial x} + \frac{\partial \breve{v}}{\partial y} + \frac{\partial \breve{w}}{\partial z} = 0 \end{cases} \tag{5}$$

where U and V are the horizontal components of a background wind, $\breve{u}$, $\breve{v}$ & $\breve{w}$ are the wind and $\breve{p}$ the pressure perturbations (*i.e.* $u=U+\breve{u}$, etc..), K is the turbulent diffusivity (product of a first order closure scheme) and $g\breve{\theta}/\langle\theta\rangle$ is the buoyancy term, being θ the potential temperature. The first three equations of set (5) are a version of the motion equation and the last one is the continuity (free-divergence) condition.

System (5) is solved by Fourier transforming the equations after specifying a vertical (exponential decaying) temperature profile depth and surface temperature. The resulting model is an extremely fast code which takes into account (till first order) not only the "neutral" (dynamical) terrain-induced effects, but also thermal effects (slope breezes and drainage winds) originated from the inclusion of the buoyancy term in (6). Details of this solving procedure can be found in Santabàrbara *et al.* (1994) where the model was run on a artificial terrain (gaussian hill). In the present paper, *L* model results on real terrain are presented for first time and compared against measured wind data.

The prognostic hydrostatic mesoscale model *PROMETEO* (Calbó and Baldasano, 1995)

As a prognostic model, *PROMETEO* solves the complete set of primitive equations. Briefly, this set includes the continuity equation (written in its anelastic form); two motion equations for each horizontal component of the wind; the equation for the energy conservation (written in terms of the potential temperature); and the hydrostatic equation, deducted from the motion equation for the vertical of the wind after some approximations. In addition, the surface energy budget is solved in order to compute the ground temperature, an important variable within a mesoscale model since most mesoscale phenomena are driven by temperature differences between different surfaces. Radiative fluxes are also computed. Turbulent fluxes, which appear in the equations due to the Reynolds decomposition, are computed by means of second order, second level closure scheme (following the hierarchy stablished by Mellor and Yamada (1974)).

The model makes use of terrain-influenced coordinates. It solves the equations by means of finite difference schemes on a staggered grid. A background (synoptic) wind, and a profile of temperature and humidity are the main initial data. Non-slip boundary conditions are defined at the bottom boundary, while constant inflow, gradient outflow conditions are applied at lateral boundaries. The goal of the model is to generate mesoscale phenomena given a specific synoptic meteorological situation and terrain. Therefore it is desingned to run in situations where the mesoscale phenomenon is stronger than the synoptic forcing.

SIMULATION SCHEME

The period chosen for the simulation was September the 15th, 1992 from 14:10 to 18:50 hours. For that day, that was the period of operation of the SODAR system which provided the wind profile needed to initialize the *MCM* model. The *L* model used data from the 30 m height tower (mast 9 in Fig.1) as input wind instead.

Observed wind speeds in this period were in the range 3 to 10 ms^{-1} with an average of 6.2 ms^{-1}. Wind came mainly from South-West (mean wind direction was 234 deg). Stability conditions could be described as changing from moderately unstable to neutral. Heat flux, takes values of the order of 100 Wm^{-2} at the central hours of the day.

A digitized topography of the terrain was used for the two diagnostic models, a rectangular grid of 90 x 75 points with a spatial step of 100 m. This was also the horizontal spatial resolution of the output wind field of the models. The resulting wind fields were linearly interpolated to get data at the meteorological masts locations.

Simulation results were obtained every 10 min. This information was later compared to the 10 min averaged available data set from the meteorological masts. By this means, a total of 29 sets of results from the models were generated.

This scheme was slightly different for the prognostic model simulation. Spatial resolution was 300 m and the output wind fields were obtained every 60 min (although the temporal step of the simulation was 2.5 s). Simulation covered a complete period of 24 hours in this case. Results between 11 and 19 h were compared with measurements.

RESULTS AND DISCUSSION

To evaluate the level of agreement between measurements and model simulation some scatter plots (Figure 2) and statistics of the comparison (Table 1) are shown. Figure 2 shows the computed *vs.* observed wind speeds and directions for the two diagnostic models. The points in the scatter diagrams are data for the studied period from 14 stations (excluding mast 9) plotted against the line of perfect prediction. Table 1 presents the following statistics of the observed/computed sets of coupled data: Root Mean Square Error (RMSE), Normalized Mean Square Error (NMSE), fractional BIAS and Deviation (FD). Figure 3 shows some of the resulting wind fields.

Table 1. Principal statistics of the computed *vs.* observed comparison for the three models.

	L		PROMETEO		MCM	
	speed	direction	speed	direction	speed	direction
RMSE	1.058	6.7	1.917	11.3	1.254	11.5
NMSE	0.037	0.001	0.091	0.002	0.05	0.002
BIAS	-0.217	0.822	-0.215	2.316	0.345	9.206
FD	-0.012	0.002	-0.015	0.005	0.034	0.019

From both table and plots it can be noted than performance is quite satisfactory for the three models, the results showing a good agreement with the measurement data, being *L* results slightly better compared against data. Best results obtained with this model give correlation factors of 0.73 (for wind speed) and 0.59 (for wind direction) between model results and data. A discriminative analysis of these last results showed than agreement with data is better for masts located at the plainer areas (for example masts 11, 12 and 13) while agreement is poorer near the more abrupt slopes. That is the case of mast 3 which gives the worst results. Excluding this tower information of the data set the comparison for model *L* would improve till correlation factors of 0.8 for both wind speed and direction.

In order to fairly compare the mass-consistent and linear model results, the *MCM* code setup was such than only a single tower measurements (in addition to the SODAR wind profile) was used as input. In that case, correlation factors between model results and data are 0.64 for wind speed and 0.52 for wind direction. Clearly, the addition of more towers as input information would improve the results. Several tests have been made with more than one tower as input information resulting in correlations as good as 0.9 for both wind speed and direction when the 15 masts were used at the same time.

Results from the prognostic model *PROMETEO* are of the same order of agreement with data than the diagnostic ones although the scenario where the models have been applied is not the best for this type of model.

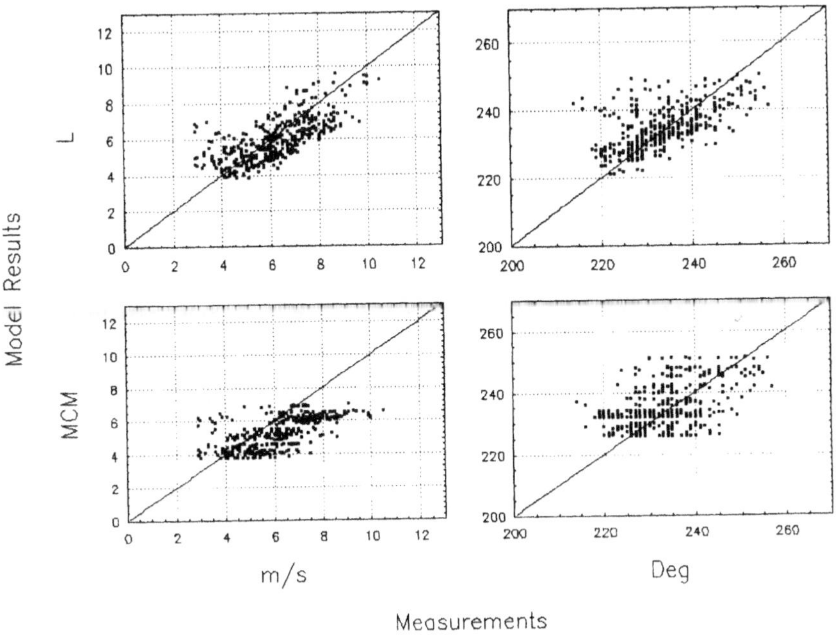

Figure 2 Scatter plots of computed *vs.* observed wind speed & direction data for the two diagnostic models.

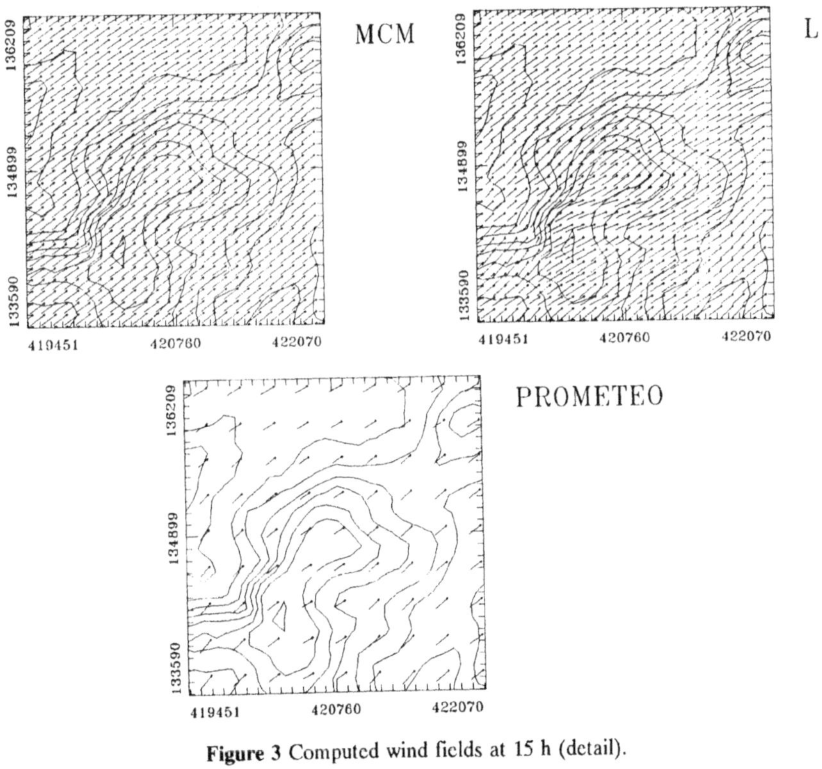

Figure 3 Computed wind fields at 15 h (detail).

REFERENCES

Calbó J. & Baldasano J.M, 1995, PROMETEO: An hydrostatic mesoscale model applied to the simulation of land-sea breeze in the Barcelona area, *Environ. Software* (in press).

Carruthers D.J., 1993, *FLOWSTAR* Model Description, available from Cambridge Environmental Research Consultants Ltd., Cambridge (UK).

Cionco R.M., aufm Kampe W., Biltotf C., Byers J.H., Collins C.G., Higgs T.J., Hin A.R.T., Johansson P.E., Jones C.D., Joeergensen H., Kimber J.F., Mikkelsen T., Nyren K., Ride D.J., Robson R., Santabárbara J.M., Streicher J., Thykier-Nielsen S., van Raden H. & Weber H., 1995, An overview of MADONA: a multi-nation field study of high resolution meteorology and diffusion over complex terrain, in the proceedings of the XI Symposium on *Boundary Layers & Turbulence*, Charlotte NC, March 1995.

Hernández J.F., Cremades L. & Baldasano J.M., 1995, Dispersion modelling of a tall stack plume in the spanish mediterranean coast by a particle model, *Atmos. Environ.*, 29, 11: 1331-1341.

Jackson P.S. & Hunt J.C.R., 1975, Turbulent wind over a low hill, *Quart.J.R.Met.Soc.*, 101: 929-955.

Jørgensen H.E., Mikkelsen T., Ride D.J., Higgs T., aufm Kampe W., Weber H., Nyrén K. & Johansson P.E., 1995, MADONA: Concentration fluctuation analysis, in the Proceedings of the XI Symposium on *Boundary Layers & Turbulence*, Charlotte NC, March 1995.

Mathur R. & Peters L.K., 1990, Adjustment of wind fields for application in air pollution modelling, *Atmos. Environ.* 124A, 5: 1095-1106.

Mellor G.L. & Yamada T., 1974. A hierarchy of turbulence closure models for planetary boundary layers, *J. Atmos. Sciences*, 31: 172-190.

Mikkelsen T. & Desiato F., 1993, Atmospheric dispersion models and pre-processing of meteorological data for real-time application, *Radiation Protection Dosimetry*, 50, 2-4: 205-218.

Mikkelsen T. & Jørgensen H.E., 1995, MADONA: on-site real-time turbulence measurements, in the Proceedings of the XI Symposium on *Boundary Layers & Turbulence*, Charlotte NC, March 1995.

Santabàrbara J.M., Sempreviva A.M., Mikkelsen T., Lai G. & Kamada R., 1994, A spectral diagnostic model for wind flow simulation: extension to thermal forcing, in *Air Pollution II*, Vol 1: Computer Simulation: Baldasano J.M., Brebbia C.A., Power H. & Zanetti P., ed., Computational Mechanichs Publications, Southampton, UK.

Sasaki Y, 1958, An objective analysis based on the variational method, *J.Met.Soc.Japan*, 36: 77-88.

Sherman C.A, 1978, A mass-consistent model for wind fields over complex terrains, *J. Appl. Met.*, 17: 312-319

Taylor P.A., Walmsley J.L. & Salmon J.R., 1983, A simple model of neutrally stratified Boundary-Layer flow over real terrain incorporating wavenumber-depending scaling, *Boundary-Layer Met.*, 26: 169-189.

Toen I. & De Baas A., 1986, A spectral diagnostic model for wind flow simulation in complex terrain. Proceedings of the *European Wind Energy Association* Conference & Exhibition. Roma, 7-9 Oct.

Wamsley J.L., Weng W., Karpik S.R., Xu D. & Taylor P.A., 1994, Applications of a mixed spectral finite-difference model and its nonlinear extension to wind flow over Blashval hill in *Air Pollution Modelling and its applications X* edited by S.E.Gryning & M.Millan, Plenum Press, New York.

DISCUSSION

J.L. WALMSLEY: Regarding the L model with thermal forcing, your equations include the buoyancy term in the vertical equation of motion. Are there any other places where thermal effects enter the model?

J.M. SANTABARBARA: No, there are not. I presume you talk about a thermal treatment of the closure scheme. By the moment, this is not included but I intended to do it in a short future.

G. SCHAYES: The typical wavelength of the topography of MADONA site seems to be of the order of 1 km. This probably explains why the prognostic model performs less well than the diagnostic ones since the hydrostatic assumption is not adequate at that size.

J.M. SANTABARBARA: The non-hydrostaticity of the flow may be one but not the only reason to explain the behaviour of the hydrostatic prognostic model PROMETEO. Indeed, the flow we are simulating can be classified approximately as a neutral advective flow, in which non-hydrostatic terms should be taken into account, even if vertical components of the wind are not very large. However, we believe that the differences among the three simulations (e.g. the coarse grid resolution used by the prognostic model) are the main factors that lead the worse behaviour of PROMETEO in front of both diagnostic models.

A COMPARISON STUDY OF AIR-QUALITY MODEL SIMULATION RESULTS WITH AIRCRAFT DATA

Gunilla Svensson[1] and Otto Klemm[2]

[1]Department of Meteorology
Uppsala University
Box 516, S-751 20 Uppsala
Sweden

[2]Fraunhofer-Institut für Atmosphärische Umweltforschung (IFU)
Kreuzeckbahnstraße 19
D-82467 Garmisch-Partenkirchen
Germany

INTRODUCTION

The air-quality in Athens, Greece, is known to be a problem for the habitants in the city. The concentrations of pollutants, both primary and secondary, are often above regulation limits. It is the combination of special meteorological conditions and high emissions in the Athens area that sometimes result in very high concentrations of photochemically produced pollutants (e.g. Güsten et al. 1988; Pilinis et al. 1993; Ziomas et al. 1995). A large field campaign, the MEDCAPHOT-TRACE (Mediterranean Campaign of Photochemical Tracers - Transport and Chemical Evolution) was carried out the 20 August - 20 September, 1994, to enlarge the knowledge about the processes behind high pollution levels in the area.

Athens is situated in a basin with mountains on three sides and the Mediterranean Sea (Saronic Gulf) on the fourth side, see Figure 1. The mountains range between 470 and 1400 m altitude above sea level. The sea surface temperature is almost always cooler than the land surface all through spring, summer and autumn. Therefore, when the synoptic flow is weak enough, conditions are favourable for sea-breezes to develop. When this occurs, the mountains act as barriers, so the pollutants can be trapped in the basin. The importance of the vertical mixing and the flow out of the valley through the gaps between the mountains increases, for the pollution concentrations within the city. These phenomena have been discussed in earlier papers, e.g. Svensson (1995b), Kallos et al. (1993) and Pilinis et al. (1993).

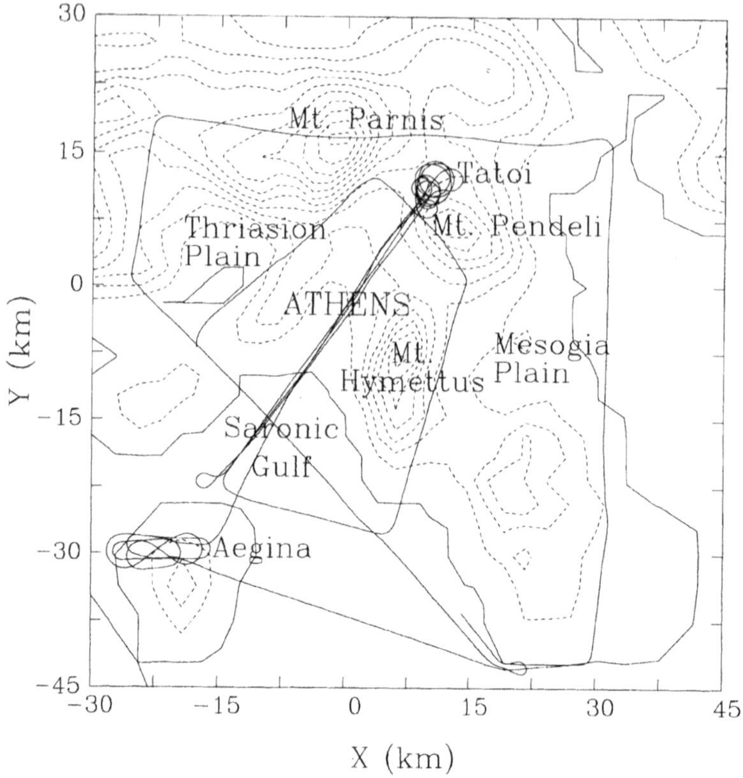

Figure 1. The Attica Peninsula and the city of Athens. The contourlines represent heights above sea level for every hundred meters. The bold line is the flight track of the research aircraft for the 15 September 1994.

The field campaign included a number of ground based measurements, soundings, tethered balloons, remote sensors and aircraft measurements. The focus here will be on the aircraft measurements on the 15 September 1994. The results of the flights are here compared with model results.

THE MODEL

The model used in this study is a three-dimensional time dependent flow model with chemical reactions. A full description of the model is found in Svensson (1995a). The main features of the model are that it has a second order closure of the turbulence scheme and that the chemical mechanism is a slightly modified version of the CBM-IV. The reaction scheme includes about 80 reactions involving 32 species. Deposition processes are calculated with a resistance model. The horizontal resolution for the simulations discussed here was 2 km in the center part, increasing towards the edges. The first grid-point in the vertical was at the roughness length (z_0) and the following levels were at 1, 3, 7, 15, 30 m etc. and the resolution decreased to about 400 m near the model top at 7 km. The center part of the model grid, with almost horizontally equally spaced grid, is shown in Figure 1. The total domain size was 150x150 km^2. The time step used in the simulations was 20 s and the model output was stored for every half hour. The model was initialized with the 00:00 UTC sounding at the Hellenicon airport right outside Athens. The actual simulation, with surface forcing, emissions and chemical reactions, started at 03:00 LT and the model was run for a total simulation time of 45 hours. Only results from the second day are analysed. After a whole diurnal cycle of emissions, chemistry, diffusion and advection, it is assumed that the initial concentrations in the model domain no longer influence the results in the polluted regions. The initial background concentrations are set very low except for ozone where 70 ppb is used. The emission data base is described in Ziomas et al. (1995).

AIRCRAFT MEASUREMENTS

The aircraft measurements were taken on a Dornier 228 twin-engine turboprop research aircraft. The measuring system includes standard meteorological measurements of location in space, horizontal wind speed and direction, temperature and humidity, and chemical measurements of O_3, NO, NO_2, NO_y, SO_2, CO, and CH_2O. The sampling time for the system is 1 Hz, and data shown here were averaged over 5 seconds. Additionally, samples of air were taken during the flights for volatile organic compound (VOC) analysis.

On the 15 September, two flights were performed, one around sunrise (6:45 – 9:00 LT) and one in the early afternoon (12:15 – 14:45 LT). The flight track is shown in Figure 1. Most of the flights were at low levels, i.e., at about 50 m height over the water and about 150 m above the land surface. In addition, several flight legs over the city were made at various heights (100, 300, 500 and 700 m) and two vertical profiles up to almost 3000 m were taken above the island Aegina and between Mt. Parnis and Mt. Pendeli (see Figure 1). The profiles are done in a spiral flight pattern which disqualify the wind measurements, however, the quality of the chemical measurements are not affected. Analysis of all flights during MEDCAPHOT-TRACE is found in Klemm et al. (1995).

RESULTS

The flow field on this day was complicated. The synoptic forcing was weak from NNW, and several sea-breeze cells developed during the day driven by the temperature difference between the air over the sea and over land, see Figure 2. The sea-breezes are very much affected by the local topography and the simulation results are sensitive to the input profile of wind and temperature. For example, in a test simulation with a change in the synoptic wind direction by only 10° towards north, the flow field was completely changed. With this northerly synoptic forcing, no sea-breeze cell on the eastern side of the Attica Peninsula was formed. However, in Figure 2, a sea-breeze is clearly seen on the eastern side of the peninsula, which is supported both by the ground based and the aircraft measurements on this day. Several model simulations with the meteorological model were thus first performed to achieve a reasonably good agreement between all the measured wind and temperature data and the model results, both near the surface and vertically. In these simulations, only the geostrophic wind speed and direction were changed. This was performed to make sure that the model captures the main features in the flow field, boundary layer height etc. This then makes it possible to run a simulation with emissions, deposition and chemical reactions and compare the results with measured concentrations.

The aircraft measurements are local in time and space, i.e. measurements at different locations are by necessity taken at different times. The model on the other hand provides volume-averaged 3-D fields at several discrete times during the flight, and in addition before and after the measurements. Thus, the model can be used as a tool to explain the processes that cause the measured concentrations. It can also be used to predict what the effect on the air-quality would be if, for example, the traffic is reduced.

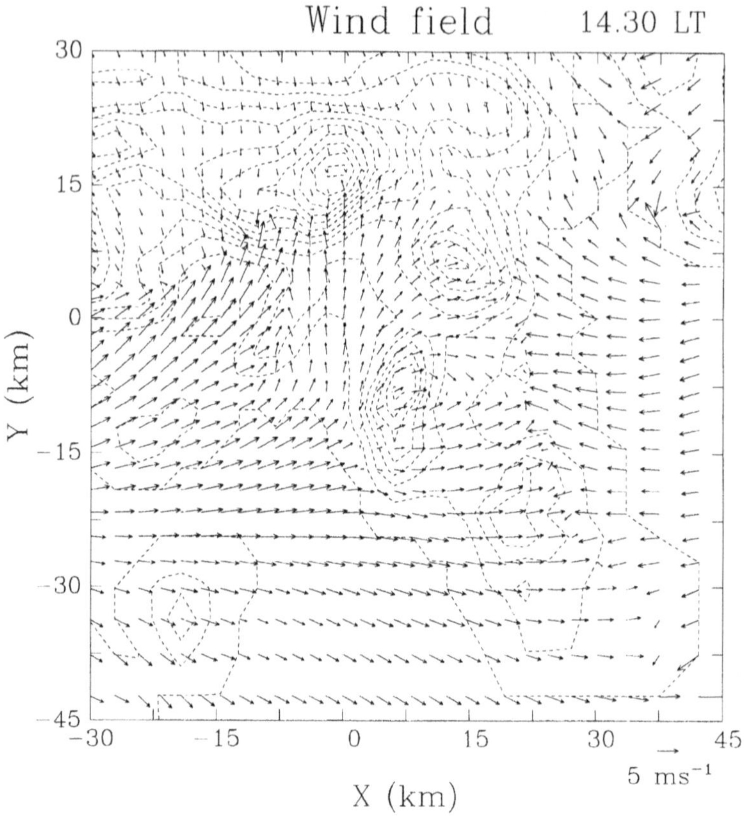

Figure 2. Modeled wind field at 14:30 LT about 100 m above ground for the 15 September 1994.

To be able to compare the parameters measured by the aircraft with the model results, some preparations of the model results were required. Considering that the flight duration was three hours, the measured variables should be compared with model output over the same time span. The space information from the aircraft measurements was used to determine which model grid volume the single aircraft data correspond to. According to this, the modeled values were extracted from the original model output files. A linear interpolation in time between the half hour outputs was then performed. The outcome of this computation are files with the same number of modeled data points as in the aircraft data sets. These data represent a 'flight in the model output', which are directly comparable to the measured aircraft data.

Figure 3 shows the wind speed and direction for the early afternoon flight, comparing the measured and the modeled values. Aircraft measurements of wind are not available if the wind speed was too low (< 2 ms^{-1}) or if the aircraft was turning. The agreement between the measured and the modeled data is generally good. However, for the eastern part of the northern flight leg (see Figure 1), the wind directions differ by about 100°. The model results give a northwesterly wind (c.f. Figure 2) at the time of this part of the flight. The measurements show easterly winds in this region, and a few hours later the model results show that too. However, the blocking of the flow out from the Athens Basin through the northern gap, which is important for the dispersion of the pollutants, is present both with the northwesterly and the easterly winds.

Three sea-breeze cells are present this day after, one in the city basin, the second at Thriassion Plain and the third over Mesogia Plain, see Figure 2. The sea-breezes in the city and at Thriassion Plain were developed at about 10:00 LT. The third cell on the eastern side of the Attica Peninsula was established later and it reached the mountains in the early after-

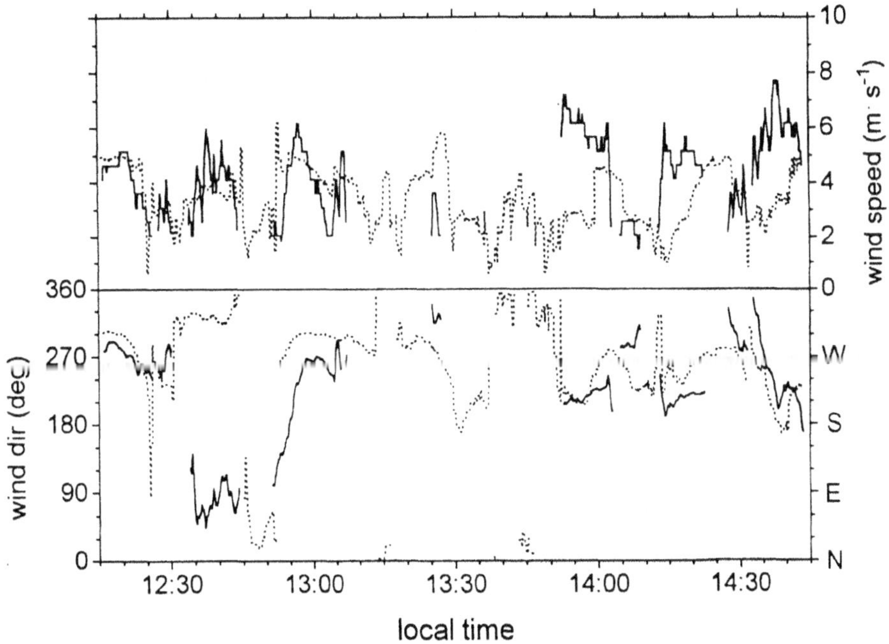

Figure 3. Measured (solid) and modeled (dashed) wind speed (ms^{-1}) and wind direction (°) for the early afternoon flight on the 15 September 1994. The flight track is shown in Figure 1. The measured wind speed and direction are only available when the wind speeds are > 2 ms^{-1} and the aircraft is not turning.

noon. They are all supported by the ground based measurements. The breezes bring cooler and stably stratified air in over land and the boundary layer heights are low, only a few hundred meters maximum heights in the afternoon. This limits the vertical mixing and the resulting pollution concentrations are high in the basin. The sea-breeze transports the pollutants emitted over the city towards the north and there they are halted by the opposing flow discussed above. Figure 4 shows measured and modeled ozone concentrations for four flight legs on various heights above the city (from about over Saronic Gulf to Tatoi in Figure 1). An increase in the ozone concentrations is seen towards north at all altitudes both in the measurements and the model results. The measured concentrations are almost always somewhat higher than the modeled ones, but the spatial pattern is the same. The concentrations are not very different at the various heights.

The top panel in Figure 4 shows the flight leg at 700 m. The modeled concentrations of ozone first increase towards north and then decreases to a level of about 60 ppb. At this level in the model, the opposing flow from NW brings air with background concentrations, which is not the case in the measurements. At the lowest level (lowest panel in Figure 4) ozone concentrations of about 100 ppb are seen out over the Saronic Gulf in the measurements. This polluted air is not seen in the model results. The origin of the high concentrations of ozone is not clear, it may be polluted morning air that is recirculated with the sea-breeze.

Figure 5, shows the measured and modeled profiles of ozone over Aegina and Tatoi (see Figure 1). The measured concentrations over Aegina were about 70 ppb and the variations with height are small, except for the lowest 500 m where some elevated ozone concentrations

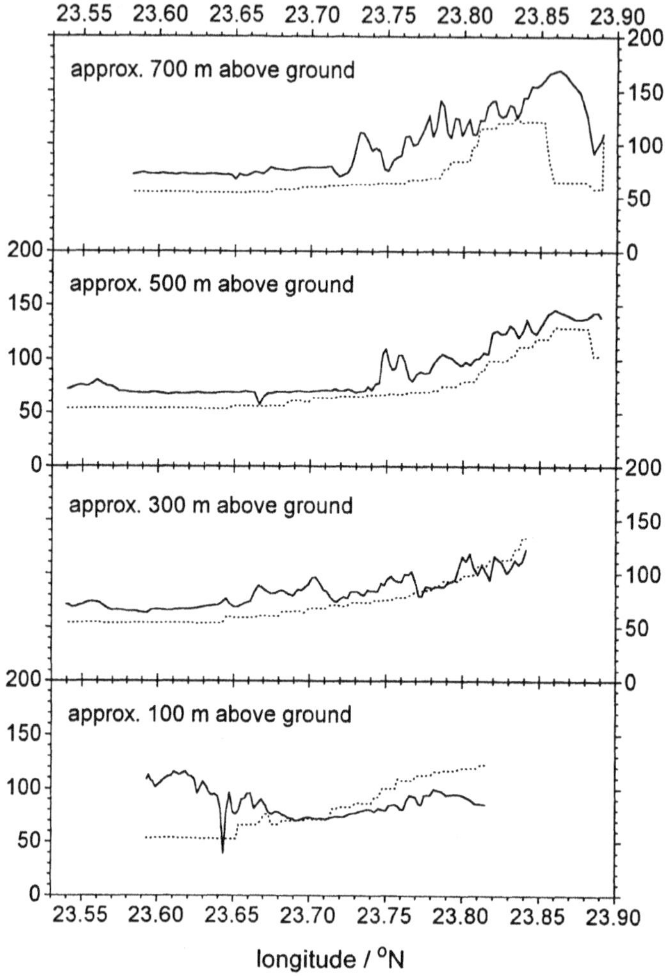

Figure 4. Ozone concentrations in ppb, measured (solid) and modeled (dashed), over the Saronic Gulf and Athens at 100, 300, 500 and 700 m above ground. The flight track in SW ↔ NE direction is shown in Figure 1. The x-axis gives the longitude and northeast is to the right in the figure. The shoreline is at 23.67°E.

are seen. This polluted air is the same that is seen over Saronic Gulf in the lowest level flight in Figure 4. The modeled profile over Aegina displays background concentrations (about 50 ppb) throughout the whole profile.

The blocking in the northern gap between Mt. Parnis and Mt. Pendeli keeps back the sea-breeze front and the resulting convergence causes vertical motion that transport pollutants up to about 2300 m. This is seen in both modeled and measured profiles over Tatoi (right panel in Figure 5). The ozone concentrations are very high both in the measurements and in the model. The maximum values measured reach almost 200 ppb, the modeled are somewhat lower, about 155 ppb. Large variations are seen in both the modeled and the measured profiles. These fluctuations do not necessarily reflect only vertical variations in the ozone con-

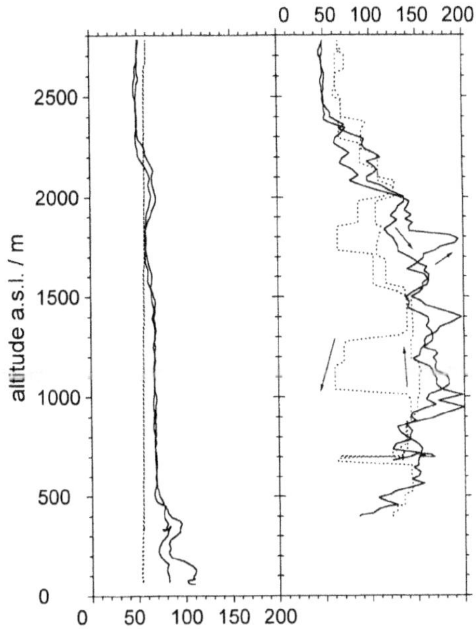

Figure 5. Measured (solid) and modeled (dashed) vertical profiles of ozone concentrations in ppb at Aegina (left) and at Tatoi (right). The arrows indicates the aircraft going up or down.

centration, but may also contain horizontal variations. The aircraft followed a spiral flight pattern during the profiles (see Figure 1), and may have flown in and out of the sea-breeze front, i.e. in and out of the polluted air.

In the northern profile, concentrations of CO, NO_y and formaldehyde are also high up to an altitude of 2300 m above sea level. The interpretation is that these pollutants are emitted over Athens, transported by the sea-breeze in a shallow layer and then lifted by the updrafts in the region between Mt. Parnis and Mt. Pendeli. The model results reveal that the updraft region, with updrafts of the order of 1-2 ms^{-1}, covers a limited area and there is a very sharp gradient in the concentrations to the background values towards north. This sharp gradient is seen in Figure 5 with much lower values when the aircraft position in the model happens to be in a grid box further north, where the model calculations gives winds from the north. The updraft region is formed around 11:00 LT in the model and lasts 5 hours. The local terrain has here created a 'chimney' that pumps up polluted air well above the boundary layer.

CONCLUSIONS

Presented here is a comparison between simulation results from a three-dimensional photochemical meso-scale model and aircraft measurements. The flow pattern of the simulated day, 15 September 1994, is very complicated. Three sea-breeze circulations are observed in the surroundings of Athens that are supported in the observations, both in the ground based and in the aircraft measurements, and the model is able to capture this intricate flow pattern. The model results for the chemical species that have undergone chemical reactions, transport and diffusion, agree well with the measured concentrations. The model results are also used to interpret the measured concentrations. Further analysis along this line is in Svensson and Klemm (1995).

On this day, the sea-breeze in the Athens region cause low boundary layers height over the city and consequently high pollution levels. Both the model simulations and the measurements show increasing concentrations of ozone and other pollutants towards northeast in the basin. In the gap between the two mountains in the north, there is an opposing wind that halts the sea-breeze. This creates a convergence zone which in turn causes vertical motions that transport polluted air up to above 2000 m, i.e. in this case into the free troposphere.

ACKNOWLEDGEMENTS

The first author was supported by the Swedish Environmental Protection Agency and the National Supercomputer Center (NSC), Linköping University. MEDCAPHOT-TRACE was supported by the European Union. All participants in the campaign, especially I. Ziomas, are recognized for their work within the project.

REFERENCES

Güsten, H., G. Heinrich, T. Cvitas, L. Klasinc, B. Ruscic and D. Lalas, 1988: Photochemical formation and transport of ozone in Athens, Greece. *Atmos. Environ.*, **22,** 1855-1861.

Kallos, G., P. Kassomenos and R. Pielke, 1993: Synoptic and mesoscale circulations associated with air pollution episodes in Athens, Greece. *Bound. Layer. Met.*, **62**, 163-184.

Klemm, O., I. Ziomas, D. Balis, P. Suppan, J. Slemr, R. Romero, L. Vyras, 1995: A 3-D summer air pollution study in Athens, Greece. *In manuscript.*

Pilinis, C., P. Kassomenos and G. Kallos, 1993: Modeling of photochemical pollution in Athens, Greece. Application of the RAMS-CALGRID modeling system. *Atmos. Environ.*, **27B**, 353-370.

Svensson, G. 1995a: A numerical model for chemical and meteorological processes in the atmospheric boundary layer – Part I. A model description and a one-dimensional parameter study. *Submitted to J. Appl. Meteor.*

Svensson, G. 1995b: A numerical model for chemical and meteorological processes in the atmospheric boundary layer – Part II. A Case Study of the Air-quality Situation in Athens, Greece. *Submitted to J. Appl. Meteor.*

Svensson, G., and O. Klemm, 1995: Aircraft measurements and model simulations of the air-quality in Athens, Greece. *In manuscript.*

Ziomas, I.C., P. Suppan, B. Rappenglück, D. Balis, P. Tzoumaka, D. Melas, A. Papayannis, P. Fabian and C.S. Zerefos, 1995: A contribution to the study of photochemical smog in the Greater Athens Area. *Contrib. Atmos. Phys.* In press.

DISCUSSION

G. KALLOS: Is the meteorological model simulations the same as the ones Dr. Ziomas presented?

G. SVENSSON: No.

G. KALLOS: What kind of surface parameterization did you use? If I understood well, you have used a heating function. If yes, is it the same for the North, South,...facing slopes? Do you believe this is an appropriate way to describe convergency zones?

G. SVENSSON: Yes, I use a function for the ground temperature. This function varies with height above sea level but not with the orientation of the slope. The reason for not using a surface energy balance equation is that the temperature prediction from this method is strongly dependent on the soil moisture, which I do not have any information about. I do not believe that the orientation of the slopes are very important for the convergency zone predicted in this case.

G. KALLOS: How did you initialize the photochemical part of the model? Did you take into account the computation from the previous day?

G. SVENSSON: Yes, the model was run for 48 hours with the same meteorological condition and the results presented here is for the second day. The assumption is that one full diurnal cycle gives the background concentrations in the Athens area.

G. KALLOS: Having all these uncertainties and shortcomings in addition to the limited domain you have used, how you can explain the good agreement with the observations you showed us.

G. SVENSSON: These results presented here indicate that the important processes, that give rise to the observed concentrations, are described properly in the model, and that the points you bring up here are of minor importance.

T. YAMADA: One of your conclusions says that the model simulations were sensitive to parameters used and topography. What kind of parameters are you talking about?

G. SVENSSON: The meteorological model is initialized with one sounding and the development of the sea-breezes are very sensitive to the geostrophic wind speed and direction given as input.

R.J. BARTHELMIE: Given presentations this week showing the potential recirculation of pollutants in coastal areas. Have you been able to use the model to describe the possible fate of pollutants once they are transported aloft?

G. SVENSSON: The model domain is too small to simulate a recirculation of pollutants from the previous day. I believe that it is very unlikely that the previous days pollutants would be transported back to the same place. However, there is some recirculation in my model results. The pollutants from the morning peak traffic are transported out over Saronic Gulf in the morning and are then brought back over the city by the sea-breeze.

A COMPARISON OF CALPUFF MODELING RESULTS WITH CAPTEX FIELD DATA RESULTS

John S. Irwin[1**], Joseph S. Scire[2], and David G. Strimaitis[2]

[1]Atmospheric Sciences Modeling Division
Air Resources Laboratory
National Oceanic Atmospheric Administration
Research Triangle Park, NC 27711
[2]Earth Tech
196 Baker Avenue
Concord, MA 01742
U.S.A.

INTRODUCTION

This paper provides a summary of results from a series of analyses of several meteorological models and puff dispersion models suitable for application to air pollution dispersion involving long-range transport. One of the primary objectives was to assess whether use of mesoscale dynamic wind fields developed using Four-Dimensional Data Assimilation (FDDA) exhibiting improved spatial and temporal resolution versus typical mesoscale wind fields determined diagnostically from the available hourly surface and twice-daily upper air observations would improve the quality of the characterization of the transport and dispersion. Results are presented for CAPTEX releases 3, 5 and 7.

The Cross-APpalachian Tracer EXperiment (CAPTEX) is a unique series of tracer releases, which besides testing a particular tracer technology, was conducted for the purpose of providing data to evaluate and improve computer models of pollutant dispersion and to provide insight into the mechanisms involved in long-range transport and dispersion (Ferber et al., 1986). A three-hour ground-level release of perfluoromonomethylcyclohexan (C_7H_{14}, PMCH) was made five times near Dayton, Ohio and twice from near Sudbury, Ontario when winds were expected to transport the tracer over the ground-level sampling network. Samplers were operated at 86 sites in Ohio, Pennsylvania, New Jersey, New York, New England and southern Canada at distances from 300 to 1100 km from the release site. Air concentrations were collected for 3 and 6 hour durations for several days following each release.

In an analysis of the CAPTEX releases 1, 2 and 5, Shi et al. (1990) used a Monte Carlo model to simulate the three-dimensional transport and dispersion of tracer, coupled with wind

**On assignment to the Office of Air Quality Planning And Standards, U.S. Environmental Protection Agency.

fields developed through objective analysis of available rawinsonde data. They illustrated that a dominate dispersion mechanism at these distances results from a combination of vertical mixing and wind shears in the vertical. Moran and Pielke (1994) describe the CAPTEX 2 episode as involving initially differential horizontal advection, with the tracer near the surface being transported northeast from Dayton while the tracer mixed initially aloft being transported nearly east. Subsequent vertical mixing during the second day of transport brings the tracer aloft down to the surface.

Davis et al. (1986) employed a Gaussian puff model to simulate CAPTEX releases 3 and 4. Results were compared in which vertical redistribution was allowed all the time or was limited to occur only from 1300 to 1600 LST. The simulation of ground-level concentration values was noticeably improved for CAPTEX 3 when the redistribution was limited to occur only during the afternoon hours. The results were less clear for CAPTEX 4, where the potential temperature difference between the surface and upper levels was considerably less than in CAPTEX 3, suggesting less potential for decoupling in CAPTEX 4 in comparison to CAPTEX 3.

Draxler (1987) employed a Gaussian puff model to simulate six of the seven CAPTEX releases, in which the potential for redistribution of mass in the vertical was computed every three hours. The boundary layer was divided into 300 m layers and new trajectories were started whenever vertical mixing distributed mass to other levels. Tests were conducted to investigate the effects of inclusion of increased temporal and spatial meteorological data in the diagnostic wind field interpolation analysis. The trajectory error was assessed using the distance between the centroid locations of observed and estimated surface concentration values. The most accurate trajectories were obtained when both spatial and temporal resolution were enhanced, such that the root-mean-squared (rms) trajectory error over all releases decreased from 180 km to 154 km for travel times of 24 to 42 hours. When vertical advection velocity was included in the trajectory estimates, the rms trajectory errors were further reduced to 144 km. The horizontal extent of the observed and estimated tracer cloud, as determined from the second moment of the surface concentration values, was typically 20% larger than observed for travel times of 12 to 36 hrs after release. The horizontal distribution of mass, which is a combination of turbulent dispersion and differential horizontal transport in the vertical, being overestimated is consistent with the result that the maximum concentrations tended to be underestimated by a factor of 2.

Godowitch (1989) presented surface concentration comparison results for six of the seven CAPTEX releases in which a Gaussian puff model (MESOPUFF II, Scire et al. 1984) was applied. Puffs within the mixed layer are advected using a wind speed and direction averaged over the vertical depth of the mixed layer. To reduce the computation burden, no provisions were made to split the puff in the vertical to allow for differential horizontal advection, and no provisions were made to limit the redistribution of mass in the vertical at nighttime. In contrast to the results of Draxler (1987) and Davis et al. (1986) which underestimated the maximum surface concentrations by as much as a factor of 2, the MESOPUFF II maximum surface concentration values were routinely overestimated by a factor of 2 or more.

METEOROLOGICAL DATA PREPARATION

Meteorological data available for use in developing the CAPTEX wind fields consisted of 122 National Weather Service (NWS) surface locations reporting hourly and 13 upper-air locations reporting twice-daily (0000 GMT and 1200 GMT) throughout the region. Furthermore, mesoscale wind fields developed using FDDA were available on an 80-km grid. Three wind field models were used to obtain a gridded field of meteorological data with a horizontal resolution of 18-km: MESOPAC II, CALMET, and CALMET using the mesoscale wind fields as STEP-1 inputs. In the discussion to follow, the latter modeled wind field is referred to as CALMET/MM4.

MM4/FDDA

Version 8 of the Penn State/NCAR Mesoscale Model - Generation 4 (MM4) was used to develop profiles of atmospheric wind, temperature, and moisture. The MM4 is a primitive equation, mesoscale, hydrostatic Eulerian framework hydrodynamic modeling system of the atmosphere (Anthes et al., 1987). The vertical layers are terrain following sigma surfaces. FDDA is incorporated as Newtonian nudging by adding to the general momentum and thermodynamic equation a "Force-Restore" term, to effectively nudge the numerical solution towards the observed data (Stauffer and Seaman, 1989; Stauffer et al., 1990). Nudging was applied for the east-west and north-south wind components, temperature and mixing ratio at all levels, except for temperature at the surface and at levels aloft within the lowest 6 layers of the analyses (top of sixth layer is typically 1500m).

The horizontal grid spacing of the MM4 simulation was 80 km in both dimensions, with a grid array size of 84 by 55 centered on 90°W longitude and 40°N latitude to cover the domain of the CAPTEX releases. The model simulation variables from the outer 2 columns and rows of grid points were not included in the data set to avoid the boundary effects typical of fixed-grid numerical models. The domain of the model's vertical coordinate system extended from the earth's surface to the 100 millibar pressure level (approximately 15 km above sea level).

MESOPAC II

MESOPAC II constructs hourly wind fields at two vertical layers: a lower layer wind field representing boundary layer flow, and an upper layer representing flow above the boundary layer. In the CAPTEX simulations, the lower layer wind field was averaged vertically from the ground to the mixing height. The upper layer winds were averaged from the mixing height to the 700 mb level. A network of 102 x 90 grid cells was used. The horizontal grid spacing was 18-km. The modeling domain included the entire CAPTEX area as well as a significant buffer zone to the west and south.

The calculation of the mixed layer averaged winds involves observational data from both surface stations reporting hourly winds and rawinsonde stations measuring profiles twice per day. MESOPAC II uses a procedure adapted from Draxler (1979) to extrapolate surface wind observations vertically through the mixed layer. The ratio of the surface wind speed to the mixed-layer averaged wind speed at the rawinsonde station is applied to adjust the surface wind observations. Similarly, wind direction adjustment factors are derived from the rawinsonde data and applied to the surface wind observations to obtain mixed-layer averaged values. A representative sounding (0000 or 1200 GMT) is selected based upon stability and time of day.

CALMET

The CALMET wind field module is based on the Diagnostic Wind Model (DWM), Douglas and Kessler (1988). A two-step procedure is involved in the computation of the gridded wind fields. In the first step, an initial guess field is adjusted for kinematic effects of terrain, slope flows, terrain blocking effects, and three-dimensional divergence minimization. The Step 2 procedure includes four substeps: inverse distance interpolation of observations into the Step 1 field, smoothing to reduce sharp gradients in the field, adjustments of vertical velocities using the O'Brien procedure (O'Brien, 1970), and divergence minimization. In the

CAPTEX simulations, the MM4 winds on the coarse (80-km horizontal grid) were interpolated to the 102 x 90 18-km CALMET grid and replaced the regular CALMET Step 1 field. Introduction of the MM4 winds by this means bypasses the CALMET adjustment of the MM4 data for fine-scale terrain effects. The assumption being made is that the coarse-grid MM4 winds already contain the most significant terrain effects of interest for the intended final grid resolution of the analysis, which in this case was 18 km. Results from a series of sensitivity analyses suggest this is a reasonable assumption for this version of MM4, at least for the eastern half of the United States going from 80 to 18 km grid resolutions. The CALMET winds were computed at 11 layers, with grid point heights of 10, 50, 100, 200, 400, 800, 1240, 1600, 2000, 2600 and 3500 m.

DISPERSION MODEL CALCULATIONS

MESOPUFF II was driven by each of the modeled wind fields to produce a set of three simulations of ground-level concentrations for each of the three CAPTEX releases (9 simulations in all). MESOPUFF II options were set as follows: 8 puffs/hour, 8 samples/hour, Gaussian vertical distribution, PG dispersion coefficients for < 10 km and Heffter formulations for > 10 km, a 2-layer wind field (below and above the mixed layer depth). For MESOPAC II wind fields the upper layer extended to 700 mb, and for CALMET wind fields the upper layer extended to 3000 m.

CALPUFF was driven by two of the modeled wind fields, CALMET and CALMET/MM4, to produce a set of two simulations of ground-level concentrations for each of the three CAPTEX releases (6 simulations in all). CALPUFF options were set as follows: the maximum puff travel distance during one sampling step (controls the puff generation rate) was 1.8 km, the integrated puff algorithm was the sampling function, Gaussian vertical distribution, dispersion coefficients were set to be the same as those in the MESOPUFF II simulations.

In both the MESOPUFF II and CALPUFF simulations, chemical transformations, dry deposition and wet removal were not modeled because PMCH was assumed to be inert and non-depositing. No attempts were made to optimize the choice of model options within CALPUFF, such as the dispersion coefficients, mode of incorporation of MM4 data, meteorological vertical layer structure. In effect, CALPUFF was run in a mode designed to make it most like MESOPUFF II, in order that the effects of different wind fields and transport characterization could be identified. MESOPUFF II and CALPUFF differ in the way transport winds are computed for each puff. MESOPUFF II uses a two-layer wind field, the lower layer for the transport of puffs within the mixed layer, and an upper layer field for puffs above the mixing height. For a surface release, as in CAPTEX, MESOPUFF II will always use the lower layer (mixed-layer averaged) wind field. CALPUFF internally computes for each sampling step, a transport wind averaged over the depth of the puff from the multi-layer winds provided to it from CALMET. As the puff grows in the vertical, the depth through which the wind is averaged is increased.

MODEL RESULT COMPARISONS

For releases 3 and 5 for those receptors closer to the release location, 3-hour concentration values were collected, whereas for all other receptor locations 6-hour average concentration values were collected. The modeling results were processed to provide

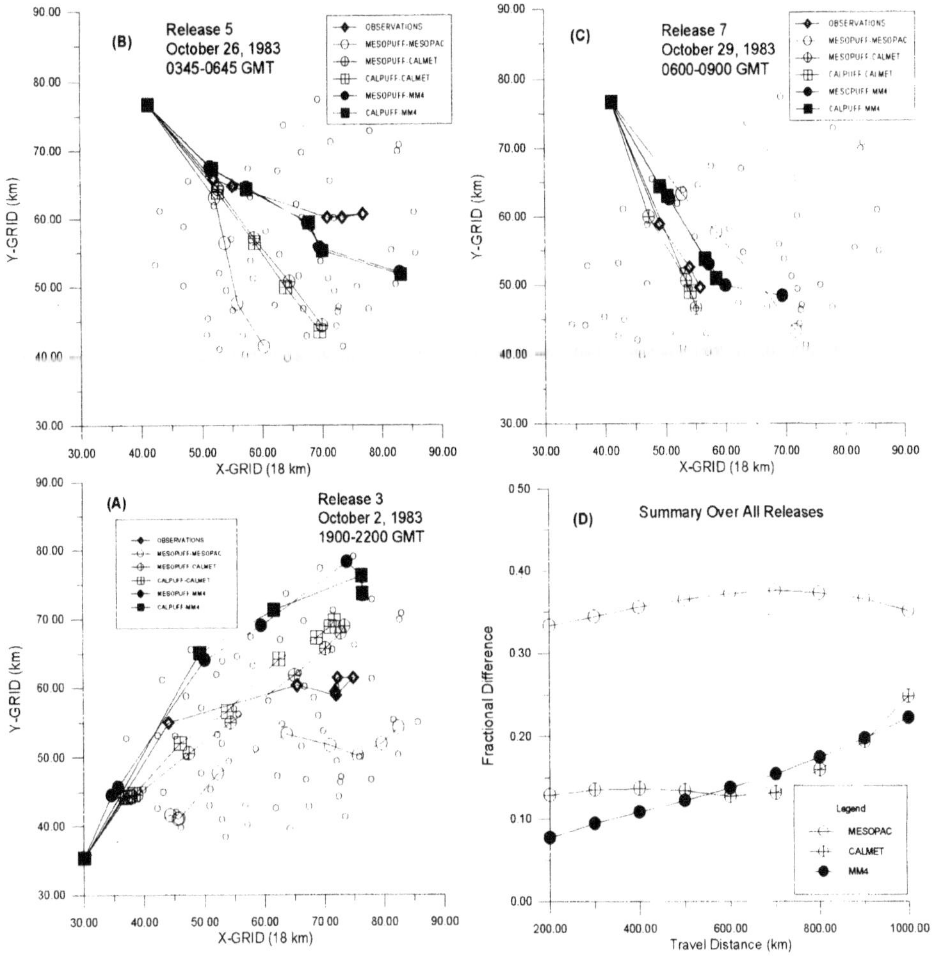

Figure 1. Summary of trajectory results for each of the releases are shown in parts (a), (b) and (c). In part (d) is shown the fractional difference computed between the observed and simulated trajectories, summarized over all releases for each of the three wind fields. The open smaller circles indicate sampling locations. The X and Y coordinates have been specified in terms of the meteorological grid, which had an 18-km grid size.

corresponding 3-hour or 6-hour concentration values for comparison with observed values. As there were far fewer 3-hour values than 6-hour values, for the analyses to be discussed, all of the concentration values were processed to provide 6-hour average concentrations for each observation site whenever possible. A background concentration of 3.4 fL/L was subtracted in the reported observed concentration values. As discussed in Ferber et al. (1986, p. 16), observed values less than twice the background level have greater analytic uncertainty, thus in the results discussed, we have only used pairs in which the observed or simulation concentration values are greater than 3 fL/L, and then only for 6-hour periods having at least 3 values above 3 fl/L for analysis.

Trajectory Errors

For each 6-hour period, the centroid position was computed for the observed and simulated tracer puff as it was transported downwind over time as,

$$X_c = \frac{\sum (X\chi)}{\sum (\chi)} \;;\; Y_c = \frac{\sum (Y\chi)}{\sum (\chi)}, \qquad (1)$$

where χ is the concentration value observed or simulated at a receptor having east-west and north-south coordinates, X and Y respectively. The results for each release are shown in Figure 1. It is concluded that the simulated trajectories are sensitive to the manner in which the wind field is characterized, but are insensitive to the model employed. This can be seen by the fact that the simulated trajectories are nearly identical for MESOPUFF and CALPUFF when the models have the same input wind fields.

In summarizing the differences seen in the simulated trajectories, we have grouped the results by the manner in which the wind fields were characterized, as the model to model differences were negligible. The fractional difference in the trajectory errors (separation between observed and simulated trajectories divided by downwind transport distance) are seen to be twice as large using the MESOPAC II winds versus using CALMET or MM4 winds. There appears to be greater divergence in the various results for Releases 3 and 5, which had larger horizontal wind direction shear in the vertical, versus that seen for Release 7.

Paired Concentrations

For each 6-hour period, the second moment of the concentration, S, about the centroid position was computed. The second moment characterizes the horizontal extent of the puff and represents a measure of the puff horizontal dispersion . Typically S was roughly 25% of the travel distance downwind. The distance from the nearest to furthest receptor was approximately 800 km and there where 86 receptor locations. The average distance between the receptors was roughly 86 km = $[\,(800)^2/86\}]^{1/2}$, which meant that typically there were 4 to 6 receptors for analysis for the shorter travel times (300 to 600 km transport) and from 8 to 16 receptors for analysis for the larger travel times (600 to 900 km transport). This suggests that the maximum concentration was likely not well characterized and is uncertain both in the observations and in the simulations.

Figure 2(a) summarizes the results obtained by dividing the simulated and observed maximum concentration values for each six-hour period. There is no noticeable trend as a function of travel time (hours after release), but clearly there is a tendency to overestimate the maximum concentration value by roughly a factor of 3.7. Some differences were seen between the various wind fields employed. For instance, the results for the maximum concentration ratio values for each pairing were (geometric average and standard deviation): MESOPUFF /MESOPAC (4.94, 3.50); MESOPUFF&CALPUFF/CALMET (4.84, 3.96); MESOPUFF &CALPUFF/MM4 (3.09, 3.01). The trend for the averages and variance to decrease in going from MESOPAC winds to MM4 winds is not statistically significant. As shown in the figure, the average geometric mean for the ratio values is 3.68, the respective geometric standard deviation is 3.54.

Figure 2(b) summarizes the results obtained by dividing the simulated and observed lateral dispersion (as determined from the second moment of the concentration values about the centroid position) for each six-hour period. As with Figure 2(a) there does not appear to be any trend with travel time following release, however there is a clear trend to underestimate the horizontal extent of the tracer puff for each six-hour period. There were only slight differences

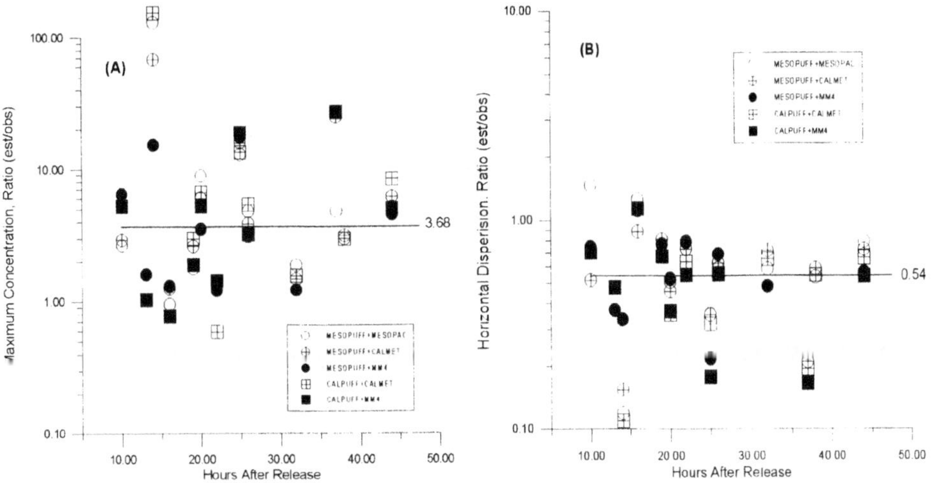

Figure 2. Summary of the comparison across all experiments of the maximum concentrations and lateral dispersion observed and simulated for each 6-hour period. In part (a) is shown as a function of hours after release the ratio formed by dividing the simulated maximum by the observed 6-hour average concentration. In part (b) is shown as a function of hours after release the ratio formed by dividing the simulated lateral dispersion, second moment of the concentration values about the centroid position, by the observed lateral dispersion.

to be seen for the different wind fields employed. The overall geometric average was 0.54, shown in Figure 2(b), and the geometric standard deviation was 1.81. If the only difficulty or "bias"in the simulation was to underestimate the horizontal extent of the puffs, the effect on the simulated maximum concentration values would be proportional to the inverse square of the bias in the horizontal dispersion, which would be $3.43 = (1/0.54)^2$. It is concluded that the underestimation seen in the simulated horizontal dispersion is able to explain the overestimates seen in the simulated maximum concentration values. The large scatter seen in Figure 2 in part may relate to the stochastic nature of atmospheric dispersion, but also is traceable to poor definition of the maximum concentration values and lateral dispersion.

CONCLUSIONS

The primary purpose of this investigation was to assess whether different methods of characterizing the wind fields affect the performance of the simulated trajectory. The comparison results presented were conducted with simplified puff dispersion model assumptions, hence the model to model differences were minimized. The results shown in Figure 1 suggest a noticeable improvement in the puff simulation results of the centroid trajectory when mesoscale dynamic wind fields developed using FDDA were used to characterize the time and space varying wind field, versus using mesoscale wind fields determined diagnostically from the available hourly surface and twice-daily upper air observations. The favorable comparison of the CALMET wind field results with the MM4 results is influenced by Release 3 results, where CALMET happened to well characterize the average transport direction of the duration. The trajectory simulations for Releases 5 and 7 derived using MM4 winds are as good or better than the results obtained with CALMET wind fields derived solely from routine National Weather Service observations. We conclude that use of MM4 is preferred and that both CALMET and MM4 wind fields provide better simulations of the trajectories than MESOPAC wind fields.

The results shown in Figure 2 suggest that the simulation assumptions employed in these results, consistently underestimate the horizontal extent of the tracer puff as it is transported downwind. The centroid maximum surface concentration was found to be correspondingly overestimated and relatively insensitive to the mesoscale wind characterization. In these simulations, no provisions were made to address delayed shear enhancement of the dispersion as described by Moran and Pielke (1994) and Shi et al. (1990). Inclusion of some sort of puff splitting is obviously warranted, but the computational demands are not trivial if one is attempting to develop an operational model for routine use. Furthermore, in those cases where the puff model dynamics have been enhanced, for example Draxler (1987) and Davis et al. (1986), there was a tendency to underestimate the surface maximum concentrations. We have left for further investigation development of the optimal simulation assumptions for best characterization of the surface distribution of concentration values.

DISCLAIMER

The information in this document has been funded wholly or in part by the United States Environmental Protection Agency under an Interagency Agreement (DW13937039-01-0) to NOAA. It has been reviewed in accordance with the Agency's peer and administrative review policies and approved for presentation and publication. Mention of trade names or commercial products does not constitute endorsement or recommendation for use.

REFERENCES

Anthes, R.A., E.Y. Hsie, and Y.H. Kuo, 1987: Description of the Penn State/NCAR mesoscale model version 4 (MM4). NCAR Tech Note, NCAR/TN-282+STR, Available from the National Center for Atmospheric Research, PO Box 3000, Boulder CO 80307, 66 pp.

Davis, W.E., A.R. Olsen, B.T. Didier and D.W. Damschen, 1986: A comparison of Meso-Alpha MLAM Model Results with Observations from CAPTEX. 5th J. Conf. On Appl. of Air Poll. Met., Am. Meteorol. Soc., Chapel Hill, NC, 69-72.

Draxler, R.R., 1979: Modeling the results of two recent mesoscale dispersion experiments. Atmos. Environ., 13, 1523-1533.

Douglas, S., and R. Kessler, 1988: User's guide to the diagnostic wind field model (Version 1.0). Systems Applications, Inc., San Rafael, CA, 48 pp.

Draxler, R.R., 1987: Sensitivity of a Trajectory Model to the Spatial and Temporal Resolution of the Meteorological data during CAPTEX. J. Clim. Appl. Meteor., 26, 1577-1588.

Ferber, G.J., J.L. Heffter, R.R. Draxler, R.J. Lagonmarsino, F.L. Thomas, R.N. Dietz, and C.M. Benkovitz, 1986: Cross-Appalachian Tracer Experiment (CAPTEX '83) Final Report. NOAA Technical Memorandum ERL ARL-142, Air Resources Laboratory, NOAA Environmental Research Laboratories, Silver Spring, MD 20910, 60 pp.

Godowitch, J.M., 1989: Evaluation and Testing of the MESOPUFF II Model System with CAPTEX Measurements. 6th J. Conf. On Appl. of Air Poll. Meteor., Am. Meteorol. Soc., Anaheim, CA, 56-59.

O'Brien, J.J., 1970: A note on the vertical structure of the eddy exchange coefficient in the planetary boundary layer. J. Atmos. Sci., 27, 1213-1215.

Moran, M.D., and R.A. Pielke, 1994: Delayed Shear Enhancement in Mesoscale Atmospheric Dispersion. 8th J. Conf. On Appl. of Air Poll. Meteor., Am. Meteorol. Soc., Nashville, TN, 96-103.

Scire, J.S., D.G. Strimatis, and M.E. Fernau, 1995: CALPUFF: A Non-Steady-State Modeling System. 21st International Technical Meeting on Air Pollution Modeling and Its Application. November 6-10, 1995, Baltimore, MD, USA.

Scire, J.S., F. Lurmann, A. Bass, S. Hanna, 1984: Development of the MESOPUFF II Dispersion Model. NTIS PB84-183753. Environmental Protection Agency, Research Triangle Park, NC. pp ??.

Shi, B., J.D. Kahl, Z.D. Christidis, and P.J. Samson, 1990: Simulation of the Three-Dimensional Distribution of Tracer During the Cross-Appalachian Tracer Experiment. J. Of Geophy. Res., 95, No. D4, 3693-3703.

Siebert, P., 1993: Convergence and Accuracy of Numerical Methods for Trajectory Calculations. J. Appl. Meteor., 32, 558-566.

Stauffer, D.R. and N.L. Seaman, 1989: Use of four dimensional data assimilation in a limited-area mesoscale model. Part I: Experiments with synoptic-scale data. Mon. Wea. Rev., 118, 1250-1277.

Stauffer, D.R., N.L. Seaman, and F.S. Binkowski, 1990: Use of four dimensional data assimilation within the planetary boundary layer. Mon. Wea. Rev., 119, 734-754.

DISCUSSION

B. BORNSTEIN: What plume rise formulations were used in the forest fire simulations ?

J. SCIRE: The plume rise equations in the CALPUFF buoyant area source algorithm are based on those in Weil (1988), except that the Boussinesq approximation is not applied. The conservation equations are numerically integrated to produce estimates of plume height from the forest fires as a function of the fire characteristics, as determined from the Emissions Production Model, and meteorological conditions.

COMPARISON OF MEASURED AND MODELED SURFACE FLUXES OF HEAT, MOISTURE, AND CHEMICAL DRY DEPOSITION

Jonathan E. Pleim[1], John F. Clarke[1], Peter L. Finkelstein[1], Ellen J. Cooter[1], Thomas G. Ellestad[2], Aijun Xiu[3], and Wayne M. Angevine[4]

[1] Atmospheric Sciences Modeling Division, Air Resources Laboratory, National Oceanic and Atmospheric Administration, Research Triangle Park, NC 27711 (on assignment to the National Exposure Research Laboratory, US Environmental Protection Agency)
[2] National Exposure Research Laboratory, US Environmental Protection Agency, Research Triangle Park, NC 27711
[3] Environmental Programs, MCNC, Information Technologies Division, Research Triangle Park, NC
[4] Cooperative Institute for Research in Environmental Sciences, University of Colorado, and NOAA Aeronomy Laboratory, Boulder, CO 80303

INTRODUCTION

Realistic air quality modeling requires accurate simulation of both meteorological and chemical processes within the planetary boundary layer (PBL). Surface energy and moisture fluxes control the temperature and humidity profiles. Similarly, chemical fluxes (dry deposition) have an important influence on PBL concentrations of trace chemical species. Therefore, accurate and consistent methods for simulation of both meteorological and chemical surface exchange processes are critical for realistic modeling of boundary layer atmospheric chemistry.

In vegetated areas, the primary pathway for surface fluxes of moisture as well as many gaseous chemicals is through vegetative transpiration. Therefore, stomatal conductance is a key element of both meteorological and chemical surface flux modeling. Typically, evapotranspiration is parameterized in the land-surface sub-model of meteorological models while dry deposition fluxes are parameterized in chemical transport models. These parameterizations are usually completely independent of each other even though they involve many of the same processes such as stomatal emission or uptake. Thus, in the course of our model development program, we are developing improved modeling techniques which will consistently simulate both meteorological and chemical surface fluxes.

Our regional and mesoscale modeling system uses the Pennsylvania State University-National Center for Atmospheric Research Mesoscale Model Version 4 (MM4) (Anthes et al., 1987), with four dimensional data assimilation (4DDA) (Stauffer and Seaman, 1990), to supply meteorology for our comprehensive atmospheric chemistry and transport model, the Regional Acid Deposition Model (RADM) (Chang et al., 1987). Evaluation of the MM4/RADM system has revealed deficiencies in the simulation of surface energy fluxes and PBL dynamics. These processes are particularly critical for air quality modeling where PBL depth largely controls ground level chemical concentrations, and surface temperatures greatly influence emission rates of many chemical species including organic biogenics. Therefore, we have been developing a more advanced land surface and PBL model for mesoscale models as described by Pleim and Xiu (1995). We are also developing a dry deposition model which uses many of the same parameters computed in the evapotranspiration model.

For evaluation of the surface energy fluxes, dry deposition velocities, and PBL evolution, we compare model simulations to field measurements made near Bondville, Illinois during the summer of 1994 (Finkelstein et al., 1995). The field data include eddy correlation measurements of fluxes of heat, momentum, water vapor, carbon dioxide, sulfur dioxide and ozone. Also, PBL heights were derived from radar wind profiler data measured at the same site. Thus, these sophisticated and comprehensive measurements of both meteorological and chemical surface fluxes, as well as PBL processes, provide an excellent test for our combined land surface and dry deposition model.

MODEL DESCRIPTION

Pleim and Xiu (1995) describe the development and initial testing of a land surface and PBL model for use in mesoscale models. Since that paper we have continued development of this model and applied it in a modified version of MM4. The model's key elements are a surface model including soil moisture and evapotranspiration based on the work of Noilhan and Planton (1989), and a non-local closure PBL model developed by Pleim and Chang (1992). The surface model includes a two layer soil model with a 1 cm surface layer and a 1 m root zone layer. Evaporation has three pathways: direct soil surface evaporation, vegetative transpiration, and evaporation from wet canopies. Ground surface (1 cm) temperature is computed from the surface energy balance using a force-restore algorithm for heat exchange within the soil. Stomatal conductance is parameterized according to root zone soil moisture, air temperature and vapor pressure deficit, leaf area index (LAI), and photosynthetically active radiation (PAR). Our modifications to the original Noilhan and Planton (1989) model include the addition of a canopy shelter factor to account for shading within denser canopies, modification of some of the stomatal functions with respect to environmental parameters, and inclusion of a data assimilation scheme similar to the technique described by Bouttier et al. (1993).

The assimilation scheme uses the errors in the predicted values of temperature and relative humidity for the lowest model layer as compared with gridded analyses of surface-based observations. These errors are used to nudge root zone and upper layer soil moisture. The concept is that errors in low level temperature and humidity may be due to erroneous partitioning of latent and sensible heat surface fluxes which in turn may be caused by unrealistic soil moisture conditions. We recognize that soil moisture, while critically important to surface exchange processes, is very difficult to initialize accurately and is very crudely modeled. Therefore, it is an obvious target for dynamic adjustment, but not by direct 4DDA since widespread operational measurements of soil moisture are not available. Indirect nudging depends on strong coupling between near surface temperature and humidity, and soil moisture through evaporation and evapotranspiration. Thus, nudging coefficients must be carefully prescribed to act only when and where this coupling is strong so that soil moisture nudging is not employed when model errors have other causes. Bouttier et al. (1993) used statistical analyses of a 1-D model to derive assimilation coefficients. However, we reason that statistical analyses of the behavior of a deterministic model should not be necessary. Therefore, we prescribe assimilation coefficients as functions of model parameters such as solar insolation, air temperature, leaf area, soil texture, vegetation coverage, and aerodynamic resistance, in order to nudge most strongly when surface-atmosphere coupling is greatest.

This assimilation technique is far from flawless since it may cause changes in soil moisture when atmospheric errors have nothing to do with surface fluxes. Therefore, the magnitude of the assimilation coefficients is small compared to physical forcings and takes several days to produce a significant change in soil moisture. Thus, short term errors due to unrelated phenomena, such as mistiming of frontal passages or erroneous predictions of cloud cover, have minimal impact on soil moisture adjustment.

Two key parameters produced by the surface model are also used in the dry deposition model. These are the aerodynamic resistance, which is a measure of turbulent transport within the atmospheric surface layer, and the stomatal resistance, which is a measure of resistance to uptake or emission through stomatal pores. Since both of these parameters depend only on physical characteristics of the atmosphere and vegetation they can be applied to any transported quantity for which these processes are important. The only adjustment needed for application to gaseous dry deposition is to weight the stomatal resistance by the ratio of molecular diffusivities of the chemical species and water vapor. Other aspects of the dry deposition model, including cuticle resistance, ground resistance and water resistance, are computed according to the scheme used in the Acid Deposition and Oxidant Model (ADOM) which was originally developed by Pleim et al. (1984) and evaluated and modified by Padro et al. (1991).

FIELD EXPERIMENT

The US Environmental Protection Agency's (EPA) nationwide network to monitor dry deposition of gases and particles, the National Dry Deposition Network (NDDN), measures concentrations of pollutants and meteorological variables relevant to deposition and then computes deposition flux as the product of the measured concentration and model computed deposition velocity. Annual and seasonal dry deposition as derived through the inferential model and concentration measurements are reported for the 50-site EPA NDDN (Clarke and Edgerton, 1993). In order to evaluate the model used for the NDDN calculation, a series of field studies making direct flux measurements and concurrent meteorology has been made at several NDDN sites which differ in terrain, climate, soil, and vegetation cover. These data are also very useful for evaluation of our land surface and dry deposition models. We have selected the 1994 Bondville, Illinois field experiment for our initial evaluation of model simulations of surface fluxes of heat, moisture, and ozone dry deposition velocity.

The measurement system at Bondville has three major instrument groups; one to measure trace gas fluxes by eddy correlation or gradient techniques, one to measure the components of the energy balance, and one to measure the variables needed by the deposition velocity models; as well as a data acquisition system. Fluxes of ozone, sulfur dioxide, carbon dioxide, water vapor, heat and momentum are measured by eddy correlation. For the gases, ambient air is pulled from close (4 cm) to a three axis sonic array ($\approx$ 7 m above ground) through a draft tube and filter to the fast response instruments that are housed in a temperature controlled box at the base of the tower. The fast ozone analyzer, based on the work of Ray et al.(1986), employs the chemiluminescent reaction of ozone with eosin-Y dye, which is borne in a carrier of ethylene glycol. Fast response data for water vapor and carbon dioxide are produced by a LICOR model 6262 infrared absorption analyzer. Details of the measurements are reported by Finkelstein et al. (1995).

The field site was at the Bondville research station (40.05°N, 88.37°W) operated by the Illinois State Water Survey. This area is very flat, with excellent fetch. The measurements were exclusively over corn. They began in August and extended past the corn's senescence in October. The period for comparison to model data was August 15-29, 1994. Since the measurement site was located near the north end of a corn field, the data were screened for acceptance only when the wind direction from all sensors (5 and 12 m vanes and 7 m sonic) was between 110° and 250°. We also checked the net energy flux, net radiation minus the fluxes of sensible and latent heat, heat fluxes into the ground, and soil heat storage, for each half-hour period. Time periods with a net energy imbalance >150 W/m^2 were excluded from the final data base.

MODEL APPLICATION

The modified version of MM4, including the land surface and dry deposition model, was run in three 5-day sections overlapping by 12 hours for initialization. Soil temperature and soil moisture in both layers were initialized in each section from the previous run. For the first run, soil moisture was initialized according to climatalogical rainfall patterns while accounting for soil texture. The model was run in hydrostatic mode with 80x80 km grid cells and 30 vertical layers up to 100 mb using a terrain following sigma-p coordinate system on a Lambert conformal map projection. In addition to the soil moisture assimilation scheme described above, direct 4DDA was used for winds at all levels, and temperature and humidity mixing ratio above the PBL. Land use related parameters were derived from a detailed inventory where the fractional coverage of 125 vegetation/land use types were determined for each model grid cell. Grid cell values of each parameter were then computed using appropriately weighted aggregation techniques. LAI and vegetation coverage are linearly averaged while roughness length is logarithmically averaged and minimum stomatal resistance is inversely averaged and weighted by LAI. The model grid cell LAI resulting from this process was 3.3 while measurements at the Bondville site were slightly greater than 3 for the study period. The model domain covers most of North America as shown in Figure 1, an example of gridded model output for dry deposition velocity of ozone.

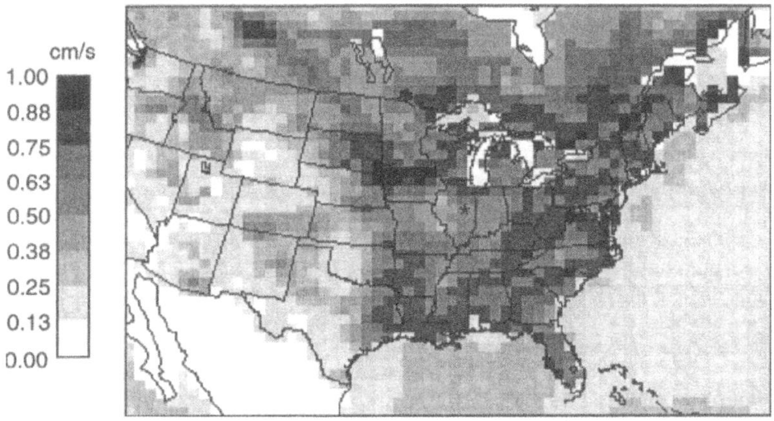

Figure 1. Example model output for dry deposition velocity of ozone at 1100 LST on August 27, 1994. The asterisk shows the location of the Bondville, IL site.

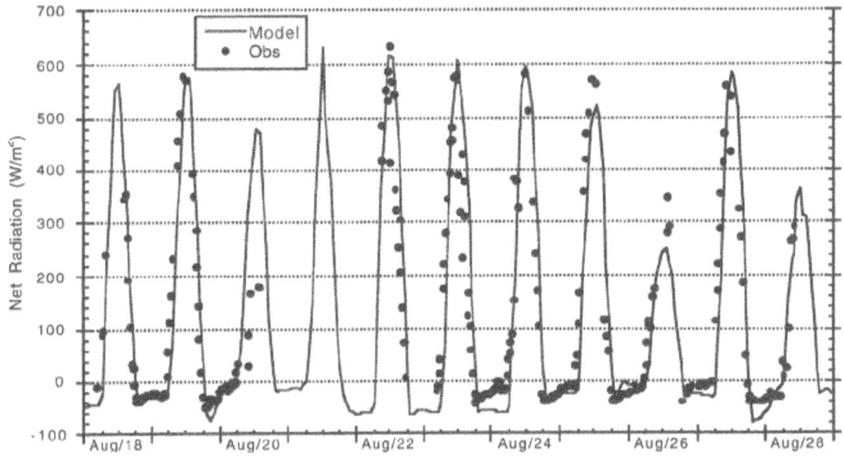

Figure 2. Modeled and observed surface net radiation at Bondville, IL for August 18-29, 1994

COMPARISON RESULTS AND DISCUSSION

Time series of model results for the grid cell which includes the measurement site were extracted for comparison to field measurements. Figure 2 shows observed and model simulated surface net radiation. Each observed data point represents a 30-minute average and is plotted at the start of the averaging interval. Gaps in the data result from the screening according to wind direction and energy balance. The model data are plotted as continuous lines from hourly output. Figure 2 shows that modeled net radiation compares quite favorably with observations particularly during clear days. The only days with substantial cloudiness were August 20, 26 and 28. On August 26 the model under-predicted the maximum with respect to the observations by almost 100 W/m^2 which is not too bad considering the coarse resolution of the model. On August 20, however, the model greatly over-predicted compared to the two observed data points near the middle of the day by under-predicting cloudiness associated with a cold front passage.

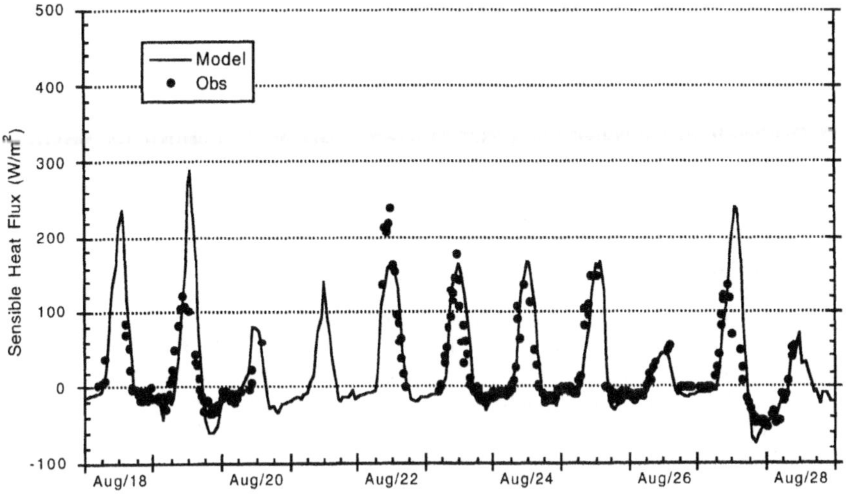

Figure 3. Modeled and observed surface sensible heat flux at Bondville, IL for August 18-29, 1994

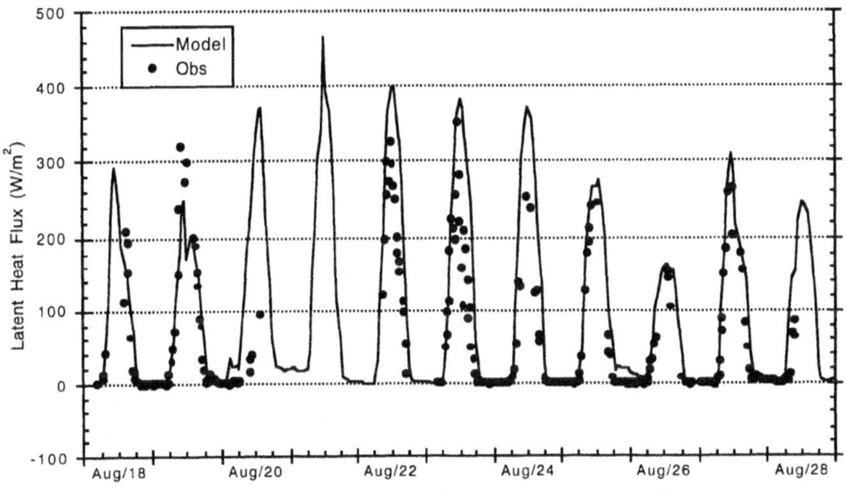

Figure 4. Modeled and observed surface latent heat flux at Bondville, IL for August 18-29, 1994

Figures 3 and 4 show time series of observed and predicted sensible and latent heat fluxes. For most days the model predictions are quite close to the observations. The greatest discrepancies occur during the early days, August 18 and 19. The 18th is difficult to evaluate since many data were missing during the middle of the day. However, the 19th clearly shows a substantial over-prediction of sensible heat flux and a lesser under-prediction of latent heat flux which indicates that the model soil moisture may be too low. August 22 shows the opposite result with sensible heat flux under-predicted and latent heat flux over-predicted suggesting that the modeled soil is too wet. Most of the other days show quite good agreement with the observations except the 27th where sensible heat is substantially over-predicted. In this case, however, the over-prediction is not balanced by under-predictions in any of the other energy components. Latent heat is very close to the measurements and soil heat flux (not shown) is also over-predicted. Unlike the model, the observed fluxes do not necessarily satisfy surface energy balance. In fact, during every day in this period there was a net energy imbalance of more than 100 W/m^2 at mid-day which leaves some uncertainty in the measurements.

The relatively poor performance of the modeled surface fluxes near the beginning and better performance later in the period suggest that the soil moisture assimilation technique is improving the estimates of soil moisture and thereby the partitioning of surface energy fluxes. Thus, the soil moisture assimilation technique seems to be able to overcome the handicap of a poor initialization in a period of a few days.

The height of the convective boundary layer (CBL) was measured for several days during this period by a 915 MHz boundary-layer profiler system. Angevine et al. (1994) describe the use of a similar system at another site. CBL height is deduced as the height of the peak signal-to-noise ratio which has been shown to occur within the capping inversion or entrainment zone. Therefore, the degree of confidence of CBL height estimates depends on the strength and depth of the inversion. Figure 5 shows time series of CBL heights derived from the profiler system along with the model simulated PBL heights. The model computes PBL height according to the technique suggested by Holtzlag et al. (1990) which uses bulk Richardson number calculated from the surface layer to the top of the PBL. This method is intended to result in a PBL height which includes the entrainment zone. Therefore, the comparability of the observed and modeled estimates depends on the depth of the entrainment zone since the model defines the top of the PBL as the top of the entrainment zone and the profiler estimates tend to correspond to the middle of the entrainment zone. The first three days (August 15-17) had the lowest uncertainty in the observed boundary layer heights, and also show the best agreement with the model. On the other days, the boundary layer was less well defined or the entrainment zone was deeper, resulting in greater uncertainty in the observations.

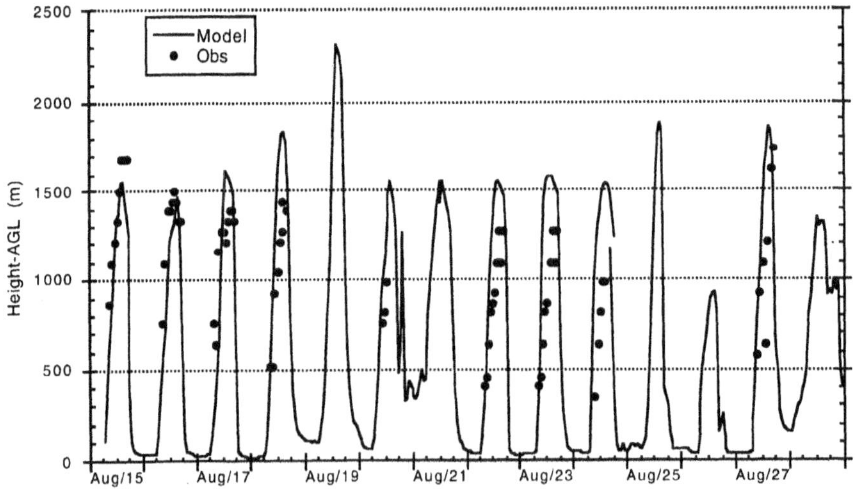

Figure 5. Modeled and observed PBL heights at Bondville, IL for August 15-29, 1994

Dry deposition fluxes were measured by eddy correlation for ozone and sulfur dioxide. Deposition velocities were computed as the fluxes divided by the concentrations. However, the SO_2 concentration measurements were usually so low that the quotients contained excessive uncertainty. Therefore, we used only the ozone dry deposition velocities for this study. Figure 6 shows time series of model estimates and observations of ozone dry deposition velocities. In general, the modeled daily peaks of dry deposition velocity agree well with observed peaks for most days, usually within 0.1-0.2 cm/s. The first two days of the comparison (4th and 5th of the simulation) show model under-predictions at the daytime peaks, afternoons, and the night of the 18th. The daytime, and particularly the afternoon, under-predictions may be due to overestimates of stomatal resistance. This is consistant with the comparisons of sensible and latent heat fluxes (Figures 3 and 4) which show tendencies for over-predictions of sensible and under-predictions of latent, both of which are indicative of underestimates of evapotranspiration. Comparison of observed and predicted friction velocity (not shown) also supports this hypothesis since the agreement was quite good for these days. Therefore, differences in dry deposition velocity were more attributable to differences in surface resistances than aerodynamic resistances. The relatively high observed dry deposition velocities during the night of the 18th remain unexplained.

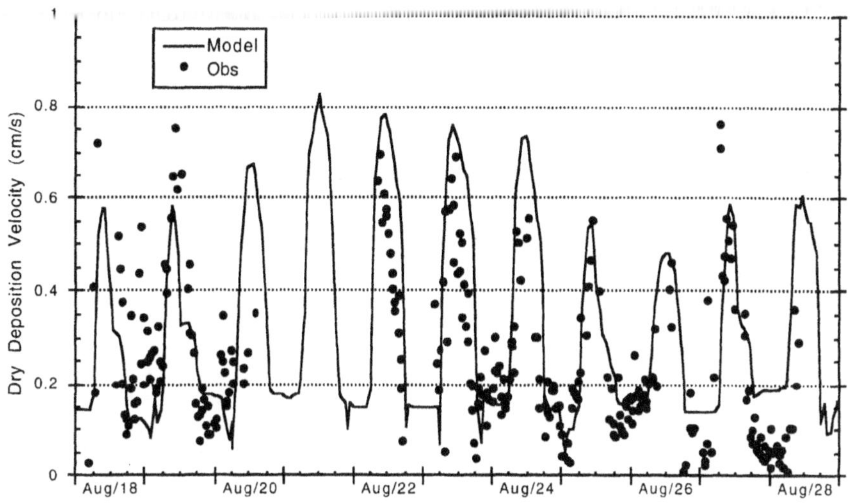

Figure 6. Modeled and observed dry deposition velocities of ozone at Bondville, IL for August 18-29, 1994

Daytime model estimates of dry deposition velocity for the rest of the period are mostly equal to or greater than the observations. The only exceptions are for two data points on the morning of August 27 at 630 and 700 CST. At these times the ozone concentrations were quite low resulting in relatively high deposition velocities. In the cases of August 22 and 23, the lower observed values in the afternoons probably result from partial afternoon cloudiness causing partial stomatal closure. Figure 2 shows reduction of observed net radiation relative to the model on both afternoons indicating partial cloudiness which the model largely missed.

CONCLUSIONS AND FUTURE WORK

Field measurements of meteorological and chemical surface fluxes as well as PBL profiler data have proven to be very valuable for testing and evaluation of surface and PBL modeling techniques. The model system used here is a prototype of an operational system for air quality research and policy analysis. The model was run on a continental scale with 80 km grid resolution and 30 vertical layers. There has been no attempt to tailor the model inputs or algorithms to this particular test case. Therefore, this comparison study, while representative of only one site, should have some general significance.

The model is based on MM4 with several important modifications. The surface and PBL portions of the model have been entirely replaced with a more sophisticated treatment of soil moisture and evapotranspiration as well as a new non-local closure PBL model. Comparison with the Bondville field measurements shows that the model well simulates surface fluxes and PBL development, particularly after a few days while the nudging scheme improves the model estimates of soil moisture. A particularly innovative aspect of this work is the use of the same parameterizations for both moisture fluxes and chemical dry deposition. Comparison of ozone dry deposition velocities demonstrate the promise of this technique for trace gases with a significant stomatal component.

Model development is continuing. We are currently making the same surface and PBL changes to MM5, which allows greater flexibility in terms of nested domains and non-hydrostatic dynamics (Grell et al., 1993). The modified MM5, along with the dry deposition model will be incorporated into the EPA's Models-3 system (Byun et al., 1995) for comprehensive multiscale air quality modeling.

Clearly, further comparison studies in different environments will add to our understanding and evaluation efforts. Also, we hope to involve other chemical species, such as SO_2, in dry deposition comparisons. Our next comparison study will be with field measurements made in southern Kentucky during the summer of 1995 in a soybean field using the same basic measurement system. These data should provide a greater challenge to the model since the period of measurement covers the entire growing season so there are tremendous changes in LAI and plant activity. There were also significant variations in moisture conditions. In addition, this experiment was coincident with the Nashville intensive field experiment of the Southern Oxidant Study which will be a major proving ground for the Models-3 system.

This paper has been reviewed in accordance with the U.S. Environmental Protection Agency's peer and administrative review policies and approved for presentation and publication. Mention of trade names or commercial products does not constitute endorsements or recommendation for use.

REFERENCES

Angevine, W.M., White, A.B., and Avery, S.K., 1994, Boundary-layer depth and entrainment zone characterization with a boundary-layer profiler, *Bound.-Layer Meteor.*, 68:375-386.

Anthes, R. A., Hsie, E.-Y., and Kuo, Y.-H., 1987, Description of the Penn State/NCAR Mesoscale Model Version 4 (MM4), NCAR Technical Note, NCAR/TN-282+STR, 66pp.

Bouttier, F., Mahfouf, J.F., and Noilhan, J., 1993, Sequential assimilation of soil moisture from atmospheric low-level parameters. Part I: sensitivity and calibration studies, *J. Appl. Meteor.*, 32:1335-1351.

Byun, D.W., Hanna, A.H., Coats, C.J., and Hwang, D., 1995, Models-3 air quality model prototype science concept development . (in press) Proceedings of the AWMA Specialty Conference on Regional Photochemical Measurement and Modeling Studies, San Diego, CA, November 7-12, 1993.

Chang, J. S., Brost, R. A., Isaksen, I. S. A., Madronich, S., Middleton, P., Stockwell, W. R. and Walcek, C. J., 1987, A three-dimensional Eulerian acid deposition model: Physical concepts and formulation, *J. Geophys. Res.*, 92:14,681-14,700.

Clarke, J.F., and Edgerton, E.S., 1993, Dry deposition flux calculations for the National Dry Deposition Network, EPA-600/R - 93/065, U.S. Environmental Protection Agency, Research Triangle Park, NC 27711, 91pp.

Finkelstein, P.L., Clarke, J.F., and Ellestad, T.G., 1995, Measurement of dry deposition for deposition velocity model evaluation, Proceedings of the 88th Annual AWMA Meeting & Exhibition, San Antonio, Texas, June 18-23, 1995.

Grell, G.A., Dudhia, J., and Stauffer, D.R., 1993, A Description of the Fifth-Generation PENN STATE/NCAR Mesoscale Model (MM5), NCAR Technical Note, NCAR/TN-398+IA.

Holtslag, A.A.M., de Bruijn, E.I.F., and Pan, H.-L., 1990, A high resolution air mass transformation model for short-range weather forecasting, *Mon. Wea. Rev.*, 118:1561-1575.

Noilhan, J. and Planton, S., 1989, A simple parameterization of land surface processes for meteorological models. *Mon. Wea. Rev.*, 117:536-549.

Padro, J., den Hartog, G., and Neumann, H.H., 1991, An investigation of the ADOM dry deposition module using summertime O_3 measurements above a deciduous forest, *Atmos. Environ.*, 25A:1689-1704.

Pleim, J. , Venkatram, A., and Yamartino, R., 1984, ADOM/TADAP Model Development Program, Volume 4, The Dry Deposition Module. ERT, Inc, Concord, MA.

Pleim, J. E. and Chang, J. S., 1992, A non-local closure model for vertical mixing in the convective boundary layer, *Atmos. Environ.*, 26A:965-981.

Pleim, J. E. and Xiu, A., 1995, Development and testing of a surface flux and planetary boundary layer model for application in mesoscale models, *J. Appl. Meteor.*, 34:16-32.

Ray, J.D., Stedman, D.H., and Wendel, G.J., 1986, Fast chemiluminescent method for measurement of ambient ozone, *Anal. Chem.*, 58:598-600.

Stauffer, D.R. and Seaman, N.L., 1990, Use of four-dimensional data assimilation in a limited area mesoscale model. Part I: Experiments with synoptic scale data., *Mon. Wea. Rev.* 118: 1250-1277.

DISCUSSION

W.A.J. van Pul: Were the flux measurements of ozone corrected for chemical reactions?

J. PLEIM: No.

L. GANZEVELD: How do you treat the uptake by the soil? For the corn, which I assume has a typical LAI of 2-3, uptake by the soil might significantly contribute in the overall uptake, especially for a high organic content of the soil, as will probably be the case for the corn field. Soil wetness probably also plays role in the soil uptake and should be considered.

J. PLEIM: The model used for this paper has a very simple algorithm for deposition to the soil. There is a pathway parallel to the vegetation uptake for deposition directly to soil with a resistance of about 533 sm^{-1} for ozone. This pathway is used only for the fractional area not covered by vegetation which was about 7% in this case. We plan to add an in-canopy pathway to the soil with an additional in-canopy resistance which we expect will increase deposition velocity values by about 0.1 cms^{-1} at most.

R. YAMARTINO: You seem to indicate some overcompensation of the soil moisture nudging technique in predicting the various heat flux terms. Did you try to adjust or optimize the nudging coefficients?

J. PLEIM: Yes. We tried to optimize these coefficients such that the time scale of the nudging is on the order of 1 day. Therefore, short term errors in predicted temperature or humidity, which may be caused by errors in cloud cover or mistiming of frontal passages, will have little effect while longer term errors, presumably caused by errors in soil moisture, will have a substantial effect. We found that it typically takes 3-5 days for the soil moisture nudging to reach an equilibrium.

AERMOD: THE DEVELOPMENTAL EVALUATION

R.F. Lee[1*], S.G. Perry[1†], A.J. Cimorelli[2], R.J. Paine[3], A. Venkatram[4], J.C. Weil[5‡], and R.B. Wilson[6]

[1]Atmospheric Sciences Modeling Division
Air Resources Laboratory
National Oceanic and Atmospheric Administration
Research Triangle Park, NC
[2]Region 3, U.S. Environmental Protection Agency
Philadelphia, PA
[3]ENSR Consulting and Engineering
Acton, MA
[4]University of California
Riverside, CA
[5]National Center for Atmospheric Research
Boulder, CO
[6]Region 10, U.S. Environmental Protection Agency
Seattle, WA

INTRODUCTION

In 1991, the American Meteorological Society (AMS) and Environmental Protection Agency (EPA) initiated a formal collaboration to accelerate the inclusion of up-to-date science into regulatory dispersion models. A working group, the **AMS/EPA Regulatory Model Improvement Committee (AERMIC)**, was established to facilitate this modeling activity (Perry et al., 1994; Weil, 1992).

AERMIC's initial model-improvement focus is on the type of regulatory models that are designed for estimating near-field impacts from a variety of industrial source types. The Industrial Source Complex Short-Term (ISCST2) (U.S. EPA, 1992) model was chosen as the platform for the new model because it is the workhorse of current regulatory tools and its recently restructured code is conducive to modification now and in the future. In updating ISCST2, AERMIC has attempted to follow certain design goals to yield a model with desirable regulatory attributes. These attributes are that the model should: 1) be robust in estimating regulatory design concentrations by providing reasonable concentration estimates under a wide range of conditions, with minimal discontinuities; 2) be easily implemented,, i.e. be user friendly and have reasonable requirements for input data and computer resources; 3) be based on up-to-date science that captures the essential physical processes

* On assignment to the Office of Air Quality Planning and Standards, U.S. Environmental Protection Agency.
† On assignment to the National Exposure Research Laboratory, U.S. Environmental Protection Agency.
‡ Permanent affiliation: Cooperative Institute for Research in Environmental Sciences, University of Colorado

Air Pollution Modeling and Its Application XI, Edited by
Gryning and Schiermeier, Plenum Press, New York, 1996

while remaining fundamentally simple; and 4) be easily revised through the use of structured computer programming. This new model is called AERMOD (**AERMIC model**).

The evaluation of AERMOD is being performed in two phases, with data from eight or more high-grade field studies. The first phase, the developmental evaluation, is being performed concurrently with the development of the model. This phase of the evaluation is primarily diagnostic, in that it is used to test whether the model represents the physics properly, and to diagnose the causes of any problems that are noted. The second phase, the final evaluation, using independent data sets, will emphasize how well the model predicts concentrations at the high end of the frequency distribution, which is important in regulatory applications.

This paper reports the results, of the developmental evaluation to date. The developmental evaluation includes comparisons of model estimates with measurements from studies that include tracer and SO_2 releases (both passive and buoyant), from surface-level and elevated sources, in flat and complex terrain, and in rural and urban areas. Only tracer comparisons are completed and reported in this paper. Comparisons of AERMOD estimates with those of ISCST2 are also included.

DATA BASES

AERMOD is being developed as a regulatory tool and thus is intended to handle non-reactive pollutants from a variety of source types in a variety of modeling situations. Therefore, it is necessary to challenge the model in as many of these situations as possible in the evaluation process. Five data bases were selected for the developmental evaluation.

1. The Prairie Grass data base (Barad, 1958) involved a near-surface, nonbuoyant SO_2 release, in a rural area, with flat terrain. Surface sampling arrays were positioned in arcs from 50m to 800m downwind of the source. Both convective and stable conditions are included.
2. The Kincaid SF_6 data base (Liu and Moore, 1984) involved an elevated, buoyant release, in a rural area, with flat terrain. Approximately 200 SF_6 monitors were placed in arcs from about 500m to 50km downwind of the single 187m stack.
3. The Indianapolis SF_6 data base (Murray and Bowne, 1988) involved an elevated, buoyant release in an urban area with flat terrain. Data were obtained from 177 SF_6 monitors in arcs from 250m to 12km downwind of an 84m stack.
4. The Lovett Power Plant SO_2 data base (Paumier et al., 1992) involved an elevated, buoyant release in a rural area with complex terrain. This one-year data set involves a 145m stack and 12 SO_2 monitoring sites on terrain features rising 250 to 330 m above stack base. The monitors are generally 2 to 3 km downwind from the stack.
5. Kincaid, SO_2 (Liu and Moore, 1984) involved an elevated, buoyant release in a rural area with flat terrain. There were 30 SO_2 monitoring stations from about 2 km to 20 km downwind of the 187m stack.

EVALUATION METHODOLOGY

The purpose of the developmental evaluation is primarily diagnostic, i.e. to identify and correct problems in the model as it is being developed. Highlights of the evaluation results are presented here using residual plots and quantile-quantile plots. In this report, all concentrations are normalized by the emission rate.

Residual (predicted/observed) plots of arcwise maximum concentrations (C_p/C_o), crosswind-integrated concentration ($CWIC_p/CWIC_o$), and plume width (σ_{yp}/σ_{yo}) as functions of several model input or calculated parameters (e.g. downwind distance, wind speed, mixing height) are used extensively in the developmental evaluation. Here, the modeled arcwise maximum concentration was taken as the concentration under plume centerline. These data are paired in time and distance. Pairing of concentration data in time and space is very demanding for a dispersion model because of the stochastic nature of turbulence and the limitations of monitoring networks. Model residuals help to identify systematic problems with model formulations (signified by a trend in the residual plot) and help to flag conditions that are associated with poor model performance. A good model would be unbiased, with some scatter about a residual (predicted/observed) of 1.0, and show no trend in the bias as a function of a parameter such as distance, stability, etc.).

Quantile–Quantile (Q–Q) plots are simple pairings of predicted concentrations, ranked highest to lowest, with observed concentrations, ranked in the same manner. One ranked set is plotted on the ordinate and one on the abscissa. If the ranked distributions are identical, then all points lie on the x = y (1:1) line. Q–Q plots are an effective method for comparing the distributions of two data sets. Since AERMOD is being designed as a possible replacement for ISCST2, comparisons between AERMOD and ISCST2 are included in the Q-Q plots.

For the tracer data bases (i.e. Prairie Grass, Kincaid SF6, and Indianapolis) the fractional bias (FB) of the mean concentrations is a useful indicator of model performance, and provides a quantitative measure with which to objectively compare the models. FB of the mean is defined as:

$$FB_{MEAN} = 2\left(\overline{C}_p - \overline{C}_o\right)/\left(\overline{C}_p + \overline{C}_o\right)$$

where $\overline{C}_p$ is the mean predicted concentration and $\overline{C}_o$ is the mean observed concentration. The average was taken over the entire data set and also over upper quartile of the concentrations. Note that with this formula, negative FB indicates underprediction and positive indicates overprediction. The fractional bias complements the Q-Q plots and allows for easy comparison of the performance of different models. The fractional bias of the mean of upper quartile of the data is presented in this paper.

The type of data taken at the field sites has also allowed investigation of specific algorithms within the models (independent of concentration prediction). For the tracer studies in which several arcs of monitors were available at increasing downwind distance, AERMIC calculated the lateral extent of the plume and compare these "observations" with the calculated σ_y.

EVALUATION RESULTS TO DATE

At the time of this writing, AERMIC had essentially completed the analyses for the three tracer data bases (Prairie Grass, Kincaid SF6 , and Indianapolis) and it is these results that are reported here. Analyses of the two full-year data bases (Kincaid SO2 and Lovett) will be completed in the next few months.

The results presented here have been stratified by data base and stability (stable or convective), and presented 1) paired in time and distance (residual plots) and 2) unpaired in time or space (Q–Q plots). Atmospheric stability, in AERMOD, is quantified with the Monin-Obukhov length, L.

Figures 1a–d display graphically the comparison of AERMOD against the Prairie Grass data during convective conditions. The residual plot in Figure 1a indicates generally little bias in the model concentration estimates with only a slight tendency toward underprediction in the near field and an equivalent overprediction at the farthest receptor arc. Figure 1b is analogous to 1a except the residuals are for the crosswind integrated concentration (CWIC) rather than concentration. The slight trend with distance found in 1a is almost indistinguishable in 1b. This points us to the lateral dispersion algorithms, and specifically to the lateral dispersion parameter, σ_y. Figure 1c suggests that AERMOD's estimates of σ_y show a slight trend toward underprediction at the farthest arc distance, which would cause a comparable trend toward overprediction of concentrations.

The Q–Q plot (Figure 1d) shows that the distribution of AERMOD predictions compare well with the distribution of observations with the highest observations being underpredicted by a small amount. However, almost the entire distribution is within a factor of two for this unpaired comparison. The fractional bias (FB) based on the mean of the top 25% of the distribution (Table 1) is -0.301 (underprediction within a factor of about 1.35). In contrast to AERMOD, ISCST2 shows a tendency to overpredict by about a factor of two or less at the high end of the distribution (Figure 1d). The FB is 0.610 (indicating an overprediction tendency by about a factor of 2).

The AERMOD and ISCST2 comparisons with the Prairie Grass data in stable conditions are shown in Figures 2a-b. AERMOD performs well as indicated in the residual plots of Figure 2a. There is no noticeable trend with distance and the scatter is very low. There is a slight bias toward underprediction. This underprediction tendency also shows in the Q–Q plot (Figure 2b) but the distribution of predictions and observations are well matched. The fractional bias based on the mean

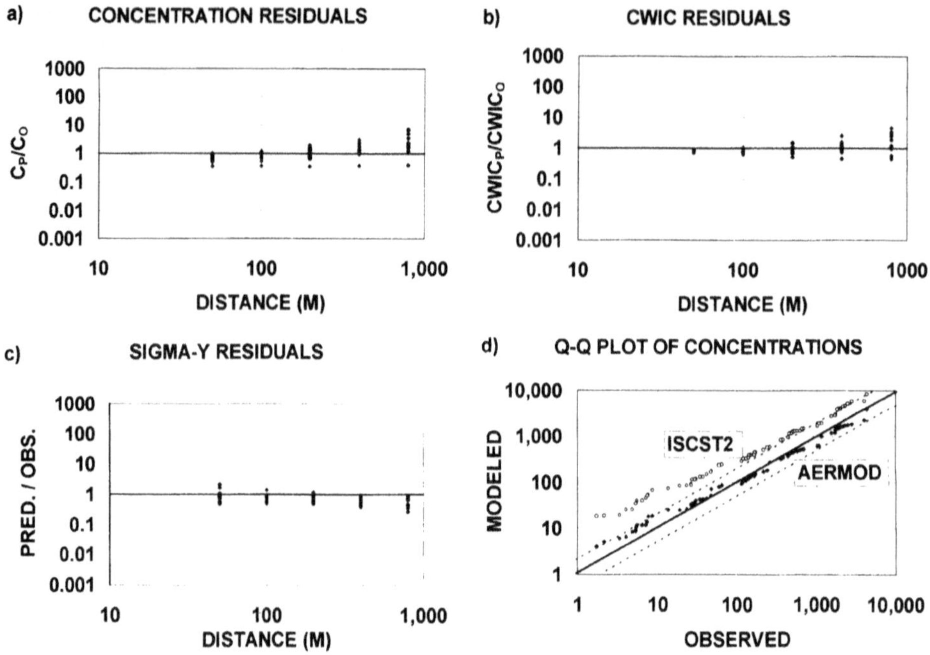

Figure 1. Prairie Grass data, convective conditions: a) residual plot (predicted/observed) of arc-maximum concentrations for AERMOD, b) residual plot of crosswind-integrated concentrations (CWIC for AERMOD), c) residual plot of plume width (σ_y for AERMOD), and quantile-quantile (Q-Q) plot of predicted versus observed arc-maximum concentrations for AERMOD and ISCST2.

of the top 25% of the distribution is -0.158, an indication of underprediction of the observations by a factor of about 1.2.

ISCST2 is performing well with data from Prairie Grass during stable conditions as shown in Figure 2b. The distribution is well matched to the observations and the highest end of the distribution is only slightly overpredicting. The fractional bias is 0.321, an overprediction of about the same amount as AERMOD's underprediction. In general, both models are performing well against the near-surface release data at Prairie Grass, with the upper end of the ISCST2 distribution containing generally higher estimates than AERMOD.

Table 1. Fractional bias of AERMOD and ISCST2 models, based on highest 25% of the data values, for Prairie Grass, Kincaid SF_6, and Indianapolis data sets.

Data Set	Stability	AERMOD Fractional Bias	ISCST2 Fractional Bias
Prairie Grass	Convective (Unstable)	-0.301	0.610
	Stable	-0.158	0.321
Kincaid SF_6	Convective (Unstable)	0.004	-1.029
	Stable	-0.100	-1.358
Indianapolis	Convective (Unstable)	0.008	0.114
	Stable	-0.143	0.117

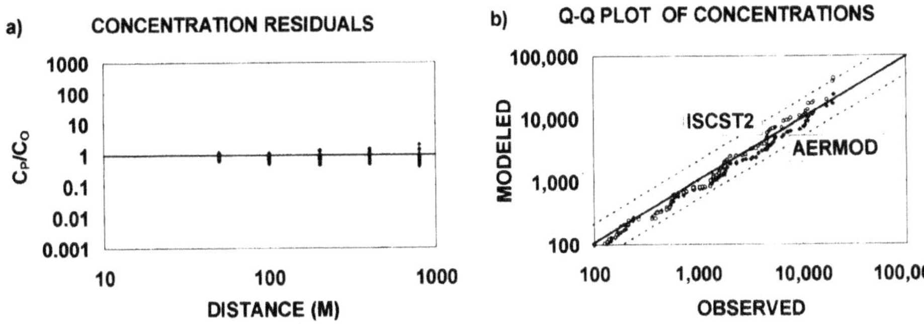

Figure 2. Prairie Grass data, stable conditions: a) residual plot of arc-maximum concentrations for AERMOD, and b) Q-Q plot of predicted versus observed arc-maximum concentrations for AERMOD and ISCST2.

The comparison of AERMOD and ISCST2 estimates to the Kincaid tracer data during convective conditions is found in Figures 3a and b. Because this is an elevated buoyant source and the observations are at the surface, the large scatter (compared to Prairie Grass where the source and monitors are both near the surface) found in the residual plots is not unexpected. In Figure 3a, the overall model seems fairly unbiased, however there is a noticeable tendency nearest to the source (at the 500 m arc) for the model to underpredict. Unfortunately there were no good estimates of observed CWIC at that monitoring arc for AERMIC to use in determining the source of this bias. For the remainder of the downwind distances, the CWIC (not shown) shows no obvious bias suggesting no systematic problems related to the vertical dispersion algorithms. The Q–Q plot (Figure 3b) for convective conditions shows a good match between the distribution of AERMOD estimates and the observations in the upper half the data. The fractional bias for this subset of data is 0.004 (Table 1). The underprediction shown by the low concentration tail of the distribution falling off (common in Q–Q plots) may be due, in part, to the increased uncertainty in the measurements with low concentrations.

ISCST2 does not perform as well as AERMOD against the Kincaid convective observations. Figure 3b suggests that ISCST2 has some problems with underprediction, however the highest few percent of the estimates are within a factor of two. The fractional bias is equal to -1.029 which represents an underprediction of about a factor of 3.

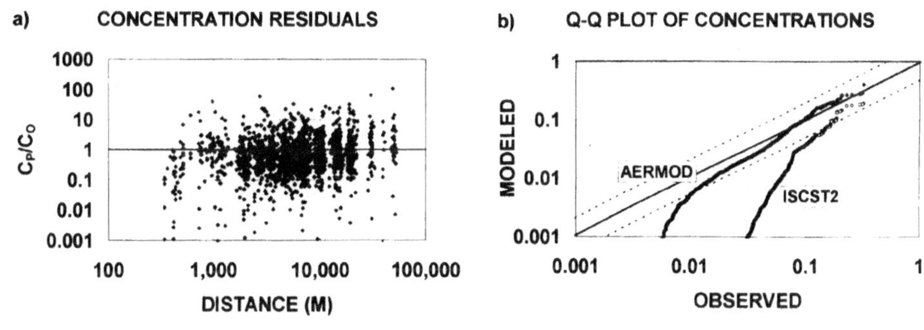

Figure 3. Kincaid SF_6 data, convective conditions: a) residual plot of arc-maximum concentrations for AERMOD, and b) Q-Q plot of predicted versus observed arc-maximum concentrations for AERMOD and ISCST2.

Figure 4 displays the AERMOD and ISCST2 comparisons to Kincaid tracer in stable conditions. The focus of the Kincaid experiments was on convective conditions so there are relatively few data points in stable conditions. Figure 4 shows that AERMOD is adequately simulating the high end of the distribution with an associated fractional bias of -0.100 (a small underprediction by a factor of about 1.1). The concentration residual plots (not shown) for stable Kincaid data show generally unbiased AERMOD estimates except at the closest monitoring arcs where the model is underpredicting these relatively low observed concentrations.

ISCST2 also has a tendency toward underprediction close to the source (based on its residuals) which shifts to unbiased out at 40 to 50 km. Figure 4 indicates a strong bias toward underprediction throughout the entire distribution. The fractional bias (-1.358) quantifies this underprediction to be by a factor of about five.

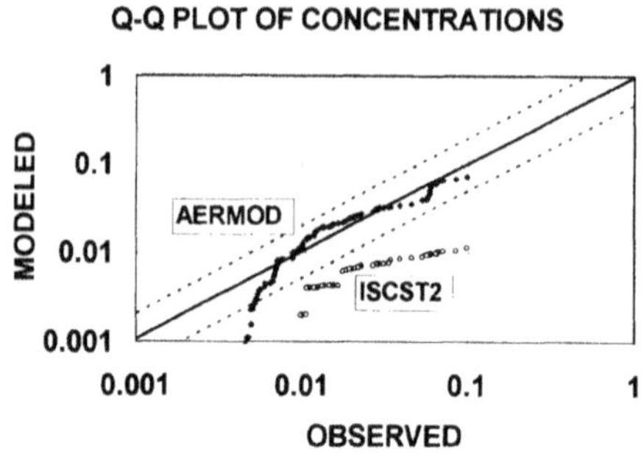

Figure 4. Kincaid SF_6 data, stable conditions: Q-Q plot of predicted versus observed arc-maximum concentrations for AERMOD and ISCST2.

The AERMOD concentration residuals (Figure 5a) for the urban, Indianapolis, convective data base, show a large amount of scatter. As with Kincaid, these are surface comparisons for an elevated buoyant plume. Overall the bias is relatively small, but there is some structure to the residuals with distance. From very near the stack to about 2000 meters downwind there may be a very slight bias toward overprediction. Beyond 2000 meters there is a small but noticeable trend toward underprediction. The Q-Q plot for this data (Figure 5b) suggests, however, that this structure in the residual plot is not significant since AERMOD is matching the upper part of the distribution remarkably well. The fractional bias of 0.008 (Table 1) confirms this. ISCST2 also shows a small fractional bias, specifically 0.114. ISCST2 also shows underprediction very near the source (not shown), overprediction at intermediate distances and a trend back toward no bias at the farthest distances of measurement.

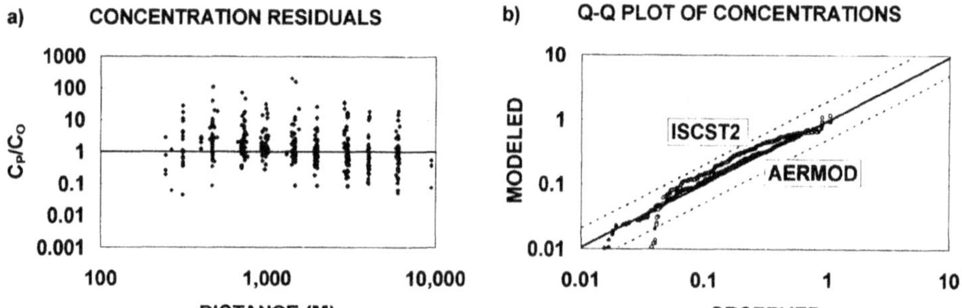

Figure 5. Indianapolis data, convective conditions: a) residual plot of arc-maximum concentrations for AERMOD, and b) Q-Q plot of predicted versus observed arc-maximum concentrations for AERMOD and ISCST2.

The Indianapolis stable data base initial residual analysis alerted AERMIC that the model needed to directly consider urban effects on the specification of stability and thus on the manner in which the dispersion and plume rise would be characterized in otherwise stable atmospheric conditions. Without consideration of anthropogenic heat flux, the model treats the plume as if it is in a stable layer and the results reflect very poor performance since the actual atmosphere is often much more neutral or even slightly unstable in nighttime urban areas. The results here include the effects of anthropogenic heat flux in AERMOD.

Figure 6 shows a good correspondence between the distributions of model estimates and observations. The FB is -0.143, an indication of underprediction by a factor of about 1.15.

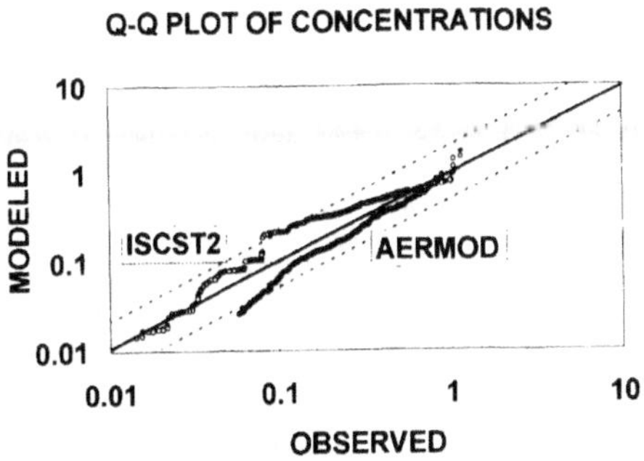

Figure 6. Indianapolis data, stable conditions: Q-Q plot of predicted versus observed arc-maximum concentrations for AERMOD and ISCST2.

ISCST2 also performs well for Indianapolis stable cases as demonstrated in Figure 6. ISCST2 accounts for urban heating by using a separate set of dispersion curves which are based on experiments made in St. Louis, MO (U.S. EPA, 1992). The distributions of model and observations are close with a slight tendency toward overprediction (FB = 0.117) as with ISCST2's performance in convective conditions at Indianapolis. ISCST2 does underpredict close to the source but overpredicts beyond about 1000 meters downwind.

SUMMARY

AERMOD and ISCST2 have been compared with observations using three of the five data bases selected for the developmental evaluation. These three data bases are all characterized as intensive tracer studies each intended to emphasize a particular source type and/or meteorological condition and/or topographic characteristic. These data have been used to identify and correct deficiencies in AERMOD during its development. The final evaluation, to be conducted after AERMOD has been completed, will be based on data bases not used during the development of AERMOD.

Table 1 best summarizes the AERMOD and ISCST2 performances in a collective and quantitative way using the fractional bias based on the average of the top 25% of the model and observed distributions. For surface releases (Prairie Grass) AERMOD has a tendency to underpredict the high concentrations by about a factor of 1.35 in convective and 1.2 in stable. ISCST2 in contrast tends to overpredict surface releases by about a factor of 2 in convective and a factor 1.4 in stable conditions. For rural elevated releases (Kincaid), AERMOD exhibits insignificant bias in convective conditions and only a slight underprediction in stable. ISCST2 significantly underpredicts in both convective and stable conditions. Finally, for the urban elevated release (Indianapolis), both AERMOD and ISCST2 perform well with only small biases for all conditions.

It should be pointed out that in the developmental evaluation, AERMIC chose to use the actual measured peak concentrations to compare with the models. However, since models simulate a smoothed, ensemble fit to the plume concentrations, use of a fitted peak in the observations may be more appropriate for model comparison. Lee and Irwin (1995, 1996) show that for the data they analyzed, the ratio of actual measured peaks to fitted peaks to typically be on the order of 1.3 to 1.4.

AERMIC will be investigating the use of the fitted peak concentrations as the developmental evaluation continues.

DISCLAIMER

This paper has been reviewed in accordance with the U.S. Environmental Protection Agency's peer and administrative review policies and approved for presentation and publication. Mention of trade names or commercial products does not constitute endorsement or recommendation for use.

REFERENCES

Barad, M.L., 1958, Project Prairie Grass, a field program in diffusion, *Geophysical Research Papers*, No. 59, Vols, I and II, Report AFCRC-TR-58-235, Air Force Cambridge Research Center.

Lee, R.F., and Irwin, J.S., 1995, Methodology for a comparative evaluation of two air quality models, Workshop on Operational Short-range Atmospheric Dispersion Models for Environmental Impact Assessment in Europe, Mol, Nov. 1994, published in *Int. J. Environment and Pollution*, Vol. 5, Nos. 4-6.

Lee, R.F., and Irwin, J.S., 1996, Improving concentration measures used for evaluating air quality models, to be presented at Ninth Joint Conference on the Applications of Air Pollution Meteorology with the Air and Waste Management Association, 28 January–2 February, 1996, at Atlanta, GA, American Meteorological Society, Boston, MA.

Liu, M.K., and Moore, G.E., 1984, Diagnostic validation of plume models at a plains site. EPRI Report No. EA-3077, Research Project 1616-9, Electric Power Research Institute, Palo Alto, CA.

Murray, D.R., and Bowne, N.E., 1988, Urban power plant plume studies. EPRI Report No. EA-5468, Research Project 2736-1, Electric Power Research Institute, Palo Alto, CA.

Paumier, J.O., Perry, S.G., Burns, D.J., 1992, CTDMPLUS, A dispersion model for sources near complex topography, Part II: Performance characteristics, *J. Appl. Meteorol.*, **31**, 646-660.

Perry, S.G., Cimorelli, A.J., Lee, R.F., Paine, R.J., Venkatram, A., Weil, J.C., 1994, AERMOD: a dispersion model for industrial source applications, *Proceedings of the 87th Annual Meeting of the Air & Waste Management Association*, 94-TA23.04, Air & Waste Management Association, Pittsburgh, PA.

U.S. EPA, 1992, *User's Guide for the Industrial Source Complex (ISC2) Dispersion Models,* Volumes I—3, EPA-450/4-92-008a-c, Office of Air Quality Planning and Standards, U.S. EPA, Research Triangle Park, NC.

Weil, J.C., 1992, Updating the ISC model through AERMIC, *Proceedings of the 85th Annual Meeting of Air & Waste Management Association*, 92-100.11, Air & Waste Management Association, Pittsburgh, PA.

DISCUSSION

R. OSA: Were tracer studies used for AERMOD development of buoyant releases?

R.F. LEE: The treatment of buoyant releases under convective conditions was based largely on work done by J.C. Weil and others, with plume rise algorithm based on G.A. Briggs (1975). Field study data taken at Kincaid (the "intensive" SF_6 study) and at the Indianapolis Perry K Power Plant were used in the developmental evaluation as a basis for checking and revising the model.

POSTER SESSION

USE OF THE UB/NMC ETA MODEL FOR THE SIMULATION OF THE LIFE-CYCLE OF CHERNOBYL RADIOACTIVE CLOUD

Milan Dacić and Boško Telenta* †

* ICSC World Laboratory; Mediterranean Research Centre
Via Guarnotta 26; I-91016 Erice; ITALY

ABSTRACT - The UB/NMC (University of Belgrade/National Meteorological Center; Washington) limited area model is employed to simulate a long-range transport of pollutant released during the Chernobyl accident. A chain of forecast sequences is used to follow the life-cycle of radioactive cloud through a contaminant mass conservation equation. A proper simulation of the wind system illustrated with a small wave-like structure near Chernobyl and cyclone over France 24 hours after the accident together with presented time-series of pollutant concentrations implies that existing numerical model may be a valuable tool in transport studies.

THE ATMOSPHERIC MODEL; UPGRADE; RESULTS; CONCLUSION

The detailed description of the UB/NMC ETA model can be found in its documentation[1] which also contains other important references. It is a hydrostatic primitive equation model with a vertical coordinate permitting a step-like design of mountains. The horizontal momentum advection has a built-in nonlinear energy cascade control. Model has a comprehensive physical package consisting of: Mellor-Yamada level 2.5 and level 2 closures; ground surface processes including surface hydrology; large scale precipitation; cumulus convection; and radiation. Three forecasting episodes were used to follow the spread of radioactive cloud: a) 25-28 April 86, b) 28 April - 1 May 86, and c) 1-4 May 86. Input fields were: geopotential, temperature, horizontal wind components and specific humidity. The model domain covers all of the Europe. The time step was 7.5 minutes. There were 16 levels in vertical direction. The grid cell containing Chernobyl is assumed to be the pollution source in the model.

The contaminant mass conservation equation has the following form:

$$\frac{\partial C}{\partial t} = -V \cdot \nabla C + \nabla \cdot K \, \nabla C + W_d + W_w - R_c + S$$

† Address all correspondence to Boško Telenta, ICSC, Erice, Italy.

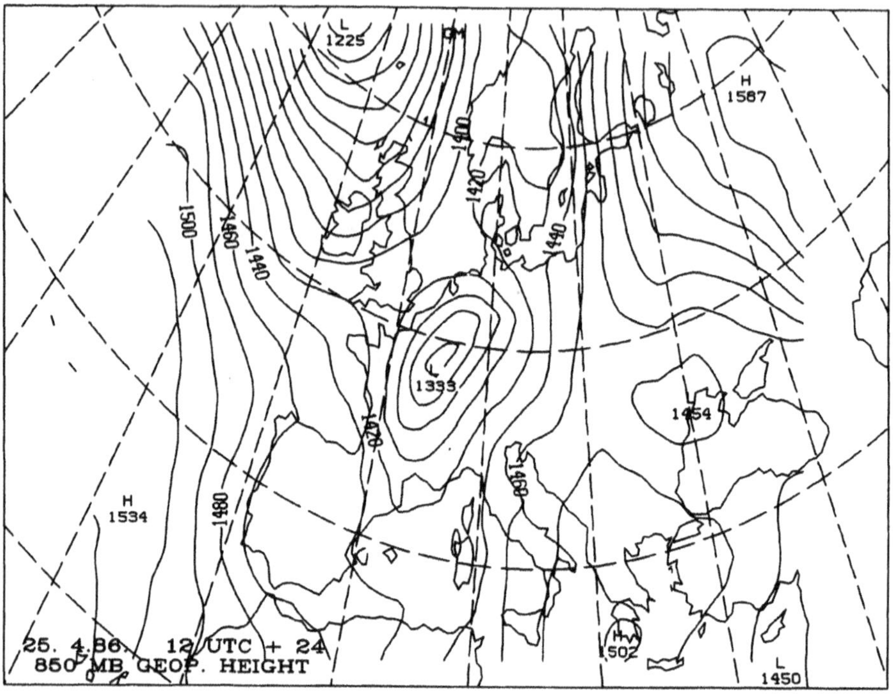

Figure 1. Forecasted AT 850 mb field for 25 April 1986 12 GMT+24 h.

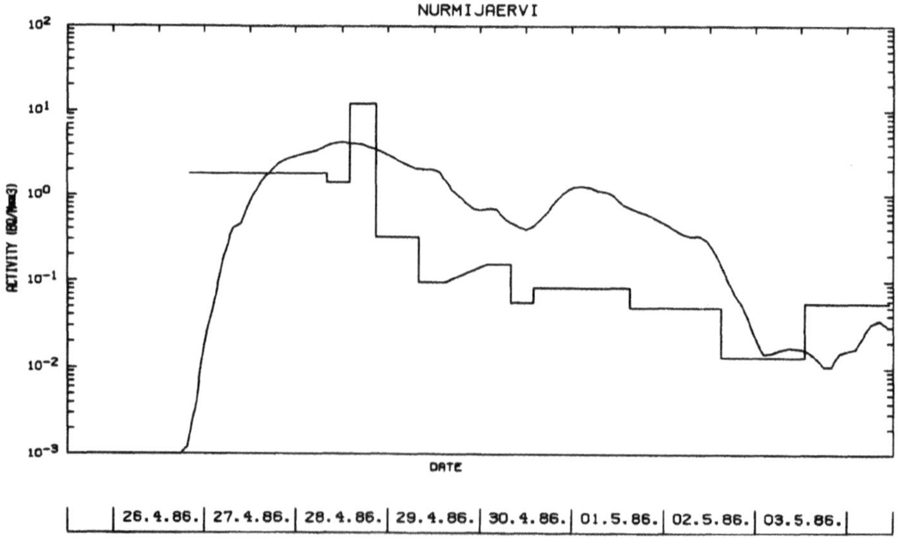

Figure 2. Forecasted and observed (polygonal line) concentrations of Cs^{137} (Bqm^{-3}) for Nurmijaervi.

where C is the specific activity in Bqm^{-3}, V the wind vector, K the turbulent diffusion coefficient, W_w and W_d depict the loss of specific activity by wet and dry deposition and S is the source term. The term R_c accounts for loss of material due to radioactive decay.

All main characteristics of the pressure field are successfully reproduced (Fig. 1.). Model predicted time series of the Cs^{137} for Nurmijaervi (Finland) presented in Fig. 2. show that the general pattern of radioactive cloud passage is captured by the model.

We may conclude that with relatively few changes in model code and modest increase in computational time the operational ETA model is able to produce reasonably good predictions of transboundary transport during ecological emergencies and therefore may serve as a predictive tool for smaller meteorological centers.

REFERENCE

1. L. Lazić and B. Telenta, Documentation of the UB/NMC ETA model,WMO/TD-No.366.(1990)

AIR POLLUTANT DIFFUSION-DEPOSITION FROM A CONTINUOUS POINT SOURCE

Ana G. Ulke and Nicolás A. Mazzeo

Dept. of Atmospheric Sciences - University of Buenos Aires
Buenos Aires - Argentina

Atmospheric diffusion models are used to estimate air pollutant's concentrations. Those considering dry deposition can be used as an alternative procedure to calculate deposition velocities and fluxes. Experimental studies show variations in a range of several orders of magnitude. The application of a diffusion-deposition model of pollutants released from a near surface continuous point source in the atmospheric boundary layer is presented. Numerical simulations of experiments carried out at Hanford (USA)[1], were done. Tracer deposition velocities are estimated from the comparison of proposed values with observational data.

The two-dimensional semiempirical advection-diffusion-deposition equation, for a continuous point source is used[2], with vertical wind and eddy diffusivity profiles obtained from an extension of the Monin-Obukhov hypothesis to the whole atmospheric boundary layer. This extension relies on the assumption of local validity of the turbulent energy equations, in a steady, horizontally homogeneous atmospheric boundary layer. The profiles compare well with other theoretical and semiempirical forms[2]. The eddy diffusivity is supposed to be equal to both eddy diffusion coefficients for momentum and heat transfer. The equation is solved numerically.

The model was run with the meteorological and emission data of the Green Glow and 30 Series experiments[1]. It was released ZnS from a near surface source. The meteorological parameters were obtained as follows: Monin-Obukhov length (L) from Richardson number calculated from wind and temperature measurements near the source[3], surface friction velocity (u_{*o}) using the near surface observed mean wind speed and the surface layer wind speed vertical profiles[3]. Mixing heights (h) were estimated with theoretical expressions[4]. The surface roughness length was z_o=0.04 m, value obtained from the Hanford-67 micrometeorological data[5]. The concentrations were normalized by emission rates. The observed normalized crosswind integrated concentrations were obtained from the peak exposures and standard deviations of the crosswind exposure distributions and total emissions[1].

First, simulations were carried out assuming no loss of material (v_d=0). The mean fractional error (MFE) between the computed and observed concentrations, and its standard deviation were calculated for each experimental run. It was found an important overestimation of concentrations. The MFE range between -0.003 and 1.3. It appears no significative difference due to the parameterization of eddy diffusivity.

Air Pollution Modeling and Its Application XI, Edited by
Gryning and Schiermeier, Plenum Press, New York, 1996

The overestimation found may be mainly due to deposition rates[2]. The model was used to obtain an effective´ deposition velocity for each experimental release. A variable quantity of numerical simulations with different deposition velocities was done. It was selected as ´effective´ that velocity for which MFE was less than 10% and the MFE obtained with no deposition. Figure 1 shows that the ´effective´ deposition velocity decreases with stability. In stable atmospheric conditions the ´effective´ deposition velocity shows an increase with surface friction velocity (Fig.2). In unstable situations, there is little correlation between the ´effective´ deposition velocity and surface friction velocity. ´Effective´ deposition velocities are within experimental ranges[6,7]. The values are in satisfactory agreement with previously obtained in the simulation of experiments known as Hanford 67-Series[5], and those done during 1983 at the same location[8].

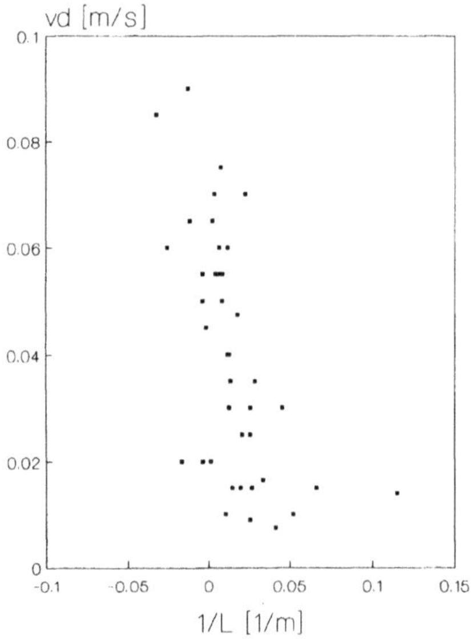

Figure 1. Variation of deposition velocity with atmospheric stability

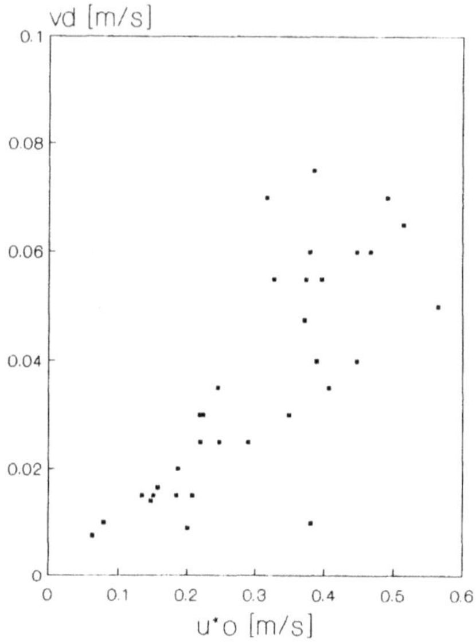

Figure 2. Variation of deposition velocity with surface friction velocity in stable conditions

REFERENCES

1.J.J. Fuquay, C.L. Simpson and W.T. Hinds, Prediction of environmental exposures from sources near the ground based on Hanford experimental data, *J. Appl. Met.*, 3:761(1964).

2.A.G. Ulke, *Diffusion-deposition of pollutants released in the atmospheric boundary layer*, PhD Thesis, University of Buenos Aires, Argentina (1992).

3.J. Wieringa, A revaluation of the Kansas mast influence on measurements of stress and cup-anemometer overspeeding, *Boundary-Layer Meteorol.*, 18, 4:411(1980).

4.S.E. Gryning, A.A.M. Holtslag, J.S. Irwin and B. Sivertsen, Applied dispersion modeling based on meteorological scaling parameters, *Atmos. Env.*, 21, 1:79(1987).

5.P.W. Nickola, *The Hanford 67-Series: A volume of atmospheric field diffusion measurements*, PNL-2433, Bat. Northwest Lab., Richland, WA, U.S.A.(1977).

6.G.A. Sehmel, Particle and gas dry deposition: A review, *Atmos. Env.*, 14, 9:983(1980).

7.T.A. McMahon and P.J.Denison, Empirical atmospheric deposition parameters - a survey, *Atmos. Env.*,13, 5:571(1979).

8.J.C. Doran and T.W. Horst, An evaluation of Gaussian plume depletion models with dual tracer field measurements, *Atmos. Env.*, 19:939(1985).

IFDM MODELLING FOR OPTIMAL SITING OF AIR QUALITY MONITORING STATIONS AROUND FIVE OIL REFINERIES

G. Cosemans[1], G. Dumont[2], E. Roekens[2], J.G. Kretzschmar[1]

[1]VITO, Division Energy, Boeretang 200, B2400 Mol, Belgium
[2]Flemish Environmental Agency (VMM), Department Networks and Research,
A.Van De Maelestraat 96
B 9320, Aalst, Belgium

Sources And Legislation

The Antwerp industrial area has among its industries 5 major oil refineries which have[1] to install ambient air quality monitors for SO_2 and NO_x. If the average immission over the last 24 hours measured in one of these monitoring sites exceeds 300 $\mu gSO_2/m^3$ or 150 $\mu gNO_x/m^3$, they have to reduce their emissions.

Calculated Immission Fields for All Sources

Modelling was done for 11 years of meteorological data. For each year, IFDM[2,3] calculated the 98 percentile of the moving 24h-averages in a 1360 receptor point grid covering the 17 km x 19 km region of interest. This yields 11 percentile values per receptor point. From those, only the minimum and maximum P98 value are considered.

Interpretation of Minimum and Maximum P98,m24h Immission Fields

Local maxima in the immission field of the minimum P98-values (Figure 1, left part) show the places where the 24h averaged immission values can be expected to be relatively high every year. A monitoring site at a such a place will be exposed quite often to the impact of the emissions of the (nearby) refineries. These are good places to monitor the environmental impact of the refineries.

Local maxima in the immission field of the maximum P98-values (Figure 1, right part) show the places where exceptionally high immission levels can occur, be it perhaps only once in 10 years. A monitor at such a place will record peak immissions to which the public can be exposed exceptionally, e.g. during worst case conditions of persistent wind direction with an upwind alignment of many sources.

Other Considerations

A similar modelling strategy is used to determine the best sitting places to monitor the impact of the emissions of the individual refineries.

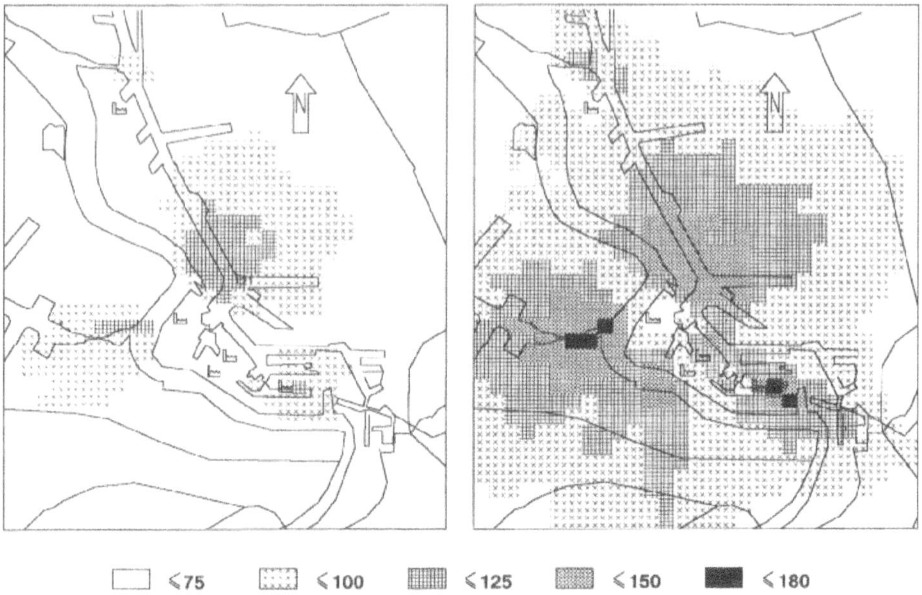

Figure 1. Immission field (µg SO_2/m^3) of minimum (left) and maximum (right) 98-percentiles of the moving 24h averages resulting from calculations for 11 different years.

Previous measuring campaigns and the existing immission monitoring network also indicate some places with high immission levels, do to emissions not present in the emission inventory used in modelling. A final aspect to be considered is the infrastructure (roads, terrain, access to electricity and telephone network) needed to install, operate and maintain the new monitoring network.

REFERENCES

1. *VLAREM II (The Flemish Environmental Legislation)*, Aminal, Brussels (1991).
2. G. Cosemans, J. Kretzschmar, and G. Maes, The Belgian immission frequency distribution model IFDM, in: *Proceedings of the DCAR Workshop on Objectives for Next Generation of Practical Short-Range Atmospheric Dispersion Models*, H.Olesen and T.Mikkelsen, eds., NERI, Riso, Denmark (1992).
3. H. Bultynck and L. Malet, Evaluation of atmospheric dilution factors for effluents diffused from an elevated continuous point source, *Tellus*, 24: 445-472 (1972).

AN ANALYTICAL FORMULATION FOR MIXING LENGTH SCALES AND EDDY DIFFUSIVITIES IN A CONVECTIVE PBL

G.A. Degrazia[1], A.P. de Oliveira[2], and O.L.L. de Moraes[1]

[1]Departamento de Física, Universidade Federal de Santa Maria - 97.119-900 , Santa Maria, RS, Brasil
[2]Departamento de Ciências Atmosféricas, Universidade de São Paulo - 01.065-970, São Paulo, SP, Brasil

INTRODUCTION

The turbulent fluxes of momentum, heat, moisture and passive scalars appear in the equations of the motion when the flow is separated into a mean and a turbulent part. This procedure leads to the first order closure problem. In general, first order closure schemes require the specification of a mixing length scale and utilize some kind of eddy diffusivity approach. In this work, a new method to evaluate mixing length scales for a turbulent PBL under convective conditions is described. An eddy diffusivity derived from the vertical mixing length scale also shows high degree of agreement with one obtained from numerical simulations using LES models.

MIXING LENGTH SCALES AND EDDY DIFFUSIVITIES IN A CONVECTIVE PBL

Recently, Degrazia et al. (1994) have shown that a general formulation for the mixing length scales in the PBL is given by the relation

$$l_i = \frac{\sigma_i \beta_i S_i(0)}{4} \tag{1}$$

where i = u,v,w, $S_i(0)$ is a turbulent velocity spectrum at the origin, β_i is defined as the ratio of the Lagrangian to the Eulerian timescales and σ_i is the turbulent velocities standard deviations. According to Degrazia (1989), in the convective PBL, the value of the spectrum at origin is of the form

$$S_i(0) = \frac{1}{(f_m)_{n,i}} \frac{Z}{Uq} \qquad (2)$$

with $(f_m)_{n,i}$ being the frequency of the spectral peak in neutral stratification, U is the mean wind speed, Z is the height above the surface and $q = (f_m)_i / (f_m)_{n,i}$, where $(f_m)_i$ is the frequency of the spectral peak regardless the stratification. The substitution of (2) in (1) results the following expression for the mixing length scales in the convective PBL

$$l_i = \frac{\sqrt{\pi}}{16} \frac{Z}{(f_m)_{n,i} q} \qquad (3)$$

In the vertical case the value of the frequency at the spectral peak was obtained from an empirical relation proposed by Caughey and Palmer (1979), so that

$$l_w = 0.2Z_i \left[1 - \exp\left(-4\frac{Z}{Z_i}\right) - 0.0003\exp\left(8\frac{Z}{Z_i}\right)\right] \qquad (4)$$

with $(f_m)_{n,w} = 0.33$. The turbulent mixing length scale (4) is similar to that proposed by Sun (1993). The proposed vertical length scale (4) multiplied by the standard deviation of the turbulent vertical velocity (Sorbjan, 1989, p.113) yields a well behaved and non-local eddy diffusivity that depends on the convective velocity scale w_*, the height Z and the convective PBL height Z_i:

$$\frac{K_{zz}}{w_* Z_i} = 0.22\left(\frac{Z}{Z_i}\right)^{1/3}\left(1 - \frac{Z}{Z_i}\right)^{1/3}\left[1 - \exp\left(-4\frac{Z}{Z_i}\right) - 0.0003\exp\left(8\frac{Z}{Z_i}\right)\right] \qquad (5)$$

The shape of the vertical profile and the magnitudes of this eddy diffusivity are similar to ones derived by Holtslag and Moeng (1991).

REFERENCES

Caughey, S.J.,and Palmer, S.G.,1979, Some aspects of turbulence structure through the depth of the convective boundary layer, *Quart.J.R.Met.Soc.* 105:811.

Degrazia, G.A., 1989, *Anwendung von Ahnlichkeitsverfahren auf die Turbulent Diffusion in der Konvektiven und Stabilen Grenzschicht*, Institut fur Meteorologie und Klimaforschung der Universitat Karlsruhe, Karlsruhe.

Degrazia, G.A., Oliveira, A.P., and Moraes, O.L.L., 1994, Multiple master length scales derived from a statistical diffusion theory in: *Air Pollution Modeling and Its Application X*, S.V.Gryning and M.M.Millan, ed., Plenum Press, New York.

Holtslag, A.A.M., and Moeng, C-H., 1991, Eddy diffusivity and countergradient transport in the convective atmospheric boundary layer, *J.Atmos.Sci.* 48:1690.

Sorbjan, Z., 1989, *Structure of the Atmospheric Boundary Layer*, Prentice Hall, New Jersey.

Sun, W.Y., 1993, Numerical simulation of a planetary boundary layer: part I, cloud-free case. *Beitr. Phys. Atmosph.* 66:3.

ON A PC-ORIENTED EULERIAN MULTY-LEVEL MODEL FOR LONG-TERM CALCULATIONS OF THE REGIONAL SULPHUR DEPOSITION

Dimiter E. Syrakov

National Institute of Meteorology and Hydrology
66 Tzarigradsko chaussee
1784 Sofia, Bulgaria

The development of NIMH dispersion model is provoked by the needs of the Environmental authorities for more detailed description of the acid loads in the region. It is realised on PC because of the broad distribution of these computers in the East European Countries and the limited possibilities for using more powerful machines. So, special attention is paid to time and space optimisation of different schemes. A new advection scheme, called TRAP, is created. It is an explicit, positively definite and conservative scheme with limited numerical dispersion and good trasportivity.

The TRAP scheme is a flux type scheme. It will be described here in one dimension. In the case of more dimensions, time splitting can easily be applied. It is well known that all flux schemes calculate the mass fluxes through the edges of the cell and the new value of the concentration is determined as

$$C_i^{n+1} = C_i^n - Fr_i + Fl_i, \qquad (1)$$

where Fr and Fl are the mass fluxes through the right and the left edges of the i-th cell. Let us suppose a staggered grid: concentrations C_i being defined in points i, and velocities U_i - in points $i+1/2$. The Courant number $c_i = U_i Dt / Dx$ is introduced for normalisation of the problem; Dt and Dx being the time and space steps.

In many schemes, the flux is determined as an area. So, Bott[1,2,3] calculates it as an integral, the curve $C(x)$ being determined through interpolation among the nearest grid points. In the slope scheme[4], $C(x)$ is a straight line starting from the nearest grid point and having a slope, determined during the previous time step, i.e. the shape of the flux area is supposed trapezoidal. To estimate this area (i.e. the flux value), the concentration in the middle of the flux area is calculated linearly and is then multiplied by the Courant number. The last approach is used in the new TRAP scheme but the concentration in the middle of the trapezium is determined through interpolation among the nearest grid points, as proposed by Bott[1]. Let $C^i(x)\geq 0$ denotes an interpolation polynomial of proper order centred at point i. Then, using the upstream differencing, we get:

Air Pollution Modeling and Its Application XI, Edited by
Gryning and Schiermeier, Plenum Press, New York, 1996

$$Fr_i = c_i.C^i(r_i), \qquad r_i = 0.5(1-c_i), \qquad c_i > 0; \qquad (2)$$
$$Fr_i = c_i.C^{i+1}(r_{i+1}), \qquad r_{i+1} = -0.5(1+c_i), \qquad c_i < 0.$$

In Eq.(2) c_i with its sign determines the sign of the flux, as well as the distance r to the nearest grid point. Interpolations of different orders have been used; best results obtained using a 5-order interpolation (Bott[1]).

The different test calculations show that the scheme after Eq.(2) is conservative and transportive, but not monotonic (positive definite). Two more versions of this scheme are developed. In the first one (TR5.1), Bott's[1] system of limiters is applied, but instead of integrating polynomials, the absolute value of Fr from Eq.(2) is included. In the second version (TR5.2), only Bott's[1] normalisation is used, right edge flux calculated by

$$FR_i = Fr_i.C_i/(A_i+\varepsilon), \qquad c_i > 0, \qquad (3)$$
$$FR_i = Fr_{i+1}.C_{i+1}/(A_{i+1}+\varepsilon), \qquad c_i > 0.$$

Here, A_i denotes the total area of the i-th cell, ε is a small value (ε~1e-15). The features of these three versions of the TRAP scheme are tested together with other schemes, results presented in [5,6]. The final impression is that Bott's scheme, TR5.1 and TR5.2 are much better than the others. These three schemes display equivalent features, but TR5.1 and TR5.2 are respectively 1.5 and 2 times faster than Bott's scheme. The TR5.2 scheme is chosen to be used in the NIMH disperion model, called EMAP[6].

The EMAP model is described in detail in [6]. It has 6-levels in PBL (non-regular) with terrain following z-coordinate system. A special z-transformation allows regularization in vertical direction. The simplest 2-order time forward explicit scheme is used for description of horizontal diffusion. The vertical diffusion block uses implicit scheme incluing the dry deposition as a bottom boundary condition. The top boundary condition is either "open boundary" or "hard lid" type. In the present version of the model, only two species - SO_2 and $SO_4^=$ - are involved and a simple transformation rate approach is used to parameterize the chemical conversion. The wet scavenging is parameterized by adequate wash-out rates, depending on precipitation intensity and type (seasonal dependence).

The present version of EMAP is realized over 48×54 horizontal grid points with 25 km step, part of EMEP standard mesh (*i*=29,36, *j*=14,22); every 150×150 km cell is divided into 36 25×25 km cells. The region is South-Eastern Europe with Bulgaria in the center (only Bulgarian SO_2-sources are processed). The 1994 sulphur deposition is calculated month by month, results presented and discussed in [6]. Using a time step of 15 min., the duration of 1 month integration on 50 Mhz IBM PC-486 occurs to be about 2.5 hours, which seems to be a good performance.

REFERENCES

1. A.Bott, A positive definite advection scheme obtained by non-linear renormalization of the advective fluxes, *Mon.Wea.Rev.*, **117**, pp.1006-1015 (1989).
2. A.Bott, Monotone flux limitation in the area preserving flux form advection algorithm, *Mon.Wea.Rev.*, **120**, pp.2592-2602 (1992).
3. A.Bott, The monotone area-preserving flux-form advection algorithm: reducing the time-splitting error in two-dimensional flow fields, *Mon.Wea.Rev.*, **121**, pp.2637-2641 (1993).
4. G.L.Russel, J.A.Lerner, A new finite-differencing scheme for the tracer transport equation, *Journal of Applied Meteorology*, December, pp 1483-1498 (1981).
5. Bulgarian contribution to EMEP, Annual Report for 1993, NIMH, Sofia, March, 1994.
6. Bulgarian contribution to EMEP, Annual Report for 1994, NIMH, Sofia, March, 1995.

AIR POLLUTION MODELING IN THE CAPITAL OF PRAGUE

Jan Pretel,[1] Josef Brechler,[2] Václav Píša,[3] and Stanislav Böhm[4]

[1]Czech Hydrometeorological Institute, Prague, CZ
[2]Faculty of Mathematics and Physics, Charles University of Prague, CZ
[3]ATEM Ltd., Prague, CZ
[4]Technical University of Prague, CZ

INTRODUCTION

A state of an ambient quality in any settled area is a matter of interest not only of habitants living there, but also of local authorities. Knowledge of this state gives local authorities an opportunity to take appropriate measures in many branches of their responsibilities. The instantaneous state of the air pollution can be obtained via network of monitoring sites located in the area of interest. On the other side, if there is necessary to answer the question how some changes in a distribution of sources could affect the air quality, than only measurement cannot support us with the sufficient information. The most complex view can be obtained if measured and computed values are put together. In this contribution we would like to introduce the modeling approach to the air pollution state assessment in the area of Prague.

APPROACH TO THE SOLUTION

The area of Prague city covers more then 500 sq. kilometers and it is located in one of the most polluted countries in the Central Europe. That is the main reason why the modeling approach is not easy. The air pollution background is "too high". The model makes the calculation of the concentration contributions from six basic categories of sources at each receptor point. The categories are following : *large sources* (thermal output higher than 5MW) - totally 359 sources in the city, *medium sources* (thermal output between 0.2 and 5MW) - totally 493 sources in the city, *small area sources* (thermal output less then 0.2MW) in the city (3498 sources), *line sources*and intersections in the city (2279 different sources), large and medium point *sources lying outside the city* on the territory of the Czech Republic (70 main sources). Additionally the model takes into the account the *influence of a transport of polluted particles* from the all districts of the Czech Republic and, in some extent, also the influence of transboundary transfer. Contribution of these sources has been parameterized via - so called - wind direction dependent values of transmission that have been added to values of level concentrations at each receptor point.

Air Pollution Modeling and Its Application XI, Edited by
Gryning and Schiermeier, Plenum Press, New York, 1996

The main data base for these calculation is the emission inventory of the main pollutants, which has a long tradition on our country and it is now quite good developed. Attention has been focused on four main pollutants: sulfur dioxide SO_2, nitrogen compounds NO_x, carbon monoxide CO and particulate matters.

A dispersion model has treated contributions from the all this set of sources. The name of the model, which has been used for the calculation is ATEM (according to the name of the company that developed this model). It is steady-state Gaussian plume model - industrial source complex model of type ISC2. The recent horizontal resolution is generally 300 by 250 meters, in some places (e.g. main streets and intersections, highways and parking lots) the resolution is even more compact. As the topography of the city is very complex and affects patterns of ambient quality concentration fields of various pollutants quite seriously, a special attention to the parameterization of orography effects has bee given. Using the model PIAPBLM, single wind roses for 4 levels (10, 40, 80 and 160 meters above the surface), five stability category and for totally 132 different places around the city has been constructed. These wind roses describe the complex topography of the city quite satisfactory. The output of the PIAPBLM model has been calibrated by the measured wind data from five meteorological stations and one aerological station in the city, the output of the ATEM model has been compared with the results of measurements on the Automatic Monitoring Network in Prague (totally 12 monitoring stations).

RESULTS

Output files contain various characteristics like annual mean ground-level concentration of the pollutants mentioned above, maximum values of concentration ("short-term concentration"), information about the main contributors to the concentration level at any receptor point both for the annual and maximum values, information about the exceedance of annual concentrations and maximum concentrations of the corresponding standard values at any receptor point ("hygiene limits"), how many times these limits are exceeded and for how long period (in per cent) in one year this exceedance could be expected.

The other output files contain data showing percentage of contribution of the six different source categories to the concentration level at any receptor point. Air pollution roses are also constructed for each receptor point. They show how ambient concentration levels depend on the wind directions both for annual mean values and for maximum values. All the results are presented both in tables and in maps and/or various types of diagrams and graphs. The system GIS has been used for the presentation of results. As a part of this presentation is a set of colour maps describing the current air quality situation in the city and most important and practical results.

There is no doubt that such a type of dispersion models can be very useful and powerful "tool" especially in planning and decision making process. I this case, dispersion model support us with an objective information about a possible impact of planned activities on the environment. On the other hand one must be very careful if short- term characteristics are desired. Meteorology is parameterized only very roughly and there are some strong limitation that restrict the use of dispersion models for such a tasks.

This contribution has been done as a part of the Prague Air Quality Project under the financial support of the Prague City Hall, Institute of the City Informatics Systems of the City Hall and the Ministry of the Environment of the Czech Republic.

MODEL PREDICTIONS ON URBAN AIR QUALITY AND THEIR COMPARISON TO MEASUREMENTS

A. Karppinen,[1] J. Kukkonen,[1] E. Valkonen,[1] J. Härkönen,[1]
E. Rantakrans,[1] L. Jalkanen,[1] S. Haarala,[1] J. Walden,[1]
T. Koskentalo,[2] T. Elolähde,[2] P. Aarnio,[2] and J. Laurikko[3]

[1]Finnish Meteorological Institute, Air Quality Department
Sahaajankatu 20 E, FIN-00810, Helsinki, Finland
[2]Helsinki Metropolitan Area Council
Opastinsilta 6 A, FIN-00520, Helsinki, Finland
[3]Technical Research Centre of Finland
PL 1601, FIN-02044 VTT, Helsinki, Finland

INTRODUCTION

We discuss studies on urban air quality in the Helsinki metropolitan area. This area includes four cities, which have a total population of 850 000. We have conducted comprehensive emission surveys comprising mobile and stationary sources, in 1990 and 1993. The emission time series were formed using the LIISA system, developed at the Technical Research Centre of Finland, and the transportation planning system EMME/2. The atmospheric dispersion was evaluated by the models UDM-FMI (The Urban Dispersion Modelling system of the FMI) and CAR-FMI (Contaminants in the Air from a Road) (Härkönen et al., 1994). Both dispersion models are connected to a meteorological pre-processing model, which is based on Monin-Obukhov type boundary layer scaling.

RESULTS

The numerical results include the time series of the spatial distributions of the CO-, NO_2-, NO- and SO_2 -concentrations in the Helsinki metropolitan area. The results show that the highest concentrations of carbon monoxide and nitrogen oxides occur in the Helsinki downtown area and at the junctions of the main streets and roads (Kukkonen et al., 1995). Figure 1 shows the annual mean CO-concentration in the Helsinki metropolitan area in 1993. As traffic is the only important source of CO in the area, the highest concentrations are found in the vicinity of the main roads and streets. The new proposed national air quality guidelines are exceeded at a few spatially limited areas in the downtown Helsinki.

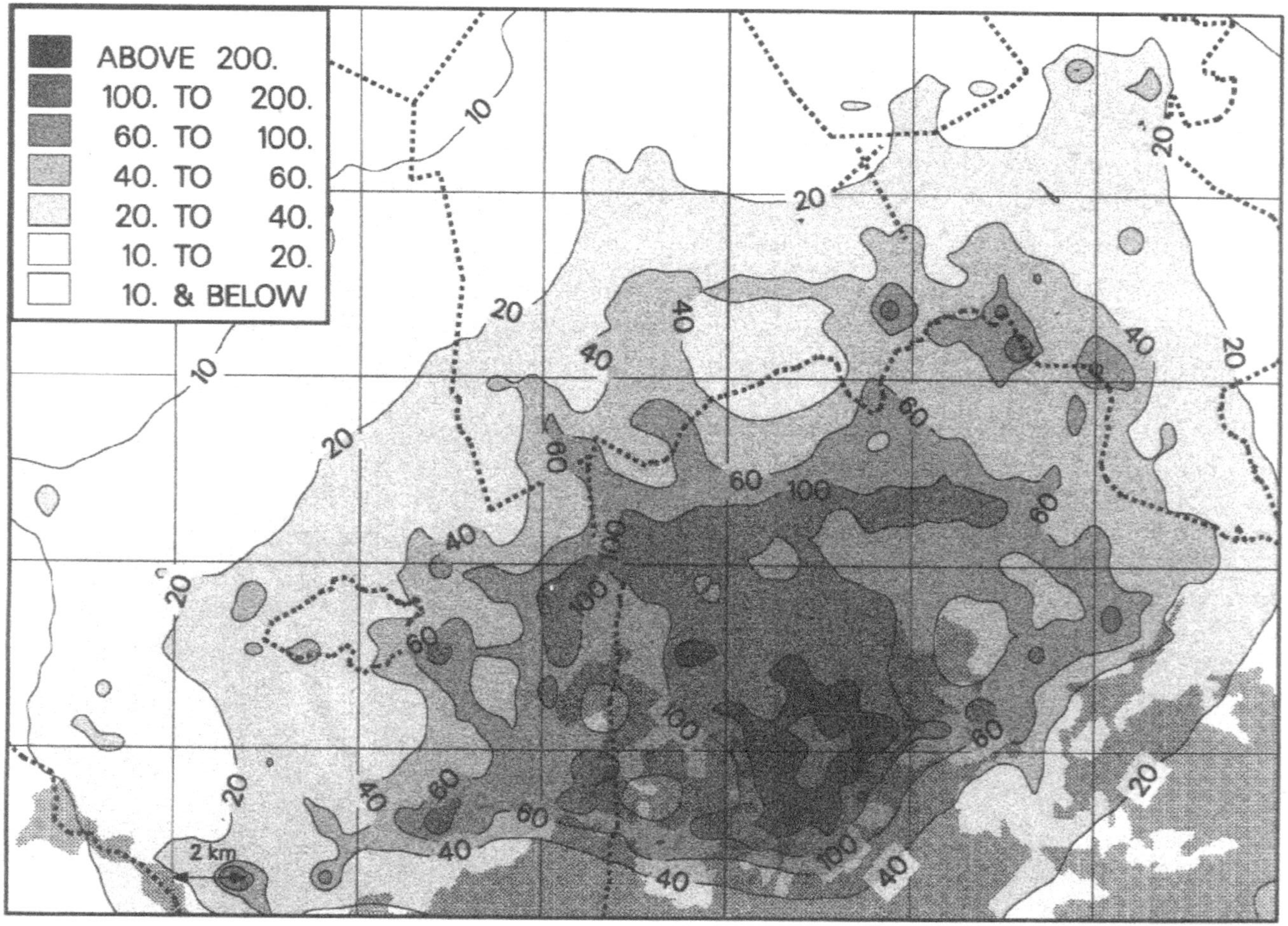

Figure 1. Distribution of the annual mean concentration of carbon monoxide (μg/m^3) in the Helsinki metropolitan area in 1993.

COMPARISON WITH MEASUREMENTS

The vertical profiles of gaseous pollutants were determined during a measurement campaign in the spring of 1994 and the measurements were compared with results of the CAR-FMI model (Walden et al., 1995). The measurements were conducted near a major road in the city of Espoo, located in the Helsinki metropolitan area. We have measured the concentrations of five gaseous pollutants (NO, NO_2, CO, SO_2 and O_3) and the meteorological parameters at the heights of 3 m, 5.5 m and 10 m, together with simultaneous measurement of traffic densities.

We have also compared the model predictions against the results of the air quality measurement network of the Helsinki Metropolitan Area Council. The models generally underpredict the measured average concentrations at the stations in the Helsinki downtown area. However, the diurnal and annual temporal variation of the measured and predicted concentrations is very similar.

REFERENCES

Härkönen, J.,Valkonen, E.,Kukkonen, J.,Rantakrans, E.,Jalkanen, L. and Lahtinen, K., 1994. An operational dispersion model for predicting pollution from a road, in: *Operational Short-range Atmospheric Dispersion Models for Environmental Impact Assessment in Europe. Vol. 1.* ,Kretzschmar, J. et al., ed., Vlaamse Instelling voor Technologisch Onderzoek, Mol, Belgium.

Kukkonen, J., Karppinen, A., Härkönen, J., Valkonen, E., Rantakrans, E., Haarala, S., Jalkanen, L., Koskentalo, T., Elolähde, T., Aarnio, P. and Laurikko, J., 1995. The influence of traffic on urban air quality - model predictions and their comparison to measurements, in: *Proceedings of the 10th World Clean Air Congress, Vol. 2*, Anttila, P. et al. ,ed., The Finnish Air Pollution Prevention Society, Helsinki, Finland.

Walden, J., Härkönen, J., Pohjola, V. and Kukkonen, J. and Kartastenpää R., 1995. Vertical concentration profiles in urban conditions - comparison of measurements and model predictions, in: *Proceedings of the 10th World Clean Air Congress, Vol. 2*, Anttila, P. et al. ,ed., The Finnish Air Pollution Prevention Society, Helsinki, Finland.

PHYSICAL MODELING OF CAR EXHAUST DISPERSION IN URBAN STREET CANYONS -- THE EFFECT OF SLANTED ROOFS

Stilianos Rafailidis and Michael Schatzmann

Meteorologisches Institut, Universität Hamburg
Bundesstraße 55, D-20 146 Hamburg, Germany

OVERVIEW

Leisen *et al.* (1982) have shown that building roof configurations adjacent to an urban street canyon influence pollution dispersion in it. The phenomenon can not be modelled numerically properly with the currently available urban pollution dispersion codes, because one needs to tackle both slanted roofs and extensive solution domains. Therefore, to elucidate the effects further, extensive physical modelling is currently underway at Hamburg University. The results form the basis of a comprehensive experimental databank for testing and validating advanced numerical dispersion models.

The phenomenon is studied in two-dimensions in a neutral stratification wind tunnel, a more extensive description of which is available elsewhere (Rafailidis *et al.*, 1995). The urban boundary layer characteristics are replicated by 20 street canyons upstream and 7 downstream of the measurement section, in the middle of which a line source is placed. The tracer concentrations are measured in the positions shown in Figure 1 and are represented as $K_i = CUHL/Q$, with C the concentration, U the free-stream velocity, H the building height and Q/L the source discharge rate. To identify differences between different canyon geometries we introduce the *canyon pollution retention parameter* $\mathbb{C}$, as defined in Table 1 in which results for some typical canyon and roof configurations are also summarised.

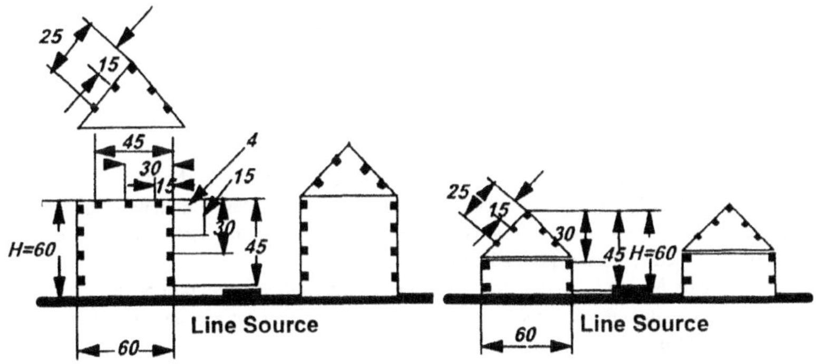

Figure 1. Sampling positions along the street canyon for the two building configurations tested.

Table 1 Total pollution $\mathbf{C}=\frac{\sum_1^n K_n}{n}\cdot B$ retained in various canyon and roof configurations; B is the street width. Because the interest is on effects on the inhabitants of neighbouring buildings, only measurements on side walls and not on the roofs are taken into account in computing **C**, as indicated by the fine dots shown on the canyon walls.

B/H=1					
C [*m*]=	3.9	3.7	1.7	1.8	3.3
B/H=1/2					
C [*m*]=	3.9	3.3	1.6	1.1	5.9

DISCUSSION

It is clear that, although street pollution levels vary, the street width has little influence on total pollution retained in the canyon, at least for the typical aspect ratios tested here. On the other hand, the configuration of adjacent roofs has a predominant effect on the canyon pollution. For instance, when the whole building with the slanted roof is confined within the urban canopy of the surrounding buildings, slanted roofs upstream of the pollution source increase dramatically canyon re-aeration, lowering total pollution retained there to less than half its original flat-roof value. Downstream roofs have, by comparison, little beneficial effect. On the contrary, when the slanted roof protrudes above the urban canopy and into the free-stream, it is roofs *downstream* of the source which improve canyon wind ventilation better, whereas upstream roofs give little improvement. It is, therefore, concluded that careful placing of roofs on buildings alongside heavy-traffic streets, and in relation to the prevailing winds in the region, may improve dramatically the long-term air quality in the polluted street canyon. This is achieved through natural wind ventilation, an environmentally-friendly and sustainable resource, which necessitates no running costs. The strategic placement of the roofs may be accomplished cheaply through improved town-planning legislation, and the resulting capital costs can be absorbed in the on-going maintenance budgets of the nearby buildings.

ACKNOWLEDGMENT

S. Rafailidis gratefully acknowledges financial support by the *Human Capital and Mobility* Programme of the European Union and the NATO CCMS Pilot Study on *Air Pollution Transport and Dispersion over Coastal Urban Areas*.

REFERENCES

Leisen, P., Jost, P., and Sonnborn, K.S., 1982, Modellierung der Schadstoffausbreitung in Straßenschluchten, Vergleich von Außenmessungen mit rechnerischer und Windkanal-Simulation, in: *Abgasbelastungen durch den Kraftfahrzeugverkehr im Nahbereich verkehrssreicher Straßen*, K Hofmann, *et al.*, eds., Verlag TÜV Rheinland, Cologne.

Rafailidis, S., Pavageau, M., and Schatzmann, M., 1995, Wind tunnel simulation of car emission dispersion in urban street canyons, in: *Annalen der Meteorologie*, Deutsche Meteorologische Gesellschaft, München. (in press)

AN IMPLICIT–EXPLICIT ALGORITHM FOR CHEMISTRY–TRANSPORT MODELS

Ralf Wolke, Oswald Knoth, and Eberhard Renner
Institut für Troposphärenforschung
Permoserstr. 15
D–04303 Leipzig, Germany

1. INTRODUCTION

The systems of ordinary differential equations (ODEs) resulting from regional atmospheric chemistry–transport models are nonlinear, highly coupled and extremely stiff. Because explicit ODE solvers require numerous short time steps in order to maintain stability, most current techniques solve stiff ODEs implicitly or semi-implicitly. In this paper we investigate a new approach based on backward differential formula (BDF) which solves the sparse linear equations during the Newton iteration by linear Gauss–Seidel iterations (Knoth, Wolke, 1994). The Jacobian is computed explicitly and not by finite differences.

In contrast to the traditional time–splitting approach vertical transport processes and chemical kinetics are treated in every horizontal grid point implicitly and simultaneously (see section 2). For the vertical advection we use a first-order upwind scheme. A step size control for the time integration is inserted in the code. For the horizontal advection a 5–point advection scheme (Verwer et. al., 1995) is implemented. The resulting fluxes are handled explicitly as "artificial" sources during the integration of columns. The time steps for horizontal advection are determined according to the CFL-criterium.

Our approach is used in the three-dimensional, non-hydrostatic mesoscale chemistry-transport model METRAS (Schlünzen, 1990). This model is applied to study an emission/immission scenario in Saxonia. The efficiency and the accuracy of the code are compared with common time–splitting algorithms.

2. SIMULTANEOUS INTEGRATION OF CHEMISTRY AND VERTICAL TRANSPORT

The dynamics of the vertical transport and the chemistry of reactive species in one column can be described by the ODE-System

$$\frac{dc_i}{dt} = R_i(c_1, \ldots, c_N, t) + Tc_i + Q_i(t) + \text{fluxes from horizontal advection}, \qquad (1)$$

where $i = 1, \ldots, N,\ t \in [t^n, t^n + \Delta_{ad}t^n], c_i \in R^{n_z}$ is the **vector** of concentrations of the i-th species in all cells of a column, the emissions are included in $Q_i(t)$, and $\Delta_{ad}t^n$ is the time step for horizontal advection. The matrix T denotes the tridiagonal block matrix from the discretization of vertical diffusion, advection, and deposition. The "initial values" $c_i(t^n)$ are known from the last integration.

In BDF methods the main computational task consists in the "corrector iteration" which is the solution of the non-linear system

$$F(c) := c - C^{n+1} - \beta \Delta t^n f(t^{n+1}, c) = 0, \tag{2}$$

where $\beta > 0$ denotes a parameter of the BDF–scheme and C^{n+1} is a linear combination of previous values. The Newton–iteration for solving (2) leads to linear systems of the form

$$(I - \beta \Delta t^n J)\, \Delta c_n = b. \tag{3}$$

Here $J = J_{\text{chem}} + T$ denotes (an approximation of) the Jacobian of $\frac{\partial f}{\partial c}$.

Equation (3) is solved by linear Gauss–Seidel iterations (2 or 3 are usually sufficient). In most cases the matrix J_{chem} from the chemical part R_i is very sparse. For instance for the CHEMSAN mechanism (Rolle et. al., 1993) only 12 % of all entries are non–zero elements. In our code the Jacobian J is generated automatically. Only the non–zero entries of this matrix have to be stored in a sparse structure. Furthermore a sparsing procedure (Nowak, 1992) for dropping small elements into the Jacobians is implemented for reducing the number of non–zero elements.

In our code we modify the original program LSODE (Hindmarsh, 1980). The step-size and order control of the code is not changed. However, the maximum order is restricted to two and the new linear algebra described above is included. The code is vectorized.

3. NUMERICAL RESULTS

Test example: CHEMSAN with 67 species and about 170 reactions, 26 vertical layers, horizontal grid size $\Delta x = 2000\,m$, itegration time: 6–72 hours, tolerance = 1 %.

advection step	average time for one step	average time for one functions
without advection	345 s	194 s
200 s	60 s	32 s
300 s	82 s	47 s

4. REFERENCES

Hindmarsh, A. C., 1980, LSODE and LSODI, two new initial value ordinary differential equation solver, ACM–SIGNUM Newsl., 15, 4, 10–11.

Knoth, O., and Wolke, R., 1994, A comparison of fast chemical kinetic solvers in a simple vertical diffusion Model, Proceedings of 20th International Meeting on Air Pollution Modeling and its Application, Valencia 1993, 147–156.

Knoth, O., and Wolke, R., 1994, Numerical methods for the solution of large kinetic problems in a simple, will be published in "Applied Numerical Mathematics".

Nowak, U., 1992, Dynamic sparsing in stiff extrapolation codes, Preprint SC 92–21, Conrad–Zuse–Zentrum für Informationstechnik, Berlin.

Rolle, W., Renner, E., and Helmig, D., 1993, Comparison of computed and measured photooxidant concentrations at a forest side, Chemosphere, 27, 881–898.

Schlünzen, K.H., 1990, Numerical studies on the inland penetration of sea breeze fronts at a coastline with tidally flooded mudflats, Beitr. Physik Atm., 63, 243–256.

Verwer, J.G., Blom, J.G., and Hundsdorfer, W., 1995, An implicit–explicit approach for atmospheric transport–chemistry problems, CWI Report NM–R9501, Centre for Mathematics and Computer Science, Amsterdam.

MODELING STUDIES ON THE ATMOSPHERIC LONG-RANGE TRANSPORT OF TOXIC METALS TO THE LAKE BALATON, HUNGARY

László Bozó,[1] Richard Artz,[2] Joseph Pinto,[3] Joseph Scudlark[4] and Krisztina Labancz[1]

[1]Institute for Atmospheric Physics
H-1675 Budapest P.O. Box 39, Hungary
[2]NOAA Air Resources Laboratory 1315 East West Highway,
Silver Spring, MD 20910
[3]U.S. EPA, Research Triangle Park, NC 27711
[4]University of Deleware, 700 Pilottown Road
Lewes, DE 19958

In absolute terms the rate of toxic metals' deposition is significantly lower than that of e.g. acidifying compounds. However, toxic metals can accumulate in soils, vegetation and surface waters after their wet and dry deposition from the atmosphere, so they can cause environmental damage over several decades. The size of aerosol particles containing toxic metals is in the fine range so they can be transported several hundred kilometers in the atmosphere from the sources where they were released.

Wet and dry deposition of lead and cadmium were measured at the Lake Balaton (Figure 1.). The results were compared with 925 hPa backward trajectory-based long-range transport model calculations where the EMEP-gridded (150 X 150 km^2) European scale emission inventories of these metals were considered. The annual total deposition of lead and cadmium over the Lake was estimated to be 4040 kg a^{-1} and 68 kg a^{-1}, respectively.

It was also computed how the different regions (western Europe, eastern Europe and Hungary contribute to the total deposition of these trace metals. The results are shown in Figure 2. It can be seen that the contribution of other eastern European sources through long-range transport is rather significant to the total atmospheric deposition over the Lake. On the basis of our model calculations the ratios of wet/total atmospheric deposition for lead and cadmium are 0.70 and 0.55, respectively.

Based on historical European emission inventories and also 925 hPa backward trajectory computations the cumulated total lead and cadmium deposition over the Lake Balaton was also calculated. Their values for lead and cadmium are 238 t and 3.6 t, respectively, considering the last fourty-years period of 1955-1995.

Air Pollution Modeling and Its Application XI, Edited by
Gryning and Schiermeier, Plenum Press, New York, 1996

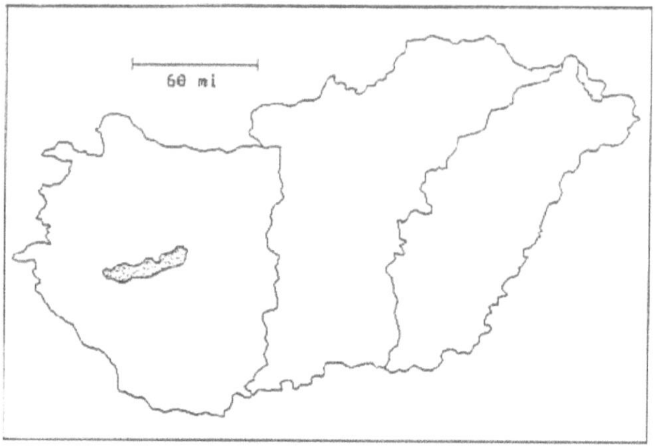

Figure 1. Geographical location of the Lake Balaton in Hungary.

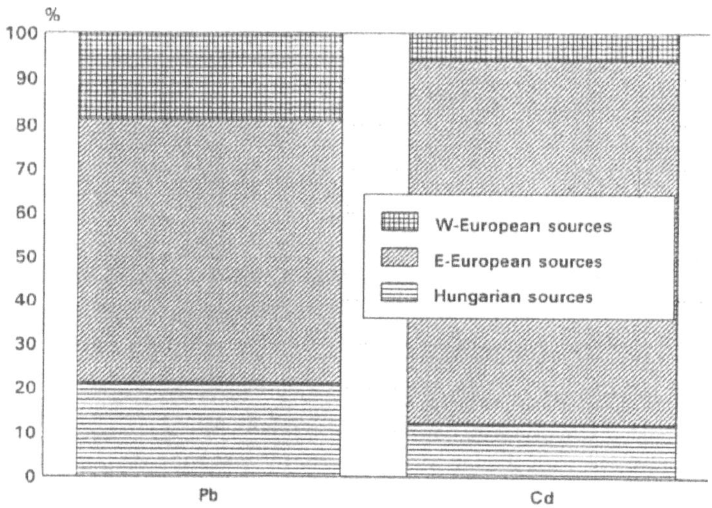

Figure 2. Relative contribution of different European regions to the total atmospheric deposition of lead and cadmium over the Lake Balaton.

Acknowledgements

This paper was sponsored by the U.S.-Hungarian Science and Technology Joint Fund in cooperation with National Oceanic and Atmospheric Administration and Hungarian Meteorological Service under Project J.F.No. 293.

LONG-TERM ATMOSPHERIC TRANSPORT AND DEPOSITION OF HEAVY METALS IN CENTRAL EUROPE

Jerzy Bartnicki[1], Krzysztof Olendrzyński[2], Józef Pacyna[3], Stefan Anderberg[4] and William Stigliani[5]

[1]Norwegian Meteorological Institute, P.O. Box 43 Blindern, N-313 Oslo, Norway
[2]International Institute for Applied Systems Analysis, Laxenburg, Austria
[3]Norwegian Institute for Air Research, Lillestrøm, Norway
[4]Lund University, Lund, Sweden
[5]University of Northern Iowa, Cedar Falls, Iowa, USA

INTRODUCTION

The so-called "Black Triangle" (BT) region covers the southern part of the former German Democratic Republic, south-western Poland and the northern part of Bohemia (Czech Republic). The BT, together with the heavily industrialized Upper Silesia region in southern Poland (BTUS), is the most polluted part of Europe, mainly due to acidic compounds and heavy metals. Continuous accumulations of cadmium, lead and zinc in soils may eventually lead to serious environmental consequences. Triggered by increased soil acidification, heavy metals may become mobilized, thus leading to plant and ground water contamination (the so-called "Chemical Time Bomb"). Acid deposition in the BTUS region has been continuously monitored in the framework of the EMEP program. However, up to now, not much is known about cumulative atmospheric deposition of heavy metals. In this study, we calculated long-term atmospheric deposition of Cd and Pb to the BTUS region (and to entire Europe) during the 1955-1987 period.

METHOD

To perform this task, we used the emissions inventory developed at IIASA's on-going Project on Regional Material Balance Approaches to Long-Term Environmental Policy Planning. The inventory includes emissions from thirteen European countries during the period 1955-1987. The grided spatial distribution of emissions for individual European countries was available only for the years 1982 (Cd) and 1985 (Pb). The same spatial distributions were assumed in the computations for the entire 33-year period.

An updated version of the Heavy Metals Eulerian Transport (HMET) model from the Norwegian Meteorological Institute (DNMI) was used for computing deposition of heavy metals. Long-term depositions were calculated indirectly since meteorological input data were not available for the entire time period. In the first step, we calculated country-to-grid source-receptor matrices for both metals based on ten year meteorological data (6-hour temporal resolution) compiled by DNMI for EMEP. In the second step, using the source-receptor matrices and emission time profiles for individual countries, we calculated 33-year cumulative deposition of Cd, Pb and Zn to the BTUS region.

RESULTS

The results, presented in the form of deposition maps for Cd - Fig. 1 and for Pb - Fig. 2, show high cumulative deposition in Western Europe (especially Benelux countries and Germany) and the BTUS region in Eastern Europe. The maximum cumulative depositions during the period 1955-1987 exceed 50 mg m^{-2} and 1900 mg m^{-2} for Pb and Cd respectively.

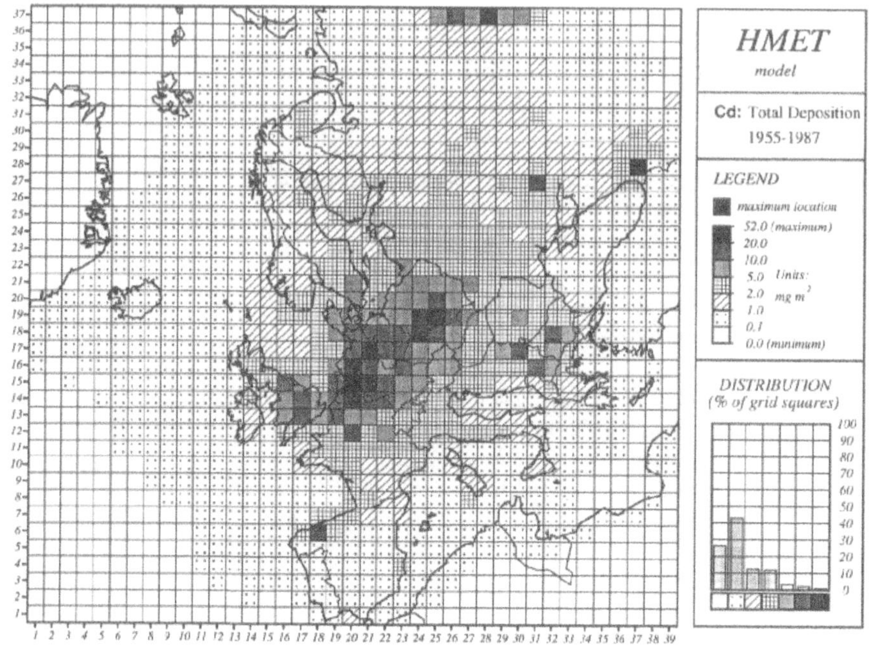

Figure 1. Computed cumulative deposition of Cd in Europe for 1955 - 1987 period. Units: mg m^{-2}.

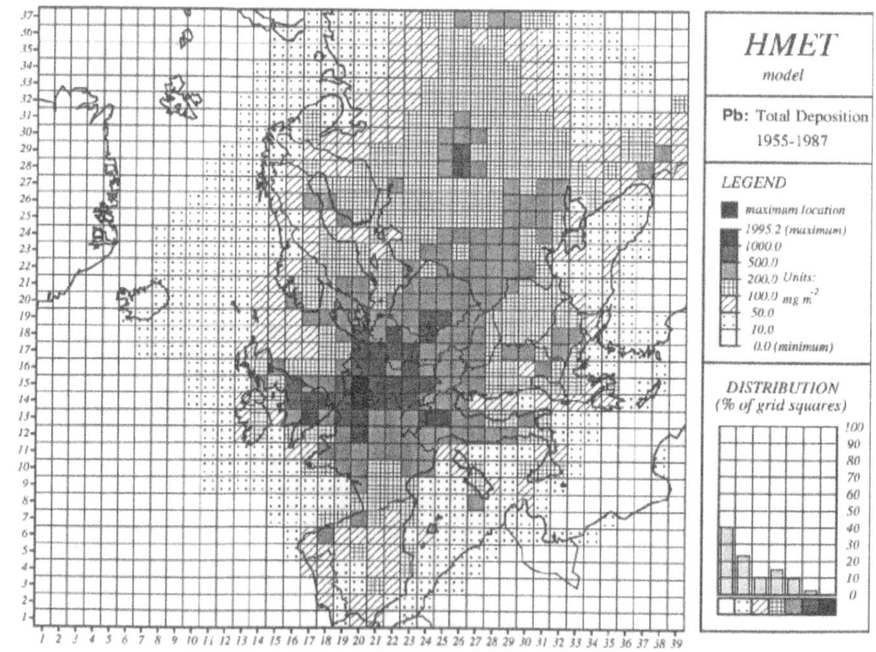

Figure 2. Computed cumulative deposition of Pb in Europe for 1955 - 1987 period. Units: mg m^{-2}.

COMPLEX MATHEMATICAL MODEL CONCERNING POLLUTANTS DISPERSION IN THE ATMOSPHERE EMITTED FROM INTERFERING SOURCES

Traian Pop [1], Livia-Mihaela Pop [1], Adrian Jelev [2] and Octavian Olteanu [1]

[1] Research & Engineering Institute for Environment
Bucharest 78, Spl. Independentei 294, Romania
[2]Simulation & Training Systems, Platf. Măgurele CP24, Bucharest, Romania

INTRODUCTION

Limited emission concentrations values of the pollutants released into the atmosphere under interference conditions, represent a topic of present interest in Romania. These limits have to join the maximum concentrations admitted by the legislation in force for protected areas. The emission limits settlement of polluting sources which belong to different industrial enterprises in a territory, has to take a zone specific into account. In this respect, a complex dispersion mathematical model was elaborated.

MATHEMATICAL MODEL

The performances of this model consist in concentration field calculation (in the air and close by the soil), exceeding concentrations areas determination and emission concentrations limiting in conditions of variation of the following parameters :
- meteorological parameters (wind, velocity and direction, air temperature, solar radiation,...)
- technological parameters of the industrial emitting sources, which determine the mass flows Q(kg/h) and the concentrations of the pollutants released into the atmosphere
- the efficiency of the facilities for pollution removal

in a long period of time and in interference conditions of the located sources in the interest territory.

The mathematical model was validated in some laboratory and field applications.

RESULTS

In order to illustrate a case of powder pollution in an interest area, some results are shown in Figs. 1, 2, and 3—as part of an application.

CONCLUSIONS

Field measurement results, in control points, validate the calculated results (Figs. 1 and 2).

The approach of the subject in this manner allows an analysis of the present state of polluting technologies and pollution removal facilities in service, as well as of the required measures to reduce the emission mass flows (Fig. 3), this representing an important parameter in an accurate selection of pollution removal facilities.

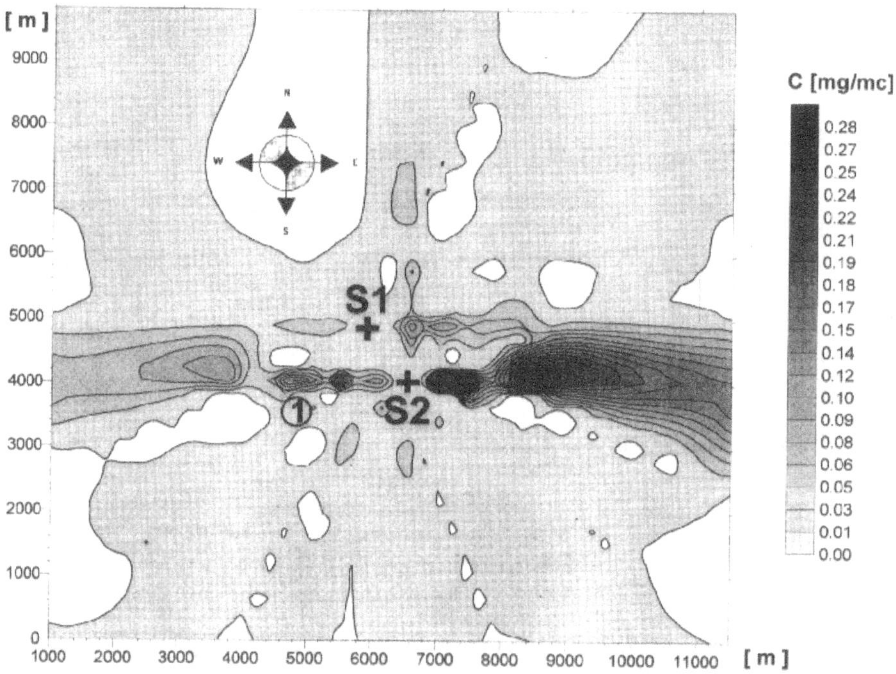

Figure 1. Average values (mg/m^3 in 24h) of suspended powders in 1994 (without pollution removal facilities). S=powder releasing sources; 1=concentration measure point.

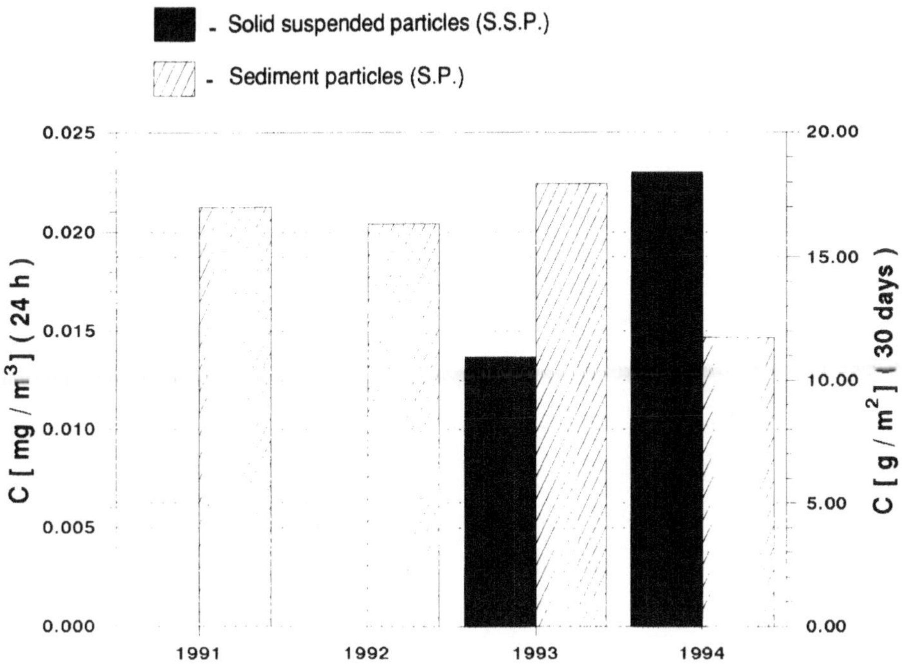

	1991	1992	1993	1994
S.S.P			0.014	0.023
S.P	17.00	16.32	17.94	11.75

Figure 2. Measured powder concentrations (control point 1).

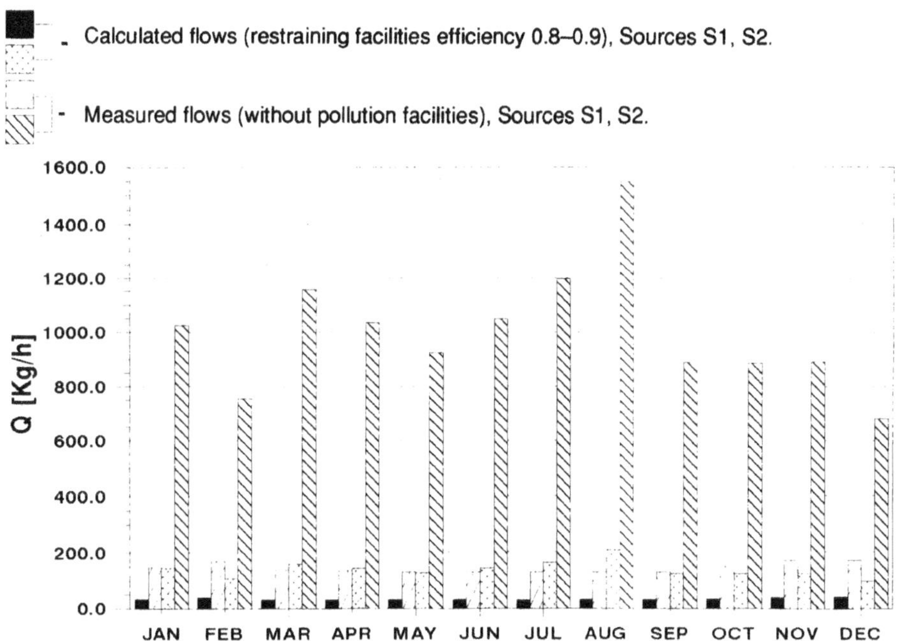

	JAN	FEB	MAR	APR	MAY	JUN	JUL	AUG	SEP	OCT	NOV	DEC
■	33.8	38.6	31.8	30.7	29.9	29.8	29.7	29.8	30.0	34.4	39.5	39.7
	147.1	167.9	138.3	133.4	130.0	129.7	129.3	129.5	130.3	149.4	171.8	172.6
	143.4	107.5	158.9	143.3	126.4	144.8	164.9	208.7	125.5	125.1	124.7	97.6
╲╲	1026.0	754.9	1157.2	1035.6	924.3	1049.8	1198.7	1548.8	889.5	886.1	890.9	681.3

Figure 3. Comparative graph concerning measured and calculated emission mass flows (Q).

NUMERICAL MODELLING OF THREE-DIMENSIONAL FLOW AND POLLUTION TRANSPORT OVER COMPLEX TERRAIN DURING STABLE STRATIFICATION

Alexander Baklanov [1,2] and Jan Burman[1]

[1]National Defence Research Establishment, FOA, S-901 82 Umeå, Sweden
[2]Institute of Northern Ecological Problems,
Kola Science Centre of RAS, 184200 Apatity, Russia

MODEL DESCRIPTION

The object of this work is to model the dynamics of the atmosphere over complex topography for neutral and stable stratification. To describe the effects of the hydro-dynamic influence of topography, for instance tear-off currents, we use the system of equations with a non-hydro-static approximation. To describe the environmental problems, in particular, while modelling for possible radiation consequences after hypothetical accidents at Nuclear Power Plants or fires, powerful sources of heat and temperature contrasts in hundred of degrees can occur. Therefor we use the equations of hydro-thermo-dynamics for the compressible atmosphere. For description of the turbulence the k-ε model is used, and it follows Rodi(1987). The k-ε model overpredicts the turbulent lengthscale for weak shear and therefor the production term in the dissipation equation is scaled with the turbulent lengthscale (l) according to Detering(1985).

$$c'_{1\varepsilon} = c_{1\varepsilon} l/h \qquad h = \frac{c_h u_*}{f_c} \qquad (1)$$

The problem is mostly intermediate between the atmospheric boundary layer problem (meteorology) and hydrodynamics. So, the hydro-thermo-dynamic CFD package PHOENICS is used for elaboration of the model. The surface layer is described by the Monin-Obuhov's theory. The friction velocity is defined by the logarithmic law. It is applicable for lower parts of the surface both for stable and neutral cases, Högström(1988). The model is applied on a district with complex relief at Monchegorsk, Russia. The area has strong pollution problems that partly are modelled here. SO_2 is released at a rate of 1140 g/s 20 meters above ground. The wind speed at the top of the domain is 15 m/s. Dispersion experiments are planned for the same area and these results should be useful to verify the model.

RESULTS

With increased stability the eddy viscosity is reduced. The mixing is reduced and the concentration gets higher at a distance from the source. Drainage flow is occurring in the stable case.

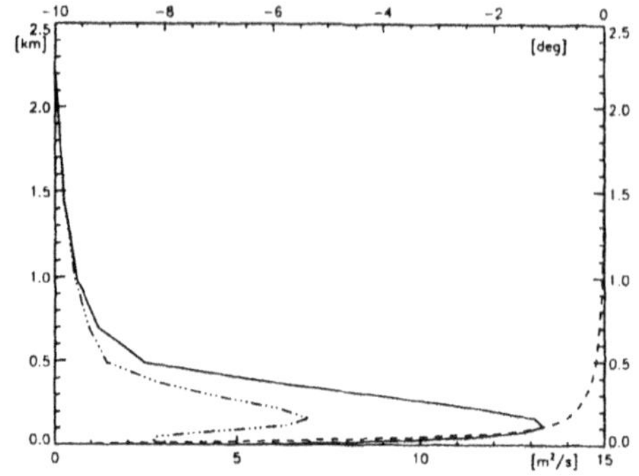

Figure 1. The eddy viscosity for the neutral (line) and the stable case (dash dot). Potential temperature for the stable case. (dashed)

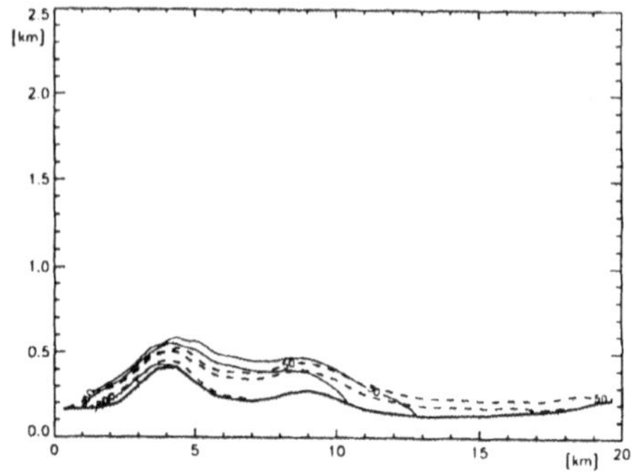

Figure 2. Isolines for concentration for stable, (dashed) and neutral case. The view is from the right, at the level of the source, (see figure 3).

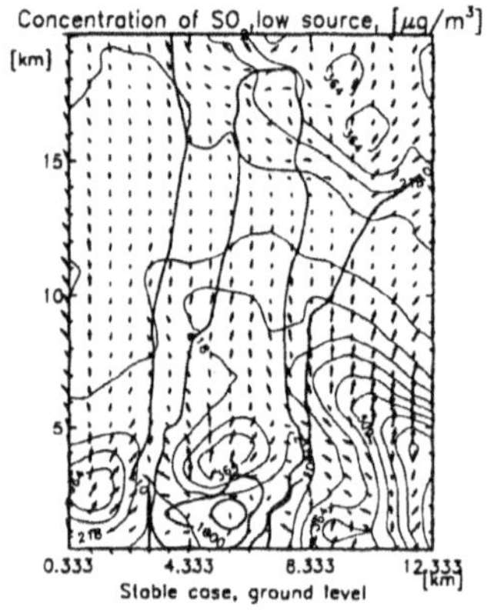

Figure 3. Topography, velocity vectors and concentration isolines for SO_2, stable case

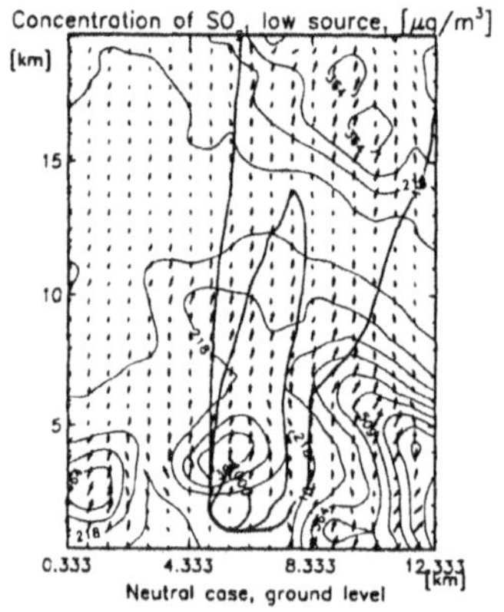

Figure 4. Topography, velocity vectors and concentration isolines for SO_2, neutral case.

DISCUSSION

The model gives reasonable good results for windflow, turbulence and concentrations but still the description of turbulence is simple. To improve the model, turbulence should be changed from scalar to vector description.

REFERENCES

Detering, H. W. and Etling, D, 1985, ´Application of the E-ε turbulence model to the atmospheric boundary layer`, *Boundary-Layer Meteorology*, **33** , 113-133

Rodi, W., 1987, ´Examples of Calculation Methods for Flow and Mixing in Stratified Fluids`, *Journal of Geophysical Research*, **92**, no C5, 5305-5328.

Högström, U. (1988): 'Non dimensional wind and temperature profiles in the atmospheric surface layer: a re-evaluation', Bound. Layer Meteor. **42**,55-78

EMERGENCY WARNING SYSTEM IN SLOVENIA FOR ACCIDENTAL RADIOACTIVE RELEASES

Marija Božnar, Rafael Martinčič and Primož Mlakar

Jožef Stefan Institute
Jamova 39
SI-61000 Ljubljana, Slovenia

ABSTRACT

A Westinghouse type nuclear power plant (NPP) has been operating for more than ten years at Krško, situated in the eastern part of Slovenia. The Chernobyl (1986) accident underlined, that severe nuclear accidents should be taken into account in emergency plans. At that time the first automated dose rate measuring stations started to operate in Slovenia. Nowdays there are two modern environmental systems around the Slovenian NPP. The first one is the Environmental Information System of the Krško NPP (EIS KNPP). It measures meteorological and dose rate data at several points in the vicinity of the NPP. In addition a Radioactive Emergency Warning System (REWS) has been built that measures only dose rate data. The design details and dispersion model applied are explained in detail.

DESCRIPTION OF THE INFORMATION SYSTEMS

Krško NPP is the only nuclear power plant in Slovenia and as such it represents a possible source of radioactive pollution. Therefore it is very important to have a sound on-line monitoring system which would also be available in the case of an emergency.

EIS KNPP consists of three automated measuring stations (a fourth one is now being built) that measure wind speed and direction, air temperature, relative humidity, precipitation and gamma dose rate. The first of these stations is a 70m meteorological tower near the NPP that measures wind on three levels and relative humidity and air temperature on four levels. It also measures global solar radiation. All the stations lie in the Krško basin. For the use of sophisticated air dispersion models this is not the best choice, because for the wind field reconstruction wind measurements at higher levels would be needed while vertical temperature profile would help to determine the thermal inversion height. For this purpose an additional automated measuring station will be built on a nearby hill. For 1996 installation of a Doppler SODAR is planned. Monitoring system also includes a station that measures indoor radioactivity at several points inside the NPP and a hydrological station on the river Sava. All the data are collected at the central unit at the NPP every half an hour and are distributed to authorised users in Slovenia.

REWS now consists of nine gamma dose rate measuring stations in a circle of five kilometres around KNPP. The central unit is sited at Administration for civil protection and disaster relief in Ljubljana. Every half an hour data are collected from these stations by means of a mobile telephone network (radio lines are planned in future). This topology is only the first stage of the planned Slovenian REWS: ten autonomous gamma dose rate measuring stations are operating in addition all over the Slovenian territory and several more are planned. On-line communications are planned for these stations. In this way the whole Slovenia will be covered.

Both information systems exchange their data automatically to ensure data availability to all authorised users.

A two level automatic control of data quality is performed in both information systems to ensure data reliability.

AIR DISPERSION MODEL

A simple Gaussian air dispersion model based on the directions given in R.G. 1.145 is implemented (automatically) using the data from the EIS KNPP. It allows dilution coefficients or actual gamma dose rates calculations based on the measured emissions.

When SODAR data will be available, more sophisticated models will be installed. Among possible models a 3D Gaussian Puff model is being considered. This kind of model has proved to perform much better than simple Gaussian models in complex orography similar to the one around Krško NPP.

DOSE ASSESSMENT

Special software is installed to estimate doses to the population due to external radiation and internal contamination following the emergency procedures of the KNPP.

CONCLUSIONS

Both information systems discussed represent the basis of the complete Slovenian Radiological Emergency Warning System. The EIS KNPP has been in operation more than ten years while REWS has been functioning slightly over one year. The operating experience shows that the systems were well designed and meet their planned performance. More on-line gamma dose rate measuring stations will be included, better air dispersion models will be used that are suitable for the complex terrain, and more reliable communications will be established in the near future.

REFERENCE

Atmospheric dispersion models for potential accident consequence assessments at nuclear power plants, U.S. Nuclear regulatory commission, Regulatory guide 1.145, 1979

A NUMERICAL STUDY ABOUT THE TRANSPORT AND DIFFUSION OF SULFER DIOXIDE DURING THE BREEZE CYCLE IN THE ISLAND OF MALLORCA (SPAIN)

Romualdo Romero, Sergio Alonso, and Clemente Ramis

Departament de Física
Universitat de les Illes Balears
07071 Palma de Mallorca, Spain

The island of Mallorca (Figure 1) is located in the centre of the Western Mediterranean. Its size is large enough to develop local circulations driven by land-sea differential heating forcing. So, during the warm half of the year, when the large scale circulation is dominated by the Azores anticyclone, the breeze circulation develops very often. As a consequence of the orography of the island, dominated by a coastal range in the west part ("Serra de Tramontana" in Figure 1), the main features of the sea breeze circulation are the blocking along the Serra de Tramontana, the strong flux from the bays of Palma and Alcudia, and the formation of a convergence line in the centre of the island.

In this work, a numerical study of the breeze circulation and the associated transport and diffusion of sulfur dioxide emitted by an electric plant located in the north bay of the island (EP1 in Figure 1) is presented. The study is extended to a future scenario when a twin electric plant (EP2 in Figure 1) which is planned to be built in the south coast is also considered. The work is based on numerical simulations starting at 0500 LST and covering a diurnal period. In the mesoscale numerical model (Nickerson et al., 1986) the horizontal diffusion is explicitally introduced by a fourth order operator, and the vertical diffusion is parameterized according to an eddy coefficient which is function of the predicted turbulent kinetic energy (Bougeault and Lacarrère, 1989).

Figure 2 shows the simulated mature breeze at 1500 LST. The model result reproduces correctly the main features of the breeze and the convergence lines are well marked (Ramis and Romero, 1995). At 0900 LST the winds are still light and restricted over the coastal areas, but the vertical diffusion over land is already appreciable. That leads to fumigation and high values of ground-level concentration not too far from the coastal sources. With the well defined breeze at 1500 LST, the colder air from the sea penetrates inland and the turbulence diminishes near the coast, but intensifies and extends vertically toward the centre of the island where strong sensible heat flux occurs. With that situation, the fumigation is translated inland, although the mixing acts in a wider layer and lower values of the concentration are reached at the ground. The convergence line limits the progression of the pollutant, deflecting the plume from EP1 toward the main mountain range, and the plume from EP2 toward the centre of the island (Figure 3). At 2100 LST the vertical diffusion has almost disappeared but horizontal diffusion and a slight transport continue acting. As a result, the pollutant derived from EP2 remains about the source level and the concentrations at ground only reach low values. The pollutant from EP1 is blocked over the eastern slope of the Serra de Tramontana. In the successive hours, vertical diffusion plays a secondary role, whereas the transport by downward air motion over the island associated with a slight land breeze becomes the dominant mechanism. The pollutant released from EP2 has sunk spectacularly, and practically all the eastern part of the island is affected (Figure 4). In this sense, the impact of the new power plant seems to be more important for the breeze case.

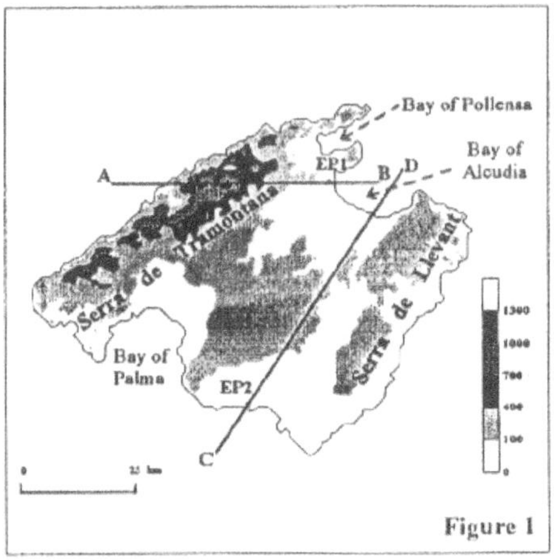

Figure 1

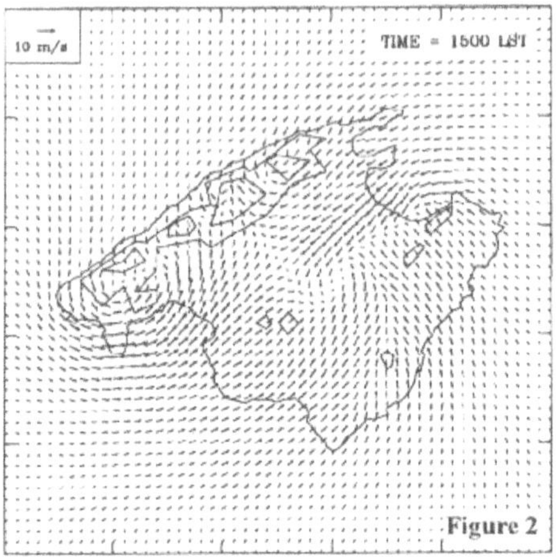

Figure 2

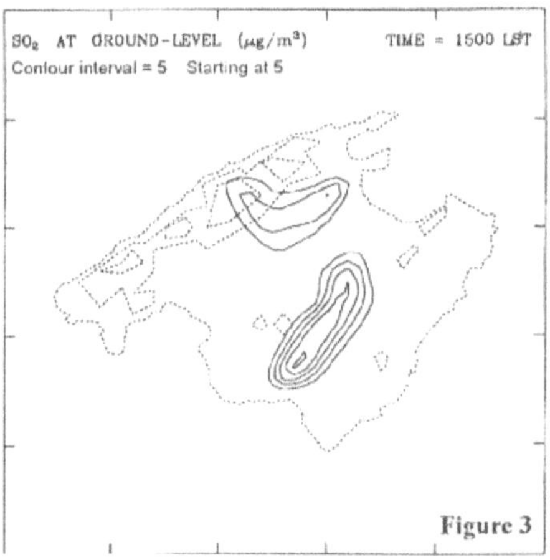

Figure 3

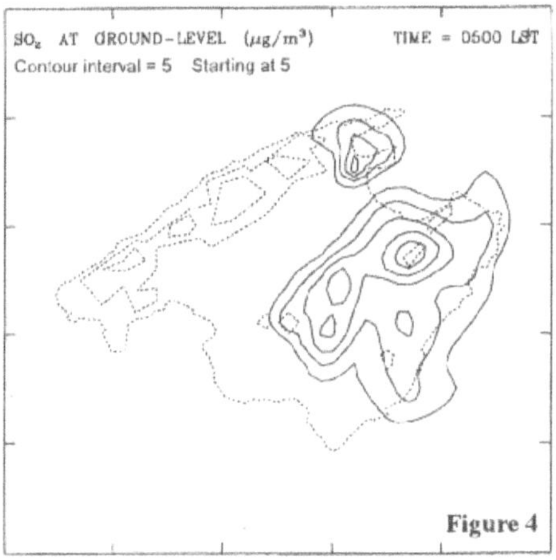

Figure 4

REFERENCES

Bougeault, P., and Lacarrère, P., 1989, Parameterization of orographic induced turbulence in a mesobeta scale model, *Mon. Wea. Rev.* 117:1872.

Nickerson, E.C., Richard, E., Rosset, R., and Smith, D.R., 1986, The numerical simulation of clouds, rain and airflow over the Vosges and Black forest mountain: a mesobeta model with parameterized microphysics, *Mon. Wea. Rev.* 114:398.

Ramis, C., and Romero, R., 1995, A first numerical simulation of the development and structure of the sea breeze in the island of Mallorca, *Ann. Geophy.*.13:981.

NUMERICAL SIMULATION OF THE MESOMETEOROLOGICAL CIRCULATIONS IN THE IBERIAN PENINSULA

R. Salvador, E. Mantilla, M.J. Salazar, and M. Millán

CEAM, Parque Tecnológico, Calle 4, Sector Oeste
46908 Paterna, Valencia Spain

INTRODUCTION

Several experimental campaigns were taken place during successive summers within the framework of EC' projects in the Iberian peninsula. Analyses of observed data served to document the structure and development of the Iberian Thermal Low (ITL), the coastal circulations and other meso-meteorological and air pollution processes involved inland and along the northern and eastern coasts of the peninsula. These processes were summarized in Millán et al.[1].

The Regional Atmospheric Modeling System (RAMS)[2] was applied to reproduce the mesometeorological processes in the peninsula and Its link with the synoptic circulation. For this purpose, two nested grids were defined: a coarse grid having 60 km grid spacing in both horizontal directions over 3720 km by 3120 km domain in the East-West and North-South direction, respectively. The finer grid was 15 km grid spacing covering a horizontal domain of 1340 Km by 1110 km in the East-West and North-South, respectively. In the vertical, twenty four with geometric grid stretching increasing with height were used and the model top was located at 20 Km.

RAMS has several options and features to initialize the model. Current simulation used the following available options: 3-D non-hydrostatic anelastic atmosphere in a terrain following coordinates. A non-homogeneous initialization with non-stationary boundary conditions was used. The model was initialized with the ECMWF global analysis data set of July 25-26 1989. The global gridded data set contain horizontal wind components, temperature, and relative humidity at the mandatory level in the standard spacing of 1 by 1 degrees in latitude and longitude. These data was ingested into RAMS isentropic analysis package as part of the model initialization procedure. Input data set was processed and the model was running for 24 hours.

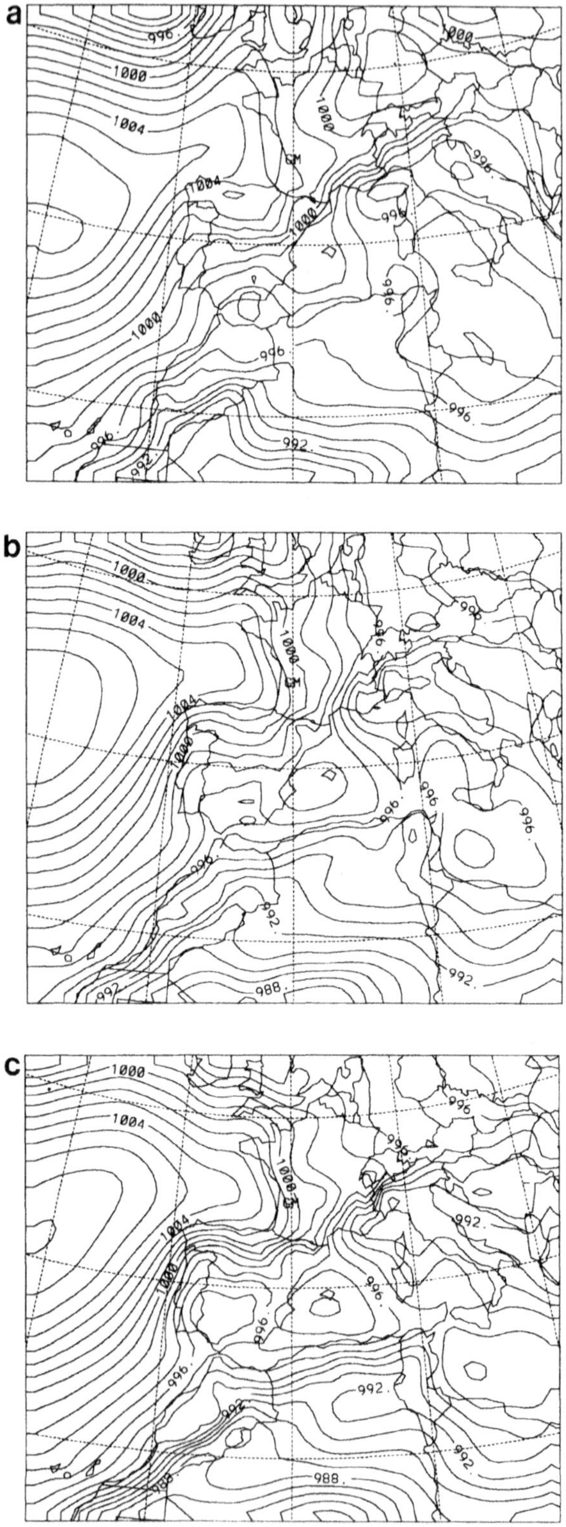

Figure 1. Modeled reduced Mean Sea Level Pressure pattern, at a height of about 58m for the coarse grid, at (a) t=06:00 UTC; (b) t=12:00 UTC; and (c) t=18:00 UTC. (MECAPIP, 25 July 1989, Grid 1.)

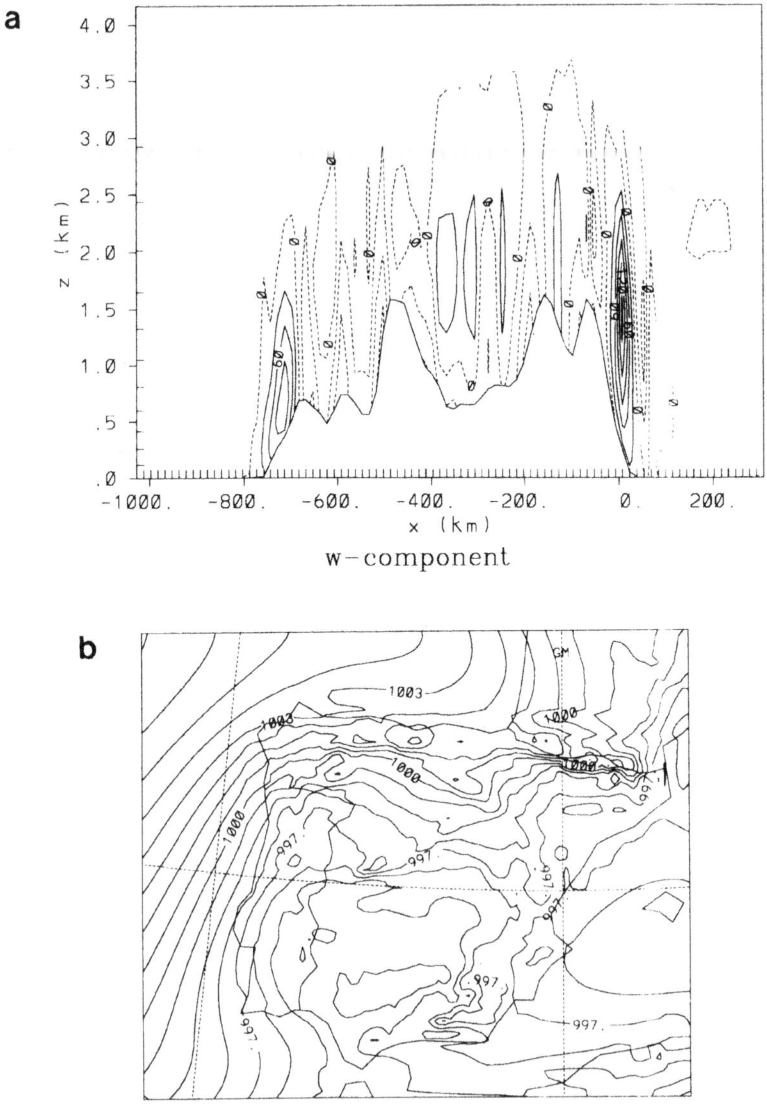

Figure 2. (a) Vertical crossection of verical wind component at 18:00 UTC, for the fine grid, along a W–E line shown in (b). (b) Pressure field at 18:00 UTC at height above the ground of about 58m for the fine grid. (MECAPIP), 25 July 1989, Grid 2.)

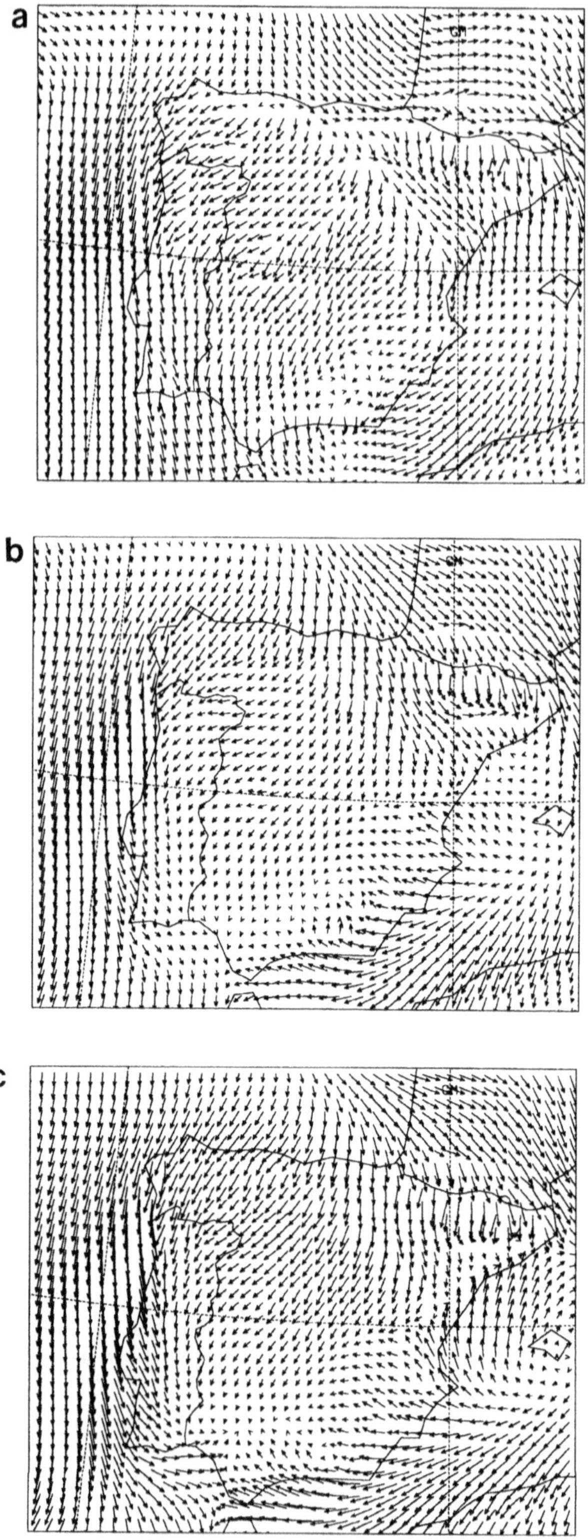

Figure 3. Modeled wind fields at a height above the ground of about 58m for the fine grid, at (a) 06:00, (b) 12:00, and (c) 18:00 UTC.→=5.0 m/s maximum vector. (MECAPIP, 25 July 1989, Grid 2.)

RESULTS AND CONCLUSIONS

Modeled reduced MSLP patterns show a typical **evolution** of the ITL (Figure 1), the Low relaxes during the night, as shown in the map at 06:00, and reaches it deepest extend around 18:00. The ITL appears as a diurnal extension of the African semi-permanent low, as it was pointed out in Millán et al.[3].

The simulated high resolution pressure plot show the **formation** of the main ITL cell over the Southern half of the peninsula by 18:00 (Figure 2a), as it was observed[3] for this day. Isobars also show a relative ridge of high pressure along the Ebro valley (aligned NW-SE in the North-east quadrant). This type of structure has been observed to accompany the formation of venting and/or drainage winds (Cierzo) along the Ebro valley. These features were observed during the experimental campaign of the MECAPIP project[3] and well predicted by the model (Figure 3).

Other interesting observed phenomena well reproduced by the model is the relative high pressure system over the cold water of the Mediterranean sea (Figure 1). This system provides a preferred area for the atmospheric subsidence, and act as a sink for some of the convective circulations which develop in the inland and coastal areas. These **compensatory** processes are seen in the vertical crossection of the vertical wind component of Figure 2b, where dashed lines denote downward motions and solid lines upward motions.

Figure 3 shows fine grid results of the modeled horizontal wind fields. Nocturnal drainage and valley flow are predicted along the main valleys. By 12:00 sea breezes combine with up-slope winds were well established along coastal areas, except in Catalonia. By 18:00, it was predicted wind **converge** toward the ITL center cell. It was also simulated the typical eastern winds across the Gibraltar Strait.

As conclusions, the main features of the area flow dynamics were well reproduced by RAMS. Further refinements of the input parameters need to be improved, such surface and soil input data. Finally, quantitative model evaluation with observed data need to be made.

ACKNOWLEDGMENTS

We acknowledge ITEMA to let us use its facilities and Dr. Kallos and Dr. Kotrini of the University of Athens for their assisstance in the use of RAMS model.

REFERENCES

1. M. Millán R. Salvador, E. Mantilla, and B. Artíñano, Meterology and Photochemical Air Pollution in Southern Europe: Experimental Results from EC Research Projects. To be published in Atmospheric Environment (1995).
2. R.A. Pieke, W.R. Cotton, R.L. Walko, C.J. Tremback, W.A. Lyons, D. Grasso, M.E. Nicholls, M.D. Moran, D.A. Wesley, T.L. Lee, and J,H. Copeland, A Comprehensive Meteorological Modeling System-RAMS, *Meteorol. Atmos. Phys.* 49, 69-91 (1992).
3. M. Millán, B. Artíñano, L. Alonso, M. Castro, F. Fernandez-Patier, J. Goberna, Mesometerological Cycles of Air Pollution in the Iberian Penninsula. *Report nº 44. Commission of the European Communities.* Brussels

THE PARTICIPATION OF THE SWISS METEOROLOGICAL INSTITUTE IN THE EUROPEAN TRACER EXPERIMENT (ETEX)

[1] D. Schneiter, [1]H. Berger and [2]S. Galmarini

[1] Swiss Meteorological Institute (SMI)
Payerne, Switzerland
[2] Institute for Marine and Atmospheric Research (IMAU)
Utrecht University, the Netherlands

In order to evaluate the state of the art in long range transport and dispersion modelling, the European Commission, Joint Research Centre, Ispra (JRC), the World Meteorological Organization (WMO), and the International Atomic Energy Agency (IAEA) have organised the European Tracer EXperiment (ETEX). This project uses long range tracer releases and measured concentration fields to evaluate the predictions of long range (2000 km) models developed by research institutes and meteorological offices around the world, in order to forecast the dispersion patterns in case of accidental releases of harmful substances.

The participation of the SMI in the two tracer experiments, which took place in autumn 1994, involved measuring constant level balloon trajectories launched at the tracer release site. The trajectories were then used to assist aircraft flights for measuring concentration profiles at great distances from the release point (500-1000 km). Wind profiles obtained at the release site by SODAR were also helpful for characterizing the flow in the vicinity of the release point.

The plume dispersion was forecasted in real time using the LORAN (Long Range Atmospheric Advection of Nuclides) model developed at the JRC and operational at the SMI. The model produced 60 hours' forecasts of trajectories and ground concentration using meteorological fields provided by the European Centre for Medium Range Weather Forecasts (ECMWF).

Preliminary results show a good agreement between trajectories measured from the release point and those computed by means of 850 hPa wind fields forecasted by the ECMWF. The ability of a model using a 850 hPa wind field to compute accurate tracer ground concentrations in complex topography is debatable considering the complex wind flows measured at ground weather stations in Switzerland during these releases. The short response time (a few minutes) of the LORAN model is of great advantage in emergency conditions.

ANALYSIS OF AIR POLLUTANTS IN ISTANBUL: A PRELIMINARY STUDY

Mehmet Karaca,[1] Arslan Saral,[2] Mete Tayanç,[3] and Ferruh Ertürk[2]

[1] Istanbul Technical University, Maden Fak., Genel Jeoloji ABD, Maslak, 80626 Istanbul-Turkey.
[2] Yıldız Technical University, Dept. of Environmental Engineering, Beşiktaş 80750, Istanbul Turkey.
[3] Marmara University, Dept. of Environmental Engineering, 81040 Göztepe, Istanbul Turkey.

INTRODUCTION

In last ten years, Istanbul faced severe air pollution problems. The occurrences of high concentrations of air pollution in the city have reached to the levels of danger for the habitants (Ertürk, 1986). The main cause of İstanbul's air pollution comes from the burning of low quality lignite and fossil fuels containing high percent of sulfur used mainly for heating purposes (Ertürk, 1986; Ertürk et al., 1995). It is the purpose of this article to gain perspective for the assessment of the spatial distribution of one of the air pollutants, SO_2, for the İstanbul municipality region by the usage of a spatial prediction scheme, Kriging (Matheron, 1971).The city's population increased exponentially at the beginning of 1980's. 1/5th of the population and 70% of the country's industry is located in İstanbul. The urbanization of the city is badly organized and unplanned.

Mean daily SO_2 and PM concentrations exceeding air quality standards have been recorded many times during the period of 1985 to 1994 at several stations of the city by the Ministry of Health.

RESULTS

The data was obtained from the State Statistical Institute. Figure 1 illustrates SO_2 distributions over the city for January 1994 based on Kriging technique (Matheron, 1971; Deniz and Karaca, 1995). Two major maxims are observed in the Golden Horn Valley of European side and Göztepe in the Asian side. We think that the topographical effects are the main factors for having these maxims with the meteorological conditions. Deniz and Karaca (1995) show that Istanbul had experienced less cyclonic activities than the climatological mean in 1993-94 winter season.

It can be seen from Figure 2 that the ratio of SO_2 emission due to coal/fuel-oil increases drastically especially after 1988 (Ertürk et al., 1995). In estimation of SO_2 emissions, EPA (1980) factors were used.

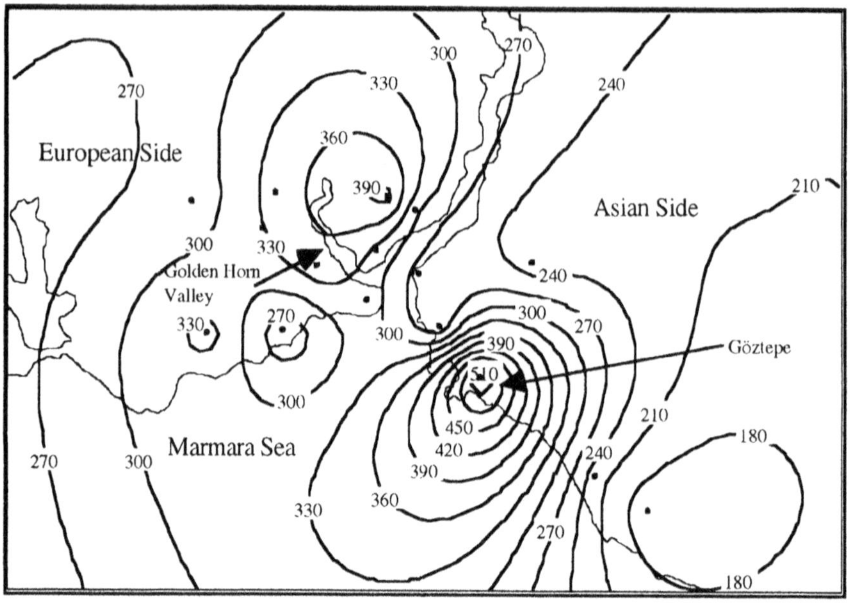

Figure 1. Distribution of monthly average SO_2 concentrations for January 1994 (μgr/m³)

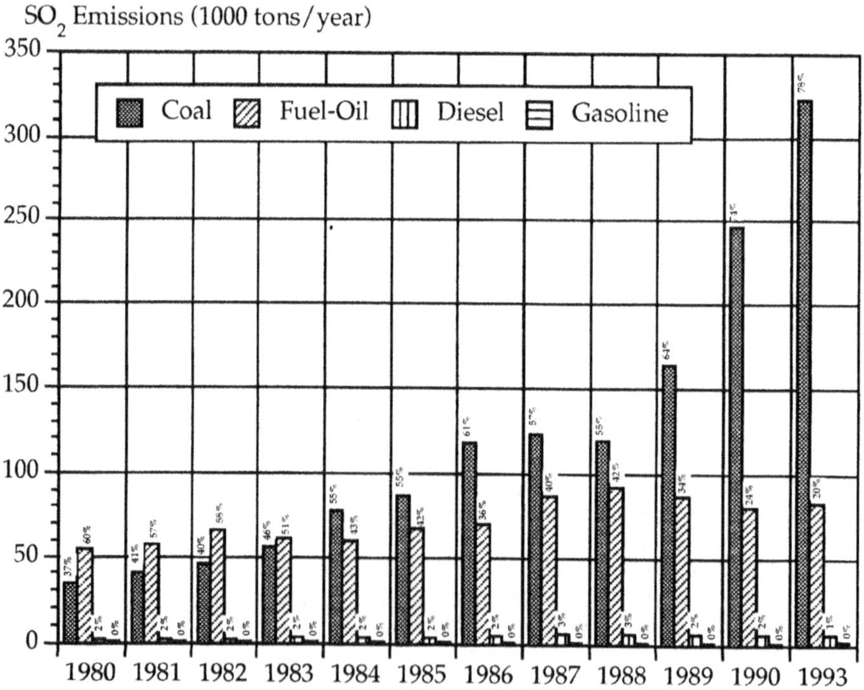

Figure 2. Estimated SO_2 emissions for different fuels from 1980 to 1993.

SUMMARY AND CONCLUSIONS

Air pollution problem in İstanbul is at alarming levels as SO_2 and PM measurements carried out at several stations since 1985 show. Both monthly and yearly average concentrations exceed the limits given by World Health Organization Air Quality Standards by several folds especially after 1987, and a definite increasing trend is observed since 1985.

The major pollution sources contributing to the high levels of SO_2 and PM is the consumption of low-quality fuels (mainly coal and fuel-oil containing high percent sulphur) in the residential areas for heating purposes in the winter season.

Previous studies has indicated that major factors that must be controlled are the sulfur content in coal and fuel-oil and particulate matter emissions from coal, and that the desired levels can be achieved by using low-sulphur fuels (coal and fuel-oil containing <1% sulphur) and high-quality coal (smokeless coal) in residential areas (Ertürk, 1986).

Air pollution problem in İstanbul is a multidimensional problem and requires a systematic approach involving all variables in order to reach a feasible solution. The determination of priorities is also considered a very important step to set the systematic solutions.

The major source for pollutants for the heating season comes from low quality coal and inefficient burning systems. The coal used for heating in the city during 1994-1995 winter season was 40% humid and had more than 2% sulfur. Beside these factors, the lack of effective heat isolation in houses causes more fuel consumption than actually needed.

Severity of the air pollution in İstanbul have already reached to the levels of danger for human beings. As a conclusion we can suggest the following short and long-term precautions:

- Usage of high quality fuels such as smokeless coal, low sulfur fuel-oil or natural gas in houses and industry,
- Urban planning,
- Preventing uncontrolled migration from rural areas,
- Controlling emissions from vehicles,
- Investing on public transportation,
- Heat isolation in houses,
- Alternative energy sources.

REFERENCES

Deniz, A., and Karaca, M., 1995, Analysis of cyclone tracks over Turkey, (in Turkish), I.T. U. Dergisi, (submitted).

EPA, 1980, *Compilation Of Air Pollution Emission Factors*", 3rd ed., OAQPS, USEPA, Research Triangle Park, NC, Publ. No: AP-42.

Ertürk, F., 1986, Investigation of strategies for the control of air pollution in the Golden Horn region, İstanbul, using a dispersion model, *Environmental Pollution (Series B)*, 11, 161-168.

Ertürk, F., Karaca M., Tayanç, M., and Saral, A., 1995, Mapping of air pollution in Istanbul and determination of regions in usage of natural gas for heating, beforehand, Istanbul Greater Municipality Research Project, 71 pp.

Matheron, G., 1971, *The Theory Of Regionalized Variables And Its Applications, Cahiers du Center de Morphologie Mathematique*, No. 5 Fontainebleau, France.

VALIDATION OF ADMS FOR DISPERSION IN CONVECTIVE CONDITIONS

Belinda M. Davies

U.K. Met. Office (APR)
London Road
Berkshire RG12 2SZ
England

INTRODUCTION

The Atmospheric Dispersion Modelling System (ADMS) is a short range dispersion model (Carruthers et al., 1994) developed jointly by the Cambridge Environmental Research Consultants (CERC), National Power and the U.K. Met. Office and is commercially available. It adopts a modern approach to boundary layer scaling based on the Monin-Obukhov length and boundary layer depth rather than on Pasquill-Gifford stability classes, which older models tend to use. R91 is an example of this older, Pasquill-Gifford type model. In this paper ADMS and R91 have been compared for passive and buoyant releases in convective conditions.

PASSIVE RELEASES

Comparisons were made with the Willis and Deardorff laboratory water tank data for a low (z_s/h=0.067, 1978) and high level source (z_s/h=0.490, 1981). The convective boundary layer was simulated using convection with zero mean wind, of a fluid layer heated from below into an overlying stable layer. Results were non-dimensionalized through the use of free-convection similarity scaling. ADMS scales according to this scaling regime and can cope with low and high level sources, two points where R91 fails. ADMS also gave better predictions of lateral and vertical spread.

BUOYANT RELEASES

A comparison of data from the Kincaid power plant in Illinois, U.S.A. (Carruthers et al., 1994), showed that ADMS (v1.06) tended to overpredict close to the source, especially so for very buoyant plumes. At this stage ADMS did not take account of plume 'lofting' (Briggs, 1993). This is the term given to very buoyant plumes which loft to the top of the boundary layer and are 'trapped' beneath the capping inversion, where

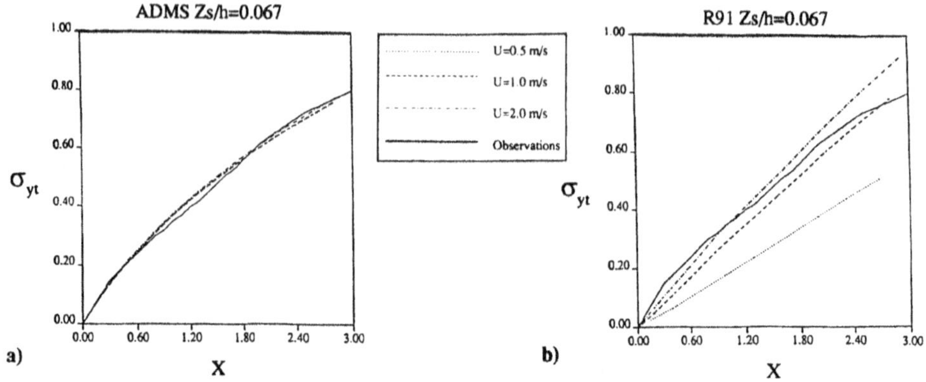

Figure 1: The dimensionless lateral standard deviation predicted by (a) ADMS (v1.06) and (b) R91 (as implemented by CERC in the code DISTARS6), compared with the Willis and Deardorff experimental data, as a function of the dimensionless downwind distance for mean wind speeds of $U = 0.5$, 1.0 and 2.0 ms^{-1} and for a source height of $0.067h$.

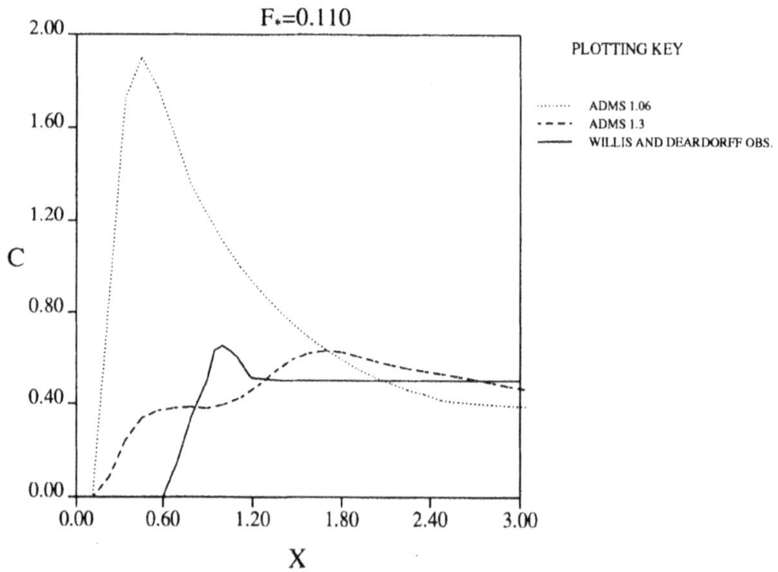

Figure 2: Comparison of the dimensionless centreline ground-level concentration versus the dimensionless distance downwind predicted by ADMS v1.06 (dotted line), and the modified version, ADMS v1.3 (dashed line) and the Willis and Deardorff water tank experimental data (1984, z_s/h=0.134) (solid line), for a strongly buoyant plume ($F_* = 0.110$, where F_* is the dimensionless buoyancy flux).

their residual buoyancy resists initial downward mixing and enhances lateral spread. Subsequently comparing R91 and ADMS, from a statistical aspect, there was found to be little difference in the performance of the two models, although R91 produced the largest over and underpredictions. Modifications were introduced in a later version of ADMS (v1.3) to represent plume lofting. At the same time the inversion penetration algorithm was improved and the effect the decorrelation of the vertical component of turbulence at large travel times was included in the plume spread model. These changes reduced the overprediction of ground level concentrations close to elevated sources and led to a more accurate fit between the Kincaid data and the experimental observations of Willis and Deardorff's water tank experimental data for buoyant releases (1984).

REFERENCES

Briggs G.A., 1993, Final results of the Condors convective diffusion experiment, *Boundary layer Meteorol.*, **62**:315.

Carruthers D.J., Holyroyd R.J., Hunt J.C.R., Weng W.S., Robins A.G., Apsley D.D., Thomson D.J. and Smith F.B., 1994, UK-ADMS: A new approach to modelling dispersion in the earth's atmospheric boundary layer, *J. Wind Eng. Ind. Aerodyn.*, **52**:139.

Carruthers D.J., Edmunds H.A., Ellis K.L. McHugh C.A., Davies B.M. and Thomson D.J., 1995, The Atmospheric Dispersion Modelling system (ADMS): comparisons with data from the Kincaid experiment, *Int. J. Environment and Pollution*, **5**.

DENSE GAS PLUME FIELD MEASUREMENTS AT THE NEVADA TEST SITE

Gary A. Briggs[1]

Atmospheric Sciences Modeling Division
Air Resources Laboratory
National Oceanic and Atmospheric Administration
Research Triangle Park, North Carolina 27711

BACKGROUND

A series of field experiments on dense gas dispersion under various meteorological and surface conditions was begun at the DOE's Spill Test Facility in Nevada in July 1993. Such investigations were authorized under the 1990 Clean Air Act Admendments. Details of planning and the execution of the tests are carried out by the Desert Research Institute[2] and the Western Research Institute[3]. We are presently preparing to return to the site in September 1995 for a second series of EPA-sponsored, continuous CO_2 releases targeting low wind speed, stable conditions over smooth desert. In addition, the EPA experiments will be immediately preceded by a series sponsored by the Petroleum Environmental Research Forum (PERF). The PERF experiments will target similar meteorological conditions, but with nearly 40,000 m^2 of desert floor covered with artificial roughness, 20-cm high baffles, to make the (scaled up) roughness length (z_0) more representative of industrial areas; $z_0 \simeq 1.1$cm, about 65 times the smooth desert z_0. There also will be a series of PERF experiments with large, 2.4-m high "flaps" added around the source position; these are intended to simulate the dispersive effect of an industrial process complex and create a local $z_0 \simeq 20$cm.

In July 1993, four CO_2 releases with SF_6 tracer were made during stabilizing conditions just before sunset at wind speeds $\bar{u} = 5.0$, 2.35, 3.0, and 5.5m/s (all measurements will be listed in order, for Tests 1, 2, 3, and 4, respectively). The purpose of this series was to test newly acquired instrumentation and to collect some baseline measurements of heavy gas dispersion over smooth, flat terrain. Details of the experimental goals, release and measurement methods, data management and QA were given by Egami et al. (1995). A fairly detailed examination of the data set was given by Briggs (1995). The intent of this manuscript and its accompanying poster is to summarize many of the results from these baseline tests.

The gases were released from a 1×1m square pit with little upward velocity, 1m/s or less. The release rates for CO_2 were about 0.55kg/s, except 1.6kg/s for Test 2. Release durations were 3 to 4½ min, except for only 70s for Test 2; nonetheless, they all could be considered continuous releases at the main, x = 38.5m arc. A key variable in dense gas models is the

[1] On assignment to the National Exposure Research Laboratory, U. S. Environmental Protection Agency.
[2] University and Community College System of Nevada.
[3] University of Wyoming Research Corporation.

buoyancy flux from the source, B_0, which is g times the specific density difference relative to air times the volumetric flow rate. $B_0 = 1.48$, 5.23, 2.12, and 2.04 m^4/s^3 for the four tests. Since the source gases were within 3°C of the ambient temperature, the buoyancy flux in the plumes could be considered conserved.

Concentrations were collected in bag samplers, most of which (43) were deployed on 9 towers at the main arc at heights z = 0.1, 0.33, 0.67, 1, and 1.4 m. This provided a fairly dense array for determining plume cross sections and checking mass balance, i. e., measured integrated flux of concentration, C, vs. measured release rate, Q. In addition, 37 continuous CO_2 sensors with about 1-s response times were deployed, mostly collocated with bag samplers to provide a test of their time-integrated C values.

The mean wind speeds have already been given for the plume centroid heights at the main arc, $\bar{z} = 0.87$, 0.55, 0.65, and 0.97m. The speeds averaged over the duration of release at z = 0.5m near the arc were 4.24, 2.75, 2.95, and 4.95m/s. The wind and temperature profiles on the 8-level meteorological tower, 112m from the source, were consistent with $z_0 = 0.17$mm, friction velocity $u^* = 0.23$, 0.115, 0.145, and 0.25 m/s, and Obukhov lengths L = 140, 20, 25, and 70 m (Högström, 1988). Thus, Tests 1 and 4 were near neutral stability, while Tests 2 and 3 were moderately stable conditions.

RESULTS

The following results are all for the main arc, at x = 38.5m. The bag samplers accounted for 0.94 ±3% of the expected mass flux of CO_2 except for Test 2 (the short test), with only 0.65. The continuous sensors measured 1.02 ±8% except for Test 2, with only 0.53. The SF_6 performance was poorer, with 0.92 ±30% and Test 2 invalidated by a release mishap. C above z = 0.33m dropped off nearly exponentially, so an exponential extrapolation was used above z = 1.4 m.

Various width and depth measures based on $\int Cdz$ and $\int Cdy$ were compared. Agreement between CO_2 sensors and bags was excellent, within ±3%, for both σ_y and width of 10% of maximum $\int Cdz$, $W_{10\%}$; however, the ratio was only 0.92 ±11% for $W_{50\%}$. The ratio $W_{10\%}/\sigma_y$ was very consistent, 1.8 ±2%, while $W_{50\%}/\sigma_y$ was 1.08 ±13%. Similarly, $h_{10\%}/\bar{z}$ was extremely consistent, 2.5 ±1%, while $h_{50\%}/\bar{z}$ was 1.03 ±12%, where h is plume depth. The ratio of SF_6 to CO_2 bag-based depth measures was consistently 1.04 ±3 or 4%; the continuous CO_2 sensors, which were arrayed differently, gave $\bar{z}$ and $h_{10\%}$ ratios near 1.12 ±5%, suggesting some deficiency in the vertical extrapolations. In sum, we found very good consistency between duplicate measures for σ_y, $W_{10\%}$, $\bar{z}$, and $h_{10\%}$, and consistent $W_{10\%}/\sigma_y$ and $h_{10\%}/\bar{z}$ ratios, but more variability for the 50% plume contours.

Briggs (1995) also tested some analytical parameterizations for plume width, depth, and centerline C, but space here is inadequate for the equations and data comparisons. The peak to average C_{max} were about 2.

Disclaimer- This paper has been reviewed in accordance with the U. S. Environmental Protection Agency's peer and administrative review policies and approved for presentation and publication. Mention of trade names or commercial products does not constitute endorsement or recommendation for use.

REFERENCES

Briggs, G.B., 1995, Field-measured dense gas plume characteristics and some parameterizations, *International Conference on Modeling & Mitigating the Consequences of Accidental Releases of Hazardous Materials*, AIChE, New Orleans, 26-29 Sept 1995.

Egami, R., Freeman, D., Bowen, J., Coulombe, W., Watson, J., Sheesley, D., King, B., Nordin, J., Routh, T., Briggs, G., and Petersen, W., 1995, Controlled experiments for dense gas diffusion- experimental design and execution, model comparison, *International Conference on Modeling & Mitigating the Consequences of Accidental Releases of Hazardous Materials*, AIChE, New Orleans, 26-29 Sept 1995.

Högström, U., 1988, Non-dimensional wind and temperature profiles in the atmospheric surface layer: a re-evaluation, *Boundary-Layer Meteor.* 42: 55-78.

PARTICIPANTS

The 21st NATO/CCMS International Technical Meeting on Air Pollution Modelling and Its Application. Baltimore, Maryland, USA, 6 -10 November 1995

ARGENTINA

Ulke A.G. — Department of Atmospheric Sciences
University of Buenos Aires
Pabellon II-Piso II
Cuidad Universitaria
1428 Buenos Aires

AUSTRALIA

Hurley P. — CSIRO
Division of Atmospheric Research
Mordialloc,
3195 Victoria

AUSTRIA

Olendrzynski K. — International Institute for Applied Systems Analysis
2361 Laxenburg

Piringer M. — Central Institute for Meteorology
Hohe Warte 38
1190 Vienna

BELGIUM

Cosemans G. — Energy Division
VITO
Boeretang 200
2400 Mol

Dutrieux A. — Universite Catholique de Louvain
Chemin du Cyclotron, 2
1348 Louvain-la-Neve

Kretzschmar J.G. — Energy Division VITO
Boeretang 200
B-2400 Mol

Schayes G.H.	Universite de Louvain 2 Chemin du Cyclotron 1348 Louvain-la-Neuve

BULGARIA

Batchvarova E.	National Institute of Meteorology and Hydrology Bulgarian Academy of Sciences 66 Tzarigradsko Chausse 1784 Sofia
Syrakov D.	National Institute of Meteorology and Hydrology Bulgarian Academy of Sciences 66 Tzarigradsko Chausse 1784 Sofia

CANADA

Conway F.	Environment Canada 4905 Dufferin Street Downsview, Ontario M3H 5T4
Dastoor A.P.	Atmospheric Environment Service 2121 Trans-Canada Highway Dorval, Quebec PQ H9P 1J3
Lin X.	Ontario Hydro Technologies, 800 Kipling Avenue, KJ135 Toronto M8Z 3N5 Ontario
Sills D.M.L.	York University Department of Earth & Atmospheric Sciences 4700 Keele St. North York, Ontario M3J 1 P3
Steyn D.G.	University of British Columbia Department of Geography 1984 West Mall Vancouver, British Columbia, V6T 1Z2
Walmsley J.L.	Atmospheric Environment Service 4905 Dufferin St. Downsview, Ontario M3H 5T4 ON

Xu D.	York University Department of Earth & Atmospheric Sciences 4700 Keele Street North York, Ontario M3J 1 P3
Zhuang Y.	Atomic Energy of Canada Limited Pinawa, MN, ROE 1L0

CZECH REPUBLIC

Bubnik J.	Czech Hydrometeorological Institute Na Sabatce 17 143 06 Prague 4
Pretel J.	Czech Hydrometeorological Institute Na Sabatce 17 143 06 Prague 4

DENMARK

Barthelmie R.J.	Risø National Laboratory P.O. Box 49 4000 Roskilde,
Brandt J.	National Environmental Research Institute Frederiksborgvej 399 P.O. Box 358 4000 Roskilde
Christensen J.	National Environmental Research Institute Frederiksborgvej 399 4000 Roskilde
Fuglsang K.	dk-TEKNIK Gladsaxe Møllevej 15 DK-2860 Soborg
Gryning S.E.	Risø National Laboratory Department of Meteorology & Wind Energy DK-4000 Roskilde

Jensen A.B.	Danish Toxicology Center Agern Alle 15 2970 Hørsholm
Jensen M.H.	Danish Meteorological Institute Lyngbyvej 100 2100 Copenhagen
Jørgensen A.M.	Danish Meteorological Institute Lyngbyvej 100 2100 Copenhagen
Løfstrøm P.	National Environmental Research Institute Frederiksborgvej 399 P.O. Box 358 4000 Roskilde
Olesen H.	National Environmental Research Institute P.O. Box 358 4000 Roskilde

FINLAND

Karppinen A.	Finnish Meteorological Institute Air Quality Department Sahaajankatu 20 E 00810 Helsinki
Nikmo J.	Finnish Meteorological Institute Sahaajankatu 20 E 00810 Helsinki
Salonoja M.	Finnish Meteorological Institute Department of Weather Research P.O. Box 503 00101 Helsinki

FRANCE

Albergel A.	ARIA Technologies 14/30 Rue de Mantes 92700 Colombes

Butet A.	METEO-FRANCE CNRM/GMEI/SPEA Avenue Coriolis 31057 Toulouse
Flossmann A.	Universite Blaise Pascal Laboratoire de Meteorologie Physique 24, Avenue des Landais 63177 Aubiere Cedex
Wendum D.	Electricite de France 6, Quai Watier 78401 Chatou Cedex

GERMANY

Eichhorn J.	Institute of Atmospheric Physics Becherweg 21 55099 Mainz
Graf J.	Deutsche Forschungsanstalt für Luft- und Raumfahrt Oberpfaffenhofen Postfach 1116 82230 Wessling
Graff A.	Federal Environmental Agency Bismarckplats 1 12303 Berlin
Heil O.	Lahmeyer International Lyoner Strasse 22 60528 Frankfurt
Klug W.	Institut für Meteorologie TH Darmstadt 64289 Darmstadt
Kunz R.	Institut für Technische Thermodynamik Universität Karlsruhe Kaiserstrasse 12 76128 Karlsruhe
Martens R.	GRS Schwertner Gasse 1 50667 Köln

Moellmann-Coers M.	Forschungszentrum Jülich Posffach 1913 52425 Jülich
Panitz H.-J.	Forschungszentrum Karlsruhe IMK, Postfach 3640 76021 Karlsruhe
Renner E.	Institute for Tropospheric Research Permoserstrasse 15 04303 Leipzig
Ries R.	Institute for Atmospheric Physics Becherweg 21 55099 Mainz
Walter H.	Federal Office for Radiation Protection Ingolstaedlen Landstrasse 1 85764 Oberschleissheim
Wichmann-Fiebig M.	Northrhine-Westphalia State Environmental Agency Wallneyer Strasse 6 45133 Essen
Wolke R.	Instute for Tropospheric Research Permoserstrasse 15 04303 Leipzig
Zilitinkevich S.	Alfred Wegener Institute for Polar & Marine Research Am Handelshafen 12 27570 Bremerhaven

GREECE

Melas D.	Aristotle University of Thessaloniki Laboratory of Atmospheric Physics 54006 Thessaloniki
Kallos G.B.	University of Athens Ippocratous 33 10680 Athens
Kambezidis H.D.	National Observatory of Athens PO Box 20048 118 10 Athens

S. Kotsopoulos	University of Patras 26110 Patra
Psyloglou B.	National Observatory of Athens I.M.P.A.E. PO Box 20048 11810 Athens
Rafailidis S.	6 Pavlou Mela 54621 Thessaloniki
Ziomas I.	Aristotle University of Thessaloniki Laboratory of Atmospheric Physics 54006 Thessaloniki

HUNGARY

Labancz K.	Institute for Atmospheric Physics P.O. Box 39 1675 Budapest

ISRAEL

Yeheskel J.	Rotem Industries Division of Environmental Technology P.O. Box 9046 84 190 Beer-Sheva

ITALY

Anfossi D.	Istituto di Cosmogeofisica CNR Corso Fiumes 10133 Torino
Bianconi R.	IDEA S.N.C. Via Volturno 80 20047 Brugherio-Milano
Brusasca G.	ENEL/CRAM Via Rubattino 54 20134 Milano
Giostra U.	Instituto Isiata CNR Via Armesano 1 73100 Lecce
Giovanni G.	Joint Research Center ISPRA 21020 Ispra

Martano P.	CNR ISTATA Via Monteroni 73100 Lecce
Mosca S.	Environment Institute Joint Research Center ISPRA 21020 Ispra
Straume A.G.	Environment Institute Joint Research Centre ISPRA 21020 Ispra

JAPAN

Kitada T.	Toyohashi University of Technology Tempaku-cho 441 Toyohashi
Kondo H.	National Institute for Resourcs & Environment 16-3 Onogawa 305 Tsukuba
Uno I.	National Institute for Environmental Studies Onogawa 16-2 305 Tsukuba, Ibaraki

KOREA

Lee T.-Y.	Yonsei University Departmentof Astronomy & Atmospheric Sciences Seodaemoon-ku 120-749 Seoul

NORWAY

Bartnicki J.	Norwegian Meteorological Institute PO Box 43 0313 Blindern
Böhler T.	Norwegian Institute for Air Research P.O. Box 100 2007 Kjeller

Iversen T.
Department of Geophysics
University of Oslo
P.O. Box 1022 - Blindern
0135 Oslo

Mylona S.
Norwegian Meteorological Institute
PO Box 43 - Blindern
0313 Oslo

PORTUGAL

Barros N.
University of Aveiro
Department of Environment &
Planning
3810 Aveiro

Borrego C.
University of Aveiro
Deparment of Environment &
Planning
3810 Aveiro

ROMANIA

Pop L.-M.
I.C.I.M.
Spl. Independentei Nr. 294
Sector 6 Cod 77703
Bucharest

Pop T.
I.C.I.M.
Spl. Independentei Nr. 294
Sector 6 Cod 77703
Bucharest

RUSSIA

Genikhovich E.L.
Main Geophysical Observatory
7 Karbysheva Street
194018 Sankt Petersburg

Kargin B.
Pr. acad. Lavrentieva, 6
Computing Center. SD RAS
630090 Novosibirsk

SLOVENIA

Boznar M.
Jozef Stefan Institute
Jamova 39
61000 Ljubljana

Mlakar P. | Jozef Stefan Institute
Jamova 39
61000 Ljubljana

SPAIN

Romero R.
University of the Balearic Islands
Department of Physics
07071 Palma De Mallorca

Salvador R.M.
Parque Tecnologico
Calle 4 - Sector Oeste
46980 Paterna (Valencia)

San Jose R.
Group of Environmental Software and Modelling
Computer Science School
Technical University of Madrid
Boadilla Del Monte 28660 Madrid

Santabarbara J.M.
ITEMA
Apt. Correus 508
08220 Terrassa (Barcelona)

SWEDEN

Karlsson E.
National Defense
Research Establishment
901 82 Umeå

Persson C.
Swedish Meteorological and
Hydrological Institute
60176 Norrköping

Svensson G.
Uppsala University
Department of Meteorology
Box 516
751 20 Uppsala

Thaning L.
National Defense Research
Establishment
901 82 Umeå

SWITZERLAND

Schneiter D. — Swiss Meteorological Institute
Les Invuardes
1530 Payerne

THE NETHERLANDS

Builtjes P.J.H. — TNO-MW
PO Box 6011
2600 JA Delft

Erbrink T.T. — KEMA
PO Box 9035
6800 Arnhem ET

Ganzeveld L. — Wageningen University
Department of Air Quality
Biotechnion, Bomenweg 2
6700 EV Wageningen

van Jaarsveld J.A. — National Institute of Public Health and Environmental Protection
PO Box 1
3720 BA Bilthoven

van Melle A. — Kon Lodewykln 140
7314 GK Apeldoorn

van Pul A. — RIVM
PO Box 1
3720 BA Bilthoven

Verver G. — Royal Netherlands Meteorological Institute
Postbus 201
3730 De Bilt

TURKEY

Yenigun O. — Bogazici University
Institute of Environmental Sciences
Bebek
80815 Istanbul

Tayanc M. — Bogazici University
Institute of Environmental Sciences
Bebek
80815 Istanbul

Ulug S.E. — Middle East Technical University
06531 Ankara

UNITED KINGDOM

Davies B. — UK Meteorological Office
MET ARR, London Road
Bracknell
RG 12 2SZ Berkshire

Pryke D.C. — Ministry of Agriculture,
Fisheries & Food
17 Smith Square
SW1P 3JR London

U.S.A

Bornstein R.D. — San Jose State University
Department of Meteorology
San Jose, CA 95192

Briggs G. — Environmental Protection
Agency/NOM
MD-80
Research Triangle Park, NC 27711

Byun D.W. — NOAA/EPA
4201 Bldg. MD-80
Research Triangle Park, NC 27711

Carmichael G.R. — University of Iowa
Center for Global &
Regional Environment Research
204 Iowa Advanced Technology
Laboratory
Iowa City, IA 52242

J.K.S. Ching — Environmental Protection
Agency/NOAA, MD-80
Research Triangle Park, NC 27711

Crist K.	Ohio University Department of Health Science 518 Tower Athens, OH 45701
Dabdub D.	California Institute of Technology 210-41 Caltech Pasadena, CA 91125
Daye R.L.	US Environmental Protection Agency 21607 NE 178th St. Holt. MO 64048
Dennis R.L.	Environmental Protection Agency/NOM MD-80 Research Triangle Park, NC 27711
Fine S.S.	North Carolina Supercomputing Center P.O. Box 12889 Research Triangle Park, NC 27709-2889
Hanna A.F.	North Carolina Supercomputing Center P.O. Box 12889 Research Triangle Park, NC 27709-2889
Hanna S.R.	Earth Tech 196 Baker Avenue Concord, MA 01742-2167
Hayami H.	University of lowa CGRER, IATL lowa City, IA 52240-1000
Hicks B.B.	NOM/ARL 1315 East-West Highway SSMC3, Rm. 3152 Silver Spring, MD 20910
Karamchandani P.	ENSR Consulting & Engineering 1220 Avenida Acaso Camarillo. CA 93012

Lehmann E.	NOM/ARL 1315 East West Highway Silver Spring, MD 20910
Meroney R.N.	Colorado State University Civil Engineering Fort Collins, CO 80523
Nappo C.J.	NOM/ARL 456 South Illinois Ave. PO Box 2456 Oak Ridge, TN 37831-2456
Odman T.	MCNC - North Carolina Supercomputering Center 3201 Cornwallis Road Research Triangle Park, NC 27709-2889
Osa R.	Science & Technology Management 2511 N. 124th Street Suite 205 Brookfield, Wl 53005-8208
Pai P.	ENSR Consulting & Engineering 1220 Avenida Acaso Camarillo, CA 93012
Petersen W.B.	Environmental Protection Agency/NOM MD-80 Research Triangle Park, NC 27711
Pleim J.E.	Environmental Protection Agency NOM MD-80 Research Triangle Park, NC 27711
Rao S.T.	SUNY - Albany Department of Atmospheric Science Albany, NY 12222

Schiermeier F.A.	Environmental Protection Agency/NOM MD-80 Research Triangle Park, NC 27711
Scire J.S.	Earth Tech 196 Baker Avenue Concord, MA 01742
Segall E.	University of Delaware 409 Smith Hall Newark, DE 19716-2586
Sykes R.I.	TITAN/ARAP 50 Washington Rd. P.O.Box 2229 Princeton NJ 08543-2229
Yamada T.	Yamada Science & Art Corp. Rt. 4, Box 81-A Santa Fe, NM 87501-9804
Yamartino R.J. Jr.	Earth Tech Sigma Research Corporation 196 Baker Avenue Concord, MA 01742
Young J.O.	Environmental Protection Agency/NOM 4201 Bldg. MD-80 Research Triangle Park, NC 27711

AUTHOR INDEX

Aarnio P., 649
Alapaty K., 433
Alonso S., 669
Anderberg S., 659
Anfossi D., 559
Angevine W.M., 613
Archer G., 181
Arndt R.L., 167
Artz R., 657
Audiffren N., 73

Baklanov A., 665
Baldasano J.M., 583
Balis D., 549
Barros N., 119
Barthelmie R.J., 529
Bartnicki J., 471, 659
Batchvarova E., 295
Berger H., 677
Binkowski F.A., 135
Biswas J., 41
Božnar M., 667
Boersen G., 213
Borrego C., 119
Bozó L., 657
Brandt J., 461
Brechler J., 647
Brenden A.K., 109
Briggs G.A., 687
Brusasca G., 559
Builtjes P.J., 213, 345
Burman J., 665
Byun D.W., 433, 491
Böhm S., 647

Calbó J., 583
Carmichael G.R., 155, 167, 239
Cautenet S., 73
Chaumerliac N., 73
Chen L.-L., 239
Ching J., 135
Christensen J., 249
Cimorelli A.J., 623
Clarke J.F., 613
Cooter E.J., 613
Cosemans G., 641
Cremades L., 583
Crist K.C., 155

Dabdub D., 491
Dacić M., 635
Dastoor A.P., 37
Davies B.M., 683
de Oliveira A.P., 643
Degrazia G.A., 643
Dennis R.L., 3
Dentener F., 239
Desiato F., 559
Dumont G., 641

Eckman R.M., 407
Eichhorn J., 539
Ellermann T., 461
Ellestad T.G., 613
Elolähde T., 649
Ertürk F., 679
Esser P., 213

Fernau M.E., 389
Ferrero E., 559
Fiedler F., 89
Finardi S., 559
Fine S., 491
Finkelstein P.L., 613
Flossmann A.I., 259
Foss A., 471

Gabruk R.S., 425
Galmarini S., 677
Ganzeveld L., 325
Genikhovich E., 399
Girardi F., 181
Gong W., 275
Graf J., 81, 443
Graff A., 443
Graziani G., 181
Gryning S.E.. 295, 363

Haarala S., 649
Hakkarinen C.,373
Hanna A.F., 491
Hanna S.R., 481
Hao W., 41
Haugen J.E.. 109
Henn D.S., 425
Hernández J.F., 583
Hertel O., 529
Hotchkiss B.T., 155
Hurley P.J., 55
Härkönen J.. 649

Irwin J.S., 603
Isogawa S., 191
Iversen T., 109

Jacobs C.M.J., 203
Jalkanen L., 649
Jelev A., 661
Jensen M.H., 109
John K., 41, 155
Jørgensen H., 285

Kallos G., 99
Karaca M.. 679

Karamchandani P., 145, 311
Karlsson E., 363
Karppinen A., 649
Kessler C., 27
Kitada T., 191
Klemm O., 593
Klug W., 181
Knoth O., 655
Kondo Y., 191
Koskentalo T., 649
Kotroni V., 99
Kretzschmar J.G., 641
Kukkonen J., 649
Kunz R., 27

Labancz K., 657
Lagouvardos K., 99
Langner J., 127
Laurikko J., 649
Lee R.F., 623
Lelieveld J., 325
Lin X., 275
Lyck E., 285, 461
Løfstrøm P., 285

Manins P.C., 55
Mantilla E., 671
Marcelo L.M., 231
Martičnič R., 667
Mathur R., 433, 491
Matthijsen J., 345
Mazzeo N.A., 639
Melas D., 549
Menard S., 275
Meroney R.N., 451
Meyer E.W., 345
Mikkelsen T., 285, 461, 583
Millán M., 671
Mlakar P., 667
Moraes O.L.L., 643
Moreno B., 231
Morselli M.G., 559
Mosca S., 181
Moussiopoulos N., 27
Mylona S., 221

Nappo C.J., 407
Nester K., 89
Nodop K., 181
Noonan J.A., 55
Näslund E., 363

Odman M.T., 491
Olendrzynski K., 659
Olesen H.R., 519
Olteanu O., 661
Osa R.H., 373

Pacyna J., 659
Pai P., 145, 311
Paine R.J., 623
Panitz H.-J., 89
Papadopoulos A., 99
Parker S.F., 425
Pavageau M., 451
Perry S.G., 379, 623
Persson C., 127, 379
Petersen W.B., 379
Pinto J., 657
Pisa V., 647
Pleim J.E., 135, 613
Pop T., 661
Pop L.-M., 661
Pretel J., 647
Pryor S.C., 529

Rafailidis S., 451, 653
Ramírez-Montesinos A., 231
Ramis C., 669
Rantakrans E., 649
Rao S.T., 41
Renner E., 655
Ries R., 303
Robertson L., 127
Rocha A., 119
Roekens E., 641
Roelofs G-J., 325
Roemer M., 213
Romero R., 669
Russell A., 491

Sacchetti D., 559
Sahm P., 27
Salazar M.J., 671
Saltbones J., 471
Salvador R., 671
San José R., 231
Santabárbara J.M., 583
Saral A., 679
Schatzmann M., 653
Schneider C., 27
Schneiter D., 677
Schoenemeyer T., 17
Scire J.S., 389, 603
Scudlark J., 657
Segall E.J., 491
Seinfeld J.H., 491
Sills D.M.L., 63
Sistla G., 41
Sivertsen B., 109
Steenkiste P., 491
Steinberg K.W., 481
Steyn D.G., 505
Stigliani. W., 659
Strimaitis D.G., 389, 443, 603
Svensson G., 593
Sykes R.I., 425
Syrakov D.E., 645

Tangirala R.S., 13
Tayanç M., 679
Taylor P.A., 63
Telenta B., 635
Thykier-Nielsen S., 461
Tinarelli G., 559
Tremmel H.G., 81
Tzoumaka P., 549

Ulke A.G., 639
Uno I., 569

Valkonen E., 649
van Jaarsveld J.A., 203, 335
van Pul W.A.J., 203
Varinou M., 99
Venkatram A., 311, 623
Verver G., 353
Vögele T., 27

Walden J., 649
Weil J.C., 623
Wichmann-Fiebig M., 17
Wilson R.B., 623
Wolke R., 655

Xiu A., 613

Yamada T., 417
Yamartino R.J., 443
Young J., 491

Zerefos C.S., 549
Zhang Y., 239
Ziomas I.C., 549
Zlatev Z., 461

SUBJECT INDEX

ADMS model, 683
ADOM, 275
Advection scheme, 311
Aegean Sea, 99
AERMOD, 623
Aerosol particles, 259
Air-soil interface, 333
Aircraft emission, 81
Aircraft, 593
Airshed models, 55
Alps, 559
Aqueous-phase chemistry, 345
Arctic, 37, 249
Asia, 167
Assessment studies, 127
Athens, 295, 549, 593
Atlantic region, 275
Atmospheric circulations, 99

Baden-Württemberg, 89
Balaton lake, 657
Bottom to top diffusion, 433

CALPUFF, 389, 603
CAPTEX experiment, 603
Car exhaust, 653
Carbon bond IV, 41
Chemical reactions, 433
Chemistry transport model, 655
Chernobyl accident, 635
Clouds, 73, 259, 345, 417
Coast line, 295
Coastal air pollution, 505
Complex airflow, 569
Complex terrain, 407, 665
Concentration fluctuations, 285, 425
Concentration intensity, 285
Constant level balloons, 677
Control policy, 41
Convective boundary layer, 353, 643, 683
CSU-MM model, 569

Danish Eulerian Model, 461
Data assimilation, 569, 613
Degradation, 203
Dense gas, 127, 167, 203, 481, 639, 645, 659, 687
Deposition, 659
Downwash, 373
Dry deposition, 325, 363, 613
Dust, 239

East Asia, 191
El Niño, 119
EMEP, 127
Emergency warning, 667
Emission control, 155
Emission inventory, 55
Environmental policy, 3
ETEX, 181, 461, 677
Eulerian, 645
EURAD, 17
Europe, 213
Evaluation methodology, 623

Gas phase, 73
Gas to particle reaction, 529
Gaussian puff, 425
General circulation model, 325
Global, 213

Hanford experiment, 639
Harmonisation, 519
Heavy metals, 659
Heilbronn experiment, 27
Hemispheric model, 249

High performance computing, 491
Historical emissions, 221
Hydrocarbons, 549
Hydrostatic model, 81

Iberian Peninsula, 671
Impaction scavenging, 259
Indoor deposition, 363
Inhomogeneous terrain, 295
Internal boundary layer, 295, 505
Istanbul, 679

Jet, 399

KAMM-model, 89
Kanto region, 569
Kincaid experiment, 425
Kinematic simulation, 443
Kola Peninsula, 109
KSP model, 443

Lagrangian, 311
Lake breeze, 63
Land/water contrast, 505
Langevin particle model, 363
Length scales, 643
Lidar, 285
Lindane, 203
Local similarity, 399
Long range tracer experiment, 181
Long range transport, 191, 333, 471, 603, 657, 659
LOTOS, 345

MADONA, 583
Madrid, 231
Mallorca, 669
MAQS, 55
MATCH-model, 127
Mathematical model, 661
MEDCAPHOT-TRACE, 549, 593
Mediterranean, 99
Mesoscale circulation, 671
Meteorological variability, 221
METRAS, 17
Microscale dispersion model, 539
Mineral aerosols, 239
MISCAM, 539
MM4 model, 603, 613
Model evaluation, 519
Montsegorsk, 109
Multi-pollutant, 3
Multi-scales, 3

Nevada test site, 481, 687
Nitrogen-oxide, 191, 231
Non-hydrostatic model, 17
Northern Pacific, 191
Nucleation scavenging, 259

Open pit, 379
Optimal siting, 641
Oxidants, 275
Ozone abatement, 17
Ozone episode, 13
Ozone formation, 27
Ozone, 41, 63, 89, 155, 191, 213, 231, 275, 345, 549

Parallel computing, 491
Particle model, 559
Particulate mass, 135
Persistent organic pollutants, 203
Photochemistry, 119, 311, 593
Photochemical module, 231
Plume rise, 373, 389
Prague, 647
Prairie Grass experiment, 363
PRIME, 373
Puff dispersion model, 389

Radioactive release, 667
Radionuclides, 471
RADM, 81
Raindrop spectrum, 259
Reactive gases, 353
Real time model, 471
RAMS, 99
Regional oxidant model, 41
Regulatory modelling, 519
REWIH3D, 81
Riming scavenging, 259
RIMPUFF, 461
Rolling terrain, 583
RPM-10, 135

Santa Barbara Channel, 417
Scenarios, 167
SCIPUFF, 425
Sea-breeze, 505, 669
SEMOS, 13, 63
Shear flow, 399
Short duration release, 481
Slanted roof, 653
Smog event, 55
SNAP, 471
SOMOS, 13, 63
SONTOS, 63
Source-receptor relationship, 145
Stable boundary layer, 407, 481, 665
Stratospheric ozone, 155
Street canyon, 451
Sulfate aerosols, 135
Sulfate, 145
Sulfur dioxide, 109, 669
Sulfur, 37, 167, 191, 221, 231, 249, 645
Sulfurdioxide, 325
Sulfurhexafluoride, 285
Suproxa, 41
Surface fluxes, 613

Top to bottom diffusion, 433
Toxic metals, 657
Trace gas species, 239
TRACT-area, 89
Traffic induced immission, 539
Transalp, 559
Tropospheric chemistry, 73
Tropospheric ozone, 155
Tunnel exit, 303
Turbulent mixing, 353

UB/NMC model, 635
Urban Airshed Model, 549, 649
Urban street canyon, 653
Vancouver, 505
Vertical diffusion, 433
Visibility, 145, 529
Volatile Organic Compounds (VOC), 89

Wavelet transform, 407
Wind tunnel, 451
Windflow pattern, 55

GPSR Compliance

The European Union's (EU) General Product Safety Regulation (GPSR) is a set of rules that requires consumer products to be safe and our obligations to ensure this.

If you have any concerns about our products, you can contact us on ProductSafety@springernature.com

In case Publisher is established outside the EU, the EU authorized representative is:

Springer Nature Customer Service Center GmbH
Europaplatz 3
69115 Heidelberg, Germany

Batch number: 10712209

Printed by Printforce, the Netherlands